THEORY AND PRACTICE OF BIOLOGICAL CONTROL

CONTRIBUTORS

Perry L. Adkisson
L. A. Andres
D. P. Annecke
E. C. Bay
Stanley D. Beck
Oscar Beingolea G.
F. D. Bennett
C. O. Berg
E. Biliotti
S. Bombosch
G. E. Bucher
L. E. Caltagirone
H. C. Chapman
P. Cochereau
Elsie Collyer
Philip S. Corbet
D. L. Dahlsten
C. J. Davis
D. E. Davis
Paul DeBach
Vittorio Delucchi
R. L. Doutt
A. T. Drooz
L. A. Falcon
J. M. Franz
M. A. Ghani
Mostafa Hafez
K. S. Hagen
Junji Hamai
P. Harris
J. B. Hoy
C. B. Huffaker
L. E. LaChance
J. E. Laing
E. F. Legner
J. A. McMurtry
A. W. MacPhee
Fowden G. Maxwell
P. S. Messenger
Kenneth Myers
L. D. Newsom
G. O. Poinar, Jr.
H. Pschorn-Walcher
R. L. Rabb
V. P. Rao
David Rosen
R. I. Sailer
E. I. Schlinger
D. Schröder
Michael F. Schuster
F. J. Simmonds
Ray F. Smith
Shirley N. Smith
William C. Snyder
Vernon M. Stern
R. E. Stinner
K. L. Taylor
E. Tremblay
W. J. Turnock
M. van de Vrie
Robert van den Bosch
G. A. Viktorov
G. W. Wallis
A. J. Wapshere
D. F. Waterhouse
W. E. Waters
J. Weiser
W. H. Whitcomb
M. J. Whitten
F. Wilson
B. J. Wood
Keizo Yasumatsu
H. Zwölfer

THEORY AND PRACTICE OF BIOLOGICAL CONTROL

EDITED BY

C. B. Huffaker

Division of Biological Control
Department of Entomological Sciences
University of California
Berkeley, California

P. S. Messenger

Division of Biological Control
Department of Entomological Sciences
University of California
Berkeley, California

Sponsored by
The International Organization for Biological Control
and
The International Center for Biological Control
University of California

ACADEMIC PRESS New York San Francisco London 1976
A Subsidiary of Harcourt Brace Jovanovich, Publishers

ACADEMIC PRESS, INC.
111 Fifth Avenue, New York, New York 10003

United Kingdom Edition published by
ACADEMIC PRESS, INC. (LONDON) LTD.
24/28 Oval Road, London NW1

Library of Congress Cataloging in Publication Data

Main entry under title:

Theory and practice of biological control.

Includes bibliographies and index.
1. Pest control–Biological control.
I. Huffaker, C. B., Date II. Messenger, P. S.
SB975.T48 632'.96 75-36648
ISBN 0–12–360350–1

PRINTED IN THE UNITED STATES OF AMERICA

To
Harry Scott Smith
who was its inspiration

CONTENTS

SECTION II BIOLOGY AND SYSTEMATICS IN BIOLOGICAL CONTROL

SECTION III METHODOLOGY

SECTION VI APPENDIX

LIST OF CONTRIBUTORS

Numbers in parentheses indicate the pages on which the authors' contributions begin.

Perry L. Adkisson (593), Department of Entomology, Texas A&M University, College Station, Texas

L. A. Andres (481), Biological Control of Weeds Laboratory, Agricultural Research Service, U. S. Department of Agriculture, Albany, California

D. P. Annecke (143), Department of Agricultural Technical Services, Plant Protection Research Institute, Pretoria, South Africa

E. C. Bay (457), Western Washington Research and Extension Center, Puyallup, Washington

Stanley D. Beck (615), Department of Entomology, University of Wisconsin, Madison, Wisconsin

Oscar Beingolea G.[1] (443, 593), Ministerio de Alimentacion, c/o Direccion General de Produccion, Lima, Peru

F. D. Bennett (359), Commonwealth Institute of Biological Control, West Indian Station, Curepe, Trinidad

C. O. Berg (457), Department of Entomology, Cornell University, Ithaca, New York

E. Biliotti (543), Départment de Zoologie, Institut National de la Recherche Agronomique, Versailles, France

S. Bombosch (93), Institut für Forstzoologie der Universität Gottingen, Gottingen-Weende, Federal Republic of Germany

G. E. Bucher[2] (169), Research Institute, Agriculture Canada, Belleville, Ontario, Canada

L. E. Caltagirone (337), Division of Biological Control, University of California, Berkeley, California

H. C. Chapman (457), Gulf Coast Mosquito Research Laboratory, Agricultural Research Service, U. S. Department of Agriculture, Lake Charles, Louisiana

P. Cochereau[3] (359), Laboratoire de Lutte Biologique, Office de la Recherche Scientifique et Technique Outre-Mer, Noumea, New Caledonia

Elsie Collyer (337), Entomology Division, Department of Scientific and Industrial Research, Nelson, New Zealand

Philip S. Corbet (661), Department of Zoology, University of Canterbury, Christchurch, New Zealand

D. L. Dahlsten (289), Division of Biological Control, University of California, Berkeley, California

C. J. Davis (481), Entomology Branch, Hawaii Department of Agriculture, Honolulu, Hawaii

D. E. Davis (501), Department of Zoology, North Carolina State University at Raleigh, Raleigh, North Carolina

Vittorio Delucchi (81), Entomologisches Institut, Eidgenössische Technische Hochschule, Zurich, Switzerland

R. L. Doutt (143), Division of Biological Control, University of California, Berkeley, California

A. T. Drooz (313), USDA Forest Service, Forestry Sciences Laboratory, Research Triangle Park, North Carolina

L. A. Falcon (443), Division of Entomology-Parasitology, University of California, Berkeley, California

[1]*Present address:* Centro de Introduccion y Cria de Insectos Utiles, c/o Direccion General de Produccion, Ministerio de Alimentacion, Lima, Peru

[2]*Present address:* Research Station, Agriculture Canada, University of Manitoba, Winnipeg, Manitoba, Canada

[3]*Present address:* Rice Pests Research Station, Office de la Recherche Scientifique et Technique Outre-Mer, Bouake, Ivory Coast

J. M. Franz (17), Institut für Biologische Schädlingsbekämpfung, Biologische Bundesanstalt für Land- und Forstwirtschaft, Darmstadt, Federal Republic of Germany

M. A. Ghani (189), Commonwealth Institute of Biological Control, Pakistan Station, Rawalpindi, Pakistan

Mostafa Hafez (443), Plant Protection Department, Ministry of Agriculture, Dokki, Egypt

K. S. Hagen (93, 397), Division of Biological Control, University of California, Berkeley, California

Junji Hamai (685), Division of Biological Control, University of California, Berkeley, California

P. Harris (481), Research Station, Agriculture Canada, Regina, Saskatchewan, Canada

J. B. Hoy[4] (501), Western Insects Affecting Man and Animals Research Laboratory, ARS, USDA, Fresno, California

C. B. Huffaker (3, 41, 255), Division of Biological Control, University of California, Berkeley, California

L. E. LaChance (637), Metabolism and Radiation Research Laboratory, Agricultural Research Service, U. S. Department of Agriculture, Fargo, North Dakota

J. E. Laing (41, 685), Department of Environmental Biology, University of Guelph, Guelph, Ontario, Canada

E. F. Legner (457), Division of Biological Control, University of California, Riverside, California

J. A. McMurtry (93), Division of Biological Control, University of California, Riverside, California.

A. W. MacPhee (255, 337), Entomology Section, Agriculture Canada, Kentville, Nova Scotia, Canada

Fowden G. Maxwell (615), Department of Entomology, Mississippi State University, State College, Mississippi

P. S. Messenger (209, 543), Department of Entomological Sciences, University of California, Berkeley, California

Kenneth Myers[5] (501), Department of Zoology, University of Guelph, Guelph, Ontario, Canada

L. D. Newsom (565), CASRD, Department of Entomology, Louisiana State University, Baton Rouge, Louisiana

G. O. Poinar, Jr. (169), Division of Entomology and Parasitology, University of California, Berkeley, California

H. Pschorn-Walcher (313), Commonwealth Institute of Biological Control, European Station, Delemont, Switzerland

R. L. Rabb (233), Department of Entomology, North Carolina State University at Raleigh, Raleigh, North Carolina

V. P. Rao (189), Commonwealth Institute of Biological Control, Indian Station, Bangalore, India

David Rosen (81, 359), Faculty of Agriculture, The Hebrew University, Rehovot, Israel

R. I. Sailer (17), Department of Entomology and Nematology, University of Florida, Gainesville, Florida

E. I. Schlinger (81), Department of Entomological Sciences, University of California, Berkeley, California

D. Schröder (289), Commonwealth Institute of Biological Control, European Station, Delemont, Switzerland

Michael F. Schuster (397), Department of Entomology, Mississippi State University, Mississippi State, Mississippi

F. J. Simmonds (17, 41), Commonwealth Institute of Biological Control, Trinidad, West Indies

Ray F. Smith (565, 661), Department of Entomological Sciences, University of California, Berkeley, California

Shirley N. Smith (521), Department of Plant Pathology, University of California, Berkeley, California

William C. Snyder (521), Department of Plant Pathology, University of California, Berkeley, California

Vernon M. Stern (593), Department of Entomology, University of California, Riverside, California

R. E. Stinner (233), Department of Entomology, North Carolina State University at Raleigh, Raleigh, North Carolina

K. L. Taylor (289), Division of Entomology, Commonwealth Scientific and Industrial Research Organization, Hobart, Tasmania, Australia

E. Tremblay (143), Istituto di Entomologia Agraria, Università di Napoli, Portici, Italy

[4]*Present address:* State University of New York, Purchase, New York

[5]*Present address:* Division of Wildlife Research, Commonwealth Scientific and Industrial Research Organization, Canberra, Australia

W. J. Turnock (289), Research Branch, Agriculture Canada, Winnipeg, Manitoba, Canada

M. van de Vrie[6] (337), Research Station for Fruitgrowing at Wilhelminadorp, Institute for Phytopathological Research, Wageningen, Netherlands

Robert van den Bosch (233, 443, 543), Division of Biological Control, University of California, Berkeley, California

G. A. Viktorov (397, 593), Institute of Evolutionary Morphology and Ecology of Animals, U. S. S. R. Academy of Sciences, Moscow, U. S. S. R.

G. W. Wallis (521), Canadian Forestry Service, Pacific Forest Research Centre, Victoria, British Columbia, Canada

A. J. Wapshere (481), Division of Entomology, Biological Control Unit, Commonwealth Scientific and Industrial Research Organization, Montepellier, France

D. F. Waterhouse (637), Division of Entomology, Commonwealth Scientific and Industrial Research Organization, Canberra, Australia

W. E. Waters (313), College of Natural Resources, University of California, Berkeley, California

J. Weiser (169), Department of Insect Pathology, Institute of Entomology, Czechoslovakia Academy of Sciences, Prague, Czechoslovakia

W. H. Whitcomb (565), Department of Entomology, University of Florida, Gainesville, Florida

M. J. Whitten (209, 637), Division of Entomology, Commonwealth Scientific and Industrial Research Organization, Canberra, Australia

F. Wilson (3,209), Department of Zoology, Imperial College of Science and Technology, London, England

B. J. Wood (359), Chemara Research Station, Kumpulan Guthrie Sendirian Berhad, Seremban, Malaysia

Keizo Yasumatsu (397), Entomological Laboratory, Faculty of Agriculture, Kyushu University, Fukuoka, Japan

H. Zwölfer[7] (189), Commonwealth Institute of Biological Control, European Station, Delemont, Switzerland

[6] *Present address:* Research Station for Floriculture, Aalsmeer, Netherlands

[7] *Present address:* Staatliches Museum für Naturkunde, Ludwigsburg, Federal Republic of Germany

PREFACE

The world's looming food shortage demands maximum food productivity. Heavy use of pesticides has played a significant role in meeting food demands as well as in alleviating insect-borne human disease. However, the extensive, often excessive, use of these powerful, broad-spectrum chemicals, some of which are nonbiodegradable, has resulted in a variety of harmful and undesirable effects on wildlife, man, and the environment. The documenting and publicizing of these effects have led to wide public concern and considerable moral and legal opposition to the continued unrestricted use of these materials. Moreover, a shortage of synthetic pesticides makes it mandatory that we use the limited supplies wisely. This must be done not only to help alleviate the pesticide-induced environmental problems, but to conserve the limited supplies of the much needed pesticides themselves, for chemicals remain the most effective immediate solution to most pest problems.

Chemicals are not the only, or indeed the best, solution to the problem. Biological control (the use of a pest's own natural enemies) has permanently solved many pest problems throughout the world. This book concerns, to a large extent, the introduction of such natural enemies (parasites, predators, and pathogens), for the purpose of suppression, from the pests' native environments to new environments in which the pests have become established as alien species. But the natural enemies of pest species often face insurmountable obstacles in accomplishing this purpose in areas where powerful, broad-spectrum chemicals are used without regard to their effects on these beneficial species. Indeed, in highly developed agricultural areas, chemicals are commonly used for the control of one or more pests of a crop. Yet existing biological control agents or potentially available ones could, if not restricted by these chemicals, control or help to control other pest species on the same crop for which these chemicals are ineffective.

There are also other biological forms of pest control which have not been conventionally classed "biological control" which, however, act in similar, ecologically nondisturbing ways to those of the natural enemies. The utilization of the varied spectrum of wild and crop plant germ plasm to develop varieties of crop plants resistant to pests is one such additional biologically based control. This tactic has been widely used in the control of plant diseases and, to a lesser extent, in insect control, but its full potential is only beginning to be tapped. The use of cultural methods of pest control, while not biological per se, is often employed in a manner to augment the effectiveness of such truly biological forms of control as the natural enemies and crop plant resistance, or to expose the pests to weather stress. Even the chemicals themselves, while inherently disruptive to conventional biological control, may be used in ways less disruptive, and sometimes as an aid to pest control by the natural enemies themselves by establishing a more favorable ratio of natural enemies to pests. Moreover, various special types of chemicals, pheromones, kairomones, juvenile hormones, etc., may be used, i.e., their special biological roles may be utilized in ways to substitute for, augment, or make more selective the conventional broad-spectrum insecticides. Finally, a number of special forms of perhaps truly biological control agents or competitors about which we know very little seem to play significant roles in the suppression of many soil-borne plant pathogens damaging to crops. These are collectively termed "antagonists."

This book was planned to supplement, not to replace, existing texts on biological control. It was designed (1) to satisfy the need for a book that covers conventional biological control achievements in the major crop types or in public health problem areas (which has not been done before), as well as to treat fully the basic features in philosophy, theory, basic biology, ecology, and practice of conventional biological control; (2) to include basic information concerning developments in other biologically based alternatives to chemical pesticides; and (3) to show how all these approaches can be combined to achieve practical integrated pest control systems that are sound both economically and ecologically.

A system of integrated pest control is dynamic and must be considered a unit in respect to the application of ecological principles in the use of various methods employed for pest suppression. Integrated control is usually centered on nature's two principal biological forces for containing excessive abundance of any species, or the severe stress of herbivores on plants, for example. These are the genetic resistance factors in the plants and control of the pests and potential pests by their own natural enemies.

This book is divided into five sections. Sections I and II deal with the philosophy, theory, scope, history, and the biological and ecological basis of biological control. Section III is concerned with methodology in biological control. Section IV details the accomplishments of conventional biological control in various types of crops, forests, and public health areas. Section V deals with various components of integrated pest control other than conventional biological control which form, along with conventional biological control, the essential tactics used in the integrated control approach.

An attempt was made to unify the contributions into a comprehensive work, each to be consistent with the others. This was not entirely possible because not all the authors have the same viewpoints. However, we asked the authors to consider the opposing viewpoints expressed in an effort to minimize the existing differences. The remaining ones should prove a challenge to other researchers.

We take this opportunity to thank the many contributors, both the authors and their invaluable assistants, without whose efforts this book could not have been produced. We wish especially to thank the Division of Biological Control, University of California, Riverside, for permitting us to refer to "Introduced Parasites and Predators of Arthropod Pests and Weeds: A World Review" by C. P. Clausen (ed.), which is being published as a U. S. Department of Agriculture Handbook. We also express our appreciation to Dr. Paul DeBach who helped us plan the book but whose health did not permit his continuing as an editor, and to Drs. Robert van den Bosch and K. S. Hagen, who also had much to do with its planning. Finally, we express our deepest appreciation to Mrs. Nettie Mackey for her tireless efforts and dedication in bringing this work to fruition, and to Dr. J. E. Laing for much of the indexing.

C. B. Huffaker
P. S. Messenger

[illegible] Stephen Wyne and Wendell A. Wolfe [illegible] contributors [illegible] [illegible] appropriate [illegible]. Paul [illegible] assisted [illegible] permissions [illegible] [illegible] who also [illegible] [illegible] Mrs. [illegible] [illegible] and [illegible] [illegible]

[illegible]

SECTION I

INTRODUCTION, HISTORY, AND ECOLOGICAL BASIS OF BIOLOGICAL CONTROL

1

THE PHILOSOPHY, SCOPE, AND IMPORTANCE OF BIOLOGICAL CONTROL

F. Wilson and C. B. Huffaker

I. INTRODUCTION AND PHILOSOPHY

The scientific basis of biological forms of pest control is very complex: possibilities for their development widen with increase in scientific knowledge, and are exploited in accordance with economic and social needs. Individual examples of the use of natural enemies to control pests have existed for centuries, but biological control emerged as a scientific method only late in the nineteenth century. This was brought about, on the one hand, by the emergence of new concepts bearing on such matters as the relationship between species, their evolution, population pressures, and the struggle for existence; and, on the other hand, the urgent need for solutions for the serious problems presented by immigrant pests in different parts of the world.

As entomology developed in the nineteenth century, an awareness grew of the importance of parasites, predators, and pathogens in the limitation of insect numbers, and suggestions were made for the practical use of such natural enemies. Understanding of population regulation at that time was very vague. Forbes (1880) wrote: "... the annihilation of all the established 'enemies' of a species would, as a rule, have no effect to increase its final numbers." However, this view was confounded within a decade by the complete and permanent control of a major pest, the cottony-cushion scale, *Icerya purchasi* Mask., in California by the introduction of the coccinellid predator *Rodolia cardinalis* (Muls.). This result quickly produced a widespread enthusiasm for biological control among agriculturists and entomologists in countries (mainly countries colonized by Europeans) in which immigrant pests occurred at population levels much above those experienced in Europe. It also initiated a long series of attempts at biological control, and each of the many successes has been a field demonstration of the intrinsic capacity of natural enemies to regulate the numbers of their hosts. Nevertheless, the doubts, felt by Forbes and others have continued, though with declining strength, to influence views on population dynamics, and to act as a brake on support for biological control and on the extension of its scope.

The term "biological control" was first used by Smith (1919) to signify the use of natural enemies (whether introduced or otherwise manipulated) to control insect pests. Its scope has expanded over the decades, and a suitable definition presents problems partly because the term has applied and academic aspects. However, some authors include in this term various other nonchemical forms of control that are biology-based. One such method, which has important applications, is the development of strains of crops that are resistant to, or tolerant of, pests or diseases; another is modification of cultural practices in a way that avoids or reduces infestation, as change of planting date, avoidance of continuity of the crop in successive seasons on the same land, or plowing, pruning, flooding, etc. A relatively new method, which aims primarily at eradication, is the release of sterile males, which has proved effective against screwworms and fruit flies. There are also such techniques as genetic, pheromonal, and other actual or potential forms of pest control that arise from new scientific knowledge. The inclusion of any such methods within the term biological control is not generally approved, and Doutt (1972) has said that such an expansion of meaning "has the damaging effect of obscuring the unique functional and ecological basis of biological control." In this treatise biological control is treated as the science that deals with the role that natural enemies play in the regulation of the numbers of their hosts, especially as it applies to animal or plant pests. Other biology-based forms of pest control are, however, discussed in order to complete the picture of using biological control in a holistic integrated control approach.

Biological control in the strict sense as defined above cannot generally be expected to solve all the problems with a crop having a complex of pests. It may provide a fully adequate solution for one or a few pest species, have a substantial but not sufficient effect on others, and be of little or no use for the remainder. However, biological control, together with plant resistance, forms nature's principal means of keeping phytophagous insects within bounds in environments otherwise favorable to them. They are the core around which pest control in crops and forests should be built. Biological control in practice, apart from special situations, is therefore often possible only within the framework of integrated control, which itself usually depends upon a core of biological control and plant resistance. Moreover, the strategy of integrated control—pest population containment, not prevention—is entirely consistent with the use of biological control. Hence, it is appropriate that this treatise should give extensive treatment to various areas of integrated control that are strongly supplemental and, indeed, essential to the utilization of biological control in many situations.

Pioneer work on biological control was based on the view that the high population level of an immigrant pest was the result of its escape from a balance of nature existing in its native habitat, and could be lowered by the reconstitution of this balance in the new habitat. This view was reflected perhaps in the "sequence theory" of Howard and Fiske (1911), which maintained that the biological control of a pest requires the establishment of a sequence of parasites on the successive stages of the host. It probably also influenced the philosophy underlying the approach to the biological control of prickly pear (*Opuntia* spp.) in Australia, which was that a useful degree of control might follow the establishment of a complex of insect species that attacked the plant's different structures (Dodd, 1940). However, experience has shown that, though the establishment of several species of natural enemies is sometimes advantageous, and seldom if ever disadvantageous (Doutt and DeBach, 1964; Huffaker *et al.*, 1971; and Chapter 3 in this volume), success in biological control has most frequently depended upon the establishment of a single well-adapted species.

The balance of nature affecting a pest in its native habitat involves an extremely complicated array of interacting biotic and abiotic factors, and it is neither possible nor desirable to reassemble such complexes in new habitats. On the other hand, such complexes commonly include species of natural enemies having specific properties that allow them to regulate host populations in suitable habitats, and research on such complexes is primarily directed toward the discovery of natural enemies that are well adapted to the habitats in which control is desired. In biological control, therefore, one is concerned not so much with what natural enemies actually do in their native habitats as with their intrinsic properties (Wilson, 1964). Each natural enemy is a single component ecological link in the complex of the original ecosystem. Such properties have evolved in the context of natural ecosystems, but the ecosystems in which natural enemies are now discovered, and the ecosystems in which they are required to provide control, are not necessarily very similar to one another or to the original ecosystem. Consequently, another research preoccupation may well be the adjustment of the new habitat to the needs of the introduced natural enemies.

Effective biological control requires that the area in which a natural enemy is to operate shall provide the conditions and requisites necessary for the expression of the intrinsic capacity to control the host; it is not required that the agroecosystem of the area shall otherwise resemble some actual or hypothetical ecosystem of the native habitat. Biological control can operate successfully in environments as "natural" as a virgin forest and as "artificial" as a glasshouse. It can accept the challenge of changing agricultural methods and the current emphasis on extensive monocultures, endeavor to discover natural enemies that are climatically preadapted, ensure that their essential requisites are available, and that cultural methods are not inimical to them.

This widened philosophical approach to biological control accords better with the experience of past research and it also gives broader perspectives for the future. Outstanding biological control successes have sometimes been achieved against native pests (as with the coconut moth, *Levuana iridescens* B. -B.), and by the use of natural enemies whose hosts belong to different species or genera from the pests they are needed to control [as with *Cactoblastis cactorum* (Berg) against prickly pears, and myxomatosis against the rabbit]. It will certainly be found that the properties of natural enemies sometimes allow them to be used effectively in unexpected ways, as, for example, in the control of the sugarcane moth borer, *Diatraea saccharalis* (F.), in Barbados by *Apanteles flavipes* Cam. (Alam *et al.*, 1971). Little was expected to result from the introduction of this Asiatic parasite of graminaceous borers of other genera. Such possibilities will receive increasing attention, though the control of immigrant pests by the introduction of natural enemies from native habitats will doubtless continue to predominate in research programs.

II. THE SCOPE OF BIOLOGICAL CONTROL

The scope of application in biological control has steadily expanded from the use of entomophagous insects to control insect pests to the use of a whole range of organisms to control insects, mites, snails, occasional vertebrates, and plants as diverse as algae, fungi, herbs, shrubs, and trees. In principle, any organism and some organic products, such as cattle dung, may be susceptible to biological control. The controlling organisms that have been used include viruses, bacteria and their toxins, fungi and other microbial pathogens, nematodes, snails, insects, mites, and vertebrates of various kinds. The range of organisms controlled biologically and of organisms providing such control may still be expected to expand considerably. While the controlling organisms usually have their effect by killing the host directly, they sometimes operate in other ways, as in antagonistic fungi, or nematodes that sterilize their female hosts (*Deladenus* in *Sirex*) (see also Chapters 7 and 21), or natural enemies that reduce the reproductive or competitive capacity of their host plants.

The number and variety of species of natural enemies available for study and use against pests are enormous. The more fully a pest is studied, the more diverse its natural enemies are found to be. An apparent single species of natural enemy not infrequently is found, on closer examination, to comprise several sibling species or geographical or biological races having different properties. Genetic diversity within species and its significance for control effectiveness have so far been comparatively little studied or exploited. Further, natural enemies are sometimes able to control the new host very effectively, and in some instances, as with certain fungi, the absence of an evolved homeostasis may render the new host especially vulnerable, as with the effect on the chestnut in America of the accidentally introduced fungus *Endothia parasitica*.

For such reasons enormous resources of natural enemies exist, and the research possibilities for pest control are virtually unlimited. Research into particular problems has often been more or less superficial, especially in earlier work, and has often been terminated without achieving adequate control. In these instances, it is probably never justifiable to conclude that such failure precludes the possibility that a renewed and more thorough investigation will not lead to success. Many of our most serious pest species, such as codling moth or cotton boll weevil, have been little investigated.

One of the most important areas in the widening scope of biological control is that involving the utilization of pathogens, of which the control of the rabbit by myxomatosis is such a notable example, though it had been preceded by various successful applications of viral, bacterial, and fungal pathogens to insect problems. The evident potential of pathogens for the control of pests has increased as a result of recent developments in this field, and there can be no doubt that microbial control will eventually attain a wide practical value. One of the most useful applications of this technique has been the commercial exploitation of *Bacillus thuringiensis* as a biological insecticide for a number of lepidopterous pests. Viruses, in particular, seem destined to serve a major role as highly specific biological insecticides when satisfactory methods have been developed for their commercial production (perhaps by exploiting tissue culture techniques), and when, as seems very probable, they have been shown to involve no human health hazard. Microbial insecticides can possess many of the desirable characteristics of chemical insecticides, such as their ease of application and their reliability within a short period of time, without many of their disadvantages (Chapter 7). It is also possible that the effectiveness of viruses as control agents may be enhanced by research aimed at elucidating the important role of stress in the manifestation of viral infection. The need for effective biological pesticides will inevitably grow as increasing restrictions are imposed on the use of chemical pesticides that involve environmental hazards. (See also Chapter 7.)

The extension of biological control into the area of weed control was a major development both academically and practically. It has had a profound effect on plant ecology by demonstrating the extraordinary potency of phytophagous insects in their effect on host abundance. While Ridley (1930) had early concluded that biological control by phytophagous insects, especially those attacking juvenile plants, is a basic reason for the great diversity of species of plants in the hot wet tropics, the significance of the insects' role was made far more convincing by the demonstrations furnished by biological control. It is fair to say that these demonstrations opened up for speculation and research a whole range of problems concerning the numerical relationships between plants and their natural enemies. The biological control of weeds has proved extremely effective, especially in view of the relatively small research effort that has been directed towards them. The great potential of biological control of weeds remains greatly underestimated by those responsible for research. This is a serious deficiency as the losses caused by weeds probably greatly exceed those attributable to both insect pests and plant diseases (see Chapter 19). The scope of research on the biological control of weeds is being extended in various ways, as, for example, the greater interest in the use of plant pathogens, the application of the method to water weeds, and its use to render the cultivation of harmful drug plants (opium poppy and hemp) commercially more difficult. So far the method has been virtually restricted to the control of immigrant weeds, but there is good reason to believe that it might prove effective also against native weedy species, particularly if mass production and releases can be undertaken economically.

III. BIOLOGICAL CONTROL AND INTEGRATED CONTROL

The integration of control measures in pest management programs for major commodities is now a characteristic aim of advanced agricultural production. The biological control component in such programs is a central or important element for reasons of economy, all-round effectiveness, and environmental harmlessness. Often enough, pesticide usage cannot be reduced without a corresponding increase in the efficiency of natural control agents. Increased effectiveness of natural enemies can be brought about by various methods in addition to the introduction of new species. Suitable manipulation of the habitat can favor natural enemy action. For example, the use in California of alfalfa interplants in cotton and strip-cutting in the periodic harvesting of alfalfa hay contributes greatly toward the conservation within the habitat of the natural enemies of pest species. It has also been shown that the provision of supplementary diets in the field results in such an augmentation of predatory insects that they bring about commercial control of *Heliothis* (van den Bosch *et al.*, 1971; and Chapter 17).

A central feature of integrated control, to establish the real need for taking *any* control action, is also central to employment of biological control. Unfortunately, we know relatively little about the real potential of many of our pests for causing economic damage, the capacity of our crops to compensate, or the capacity of other later-acting factors to prevent threatened economic loss. Integrated control practitioners are now intensively exploring this area (Chapter 27). Moreover, if the maximum is to be made of the biological control potentials for control of any of a complex of pests, we must learn more about the whole agro- or forest ecosystem, as the case may be. We must investigate the pertinent factor interactions among crops, pests, natural enemies, soil and climatic factors, and cultural practices, and how to structure these inputs to permit predictive modeling for checking against real outcome, and for use in decision making.

A number of other prime components in integrated control programs are also seen as having real potential for supplementing, or even supporting, better action of natural enemies. One of these components is the development of plant varieties offering tolerance of or resistance to insect attack. This has been widely and successfully used in plant pathology but rather neglected until recently by entomologists in spite of some notable early successes (e.g., grape rootstocks resistant to Phylloxera and wheat cultigens resistant to Hessian fly) (Chapter 25). Various elaborations of cultural measures, in addition to those referred to above, can provide refuges for natural enemies, resources for them or their alternate prey or hosts, and various supplementive as well as supporting advantages (Chapter 24). Advancing knowledge in the use of insect behavioral attractants and repellents, especially pheromones (Wood *et al.*, 1970), offer marked advantages in integrated control efforts, especially now for use in monitoring populations, but perhaps soon as a direct means of insect control. However, for this promising horizon, as well as that of the use of insect hormones or analogs as insecticides, the utmost caution and extensive prior study will be required before their commercial employment can be foreseen (Huffaker, 1972; Chapter 23). Otherwise these "third generation insecticides" could have a sequel not unlike the current sad situation. Finally insecticides in general can be more selectively and compatibly used than has been the common practice, and more selective kinds of insecticides should be developed, thus offering a greater compatibility with biological control and the environment (Chapter 23).

Another means of increasing natural enemy effectiveness is periodic colonization of laboratory-bred natural enemies. For example, such colonization of *Metaphycus helvolus* (Compere) on black scale and of *Aphytis melinus* DeBach on citrus red scale in the Fillmore-Piru area of southern California provided strikingly effective commercial control of these major pests (Lorbeer, 1971). Such methods are being employed also in the control of glasshouse pests (Hussey and Bravenboer, 1971).

It often happens that serious pest damage to crops arises because of the seasonal delay between initial pest infestation and the appearance of effective natural enemies. An interesting technique for avoiding this problem is to ensure that both pest and natural enemy are present in appropriate numbers and suitably dispersed at the time of planting. This method of control was first tested by Huffaker and Kennett (1956) for the cyclamen mite on strawberries, and subsequent work along similar lines with the cabbage butterfly on cruciferous crops has been carried out by Parker (1971). This technique should appeal to agriculturists for use against pests that have highly effective natural enemies that can be relied on to provide a high level of control.

IV. THE IMPORTANCE, LIMITATIONS, AND FUTURE OF BIOLOGICAL CONTROL

Considerable responsibility is involved in the deliberate extension of the geographic ranges of organisms for the purposes of biological control, and caution is essential, particularly in dealing with categories of natural enemies, and also pests and habitats, that have been little studied, as with the use of pathogens and work in aquatic habitats. The field is essentially one for experts. Nevertheless, a notable feature of biological control has been its freedom from harmful side effects, especially compared to chemical control. The safety with which biological control has been conducted has depended on a number of factors. The biotic agents used in biological control are selected from naturally occurring, self-balancing population systems, in which many of the natural enemies, such as entomophagous and phytophagous insects, often show a high degree of host specificity. This allows them to be used with considerable confidence that undesirable complications will not arise. Also, the preliberation research now undertaken is being pursued at deeper levels in respect of the properties of natural enemies and the characteristics of the pest and its habitat, and with the increasing expertise that arises from greater ecological understanding and scientific background. In no field perhaps has caution been more necessary than with research on the biological control of weeds, because of the risk that natural enemies introduced may attack crops or other useful plants. Although this risk has hampered research in some respects, as with the attempt to control *Xanthium* in Australia, the instances of introduced insects attacking any useful species have been trifling.

Care is particularly necessary when polyphagous species are being considered for use, as when fish are to be employed to control mosquitoes or weeds, or omnivorous mammals for forest insect control. Useful fish, such as *Gambusia*, *Tilapia*, or *Lebistes*, have sometimes been considered (though actual proof may well be lacking) to reduce the abundance of game or food fish. Special caution is also required when pathogens (or antagonists or competing species) are under consideration for use against plant or animals pests. The range of pathogens available as agents of pest control is wide, and the safety with which they may be employed is variable and often uncertain. Fortunately, the insect viruses, which are potentially so important, appear to be highly specific; their safety for pest control is at present under consideration at the international level.

The advantages of biological control are numerous. They often include a high level of control at low cost, self-perpetuation at little or no cost following the initial effort, absence of harmful effects on man, his cultivated plants, and domesticated animals, wildlife, and other beneficial organisms on land or in the sea, and the utility of some types as biotic insecticides. The ability of natural enemies to reproduce rapidly and to search out their hosts and survive at relatively low host densities makes outstanding advantages possible. The development of host resistance to introduced biological control agents, with the consequence of imperiling a whole program, is virtually unknown, although host resistance to insect parasitoids and other types of parasites is common. Initially successful biological control programs have continued to be so, in many instances, over long periods of time. Continued effort and new introductions have apparently solved the problem of development of resistance in the larch sawfly in Canada to the parasite *Mesoleius tenthredinis* Morley (Chapter 12). Moreover, the degree of population control of the rabbit by myxoma virus in Australia is still very substantial, contrary to many misleading implications in the literature (Chapter 20).

The limitations of the biological control method are difficult to indicate or forecast; they will be properly known only after greatly expanded research over a long period. However, one basic limitation is that the host population will continue to exist at a level determined by the properties of the host and its natural enemies and of the habitat; when this level is still economically troublesome after the establishment of natural enemies, it must be lowered by some form of integrated control, or the habitat or natural enemies must be manipulated in some way to make them more efficient, until more effective natural enemies can be discovered. Second, the attainment of biological control of one major pest on a crop necessitates the elaboration of a system of integrated control for other pests of the crop, if any. This often requires a highly complex technology which we are just beginning to explore adequately.

It is also true that the research necessary in seeking a biological control solution to a problem is often demanding in terms of scientific and technical staff, funds, and time, and that a solution cannot be guaranteed in advance. Research may provide a complete solution in a few years, but it may remain unproductive for many years. While work on a pest of a promising type may be initiated with justifiable optimism (and this applies to a wide range of pests), there are many kinds of problems of great importance for which the prospects of success are unknown. These problems fall largely into categories that because of their complexity or obscurity, and lack of information arising from inadequate research support, are too little understood to assess the prospects of successful practical results. Biological control successes have so far been rare in some fields, including the control of plant pathogens, nematodes, insect vectors of plant diseases, medical, veterinary, and stored products pests, and mammals.

In these little-known fields some valuable and encouraging results have been obtained, as with the rabbit in Australia (Chapter 20), the use of antagonistic fungi to control fungal pests (Chapter 21), the control of mosquitoes by fish (Chapter 18), and so on. A considerable expansion of research in such fields is highly desirable. The World Health Organization is fostering research into biological and genetic controls for disease vectors, such work having been rendered urgent by the widespread development of resistance to pesticides among these pests. A measure of optimism about the prospects for the biological control of plant pathogens has been expressed by Garrett (1965): ". . . I now believe that eventual prospects for biological control are brighter than ever before, partly because we now comprehend the magnitude of the problem to be solved." (See Chapter 21.)

Most criticisms of biological control have been concerned with its supposedly narrow limitations as a technique of pest suppression. In general, events have invalidated such criticisms, but adverse comment, often grossly distorted, has played a significant role over the decades in reducing support for biological control research and in retarding progress. The real basis of such criticism lies in inadequate appreciation of the important role of natural enemies in the determination of the abundance of plants and animals, and a failure to grasp imaginatively the immense possibilities for pest control implicit in the ability to expand the geographical ranges of natural enemies of species noxious to man. As a consequence, the scope of biological control has often tended to be equated with the area of already proved effectiveness, and attempts to extend its scope have seemed to some to lack a sound ecological basis.

The many successes of the kinds secured in Hawaii and other oceanic islands, and in California and elsewhere, have been used as the basis for various hypotheses concerning the limited applicability of biological control. Favorable subjects or circumstances have been held to be limited to introduced pests, permanent crops or habitats such as orchards or forests, islands or "ecological islands," a restricted biota, and a tropical or subtropical climate. More recently, it has been contended that the method is applicable only to "indirect" pests, i.e., those that do not attack the marketable commodity itself. Pests or circumstances of a contrary kind were, according to such hypotheses, unsuitable or at least unfavorable for a successful outcome. It has also been argued that there is little evidence that natural enemies have any important effect on plant numbers, and consequently that the scope for the biological control of weeds is very restricted. After various remarkable successes against perennial weeds of range lands, the criticism was modified to have relevance mainly to particular weed categories, such as annuals or weeds of crops.

Of the various critical comments on biological control's scope and prospects, one of the more sweeping and temporarily influential was that of Taylor (1955) who, doubtless influenced by the then current pesticide euphoria, concluded that biological control seldom worked, most susceptible pests had already been so controlled, few possibilities remained for future successes, the propects for success in continental areas was particularly slight, and that biological and chemical controls were incompatible.

For the most part these restrictive hypotheses and forecasts have been contradicted by subsequent research results. There are well over 200 examples of completely or partially successful classical biological control. These have been obtained with the most diverse kinds of pests, crops, habitats, and climates, in continental countries no less than in islands, in temperate and cold climates as well as tropical and subtropical areas, with some native as well as many immigrant pests, against pests of annual crops in addition to those of orchards and forests, and with direct as well as indirect pests, and with annual, perennial, and aquatic weeds, etc. Biological control successes are becoming more frequent, not less, and research on it is rapidly increasing, not decreasing. Effective integrated control has been secured against the complex of pests of various major crops. Far from chemical and biological controls being quite incompatible, integrated control, utilizing biological control as an important component, is seen by many entomologists as the most promising approach to pest control under modern forms of crop production.

It may well be found in time that biological control has important limitations, but these will need to be established by research exploration of each problem. Ill-founded hypotheses, like excessive caution (Huffaker and Kennett, 1969), can paralyze research action and scientific progress. An examination of the results provided by biological control research suggests that intensive, well-supported programs are usually successful (Bartlett and van den Bosch, 1964), that in every area where biological control has been seriously applied there have been major successes (Munroe, 1971), and that the biological control success achieved in different countries has been related to the input of research effort, rather than to the kind of pest, crop, or climate (DeBach, 1964). The main conclusion to be drawn from the general experience is that the possibility of a biological control solution should seldom, if ever, be excluded on principle when considering a suitable research approach.

It can readily be demonstrated that biological control research has been an extremely profitable community investment (DeBach, 1964; Simmonds, 1968; and Chapter 3). This research must, in general, be funded by governments or international agencies because safety demands the official sanction and the general benefit conferred is not condusive to commercial exploitation, except for periodic colonization procedures and the production of microbial insecticides. The savings resulting from effective biological control, which stem from the reduced need for pesticide treatment, and increased crop yields and land values are rendered the larger because biological control successes frequently involve major pests of high value crops in high production, technically advanced areas.

Historically, there has been a sharp divergence in the thinking of those engaged in biological control research and those concerned with the toxicological control of pests. In an era dominated by the chemical approach to pest control, biological control research, broadly speaking, has been provided with minimal funds mainly to find solutions for problems with which chemical control could not cope. Under these conditions, biological control scientists developed a tough-minded individualism and succeeded, with gradually increasing authority, to demonstrate that a wide range of natural enemies and biological control procedures could be effectively used to provide efficient control of a wide range of animal and plant pests. At the same time, they took a deep interest in the ecological foundations of the applied side of the science, and contributed both directly, and indirectly by influencing academic ecologists, in the great expansion in ecological understanding that has occurred during the last 40 years.

The cleavage between those concerned with biological and with chemical approaches to pest control was greatest in the period following the development of persistent, broad-spectrum, organosynthetic insecticides, when many believed that chemistry was on the point of providing a complete solution to entomological problems. This opinion was not shared by those engaged on biological control problems who observed some of the adverse effects of these pesticides (e.g., Smith, 1941, 1944), and by some purely academic entomologists (e.g., Wigglesworth, 1945). Indeed, Smith (1941) had warned of the insect's potential to develop resistance to insecticides and of the probable consequences of an insect control strategy based primarily on chemical insecticides and ignoring ecological principles. The correctness of his prophetic counsel is now evident. Over 200 examples of resistance are now known, and it is becoming impossible to control some species with any insecticides. In the last decade, the awareness of all these problems has become general (various chapters herein), and the attitude of many contemporary entomologists is summed up in the following passage from a recent report of the International Biological Program (1971): "The timing of the IBP effort has been coincident with increasing realization that the era of nearly sole reliance on pesticides for insect control has been a major failure—due to the attendant problems of pesticide resistance, inducement of secondary pest species, the ever increasing need for more pesticides (pesticide 'addiction'), and mounting costs yet poor insect control, and harm to non-target species and the environment. In short a more ecological approach has become a recognized necessity."

That pest control is entering a new phase is now generally appreciated. Increased scientific knowledge bearing on the principles of population regulation, the development of other biological forms of pest control, strongly supplemental or supportive of control by natural enemies, and research into the integration of biological and chemical agents of control are providing an ever-widening basis for developing modes of pest control that are free from substantial harmful side effects. At the same time, worldwide concern about environmental pollution is providing a social climate in which it will increasingly be required that pest control be studied, not from the viewpoint of short-term expediency, but with the perspective required for satisfactory long-term solutions. This concern has been reflected in the recent reconstruction of the International Organization for Biological Control, and its formation in different parts of the world of regional sections concerned to promote the development and application of biological and integrated forms of control capable of dealing with regional pest problems.

It is also being recognized that modes of land use and technical developments in crop production, variety improvement, and especially pesticide use are having, and threaten to increasingly have, a serious effect on the diversity of organisms. There is concern, for example, about the loss of genetic resources of useful plants, and this includes the pool of factors for resistance to pests. Comparatively few insects or plants are pests. Vast numbers of insect species serve important biological functions in the cycling of organic materials, the increase in soil fertility, and the population regulation of their hosts, many serving this role for insect and plant pests. The natural enemies of pests constitute a self-renewing natural resource of inestimable value, and the preservation and rational utilization of this resource is of the greatest importance (Wilson, 1971). The numerous biological control successes that have been obtained in a period dominated by chemical control give some measure of the benefits that will accrue to agriculture when research into biological forms of pest control are supported by governments on a scale that allows their fullest exploitation.

REFERENCES

Alam, M. M., Bennett, F. D., and Carl, K. P. (1971). Biological control of *Diatraea saccharalis* (F.) in Barbados by *Apanteles flavipes* Cam. and *Lixophaga diatraeae* T. T. *Entomophaga* **16**, 151–158.

Bartlett, B. R., and van den Bosch, R. (1964). Foregin exploration for beneficial organisms. *In* "Biological Control of Insect Pests and Weeds" (P. DeBach, ed.), pp. 283–304. Reinhold, New York.

DeBach, P. (1964). Successes, trends, and future possibilities. *In* "Biological Control of Insect Pests and Weeds" (P. DeBach, ed.), pp. 673–713. Reinhold, New York.

Dodd, A. P. (1940). "The Biological Campaign Against Prickly Pear," 177 pp. Commonwealth Prickly Pear Board, Brisbane, Australia.

Doutt, R. L. (1972). Biological control: parasites and predators. *In* "Pest Control Strategies for the Future," pp. 228–297. Nat. Acad. Sci. Nat. Res. Council, Washington, D.C.

Doutt, R. L., and DeBach, P. (1964). Some biological control concepts and questions. *In* "Biological Control of Insect Pests and Weeds" (P. DeBach, ed.), pp. 118–142. Reinhold, New York.

Forbes, S. A. (1880). On some interactions of organisms. *Bull. Ill. Natur. Hist. Surv.* **1**, 1–17.

Garrett, S. D. (1965). Toward biological control of soil-borne plant pathogens. *In* "Ecology of Soil-Borne Plant Pathogens. Prelude to Biological Control" (K. F. Baker and W. O. Snyder, eds.), pp. 4–17. John Murray, London.

Howard, L. O., and Fiske, W. F. (1911). The importation into the United States of the gypsy moth and the brown-tail moth. *U.S. Dep. Agr. Bur. Entomol. Bull.* **91**, 312 pp.

Huffaker, C. B. (1972). Ecological management of pest systems. *In* "Challenging Biological Problems: Directions toward Their Solution" (John A. Behnke, ed.), pp. 313–342. Oxford Univ. Press, New York.

Huffaker, C. B., and Kennett, C. E. (1956). Experimental studies on predation: Predation and cyclamen mite populations on strawberries in California. *Hilgardia* **26**, 191–222.

Huffaker, C. B., and Kennett, C. E. (1969). Some aspects of assessing the efficiency of natural enemies. *Can. Entomol.* **101**, 425–447.

Huffaker, C. B., Messenger, P. S., and DeBach, P. (1971). The natural enemy component in natural control and the theory of biological control. *In* "Biological Control" (C. B. Huffaker, ed.), pp. 16–67. Plenum, New York.

Hussey, N. W., and Bravenboer, L. (1971). Control of pests in glasshouse culture by the introduction of natural enemies. *In* "Biological Control" (C. B. Huffaker, ed.), pp. 195–216. Plenum, New York.

International Biological Program. (1971). *Rep. 4th Meet. UM/IBP Committee Biol. Control, Canberra (Australia), August 23–25, 1971.*

Lorbeer, H. B. (1971). *49th Annu. Rep. Fillmore Citrus Protective District (Calif.), Jan. 1 to Dec. 31, 1970.*

Munroe, E. G. (1971). Status and potential of biological control in Canada. *Commonw. Inst. Biol. Control Tech. Commun.* **4**, 213–255.

Parker, F. D. (1971). Management of pest populations by manipulating densities of both hosts and parasites through periodic releases. *In* "Biological Control" (C. B. Huffaker, ed.), pp. 365–376. Plenum, New York.

Ridley, H. N. (1930). "The Dispersal of Plants throughout the World," 744 pp. L. Reeve and Co., Ashford, England.

Simmonds, F. J. (1968). Economics of biological control. *PANS* **14**, 207–215.

Smith, H. S. (1919). On some phases of insect control by the biological method. *J. Econ. Entomol.* **12**, 288–292.

Smith, H. S. (1941). Racial segregation in insect populations and its significance in applied entomology. *J. Econ. Entomol.* **34**, 1–13.

Smith, H. S. (1944). Julian Huxley on evolution. *Ecology* **25**, 477–479.

Taylor, T. H. C. (1955). Biological control of insect pests. *Ann. Appl. Biol.* **42**, 190–196.

van den Bosch, R., Leigh, T. F., Falcon, L. A., Stern, V. M., Gonzales, D. and Hagen, K. S. (1971). The developing program of integrated control of cotton pests in California. *In* "Biological Control" (C. B. Huffaker, ed.), pp. 377–394. Plenum, New York.

Wigglesworth, W. B. (1945). DDT and the balance of nature. *Atl. Mon.* **176**, 107.

Wilson, F. (1964). The biological control of weeds. *Annu. Rev. Entomol.* **9**, 225–244.

Wilson, F. (1971). Biotic agents of pest control as an important natural resource. *4th Gooding Memorial Lect. Cent. Ass. Bee-Keepers.* 12 pp.

Wood, D. L., Silverstein, R. M., and Nakajima, M. (1970). "Control of Insect Behavior by Natural Products," 345 pp. Academic Press, New York.

2

HISTORY OF BIOLOGICAL CONTROL

F. J. Simmonds, J. M. Franz, and R. I. Sailer

I. INTRODUCTION

This chapter is concerned primarily with the history and foundations of classical biological control which involves planned relocation of natural enemies of insect pests and weeds from one locality to another. Normally the method has been used to combat pests that have invaded agroecosystems geographically or ecologically isolated from the ecosystems in which the invading organisms evolved. For the most part, the insect pests against which natural enemies have been moved from one geographical area to another are species inadvertently carried in commerce, while weeds have often been purposely introduced to new areas as ornamental plants or for other reasons.

When such immigrant organisms encounter a climatically and edaphically favorable environment in which there is an abundance of food and space, their populations commonly increase to the carrying capacity of the available food or space. Such explosive increases in the abundance of an invading organism are usually the result of escape from regulation imposed in native ecosystems by coevolving organisms highly adapted to utilize the invader as a food resource. Obviously such specialized natural enemies cannot colonize a new area until their host attains a critical population level. By breaching geographical and ecological barriers, commerce has served as a selective filter through which insect pests and weeds have passed more readily than their enemies. The objective of "classical" biological control is, therefore, that of finding the most effective enemy species and colonizing them in the invaded areas.

In addition to introduction of natural enemies from one area to another, the biological control concept includes other methods of manipulating parasites, predators, and pathogens for suppression of pests. These methods include measures taken to augment natural enemy action through inoculative or inundative releases. The recent trend in pest control practices away from pesticides toward integrated management in great part involves changes in agricultural and silvicultural practices designed to maximize the effect of such enemy organisms.

In recent years the biological control concept has been enlarged by some to include research on or operational use of the autocidal approach utilizing sterile males or genetic manipulation designed to introduce lethal or deleterious alleles into pest populations. The use of key physical or chemical stimuli which release intensive and rather specific reactions in pest organisms, summarized as "biotechnical control," has proved to be compatible with natural enemy action. Stimuli such as flashes, sound, pheromones, hormones, attractants, or deterrents cause disturbances of normal behavior and can be detrimental to target populations. Autocidal and biotechnical control might be considered as specific subsidiary areas for a broader field of biological control and are covered in other chapters within this treatise.

One may divide the history of biological control into three sections: (1) preliminary efforts, representing really early manifestations of biological control when living agents were released rather haphazardly, with little or no scientific approach, and while to a certain degree the results are known, there is certainly a lack of precise information; (2) the intermediate period of more discriminating biological control which might, possibly, be said to have started in 1888 when, with the introduction of *Rodolia cardinalis* Mulsant from Australia into California against the cottony-cushion scale, *Icerya purchasi* Maskell, the first really planned, and successful, biological control attempt took place; and (3) the modern period, which we think could be taken to cover the past 20 years. This period is characterized by more careful planning and more precise evaluation of results as well as by the development of various methods, other than the release of natural enemies, which are now commonly classed as biological control. Overall, there are various facets of the subject involving, for example, insect pathogens, strategic and inundative releases, and relationships with integrated control, which together with development differences in different geographical regions combine to make the history of biological control multidimensional.

II. EARLY HISTORY

As with all good histories we may start at the nearest known approach to the beginning, where we rely somewhat on hearsay. It is said that from early days—a method of intimating that the exact time is unknown—the Chinese used Pharaoh's ant, *Monomorium pharaonis* (L.), by introducing nests into barns to combat stored product insects. Perhaps this is the first example of an inundative or inoculative release of a native species for biological control. Whether such introductions were successful seems not to have been recorded. Bearing in mind the state of entomological research (the word was nonexistent at that time) and that of taxonomy, it is not surprising that details are lacking. Much of this historical matter has, naturally, been given in other textbooks on biological control, but a few other examples less readily available in the literature are added here.

Before 1859 the giant toad, *Bufo marinus* (L.), was introduced from Cayenne in northern South America into Martinique in an attempt to control pests, particularly white grubs in sugarcane. From there it was introduced into Barbados, Jamaica, and, subsequently, into other areas.

Rats were a very serious pest of sugarcane in Jamaica—a virtual plague in 1764—and in 1789 it was estimated that 25% of the crop was lost to this pest. In 1750, in an attempt at biological control, Sir Charles Price imported unsuccessfully an unknown carnivore from South America, after ferrets brought from England had been incapacitated by Chigoe fleas. In 1762 Thomas Raffles introduced into Jamaica an ant, "*Formica omnivora*," from Cuba to control "insects and vermin."

In 1872, the mongoose *Mungos birmanicus* (Thom.) was introduced into Trinidad against rats; four pairs being imported from Calcutta. In the next 20 years this species was introduced into other West Indian islands and in 10 years was said to have saved sugar losses of £45,000—an enormous sum in those days. Now the mongoose is considered to be a pest, killing chickens and decimating ground lizards as well as ground-nesting birds. While it is often cited as an example illustrating the consequence of well-intentioned but ill-conceived biological control work, the mongoose remains a highly effective enemy of rats. Although it is true that it also destroys chickens, lizards, and ground birds, what are the consequences of an unchecked rat population?

In Swan Island in the Caribbean, accidentally introduced rats are believed to have exterminated a species of lizard and a thrush. The biota of many other oceanic islands have been decimated by rats. In New Zealand, cats and dogs caused an alarming reduction in the number of kiwis (Myers, 1931). Also, rats as a potential reservoir of plague constitute a threat to human populations. Hence, it might be argued that once the rat gained entry to these insular ecosystems, the mongoose has tended to restore a balance more favorable to the other animals and, at worst, is the lesser of two evils.

The lizard, *Anolis grahami*, was introduced into Bermuda about 1900 to control mosquitoes.

Sweetman (1958), DeBach (1964), Ordish (1967), van den Bosch (1971), and Hagen and Franz (1973) give accounts of many other introductions attempted prior to 1888 and of their subsequent history. We will not repeat these accounts here, but many are summarized in Table 1.

A common characteristic of these early examples of biological control is the haphazard way in which various introductions were made. That many were unsuccessful and that

TABLE 1

Pertinent Dates in the Early History of Biological Control

Year	Natural enemy	Pest species	Significant feature	Reference
1200	*Oecophylla smaragdina*	*Tessarotoma papillosa*	Nests put in citrus and litchi trees by farmers in China	Liu (1939)
1200	Ants	Date palm pests	Nests brought from hills to Yemen and put in trees	Forskal (1775)
1200	Ladybird beetles	Aphids and scales	Usefulness at least recognized	Kirby and Spence (1815, 1856)
1602	*Apanteles glomeratus*	*Pieris rapae*	Nature of parasitism recognized	van Leuwenhoek (1701); Aldrovandi (in Silvestri, 1909)
1718	"Ichneumonids"	"Caterpillars"	Recorded from caterpillars	Bradley (1718)
1734	"Aphidivorous fly"	Aphids	Suggested to collect eggs and put them in greenhouses to contain aphids	de Réaumur (1734)
1762	Mynah, *Acridotheres tristis*	Red locust	Bird successfully introduced from India to Mauritius	Moutia and Mamet (1946)
1763	*Calosoma sycophanta*	"Caterpillars"	Suggested for use, but not used	Linnaeus (1756)
1776	*Reduvius personatus; Picromeris bidens*	*Cimex lectularius*	Used as predators against bedbugs	Kirby and Spence (1815); Clausen (1940)
1800	"Ichneumonids"	"Cabbage caterpillars"	Discussed as natural control factor	Darwin (1800)
1837	—	—	Concept of natural control	Kollar (1837)
1840	*Calosoma sycophanta; Staphylinus olens*	*Porthetria dispar; Forficula auricularia*	Used in village by Boisgiraud to control gypsy moth larvae and earwigs	Joly (1842)
1844	Staphylinids and carabids	Garden pests	Villa used these experimentally in Milan to control pests	Villa (1845)

1856	Parasites	*Contarinia tritici*	Suggested bringing parasites into United States from Europe	Fitch (1856)
1860	Parasites	*Contarinia tritici*	Parasites requested, but nothing developed	Curtis (1860)
1870	Parasites	Weevils	Parasites moved from one area of Missouri to others	Riley (1893)
1870	*Aphytis mytilaspidis*	"Scales"	Moved apple branches with parasitized scales from one orchard to others (Geneva, Illinois)	LeBaron (1870)
1873	*Tyroglyphus phylloxerae*	*Phylloxera vitifoliae*	Mite sent from United States to France. Established, but of no value	Riley (1893)
1874	*Coccinella undecimpunctata*	Aphids	Sent from United Kingdom to New Zealand. No results	
1880	*Stegodyphus mimosarum*	Houseflies	South African social spider used by Zulus, etc., to control flies and Vortrekkers in Transvaal and later to control flies in dairies, etc.	Steyn (1959)
1882	*Trichogramma* sp.	*Nematus ribesii*	Saunders shipped the parasite from United States to Canada	Baird (1956)
1883	*Apanteles glomeratus*	*Pieris rapae*	Parasite from United Kingdom to United States. Successful	Riley (1893)
1888	*Rodolia cardinalis*	*Icerya purchasi*	Shipped from Australia to California. Spectacular success. Ushered in modern era of introductions	Numerous

others produced definite harmful side effects could have been predicted had present knowledge been available. Even so, the fact that the predators mostly concerned in these early efforts normally attacked a wide range of prey species must have been known to the people handling the introductions and they might have foreseen that some undesirable effects would result. However, it is evident that they concerned themselves only with the immediate economic problem, an attitude reminiscent of a more modern era when chemical toxicants appeared to be the panacea of pest control.

III. INTENSIVE PERIOD

Although it will be seen that there were several successful attempts at biological control prior to 1888, notably the introduction of *Apanteles glomeratus* (L.) from England into the United States, and that attention to biological control had increased considerably in the previous 50 years, it was the spectacular success of *Rodolia cardinalis* that initiated development of biological control introductions as subsequently practiced. This example has often been related in some detail (e.g., DeBach, 1964; van den Bosch and Messenger, 1973). Briefly, the cottony-cushion scale *Icerya purchasi* became established in the Los Angeles area some time prior to 1876 and by 1885 was threatening to destroy the young citrus industry of southern California. The pest had actually been first observed in 1868 at Menlo Park, 400 miles north of Los Angeles. In 1872, C. V. Riley, then State Entomologist for Missouri, after examining specimens sent to him from California, suggested that the new pest may have come from Australia. Soon after becoming Chief Entomologist for the United States Department of Agriculture (USDA) in 1878, he again interested himself in the cottony-cushion scale problem. However, it was not until 1888 that through devious means and with the moral support of the California State Board of Horticulture, he was able to send Albert Koebele to Australia.

Soon after his arrival, Koebele found two enemies attacking cottony-cushion scale on citrus. One was a dipterous parasite, *Cryptochaetum iceryae* (Williston), and the other a coccinellid predator, *Rodolia cardinalis*, now commonly known as the vedalia beetle. Stock of both species were shipped to San Francisco where they were examined, reared, and then sent to Los Angeles. There they were released on scale-infested trees enclosed in canvas tents. The vedalia beetles increased rapidly and were then allowed to spread into adjacent trees. With subsequent help from citrus grove owners and the Los Angeles County Board of Horticultural Commissioners the beetles soon spread through the county. Within 1 year, shipments of oranges increased from 700 to 2000 cars. The California citrus industry was saved and $500 of the original $2000 allocated for Koebele's trip to Australia remained unspent (Doutt, 1958).

The spectacular success of this venture and its successful extension to other parts of the world where the cottony-cushion scale was also a serious economic problem, coupled with its permanency, simplicity, and low cost generated enthusiastic support for biological control as a solution for other agricultural pest problems. Many successful introductions were undertaken against pests in different areas of the world. Examples of the more important projects were against the white grub, *Anomala orientalis* Waterh., and the sugarcane weevil, *Rhabdoscelus obscurus* (Boisduval), in Hawaii; the rhinoceros beetle, *Oryctes rhinoceros* (L.), in Mauritius, and the gypsy moth, *Porthetria dispar* (L.), the brown-tail moth, *Nygmia phaeorrhoea* (Don.), and the alfalfa weevil, *Hypera postica* (Gyll.), in the United States. In Kansas there was a campaign in 1907 to increase parasitism of the greenbug, *Schizaphis graminum* Rondani, by collecting parasitized aphids in the south of the state and releasing large numbers in the north (Hunter, 1909). These and many other efforts to exploit natural enemies are listed by DeBach (1964), indicating that over these years there was a steadily increasing number of introduction efforts, with 28 recorded in 1920–1930 and 57 in 1930–1940.

During this period the USDA initially had a minor role in biological control research. In fact, L. O. Howard, who became Chief of the Division of Entomology in 1894, was most critical of the biological control work as practiced at that time in California and elsewhere. However, in 1905 the gypsy moth was causing a severe problem in Massachussetts. The USDA agreed, using funds from Massachussetts, to obtain parasites from Europe, where the gypsy moth had originated. W. F. Fiske was in charge of the operations in Massachussetts while Howard traveled extensively in Europe establishing contacts and facilitating the later work of others who were sent to collect parasites of both the gypsy and brown-tail moths. Both H. S. Smith and W. R. Thompson, who were to gain international recognition as leaders in biological control, gained their early experience in this project. The latter became the first Director of the Commonwealth Institute of Biological Control (CIBC). World War I interrupted the work in Europe but it was renewed from 1922 to 1933. Overall, the parasites successfully introduced into the United States did much to reduce the general populations of gypsy moth. However, in the 1950's its range began to increase alarmingly.

Over the years the United States has, like many other countries, carried out biological control programs on a number of important pests, with a portion of them resulting in outstanding successes. Much of this work has been done in close collaboration of the USDA with various State and University entomologists, and also with Canada since, of course, many of the pest problems are of mutual interest to the two countries.

During the early period of parasite introduction efforts the search for and collection of material for shipment to other countries was carried out either with the help of cooperators in foreign countries or by entomologists sent on temporary assignments to explore for entomophagous insects. Even at this time in California, little care was taken to discriminate between useful primary parasites and harmful secondaries. That only one of the latter, *Quaylea whittieri* (Gir.), is known to have been introduced during this period was indeed fortunate.

Also, the hazards of allowing untrained or careless people to move natural enemies from one place to another became apparent when the cottony-cushion scale invaded

Florida as a result of a misguided attempt to use the vedalia beetle for control of other kinds of scale insects. Aware of the successful use of the beetle against the cottony-cushion scale, a Florida citrus grower obtained stock from a California grower. The beetles arrived accompanied by a food supply of cottony-cushion scale and were released on a tree. The box in which they were received was discarded nearby. The beetles, unable to feed on the unaccustomed host, failed to survive but the inoculum of cottony-cushion scale precipitated one of the first of many unsuccessful attempts to eradicate an invading insect pest (Berger, 1915). Control was eventually obtained through a later introduction of the vedalia beetle.

It became apparent that training and experience were essential in conducting successful biological control work. A well-staffed organization was developed by the State of California over a period from 1913 when H. S. Smith took charge of its biological control activities. In 1923 the operation was moved to the University of California's Citrus Experiment Station at Riverside, with Smith still in charge and with steadily improving facilities. In 1945 a second subunit was established, headed by J. K. Holloway at Albany, where the first attempt at biological control of a weed on the North American continent was initiated. Subsequently, a separate insect pathology laboratory was developed by E. A. Steinhaus in Berkeley, which became the leading center for research in insect diseases.

Not only were there many programs of research into biological control of specific pests, a number of them successful (see following chapters and Appendix), together with development of mass culture techniques for both hosts and parasites, but studies were carried out on the underlying principles of biological control. Important contributions were made on host/parasite relationships and what is now termed population dynamics with the establishment of the concepts of density-dependent, inverse density-dependent, and density-independent factors (Smith, 1935).

An important step forward in the development of biological control was the establishment in 1919 by the USDA of a laboratory in France. First located at Auch for research on parasites of the European corn-borer, *Ostrinia nubilalis* (Hubn.), it was soon moved to Hyères and since 1934 has been known as the European Parasite Laboratory. In 1936 it was moved to a suburb of Paris where it has continued to operate except for a 7-year period during World War II. This was followed in 1927 by the establishment of the Farnham House Laboratory (see below) to work on problems of British Commonwealth interest, both in Europe and the West Indies.

Although biological control work declined, not unnaturally, to a comparatively low level during World War II, some projects were continued even then. For example, some were conducted by the CIBC. Following the war this field was actively developed in a number of countries including: the United States (California and Hawaii, in particular), Canada, Mauritius, and Bermuda, to name a few (where the overseas work was carried out by CIBC), and in Australia and South Africa. A substantial number of complete and partial successes were obtained during this period (see Appendix of Successful Examples).

During the war, DDT had been used very extensively for control of mosquitoes and other insect vectors of diseases. Because of the ease of application and degree of control achieved through its use, and that of other new insecticides which were being developed, biological control methods were thought to be obsolete and in some quarters held in disrepute. However, within less than 10 years, predicted by H. S. Smith in 1941 (Smith, 1941), economic entomologists were finding themselves confronted by a new class of problems resulting from excessive reliance on insecticides. Pests became resistant to DDT and successively to other, usually increasingly toxic, chemicals. Previously innocuous species became major pests. Pest populations decimated by chemical applications commonly rebounded with increasing violence. Residue problems became a major concern and chemists were required to perfect more sensitive analytical methods. Now, in the search for means to reduce dependence on chemicals for pest control, interest in biological control has revived and prospects are good that it will play an increasingly important role in the pest management strategies currently being developed.

At this time greater interest began to be shown in the importance of ecological considerations in biological control.

C. P. Clausen, previously head of the USDA's foreign parasite introductions, succeeded H. S. Smith at California from 1951 to 1959 when he retired. In this period, increasing attention was paid to natural enemies of scale insects, aphids, spider mites, and weeds. It was also at this time that the concept of integrated control was taking on a formal entity through the work of the two biological control groups at Riverside and Berkeley. Also involved were A. E. Michelbacher and co-workers and A. D. Pickett and his colleagues in Nova Scotia.

Canada has engaged in biological control work for a long time, principally against pests introduced from Europe, more particularly forest insects. In 1882, W. Saunders imported and released a *Trichogramma* species against the European currant sawfly, *Nematus ribesii* (Scop.). In 1915, a small biological control laboratory was built at Fredericton. In 1923 the operations were moved to Ontario and in 1929 to Belleville, with the establishment of the Dominion Parasite Laboratory. Modern facilities and quarantine rooms were provided in a new building in 1936. Cooperation developed with similar projects in the United States. Later this Canadian laboratory was again expanded. It was afforded greater facilities in 1956, being renamed the Entomology Research Institute. In later years, basic research in various aspects of physiology, behavior, and population ecology of entomophagous insects was emphasized. In 1972, this Institute was disbanded and its staff dispersed to other laboratories in Canada where biological control in interdisciplinary, commodity-oriented research is being continued.

Actual research in biological control was rather late in developing in Europe compared with that in some other areas, principally because of the comparatively few introduced pest species which warranted attention. However, there is a notable early exception. The grape phylloxera, *Phylloxera vitifoliae* Fitch, of American origin, entered France prior to 1867. Although the damage to vines was first thought to be due to a disease, its causative organism and the fact that both leaf and root forms were involved was soon elucidated by cooperation between French and American entomologists. By 1884 over a million hectares of vines had been destroyed by the pest. The damage, together with the cost of importing wine and dried grapes for making it, produced a staggering total economic impact of *Phylloxera* of some 2 billion dollars (and this in pre-World War I values). The pest spread to most of the European grape-producing areas and was found in Australia in 1875.

Exchange of experts between France and America took place from 1870. It was found that American vines that had been planted in 1862 in France were immune to attack. Experiments were then carried out grafting European varieties onto American rootstocks, more of which were imported from the United States. In 1873 C. V. Riley sent a predatory mite, *Tyroglyphus phylloxerae* Riley, from the United States to France. It became established but did not appreciably affect *Phylloxera* populations. However, with the general use of American vine rootstocks the problem of *Phylloxera* in Europe was virtually eliminated. However, many repercussions resulted. Riley's involvement stimulated his interest in the importation of natural enemies to combat introduced foreign pests. This may well have been a factor in the search for and importation of *Rodolia* against *Icerya* in California in 1888, which in turn, as discussed above, sparked a worldwide interest in biological control.

In Europe, the emergence of ideas fundamental to the practical use of natural enemies in pest control was more important in the early periods than the execution of projects of the classical type (Hagen and Franz, 1973). However, three introductions of entomophagous insects that originated in Europe are of special historical interest. One is the colonization in Italy and other Mediterranean countries of *Prospaltella berlesei* (How.) from North America which successfully suppressed the white peach scale, *Pseudaulacaspis pentagona* (Targ.-Tozzetti). Second is the discovery and mass production of the eulophid *Encarsia formosa* Gahan, the parasite of the greenhouse whitefly, *Trialeurodes vaporariorum* (Westwood), in England, from which country it has subsequently been distributed to all other continents. Third, importations have been made of various races of *Prospaltella perniciosi* Tow. from several faunal regions to reduce damage caused by the San Jose scale, accidentally introduced into Europe after World War II. The cooperation of European countries in this successful project under the guidance of W. Klett in Germany was one of the focal points for the early functioning of the Organisation Internationale de Lutte Biologique (OILB).

The main contribution, though, of European researchers to the development of biological control was the utilization of native arthropods and vertebrates. This began with the artificial concentration of carabids and staphylinids by Boisgiraud around 1840 (Joly, 1842). Important additional steps were the breeding of parasitized pest insects in cages allowing only the parasites and not the hosts to escape (France, Germany) and the mass production of *Trichogramma* species, originating in England and Russia (Hagen and Franz, 1973). This tradition continues and the practical utilization of *Trichogramma* against Lepidoptera in orchards and field crops has developed to huge operations, particularly in the U.S.S.R. and China. Finally, the method of artificial colonization of forest ants of the *Formica rufa* group, as well as of insectivorous birds by providing nesting facilities, is noteworthy because these activities are specific elements in the European pattern of biological control (Franz, 1961b). At present, modern laboratories for the various aspects of biocontrol exist in most countries of Europe, indicating the general interest in the development of pesticide-free methods of pest control.

Australia has played an important role in the development of biological control, beginning in 1903 when a *Dactylopius* sp. was introduced from India in an attempt to control *Opuntia* cacti, of South American origin, which eventually rendered useless some 60,000,000 acres of land in Queensland and northern New South Wales. Additional natural enemies of *Opuntia* spp. were introduced in 1913, but it was in 1921–1935 that the most intensive and successful work on this problem was carried out. While about 50 insect species were imported into Australia for trial it was the phycitid moth, *Cactoblastis cactorum* (Berg.), imported from South America in 1925, that was effective in destroying the two most important species of *Opuntia*. By the late 1920's they were no longer a serious problem in most areas. It may be noted here that as with a number of problems which may well be amenable to "classical" biological control, there was a long time lag between the initiation of the idea and the really intensive investigations which led to a successful conclusion.

Since then Australia has placed strong emphasis on biological control, not only in its purely practical application, but also on research into the principles underlying biological control and its ecological implications. Noteworthy are the contributions of Nicholson (1933) on the concept of population regulation and mathematical theories relating to the action of parasites on their hosts. These subjects were also discussed in depth by W. R. Thompson some years before (see Thompson, 1930b) and became a topic that engendered violent controversies and even personal animosities (see Wilson, 1960, 1963).

It is perhaps rather invidious to describe in some detail the gradual progress in biological control in a few countries while failing to describe that which took place in others. In New Zealand, South Africa, Fiji, and Mauritius there has been considerable emphasis on biological control for quite some time, and with other examples certainly of historical interest in, for example, Ceylon, Indonesia, Samoa, and the Seychelles. It is simply impossible in a limited space to describe developments in all the different areas where biological control has been conducted. However, we have tried to highlight some of the main centers where biological control has been actively pursued.

No history of biological control would be complete without a specific reference to the work done in Hawaii, where possibly more attention has been given to it and more successes achieved than anywhere else in the world (see DeBach, 1964). This should occasion no great surprise since all the major crop plants and their pests are introduced ones, the native flora and fauna were comparatively poor, and hence these introduced pests had virtually no natural enemies (Timberlake, 1927). Following the success of

Rodolia against *Icerya* in California, immediate steps were taken in Hawaii to initiate biological control. In 1893, Albert Koebele was sent to Australia and the Orient to collect material for Hawaii. Of the material he sent between 1894—1896, six coccinellids and two hymenopterous parasites were established.

In 1900 the sugarcane leafhopper, *Perkinsiella saccharicida* Kirkaldy, of Australian origin, was found attacking sugarcane and by 1904 losses were such as to threaten the entire industry. The Division of Entomology of the Hawaiian Sugar Planters Experiment Station, with R. E. L. Perkins in charge, was formed and as a result of their importations of natural enemies the leafhopper was completely controlled. Research in Hawaii contributed notably to the practical development of biological control as a whole, including that of weeds, in which Hawaii was a pioneer, particularly with respect to *Lantana camara* L. and later *Eupatorium adenophorum* Spreng. In fact, the history of entomology in Hawaii has been largely that of biological control (Flanders, 1955; Howard, 1930; Pemberton, 1953, 1964).

In viewing the origins and the development of concepts and principles that serve to guide practical applications of biological control the contributions of scientists outside the field must be considered. Biological control is, of course, no more than a sector of applied ecology. Therefore, it might be expected that many of the basic principles of biological control would have derived from the research of ecologists or ecological research of scientiest in fields quite removed from pest control. To an important degree this is true. Scientists of many universities have made fundamental contributions applicable to biological control particularly in the fields of behavior, nutrition, genetics, and population dynamics. Others engaged in applied research relating to forestry and agriculture have also contributed. Not the least have been the taxonomists who provide the basic inventory tool and key to communication of knowledge about organisms. In this sense the history of biological control, ecology, and in fact that of biology, is a continuum.

The operational level of classical biological control, involving the movement of natural enemies differs markedly from most other fields of basic and applied science. While science in general is international in the sense of communicating knowledge and cooperation between individuals and institutions, the success of many biological control programs is dependent on research that, of necessity, must be conducted internationally. Success is also dependent on activities that are sequential in nature, with the phases in different countries carefully coordinated. Thus organization is an essential element of biological control. While institutional organizations such as the University of California's Divisions of Biological Control and the United States Department of Agriculture's now sadly defunct Insect Identification and Parasite Introduction Research Branch have functioned internationally, the much broader need for international cooperation and coordination has resulted in the development of two international organizations devoted to advancement of biological control. One of these is the Commonwealth Institute of Biological Control, which maintains staffed laboratories in many countries. The second is the International Organization for Biological

Control/Organisation Internationale de Lutte Biologique (IOBC/OILB), which differs from the first in that it is an association of individual and institutional members. The history, composition, and objectives of these organizations are discussed in Section V.

IV. MICROBIAL CONTROL

The term "microbial control" refers to that type of biological control involving the utilization of microorganisms — including viruses — for the control of pests (Steinhaus, 1949). The idea has long been held, but its more fruitful approach began following an accumulation of basic knowledge of the various pathogens of potential use.

A. Microbial Control of Arthropods

That domesticated insects, bees, and silkworms suffer from disease was well recognized by naturalists in the Middle Ages. The recorded history of insect pathology, in fact, begins with Aristotle's *Historia Animalium* in which he described certain diseases of honeybees (335–322 B.C.). Still older are reports of muscardined silkworms which were used for treatment of human paralysis as early as 900 B.C. The emergence of the idea of microbial control of insects has been traced in detail by Steinhaus (1956). Other historical sketches are also available (e.g., Sachtleben, 1941; Franz, 1961a; Steinhaus, 1963; Cameron, 1973; see also Chapter 7); a much condensed treatment will suffice here.

Subsequent to the Middle Ages, de Réaumur (1726) described *Cordyceps* on a noctuid larva. However, the beginning of a scientific insect pathology is seen in the chapter, "Diseases of Insects," in the famous book by Kirby and Spence (1826), "An Introduction to Entomology." Steinhaus (1956) concluded that the idea of using microorganisms to control pest insects had its roots in studies of the diseases of the silkworm [*Bombyx mori* (L.)]. Experimental work began with Agostino Bassi (1835) who first showed that a microorganism, the fungus *Beauveria bassiana* (Balsamo) Vuillemin, caused an infectious disease in an insect (the silkworm). Bassi suggested the use of microorganisms to kill harmful insects. Audouin (1839) transmitted the muscardine fungus to several species of pest insects. He also reported that a sericulturist emptied fungus-contaminated trays used in rearing silkworms out of a window onto trees infested with pest insects, some of which died of muscardine 4 days later (Steinhaus, 1956). This is apparently the first reported instance of pest insects in the field being killed through the artificial dissemination of a microbial pathogen. Although Louis Pasteur conducted his famous experiments to protect insect life (that of the silkworm) from disease by selection of offspring from healthy parents, his work greatly enhanced our insight into the nature of infectious diseases (Pasteur, 1870), which has paved the way for the use of pathogens to control pest insects.

The American entomologist J. L. Leconte (1873, publ. 1874) was the first definitely to recommend Bassi's visionary suggestion to study and use pathogenic microorganisms to combat pest insects. Finally, in 1879, Elie Metchnikoff carried out the first convincing experimental tests using the fungus *Metarrhizium anisopliae* (Metch.) Sorokin to control larvae of the wheat grain beetle, *Anisoplia austriaca* Herbst. He advocated mass production of entomopathogenic fungi on artificial media for use in the field and succeeded in artificial propagation of the green muscardine fungus (Metchnikoff, 1880). Krassilstschik (1888) followed with apparently successful experiments using *M. anisopliae* to control the sugar beet curculio, *Cleonus punctiventris* Germ., near Kiev, U.S.S.R. Almost simultaneously, several American workers (e.g., Snow, 1891) produced and distributed *Beauveria globulifera* (syn. *B. bassiana*) to control the chinch bug, *Blissus leucopterus* (Say). Although Billings and Glenn (1911) showed that this work had been without immediate practical results, it provided an important stimulus to the development of insect pathology and microbial control. Fungi still dominated the field at the beginning of the 20th century when H. S. Fawcett, E. W. Berger, J. R. Watson, and others at the Florida Experiment Station (1908 and later) studied their role in the control of scales and other citrus insects (Steinhaus, 1956). (See further Chapter 7.)

Bacteria were used in practical tests for microbial control on two early occasions. D'Hérelle gave optimistic reports between 1911 and 1914 on the use and effectiveness of a bacterium which he called *Coccobacillus acridiorum* against locusts (*Schistocerca paranensis* Burm.) (e.g., d'Hérelle, 1911–1914). He increased the virulence of the strain by repeated passages and observed transmission and epizootic effects. However, neither his own experiments nor those of others such as Beguet, Musso, and Sergent in Algeria (reviewed extensively in Sachtleben, 1941) against *Schistocerca gregaria* Forsk. and other locusts could be repeated successfully. Apparently, fundamental knowledge of the identification, character, and handling and storage requirements of bacteria was not yet far enough advanced. The second instance of the use of bacteria was more successful. Berliner (1911) described the causative agent of a bacterial disease of the Mediterranean flourmoth, *Anagasta kuehniella* (Zeller), as *Bacillus thuringiensis*. Practical tests were initiated, first in Hungary, for the microbial control of the European corn borer (Husz, 1928, 1929). Further tests followed against the pink bollworm [*Pectinophora gossypiella* (Saunders)], cabbageworms (*Pieris* spp.), and other lepidopterous larvae in Europe (see Krieg, 1961). The first commercial product of *Bacillus thuringiensis*, named Sporeine, was produced in France before 1938 (Jacobs, 1951). Experiments with *B. thuringiensis* were resumed after World War II almost simultaneously in Europe against the fall webworm (Klement, 1953) and in America against the alfalfa caterpillar (*Colias philodice eurytheme* Boisd.) (Steinhaus, 1951). The impact of the work of Steinhaus and co-workers and some commercial firms in the United States led to a worldwide development of this microbial material (see further Chapter 7). In between the earlier work on *B. thuringiensis* and its later development was the highly successful, permanent type of control of the Japanese beetle (*Popillia japonica* Newm.) afforded by spread of the milky disease bacterium in the United States beginning in 1940. This constituted a powerful stimulus to the developments in this field by Steinhaus and others throughout the world.

Virus diseases of insects had also been known for centuries, again particularly that of the silkworm under the name of "jaundice." Early descriptions of the symptomatology and pathology were given, e.g., by Cornalia (1856) and Maestri (1856). Bolle (1898) correctly considered the polyhedral bodies observed to contain the causative agents of the disease and found that the polyhedra were soluble in weak alkali. The viral nature of the pathogen was concluded but not shown by von Prowazek (1907).

In these first decades, control experiments against several forest Lepidoptera such as *Lymantria monacha* L. and *Porthetria dispar* (L.) were carried out in Europe and in North America. Macerated, virus-killed larvae, or contaminated forest litter were used, but these trials produced no practical results (reviewed by Sachtleben, 1941). Only when the nuclear polyhedrosis of the European spruce sawfly [*Diprion hercyniae* (Htg.)] was introduced into Canada around 1943, probably along with parasites in cocoons shipped from Europe, was mass production and modern use of virus application initiated against this forest pest (Balch and Bird, 1944). The first success using a virus in agriculture was directed against the alfalfa caterpillar (Steinhaus and Thompson, 1949). The history of virus applications provides another example of the necessity of having sound fundamental knowledge in order to develop field programs. Reciprocally, it shows that a favorable result in practical control is able to stimulate public interest necessary for the further development of such a field. Examples of this are the creation of the first laboratories of insect pathology, at the University of California, Berkeley, in 1945, and at Sault Ste. Marie, Ontario, Canada, in 1946, following the unrivaled successful microbial control campaign using the milky disease pathogen against the Japanese beetle in northeastern United States (Dutky, 1959; Hall, 1964; Burges and Hussey, 1971) and the successful results with the inadvertently introduced virus disease of the European spruce sawfly in Canada (above). (See further this volume, Chapters 7 and 12.)

B. Microbial Control of Mammals

The first experiments using bacteria against mammals were carried out under the supervision of Louis Pasteur in France in 1887 and 1888. The fowl cholera bacillus, *Pasteurella multocida* (L. & N.), proved capable of killing wild rabbits when contaminated cut alfalfa was eaten. Pasteur then sent an assistant to Australia to organize an anti-rabbit campaign there. However, no project was initiated because the necessary permission was not granted by the Australian Department of Agriculture. Shortly afterward, Loeffler isolated a bacillus from laboratory mice, which was described as *Bacillus* (now *Salmonella*) *typhimurium* (Loeffler, 1892a). It proved to be highly infectious to mice (*Mus*) and some voles (*Microtus*) and was tested in large-scale field experiments in Thessalia (Greece) (Loeffler, 1892b). The success achieved led to a wide acceptance of this method, particularly in the control of *Microtus arvalis* Pall. and

Rattus norvegicus Berkenh. Other slightly different *Salmonella* species were isolated, for instance, by Danysz (1893), and widely used in Europe (review Regnier and Pussard, 1926). However, because of their unfortunate close relationship to the bacilli causing paratyphus in man, and certain accidents that occurred, some countries prohibited or greatly restricted the use of bacteria in pest control (e.g., in Germany and the United States); other countries did continue their use (e.g., France, Denmark, Poland, and, particularly, the U.S.S.R.).

The recent use of the myxomatosis virus against the rabbit, *Oryctolagus cuniculus*, caused more general interest than the above-mentioned use of bacteria because it resulted in a high degree of control and some interesting fluctuating epizootic consequences (see this volume, Chapter 20). It began with the observation that European rabbits in South America caught a disease called myxomatosis (Sanarelli, 1898). The causative agent was later discovered to be a virus of endemic occurrence in local South American cottontail rabbits (*Sylvilagus*). Usually transmitted by insect vectors, it caused originally a very high mortality in European rabbit populations (Aragão, 1927). Early trials using this virus to control rabbits in Europe, Denmark, England, and Sweden failed, probably because of insufficient vector populations (Lockley, 1955). The final breakthrough was the well-prepared series of field experiments in Australia along the Murray and Darling Rivers in 1951 and 1952 where artificially infected rabbits, when released, triggered an impressive epizootic. The results of this experiment which led to exceptional, but not total, reduction of the rabbit population in Australia (but with some rebound in its numbers later), and later work in England and the European continent as well, are reviewed by Fenner and Ratcliffe (1965) and the more up-to-date Chapter 20 in this treatise. The historic merit of this operation was the successful effort to use a virus disease of low morbidity for its native hosts to control a more susceptible, and in this case an exotic, pest species new to it. This type of microbial control might also prove useful in solving other problems in the future. (See especially Chapter 20.)

C. Microbial Control of Weeds

In addition to the use of insect and mammalian pathogens in biological control, plant pathogens have also been used, although the history of their exploitation has for the most part been quite recent. These developments are given in Table 2, and in Chapter 19.

V. INTERNATIONAL ORGANIZATIONS

A. Commonwealth Institute of Biological Control (CIBC)

This institute was formed in 1927 as the Farnham House Laboratory, a part of the then Imperial (now Commonwealth) Institute of Entomology, to develop practical biological control work. Originally, work was conducted in the United Kingdom, and in

TABLE 2
Plant Pathogens Used in Weed Control

Year		Reference
1893	The dissemination of the rust fungus *Puccinia suaveolens* against Canada thistle (*Cirsium arvense*) in New Jersey is suggested	Halsted, 1894
1927	Success in controlling *Opuntia* in Australia results from extensive study and introduction of a number of insects and associated microbes	Dodd, 1927
1927	Weir describes *Ustulina zonata*, a fungus attacking the dangerous weed tree *Dichrostachys nutans* in Cuba, and discusses its value as a possible biological control agent	Weir, 1927
1946	Leach suggests that the disease of dodder caused by *Colletotrichum destructivum* may be used as a control agent against *Cuscuta*	Leach, 1946
1951	Artificial dissemination of the pathogen *Colletotrichum xanthii* Halst. against *Xanthium spinosum* in New South Wales	Butler, 1951
1954	Unsuccessful attempt to use *Fusarium oxysporum* to control the white form of *Opuntia megacantha* in Hawaii	Fullaway, 1954
1960	Biological control of *Cuscuta* with the fungus *Alternaria cuscutacidae* in the U.S.S.R.	Rudakov, 1960
1963	Discovery of a viral disease of blue-green algae stimulates intensive studies with the view to use the virus (Lpp-1) to control undesirable algal "blooms" in sewage	Saffarman and Morris, 1963
1971	Successful release of *Puccinia chondrillina* for control of the skeleton weed *Chondrilla juncea* in Australia	CSIRO, Div. of Entomol., Annu. Rep. 1971/72

Europe (for the United Kingdom, New Zealand, and Canada), and in the West Indies, particularly on the sugarcane borer, *Diatraea saccharalis* (F.), with a number of notable successes being achieved (Thompson, 1930a; Myers, 1931). This work developed well during the 1930's, the later work taking place only in Europe. With the start of World War II work ceased. However, in 1940 operations were transferred to Canada, under the name Imperial Parasite Service, centered at the Canadian Biological Control Laboratory at Belleville, Ontario. Here considerable work was carried on during the next 10 years on projects for the United Kingdom, South Africa, West Indies, Bermuda, and other Commonwealth countries, with stations in California (1941—1965) and later in Uruguay (1942—1949) and Argentina. The forerunner of the present West Indian Station was started in 1946 with work on what was to prove the very successful biological control in Mauritius of the shrub *Cordia currasavica* (Jacq.) R & S [= *macrostachya* (Jacq.)] of South American origin. Work in Europe was started again on Canadian projects in 1949 with the development of the European Station, first at Feldmeilen near Zurich, later transferred to Delémont near Basle, and now with an Austrian substation at Neulengbach, near Vienna.

In 1956, the Indian and Pakistan Stations were established with the aid of Canadian Colombo Plan funds, now with substations in various parts of these countries.

A Station was organized at Kawanda, near Kampala, Uganda, in 1961 for work in the general East African area. In 1969 work was started in Malaysia with development of the Sabah Substation working successfully on the manipulation of natural enemies of oil palm bagworms. This Substation was closed at the end of 1974. The initiation of work on graminaceous stem borers at the West African Substation at Kumasi, Ghana was begun in 1969.

At the present time at its various stations, CIBC is carrying out work on some 50 projects for over 30 countries, involving in 1974, a total of 615 shipments of 137 species of parasites and predators. Naturally some of these projects are of a long-term nature and entail considerable research into various aspects of the biology and ecology of both pests and natural enemies. Other services involve only shipment of known biotic agents to areas where they are required. In addition, CIBC has, since 1946, published a catalog of parasites and predators, and also publishes Technical Bulletins comprising papers on specific items of research and Technical Communications reviewing biological control in different regions of the world. Since 1958, work on the Institute has been summarized in annual reports and other documents.

B. International Organization for Biological Control/Organisation Internationale de Lutte Biologique (IOBC/OILB)

In 1956, based on the efforts of French and other European entomologists, the Commission (changed in 1965 to Organisation) Internationale de Lutte Biologique was established under the auspices of the International Union of Biological Sciences. By 1967 it had attained membership from 16 countries, mostly in western Europe and the Mediterranean area. Within OILB various commissions dealing with different aspects of biological control were established, working groups for specific pests were created, and the journal *Entomophaga* was initiated. Although definitely international, OILB remained for some years restricted in geographical scope, during which time it was felt that a more truly global organization was needed. This was discussed by interested individuals from many parts of the world in the 1960's, and in 1964 at the time of the International Congress of Entomology in London an International Advisory Committee for Biological Control was established with the view to creating a truly worldwide biological control organization. The advisory Committee's proposals were discussed with representatives of the International Union of Biological Sciences (IUBS), and also publicly at the International Congress of Entomology in Moscow in 1968, as a result of which IUBS became interested in developing a truly worldwide organization to evolve from the OILB into an appropriate form consisting of a central organization with largely independent regional sections covering different biogeographical areas affiliated to it. There followed a meeting of 34 biological control specialists from 23 countries convened by IUBS in Amsterdam in November, 1969. This resulted in formal proposals of an enlarged International Organization for Biological Control, covering all aspects of biological control. The new organization was formally created in March, 1971. At present a Western Palaearctic Regional Section (comprising the old OILB), an Eastern Palaearctic Regional Section, a Western Hemisphere Regional Section, and a South and East Asian Regional Section have already been formed, with others covering Africa and the Pacific areas proposed.

VI. CONCLUSIONS

While the foregoing gives a general history of biological control, its development in different parts of the world has occurred at different times and at different rates of progress. Some areas have been notably successful with biological control in the early attempts and hence have naturally devoted more effort and expenditure to this field and have better documented both the methodology and results. Accounts of such efforts should be consulted; e.g., Australia (Wilson, 1960, 1963); New Zealand (Miller, 1970); Canada (McLeod *et al.*, 1962; Research Institute of Canada Department of Agriculture, 1971); California (DeBach, 1964); Hawaii (Pemberton, 1953); the United States (Clausen, 1956; Dowden, 1962; Sailer, 1973); South Africa (Greathead, 1971); Mauritius (Greathead, 1971); Fiji (Rao *et al.*, 1971).

In other world areas a number of problems have been explored and in many cases parasites or predators found to be effective in another area have been successfully imported. Examples are *Aphelinus mali* (Hald.) introduced into 50 countries for control of the woolly apple aphid, *Eriosoma lanigerum* (Hausm.), and the vedalia beetle, *Rodolia cardinalis*, that has been successfully used in at least 50 citrus-producing countries for control of cottony-cushion scale, *Icerya purchasi*. Such work as has been done in Europe was reviewed by Franz (1961b) and Franz and Krieg (1972), in most of Asia by Rao *et al.* (1971), and in Africa by Greathead (1971). The seeming lack of interest or activity in some world areas may be attributed to the comparative absence of severe introduced pest problems. In other areas it is best explained by the absence of competent or highly motivated personnel or other resources needed to undertake such work. More often than not absence of organizational capabilities rather than resources is the basic factor. Hopefully, as these deficiencies are recognized and corrected the full potential of biological control will become more nearly realized.

REFERENCES

Aragão, H. de Beaurepaire (1927). Myxoma dos coelhas. *Mem. Inst. Oswaldo Cruz* **20**, 225–235.

Audouin, V. (1839). Quelques remarques sur de la muscardine, à l'occasion d'une lettre de M. de Bonafous, faisant connaître les heureux resultats obtenus par M. Podebard, dans la magnanerie de M. le comte de Demidoff. *C. R. Acad. Sci. (Paris)* **8**, 622–625.

Baird, A. B. (1956). Biological control. *In* "Entomology in Canada up to 1956. A Review of Developments and Accomplishments" (R. Glenn, ed.), pp. 363–367. (*Can. Entomol.* **88**, 290–371.).
Balch, R. E., and Bird, F. T. (1944). A disease of the European spruce sawfly, *Gilpinia hercyniae* (Htg.) and its place in natural control. *Sci. Agr.* **25**, 65–80.
Bassi, A. (1835). "Del mal del segno, calcinaccio o moscardino, malattia che affligge i bachi da seta e sul modo di liberarne le bigattaie anche le piu infestate." Parte Prima, XI + 67 pp. Teoria. Orcesi, Lodi.
Berger, E. W. (1915). Cottony cushion scale. *Fla. Grower*, Dec. 18, 1915, pp. 10–11.
Berliner, E. (1911). Uber die Schlaffsucht der Mehlmottenraupe. *Z. Gesamtes Getreidewesen* **3**, 63–70.
Billings, F. H., and Glenn, P. A. (1911). Results of the artificial use of the white-fungus disease in Kansas. *U.S. Dep. Agr. Bur. Entomol. Bull.* **107**, 58 pp.
Bolle, G. (1898). "Die Gelb- oder Fettsucht des Seidenspinners, eine Schmarotzerkrankheit. Suppl. Der Seidenbau in Japan," 141 pp. A. Hartleben, Budapest, Wien, Leipzig.
Bradley, R. (1781). "New Improvements of Planting and Gardening." London.
Burges, H. D., and Hussey, N. W. (eds.). (1971). "Microbial Control of Insects and Mites," 861 pp. Academic Press, New York.
Butler, F. C. (1951). Anthracnose and seedling blight of Bathurst burr caused by *Colletotrichum xanthii* Halst. *Aust. J. Agric. Res.* **2**, 401–410.
Cameron, J. W. MacBain. (1973). Insect pathology. *In* "A History of Entomology" (R. F. Smith, T. E. Mittler, and C. N. Smith, eds.), pp. 285–306. Annu. Rev. Inc., Palo Alto, California.
Clausen, C. P. (1940). "Entomophagous Insects," 688 pp. McGraw-Hill, New York.
Clausen, C. P. (1956). Biological control of insects in the Continental United States. *U.S. Dep. Agr. Tech. Bull.* 1139, 151 pp.
Cornalia, E. (1856). Monografia del bombice del gelso (*Bombyx mori* L.). *Mem. I. R. Inst. Lombardo Sci. Lett. Arti* **6**, 1–387.
Curtis, J. (1860). "Farm Insects." London.
Danysz, J. (1893). La destruction des mulots. *Bull. Soc. Nat. Agr.* **10**, 681.
Darwin, E. (1800). "Phytologia." London.
DeBach, P. (ed.). (1964). "Biological Control of Insect Pests and Weeds," 844 pp. Chapman and Hall, London.
d'Hérelle, F. (1911). Sur une épizootie de nature bacterienne sévissant sur les sauterelles au Mexique. *C. R. Acad. Sci. (Paris)* **152**, 1413–1415.
d'Hérelle, F. (1914). Le coccobacille des sauterelles. *Ann. Inst. Pasteur* **28**, 280–328, 387–407.
de Réaumur, R. A. (1726). Remarques sur la plante appellée à la Chine Hia Tsao Tom Tchom, ou plante verte. *Mem. Acad. Roy. Sci.*, pp. 302–305.
de Réaumur, R. A. (1734). "Memoires pour servir à l'Histoire des Insectes." Paris.
Dodd, A. P. (1927). The biological control of prickly pear in Australia. *Aust. Commonw. Counc. Sci. Ind. Res. Bull.* **34**, 1–44.
Doutt, R. L. (1958). Vice, virtue and the vedalia. *Bull. Entomol. Soc. Amer.* **4**, 119–123.
Dowden, P. B. (1962). Parasites and predators of forest insects liberated in the United States through 1960. *U.S. Dep. Agr. Handb.* **226**, 70 pp.
Dutky, S. R. (1959). Insect microbiology. *Advan. Appl. Microbiol.* **1**, 175–200.
Fenner, F., and Ratcliffe, F. N. (1965). "Myxomatosis," 348 pp. Cambridge Univ. Press, New York.
Fitch, A. (1856). "Sixth, Seventh, Eighth, and Ninth Reports On The Noxious, Beneficial and Other Insects of the State of New York," 259 pp. Albany, New York.
Flanders, S. E. (1955). Principles and practices of biological control utilizing entomophagous

insects. (Extracts from series of lectures presented at the University of Naples, Italy.) In Univ. Calif. libraries at Davis and Riverside, California.

Forskal, P. (1775). "Descriptiones Animalium, Avium, Amphibiorum, Piacium, Insectorum, Vermium; quae in Itinere Orientali Observavit P. Forskal, Postmortem Auctoris Edidit, Carsten Niebuhr," part 3. Hauniae, Moeller.

Franz, J. M. (1961a). Biologische Schädlingsbekämpfung. *In* "Handbuch der Pflanzen Krankheiten" (P. Soraurer, ed.), Vol. VI, pp. 1–302. Parey, Berlin.

Franz, J. M. (1961b). Biological control of pest insects in Europe. *Annu. Rev. Entomol.* **6**, 183–200.

Franz, J. M., and Krieg, A. (1972). "Biologische Schädlingsbekämpfung," 208 pp. Parey, Berlin.

Fullaway, D. T. (1954). Biological control of cactus in Hawaii. *J. Econ. Entomol.* **47**, 696–700.

Greathead, D. J. (1971). A review of biological control in the Ethiopian Region. *Commonw. Inst. Biol. Control. Tech. Commun.* **5**, 162 pp.

Hagen, K. S., and Franz, J. M. (1973). A history of biological control. *In* "A History of Entomology" (R. F. Smith, T. E. Mittler, and C. N. Smith, eds.), pp. 433–476. Annu. Rev. Inc., Palo Alto, California.

Hall, I. M. (1964). Use of micro-organisms in biological control. *In* "Biological Control of Insect Pests and Weeds" (P. DeBach, ed.), pp. 610–628. Chapman and Hall, London.

Halsted, B. D. (1894). Weeds and their most common fungi. *N. Jersey Exp. Sta. Rep.*, pp. 379–381.

Howard, L. O. (1930). A history of applied entomology. *Smithsonian Misc. Collect. (Publ. 3065)* **84**, 564 pp.

Hunter, S. J. (1909). The green bug and its natural enemies. *Kans. State Univ. Bull.* **9**, 1–163.

Husz, B. (1928). *Bacillus thuringiensis* Berl., a bacterium pathogenic to corn borer larvae. *Int. Corn Borer Invest. Sci. Rep.* **1**, 191–193.

Husz, B. (1929). On the use of *Bacillus thuringiensis* in the fight against the corn borer. *Int. Corn Borer Invest. Sci. Rep.* **2**, 99–110.

Jacobs, S. E. (1951). Bacteriological control of the flour moth, *Ephestia kuehniella* Z. *Proc. Soc. Appl. Bacteriol.* **13**, 83–91.

Joly, N. (1842). Notice sur les ravages que la *Liparis dispar* (*Bombyx dispar* Latr.). a exercés aux environs de Toulouse, suivie de quelques réflexions sur un nouveau moyen de détruire certains insectes nuisibles. *Rev. Zool. Soc. Cuvierienne*, pp. 115–119.

Klement, Z. (1953). Further experiments on the use of *Bacillus thuringiensis* for biological control of *Hyphantria cunea* Drury. *Ann. Hung. Plant Prot. Inst* (1951) **6**, 177–183 (Hung.).

Kirby, W. and Spence, W. (1815). "An Introduction to Entomology," 285 pp. Longman, Brown, Green and Longmans, London.

Kirby, W. and Spence, W. (1826). Diseases of insects. *In* "An Introduction to Entomology: or Elements of the Natural History of Insects," Chapt. XLIV, Vol. 4, pp. 197–232. Longman, London.

Kirby, W., and Spence, W. (1856). "An Introduction to Entomology," 7th ed., 607 pp. Longman, London.

Kollar, V. (1837). *In* "London's Gardener's Magazine." (Engl. Transl.)

Krassilstschik, J. (1888). La production industrielle des parasites végétaux pour la destruction des insectes nuisibles. *Bull. Sci. France, Belg.* **19**, 461–472.

Krieg, A. (1961). *Bacillus thuringiensis* Berliner. Über seine Biologie, Pathogenie und Anwendung in der biologischen Schädlingsbekämpfung. [In memoriam Dr. Ernst Berliner (1880–1957).] *Mitt. Biol. Bundesanst. Land-Forstwirt. Berlin-Dahlem.* No. 103, 79 pp.

Leach, C. M. (1946). A disease of dodder caused by the fungus *Colleotrotrichum destructivum*. *Plant Dis. Rep.* **42**, 827–829.

LeBaron, W. (1870). The chalcideous parasite of the apple-tree bark-louse (*Chalcis mytilaspidus*, n. sp.). *Amer. Entomol. Bot.* **2**, 360–362.

LeConte, J. L. (1874). Hints for the promotion of economic entomology. *Proc. Amer. Ass. Advan. Sci.* **22**, 10–22.

Linnaeus, C. (1756). "Reisen durch das Königreich Schweden, welche auf Befehl der hohen Obrigkeit zur Verbesserung der Naturkunde, Haushaltungs- und Arzneykunst von ihm angestellt". A.d.Schwed. übers. von Klein, C. E. 1. Teil. Leipzig, Gottfried Kiesewetter, 336 pp., 2 Taf. 1 Faltkarte.

Liu, G. (1939). Some extracts from the history of entomology in China. *Psyche* **46**, 23–28.

Lockley, R. M. (1955). Failure of myxomatosis on Skokholm Islands. *Nature* (*London*) **175**, 906–907.

Loeffler, F. (1892a). Ueber Epidemien unter den im hygienischen Institute zu Greifswald gehaltenen Mäusen und über die Bekämpfung der Feldmausplage. *Zentralbl. Bakteriol. Parasitenk. Infektionsk. Abtäl* **11**, 129–141.

Loeffler, F. (1892b). Die Feldmausplage in Thessalien und ihre erfolgreiche Bekämpfung mittels des *Bacillus typhi murium*. *Zentralbl. Bakteriol. Parasitenk. Infektionsk. Abt. 1* **12**, 1–17.

McLeod, J. H., McGugan, B. M., and Coppel, H. C. (1962). A review of the biological control attempts against insects and weeds in Canada. *Commonw. Inst. Biol. Control Tech. Commun.* **2**, 216 pp.

Maestri, A. (1856). "Frammenti Anatomici, Fisiologici e Patologici sul Baco da Seta (*Bombyx mori* Linn.)," 172 pp. Fratelli Fusi, Pavia.

Metchnikoff, E. (1879). Diseases of the larvae of the grain weevil. Insects harmful to agriculture (series). Issue III. The grain weevil. 32 pp. Published by the Commission attached to the Odessa Zemstvo Office for the investigation of the problem of insects harmful to agriculture. Odessa. (In Russ.)

Metchnikoff, E. (1880). Zur Lehre über Insectenkrankheiten. *Zool. Anz.* **3**, 44–47.

Miller, D. (1970). Biological control of weeds in New Zealand 1927–48. *N. Z. Dep. Sci. Ind. Res. Mat. Ser. No.* **74**, 1–104.

Moutia, A. L., and Mamet, R. (1946). A review of 25 years of economic entomology in the Island of Mauritius. *Bull. Entomol. Res.* **36**, 439–472.

Myers, J. G. (1931). "A Preliminary Report on an Investigation into the Biological Control of West Indian Insect Pests," 178 pp. H.M. Stationery Office, London.

Nicholson, A. J. (1933). The balance of animal populations. *J. Anim. Ecol.* **2**, 132–178.

Ordish, G. (1967). "Biological Methods in Crop Pest Control," 242 pp. Constable, London.

Pasteur, L. (1870). "Etudes sur la Maladie des vers à Soie," Vol. I, 332 pp.; Vol. II, 327 pp. Gauthier-Villars, Paris.

Pemberton, C. E. (1953). The biological control of insects in Hawaii 1778–1963. *Proc. 7th Pac. Sci. Congr.* **4**, 220–223.

Pemberton, C. E. (1964). Highlights in the history of entomology in Hawaii. *Pac. Insect.* **6**, 689–729.

Rao, V. P. *et al.* (1971). A review of the biological control of insects and other pests in South-east Asia and the Pacific region. *Commonw. Inst. Biol. Control. Tech. Commun.* **6**, 149 pp.

Regnier, R., and Pussard, R. (1926). Le campagnol des champs (*Microtus arvalis* Pall.) et sa destruction. *An. Serv. Epiph.* **12**, 470–514.

Research Institute Canada Department of Agriculture (1971). Biological control programmes against insects and weeds in Canada 1959–1968. *Commonw. Inst. Biol. Control Tech. Commun.* **4**, 266 pp. (Multiple authorship.)

Riley, C. V. (1893). Parasitic and predaceous insects in applied entomology. *Insect Life* **6**, 130–141.

Rudakov, O. L. (1960). Grib alternarii—vrag poviliki *Sel. Khoz. Kirgizii SSR* **6**, 42–43.

Sachtleben, H. (1941). Biologische Bekämpfungsmassnahmen. *In* "Handbuch der Pflanzenkrankheiten" (P. Sorauer, ed.), Vol. VI, Part 2, pp. 1–120 Parey, Berlin.

Safferman, R. S., and Morris, M. E. (1963). Algal virus: isolation. *Science* **140**, 679–680.

Sailer, R. I. (1973). A look at USDA's biological control of insects pests: 1888 to present. *Agri. Sci. Rev.* **10** (4), 15–27.

Sanarelli, G. (1898). Das myxomatogene Virus. Beitrag zum Studium der Krankheitserreger ausserhalb des Sichtbaren. *Zentralbl. Bakteriol. Parasitenk. Infektionsk. Abt. 1* **23**, 865–873.

Silvestri, F. (1909). Squardo allo stato attuale dell' entomologica agraria negli statiuniti del Nord America e ammastramenti che possono derivarne per l'agricoltura Italiana. *Boll. Soc. Agr. Ital. Roma* **14**, 1–65.

Smith, H. S. (1935). The role of biotic factors in determination of population densities. *J. Econ. Entomol.* **28**, 873–898.

Smith, H. S. (1941). Racial segregation in insect populations and its significance in applied entomology. *J. Econ. Entomol.* **34**, 1–13.

Snow, F. H. (1891). Chinch-bugs. Experiments in 1890 for their destruction in the field by the artificial introduction of contagious diseases. *7th Bienn. Rep. Kans. State Board Agr.* **12**, 184–188.

Steinhous, E. A. (1949). "Principles of Insect Pathology," 757 pp. McGraw-Hill, New York.

Steinhaus, E. A. (1951). Possible use of *Bacillus thuringiensis* Berliner as an aid in the biological control of the alfalfa caterpillar. *Hilgardia* **20**, 359–381.

Steinhaus, E. A. (1956). Microbial control—the emergence of an idea. A brief history of insect pathology through the nineteenth century. *Hilgardia* **26**, 107–160.

Steinhaus, E. A. (ed.). (1963). "Insect Pathology. An Advanced Treatise," Vol. I, 661 pp.; Vol. II, 689 pp. Academic Press, New York.

Steinhaus, E. A., and Thompson, C. G. (1949). Preliminary field tests using a polyhedrosis virus to control the alfalfa caterpillar. *J. Econ. Entomol.* **42**, 301–305.

Steyn, J. J. (1959). Use of social spiders against gastrointestinal infections spread by house-flies. *S. Afr. Med. J.* **33**, 730–731.

Sweetman, H. L. (1958). "The Principles of Biological Control," 560 pp. Wm. C. Brown, Dubuque, Iowa.

Thompson, W. R. (1930a). The biological control of insect and plant pests. *Publ. Empire Marketing Board No.* **29**, 124 pp.

Thompson, W. R. (1930b). The utility of mathematical methods in relation to work on biological control. *Ann. Appl. Biol.* **17**, 641–648.

Timberlake, P. H. (1927). Biological control of insect pests in the Hawaiian islands. *Proc. Hawaii. Entomol. Soc.* **6**(3), 529–556.

van den Bosch, R. (1971). Biological control of insects. *Annu. Rev. Ecol. Syst.* **2**, 45–66.

van den Bosch, R., and Messenger, P. S. (1973). "Biological Control," 180 pp. Intext Educ. Publ., New York.

Villa, A. (1845). Degli insetti carnivori adoperati a distruggere le specie dannose all'agricoltura. *Spettatore* **3**(19), 359 pp.

von Leuwenhoek, A. (1701). Part of a letter concerning excrescenses growing on willow leaves, etc. *Phil. Trans. Roy. Soc. London* **22**, 786–799.

von Prowazek, S. (1907). Chlamydozoa. II. Gelbsucht der Seidenraupen. *Arch. Protistenkde.* **10**, 358–364.

Weir, J. R. (1927). The problem of *Dichrostachys nutans*, a weed tree, in Cuba with remarks on its pathology. *Phytopathology* **17**, 137–146.

Wilson, F. (1960). A review of the biological control of insects and weeds in Australia and Australian New Guinea. *Commonw. Inst. Biol. Control Tech. Commun.* **1**, 102 pp.

Wilson, F. (1963). The results of biological control investigations in Australia and New Guinea. *Proc. 9th Pac. Sci. Congr.* (1957) **9**, 112–123.

3

THE THEORETICAL AND EMPIRICAL BASIS OF BIOLOGICAL CONTROL

C. B. Huffaker, F. J. Simmonds, and J. E. Laing

I. INTRODUCTION

The practice of biological pest control has had essentially an empirical, not a theoretical, origin. The striking results following the introduction of the vedalia beetle for control of cottony-cushion scale in California, and similar results from other introductions worldwide stand as the basis of this practice. Theory must conform to fact. In endeavoring to make law out of hypothesis or theory we often put the cart before the horse. In this chapter we endeavor to use ecological theory as a means of better understanding the empirical events established through biological control, and we try to relate the sounder elements of theory to the empirical facts. The principal basis of the practice of biological control derives from the fact that it has worked on many occasions. The degree of this success and the probability of achieving new successes is a matter also of economics. Thus, the economics of biological control is also dealt with in this chapter, and a table listing the results on a worldwide basis is given as an Appendix as a suggestion of further possibilities.

II. NATURAL CONTROL AND THE BALANCE OF NATURE

The concept of natural control derives from the observation that, in general, populations of organisms in nature commonly exhibit a considerable stability in their space—time relationships. MacFadyen (1957) stated, "It is generally agreed that the same species are found in the same habitats at the same seasons for many years in succession and that they occur in numbers which are of the same order of magnitude." The term "balance of nature" implies a stability of numbers, although it is obvious that populations continually undergo changes in density, e.g., as related to mortality factors and to reproductive and nonreproductive periods. Thus, insect populations are characterized by inherent stability, on the one hand, and by recurrent fluctuations, on the other. Smith (1935) characterized this in the idea that population densities are continually changing but that they tend to vary about a mean which is comparatively stable although itself subject to change.

The more diverse the natural communities, the greater the range of compensations in preventing drastic upsets when controlling agents which are specific over certain ranges of host density may temporarily lose their role. Here we wish to explore this process of natural control. The subject has long been one of considerable debate, but in the last decade rather more agreement has come about. Population ecologists now view the system at several levels and they present a reasonably integrated concept of how genetic features, including behavior, limitations in food and space, natural enemies, the physical factors, etc., interplay to give us varying degrees of stability or relative instability of populations and variations in species diversity. Our purpose here is to place the roles of all these factors in perspective.

Natural control has been defined as the maintenance of a population's density (or biomass) within certain upper and lower limits (*characteristic* abundance) by the combined actions of the whole environment, including other population members and necessarily a density-induced element which is regulating in relation to the conditions of the environment and the properties of the species. Some populations are regulated at very stable levels [e.g., magpies in Australia (Carrick, 1963)] while the range of variation in numbers is much greater for others, especially for those inhabiting very unstable environments such as occur in many of our oversimplified agroecosystems (see Huffaker and Messenger, 1964; Huffaker *et al.*, 1971).

Two basic views of population limitation in nature have been proposed; the first conforms to the above definition of natural control, the other does not. Briefly, the latter, or second, view sees population limitation as control by the restrictive forces of the environment such that, by chance, the favorability and unfavorability of the environment balances out and there is only a fluctuation in numbers accordingly as the conditions wax and wane (e.g., Uvarov, 1931; Thompson, 1939, 1956; Andrewartha and Birch, 1954).

The first view takes as a premise the stipulation in the above definition that at least one element must act through density-induced feedback for long-term natural control to occur. Thus, a population increases in numbers in a favorable environment, but this increase itself induces an increase in the repressive forces such as natural enemies or competition for food and shelter. [See Nicholson's (1933) competition curve.] Repressive forces eventually prevent further increase and usually, due to overreaction, lead to a decline in numbers. As the population declines these repressive forces relax in intensity, although some do so with a lag, and permit the population to increase again. Hence, one view sees no essential connection between density and mortality or natality, or losses or gains through movements, whereas the other view holds this connection to be critical. Only in this latter view, the view of density-dependent control, is the term "regulation" appropriate.

A. The Concept of Natural Control by Density Unrelated Factors

While a host of authors (e.g., Nicholson, 1954; Solomon, 1957; Milne, 1962) have objected to this view, we give here only the summary objections stated by Huffaker *et al.*, 1971 (see also Huffaker and Messenger, 1964):

> Control in the sense of temporary suppression or limitation can of course occur from any adverse action. The view of real long-term natural control unrelated to density . . . implies that populations may be held within the long-term ranges of densities observed by the ceaseless change in the favorableness or magnitude of conditions and events unrelated to the population's own density, the latter exerting no necessary or special influence. Density-independent resistance factors, the heterogeneity of the environment in time and place, and the play of chance in the action of such factors, is considered sufficient.
>
> We reject this hypothesis because: (1) it concerns itself mainly with *changes* in density, largely leaving out the causes of the magnitude of mean density; (2) being

concerned mainly with changes in density, the view ignores or denies the fact of characteristic abundance, and for example provides no logical explanation of why some species are always rare, others common and still others abundant, even though each may respond similarly to changes in the weather; (3) to accomplish such long-term natural 'control,' it is presumed that the ceaseless change from favorability to unfavorability of the environment, in terms of a species' tolerances, can be so delicately balanced (a knife-edged balance) over long periods of time as to keep a population in being without density-related stresses coming into play to stop population increase at high densities, and without any lessening of these density-related stresses coming into play at low densities to reduce the likelihood of extinction; (4) such knife-edged balance relative to a species' adaptations, i.e., a balanced favorability and unfavorability, is essentially control by chance, and is incompatible with the view of adaptive improvement resulting through evolution. If, as Birch (1960) claims, there is a tendency for adaptive increase in 'r' or any adaptive tendency to overcome environmental stresses (increased fecundity, better protective behavior, resistance to natural enemies, all assumed unrelated to density), this would immediately destroy this precise balance unless a corresponding repressive change in environment were at the same time automatically elicited. Otherwise, the improved population would increase without limit; (5) last, and most important, the special significance of density-dependent processes (which the view denies as essential) can be easily demonstrated except for the most unmanageable and violently fluctuating examples by applying a heavy mortality or adding enormous numbers to the populations, or by removing mortality factors, or adding additional resources and then following the population trends afterwards.

B. The Concept of Density-Dependent Regulation

Again we quote from Huffaker *et al.* (1971) for this account of the concept of density-dependent regulation of population size, but refer the reader to their work for aspects not treated here:

Regarding the concept of a knife-edged balance hypothesis, as Huffaker and Messenger (1964) stated, "Hence it appears to us mandatory that a suitable theory of natural control of animal abundance should include some provision whereby the tendency of the animals to adapt to, and hence in numbers to overcome, stresses from the environment, in particular the physical factors in the environment, be met by reaction of some sort or other. To us, again, the mechanism of reactive change in the intensity of action of a given controlling factor elicited by change in abundance, i.e., density-dependent control factors, provides the simplest and most elegant explanation of how such adaptational possibilities are counter-balanced." The objections to the view of environmental balance by chance do not apply to the concept of density-dependent regulation of numbers. This view offers an explanation of the fact of characteristic abundance and, as stated above, its reality can be demonstrated rather easily. It is consistent with the observation that extreme densities are inevitably met by starvation or lack of space, if not by predation, disease or other catastrophe associated with such abundance, and with many studies showing that such extreme densities are often prevented by density-related processes operating at much lower densities.

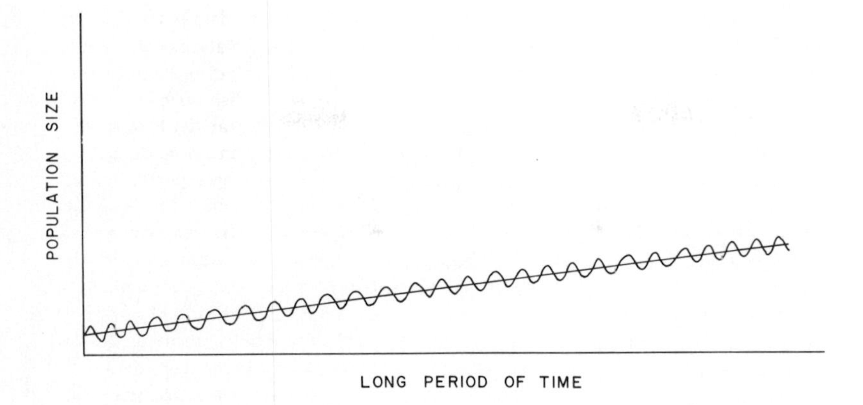

Fig. 1. The relationship of "control" in the relative sense (as through competition for a resource) (see oscillations), and "control" in the absolute sense (here no control at all but, rather, a steady increase in mean density), as the resource, e.g., food or the efficiency of its use, is hypothesized here as being indefinitely increased. (From Huffaker *et al.*, 1971.)

C. The Interrelation of Conditioning Forces and Regulatory Actions

Figure 1 taken from Huffaker *et al.* (1971, p. 22) and discussed therein shows the interrelatedness of the density unrelated *conditioning forces*, the density-dependent, or *regulating actions*, and the *properties* of the species. Here we also present additional illustrations, taken mostly from Huffaker and Messenger (1964), to help explain this interrelatedness and the nature of the roles of each factor.

Figure 2 is used to explain the relationship of the favorability of the environment to the *changes* in density that occur and to the process of regulation as well. In a salubrious tropical environment, represented by zone 1, changes in conditions are relatively mild and the conditions remain continuously conducive to potential increase except for the biological checks or processes related to density rather than density-unrelated factors, such as the weather. Thus, the biotic factors predominate in causing changes in density in this zone and in addition operate in their usual regulating way, *in relation to* the environmental conditions, in determining mean population size. The space between the wavy lines and the enclosing arbitrary center circle indicates the limited potential for population densities to fluctuate purely from changes in the meteorological conditions. Zone 3 represents the most rigorous conditions where permanent existence is possible. These areas of permanent existence are small and widely scattered. There is much potential for the population to fluctuate in density due to the marked changes in environmental favorability independent of density itself. In the small permanent breeding areas the regulating actions and the conditioning factors again act together in determining long-term mean population size, and thus the numbers of individuals that are available to move to other areas. In zone 4 only

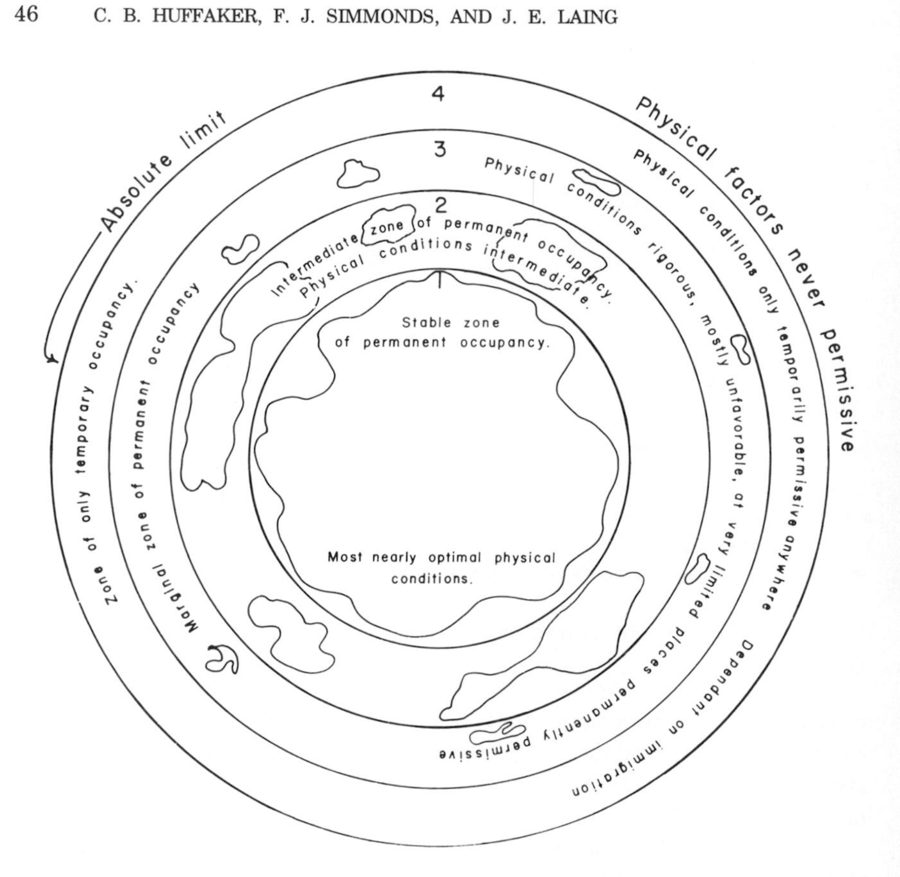

Fig. 2. The geographical distribution of a species population and the interrelation of conditioning and regulating forces. Irregular patches in each zone represent localized areas of relatively permanent favorability and the interspaces indicate room for waxing and waning of such areas in time. (From Huffaker and Messenger, 1964.)

temporary existence is possible over vast areas, and existence is dependent upon immigrations from the permanent breeding areas of zones 1, 2, and 3.

To illustrate more precisely our description of the interrelated and inseparableness of the density-dependent and conditioning forces relative to determination of long-term mean population size, Figure 3 (from Terao and Tanaka, 1928) shows that a density-dependent process is in operation at each temperature employed in causing the populations of the water flea, *Moina macrocarpa* Strauss, to decline in rate of growth and subsequently to stop growth, within each system for any changes were associated with increasing density itself. On the other hand, by comparing the levels of density at which growth stopped under the different temperatures, the role of the density-unrelated conditioning environment is evident [see Huffaker and Messenger (1964) for additional illustrations].

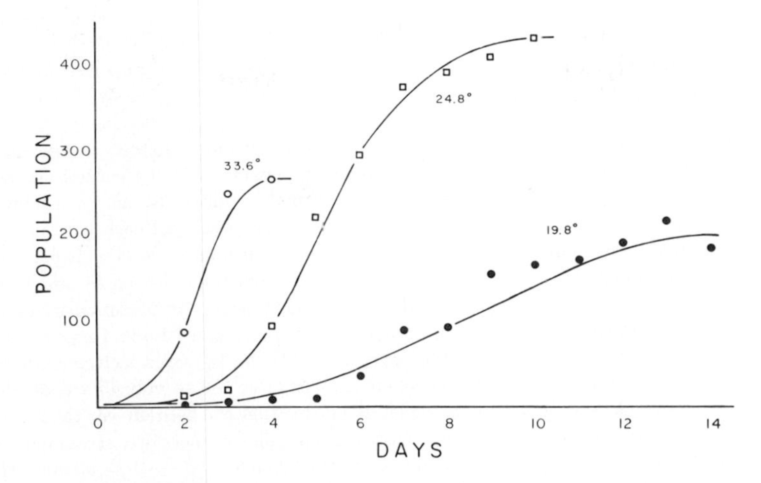

Fig. 3. The logistic growth of three laboratory populations of the water flea, *Moina macrocarpa*, at three temperatures. (Redrawn after Terao and Tanaka, 1928.)

D. Genetic Feedback in Natural Control

In their space-time relationships plants and animals present considerable homeostasis. The properties leading to this have arisen through genetic feedback over vast periods of time. Some genetic adaptability also exists currently and allows for adjustments enabling homeostatic response over short periods of time (dynamics).

David Pimentel and associates have emphasized this aspect in interspecies relations. They consider that the selective pressure exerted by predators or parasites in time leads to genetic changes in the prey population which make it better adapted to resist or tolerate attacks. Like Huffaker *et al.* (1971), we do not challenge this explanation of the evolutionary and genetic process, but we differ with Pimentel regarding two of the basic claimed consequences of the process: (1) that this process represents a distinct mechanism of regulation of population size, and (2) that because of such genetic feedback processes, a natural enemy will in time usually lose its capacity to control the host population at the former low density level, and corollary to this, that natural enemies coevolved with a given host species will be less efficient control agents of that host than ones coevolved elsewhere with a related host species and introduced against the given host.

Regarding the first point, Huffaker *et al.* (1971) claim that the genetic feedback mechanism only changes the quantitative rules of the game.

To illustrate the unacceptability of Pimentel's concept of "regulation" we will consider some of his experiments and conclusions at this point.

A comparison was made (Pimentel, 1964; Pimentel and Al-Hafidh, 1965) between populations of a control or newly associated experimental parasite—host system and an older one in which some homeostasis had evolved during a 2-year interaction. After noting the changes in genetic characteristics and differences in population size and stability, Pimentel (1964) stated:

> Different regulatory [sic] mechanisms were apparently dominant in these two population systems. The average density of the parasite population in the control system was 3,589; progeny production was 138 per female [side tests on reproductive capability], thus equilibrium would require 136 to be destroyed each generation. Competition, environmental discontinuity and other factors were responsible for limiting the numbers of the control population. To illustrate, let us assume that competition was the only limiting factor and it eliminated the 136 progeny each generation. Parasite density of the experimental system averaged 1,436. Progeny production was 45 per female; and equilibrium requires 43 to be destroyed each generation. The decline from the original reproduction [capacity] of 138 to only 45 was due to genetic feedback. For simplicity, again let us assume that competition was the factor which eliminated the 43. The genetic feed-back mechanism, therefore, accounted for almost twice this number or 83 [actually, more than double: 138 − 45 = 93, not 83].

Pimentel and Al-Hafidh (1965) add, relative presumably to extended data: "Thus, there was evolution from dominant regulation by the competitive mechanism to dominant regulation [sic] by the genetic feed-back mechanism."

Let us suppose that their "co-adapted" system were to continue some 50 years more, with no further genetic adjustments. Would those authors then claim genetic feedback to be still the main density *regulator*? Surely some related Nicholson-type of mechanism would then be *doing* the regulating (as it is here). The reproductive capacity of the wild stock (138) must have derived historically from some other value. How then could Pimentel, consistent with his view, have assumed that at that time competition (and environmental discontinuity) was the responsible regulating mechanism, rather than the genetic feedback mechanism which must have been somewhere in past history responsible for change to *that* particular level of reproductive capacity? Such a change in their experimental populations forms the basis of their claim that genetic feedback *had taken over* the role of regulation. In Pimental's earlier concepts, at least, *control* (density determination) and *regulation* are not separated. Control of density must include a true negative feedback feature geared to density (a regulator) but it also includes the *properties* of the population and the *conditions* of the environment. The properties may, as Pimentel and associates have shown, also be subject to density-induced alterations. Thus, the density at which the population is controlled by the density-geared regulation is as much a function of the conditions and the properties of the populations involved as it is of the regulating mechanism itself. Genetic feedback, (1) may permanently alter properties (long-term evolutionary aspect), this being reflected in the density at which the population is regulated in time, and (2) it may produce oscillations or fluctuations in the properties (the dynamic, functional aspect), and this too is reflected in an oscillating or fluctuating population size under whatever competitive regulating process as is in operation. This was seen in the experiments of Pimentel and Al-Hafidh (1965) wherein coexistence of housefly and blowfly populations under regulation by competition for food was accounted for through density-geared alternating shifts in the properties of the two species in time. Thus, we consider that the genetic factor is an inherent, inseparable component that modifies the level at which competition does the regulating, but it does not itself *do* the regulating. Genetic feedback can, however, result in characteristics that are in "course balance" with needs, e.g., fecundity and the risk of survival to maturity, and Pimentel and Soans (1971) are now correct in noting that genetic feedback can only lead to a "course" balance.

Regarding the second objection to Pimentel's major conclusions, Huffaker *et al.* (1971) present countering arguments, including a number of examples of biological control agents, which have not lessened in efficiency during the relatively long periods since their initial success. In fact, contrary to Pimentel's view, most workers consider that entomophagous insects generally possess plasticity and become better adapted to control a host after associating with it for some time in a new environment (Messenger and van den Bosch, 1971 and Chapter 9).

Some natural enemy which coevolved with a natural host may well be a more effective biological control agent of a close relative of its host than of its normal host. This especially applies to true parasitism, but rarely at least to entomophagous parasitoids and predators, wherein there is no evolution of a tolerable relationship at the individual contact level. However, a harmonious relation can result under partial or complete reciprocal, density-dependent interaction at the population level.

It is extremely interesting that the entomophagous parasitoids have not evolved examples illustrating the trend in true parasitism from early severity, to toleration, to mutualism, in the attacked-attacker relationship at the individual level. Even when, in the rare examples, the parasitoid does not kill the host (or die itself), the host is rendered sterile, which accomplishes the same result—a population control effect.

III. THE ROLE OF NATURAL ENEMIES IN NATURAL CONTROL

A. General Aspects of Predation and the Kinds of Natural Enemies

The general role of natural enemies in the regulation of populations of their hosts is seen in the results of natural enemy introductions (e.g., Chapters 12–21 and Appendix) and by analyses of the effects of indigenous natural enemies in the control of their indigenous hosts, as derived from deliberate experimentation. There is limited evidence concerning the latter, but recently Hagen *et al.* (1971), MacPhee and MacLellan (1971), and Rabb (1971) summarized records of some significant examples in the United States and Canada. Also, the extensive record of unusual outbreaks of secondary pests following the introduction of synthetic insecticides after World War II is generally considered due in part to disruption of resident natural enemies (e.g., Ripper, 1956; Huffaker *et al.*, 1971; Harpaz and Rosen, 1971; Wood, 1971; see also Chapters 14, 15, 17, and 22). We need also to look at the questions posed somewhat more theoretically and in more detail than those presented in terms of a release from control due to chemical disruptions or the degree of control exerted as indicated by examples of "before" and "after" results, or through use of "check methods" (Chapter 11).

To understand the phenomenon of "predation" in the broadest sense we may look at predation as the exploitative feeding of one organism on another, thus including herbivores, carnivores, true parasites, and the entomophagous parasitoids and predators. There are many levels of such predation and its functions are not always the same. Thus predation may shape the morphology, physiology, and behavior of many species (e.g., remarkable examples of mimicry); it may cause complete or merely local extinction, thus affecting species distribution; predation may cause reductions in numbers little related to population regulation or may actually *do* the regulating; finally, predation may enrich or deplete the diversity and structure of natural communities.

Differences in degree of host specificity, behavior, and immunity of the host, and the changes in the environment wrought by man relative to the roles of classical vertebrate predators, versus their counterparts among the invertebrates, may account for apparently quite different potentials for their use in biological control (Chapter 20). The first point is most significant. We simply cannot introduce natural enemies lacking adequate host or prey specificity and few vertebrate predators are sufficiently restricted in their diet. Usually, the most efficient and reliable natural enemies are the relatively highly host-specific ones and these abound mainly among the insects. In permanent habitats these latter commonly have a reciprocal density-dependent relation with their host (Huffaker and Messenger, 1964). Such a natural enemy limits the numbers of its host and in turn is itself food-limited. This seems to be the general rule for invertebrate predators except in unstable situations (Ehler *et al.*, 1973). This may be true for some vertebrates as well, even though their evolution of complex systems of social hierarchy and territorialism may have modified this (Huffaker, 1971a).

It may also be true that the development of immunity in vertebrate hosts which are the main "targets" of specific true parasites would be more apt to nullify a biological control effort than is the case for biological control of insects because immunity reactions in insect hosts to entomophagous parasites (e.g., encapsulation) are not substantial and immunity reactions in insects to pathogenic microbes are weak in contrast to those in higher vertebrates (Stephens, 1963).

Pathogens often decimate host populations and may exert a balancing role through mortality or chronic debilitating effects. Here we are interested in entomopathogenic species (but see also Chapters 20 and 21). We know a great deal about the kinds of insect pathogens, their virulence, infectivity, host specificity, and means of identification and their production in the laboratory. We know little about their ecological roles in nature. They appear to be density-dependent in nature, but only rather loosely so because of the complex of events (other than the host population's density) that is often required to trigger epizootics (see Tanada, 1964; Huffaker *et al.*, 1971). As Tanada noted, an epizootic may cause prolonged control through its near annihilative effects. Pathogens also present promising advantages as selective insecticides (Chapter 23). Thus investigation of the roles of entomopathogenic microbes in population suppression and as regulating factors in the field, and their interactions with other control factors, needs to be intensified (Chapter 7).

B. The Utility and Attributes of Effective Natural Enemies

The two main avenues to greater use of natural enemies are (1) to introduce more natural enemies and (2) to make better use of resident ones. Evidence of the role of natural enemies is gained through such empirical events. Extensive examples suggest, and some well documented ones prove, that adverse influences of insecticides on natural enemies have occasioned many rises to pest status of species formerly innocuous or of minor importance (e.g., Huffaker, 1971b; see also Chapters 14–16, 22, 27 herein).

Natural enemies are generally recognized as a significant means of control of pest populations even by those who deny that their action is density-dependent or truly regulating. In fact, it is far easier to establish their *control potential* (as by use of check methods, Chapter 11) than to prove that this effect is achieved through true density-dependent action. We contend, however, that their action is often regulatory even when this is difficult to demonstrate. We contend also that they often act to *stabilize* their host's densities, and their own, at relatively low levels (e.g., Huffaker *et al.*, 1971, pp. 38–40).

The intensity of action of nonterritorial natural enemies may be gauged mainly by the ability of the most efficient species to lower the prey's numbers to the point where other natural enemies are displaced (Huffaker *et al.*, 1971). Searching capacity is the main criterion in relatively stable environments, while intrinsic rate of increase is a very important *additional* feature in unstable environments (e.g., Huffaker and Laing, 1972). Thus, among the attributes of an effective natural enemy, searching capacity comes first. These attributes are summarized by Doutt and DeBach (1964), and Huffaker (1974) lists the five main characteristics as: (1) adaptability to the varying physical conditions; (2) searching capacity, including general mobility (or capacity for dispersal); (3) power of increase relative to that of its prey (or host); (4) power of prey consumption; and (5) other intrinsic properties such as synchronization with host life history, host specificity, degree of discrimination, ability to survive host-free periods, and behavior which may alter performance as related to density or dispersion of hosts or its own population.

In Chapters 8 and 9 the desirable characteristics of a natural enemy are considered further. In addition to these basic properties, five somewhat related factors are discussed in Chapter 8, i.e., (1) host range, (2) host preference, (3) constancy of association, (4) abundance, and (5) intrinsic and extrinsic competitive capacities. Clearly, if a strictly host-specific parasite is intrinsically inferior to another strictly host-specific parasite, and if they occupy the same niche, the intrinsically inferior species must be extrinsically superior if it is to coexist. Mere existence requires it to have superior searching capacity; thus, it is capable of controlling the host at lower densities. High searching efficiency itself also requires high adaptation to the host and habitat. Where the above strictures apply to potential competitors it is likely that those possessing extrinsic superiority would normally succeed over and displace the extrinsically inferior but intrinsically superior competitors. However, host specificity is not commonly so strict, and the adaptations of parasites are not equally suited over an entire range of the host. Hence, closer evaluation is required concerning any intrinsically superior hosts of somewhat unspecialized host adaptations, especially ones exhibiting an hyperparasitic habit of cleptoparasitic nature (see further Smith, 1929 and Chapter 8).

Natural enemies are the main natural control factor which we can conveniently manipulate. Pervading weather, another major natural control factor, cannot be so manipulated. (We can alter the microweather by manipulation.) We do effectively develop and use cultural practices, resistant host stocks, pesticides, etc., all of which are artificial measures, though perhaps acting in conjunction with climatic and edaphic factors.

C. Models of Host–Parasite and Predator–Prey Interactions—The Functional, Numerical, and Overall Responses

Models that have been developed to simulate host–parasite and predator–prey interactions in agroecosystems are as yet inadequate representations for the purpose of optimum pest control decision making (Chapter 27). However, various concepts concerning modeling such relationships aid our understanding of host–parasite and predator–prey interactions.

Huffaker *et al.* (1968) graphically presented the four postulated types of functional response (to host or prey density) of predators (including parasitoids). These are shown in Figure 4. As noted by Hassell (1966) and the aforementioned authors, only the sigmoid type of functional response, uncommonly encountered in entomophagous insects, presents the possibility of prey population control or regulation inherent to the functional response itself (Fig. 4). Hence, the major avenue of host (or prey) population control of insect pests by entomophagous insects is through the numerical response, although an increased functional performance (prey consumption) tends to result in a numerical response. In Thompson's (1924, 1939) conception (Fig. 4a) no difficulty is encountered in finding hosts or prey; the number attacked is unrelated to density, being limited only by satiation in appetite (predators) or exhaustion of egg supply (parasitoids). In Nicholson's (1933, 1954) conception, the number attacked is dependent only upon searching capacity, taken as a constant, and the number attacked (Fig. 4a) is directly proportional to the density of the attacked species, with the *proportion* attacked (Fig. 4b) being a constant. The established functional response of entomophagous insects is better represented, however, by the disc equation of Holling (1959) (see also Burnett, 1954, 1958; Morris, 1963; Messenger, 1969). Numbers attacked increase at a

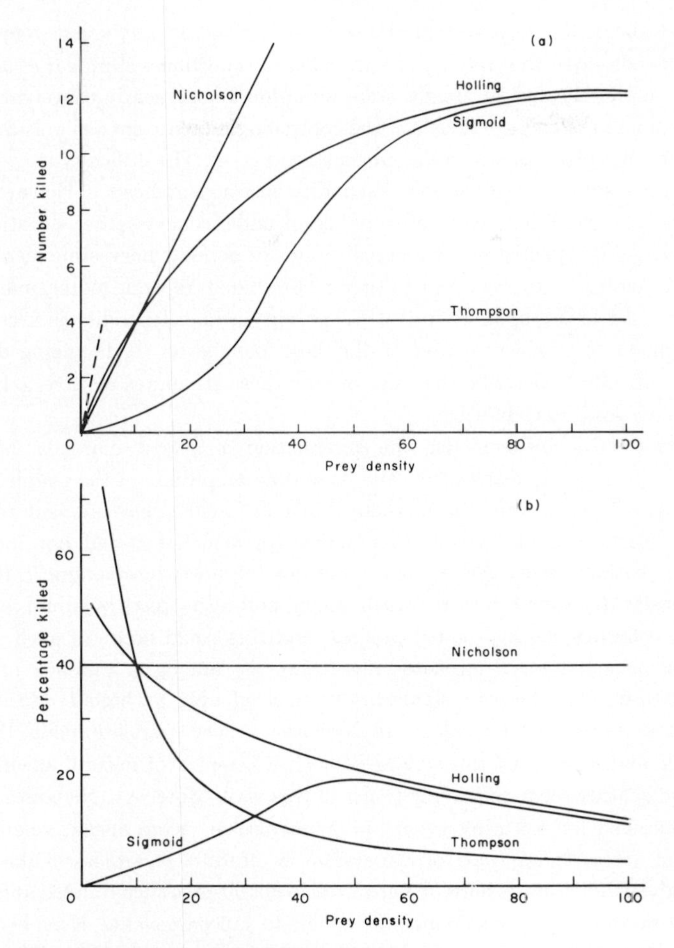

Fig. 4. (a) The four postulated types of functional response of predators (including entomophagous parasites) to prey density. (b) The same types of functional response expressed in terms of proportionate kill. (After Huffaker *et al.*, 1968.)

progressively decreasing rate with increase in host density while the proportion attacked decreases (Fig. 4) (see Huffaker *et al.*, 1971).

The numerical response of entomophagous insects is the customary avenue to their control and regulatory power. Success following introduction of good natural enemies customarily requires time for several generations of numerical buildup. This is a basic reason, in addition to statistical ones, why Morris' (1959, 1963) method of evaluating the role of natural enemies is faulty (e.g., Hassell and Huffaker, 1969; Luck, 1971; Varley and Gradwell, 1971). His methods do not assess the numerical response, which is a generation to generation phenomenon, but relate only to within-generation phenomena.

As stated above, the relative importance of searching capacity versus reproductive capacity depends upon the stability or variability of conditions. Huffaker *et al.* (1968) noted that in the theoretical steady state situation (most nearly conceivable under stable conditions) the average female of the controlling parasite species only has to find enough hosts to replace herself in the progeny generation. The difficulty is in finding a host, not in a shortage of egg supply. Fecundity is of no significance. However, under widely fluctuating conditions (such as would occur under some weather situations or in an area undergoing repeated pesticide applications or periodic harvesting disruptions), a controlling parasite may from time to time be precluded from adequately maintaining its numbers and hence upon return of favorability a high fecundity is a real asset, serving perhaps to prevent escape of the host population to damaging densities. Huffaker *et al.* (1971) describe the value of both these attributes relative to biological control of olive scale in California.

Based on the Nicholson model, the annihilation of a host—parasite interaction (which, it is postulated, results from the increasing amplitude of the coupled oscillations with time) means that the parasite is not in itself a self-sufficient regulating agent. In nature it seems evident that successive densities would not increase in this manner without other factors (e.g., food for the host) intervening in the interaction. Density-dependent factors which might enter the picture would become a part of the *collective* density-geared process, and this could occur in such a way as to damp a principal host—parasite interaction. If, among a complex of natural enemies, certain ones become significantly involved only at higher densities, the overall process is density-dependent. In Nicholson's host—parasite model the interaction would be damped by the actions of such a complex of natural enemies. Less host-specific species seem commonly to act in this way. Moreover, Nicholson's model itself is conceived as picturing events in a microcosm, so to speak, wherein completely equal access to any part of the system is afforded regardless of the position at which a given individual is born. In nature we do not observe such coupled annihilative interactions pertaining to whole populations, only to subcomponents. Even here, movements between subcomponents appear to damp the interactions considerably. Huffaker *et al.* (1971) considered this question, and we quote from them:

> In the case of entomophagous parasites and predators, we now consider the possibilities by which the theoretical, inherently annihilative, predominantly delayed density-dependent action may be damped. An explanation of these features seems essential in bringing theory in line with what appears overwhelmingly implied in the empirical results of biological control.
>
> In the Nicholson-Bailey model, or some other model incorporating lag effects inherently associated with such numerical response phenomena, the increasing amplitude that is a direct consequence of these lag effects* must be damped if stability is to be

* "Introducing a lag effect into the Lotka-Volterra model also produces this result (Wangersky and Cunningham, 1957)."

> achieved. Thus, if theory and the status of a rather persistent, stable control of many pest species by their natural enemies are to be brought into accord, damping processes tending to stabilize must be built into theory and used in modeling.

After discussing the works of a number of investigators, Huffaker, *et al.* (1971) further concluded:

> Summarizing, it would appear that any action that results in a sufficiently greater adverse effect on the predator than on the prey during the crash phases of the interactions when there are "too many" predators would add a damping tendency to the interaction, and the degree of this effect would determine its sufficiency or insufficiency for stabilization. Extrinsic damping tendencies include, (1) competition for food or shelter on the part of the host (or prey), (2) physical refuges affording some protection for the host, and (3) actions of other natural enemies superimposed on the given host/parasite system where the above stricture is satisfied. Intrinsic damping tendencies include, (1) a functional response that is directly density-dependent in effect—i.e., sigmoid (even if insufficient alone to regulate), (2) intrinsic behavior or limitations leading to partial host protection through imperfect temporal or spatial synchrony, (3) mutually interfering contacts, cannibalism, superparasitism, wound-killing or host-feeding (by parasite adults) that may satisfy the above strictures, (4) an appropriate contagion of attacks, and (5) general density-related movements into and away from sub-population units of the habitat. (See also Hassell and Varley, 1969.)

Auslander *et al.* (1974) developed a model showing that where the natural enemy exhibits a preference for an age class of the host, stability may result quite independent of the aforestated intrinsic damping processes.

IV. SOME CONTROVERSIAL QUESTIONS CONCERNING BIOLOGICAL CONTROL

A. The Premise of Biological Control

The premise of biological control is that organisms have natural enemies, that in certain circumstances of place, time, and combination of species, strains, etc., many organisms are held at low, noninjurious levels by these natural enemies, and that in the cases of specific pest species or potential pest species these natural enemies, or natural enemies of their relatives, may provide a control solution. Experience has proved the soundness of the premise. Applied biological control has had many successes, mainly through the introduction of exotic natural enemies to control exotic pests, but more recently by manipulating indigenous natural enemies. Hagen *et al.* (1971), Rabb (1971), and MacPhee and MacLellan (1971) have recently detailed the value that can result from indigenous natural enemies. Naturally occurring biological control exists all around us and the possibilities of its conservation and augmentation are many. Yet the spectacular examples of manipulated or applied biological control as achieved through introductions of exotics to control exotics have constituted most of the world effort until recent years.

A common misconception is that applied biological control rests mainly on the concept of community stability, with such stability being dependent upon species diversity; thus, natural enemies are introduced mainly to increase the complex of natural enemies (Turnbull, 1967). This idea, plus the additional idea that the support of two or more competing enemy species will necessarily lead to an increase in the density of the host species, seems to have led some authors (e.g., Turnbull, 1967; Watt, 1965) to conclude that introduction of any natural enemies other than the single "best" species would result in poorer biological control than that possible by the "best" species alone (see below and Chapter 9). Thus, in their view the community would be rendered more stable by the larger complex of enemies but at a higher density of the target pest species. Since, however, the objective of classical biological control is to achieve reduction in the density of the target species by introduction of a better natural enemy, or one which will supplement others, the objective may be achieved, even though the result is a reduction in species diversity. Such is not uncommon. Most of the examples of excellent biological control by introductions have involved rather highly host-specific natural enemies, and in such cases a superior natural enemy often so reduces the target species that several competing species previously present are displaced entirely from the habitat (e.g. DeBach *et al.*, 1971; DeBach and Sundby, 1963; Huffaker and Kennett, 1969; Huffaker *et al.*, 1971; and Chapter 15).

However, a complex of natural enemies of broader host range can serve as community stabilizers and control various pest species, acting primarily upon any of the phytophagous species that may reach densities high enough to be attractive at the time, including species which commonly reach high densities and ones normally held at low densities by their usually more efficient highly specific natural enemies. Huffaker *et al.* (1969) noted, "Theoretically, general predators tend to serve as regulators of community stability while the specialists tend to regulate single species stability." (See further Huffaker *et al.*, 1971.) Such predators, commonly also possessing considerable mobility, may be mainly responsible for control in frequently disturbed habitats, as in annual crops (Ehler *et al.*, 1973).

B. Multiple Parasitism and Multiple Introductions

Two or more relatively host-specific species of natural enemies which attack the same species obviously compete with one another when they occur in the same habitat, *extrinsically* (outside of the host) and also *intrinsically* (inside or on the host) as larvae if they occur together with the same host. Typically, in the latter case, the individual of only the intrinsically superior species will survive.

As a result of early work in Hawaii on the Mediterranean fruit fly, Pemberton and Willard (1918) advanced the hypothesis that due to the possibility that the best parasite may be decimated through the effects of intrinsic competition from an extrinsically inferior parasite, only the best species should be introduced. No one has yet been able to say just how such a best species can be predetermined, or in fact if a generally "best" species even exists. This serious question involving multiple parasitism was analyzed by Smith (1929). Using data of Pemberton and Willard (1918) and Willard and Bissell (1926), and logic, Smith supported the continued practice of introducing any primary parasites of general promise.

Since there are other types of competition, the problem is much broader, however, than that involving multiple parasitism (Doutt and DeBach, 1964). Nicholson (1933), in a rare example wherein he and Smith differed, came to contrary conclusions. His method was quantitative and entirely theoretical, embracing the use of a model now shown to have at least two basic flaws (Hassel and Varley, 1969; Varley and Gradwell, 1971). His results suggest that an earlier acting parasite, once established, could continue to exist with a later acting regulating parasite even though the earlier acting parasite possessed (within narrowly restricted limits) a lower searching power than the later acting parasite, and that the density of the host stage which it attacks would be greater than if the later acting superior searching parasite were present alone. [Nicholson, however, never used this theoretical conclusion to deny introduction into Australia of a complex of appropriate primary parasites (personal communication)].

The practice of introducing a complex of enemy species went largely unchallenged until Turnbull and Chant (1961) concluded that because of the effects of competition the establishment of any species except the best one would reduce the overall degree of control, a thesis supported either directly or by implication by Zwölfer (1963), Watt (1965), and Turnbull (1967).

A variety of objections have been raised to the view that under no circumstances could the introduction of a complex of enemy species reduce the level of control. Certain of these objections seem valid conceptually and are supported to a degree by the empirical record (e.g., Zwölfer, 1971), while others appear to have little theoretical foundation and are inconsistent with the empirical record. The essential feature is to distinguish between rather highly host-specific natural enemies and the more general ones, and between ones which act more catastrophically and those which act in a more stable, reliable manner. There are, of course, two aspects of this question. One has to do with the geographical coverage of the hosts' environment. It is unlikely that a single enemy species would be the superior one over an extensive range. The value to be gained from a complex of enemies, such that better control would be obtained over wide geographical areas (with different enemies superior in different situations), is largely ignored in criticisms leveled against current practices.

The empirical record itself furnishes the clearest basis for an evaluation of this question. This record overwhelmingly supports the practice of introducing a complex of primary highly specific parasites and predators (e.g., California red scale, De Bach *et al.*, 1971; spotted alfalfa aphid, van den Bosch *et al.*, 1964; Klamath weed, Huffaker, 1967; olive scale, Huffaker and Kennett, 1966). Moreover, there has been no general indication of detrimental consequences that might ensue from introducing the more euryphagous species. Yet this record is not entirely free of warnings relative to less host-specific parasites and predators, and especially pathogens. The experience with pathogens has been much less extensive, and by their very nature they are more "catastrophic" and erratic in action. (See further Huffaker *et al.*, 1971.)

Ullyett (1947) considered that "catastrophic" factors such as disease (or, we add, general predators, above) might cause such a seasonal or periodic annihilation of a given pest as to cause an even-brooded condition or inability of a more reliable and stable regulating enemy species to exert its potential. The pest species would outbreak frequently. Ullyett noted that disease epizootics of the diamondback moth, *Plutella xylostella* (L.), in South Africa acted in this way. C. E. Kennett (personal communication) considers that the catastrophic but erratic occurrence of virus epizootics in citrus red mite [*Panonychus citri* (McG.)] populations in central California may be the reason why more reliable and better control by phytoseiids and other predators is not exerted. In fact, the only local phytoseiid that can survive these conditions is *Amblyseius hibisci* (Chant), a rather poor predator which prefers pollen to prey as food but can, on occasion, exert a good suppressive effect on the citrus red mites.

Another example involves the disruption of biological control by indigenous parasites of the hispid beetle *Promecotheca reichei* Baly in Fiji with the inadvertent establishment of the general predatory mite *Pediculoides ventricosus* (Newp.) This situation was corrected, however, by adding still further to the complex of natural enemies with the introduction of a highly specific parasite *Pediobius parvulus* (Ferr.) which attacks all the larval and pupal stages (Taylor, 1937).

Huffaker and Kennett (1966) give a synopsis of the empirical record which appears to justify continuation of current practices, but with greater care:

> While the gross record of deliberate introductions of enemies to control given pests is largely undocumented as to the degrees of control achieved by the first introduced enemy versus the first two, versus the first three or four, *et cetera*, two facts seem abundantly clear from experience alone: (1) introduction of an additional enemy has invariably improved upon the degree of overall control or else it has made little difference, and (2) unless we can make the highly unlikely (ridiculous) assumption that in all cases the first enemy was an inferior one and that each later one added could have done a better job if introduced alone, the suggestion is disproved that the more enemies a species has, the poorer and more unstable will be the degree of control.

By chance, we would expect that a second species would be inferior to the first, half of the time, and according to the general hypothesis, there should be as many cases wherein the degree of control was noticeably reduced as there would be cases of noticeable improvement. This is patently not true!

Several parasites were introduced against the Mediterranean fruit fly in Hawaii early in this century, including *Opius humilis* Silv., and *O. tryoni* (Cameron). In 1915, *O. humilis* was virtually acting alone, and the total parasitization was less than it was in later years after *O. tryoni* and other species were established. Also, the records for coffee over the 20-year period, 1914–1933, show that intense competition between the three main species of parasites, and others, was not detrimental to overall biological control as reflected by the degree of infestation of coffee berries by the fruit fly *before* (1916–1924, 0.4 larva per coffee berry) and *after* (1925–1933, 0.1 larva per berry) the time when *O. humilis* was virtually completely eliminated by competition (Willard and Mason, 1937).

Van den Bosch and Haramoto (1953) and Bess and Haramoto (1958) relate the consequencies of another similar case of introductions against another fruit fly, *Dacus dorsalis* Hend., in Hawaii. *Opius longicaudatus* (Ashm.) was released about the same time as *O. vandenboschi* Full. and the former, being better adapted for rapid multiplication, first became heavily and widely established in 1948. However, by late 1949 the intrinsically superior *O. vandenboschi* had become the dominant species. This species in turn lost its position to a later established, and again intrinsically superior species, *O. oophilus* Full. in 1950. Each of the near replacements of one species by another was accompanied by a higher total parasitization and a decrease in the fruit fly infestation. Currently, *O. oophilus* is nearly the sole dominant with about 70% parasitization occurring and infestations greatly reduced from their peak before introductions were initiated.

Several well-documented cases are related by Huffaker *et al.* (1971) and DeBach *et al.* (1971), one of the best being the biological control in California of olive scale, *Parlatoria oleae* (Colvée), following a sequence of introductions of parasites and predators, and the successive improvement in biological control of California red scale by natural enemy introductions over a period of 75 years. (See Chapters 14 and 15.) Other examples can be gleaned from accounts in the book "Introduced Parasites and Predators of Arthropod Pests and Weeds—A World Review" edited by C. P. Clausen (in press).

Thus, so far as the documentation permits, the statement given previously from Huffaker and Kennett (1966) is valid. Doutt and DeBach (1964, p. 127) similarly comment: "Multiple introductions whether simultaneous or spread over a period of time, have apparently nearly always added something to the effectiveness of host population regulation, either in time or space, regardless of the intense competition which occurs. . . ."

The interrelationships between this whole question and that of competitive displacement is fundamental, and is treated in detail by Huffaker *et al.* (1971) and Huffaker and Laing (1972). (See also DeBach and Sundby, 1963; DeBach, 1966; and Chapter 8 in this treatise.)

C. The Relative Value of Parasites and Predators

Usually a predatory insect can be distinguished from a parasitic one (parasitoid) by the fact that the predator must consume more than one host (prey) individual to reach maturity whereas the parasite develops in or on a single host individual. Here we do not attempt to characterize these two types of natural enemies (however, see Doutt and

DeBach, 1964; Doutt, 1964; and Chapters 5 and 6 in this treatise). We deal here only with their relative importance in biological control. Insights are derived from both theoretical and empirical results, primarily the latter.

Briefly, since an individual predator destroys more prey than the parasite, it would appear superficially that the predator is the better biological control agent. However, closer theoretical consideration of results at the *population* level, and the empirical record itself, seem to favor parasites over predators, usually, for the following reasons: in general, parasites are more host-specific* (see below and Chapters 5 and 6), better adapted and synchronized in interrelationships, have lower food requirements per individual thereby maintaining a balance with their host species at a lower host density (better control), and the young (the less mobile stages) do not need to search for food (Doutt and DeBach, 1964). Huffaker *et al.* (1969) noted that for natural populations to be equally as good in maintaining a low prey density, a more voracious species would have to have a compensating advantage in higher searching capacity. Yet, as Doutt and DeBach cautioned, there are many examples of excellent biological control by predators and many more will probably be shown to exist. It would be quite wrong to deprecate the value of predators. Each situation must be independently determined. Some of the most striking successes from introductions involve predators, and predators stand essentially alone (there are no known parasitoids) in the substantial degree of naturally occurring biological control of spider mites throughout the world (Huffaker *et al.*, 1969). Doutt and DeBach (1964) further comment that in some enemy complexes the predators are clearly dominant, and although there has been a tendency to play down the role of predators, there is now an awareness of the stabilizing effect of this massive reservoir of naturally occurring predators. However, about two-thirds of the successes from biological control introductions up to about 1963 have resulted from parasites (DeBach, 1964a), and this proportion is higher when the more recent examples are included (Appendix).

D. Specific Versus General Parasites or Predators

The great majority of successful biological control introductions have involved rather highly host-specific enemy species. We caution that some rather stenophagous species may be practically monophagous in given habitats. The more general parasites or predators (euryphagous species) tend to serve in overall community balance and as rearguard control factors when the more effective highly host-specific species (in stable habitats) may have failed locally in the control of given host species. Host-specific species are more commonly the main reliable control factors holding given species at low densities in stable habitats. Huffaker *et al.* (1971) state:

* Many predators, however, are far more restricted in their range of acceptable prey than commonly supposed and many are relatively prey-specific in particular habitats (Thompson, 1951b).

> Evolution has served both to promote monophagy and to promote and maintain polyphagy in relation to the organisms and the environment. The value of a broad diet is obvious: if one food (prey) is scarce another can be substituted. By their very nature, specialists are better adapted to utilize a specific prey at low prey densities in maintaining their own populations. They are more closely synchronized in their habits, haunts and seasonal life phases, and are normally better attuned in nutritional needs, reproductive potential and searching behavior to effectively utilize their prey at a minimal prey density. Thus, they are more effective and reliable biological control agents.
>
> Since a polyphagous predator (or parasite) can hardly be equally well adapted to utilization of each of its many acceptable prey species, it could hardly be as efficient or as reliable as a control or regulating agent, relative to a prey species also attacked by a monophagous predator (other qualities being equal).

Doutt and DeBach (1964) also note that in disturbed situations as in annual crop situations where the host species is frequently much depressed by other factors (Ehler *et al.*, 1973), the highly host-specific natural enemy will suffer most, whereas the more general feeder will maintain itself nearby on other species, and be at hand to retain control of the given species when it starts to increase again. Moreover, they note that in any area of great ecological diversity and steep climatic gradients, a given natural enemy may not be equally effective (or even able to exist) everywhere. Thus, the high degree of specificity in host relations also has its counterpart in a high degree of specificity in adaptation to the environment. The complex additional demands for survival seem to mean that a high degree of host specificity is correlated with less adaptability to environmental differences. We tend to find, then, that excellent biological control of a widely distributed host species throughout most of its range by parasitoids will be the result of a *complex* of often highly specific species of a single genus or perhaps different strains of the species. We are just beginning to tap this reservoir of expanded possibilities. Clausen (1936) had explored the idea, then commonly held, that "... a given species had a definite and fixed capacity in relation to its host and that this applied throughout its range of distribution." This idea led to the practice of obtaining a species from a single place and often where it could be most easily obtained. Clausen urged a broader search for genetic material. (See, in addition, Doutt and DeBach, 1964; Messenger and van den Bosch, 1971; and Chapter 9 in this treatise.)

E. The Role of Hyperparasites

Very little is known about the role of secondary natural enemies, including parasitic ones (hyperparasites). Hyperparasitism is the phenomenon of parasitism on another parasite. The major question posed is whether or not secondary parasites are a significant factor in reducing the efficiency of primary parasites, and on the contrary, whether or not such a parasite may be beneficial to a more stable and better biological control by its host, the primary parasite, possibly through damping the primary host—parasite interaction. Muesebeck and Dohanian (1927) considered, on the one hand, that resident hyperparasites may overwhelm and prevent the establishment of primary parasites being introduced [there are no data showing that this has occurred (Doutt and DeBach, 1964)], and on the other, that hyperparasites may have a beneficial, stabilizing role. On theoretical grounds, Nicholson (1933) and Flanders (1943) proposed that the concern over the possible adverse effect of hyperparasites has been overstated.

It has long been established policy in the quarantine operation to eliminate all hyperparasites. While there is little solid data to support the practice, prudence dictates it in the absence of countermanding evidence. Biological control can operate at various trophic levels. The phytophagous insect may, as a primary natural enemy of the plant, control its host plant's abundance (biological control of weeds, Chapter 19) or it may be prevented from exerting this role by the action of a primary parasite (classical or natural biological control), or the primary parasite may be circumvented in its role by a secondary species, or hyperparasite. It seems logical that natural selection will have established each of these forms of biological control under the multitudinous variety of conditions found in nature. Hairston *et al.* (1960) suggested a structured heirarchy of biological control: since vegetation is seldom severely utilized by herbivores, the carnivores (here the primary parasites of phytophagous insects) are responsible as biological control agents, and furthermore the secondary parasites are ineffective, or else the vegetation would not be protected from the phytophagous species by the primary parasites. Huffaker (1962) clarified certain aspects of this proposal, in general agreed with the main thesis, but noted that a great many more individual species of plants among the mass of vegetation referred to may actually be under good but inconspicuous biological control.

Telford (1961) has suggested that hyperparasites may have a real role in triggering the rare and unpredictable outbreaks of some insects, e.g., lodgepole needle miner, *Coleotechnites* (*Recurvaria*) *milleri* (Busck). Also Clark (1964) found a very high level of parasitization of the primary parasites of the eucalyptus psyllid, *Cardiaspina albitextura* Taylor. Perhaps in part because of this, parasites were ineffective as regulating factors. Also, DeBach (1949) used an insecticide check method of evaluation which selectively retarded the hyperparasite, *Lygocerus* sp., which attacks the mealybug parasite, *Anarhopus sydneyensis* (Timb.). The results suggested that the hyperparasitic species effectively reduced the degree of biological control of the mealybug.

Doutt and DeBach (1964) noted, "It is perhaps significant that the introduced primary parasites that have brought about control of a pest are species that are equally successful in their native habitats, in spite of attack by their natural enemies (the secondary parasites and predators)." It appears that those secondary species are not very effective, and it is illogical to presume that the tertiary species are the reason for this, for they would not in general be expected to be more effective than the secondaries. Rather, the farther up the trophic pyramid, the more generalized are the natural enemies in their host or prey acceptance. They would be expected to act more as opportunists or community stabilizers and as damping factors relative to lower level interactions.

In summary, the burden must lie in *proving* that a hyperparasite would be beneficial in a new situation. It is not likely that we can prove in advance that a hyperparasite would have a beneficial effect in a new environment even if this did occur in the native home region or in laboratory populations. The total faunal, floral, and meteorological complex is never duplicated exactly. Thus, there are as yet no grounds for relaxing the rigid practice of excluding the hyperparasites, but the matter should be investigated.

F. The Time Factor in Biological Control

An analysis of many of the successful cases of biological control by Clausen (1951) led him to conclude that an effective introduced natural enemy should show evidence of commercial control of a pest at the point of release within a period of three host generations or 3 years. Clausen stated that the attempted "colonization of an imported parasite or predator may well be discontinued after 3 years if there is still no evidence of establishment," provided: "(a) that colonization has been effected in each distinct climatic zone occupied by the host; (b) that the colonies were adequate in size and number; (c) that releases were synchronized with the time of abundance of the preferred host stages; (d) that recovery collections were adequate; and (e) that no biological factor directly affecting continued reproduction is involved (i.e., sex differentiation in host relationships, as in *Coccophagus*)." He further stated that a parasite or predator that requires colonization beyond this time may become established but will be of little real value and will not compensate for the additional costs involved in its establishment.

Acceptance of Clausen's hypothesis would seem to preclude from consideration parasites and predators which by themselves would give little "commercial control" but which in combination might reduce their host to nonpest status. Sellers (1953) concluded that the time element for evaluation of partially effective parasites which are to become useful may extend beyond 3 years and to adopt Clausen's hypothesis in such cases would involve a danger of prematurely discontinuing projects on potentially valuable species. Thompson (1951a) remarked that the qualities which a parasite must possess to be successful in Clausen's view are very restrictive, and he (1927) had earlier shown theoretically that a parasite with a low rate of multiplication and a small initial population in relation to the host population may take much longer than Clausen allows to overtake and reduce its host population.

Clausen (1951) supported his hypothesis with many examples of successful biological control and pointed out that few successful examples exist of introduced parasites which have taken more than 3 years to become successful. However, Tooke (1953) reported that the eucalyptus snout beetle *Gonipterus scutellatus* Gyll. was controlled by the mymarid egg parasite *Patasson nitens* (Grlt.) within 3 years in the coastal area of the southwest Cape of South Africa but was not controlled at high altitudes until 5—15 years after the introduction of *P. nitens*. Also, in the most favorable areas (e.g., Humboldt County, California), the control of St. Johnswort by *Chrysolina quadrigemina* Suffrian was accomplished in 3 years, but not until 5 or 6 years in some areas, e.g., parts of Shasta County, and even longer in British Columbia, Canada (below).

The possibility exists that an exotic natural enemy which has the genetic variability to adapt to a climate which differs to some degree from that of its native area would not exert a significant reduction in its host within three generations or 3 years yet could exert "commercial control" once selection has occurred in the release area. This has happened over a period of years with respect to *C. quadrigemina* in the control of St. Johnswort in parts of British Columbia (Harris *et al.*, 1969). According to DeBach (1965), "No case is known, however, where an established colonized species which was originally obviously ineffective in host-population regulation later produced completely successful biological control." However, DeBach *et al.* (1962) cite two cases of pest colonization where adaptation may have occurred for less than completely efficient natural enemies. *Comperiella bifasciata* How., a parasite of California red scale, *Aonidiella aurantii* (Mask.), was imported in 1941 yet did not become widely established until 1951, continuing to spread until 1965. Clausen (1956) and DeBach and Sundby (1963) reported that *Agathis diversus* (Mues.), a parasite of the Oriental fruit moth, *Grapholitha molesta* (Busck), was imported from Japan to New Jersey from 1933 to 1936, yet declined after initial releases and did not become firmly established until 1943. In Japan this parasite attacks larvae in peach twigs early in the season yet in the United States it changed its habits to attack larvae in mature peach fruits. (See also Chapter 9.)

G. Salubrious and Rigorous Environments

Climate is the most important factor limiting the worldwide distribution of insects. This applies to natural enemies as well as pest species. Thus salubrious environments are felt to be more conducive to successful biological control than rigorous environments. This idea was reinforced by the numbers of early successful projects in tropical areas, principally islands. However, DeBach (1965) (and the Appendix) show that many successes (more than 50% of the total) have occurred above or below the 30th parallel: the cool temperate climates of Australia (Tasmania), Canada (British Columbia), New Zealand, and the United States (outside of California and Hawaii).

Also, the large number of successes which have occurred in warm climates is partially due to the relatively greater abundance of homodynamic species in tropical versus temperate climates. These are species that reproduce continuously throughout the year and whose development is arrested only by adverse conditions such as cold or drought (Wigglesworth, 1965). In temperate climates most insects enter true diapause with a prolonged arrest of growth during which synchrony of natural enemies and their hosts becomes a critical factor for success of the natural enemies. Homodynamic species usually do not have discrete generations; each stage is continuously available for attack by its natural enemies. This reduces the importance of precise parasite—host or predator—prey synchrony. Other obvious advantages of salubrious climates which contribute to successful colonization of natural enemies are the numbers of potential alternate hosts available and the avoidance of the problem of adaptation to a rigorous environment.

H. The Island Theory in Biological Control

Many of the successes in biological control have occurred in such areas as Hawaii, Bermuda, Fiji, Celebes, Guam, Mauritius, Seychelles, and Tasmania. Because early successes often were concentrated in insular areas the hypothesis was advanced that biological control could be effective only on islands (Imms, 1931). However, this concentration was partly because initial efforts in biological control were heavily focused on these islands.

Many of the early successes in such areas as Tasmania, Hawaii, Fiji, and Madagascar have since been repeated in continental areas such as Australia, California, Portuguese West Africa, and South Africa. Many more recent projects have been successful on continents as well as islands, as shown by the data in Tables A-2 and A-4 of the Appendix. Of the 102 projects for insects which are regarded as complete successes, 67 or 66% occur on continental areas. In the case of substantial but not complete successes, 86 out of 144 or 59% have been reported for continental areas and 56% of 81 partial successes are recorded from continents. For weeds, 37 of 57 rated successes (65%) have occurred on continental areas, thus making the hypothesis untenable.

V. THE ECONOMICS OF BIOLOGICAL CONTROL

It is relatively simple to obtain data on the costs and benefits for chemical methods of pest control. The costs of research and development of new effective pesticides, and expenditures on research which result in little commercial value can be readily estimated by chemical pesticide manufacturers, as can the costs of promoting and advertising a new pesticide to the stage where it is available for purchase and field application. This is one aspect of the costs of pesticides. The second stage, again where cost/benefit figures are comparatively easy to obtain, is the actual application of the chemical. Costs of pesticide, equipment, and labor for any given operation can be calculated, as can the cash benefits due to diminution of losses (both in quantity and of produce) from the pests, if there are untreated control plots. What is not put on any balance sheet of this sort is the "cost" (not necessarily in terms of cash) of side effects such as reduction in natural enemies (or development of resistance) which result in the necessity for additional recurrent costs of pesticide applications because natural control has been eliminated. All these effects, which are now grouped together as pollution of the environment, may well in the long term be extremely important and costly to remedy. Thus, apart from any accurate evaluation of the undesirable side effects resulting from chemical pesticide applications, the cost/benefit figures are fairly readily calculable.

Biological control is in a somewhat different and less satisfactory position because cost/benefit figures are often very difficult to obtain and assess. These figures comprise the costs of research into a particular problem with a view to finding suitable natural enemies for introduction into areas where control of the pest is required, followed by the collection, quarantine handling, and breeding of these beneficial insects, their release in the target area, and an evaluation of their effectiveness against the pest. This phase corresponds with the research, development, and marketing of a given chemical pesticide. However, from this stage on there is a complete difference in the nature of the economic costs and benefits. Another major distinction is that the pesticide developer and promoter is a private manufacturer while the biological control developer is usually a tax-supported institution, federal or state.

The user of chemicals has to purchase and reapply them and absorb the recurring annual costs. Classical biological control entails virtually no further costs, except possibly that of distributing the successful natural enemy widely, rather than wait for it to spread naturally. Thus, in general, the economic advantage is overwhelmingly with biological control in the case of its classical use. It is, however, obvious that while the costs of research, development, and marketing of a chemical pesticide are considerable [estimated in 1970 to be $8 million in the United States (Wierenga, 1971)], the resulting pesticide is usually capable of dealing with a number of pests, whereas biological control in each instance is aimed at only one or at the most a very few closely related pest species. However, as already pointed out the recurrent costs of chemical applications due to resistance, resurgence, and secondary pest outbreaks have no counterpart in classical biological control, nor do the possible long-term costs involved in dealing with any resulting environmental pollution such as chemical residues or effects on applicators or nontarget species.

We will now give some definite cost/benefit figures for specific biological control projects. However, before doing so it may be worthwhile to cite some comparative figures for the funds expended on research in the field of biological control and that spent by the chemical industry to develop pesticides. We stress again that the cost of biological control *includes* the application of the control methods, whereas the cost of research and development in pesticides only involves bringing the product onto the market. From then on the chemical user expends additional funds per unit area treated every time he wishes to control a pest. The total amounts by all users are enormous. Bearing in mind the known costs of biological control work in several areas and approximately the number of workers in this field, it was estimated in 1967 that the worldwide expenditure on biological control in its broad sense was some $10,000,000 per year (Simmonds, 1967) compared with research to *develop* chemical pesticides, not including the cost of their use, of $84,000,000 per year (Galley and Stevens, 1966). These figures, because of both currency inflation and increased stress on research, are now in 1974 undoubtedly higher, possibly proportionately so in both cases.

A. Some Successes by CIBC

The total costs of the operation of the Commonwealth Institute of Biological Control (CIBC), including the costs of building laboratories, equipment, staff, the conduct of the specific projects, and the publication of results and reports, had been since its inception in 1927 up to 1966 approximately $3,700,000. In the next 5 years the costs rose from $542,500 to $657,000 per annum with a total from 1966 to 1971 of approximately $3,040,000, for a grand total of some $6,750,000. Because of the continuous decrease in the value of money a simple addition of costs such as this gives a wrong impression. The large steady increase in annual expenditures over the years does, however, indicate not only this devaluation but also a considerable increase in the work itself by CIBC.

As a result of this work of CIBC, i.e., since 1927, there have been a number of major successes where the actual cost of the control achieved can be assessed fairly accurately. Some were detailed by Simmonds (1967). In addition there is the very recent spectacular control of *Diatraea saccharalis* (Fab.) in Barbados (see Alam *et al.*, 1971) about which Simmonds (1967) had stated, "For years attempts have been made, and are continuing, to control the sugarcane borer, *D. saccharalis*. in Barbados, so far without success. . . . " With perserverance, this situation changed markedly. These successes, where cost/benefit figures are available, are given in Table 1. Thus the eight CIBC projects alone, which cost a total of approximately $610,000, have accrued a total benefit to date of some $29,472,000, with a continuing annual benefit of some $2,175,000 or about 350% yearly on the original investment, or 330% of the present total annual costs of CIBC.

However, these are by no means the only successes achieved from CIBC's work over this 40-year period, and with some of these additional successes a rough idea of the financial implications can be given.

In Nevis, West Indies, the cactus *Opuntia triacantha* was a very serious pest of pastures and rendered thousands of acres of grassland unusable. This cactus was spectacularly controlled during 1957–1960 by means of *Cactoblastis cactorum* (Berg)—at a total cost of some $1360 (including visits to make releases and check on results). To achieve the same results with herbicides—not that this would in fact have been carried out—would have continuously cost between $25,000 and $50,000 per year (see Simmonds and Bennett, 1966; Simmonds, 1967).

The 1888 success of *Rodolia cardinalis* (Muls.) against *Icerya purchasi* (Mask.) in California has been repeated many times since, almost wherever the pest has been introduced (DeBach, 1964a). In the West Indian Islands of St. Kitts, Nevis, and Montserrat this scale only became established about 1964, and was very seriously

TABLE 1

Cost/Benefit Figures for Several Biological Control Successes Resulting from CIBC Work Where Fairly Accurate Data Are Available

Pest species controlled	Area	Date of control	Total cost (in dollars)	Benefit per annum (in dollars)	Percentage of cost	Total value of benefit to 1973 (in dollars)	Reference
Aspidiotus destructor Sign. Coconut scale	Principe, West Africa	1956	10,000	180,000	1800	3,060,000	Simmonds (1960)
Diatraea saccharalis (Fab.) Sugarcane borer	Antiqua, West Indies	1931–1945	21,250	41,250	194	1,562,500	Box (1960) (to 1968 when sugar production ceased)
Sugarcane borer	St. Kitts, West Indies	1934	500	125,000	25,000	4,875,000	Box (1960)
Sugarcane borer	St. Lucia, West Indies	1933	2500	30,000	1200	1,050,000	Box (1960) (to 1966 when sugar production ceased)
Sugarcane borer	Barbados	1967	150,000	1,000,000	667	6,000,000	Alam *et al.* (1971)
Cordia macrostachya (Jacquis) Roemer and Schultes Black sage	Mauritius	1952	25,000	250,000	1000	5,250,000	Simmonds (1967)
Operophtera brumata (L.) Winter moth	Nova Scotia, Canada	1954–1962	150,000	175,000	117	1,925,000	CIBC (1971)
Diprion hercyniae (Hart.) European spruce sawfly	Canada	1932–1946	250,000	375,000	150	5,750,000	CIBC (1971)
Opuntia megacantha Salm-Dyck. Cactus	South Africa	1950	42,500	237,500	560	5,475,000	Pettey (1950)[a]
Planococcus kenyae (LePelley) Coffee mealybug	Kenya, East Africa	1939	75,000	1,250,000	1667	42,500,000	Melville (1959)[a]

[a] Not from CIBC work.

affecting citrus and pigeon peas when *R. cardinalis* was introduced. It is impossible to give accurate cost/benefit figures, except in terms of what control by chemical insecticides would cost. Introductions of *Rodolia* into all these islands (again with visits to make releases and to check results) cost under $1250 per year. If control had been by chemical applications, it would have cost some $6250 per year. Assuming that $625 per year will still have to be spent against scale insects in Montserrat, particularly on limes, then the net annual saving from the $6250 total invested is $5625.

It is quite obvious that considerable economic benefits have accrued from other CIBC efforts but it is not possible to assess their costs and benefits. Rhodesgrass scale *Antonina graminis* (Mask.) was successfully controlled at little cost in the southwest United States following introduction of a parasite, *Neodusmetia sangwani* (Rao), into Texas from the CIBC Indian Station in 1959. The scale population was reduced by some 36–70%, with a 30% increase in forage grasses over a vast area (Dean *et al.*, 1961). No accurate cost/benefit figures for this project are available but the benefit is obviously great.

The mealybug parasite, *Coccophagus gurneyi* Comp., obtained from the CIBC California Station in 1960, was reared and released in Uzbekistan, U.S.S.R., against the mealybug *Pseudococcus gahani* Green, a serious pest of citrus. By 1966, some 200,000 had been released over 180 hectares. Parasitism was 95–98% with a resultant negligible damage by the mealybug, where 50–70% loss of fruits had been usual before introduction of the parasite. It is now difficult to find the mealybug, and *C. gurneyi* is as effective in the U.S.S.R. as it is in California. Again no accurate cost/benefit figures are available.

The citrus blackfly *Aleurocanthus woglumi* Ashby has been very successfully controlled in the past 10 years in the Seychelles, South Africa (in 1965), East Africa (in 1964), and Barbados (in 1967) following similar control earlier in Cuba, Jamaica, the Bahamas, and Mexico; pigeon pea pod borers have been controlled in Mauritius (in 1958) by parasites from the West Indian Station; oak leaf miner, *Lithocolletis messaniella* Zell., in New Zealand has been controlled with parasites from the European Station; larch sawflies (in 1965) and pine sawflies (in 1964) in some areas of Canada have been controlled with parasites from the same station (Muldrew, 1967); clover casebearer in New Zealand, potato tubermoth in Australia, Cyprus, Zambia, etc. It has not been possible as yet to obtain monetary assessments of the benefits derived from these projects, but obviously in aggregate they are considerable.

Thus, when the additional actual and potential future benefits mentioned above are also considered it is very clear that the economic return from funds invested in biological control by CIBC is far in excess of the 330% of the present annual expenditure as indicated above.

However, it must be remembered that this total overall return does not mean that every project undertaken will result in a similar economic success. Many projects, and this includes some which are very well financed, are unfortunately complete failures, and it is virtually impossible to predict a priori which projects will be successful, even marginally. If this were true the benefits related to the costs of the CIBC operation would be even better.

B. Successes in California

California has strongly supported biological control of pest insects for many years. This research support, as well as the varied agriculture and the salubrious climate throughout much of the state, have led to many successful biological control projects and considerable savings to agriculture. DeBach (1964a) presented the general picture of costs/benefits to California due to biological control from 1923 to 1959. These data are used here and supplemented with results subsequent to 1959 (Table 2). The estimates of Table 2 have been purposefully evaluated conservatively. Many factors, such as discussed above, complicate this type of cost/benefit analysis (for instance, beef relative to rangeland formerly occupied by Klamath weed), including especially the likelihood of a wider distribution of former pests now under successful biological control, the increase in costs of chemical controls, and the value to society

TABLE 2
Estimates of Savings to the Agricultural Industry in California from 1928–1973 through Major Successful Biological Control Projects[a]

Biological control project	Degree of success	Yearly savings over previous losses plus pest control costs (in dollars)	Total savings to 1973 (in dollars)
Black scale on citrus	Partial to complete	1,684,000 (1940–1959) 2,100,000 (1959–1973)[b]	59,712,000
Citrophilus mealybug	Complete	2,000,000 (1930–1959) 2,500,000 (1959–1973)[b]	91,000,000
Grape leaf skeletonizer	Partial to complete	75,000 (1945–1956) 1,000,000 (1956–1973)[b]	2,050,000
Klamath weed	Complete	19,200,000 (1953–1973)[c] 2,000,000—weight gain in cattle (1953–1959) 2,500,000—weight gain in cattle (1959–1973)[b]	66,200,000
Olive parlatoria scale	Complete	465,000 (1962–1966) 725,000 (1967–1973)	7,400,000
Spotted alfalfa aphid	Substantial	5,580,000 (1958–1959) 3,000,000 (1959–1973)	47,580,000
Walnut aphid	Substantial	250,000 (1970–1973)	1,000,000
			274,942,000

[a] After DeBach (1964a) with additions by the authors.
[b] Projected at a higher rate to account for inflation.
[c] Savings from increased land values plus cessation of chemical control, adjusted for inflation.

through the reduction in use of insecticides. However, a rough balance sheet can be presented.

The Divisions of Biological Control at the Riverside and Berkeley campuses (formerly a single department) of the University of California are responsible for conducting research in biological control for the state of California. The total annual budget ($1,005,700) does not include grants which support some graduate students and postdoctorates who may be peripherally involved on some biological control projects. However, the work of these nonbudgeted individuals would be more than offset by the efforts of regular staff in both divisions that is not devoted to classical biological control work. Many academic staff, at least in Berkeley, have teaching or administrative duties which often leave them little time for research and much of the research they do is focused on integrated control, with emphasis on indigenous enemies and somewhat to the neglect of biological control, especially classical biological control.

One important factor limiting the importation of natural enemies is the small annual expenditure for foreign exploration which is the very heart of biological control. Currently, the budgeted amount, $8400 per year, is shared by the Riverside and Berkeley divisions. This severely limits importations and thus reduces the successes.

Nevertheless, as shown in Table 2 the successes for which we can estimate costs/benefits indicate the enormous savings which can be attributed to biological control. Although the figures are only estimates and have been calculated conservatively, for each dollar spent by these divisions of biological control on research agribusiness in California, $30.00 has been saved to the industry. The list given in Table 2 is by no means complete. Many other successes after 1928 could be cited (Chapter 28) but the lack of cost/benefit information prevents listing such examples as nigra scale, *Saissetia nigra* (Nietn.), and black scale, *Saissetia oleae* (Olivier) and olive parlatoria scale, *Parlatoria oleae* (Colvée), on ornamentals, alfalfa weevil, *Hypera postica* (Gyll.), and pea aphid, *Acyrthosiphon pisum* (Harris), on alfalfa in northern California, and California red scale, *Aonidiella aurantii* (Maskell), and yellow scale, *A. citrina* (Coq.), on citrus.

The figures given by DeBach (1964a) have been updated for the next 14 years and savings have been projected at a higher rate due to inflation. The reader is referred to DeBach (1964b, Chapter 1) for details of the calculations of savings to California agriculture due to biological control of black scale, citrophilus mealybug, grape leaf skeletonizer, Klamath weed, and spotted alfalfa aphid. Two projects, those of olive parlatoria scale and walnut aphid, have been added.

Data for olive scale were obtained with the help of C. E. Kennett.* The olive scale was a severe pest on approximately 12,000 acres of olives in the state when the project was initiated and the scale had the potential to spread (and since has spread) to the remainder of the more than 27,500 bearing acres in the state. Although reduc-

* Division of Biological Control, University of California, Berkeley, California.

tion of the scale population began before 1962 and considerable savings accrued prior to this time, this year was taken as the initial year of significant biological control of this pest. It was estimated that $30 per acre per year was saved through reduction in chemical control costs. Because the spread of this pest to other acreage was slowed greatly and this natural control tends to follow (spread) quickly with the spread of the pest there is thus a potential savings by this project of approximately $825,000 per year in control costs. The total savings to date are estimated at $7,400,000.

The savings to agriculture for the walnut aphid project are more difficult to estimate because the project was initiated too recently for much impact to be felt by the walnut industry. However, R. van den Bosch (personal communication) conservatively estimates that one-half of the losses due to this pest have been saved annually from 1970 to 1973. Greater savings will accrue in the future as more walnut growers turn to biological control. Estimates of crop losses due to walnut aphid (California State Department of Agriculture Annual Crop Loss Reports) averaged $746,000 per year from 1963 to 1969. Thus savings would be approximately $350,000 per year. (The estimate of crop losses due to walnut aphid in 1970 was $328,000, which agrees with our estimate.) Yet, to be conservative, we have taken the annual savings to be $250,000 for a total savings of $1,000,000 for the 1970 to 1973 period.

C. Inundative Releases and Successive Introductions

As an example of an inundative release where cost/benefits can be assessed indirectly, one may cite the mass breeding of the parasite *Opius concolor* Szel. on the factitious host *Ceratitis capitata* (Weid.) and its release by helicopter over olive groves in Sicily for the control of the olive fly, *Dacus oleae* (Gmel.). In 1965–1967, the cost of treatment and control by parasites was 14¢ per tree as against 33¢ per tree with chemicals. This was only an experimental effort and it must be pointed out that while the chemicals prevented attack, parasites did not. The latter, however, ensured that the next generation was of no economic importance, with the added advantage of no secondary hazards (Delangoue, 1970).

One example of reduction of damage serially with successive introductions of parasites is that of the sugarcane borer, *Diatraea saccharalis* (Fab.), in Guadeloupe (Simmonds, 1959). Unfortunately, there are no accurate "joint infestation counts," the standard method of assessing borer damage and sugar loss, prior to the introduction of *Metagonistylum minense* Tns. into Guadeloupe about 1940. However, planters contend there was a "considerable reduction" in damage following this introduction. With increasing use of susceptible varieties of sugarcane, borer damage increased to around 20–25% joint infestation by 1947–1948 when introductions of *Lixophaga diatraea* (Tns.) were made. Following its successful establishment, joint infestation (and hence damage) decreased to about 10 to 15%. In 1954, *Paratheresia claripalpis* (Wulp) was introduced from Trinidad and became established and this resulted in a further reduction in sugarcane borer damage. Joint infestation at Beauport has been assessed as follows:

	(1)			(2)			
Pre 1949	1949	1952	1953	1954	1955	1956	1957
10–15%	12.81	7.32	7.77	5.90	5.07	3.07	2.53

At (1) the effect of the *Lixophaga* release was becoming apparently felt, with a reduction from 13 to 5–6% in joint infestation. From (2), following the introduction of *Paratheresia*, a further reduction occurred over 3 years. The records indicate *Metagonistylum* had a considerable initial effect in reducing borer attack, that of *Lixophaga* decreased the attack further, and that of *Paratheresia* caused a still further reduction to a satisfactory level of 3%. Each parasite contributed toward the final result, and although in competition for hosts the net result was supplemental.

VI. CONCLUSION

With the available evidence of the marked economic benefits derivable from the use of biological control (this chapter and Chapters 12–20), the practice of biological control is seen to be largely empirically based. It is, nevertheless, theoretically sound, and there is little reason for the hesitancy with which decisions to engage in biological control are approached. With a more solid financial and philosophical commitment there is no reason why biological control, in both its classical and other aspects, should not make much greater progress than it has.

REFERENCES

Alam, M. M., Bennett, F. D., and Carl, K. P. (1971). Biological control of *Diatraea saccharalis* (F.) in Barbados by *Apanteles flavipes* Cam. and *Lixophaga diatraeae* T. *Entomophaga* **16**, 151–158.

Andrewartha, H. G., and Birch, L. C. (1954). "The Distribution and Abundance of Animals," 782 pp. Univ. Chicago Press, Chicago, Illinois.

Auslander, D. M., Oster, G. F., and Huffaker, C. B. (1974). Dynamics of interacting populations. *J. Franklin Inst.* **297**, 345–376.

Bess, H. A., and Haramoto, F. H. (1958). Biological control of the oriental fruit fly in Hawaii. *Proc. 10th Int. Congr. Entomol.* **4**, 835–840.

Birch, L. C. (1960). Stability and instability in natural populations. *N. Z. Sci. Rev.* **20**, 9–14.

Box, H. E. (1960). Status of the moth borer, *Diatraea saccharalis* (F.), and its parasites in St. Kitts, Antigua and St. Lucia, with observations on Guadeloupe and an account of the situation in Haiti. *Proc. 10th Congr. Int. Soc. Sugarcane Technol.*, pp. 901–914.

Burnett, T. (1954). Influences of natural temperatures and controlled host densities on oviposition of an insect parasite. *Physiol. Zool.* **27**, 239–248.

Burnett, T. (1958). Effect of host distribution on the reproduction of *Encarsia formosa* Gahan (Hymenoptera: Chalcidoidea). *Can. Entomol.* **90**, 179–191.

Carrick, R. (1963). Ecological significance of territory in the Australian magpie *Gymporhina tibicens*. *Proc. 13th Int. Ornithol. Congr.*, pp. 740–753.

Clark, L. R. (1964). The population dynamics of *Cardiaspina albitextura* (Psyllidae). *Aust. J. Zool.* **12**, 362–380.

Clausen, C. P. (1936). Insect parasitism and biological control. *Ann. Entomol. Soc. Amer.* **29**, 201–223.

Clausen, C. P. (1951). The time factor in biological control. *J. Econ. Entomol.* **44**, 1–9.

Clausen, C. P. (1956). Biological control of insect pests in the continental United States. *U.S. Dept. Agr. Tech. Bull.* **1139**, 151 pp.

Clausen, C. P. Introduced Parasites and Predators of Arthropod Pests and Weeds: A World Review, *U.S. Dept. Agr. Handbk.* **480**, in press.

Commonwealth Institute of Biological Control. (1971). Biological programmes against insects and weeds in Canada 1959–1968. *Commonw. Agr. Bur. Tech. Commun. No.* **4**, 266 pp.

Dean, H. A., Schuster, M. P., and Bailey, J. C. (1961). The introduction and establishment of *Dusmetia sangwani* on *Antonina graminis* in South Texas. *J. Econ. Entomol.* **54**, 952–954.

DeBach, P. (1949). Population studies of the long-tailed mealybug and its natural enemies on citrus trees in southern California, 1946. *Ecology* **30**, 14–25.

DeBach, P. (1964a). Successes, trends and future possibilities. *In* "Biological Control of Insect Pests and Weeds" (P. DeBach, ed.), pp. 673–713. Reinhold, New York.

DeBach, P. (ed.). (1964b). "Biological Control of Insect Pests and Weeds," 844 pp. Reinhold, New York.

DeBach, P. (1965). Some biological and ecological phenomena associated with colonizing entomophagous insects. *In* "Genetics of Colonizing Species" (H. G. Baker and G. L. Stebbins, eds.), pp. 287–306. Academic Press, New York.

DeBach, P. (1966). The competitive displacement and coexistence principles. *Annu. Rev. Entomol.* **11**, 183–212.

DeBach, P., Landi, P. J., and White, E. B. (1962). Parasites are controlling red scale in southern California citrus. *Calif. Agr.* **16** (12), 2–3.

DeBach, P., Rosen, D., and Kennett, C. E. (1971). Biological control of coccids by introduced natural enemies. *In* "Biological Control" (C. B. Huffaker, ed.), pp. 165–194. Plenum, New York.

DeBach, P., and Sundby, R. A. (1963). Competitive displacement between ecological homologues. *Hilgardia* **34**, 105–166.

Delangue, P. (1970). Utilisation d'*Opius concolor* Szepl. en vue de la lutte contre *Dacus oleae* Gmel. *Colloq. Franco-Sovietique l'Utilization des Entomophages, Antibes, France, May 13–18, 1968*, pp. 63–69.

Doutt, R. L. (1964). Biological characteristics of entomophagous adults. *In* "Biological Control of Insect Pests and Weeds" (P. DeBach, ed.), pp. 145–167. Reinhold, New York.

Doutt, R. L., and DeBach, P. (1964). Some biological control concepts and questions. *In* "Biological Control of Insect Pests and Weeds" (P. DeBach, ed.), pp. 118–142. Reinhold, New York.

Ehler, L. E., Eveleens, K. G., van den Bosch, R., (1973). An evaluation of some natural enemies of cabbage looper on cotton in California. *Environ. Entomol.* **2**, 1009–1015.

Ehler, L. E., and van den Bosch, R. (1974). An analysis of the natural biological control of *Trichoplusia ni* (Lepidoptera: Noctuidae) on cotton in California. *Can. Entomol.* **106**, 1067–1073.

Flanders, S. E. (1943). Indirect hyperparastism and observations on three species of indirect hyperparasites. *J. Econ. Entomol.* **36**, 921–926.

Franz, J. M. (1961). Biological control of pest insects in Europe. *Annu. Rev. Entomol.* **6**, 183–200.

Galley, R. A. E., and Stevans, J. G. R. (1966). Pesticide research today. *Span* **9**(2), 107–109.

Hagen, K. S., van den Bosch, R., and Dahlsten, D. L. (1971). The importance of naturally occurring biological control in the western United States. *In* "Biological Control" (C. B. Huffaker, ed.), pp. 253–293. Plenum, New York.

Hairston, N. A., Smith, F. E., and Slobodkin, L. B. (1960). Community structure, population control, and competition. *Amer. Natur.* **94**(897), 421–425.

Harpaz, I., and Rosen, D. (1971). Developments of integrated control programs for crop pests in Israel. *In* "Biological Control" (C. B. Huffaker, ed.), pp. 458–468. Plenum, New York.

Harris, P., Peschken, D., and Milroy, J. (1969). The status of biological control of the weed *Hypericum perforatum* in British Columbia. *Can. Entomol.* **101**, 1–15.

Hassell, M. P. (1966). Evaluation of parasite and predator responses. *J. Anim. Ecol.* **35**, 65–75.

Hassell, M. P., and Huffaker, C. B. (1969). The appraisal of delayed and direct density dependence. *Can. Entomol.* **101**, 353–361.

Hassell, M. P., and Varley, G. C. (1969). New inductive population model for insect parasites and its bearing on biological control. *Nature (London)* **223**, 1133–1137.

Holling, C. S. (1959). The components of predation as revealed by a study of small mammal predation of the European pine sawfly. *Can. Entomol.* **91**, 293–320.

Huffaker, C. B. (1962). Some concepts on the ecological basis of biological control of weeds. *Can. Entomol.* **94**, 507–514.

Huffaker, C. B. (1967). A comparison of the status of biological control of St. Johnswort in California and Australia. *Mushi (Suppl.)* **39**, 51–73.

Huffaker, C. B. (1971a). The phenomenon of predation and its roles in nature. *In* "Dynamics of Populations" (P. J. den Boer and G. R. Gradwell, eds.), pp. 327–341. Centre for Agr. Publ. Doc., Wageningen.

Huffaker, C. B. (ed.). (1971b). "Biological Control," 511 pp. Plenum, New York.

Huffaker, C. B. (1974). Some ecological roots of pest control. *Entomophaga* **19**, 371–389.

Huffaker, C. B., and Kennett, C. E. (1966). Studies of two parasites of olive scale, *Parlatoria oleae* (Colvée). IV. Biological control of *Parlatoria oleae* (Colvée) through the compensatory action of two introduced parasites. *Hilgardia* **37**, 283–335.

Huffaker, C. B., and Kennett, C. E. (1969). Some aspects of assessing efficiency of natural enemies. *Can. Entomol.* **101**, 425–447.

Huffaker, C. B., and Laing, J. E. (1972). "Competitive displacement" without a shortage of resources? *Res. Population Ecol.* **14**, 1–17.

Huffaker, C. B., and Messenger, P. S. (1964). The concept and significance of natural control. *In* "Biological Control of Insect Pests and Weeds" (P. DeBach, ed.), pp. 74–117. Reinhold, New York.

Huffaker, C. B., Kennett, C. E., Matsumoto, B., and White, E. G. (1968). Some parameters in the role of enemies in the natural control of insect abundance. *In* "Insect Abundance" (T. R. E. Southwood, ed.), pp. 59–75. Blackwell, Oxford.

Huffaker, C. B., van de Vrie, M., and McMurtry, J. A. (1969). The ecology of tetranychid mites and their natural control. *Annu. Rev. Entomol.* **14**, 125–174.

Huffaker, C. B., Messenger, P. S., and DeBach, P. (1971). The natural enemy component in natural control and the theory of biological control. *In* "Biological Control" (C. B. Huffaker, ed.), pp. 16–67. Plenum, New York.

Imms, A. D. (1931). "Recent Advances in Entomology," 374 pp. Churchill, London.

Luck, R. F. (1971). An appraisal of two methods of analyzing insect life tables. *Can. Entomol.* **103**, 1261–1271.

MacFadyen, A. (1957). "Animal Ecology: Aims and Methods," 255 pp. Pitman and Sons, London.

MacPhee, A. W., and MacLellan, C. R. (1971). Cases of naturally occurring biological control in Canada. *In* "Biological Control" (C. B. Huffaker, ed.), pp. 312–328. Plenum, New York.

Melville, A. R. (1959). The place of biological control in the modern science of entomology. *Kenya Coffee* **24**, 81–85.

Messenger, P. S. (1969). Bioclimatic studies of the aphid parasite *Praon exsoletum*. 2. Thermal limits to development and effects of temperature on rate of development and occurrence of diapause. *Ann. Entomol. Soc. Amer.* **62**, 1026–1031.

Messenger, P. S., and van den Bosch, R. (1971). The adaptability of introduced biological control agents. *In* "Biological Control" (C. B. Huffaker, ed.), pp. 68–92. Plenum, New York.

Milne, A. (1962). On the theory of natural control of insect populations. *J. Theoret. Biol.* **3**, 19–50.

Morris, R. F. (1959). Single-factor analysis in population dynamics. *Ecology* **40**, 580–588.

Morris, R. F. (1963). Predictive population equations based on key factors. *Mem. Entomol. Soc. Can.* **32**, 16–21.

Muesebeck, C. F. W., and Dohanian, S. M. (1927). A study of hyperparasitism with particular reference to the parasites of *Apanteles melanoscelus* (Ratzeburg). *U.S. Dept. Agr. Bull.* **1487**, 35 pp.

Muldrew, J. A. (1967). Biology and initial dispersal of *Olesicampe* (*Holocremnus*) sp. nr. *nematorum* (Hymenoptera: Ichneumonidae), a parasite of the larch sawfly recently established in Manitoba. *Can. Entomol.* **99**, 312–321.

Nicholson, A. J. (1933). The balance of animal populations. *J. Anim. Ecol. Suppl.* **2**, 132–178.

Nicholson, A. J. (1954). An outline of the dynamics of animal populations. *Aust. J. Zool.* **2**, 9–65.

Pemberton, C. E., and Willard, H. F. (1918). A contribution to the biology of fruit-fly parasites in Hawaii. *J. Agr. Res.* **15**, 419–465.

Pettey, F. W. (1950). The cochineal (*Dactylopius opuntiae*) and the problem of its control in spineless cactus plantations. Pt. I. Its history, distribution, biology and what it has accomplished in the control of prickly pear in South Africa. Pt. II. The control of cochineal in spineless cactus plantations. *Union. S. Afr. Dept. Agr. Sci. Bull.* **296**, 34 pp.

Pimentel, D. (1964). Population ecology and the genetic feed-back mechanism. *In* "Genetics Today" (S. J. Geerts, ed.), Vol. 2, pp. 483–488. Pergamon, New York.

Pimentel, D., and Al-Hafidh, R. (1965). Ecological control of a parasite population by genetic evolution in the parasite-host system. *Ann. Entomol. Soc. Amer.* **58**, 1–6.

Pimentel, D., and Soans, A. B. (1971). Animal populations regulated to carrying capacity of plant host by genetic feedback. *In* "Dynamics of Populations" (P. J. den Boer and G. R. Gradwell, eds.), pp. 313–326. Centre for Agr. Publ. Doc., Wageningen.

Rabb, R. L. (1971). Naturally occurring biological control in the eastern United States, with particular reference to tobacco insects. *In* "Biological Control" (C. B. Huffaker, ed.), pp. 294–311. Plenum, New York.

Ripper, W. E. (1956). Effect of pesticides on balance of arthropod populations. *Annu. Rev. Entomol.* **1**, 403–438.

Sellers, W. F. (1953). A critique on the time factor in biological control. *Bull. Entomol. Res.* **83**, 230–240.

Simmonds, F. J. (1959). The successful biological control of the sugarcane moth-borer, *Diatraea saccharalis* (F.) (Lepidoptera: Pyralidae), in Guadeloupe, F.W.I. *Proc. 10th Congr. Int. Soc. Sugarcane Technol.*, pp. 914–918.

Simmonds, F. J. (1960). Biological control of the coconut scale *Aspidiotus destructor* Sign. in Principe, Portuguese West Africa. *Bull. Entomol. Res.* **51**, 223–237.

Simmonds, F. J. (1967). The economics of biological control. *J. Roy. Soc. Arts* **115**, 880–898.

Simmonds, F. J., and Bennett, F. D. (1966). Biological control of *Opuntia* spp. by *Cactoblastis cactorum* in the Leeward Islands (West Indies). *Entomophaga* **11**, 183–189.

Smith, H. S. (1929). Multiple parasitism: its relation to the biological control of insect pests. *Bull. Entomol. Res.* **20**, 141–149.

Smith, H. S. (1935). The role of biotic factors in the determination of population densities. *J. Econ. Entomol.* **28**, 873–898.

Solomon, M. E. (1957). Dynamics of insect populations. *Annu. Rev. Entomol.* **2**, 121–142.

Stephens, J. M. (1963). Immunity in insects. *In* "Insect Pathology: An Advanced Treatise" E. A. Steinhaus, ed.), Vol. 1, pp. 273–297. Academic Press, New York.

Tanada, Y. (1964). Epizootiology of insect diseases. *In* "Biological Control of Insect Pests and Weeds" (P. DeBach, ed.), pp. 548–578. Reinhold, New York.

Taylor, T. H. C. (1937). "The Biological Control of an Insect in Fiji," 239 pp. Imp. Inst. Entomol., London.

Telford, A. D. (1961). Feature of the lodgepole needle miner parasite complex in California. *Can. Entomol.* **93**, 394–402.

Terao, A., and Tanaka, T. (1928). Population growth of the water flea *Moina macrocarpa* Strauss. *Proc. Imp. Acad. (Tokyo)* **4**, 550–552.

Thompson, W. R. (1924). La theorie mathematique de l'action des parasites entomophages et le facteur du hasard. *Ann. Fac. Sci. Marseille Ser. 2* **2**, 69–89.

Thompson, W. R. (1927). On the effects of methods of mechanical control on the progress of introduced parasites of insect pests. *Bull. Entomol. Res.* **18**, 13–16.

Thompson, W. R. (1939). Biological control and the theories of the interactions of populations. *Parasitology* **31**, 299–388.

Thompson, W. R. (1951a). The time factor in biological control. *Can. Entomol.* **83**, 230–240.

Thompson, W. R. (1951b). The specificity of host relations in predaceous insects. *Can. Entomol.* **83**, 262–269.

Thompson, W. R. (1956). The fundamental theory of natural and biological control. *Annu. Rev. Entomol.* **1**, 379–402.

Tooke, F. G. C. (1953). The eucalyptus snout-beetle, *Gonipterus scutellatus* Gyll. A study of its ecology and control by biological means. *Union S. Afr. Dept. Agr. Entomol. Mem.* **3**, 282 pp.

Turnbull, A. L. (1967). Population dynamics of exotic insects. *Bull. Entomol. Soc. Amer.* **13**, 333–337.

Turnbull, A. L., and Chant, D. A. (1961). The practice and theory of biological control of insects in Canada. *Can. J. Zool.* **39**, 697–753.

Ullyett, G. C. (1947). Mortality factors in populations of *Plutella maculipennis* (Curtis) (Lep. Tineidae) and their relation to the problem of control. *Union S. Afr. Dept. Agr. Entomol. Mem.* **2**, 77–202.

Uvarov, B. P. (1931). Insects and climate. *Trans. Entomol. Soc. London* **79**, 1–247.

van den Bosch, R., and Haramoto, F. H. (1953). Competition among parasites of the oriental fruit fly. *Proc. Hawaii Entomol. Soc.* **15**, 201–206.

van den Bosch, R., Schlinger, E. I., Dietrick, E. J., Hall, J. S., and Puttler, B. (1964). Studies on succession, distribution, and phenology of imported parasites of *Therioaphis trifolii* (Monell) in southern California. *Ecology* **45**, 602–621.

Varley, G. C., and Gradwell, G. R. (1971). The use of models and life tables in assessing the role of natural enemies. *In* "Biological Control" (C. B. Huffaker, ed.), pp. 93–112. Plenum, New York.

Wangersky, P. J., and Cunningham, W. J. (1957). Time lag in prey–predator population models. *Ecology* **38**, 136–139.

Watt, K. E. F. (1965). Community stability and the strategy of biological control. *Can. Entomol.* **97**, 887–895.

Wierenga, M. E. (1971). What is being done by the pesticide chemical industry. *In* "Agricultural Chemicals—Harmony or Discord for Food, People, Environment" (J. E. Swift, ed.), pp. 130–132. Univ. Calif., Div. Agr. Sci., Berkeley, California.

Wigglesworth, V. B. (1965). "The Principles of Insect Physiology," 6th ed., 741 pp. Methuen, London.

Willard, H. F., and Bissell, T. L. (1926). Work and parasitism of the Mediterranean fruit fly in Hawaii in 1921. *J. Agr. Res.* **33**, 9–15.

Willard, H. F., and Mason, A. C. (1937). Parasitization of the Mediterranean fruit fly in Hawaii, 1914–33. *U.S. Dept. Agr. Circ.* **439**, 17 pp.

Wood, B. J. (1971). Development of integrated control programs for pests of tropical perennial crops in Malaysia. *In* "Biological Control" (C. B. Huffaker, ed.), pp. 422–457. Plenum, New York.

Zwölfer, H. (1963). The structure of the parasite complexes of some Lepidoptera. *Z. Angew. Entomol.* **51**, 346–357.

Zwölfer, H. (1971). The structure and effect of parasite complexes attacking phytophagous host insects. *In* "Dynamics of Populations" (P. J. den Boer and G. R. Gradwell, eds.), pp. 405–418. Centre for Agr. Publ. Doc., Wageningen.

SECTION II

BIOLOGY AND SYSTEMATICS IN BIOLOGICAL CONTROL

4

RELATIONSHIP OF SYSTEMATICS TO BIOLOGICAL CONTROL

Vittorio Delucchi, David Rosen, and E. I. Schlinger

I. THE PROBLEM OF IDENTIFICATION

Like any method of control, biological control agents are (or should be) utilized on the basis of their reliability. Failure to apply a biological method adequately has, therefore, economic implications. In its original, conventional form biological control takes advantage mainly of the relationship between natural enemy and victim (host or prey). Failure may, therefore, result from incorrect identification of any of the species involved. Among the interdependencies of taxonomy and biological control, the consequences of an incorrect identification come first to mind which generally result in a considerable delay (sometimes many years) in the application of the method.

Identification is the first step in a biological control effort. Characteristically, this identification is made by a specialist located far from the project area and it is made indifferent to the potential impact the project may have. The classical approach of the professional taxonomist* to this problem does not go beyond a morphological analysis of

*The term "professional" is used here to designate a taxonomist whose main activity concerns identification and classification of organisms. In general, the professional taxonomist works in a museum, is a specialist in a relatively high systematic category, and has a good knowledge of the order containing that category.

dead specimens, the results of which are partly influenced by the adequacy or inadequacy of his reference material, and the methods used have in general serious limitations despite careful studies on the variability of the systematic category considered. To cite just one recent example, specimens of *Aphytis chilensis* Howard (Hymenoptera: Aphelinidae) obtained from an unusual host, *Parlatoria pergandei* Comstock (Homoptera: Diaspididae), have been attributed to a new species on the basis of morphological differences determined by the host (Delucchi, 1964).

The possible influence of the host on relevant characters used for species segregation is an additional difficulty for which a very limited amount of information is available. Working groups of the International Organization for Biological Control (IOBC) have often included taxonomists in project teams so as to increase mutual information exchange and create a bridge of common interests. However, tangible results from this procedure have so far been the exception. On the other hand, the same Organization has often contributed to a better understanding of morphological species and to a more careful evaluation of taxonomic entities by continuously supplying systematists with material of a given taxonomic category from all areas of its distribution and by publishing identifications.

A. Identification of Target Species and the Determination of Native Habitats

Search for natural enemies of a pest species in the latter's native habitat is a first priority matter in biological control. Whereas the identification of an exotic organism may be rather easy in a new area of invasion, in its area of origin identification is often complicated by the presence of closely related species. As a rule, there is little information on this complex of forms in an area where they are commonly of no economic importance. The most efficient natural enemies are commonly nearly monophagous or narrowly oligophagous. Therefore, misidentification of the target organism in the native habitat may lead to introduction of the wrong natural enemies (which may also fail to become established in the new area). Such misidentification was one of the main causes of repeated failures in the long and involved history of biological control of the California red scale, *Aonidiella aurantii* (Maskell) (Homoptera: Diaspididae) (Compere, 1961). During the first one-third of this century, several attempts were made to introduce natural enemies from the Far East into California. However, various species of *Aonidiella* were at that time misidentified as California red scale and their parasites failed to become established in California on that host. These repeated importation failures led to the erroneous conclusion in the 1930's that no effective parasites of the California red scale existed in the Orient. (See also Chapter 15.)

Correct identification may help direct biological control workers to the area of origin of the pest. Thus, misclassification of California red scale as a species of *Chrysomphalus*, a genus believed to be of South American origin, misdirected the search for natural enemies toward that continent in 1934–1935, with negative results. Only after thorough taxonomic revision by McKenzie (1937) had established the specific identity and generic affinities of California red scale, were researchers able to discover efficient parasites of that pest in the Far East (Compere, 1961; DeBach *et al.*, 1971; Rosen and DeBach, 1976a).

The crucial importance of correct specific identification of the host is also provided by the successful biological control of the coffee mealybug, *Planococcus kenyae* (LePelley) (Homoptera: Pseudococcidae) in Kenya. During the time when this important pest was misidentified, numerous unsuccessful attempts were made to introduce natural enemies against it from four different continents. Only after its correct determination as an undescribed species, known only from East Africa, was the search for its parasites conducted in neighboring Uganda and Tanganyika. Several effective parasites were then discovered and successfully introduced into Kenya (Le Pelley, 1943).

Determination of the native home area and habitat of a pest is not always easy. Preserved museum specimens and taxonomic data from technical publications often provide crucial information; the interpretation(s) and opinion(s) of a professional taxonomist(s) is essential. According to van den Bosch and Messenger (1973), many ineffective natural enemies were introduced into California against the black scale, *Saissetia oleae* (Olivier) (Homoptera: Coccidae) from several parts of the world, before an outstanding parasite, *Metaphycus helvolus* (Compere) (Hymenoptera: Encyrtidae), was imported from South Africa, which appears to be the native habitat of the scale. For several species of economic importance over extensive areas, e.g., the codling moth [*Laspeyresia pomonella* (L.)], olive fly (*Dacus oleae* Gmelin), and coconut rhinoceros beetle [*Oryctes rhinoceros* (L.)], the determination of the native home might lead to the discovery of more effective natural enemies. The problem is more complicated when the invading species is new, as in the case of *Agathodes thomensis* Castel-branco (Lep.), accidentally introduced into the Portuguese island of S. Tomé. It was first noted in 1957 defoliating *Erythrina* shade trees, thus causing increases in densities of otherwise unimportant arthropods associated with the cacao trees, creating indirectly a menace to the cacao plantations. Since chemical control is the cause of undesirable side effects and must be avoided, the present natural balance in the biocoenosis could be maintained through the introduction of natural enemies. Determination of the native range of *Agathodes* is primarily a taxonomic problem. Interesting examples of the importance of knowing the center of origin of a pest species are given by Doutt (1967) and van den Bosch and Messenger (1973).

B. Identification of Natural Enemies

Taxonomists of entomophagous species are rather rare. Most natural enemies belong to systematic categories which have not been modernized. With initiation of almost every conventional biological control project, there is a taxonomic problem extending beyond the pest and natural enemy species concerned. The history of the systematics of important parasite genera like *Aphytis* (Hymenoptera: Aphelinidae) or *Trichogramma*

(Hymenoptera: Trichogrammatidae) offers a panoramic view of taxonomic difficulties and of their implications for biological control.

Potential success in biological control of California red scale in California was delayed for over half a century because the species of *Aphytis* attacking it in the Orient were all misidentified as *A. chrysomphali* (Mercet), an inefficient parasite already present in California. The two most effective natural enemies of the scale, *A. lingnanensis* Compere and *A. melinus* DeBach, were not recognized as distinct species until 1948 and 1956, respectively. Another example is *A. holoxanthus* DeBach, currently known as the most efficient parasite of the Florida red scale, *Chrysomphalus aonidum* (L.). It was apparently first encountered near the turn of the century, but was ignored because it was misidentified as another species. It became available for biological control in 1960 when its identity was definitely established (DeBach, 1960; DeBach *et al.*, 1971). (See also Chapter 15.)

Correct identification of natural enemies is not only important during the importation phase of a project but also in subsequent manipulations of (or evaluations of) the organisms. Mass culture of entomophagous forms, especially minute parasitic Hymenoptera, may be readily contaminated by unwanted species. Unless continuous collaboration with a taxonomist is maintained, such contamination may not be noticed for long periods resulting in useless waste of effort. This apparently occurred in the intensive project on biological control of San Jose scale, *Quadraspidiotus perniciosus* (Comstock) (Homoptera: Diaspididae), in Europe. Mass cultures of the imported parasite *Prospaltella perniciosi* Tower (Hymenoptera: Aphelinidae) were apparently invaded by the related, ineffective species, *P. fasciata* Malenotti. Some 4 millions of this species were released in the Heidelberg region in 1956–1958 before its identity was determined by taxonomists. *Prospaltella fasciata* never became established in West Germany (Rosen and DeBach, 1976a).

II. THE LIMITATIONS OF THE MORPHOLOGICAL SPECIES CONCEPT

The aforementioned examples are sufficient to demonstrate the basic importance of taxonomic information in biological control. Unfortunately, the importance of taxonomy is expressed by the negative implications of incorrect identification or of incorrect classification of the taxa concerned, for which the professional taxonomist does not feel a direct responsibility. He knows, of course, that the correct name provides a link with work carried out in the past and therefore represents a key to available information on faunistics, ethology, and ecology of the species. He also knows that the classification of taxonomic categories according to sound phylogenetic principles constitutes a basis of predictability and is essential for the exploitation of all biological control potentialities. Recent contributions to the significance of taxonomy emphasize these points, e.g., DeBach (1960), Slater (1960), Labeyrie (1961, 1964), Schlinger and Doutt (1964),

Oehlke (1966), Delucchi (1967), Sailer (1969), and Sabrosky (1970). The taxonomist evaluates the accuracy of earlier work and synthesizes the accumulated knowledge on a given species. This is what has been attempted in the "Index of Entomophagous Insects" (Delucchi and Remaudière, 1966 and following).

However, the professional insect taxonomist deals mainly on the basis of reference collections and descriptions of dead adult stages. Morphological and chromatic characters remain his principal criteria. The result is largely influenced by this restriction of criteria as well as the limitations of the tools employed to gain insight into these characters. Until relatively recently, he had to contend with generally poor optical equipment, and the knowledge of arthropod morphology was often rather inadequate.

With the introduction of better equipment and the adoption of more sophisticated procedures for the preparation of specimens, the quality of morphological research has been greatly improved. Whereas in the past, taxonomic investigation of insects was usually confined to examination of pinned museum specimens under a low-power stereoscopic microscope, taxonomists have available today high-power phase-contrast microscopes for slide-mounted organs, the Ultropak incident light illuminator systems for examination of integumental structures, and, as well, the scanning electron microscope, therefore providing far better results. The scanning electron microscope has placed a revolutionary tool in the hands of morphologists and opened new dimensions for taxonomic research (Taylor and Beaton, 1970). As a result, closely related species, previously considered morphologically identical in the adult stage, and a single species, can now be separated on the basis of new morphological criteria, thus approaching the needs of biological control. Recent studies of *Muscidifurax* (Hymenoptera: Pteromalidae) by Kogan and Legner (1970) and of *Signiphora* (Hymenoptera: Signiphoridae) by Quezada *et al.* (1973) serve to illustrate the great potential of these modern techniques for systematic research on entomophagous insects.

Systematic research has often suffered from lack of appreciation of intraspecific variation in morphological characters. When species are narrowly viewed as fixed entities, identical with certain "type" specimens, and no allowance is made for the natural range of variation, specimens exhibiting insignificant morphological variation may be inadvertently described as distinct species. On the other hand, if intraspecific variation is not thoroughly understood, valid diagnostic characters may be overlooked. Serious revisional work in the taxonomy of any group of organisms should, therefore, be preceded by a comparative biometric investigation of large series of specimens, in order to evaluate the validity of characters used for the separation of species.

A recent revision of the genus *Aphytis*, for instance, has revealed that some of the characters that had been commonly used in classification, such as the number and relative lengths of certain setae, tend to vary with the size of specimens, and hence are of little diagnostic value. On the other hand, an important valid character such as the pattern of pigmentation was often previously ignored. Thus, numerous species had to be redefined, and many have been recognized as new (Rosen and DeBach, 1976b, in press).

Even though the research of a professional taxonomist ends at the morphological level, this contribution to biological control is enormous. The mere indication that a biologically unknown species belongs to a taxonomic category where representatives show mainly hyperparasitic behavior may avoid a serious error of which biological control has not been completely free.

The extension of the morphological criteria to the preimaginal stages has met with success, although little use has so far been made of its potentialities in classifying and identifying natural enemies. There are technical problems (preparation and preservation of material, stage differentiation, etc.) for both the taxonomist and the entomologist requesting the identification, which are difficult to overcome. Yet, attempts have been made to utilize larval characters for the classification of Braconidae and Ichneumonidae (Short, 1952, 1959). Such studies on the comparative morphology of preimaginal stages have generally been conducted by entomologists closely connected with a biological control project and with a view to differentiating sibling species. Cocoons and puparia, and their contents, have been used for identification of parasites of Diprionidae (Finlayson, 1963). Pupal pigmentation has been considered for the separation of *Aphytis* species (DeBach, 1959).

III. THE USE OF OTHER MORPHOLOGICAL CHARACTERISTICS

Additionally interesting morphological features connected with behavior may prove useful in population studies of natural enemies. The different emergence holes of the egg parasites *Asolcus* and *Gryon* (Hymenoptera: Scelionidae) and *Ooencyrtus* (Hymenoptera: Encyrtidae) in eggs of *Eurygaster* species (Heteroptera: Pentatomidae) serve for generic identification (Voegelé, 1961). Moreover, differences in pigmentation of the membrane which contains *Asolcus* pupae in parasitized *Eurygaster* eggs is, in many cases, a fine character for the separation of species (Voegelé, 1962). As this character persists after emergence of the parasite, it is useful in field population studies. Viggiani (1964) has indicated that eulophid species parasitizing leaf miners may be distinguished by the type and arrangement of the meconium excreted by the parasite larvae. The meconium particles are arranged by the larva to form columns which maintain a free space for the pupating parasite between the upper and lower leaf epidermal layers. The distinctness of the respective columns serves to identify a species, or group of species, even after emergence of the adults. Thus, closer study of these and similar morphological characters associated with parasitism and predation could supply taxonomists and biological control specialists with important characters for species identification.

IV. THE INTERPRETATION OF SIBLING SPECIES

While supraspecific systematic categories may be rather easily differentiated according to morphological characters, at the species level these differences may be so indistinct that conventional equipment does not reveal them or the taxonomist cannot interpret them. This is the case with many so-called strains or races which are coming more and more to be recognized as valid taxa designated as cryptic or sibling species. Their detection is possible on the basis of biological parameters such as host and habitat specificity, number of generations in a given area, diapause habitat, and different tolerances to climatic factors, although the validity of a species must be ascertained by establishing reproductive isolation. The significance of sibling species in biological control has been illustrated by DeBach (1959, 1960, 1969), and Schlinger and Doutt (1964).

Recognition of biological differences between morphologically indistinguishable "strains" of *Aphytis maculicornis* Masi was the key to the outstanding success in control of olive scale, *Parlatoria oleae* (Colvée) (Homoptera: Diaspididae), in California. A biparental "strain" of *A. maculicornis*, introduced from Iran, was found to be superior to several other "strains" or sibling species imported from other countries, including one already present in California (Hafez and Doutt, 1954). When released in California the "Persian *Aphytis*" became a major factor in the control of the olive scale (Huffaker and Kennett, 1966; DeBach *et al.*, 1971).

Another taxonomic problem of great interest to biological control concerns semispecies, which represent conspecific populations exhibiting only partial reproductive isolation (DeBach, 1969). Semispecies can only be evaluated by crossing tests between various populations. Crossing tests performed by Rao and DeBach (1969a,b) with populations of *Aphytis* morphologically identical to *A. lingnanensis* Compere revealed existence, not only of three sibling species, but also of two semi-species having various degrees of reproductive isolation. Some of these morphologically indistinguishable forms possess distinct host preferences and other biological characteristics which may be of utmost importance in biological control projects.

The development of taxonomy in relation to biological control is perhaps best exemplified by tracing the history of its application to the genus *Aphytis* in California. This development has been possible because of the extensive period of research on this genus both in the laboratory and field. This example indicates the extent of progress still required in taxonomy to serve the needs of biological control.

From the studies on sibling species during the last decade, it seems clear that the identification of a species according to the morphological concept, i.e., in the traditional sense, represents a necessary step to the final determination of a true specific entity. Whether or not the morphological criteria are sufficient or other biological parameters are needed, will be decided case after case by taxonomist-biological control specialists. Such a specialist must become a highly qualified taxonomist for a small group of taxa. However, the more essential information that can be obtained through the morphological approach, the easier will be the task of the biological control specialist in establishing other distinguishing parameters. An enormous amount of data is stored in museums, institutes, and private collections and is not readily available to ecologists. On the other hand, annotations concerning the capture or the rearing of an entomophagous species are often completely neglected.

V. TAXONOMIC STUDIES AND ECOSYSTEM SAMPLING

While most of the earlier discussion related the need to precisely discriminate taxonomically both the host species and its parasitoid(s) or predator(s), it is becoming increasingly more important in biological control projects to integrate this knowledge into the study of the ecosystem(s) in which they occur. For biological control projects, this need is more acute in the local area in which importations are made than in the original exotic ecosystem(s). It is in the "new" ecosystem that critical ecological studies are made to evaluate the effectiveness of the imported beneficial organisms, and it is here where knowledge of the newly interrelated organisms may have a profound effect on the success or failure of the project.

Knowledge of the arthropods in a given ecosystem starts with proper identification of the various taxa as they are sampled and shown to be related in some way ecologically to the pest/natural enemy complex. In some cases, e.g., alfalfa in California, it is now known that nearly 1000 taxa exist, and have some ecological relationship to alfalfa during the year (E. I. Schlinger, unpublished data). In this case, only about 30% of the taxa are well enough known taxonomically to carry specific names, even though more than 70 professional taxonomists have studied this material during the years 1958–1974.

If we wish to sample an ecosystem in order to gain a full, proper, or better understanding of the complex arthropod interactions involved with a biological control project, then we should be prepared to (1) learn how to sample the ecosystem, (2) be sure to get good seasonal data for all stages of the various arthropods present, (3) properly preserve enough specimens (all stages) for later taxonomic studies, and especially (4) realize and be prepared to sponsor the great need for many more well-trained taxonomists to aid the endeavors than are presently available. Unless the taxonomists are specifically paid by the project for their efforts in behalf of these biological control-taxonomic projects, means must be found to pay for the services of the few available worldwide taxonomists who may be asked to perform these tasks, while carrying on their normal official duties.

VI. NEED FOR ARTHROPOD SPECIMEN DATA-MANAGEMENT SYSTEMS

The rate of accumulation of biological data and of specimens from various ecological studies throughout the world is staggering, and the rate of accrual appears to be increasing each year. It has been estimated that in the United States National Museum alone there are over 200 million specimens of insects already preserved, and that the number is increasing at the rate of about 1% per year, or 2 million specimens per year. If each specimen carried only three ecological parameters of data, i.e., locality, date, and collector, we find that, together with the scientific name of the specimen, we have about 200 million bits of taxonomic and geographic data already on hand in the United States National Museum. Hopefully, however, many parasitoids and predator specimens will also carry data on host plant association, host insect information, climatic data, rearing conditions and dates, breeding cites, etc., which further complicates the utilization (the extraction) of these data except through the ordinary method of taxonomic categories.

Since the ecologist could use these stored data in a variety of ways if it could be made available to him, it seems appropriate to ask this question: How can specimens and specimen data, at least from this point in time forward, be preserved and managed to ensure ecologists the use of such data? An obvious answer to this problem is to use computers to store all the specimen data, and still use the traditional museum storage method of handling specimens in a taxonomic way.

Starting in 1970, E. I. Schlinger and Peter Rauch (unpublished) have been developing such a computer storage system for a Field Entomology-Ecology course at the University of California at Berkeley. To date, they (and their students) have amassed a collection of about 50,000 insects and spiders from seven different ecological zones in California. Each specimen carries a unique specimen number, together with traditional data such as locality, date, and collector. Each ecological site collection, which may represent one to many specimens, is associated in a specific way as the specimens were collected, and they are recorded on a small, preprinted, field notebook page. Each item on this page, whether an activity pattern, a host-plant feeding record, or a parasitoid rearing association, is numbered for later key punch use. As the collected specimens are prepared and numbered, each number referring to that page of data is recorded on that page.

In this way, all ecological data and all taxonomic specimen data are interlocked for computer storage. As determinations are made either for host plants or insects at any later date, the taxonomic name, together with the unique number, is key punched and stored in the same computer bank. The further development of several computer programs will allow the ecologist to have access to any and all kinds of data stored in the computer bank by simply asking that the specific data needs to be prepared on a computer print-out sheet. This system at Berkeley is in partial operation now and it illustrates the feasibility of such an undertaking, but the development of still other computer programs is needed before complete use of these stored data can be made available.

There can be little doubt that some form of computerization of ecological specimen data is needed and long overdue. Just what form this system eventually will take is unknown, but those of us dealing with biological control projects throughout the world would likely benefit the most from such a specimen data-management system, and we should be making a strong effort toward achieving this goal.

REFERENCES

Compere, H. (1961). The red scale and its insect enemies. *Hilgardia* **31**, 173–278.

DeBach, P. (1959). New species and strains of *Aphytis* (*Hymenoptera, Eulophidae*) parasitic on the California red scale, *Aonidiella aurantii* (Mask.), in the Orient. *Ann. Entomol. Soc. Amer.* **52**, 354–362.

DeBach, P. (1960). The importance of taxonomy to biological control as illustrated by the cryptic history of *Aphytis holoxanthus* n. sp. (*Hymenoptera: Aphelinidae*), a parasite of *Chrysomphalus aonidum*, and *Aphytis coheni* n. sp., a parasite of *Aonidiella aurantii*. *Ann. Entomol. Soc. Amer.* **53**, 701–705.

DeBach, P. (1969). Uniparental, sibling and semi-species in relation to taxonomy and biological control. *Isr. J. Entomol.* **4**, 11–28.

DeBach, P., Rosen, D., and Kennett, C. E. (1971). Biological control of coccids by intoduced natural enemies. *In* "Biological Control" (C. B. Huffaker, ed.), pp. 165–194. Plenum, New York.

Delucchi, V. (1964). Une nouvelle espèce d'*Aphytis* du groupe *chilensis* Howard (Hym., Chalcidoidea, Aphelinidae). *Rev. Path. Vég. et Ent. Agr. Fr.* **43**, 135–140.

Delucchi, V. (1967). The significance of biotaxonomy to biological control. *Mushi Suppl.* **39**, 119–125.

Delucchi, V., and Remaudière, G. (eds.). (1966). "Index of Entomophagous Insects." Librairie Le Francois, Paris. (Five volumes issued).

Doutt, R. L. (1967). Biological control. *In* "Pest Control: Biological, Physical and Selected Chemical Methods" (W. W. Kilgore and R. L. Doutt, eds.), pp. 3–30. Academic Press, New York.

Finlayson, T. (1963). Taxonomy of coccoons and puparia, and their contents, of Canadian parasites of some native *Diprionidae* (*Hymenoptera*). *Can. Entomol.* **95**, 475–507.

Hafez, M., and Doutt, R. L. (1954). Biological evidence of sibling species in *Aphytis maculicornis* (Masi) (*Hymenoptera, Aphelinidae*). *Can. Entomol.* **86**, 90–96.

Huffaker, C. B., and Kennett, C. E. (1966). Biological control of *Parlatoria oleae* (Colvée) through the compensatory action of two introduced parasites. *Hilgardia* **37**, 283–335.

Kogan, M., and Legner, E. F. (1970). A biosystematic revision of the genus *Muscidifurax* (*Hymenoptera: Pteromalidae*) with descriptions of four new species. *Can. Entomol.* **102**, 1268–1290.

Labeyrie, V. (1961). Taxonomie, écologie et lutte biologique. *Entomophaga* **6**, 125–131.

Labeyrie, V. (1964). Taxonomie et écologie. E.N.S.A.T., *Bull. Trim. Ass. ing. Anciens Elèves*, janvier, pp. 11–14.

Le Pelley, R. H. (1943). The biological control of a mealybug on coffee and other crops in Kenya. *Emp. J. Exp. Agr.* **11**, 78–88.

McKenzie, H. L. (1937). Morphological differences distinguishing California red scale, yellow scale, and related species. *Univ. Calif. Berkeley Publ. Entomol.* **6**, 323–336.

Oehlke, J. (1966). Moderne taxonomische Forschung. Grundlage für Prognose und Methoden der Bekämpfung von Schadinsekten, dargestellt am Beispiel der Kiefernbuschhornblattwespen (*Diprioninae*) und ihre Parasiten (*Ichneumonidae*). *Arch. Forstw.* **15**, 953–958.

Quezada, J. R., DeBach P., and Rosen, D. (1973). Biological studies of *Signiphora borinquensis*, new species (*Hymenoptera: Signiphoridae*), a primary parasite of diaspine scales. *Hilgardia* **41**, 543–603.

Rao, S. V., and DeBach, P. (1969a). Experimental studies on hybridization and sexual isolation between some *Aphytis* species (*Hymenoptera: Aphelinidae*). I. Experimental hybridization and an interpretation of evolutionary relationships among the species. *Hilgardia* **39**, 515–553.

Rao, S. V., and DeBach, P. (1969b). Experimental studies on hybridization and sexual isolation. between some *Aphytis* species (*Hymenoptera: Aphelinidae*). II. Experiments on sexual isolation. *Hilgardia* **39**, 555–567.

Rosen, D., and DeBach, P. (1976a). Diaspididae. *In* "Introduced Parasites and Predators of Arthropod Pests and Weeds—A World Review" (C. P. Clausen, ed.), in press. U.S. Dept. Agr., Washington D.C.

Rosen, D., and DeBach, P. (1976b). Biosystematic studies on the species of *Aphytis* (*Hymenoptera, Aphelinidae*). *Mushi* (in press).

Sabrosky, C. W. (1970). Quo vadis taxonomy? *Bull. Entomol. Soc. Amer.* **16**, 3–7.

Sailer, R. I. (1969). A taxonomist's view of environmental research and habitat manipulation. *Proc. Tall Timbers Conf. Ecol. Anim. Control Habitat Manage.*, **1**, 37–45.

Schlinger, E. I., and Doutt, R. L. (1964). Systematics in relation to biological control. *In* "Biological Control of Insect Pests and Weeds" (P. DeBach, ed.), pp. 247–280, Chapman and Hall, London.

Short, J. R. T. (1952). The morphology of the head of larval Hymenoptera with special reference to the head of *Ichneumonoidea*, including a classification of the final instar larvae of the Braconidae. *Trans. Roy. Entomol. Soc. London* **103**, 27–84.

Short, J. R. T. (1959). A description and classification of the final instar larvae of the *Ichneumonidae (Insecta, Hymenoptera)*. *Proc. U.S. Nat. Mus.* **110**, 391–511.

Slater, J. A. (1960). The responsibility of the insect taxonomist. *Bull. Entomol. Soc. Amer.* **6**, 17–19.

Taylor, R. W., and Beaton, C. D. (1970). Insect systematics and the scanning electron microscope. *Search* **1**, 347–348.

van den Bosch, R., and Messenger, P. S. (1973). Biological control. Intext Educ. Publ., New York, pp. 180.

Viggiani, G. (1964). La specializzazione entomoparassitica in alcuni eulofidi (*Hym., Chalcidoidea*). *Entomophaga* **9**, 111–118.

Voegelé, J. (1961). Contribution à l'étude de la biologie des Hyménoptères oophages des punaises des céréales au Maroc. *Cah. Rech. Agron.* **14**, 69–90.

Voegelé, J. (1962). Reconnaissance des espèces *Asolcus tumidus* Mayr et *A. basalis* Wollaston (*Hymenoptera, Proctotrupoidea*) d'après les caractères externes de l'oeuf hôte. *Al Awamia* **4**, 147–153.

5

THE BIOLOGY AND IMPACT OF PREDATORS

K. S. Hagen, S. Bombosch, and J. A. McMurtry

I. INTRODUCTION

Up to 1963 only 15 exotic predaceous insect species had been successfully introduced against insect pests and exercised a distinct degree of control as contrasted to over 115 parasites (DeBach, 1964). One is naturally concerned about the reasons for this difference. It should be noted, however, that a number of pest species have been outstandingly controlled by the few successfully introduced predators. These include certain scale insects, mealybugs, and eggs of leafhoppers. These types of prey are essentially sessile, nondiapausing, nonmigratory, and are associated with evergreen perennial plants or crops.

The reasons for the relatively poor success in establishing predaceous insects compared to parasitic species are no doubt many. The "shot gun" approach of haphazardly releasing predators and failure in the culture of many predators due to technical problems have accounted for some failures. There are, however, more subtle reasons that may have accounted for some unsuccessful introductions. When the biological attributes of the predators that have been successfully introduced are compared to those that have failed, as well as to the characteristics of their prey, there are a number of attributes common to the most effective introduced predaceous insects.

These attributes are multivoltinism (nondiapausing), narrow prey specificity, and high searching efficiency by long-lived adults. Furthermore, their temperature thresholds for activity are quite close to those of their prey, and generally the effective predators have more generations than their prey. These qualities approximate those of the more effective parasitic insects (see Chapters 3, 6, and 9).

Most of these properties are possessed by *Rodolia cardinalis* (Mulsant) (Thorpe, 1930; Quezada and DeBach, 1973) and by *Cryptognatha nodiceps* Marshall (Taylor, 1935), two introduced coccinellids which have had spectacular success. In the few locations where these predators were released and failed to give complete biological control of their respective prey, the cottony-cushion scale and the coconut scale, the climate was rather too cool. In these cooler areas biological control of both pests, however, was achieved by introducing parasitic species that were active at cooler temperatures. For example, the parasitic fly, *Cryptochaetum iceryae* Coq. dominates over *R. cardinalis* in controlling cottony-cushion scale in cooler coastal areas and in cooler seasons in California (Quezada and DeBach, 1973) and at cooler higher elevations in Bermuda (Bennett and Hughes, 1959). The introduction of *C. nodiceps* from Fiji to the New Hebrides failed to control the coconut scale, but biological control of the scale was achieved by introducing another predatory coccinellid, *Rhizobius pulchellus* Montrouzier, from New Caledonia. Cochereau (1969) believes that the climate in New Hebrides being somewhat cooler than in Fiji accounted for *C. nodiceps* not becoming established.

Thus some failures of past predator introductions may be attributed to poor climatic fit between the origin of the predators and the release areas. The limitations of the physical environment for a predator (or a parasite) can be much more restricted than for their prey. The following coccid-eating coccinellids clearly demonstrate this. *Orcus chalybeus* (Boisd.) from Australia (Essig, 1930), *Chilocorus similis* (Rossi) from Asia (Smith, 1965), and *Exochomus metalicus* Korchefsky from Eritrea (Bartlett and Lloyd, 1958) were released widely in California but became established and persisted only in the coastal Santa Barbara area. Similarly, *Cryptolaemus montrouzieri* Mulsant established in coastal San Diego and Santa Barbara counties (Essig, 1930). The prey of these coccinellids have much wider distributions, not only in California but in other areas of the United States where they have invaded.

Prey specificity cannot be discounted as being another reason for some of the failures in establishing introduced predators. C. B. Huffaker (personal communication) thinks that the lack of high prey specialization in a predator means that it would have to compete, without the advantages such specialization confers, against a complex of indigenous predators, making establishment especially difficult. Thompson (1951) discussed the question of prey specialization concerning thirteen species of coccinellids which he attempted to introduce into Bermuda to control a scale insect. Only one of the thirteen became established even though some of the others did develop a generation when enclosed in field sleeve cages. Hodek (1973) reviewed the food relations of the Coccinellidae and reaffirmed Thompson's conclusion that the specificity of many coccinellids is more narrow than generally believed.

Simply observing a predator feeding on certain prey species is not sufficient to indicate that such species is a preferred or normal prey. A candidate species should be cultured for several generations to determine if the target prey is suitable. Also, its developmental periods, fecundity, and mortality should be studied. *Adalia bipunctata* (L.) eats many different kinds of aphids, but larvae fed different aphid species varied in their developmental times, weight of adults produced, and amount of larval mortality (Blackman, 1965). The preoviposition period of *Propylea quatuordecimpunctata* varied depending upon the species of aphid eaten (Rogers *et al.*, 1972). Some such differences have also been reported concerning predaceous mites (McMurtry *et al.*, 1970). Fecundity can also be influenced. When fed *Macrosiphum rosae* (L.), *Coccinella septempunctata* L. deposited over twice as many eggs than when fed *Myzus persicae* (Sulzer) (Hämäläinen and Markula, 1972). Furthermore, an aphid species on one host plant may be unsuitable food for a coccinellid species, while the same aphid on another host plant is suitable (Okamoto, 1966). Also, one species of aphid may be unsuitable for growth and reproduction for certain coccinellids during one season but not in another season (Takeda *et al.*, 1964).

In the early 1900's many coccinellid species were introduced into western United States, the Hawaiian Islands, and South Africa with only a few species becoming established. Among the species introduced from Australia into the Hawaiian Islands, was *Harmonia* (*Leis*) *conformis* (Boisd). It was released against various aphids but failed to become established. Recently it was reintroduced into Hawaii against an accidentally introduced psyllid pest, *Psylla uncatoides* Ferris and Klyver from Australia, and it not only became established but has now controlled the psyllid in one location in Hawaii (J. Beardsley, personal communication). Furthermore, Beardsley found that although *Harmonia* would eat aphids it did not completely develop on such a diet. Thus, their prey specificity is indeed an important consideration which has often been considered little when introducing predators.

The absence of symbiotic microorganisms in certain predaceous species or in the target environments may preclude establishment. Although symbiotic microorganisms are not involved in strictly predaceous insects, this must be considered when introducing at least some *Chrysopa* spp. Of the 26 *Chrysopa* species for which the adult feeding habits are known, about one-half are not predaceous as adults but are suspected to feed mainly upon honeydews and perhaps nectar and pollen (Hagen *et al.*, 1970; Sheldon and MacLeod, 1971). *Chrysopa carnea* Stephens (Fig. 1) is an example of one of these species that is not predaceous in the adult stage. Its crop is adapted for a mutualistic relationship with the yeast *Torulopsis*. The yeast apparently synthesizes the missing essential amino acids that are usually lacking in honeydews. In the laboratory when

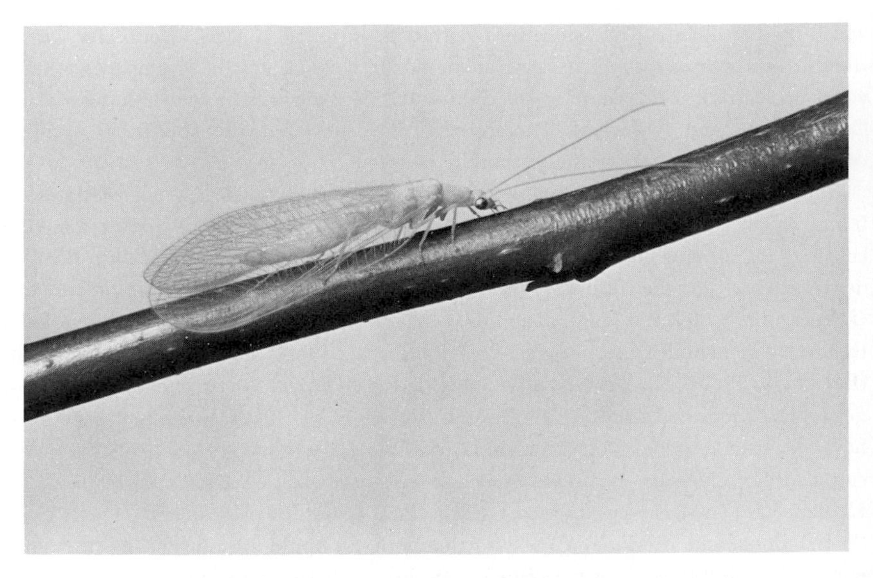

Fig. 1. *Chrysopa carnea* adult. Photo by F. E. Skinner, Univ. of Calif., Berkeley, California.

complete diets containing the 10 essential amino acids are fed to the adults the symbiotes are not necessary (Hagen and Tassan, 1972).

Recently, eggs of *C. carnea* were sent to New Zealand for introduction since no *Chrysopa* spp. now exist there. The eggs were reared to adults and released, but they did not become established (E. W. Valentine, personal communication). Since the yeast symbiote is not transferred from generation to generation via the egg, the *C. carnea* adults must pick up the yeasts from honeydew and/or possibly nectar in the field. The *C. carnea* reared in the insectary from eggs would not contain the yeast, and if the *Torulopsis* is not present in the new environment, the released *C. carnea* would not be able to produce many eggs; in any event the progeny adults would lack *Torulopsis*. Establishment would be greatly hindered in any such situation.

The most numerous failures involving predator introductions have been the attempts to establish predators against aphids. Poor phenological synchrony, low numbers released, and missing habitat requirements in the target area probably account for many of these failures. The ephemeral nature of aphid populations, with their prolonged periods of absence in many crops, causes disengagement of a colonizing predator and requires long periods of dormancy which may be either aestival or hibernal or a combination of both. If the predator is unadapted in this respect it is likely to fail. Even if so adapted there must be suitable refuges available to permit survival during dormant periods. Adequate refuges are absent or are becoming scarce in many areas of the world where monoculture is practiced. Many monocultures are annual crops, with periods of fallow intervening, and also in these areas, weeds and natural border vegetation are greatly restricted. Furthermore, these "sterile" environments do not provide alternate sources of foods for predators which might help to bridge gaps when target aphids are absent, or provide foods for building up the necessary fat for surviving dormancy periods (Hagen, 1962; Hodek, 1973).

It is difficult even for many naturally occurring species of predators to exert their potential effectiveness today in monoculture areas (DeLoach, 1971). It is only because of their polyphagy, their wide aerial distribution, and migration capacities that permit them to survive as populations and reinvade crops in effective numbers. Some once common predators have actually disappeared in vast monoculture areas. Some 20 years ago, *Paranaemia vittegera* (Mann.) was a common predator in alfalfa and cotton in the San Joaquin Valley of California, but today this coccinellid is rarely seen in these or other field crops. It spends its aestival and hibernal dormancy periods as aggregations at bases of perennial plants and weeds in and near their feeding areas. Such refuge sites rarely are allowed to persist today under monoculture practices. The coccinellids having similar habits of aggregating for the dormant period, and which persist today, are species that migrate considerable distances and are able to find more or less suitable dormancy sites, often far removed from the monocultures.

In order to establish aphidophagous predators for field crop pest control, it may be necessary to import those species that do migrate, and it may be necessary to release great numbers in different habitats even far removed from the target aphid to be attacked. *Hippodamia convergens* Guerin is a long-range migrating aphid predator that usually aggregates in mountains (Hagen, 1962). It has been released in many areas of the world. It was finally established in Peru (Wille, 1958) after millions of *H. convergens*, collected from California's mountain aggregations, were released over many years. The Andes mountains and valleys to the west in Peru appear to provide the habitats and appropriate seasonal winds similar to mountains and valleys of California. *Hippodamia convergens* is also now established in Chile (Caltagirone, personal communication) and apparently in Venezuela (Guagliumi, 1966). Large numbers of *H. convergens* were sent to the Hawaiian Islands from California aggregations in 1896, 1905, 1910, and 1952 (Davis and Chong, 1968), but only recently (1964) has it been found established on the high mountain slopes on the Islands of Hawaii and Maui (Beardsley, personal communication). It is suspected that in Hawaii great numbers of *H. convergens* are lost by being blown out over the ocean when they are migrating. Thus, it appears possible to establish aphid predators that migrate, aggregate, and diapause if great numbers are released in suitable regions similar to where they originated.

II. IMPACT OF PREDATORS

The impact of predaceous arthropods in natural communities and agricultural crops is receiving more attention in recent years, and the importance of predators as controlling agents is coming into closer focus, based on modern ecological investigations and experiments. Chiang (1974), Doutt and DeBach (1964), DeBach (1974), Ehler and van den Bosch (1974), Huffaker (1970, 1971), and van Emden (1972) are among many authors that discuss the role of predators in controlling agricultural pests. Analysis of the impact of indigenous predators in natural communities is most difficult compared to that of indigenous predators in agricultural crops even though the latter is far from easy. However, use of releases and exclusion of predators is experimentally demonstrating the importance of such predation (see Chapter 11). The importation and establishment of exotic predators against introduced pests have provided the most obvious tests of importance of some predators in regulating their prey populations. The use of selective pesticides sparing predators has also shown the importance of some predators. A few examples of each of these above situations are offered below.

A. Predators in Natural Communities

There are several examples that illustrate the importance of predators in natural communities. The factors that regulate the winter moth, *Operophtera brumata* (L.), in a natural oak community have been studied for over 20 years near Oxford, England by G. C. Varley and colleagues (Varley, 1971). Using the key factor analysis technique, Varley and Gradwell (1968) found a statistically significant density-dependent relationship between the aggregate effect of all predators and the logarithm of the winter moth population density on which they acted. This density-dependent effect contributed significantly to the stability of the population density of the winter moth. East (1974) found that the pupal predation appeared to be due mainly to carabids (which destroyed 38%), staphylinids (30%), and small mammals (4%). The predaceous insects responded behaviorally to pupal density.

The importance of naturally occurring predators in a California pine forest came to light when natural enemies of diaspine scales on pines were inadvertently killed by insecticide. A lightly developed residential area in the pine forest at South Lake Tahoe, California had received Malathion fogging treatments to control adult mosquitoes for five successive summers. In precisely the "fogged" area, several scale insects on the pines appeared in outbreak proportions. When the insecticide applications were stopped in July, 1969, predation and parasitization of the scale insects increased (Dahlsten *et al.*, 1969) and by 1970 their population densities had declined substantially (Luck and Dahlsten, 1975).

Luck and Dahlsten (1975) studied the population dynamics of one of these scales, *Chionaspis pinifoliae* (Fitch), during the decline period on two hosts, lodge pole (*Pinus contorta* Dougl.) and Jeffrey pine (*P. jeffreyi* Grev. and Balf.), and on the basis of life table data and *k*-factor analysis, Luck concluded that predation, mainly by *Chilocorus orbus* Casey, on immature scales was responsible for only a slight amount of mortality, but that this amount coupled with other losses in the egg and crawler stages was sufficient to account for the population decline. However, at low densities, Luck believes that parasites are the main regulators of the scale.

Another predator, a tiny coccinellid, *Cryptowesia atronitens* (Casey), was also destroying scales in the South Lake Tahoe pine plots, but was not considered as important as *C. orbus*. However, on pines in Canada, *C. atronitens* was found to be an important predator of the same scale species (Martel and Sharma, 1970).

The impact of a predator of a coccid on pines was demonstrated by use of tree bands to prevent ants from interfering with the coccinellid, *Hyperaspis congressis* Watson. The undisturbed predators greatly increased and rapidly reduced the scale (Bradley, 1973). Way (1963) has reviewed similar cases with other honeydew-producing Homoptera which increased in abundance when the ants attending them were removed. The role of ants as predators is highly developed in the coevolution of the mutualistic association of certain ants and acacias, and certainly points to the effectiveness of the ants as predators, for the removal of ants allows defoliation by herbivorus insects (Janzen, 1966).

B. Indigenous Predators of Agricultural Pests

Impact studies of indigenous predators on various agricultural pests have been largely based on observed correlations between abundance of predators and population changes of the prey. Since pathogens and parasites are also commonly involved in the same pest population changes, only estimates of predator impact are usually presented. As seen in the reviews of the impact of natural enemies on aphids by Hagen and van den Bosch (1968), Hodek (1973), van Emden *et al.* (1969), and van Emden (1972) very few studies clearly show a significance of predators because of the difficulty in isolating their action among a complex of natural control factors in field situations.

A unique situation occurred in California when the spotted alfalfa aphid (SAA) invaded the state, permitting an evaluation of mainly predation on both SAA and pea aphid (PA) populations before any parasites were firmly established and before aphid-resistant alfalfas were widely planted. Alfalfa fields in two climatically different regions were sampled throughout the year from 1957 to 1959 to determine the factors that influenced the aphids' populations. It was found that coccinellids were mainly responsible for aphid control in the spring and fall (Smith and Hagen, 1966).

Neuenschwander *et al.* (1975) pooled all data from the two California regions studied, allowing more accurate determination of the changes in the population of these mobile predators than is possible from counts taken in single alfalfa fields. Coccinellids and hemerobiids, two rather oligophagous groups, were closely linked numerically to the aphid populations, and showed density-dependent relationship to aphid abundance. This density dependence, particularly of the lady beetles, is a good indication of their ability to control aphids. The more polyphagous predators, the chrysopids and Hemiptera, became important in the control of the aphids only when the dominant coccinellids were inactive because of diapause or when lack of aphids prompted coccinellids to leave the fields.

On the other hand, when aphid numbers exceed the momentary capacity for compensating increase in lady beetle numbers, syrphids often take a large toll. Furthermore, several species having similar biologies were found to replace each other at different times of the year.

Other studies confirm the importance of indigenous predators in the control of aphid populations. *Coccinella septempunctata* L. and *Adonia variegata* (Goeze) were mainly responsible for controlling *Aphis fabae* infesting sugar beets in southern European U.S.S.R (Minoranskii, 1967). Native syrphids effectively suppressed fall populations of *Myzus persicae* on peach trees in Washington (United States) (Tamaki *et al.*, 1967). Syrphids are also considered to play the dominant role in controlling aphids on peach in the Middle Rhone Valley (Remaudière and Leclant, 1971). During the summer, coccinellids and syrphids reduced the production of alate aphids on peach trees by 95% (Tamaki, 1973).

In Maine, Shands *et al.* (1972) found a correlation between predation and aphid declines on potatoes in 14 cases out of 31, while in 17 cases, fungi were associated as causing aphid population crashes. However, in most instances the aphid decline occurred too late to prevent some damage.

The role of general predators (polyphagous species) appears to be important in preventing some pest outbreaks, particularly in monocultures. DeBach (1951) considered that "general predators may act as sort of a balance wheel in the pest-natural enemy complex" tending to feed upon whatever pest is present in abundance. Even in situations where they may be in themselves incapable of achieving natural control below economic levels, they slow down the rate of increase of potential pests or reduce peak infestations when more host-specific natural enemies are ineffective.

In annual monoculture crops like cotton, general predators often prevent damaging outbreaks of certain noctuids. In California, *Spodoptera exigua* (Hubner) and *Trichoplusia ni* (Hubner) are secondary pests of cotton and are commonly becoming abundant following insecticide applications against lygus bugs (van den Bosch *et al.*, 1971). Eveleens *et al.* (1973) and Ehler and van den Bosch (1974) believe it is the insecticide destruction of the general predators, *Geocoris*, *Orius*, *Nabis*, and *Chrysopa*, that allows the increase of the noctuids. Ehler and van den Bosch (1974) consider these predators to act in a density-independent fashion relative to the intrageneration abundance of their lepidopterous larval prey in cotton. They classify these general predators as r-strategists, a strategy of expected value to unstable situations. They believe that consideration should be given to the importation of such predators against pests of short-duration crops.

Wilson (1974) also suggests that introductions be made into highly modified agroecosystems, for often the original habitat of a natural enemy may have been changed yet a parasite or predator remains effective in spite of the changes in its native habitat. If importation of exotic general predators is contemplated, it must be kept in mind that direct introduction into short-cycle crops may not lead to establishment. There also must be in the general area of release other crops or wild vegetation to serve as refuges and habitats for alternate prey when the target crop is removed. The population densities of such general predators are density-dependent (in a general sense) on the availability of acceptable prey as a collective biomass. Also there may well be a greater degree of food and habitat specificity among the "general" predators than is suspected.

C. Manipulation of Indigenous Predators

By removing or adding predators, their impact can be demonstrated (see also Chapter 11).

1. Removing Predators

Excluding predators has been done by hand, by use of exclusion, and by the use of insecticides. Hand removal of spider mite predators has resulted in sharp increases in mite populations on avocados (Fleschner *et al.*, 1955) and on strawberries (Huffaker and Kennett, 1956). When the egg predator *Cyrtorhinus fulvus* Knight was removed from taro plots every other day for 3 months, the taro leafhopper significantly increased by the third month (Matsumoto and Nishida, 1966). Pollard (1969) differentially removed predators of the cabbage aphid on brussels sprouts in England which revealed the importance of two syrphid spp. and of a cecidomyid in causing declines in aphid abundance in August. Tamaki and Weeks (1973) also hand-removed aphidophagous predators in field-grown sugar beets and the aphids increased.

2. Adding Predators

Various predaceous insects and mites have been released or added for use against prey in cages or open field plots, not only to determine the impact of specific predators but also to evaluate the practicality of such releases or manipulations. Besides the release or augmentation of indigenous predators, other techniques that attract, aggregate, and increase oviposition by wild predators have been used to increase the natural predation.

Releases of predators have involved mainly *Chrysopa carnea*, various coccinellids, *Geocoris*, *Nabis*, and phytoseiid mites.

Chrysopa carnea releases have depressed mealybugs, moth eggs and larvae, and aphids. Control of mealybugs infesting DDT-treated pears in California was achieved by placing 250 *C. carnea* eggs per tree three times at 2-week intervals (Doutt and Hagen, 1950). Eggs and larvae of *C. carnea* released on cotton in Texas accounted for suppression of *Heliothis* eggs and larval populations (Ridgway and Kinzer, 1974).

In England, the green peach aphid infesting chrysanthemums in greenhouses was controlled by releasing *C. carnea* larvae (Scopes, 1969). In Maine (United States), Shands and Simpson (1972) sprayed a total of 55,000 *Chrysopa* eggs per acre onto potato plants in four different spacial arrangements. Eggs added to circular areas (9-foot diameters) spaced 45 feet apart reduced the aphids to lower levels than where eggs were added at the same overall rate but on every row. In Washington (United States), Tamaki and Weeks (1973) released several aphidophagous predators, including larvae of *C. carnea*, against *M. persicae*. Twelve releases of *C. carnea* during June and July resulted in an average of 4.65 larvae being released per sugar beet plant. The aphid population was reduced by two-thirds, compared to plots where the natural predators were removed.

Native coccinellids that have been released mainly against aphids have given variable results. In the early 1900's many millions of *Hippodamia convergens* were collected from their California mountain aestival-hibernal aggregations and released against aphids infesting lowland crops. Control of aphids in the fields where they were placed was not achieved because they either dispersed too widely upon release in the spring (Davidson, 1924) or remained dormant and ineffective when released in the summer (Hagen, 1962). However, releases of *H. convergens* and *Coccinella transversoguttata* Fald. collected from alfalfa fields in Washington (United States) that were treated so they could not fly (by gluing their elytra) were effective. Eight releases totaling 2.25 beetles and larvae per plant reduced *M. persicae* populations to less than one-third the population level that occurred in sugar beet plots where predators were removed. The aphids were also more depressed than in those plots where either *Geocoris*, *Nabis*, or *Chrysopa* larvae were released (Tamaki and Weeks, 1972b). *Dysaphis plantaginea* (Pass.) on apple in southern France was suppressed by adding 20 *Adalia bipunctata* (L.) adults per five trees at 15-day intervals (Remaudière *et al.*, 1973). Gurney and Hussey (1970) released larvae of three species of coccinellids against aphids infesting greenhouse-grown cucumbers and chrysanthemums and obtained different degrees of reduction of aphids. Ten *Cycloneda sanguinea* (L.) larvae suppressed initial populations of up to 400 *Aphis gossypii* Glover per cucumber plant within 10 days.

Since tetranychid mites do not have any recorded insect parasites, predators are their main natural enemies. Many kinds of predators feed on spider mites (McMurtry *et al.*, 1970). Among the main predators are phytoseiid mites, *Stethorus* spp. and *Scolothrips* spp. Release of phytoseiids against spider mites is commonly used in greenhouses but they and *Stethorus* have also been used in strawberry plantings, in various tree crops, and in grapes. Huffaker and Kennett (1956) demonstrated the importance of phytoseiid mites in controlling cyclamen mite in 2nd-, 3rd-, and 4th-year strawberry plantings by introducing both phytoseiids and their prey into 1st-year fields (and by use of a chemical check method). In 1st-year strawberry plantings, Oatman *et al.* (1968) obtained significant reduction of the two-spotted mite by releasing 320,000 *Phytoseiulus persimilis* Athias-Henriot per acre at weekly intervals for 10 weeks. Such a rate of application is not considered practical, however. *Phytoseiulus persimilis* is available commercially and is used commonly in greenhouses in England and Europe to control the two-spotted mite on cucumbers (many authors in OILB bull. SCROP 1973/1974). A release of 400 to 500 *Stethorus picipes* Casey adults per avocado tree reduced the avocado brown mite (McMurtry *et al.*, 1969).

3. *Increasing Indigenous Predators by Addition of Subsidiary Food*

The importance of predaceous insects and mites has also been demonstrated by providing supplementary foods to retain, arrest, or attract them, and for some predators to increase their oviposition (Chapter 10; Hagen and Hale, 1974). By spraying sucrose solutions in corn fields, adult coccinellid populations were increased on the plants, and higher adult coccinellid and chrysopid ratios to their aphid prey were achieved, resulting in a lower aphid population (Schiefelbein and Chiang, 1966). *Hippodamia* and *Chrysopa carnea* are not attracted to sucrose-sprayed plots, but they are arrested when they land in such plots (Hagen *et al.*, 1971). By adding yeast products with the sugar, honeydew is simulated and *C. carnea* is then attracted to such food sprays and its oviposition is increased. The importance of abnormally early oviposition, induced by use of "artificial honeydew," has resulted in reductions of aphids in alfalfa, bollworms in cotton (Hagen *et al.*, 1971), and aphids in peppers (Hagen and Hale, 1974).

Pollen dusted on plants has increased the effectiveness of certain phytoseiid mites which eat pollen when mites are scarce. Also predator effectiveness has been increased by the increase of nonpest mites which develop on the pollen and which serve as alternate prey for nonpollen-feeding phytoseiids when target mites are scarce (McMurtry and Scriven, 1966; Flaherty and Hoy, 1971).

D. Predators Introduced for Crop Pest Control

The clearest evidence of the impact of predators in controlling insect pests comes from the introductions of several coccinellids, and a mirid. Ever since the establishment of the coccinellids *Rodolia cardinalis*, *Cryptognatha nodiceps*, *Rhizobius pulchellus*, and the mirid *Tytthus mundulus* (Breddin), these predators have controlled their respective target pests except where disturbed by use of pesticides, and the outbreaks following such disturbances only dramatizes their lasting effectiveness over the years.

The spectacular control of *Icerya purchasi* within a year after introducing *R. cardinalis* in California in 1888 has been repeated in 29 countries (DeBach, 1964; also Chapter 15). In 25 of these countries its introduction resulted in complete control, and in the remaining 4 substantial control was achieved. Using a modified life table approach, Quezada and DeBach (1973) determined that 74–91% of tagged scales on field citrus were destroyed by the vedalia with only 0.01 to 0.12% of scales surviving from generation to generation.

The coconut scale in Fiji was completely suppressed by 1929 with the introduction of *C. nodiceps* from Trinidad in 1928 (Taylor, 1935; Chapter 15). In 1952 *Aspidiotus destructor* invaded the Island of Principe and the introduction of *C. nodiceps* in 1955 accounted for nearly complete control by 1956, and by mid-1958, copra production was back to normal, indicating that biological control of the scale had been successful (Simmonds, 1960; Chapter 15). Simmonds (Chapter 3) has estimated a potential copra savings of $165,000 per year (based on former heavy infestations). This was achieved at a cost of less than $10,000 for importation and release of *C. nodiceps*. Furthermore, the probably continuing control and savings with no further cost is the bonus gained by a successful biological control program.

This coconut scale was not controlled, however, in the New Hebrides by the introduction of *C. nodiceps*, although the importation of another coccinellid, *Rhizobius pulchellus*, from New Caledonia rapidly suppressed the outbreaks of this scale on Vate Island. Apparently the cooler climate in the New Hebrides precluded the effectiveness of *C. nodiceps* as mentioned earlier (Cochereau, 1969).

The date scale *Parlatoria blanchardi* Berg. has been controlled by the introduction of *Chilocorus bipustulatus* L. from Iran into Mauritania (Iperti *et al.*, 1970).

The mirid *Tytthus mundulus* is an egg predator of the sugarcane leafhopper, and was one of the very early successful introductions for biological control. It has achieved continued biological control of the leafhopper in the Hawaiian Islands since its introduction from Australia and Fiji in 1920 (Swezey, 1936; Chapter 15). Another egg-eating mirid, *Cyrtorhinus fulvus* Knight, was imported from the Philippines and introduced against taro leafhopper in Hawaii in 1938 and in Guam in 1947 and has effectively controlled the leafhopper (Matsumoto and Nishida, 1966).

Cryptolaemus montrouzieri, introduced from Australia into California in 1892, became established against citrus mealybugs in southern California, but only successfully overwintered along the coast. This situation, coupled with the failure of the beetles to disperse readily throughout a grove to contain early season mealybug reproduction, prompted Smith and Armitage (1926) to develop a technique of mass culturing *Cryptolaemus on* mealybugs grown on potato sprouts, and to release the cultured beetles each spring in citrus trees. These first inoculative-type releases (10 cultured beetles per tree) were effective in controlling mealybugs through the predation achieved by the progeny of the released beetles. This technique was adopted by private companies and citrus grower groups which built their own insectaries. In the 1960's, one large California insectary produced over 30 million *Cryptolaemus* a year (Fisher, 1963), and in Sicily, more than 1000 citrus orchards received insectary-produced *Cryptolaemus* each year, with satisfactory control of the citrus mealybug being achieved (Liotta and Mineo, 1963).

Cryptolaemus has also been periodically released in greenhouses to control mealybugs. Whitcomb (1940) obtained control of *Planococcus citri* on gardenias and *Phenococcus gossypii* Town. and Ckll. on chrysanthemums. Doutt (1951) found that if greenhouses in California were warm enough, control of mealybugs on gardenias could be obtained from periodic release of the beetles.

E. Use of Predators in Integrated Control

The efficiency of various predators in several agroecosystems has been a significant feature to development of integrated control programs—e.g., in orchards (Asquith and

Colburn, 1971; Mathys and Baggiolini, 1965; Hoyt and Caltagirone, 1971; Hukusima, 1970; Kennett and Flaherty, 1974; Niemczyk *et al.*, 1974; Pickett and Patterson, 1953; Remaudière *et al.*, 1973; Steiner, 1964; and Chapter 14); in alfalfa (Stern *et al.*, 1959; Chapter 16); in cotton (van den Bosch *et al.*, 1971; Chapter 17); in rice (Itô *et al.*, 1962; Kiritani *et al.*, 1971; Chapter 16); in tobacco (Rabb *et al.*, 1974); in strawberries (Oatman *et al.*, 1967); in vegetables (Bombosch, 1965; Hodek, 1966; Oatman, 1970); in greenhouses (Gould, 1970; Hussey and Bravenboer, 1971; Helgesen and Tauber, 1974; and numerous examples reported in the 1973—1974 bulletin of the West Palearctic Regional Section of the International Organization for Biological Control, IOBC—OILB). Newsom (1974) reviewed the relationships of insecticide use and predation. He and others have stressed that predators may be more adversely affected than the pests by use of an insecticide because not only is the predator exposed to the toxic material but it also suffers the additional hazard of an insecticide-induced shortage of its food.

Methods used in the field to determine the impact of predators on insect populations are more difficult than those used to evaluate the role of parasites, for parasites can be easily detected by their presence in or on a host but many predators leave no sign of their attack. However, there are both direct and indirect methods of determining their impact. These are (1) by direct inspection of the relationship between effect and density, (2) by key factor analysis, (3) by life table analysis, and (4) by artificial reduction or exclusion of a regulatory factor (DeBach, 1964; Huffaker, 1971; Hughes, 1972; Kiritani and Dempster, 1973; Solomon, 1964; Southwood, 1966; van Emden, 1972; Varley *et al.*, 1973; and Chapter 11).

Studies on energetics of predatory insects are few indeed. Mukerji and LeRoux (1969) made such a study involving *Podisus maculiventris* (Say) fed different levels of *Galleria mellonella* (L.). For all predator nymphal stages combined, from 38 to 55% of biomass in calories of the prey they consumed was converted to their biomass in calories, and efficiency of converting prey biomass consumed by females into eggs ranged from 37 to 51% over 55 days.

F. Predator Models

Mathematical models of true predation thus far have involved mainly intrageneration responses of predators. Solomon (1949) classified predator responses into main divisions: (1) the functional response, which involves prey consumption, and (2) the numerical response, which is concerned mainly with reproduction of predators. Hassell (1966) proposed the terms behavioral and reproductive responses to describe these. He stressed that several consecutive generations of the parasite or predator should be followed to determine their responses to host or prey densities. These responses are expressed in terms of changes in the percentage parasitization or predation.

Holling (1961, 1966) identified and quantified the components of the functional responses, largely utilizing the preying mantid as his experimental predator. He reasoned that since the numerical responses are largely the outcome of effects of consumption on reproduction, immigration, emigration, and mortality, it follows then that the functional responses measured in terms of prey consumed are of basic importance. Yet it is the numerical response, and not the functional response itself, that is usually density-dependent and regulatory in nature (see Chapter 3). Although Holling neglected the more important numerical response, he developed a detailed model, primarily of the functional response, using the systems analysis approach, but his model does not embrace a response to changes in predator density. To develop a complete model of predation, the functional responses to both prey and predator density and the numerical response in various aspects must be dealt with.

Three submodels of competition involving parasites and predators (Griffiths and Holling, 1969) were combined and coupled with Holling's earlier model which introduced the full consequences of simulated different prey and predator densities. From this model, Holling (1968) identified four specific tactics of predators based on the assumption that the strategy of the organism is to produce stability in the existing conditions.

The combined functional and numerical responses exhibited by some phytoseiids are sufficient to suppress and then rapidly contain increasing spider mite populations, and the regulation of such populations has been demonstrated in both laboratory and field (e.g., review by Huffaker *et al.*, 1970). Recently, Laing and Osborn (1974) described the combined effective functional and numerical responses of three phytoseiid species.

Some of the first mathematical models describing insect predation specifically dealt with predators of aphid populations. The components of these models included the potential rate of increase of the aphid population without predation, the initial aphid population, the aphid population density on a subsequent day, and the number of aphids consumed by the predators present (Bombosch, 1963). van Emden (1966) viewed the predator/aphid relationship as an interaction of three major components: (1) *voracity* (a function of appetite, activity, and abundance of predators), (2) *synchronization* (the time relationship of the attack by aphidophagous insects with the aphid population and the life span of the individual aphid), and (3) aphid *reproductive rate*. By using Bombosch's model, van Emden found that environmental factors influenced the responses of both the aphids and predators and he related the model to the efficiency of predators in the field.

Tamaki *et al.* (1974) modified Bombosch's predator model in an attempt to formulate a predictive one for field evaluation. The major innovation was dividing the aphid predator complex into three classes depending upon their voracity, based on similar feeding capacities. Each class is given a value called "predator power." The polyphagous predators (class 1), e.g., nabids and anthocorids, are rated with a "predator power" of 1, the aphidophagous larvae of large coccinellids, syrphids, chrysopids, and *Scymnus* adults were rated 4 (class 2), and the large aphidophagous adult coccinellids rated 8 (class 3). Total predator power is, therefore, the sum of all predators monitored in class 1 plus 4 times the sum in class 2 plus 8 times the sum in class 3.

The total predator power is multiplied by predator efficacy to give projected reduction of aphids. According to Tamaki *et al.*, predator efficacy is affected by such external factors as temperature, prey density, seasonal variation, and host plant condition that affect predator power by reducing the population of prey. To obtain a value for predator efficacy they modified Bombosch's (1963) original equation. Therefore, predator power multiplied by predator efficacy yielded an estimate of the impact of the predator complex, with the resulting product giving the projected reduction of a population of aphids both with respect to the number of aphids consumed and the potential growth of the aphid population in the absence of predators. The calculated population curves of aphids on potato plants and sugar beets exposed to various predator population densities, using the model, were similar to those observed in field plots where aphid and predator densities were known.

Hughes and Gilbert (1968) made a computer model of aphids, their predators, parasites, and hyperparasites. The biological information came from experience gained from studying the cabbage aphid and its natural enemies in Australia. The model uses a physiological time scale with aphid instar periods as units. Input for the model consists of 16 independent parameters. The model predicts the seasonal behavior of 9 observable life system characteristics. The model output indicates that the above authors' knowledge of kale-cabbage aphid-parasite life systems are reasonably complete and suggest how aphid numbers were controlled in their situation and predicts the effects of various management practices. However, no generalizations concerning population dynamics were expressed. Gilbert and Hughes (1971) then used the model to examine (1) stochastic variation of population numbers, (2) the parasite's strategy, and (3) a field test of the model's predictions for biological control. From these studies, it was concluded, however, that the model appears to have a general value in the insight it gives on the relation between the two most accurately modeled components—the aphid and its parasite.

III. PREDATORY INSECTS

Insects that prey upon other insects and spider mites occur in most of the orders of insects, the exceptions being the Protura, Embioptera, Zoraptera, Isoptera, Mallophaga, Anoplura, Homoptera, and Siphonaptera. Space limitations preclude discussion of the biologies of all predaceous Insecta. Only a list of families containing predaceous insects (Table 1) is presented, followed by some generalizations concerning their characteristic morphology and some biology. The list also shows the predatory stage(s) and the prey utilized (based primarily on data from Adashkevitch and Popov, 1972; Balduf, 1935, 1939; Bänsch, 1964, 1966; Bay, 1974; Bequaert, 1922; Blumberg and Swirski, 1974; Bombosch, 1963; Brues, 1946; Carayon, 1961; Ceianu, *et al.*, 1965; Clausen, 1940; Collyer, 1953; Collyer and Massee, 1958; Corbet, 1962; Cullen, 1969; Davis, 1919; DeBach, 1964; Escherich, 1942; Evans and Eberhard, 1970; Finnegan, 1974; Franz, 1958, 1961; Grandi, 1961; Grasse, 1951; Greathead, 1963; Hagen, 1974; Hagen and van den Bosch, 1968; Hinton, 1945; Hodek, 1966, 1973; Holling, 1961, 1965, 1966; Huffaker, 1970; Iwata, 1964; Jenkins, 1964; Kline

TABLE 1

Predaceous Insects, Their Predatory Stages, and Their Prey

Taxa	Predatory stage*		Prey
	Immature	Adult	
Thysanura			
Lepismatidae[a,b]	+	+	Miscellaneous moribund insects
Diplura			
Japygidae	+	+	Soil arthropoda
Collembola			
Entomobryidae[a]	+	+	Nematodes
Ephemeroptera			
Siphlonuridae	+	–	Aquatic fauna
Odonata (16 families)	+	+	Aquatic fauna, miscellaneous insects
Orthoptera			
Acrididae[a]	+	+	Grasshoppers
Gryllidae[a]	+	+	Miscellaneous insects
Gryllacrididae[a]	+	+	Miscellaneous insects
Mantodea			
Amorphoscelidae	+	+	Miscellaneous insects
Mantidae	+	+	Miscellaneous insects
Dermaptera			
Arixeniidae	+	+	Insects in bat dung
Chelisochidae	+	+	Miscellaneous insects
Forficulidae[a]	+	+	Miscellaneous insects
Labiduridae[a]	+	+	Miscellaneous insects
Plecoptera			
Chloroperlidae[a]	+	–	Aquatic insects
Eustheniidae	+	–	Aquatic insects
Perlidae	+	–	Aquatic insects
Perlodidae	+	–	Aquatic insects
Psocoptera			
Atropidae[a]	+	+	Miscellaneous insects
Caeciliidae[a]	+	+	Miscellaneous insects
Liposcelidae[a]	+	+	Lepidoptera eggs
Thysanoptera			
Aeolothripidae	+	+	Small insects, mites
Phloeothripidae[a]	+	+	Small insects, eggs, mites
Thripidae[a]	+	+	Small insects, mites
Hemiptera			
Heteroptera			
Anthocoridae[a]	+ –	+ –	Homoptera, Thysanoptera, Lepidoptera, Coleoptera, mites
Belostomatidae	+	+	Aquatic fauna
Berytidae (Neididae)	+	+	Lepidoptera eggs
Corixidae[a]	+	+	Aquatic fauna
Enicocephalidae[a]	+	+	Ants
Gelastocoridae	+	+	Shore Arthropoda
Gerridae	+	+	Aquatic and surface Arthropoda

TABLE 1 *(continued)*

Taxa	Predatory stage		Prey
	Immature	Adult	
Hebridae	+	+	Shore insects
Lygaeidae[a]	+−	+−	Homoptera, Hemiptera, Lepidoptera, eggs, mites
Mesoveliidae	+−	+−	Dead or feeble insects
Miridae[a]	+−	+−	Homoptera, Hemiptera, Lepidoptera, mites
Nabidae	+−	+−	Homoptera, Lepidoptera, mites
Naucoridae	+	+	Aquatic fauna
Nepidae	+	+	Aquatic fauna
Notonectidae	+	+	Aquatic Arthropoda
Ochteridae	+	+	Aquatic and shore insects
Pentatomidae[a]	+	+	Lepidoptera, Coleoptera
Phymatidae	+	+	Small insects
Pyrrhocoridae[a]	+	+	Myriapoda, Lepidoptera, Coleoptera
Reduviidae[a]	+−	+−	Miscellaneous insects
Saldidae	+	+	Shore insects
Veliidae	+	+	Diptera, Collembola
Coleoptera			
Adephaga			
Amphizoidae	+	+	Plecoptera nymphs
Carabidae[a]	+−	+−	Miscellaneous insects, snails, Myriapoda
Cicindelidae	+	+	Miscellaneous Arthropoda
Dytiscidae	+	+	Aquatic fauna
Gyrinidae	+	+−	Aquatic fauna
Haliplidae[a]	+	+	Miscellaneous aquatic insects, nematodes
Hygrobiidae	+	+	Tubifex worms, aquatic fauna
Noteridae	+	+	Aquatic fauna
Paussidae	+	+	Ant adults and larvae
Polyphaga			
Anthicidae[a]	+−	+−	Miscellaneous insects, mites
Anthribidae[a]	+	+	Coccids
Brentidae[a]	+	+−	Wood-boring Coleoptera, ants
Cantharidae[a]	+	+	Soil Arthropoda, aphids
Cleridae[a]	+	+−	Wood-boring Coleoptera, Lepidoptera, Hymenoptera larvae, Orthoptera eggs
Coccinellidae[a]	+−	+−	Homoptera, Lepidoptera, Coleoptera, mites
Colydiidae[a]	+	+−	Wood-boring Coleoptera
Cryptophagidae	+	+	Bark beetles
Cucujidae[a]	+	+	Bark beetles, Coleoptera larvae
Cybocephalidae	+	+	Diaspine scales
Dermestidae[a]	+	−	Mantid and blattid eggs, Lepidoptera larvae
Derodontidae[a]	+	+	Chermidae

TABLE 1 *(continued)*

Taxa	Predatory stage		Prey
	Immature	Adult	
Drilidae	+	−	Snails
Endomycidae[a]	+	+ −	Coccids
Elateridae[a]	+ −	+ −	Larvae of Coleoptera, sawflies, Lepidoptera
Histeridae[a]	+	+	Miscellaneous insects in dung, carrion, logs, ants
Hydrophilidae[a]	+	+ −	Aquatic fauna, Lepidoptera larvae
Lampyridae	+	+ −	Snails, anneillids, Coleoptera adults
Lathridiidae[a]	−	+ −	Coccids
Lycidae[a]	+	−	Wood-boring Coleoptera
Meloidae	+	−	Orthoptera and bee eggs
Melyridae[a]	+	+ −	Homoptera, Lepidoptera larvae, miscellaneous Arthropoda
Melandryidae[a]	+	−	Wood insects
Mordellidae[a]	+	−	Wood-boring Coleoptera
Nitidulidae[a]	+	+	Bark beetles
Nosodendridae	+	+	Diptera larvae
Orthoperidae	+	+	Coccids, scolytids
Othniidae	?	+	Leaf insects
Passandridae	+	?	Wood-boring Coleoptera
Pedilidae[a]	?	+ −	Scarab and meloid adults
Phengodidae	+	+	Snails, Myriapoda
Pselaphidae[a]	+	+ −	Soil insects, mites, ants
Pyrochroidae[a]	+?	+ −?	Bark insects
Pythidae[a]	+	+	Bark beetles
Rhipiphoridae[a]	+	−	Wasp larvae
Rhizophagidae[a]	+	+	Bark beetles
Scarabaeidae[a]	+	+ −	Ants, termites, Orthoptera eggs
Salpingidae	+	+	Intertidal fauna, wood-boring Coleoptera
Silphidae[a]	+ −	+ −	Diptera and Lepidoptera larvae, snails
Staphylinidae[a]	+	+	Miscellaneous arthropods, snails
Tenebrionidae[a]	+	+	Wood-boring Coleoptera, flea larvae, coccids
Trogositidae[a]	+	+	Wood-boring Coleoptera, Lepidoptera larvae
Mecoptera			
Bittacidae	?	+	Miscellaneous small insects
Boreidae[a]	−	+?	Miscellaneous insects, mites
Panorpidae[a]	−	+?	Miscellaneous insects
Neuroptera			
Megaloptera			
Sialidae	+	−	Benthic fauna
Corydalidae	+	−	Benthic fauna
Raphidiodea			
Raphidiidae	+	+	Insects under bark, miscellaneous insects

TABLE 1 *(continued)*

Taxa	Predatory stage		Prey
	Immature	Adult	
Inocelliidae	+	+	Insects under bark, miscellaneous insects
Planipennia			
Ascalaphidae	+	+	Miscellaneous insects, mites
Berothidae	+	?	Associated with ants, termites
Chrysopidae	+ −	+ −	Homoptera, Lepidoptera eggs, larvae, thrips, spider mites
Coniopterygidae	+	+ −	Homoptera, spider mites
Dilaridae	+	?	Insects in wood galleries
Hemerobiidae	+	+	Homoptera, Lepidoptera eggs, larvae, spider mites
Ithonidae	+	?	Associated with scarab larvae
Mantispidae	+	+	Spider eggs, miscellaneous insects
Myrmeleontidae	+	+ −	Ants, soil surface insects
Nemopteridae	+	?	Miscellaneous soil surface insects
Nymphidae	+	?	Miscellaneous insects in logs
Osmylidae	+	+ −	Miscellaneous insects under bark, Diptera at edge of water
Polystoechotidae	+	+	Miscellaneous insects
Psychopsidae	+	?	Miscellaneous insects under bark
Sisyridae	+	−	Sponges
Sympherobiidae	+	+	Pseudococcidae, Chermidae
Trichoptera			
Annulipalpia[c]			
Bottom dwelling spp.	+	−	Benthic fauna
Lepidoptera			
Blastobastidae[a]	+	−	Coccids
Cyclotornidae	+	−	Leafhoppers, ants
Heliodinidae[a]	+	−	Coccids, spider eggs
Cosmopteryidae[a]	+	−	Coccids, ant brood
Lycaenidae[a]	+	−	Associated with ants, Homoptera
Noctuidae[a]	+	−	Coccids
Oinophilidae[a]	+	−	Coccids
Psychidae[a]	+	−	Coccids
Pyralidae[a]	+	−	Homoptera, Lepidoptera, ants
Tineidae[a]	+	−	Coccids, ant pupae
Tortricidae[a]	+	−	Homoptera, mites
Diptera			
Nematocera			
Cecidomyiidae[a]	+	−	Homoptera, thrips, mites
Ceratopogonidae	+	−	Chironomid eggs and larvae
Chaboridae[a]	+	−	Aquatic Arthropoda
Chironomidae[a]	+	−	Aquatic invertebrates (Chironomid immatures), ant larvae
Culicidae[a]	+	−	Aquatic Diptera, miscellaneous insects
Mycetophilidae[a]	+	−	Adult Diptera, miscellaneous insects

TABLE 1 *(continued)*

Taxa	Predatory stage		Prey
	Immature	Adult	
Tipulidae[a]	+	—	Aquatic Diptera, worms
Brachycera			
Asilidae	+	+	Soil insects, miscellaneous insects
Bomblyliidae[a]	+	—	Orthoptera eggs
Dolichopodidae	+ —	+	Bark beetle, miscellaneous insects, mites
Empididae[a]	+ —	+	Miscellaneous insects, mites
Mydidae	+	+	Wood-boring Coleoptera, miscellaneous insects
Rhagionidae	+	—	Orthoptera eggs, miscellaneous insects
Tabanidae[a]	+	—	Scarab larvae, miscellaneous soil fauna
Therevidae	+	—	Coleoptera and Diptera larvae in soil
Xylophagidae	+	—	Bark beetles
Cyclorrhapha			
Anthomyiidae[a]	+	+	Orthoptera eggs, miscellaneous insects
Calliphoridae[a]	+	+	Orthoptera eggs, termite adults, ants
Chamaemyiidae	+	—	Homoptera
Chloropidae	+	—	Aphids, miscellaneous insects, spider eggs
Drosophilidae[a]	+	—	Homoptera, spider eggs, scolytids
Ephydridae[a]	+	—	Aquatic larvae, terrestrial insects
Lonchaeidae[a]	+	—	Wood-boring Coleoptera
Otitidae[a]	+	—	Orthoptera eggs
Pallopteridae	+	—	Bark beetles
Phoridae[a]	+	—	Eggs of Orthoptera, coccids and spiders, Lepidoptera
Sarcophagidae[a]	+	—	Lepidoptera pupae, locust and spider eggs
Scatophagidae[a]	+	—	Miscellaneous insects
Sciomyzidae	+	—	Snails
Syrphidae[a]	+ —	—	Homoptera, Coleoptera, Lepidoptera, thrips
Hymenoptera			
Symphyta[d]			
Tenthredinidae[a]	—	+	Coleoptera, Diptera, Homoptera
Apocrita			
Ichneumonoidea[e]			
Braconidae[b]	+	+	Diptera and Lepidoptera larvae
Ichneumonidae[b]	+	+ —	Spider eggs, bee eggs, larvae
Chalcidoidea[f]			
Encyrtidae[b]	+	+ —	Coccids and diaspine eggs
Eulophidae[b]	+	+ —	Eggs, miscellaneous insects
Eupelmidae[b]	+	+ —	Eggs, miscellaneous insects, Hymenoptera larvae
Eurytomidae[a,b]	+	—	Orthoptera and Homoptera eggs

TABLE 1 *(continued)*

Taxa	Predatory stage		Prey
	Immature	Adult	
Pteromalidae[b]	+	+ −	Coccids eggs
Torymidae[b]	+	−	Mantid eggs
Proctotrupoidea			
Evaniidae	+	−	Blattid eggs
Gasteruptiidae[b]	+	−?	Hymenoptera larvae
Chrysidoidea			
Chrysididae[b]	+	+ −	Hymenoptera and stored prey
Bethyloidea[g]			
Bethylidae[b]	−	+	Coleoptera and Lepidoptera larvae
Dryinidae[b]	−	+ −	Leafhoppers
Scolioidea			
Mutillidae[b]	−	+ −	Hymenoptera larvae in cells
Tiphiidae[b]	−	+ −	Scarab larvae
Thynnidae[b]	−	+ −	Scarab larvae
Pompiloidea			
Popilidae[b]	−	+ −	Spiders
Formicoidea			
Formicidae	+ −	+ −	Miscellaneous insects
Vespoidea			
Eumenidae	+	+ −	Lepidoptera and Coleoptera larvae
Masaridae[a]	+	+ −	Few species predaceous
Vespidae	+	+ −	Lepidoptera and Coleoptera larvae
Sphecoidea			
Sphecidae			
Ampulicinae[b]	+	+ −	Orthoptera (Blattids)
Astatinae	+	−	Hemiptera
Crabroninae	+	−	Miscellaneous insects
Larrinae[b]	+	+ −	Orthoptera, Hemiptera, spiders
Mellininae	+	−	Diptera
Nyssoninae	+	−	Orthoptera, Hemiptera, Lepidoptera, Diptera
Pemphredoninae	+	−	Collembola, Hemiptera, Thysanoptera
Philanthinae	+	−	Coleoptera, Hymenoptera
Sphecinae	+	−	Orthoptera, spiders

*+, Individual eats only prey; −, individual not predatory; + −, individual eats both prey and non-insect food.

[a]Not all species in the taxon are predaceous, or species that are predaceous may also have other feeding habits.

[b]Species in the family are mostly parasitic.

[c]Most families.

[d]Probably all sawfly families predaceous in adult stage.

[e]Some adults host feed.

[f]Many adults host feed.

[g]Adults host feed.

and Rudinsky, 1964; Krombein, 1967; Kulman, 1974; Ler, 1964; Lewis, 1973; Lindner, 1949; Lord, 1971; Miller, 1971; Muma, 1955; Nuorteva, 1971; Oldroyd, 1964; Seguy, 1950; Shurovenkov, 1962; Stark and Dahlsten, 1970; Swan, 1964; Sweetman, 1958; Tao and Chiu, 1971; Thompson, 1951; Thompson and Simmonds, 1964–1965; Sundby, 1966; Usinger, 1963; van den Bosch and Hagen, 1966; van den Bosch and Messenger, 1973; van Emden, 1966, 1972; Whitcomb and Bell, 1964; Wilson, 1971; and a series of papers in OILB, Sect. Region Ouest Palearct. Brochure No. 3, 1974).

A. Stages

1. Eggs

The diversity in egg form exhibited by predaceous insects carries more genotypic characteristics than the phenotypic adaptive modifications found among parasitic insects (Fig. 2). There appears to be some uniformity in egg form in many families, at least at the generic level. In general, eggs are proportionate in size to the insects

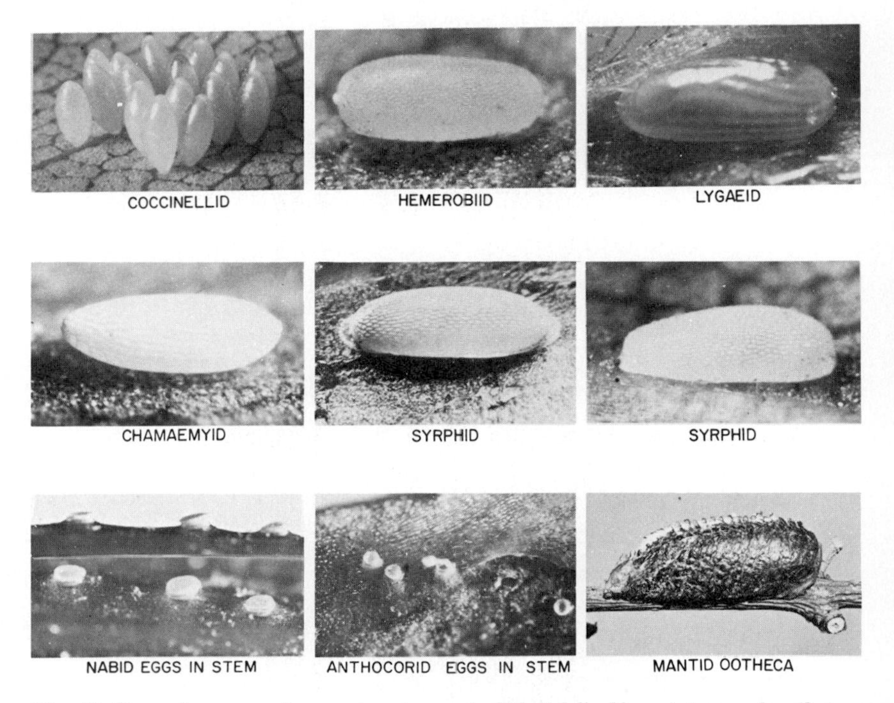

Fig. 2. Eggs of some predaceous insects: coccinellid, *Adalia bipunctata*, egg length 1 mm; hemerobiid, *Hemerobius* sp., egg length 0.87 mm; lygaeid, *Geocoris* sp., egg length 0.87 mm; chamaemyid, *Leucopis* sp., egg length 0.47 mm; syrphid eggs, egg lengths 0.87 mm; nabid, *Nabis* sp., inserted into alfalfa stem, egg cap diam. 0.3–0.4 mm; anthocorid, *Orius tristicolor*, inserted in alfalfa stem, egg cap diam. 0.25 mm; mantid egg case, 27 mm long. Photos by F. E. Skinner, Univ. of Calif., Berkeley, California.

depositing them. In structure, some characteristics are common to many families of certain orders.

Eggs with conspicuous stalks and distinct disklike micropyles are those of Chrysopidae or Mantispidae. The eggs of most Hemiptera possess an operculum, or at least a false operculum which is surrounded by chorionic processes. The eggs of most predaceous Diptera and all Neuroptera have perceptible, often projecting micropyles. The eggs of predaceous Coleoptera do not seem to possess conspicuous micropyles.

Eggs of predators are usually deposited near their prey, i.e., exposed upon plants, or on or in the soil. Those inserted into plants are mostly restricted to a few families of Hemiptera and thrips.

2. *Nymphs or Larvae*

In the family to which a predator belongs there often are other genera or subfamilies which are phytophagous or omnivorous (Fig. 3). The structure of the mouthparts and

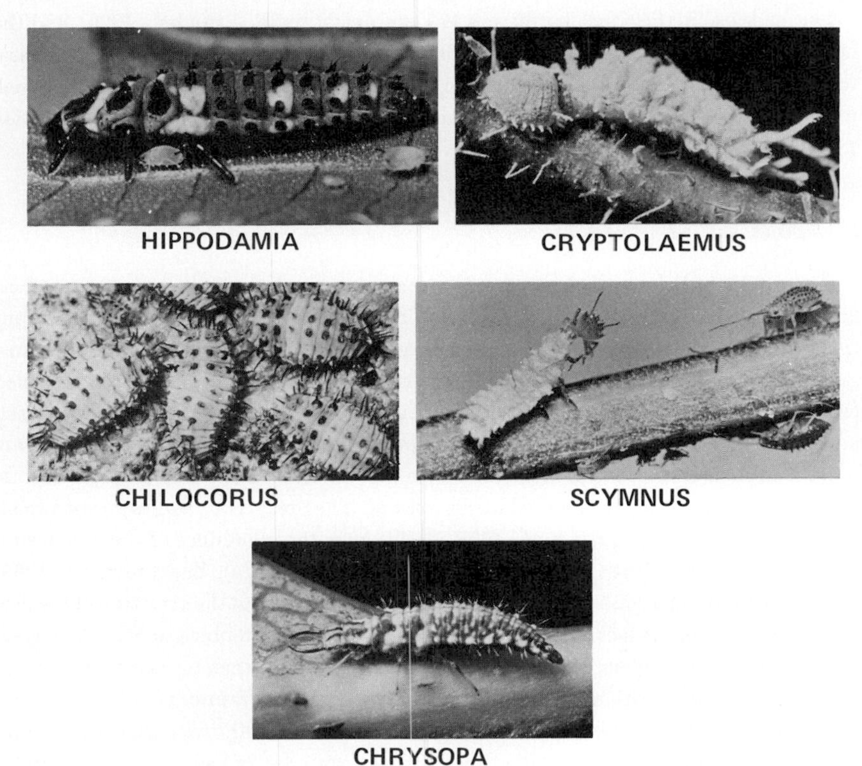

Fig. 3. Mature larvae of some predaceous insects. *Hippodamia quinquesignata*, larva length 8.3 mm; *Cryptolaemus montrouzieri* feeding on mealybug, larva length 6.5 mm; *Chilocorus* sp., larva length 7 mm; *Scymnus loewii*, feeding on an aphid, larva length 2.8 mm; *Chrysopa carnea*, larva length 7 mm. Photos by F. E. Skinner, Univ. of Calif., Berkeley, California.

front legs often presents clues to the predaceous habit. Thus, there are ecological inferences from morphological data (Hespenheide, 1973). In the Hemiptera the rostrum is shorter, and stouter, and angular, for the rostrum is usually held in a horizontal plane to the body when feeding, while the plant feeders have the rostrum projecting at a right angle to the body when feeding. The anterior legs are modified or smeared with resin for capturing prey, at least in those predaceous Hemiptera that rely on the ambush strategy of prey capture (Miller, 1971). Internally, the alimentary tract is longer and the salivary glands produce an alkaline secretion which is injected into the prey causing paralysis (Elson, 1937). In the Orthoptera, Isely (1944) described the carnivorous mandibles found among some Decticinae, but some predaceous Conocephalinae which also eat grass seeds have mandibles of the carnivorous type. In the Neuroptera, the larvae are all predaceous and distinguished by their sucking, tonglike mouthparts, composed of closely fitting side-by-side maxillae and mandibles. In the Coccinellidae both the larvae and adults of the predaceous genera have mandibles of the incisor type, having one or two apical teeth and a basal tooth, while the fungus-eating species in the Psylloborini possess six apical teeth (Strouhal, 1926) and the phytophagous Epilachninae have mandibles with several apical teeth, without a basal tooth (Boving, 1917). The larvae of Scymnini and Stethorini (all predaceous) have small mandibles with an internal duct for sucking liquid food and for extraoral digestion (Sovoiskaya, 1960).

3. Adults

The adults of holometabolous predators may have similar mouth structures as their larvae, but usually they are quite different. Here again, the mouthparts are different from those of closely related phytophagous groups, sometimes even within the same genus. About one-half of the species of *Chrysopa* adults studied are honeydew and pollen feeders while the other one-half are predaceous. A comparison of a few species of each adult type in *Chrysopa* reveals that in the nonpredaceous species, yeast symbiotes are present in their crops and the tracheal trunks associated with the crop are significantly larger than those found in the predaceous species. The predaceous adults do not appear to harbor symbiotes (Hagen *et al.*, 1970, 1972). Also, the mandibles of the predaceous adults are longer (Ickert, 1968). In the Carabidae (Coleoptera), Zhavoronkova (1969) found a definite relationship between the nature of feeding and the structure of the proventriculus and mandibles. He divided the family into three trophic groups, (1) obligate zoophages: mandibles long, pointed, falcate, their length more than twice their width; the proventriculus lined with dense long hairs, and practicing extraintestinal digestion, (2) predominantly zoophages: mandibles as above but their length twice their width, proventriculus anteriorally having either teeth, spines, spinules, or long acicular structures, and (3) predominantly phytophages: mandibles broad, triangular, apically blunt with length slightly greater than width, and the proventriculus similar to group 2. Many Carabidae are omnivorous and phytophagous (Johnson and Cameron, 1969). In the

Diptera, adults of many species, scattered in various families have the proboscis modified for feeding on insects by increased rigidity and, often, elongation. The most remarkable development is in some Dolichopodidae in which the mouthparts appear to be mandibulate. Here, the labella (or paraglossae) has developed a sharp sclerotized lobe on the inner side which simulates the pinching jaws of mandibulate insects (Aldrich, 1922; Brues, 1946). In the entomophagous Hymenoptera, the mouthparts are mainly mandibulate but in some species are lengthened, enabling them to feed at flowers. The most distinctive feature is the modified ovipositor, used by species feeding as predators as adults, for either paralyzing their prey or building feeding tubes to obtain prey hemolymph coming from "protected" prey by capillary action, or to open wounds for simply feeding on exposed prey (Clausen, 1940; Doutt, 1959).

B. Phenology

1. Diapause

As in other insects, the synchronization of life cycles of predators with their environment is largely based on entering a dormant state when conditions are unfavorable. The induction and termination of diapause is likewise similar to that in other insects, with photoperiod, temperature, humidity, and nutrition acting as the environmental triggers (Tauber and Tauber, 1973). The role of these factors varies with the species, even in the same genus. In *Chrysopa*, some species overwinter as adults, while others as third instar larvae within the cocoons (Balduf, 1939; Principi, 1956; Toschi, 1965; Tauber and Tauber, 1970). The length of changes in photoperiod induced diapause in the species investigated (MacLeod, 1967; Sheldon and MacLeod, 1974; Tauber and Tauber, 1970, 1972). *Chrysopa carnea* adults can enter diapause in response to the direction of change in day length that does not encroach on the critical photoperiod (Tauber and Tauber, 1970). Many Odonata in temperate regions apparently diapause in the final larval instar. Corbet (1962) found that, at least in *Anax imperator* Leach, the final instar nymphs can discern the rate of increase in day length; they forego diapause if this amount exceeds 2 minutes/day.

Predaceous Heteroptera appear to overwinter mainly as adults or in the egg stage. The stimuli that induce diapause in this group apparently has not been determined. Some species of *Geocoris* overwinter in the adult stage, but Tamaki and Weeks (1972a) observed *G. bullatus* to overwinter in the egg stage. *Nabis* spp. appear to overwinter in the adult stage, but the nabid *Himacerus apterus* F. hibernates in the egg stage in Europe (von Koschel, 1971). The predaceous Pentatomidae, at least in *Podisus* spp., overwinter as adults (Coppel and Jones, 1962). Among the anthocorids, species of *Anthocoris*, *Orius*, *Tetraphleps*, and *Malacoris* overwinter as adults, but some species, at least in *Xylocoris* and *Elatophilus*, overwinter as large larvae (Anderson, 1962a,b). Of 12 species representing 10 predaceous mirid genera all hibernate in the egg stage, but *Camptobrachis lutescens* (Schilling) overwinters as an adult (Collyer, 1953).

In the Coccinellidae, photoperiod, temperature, and/or nutrition are involved in induction of reproductive diapause which is induced in the young adults (Hagen, 1962; Hodek, 1973). Hibernal diapause is induced under short photoperiod (10—12 hours) in the bivoltine *Coccinella novemnotata* Herbst and aestival diapause is induced under an exceedingly long photoperiod of 18 hours (McMullen, 1967).

The Carabidae are mostly univoltine, either spring breeders whose summer larvae emerge as new generation adults later the same summer, or autumn breeders whose larvae emerge as new generation adults the following spring (Mitchell, 1963; Skuhravy, 1959). One species studied by Penney (1969) underwent an obligatory summer diapause; various regimes of photoperiod, temperature, and humidity did not prevent induction of diapause, although starving the adults prevented diapause. Photoperiod was involved in termination of diapause. The males of three *Pterostichus* spp., mature fully under short-day conditions, and cessation of reproduction is induced by long-day photoperiod which can be overcome by application of corpus allatum hormone (Ferenz and Holters, 1975).

The aphidophagous Syrphidae overwinter in the last larval instar or in the adult stage; the diapause-inducing factors apparently are unknown but desiccation may play a role. Diapausing larvae in the soil can take in water through their anal tubes (Schneider, 1969).

2. Thermal Thresholds

The thermal thresholds for development of aphidophagous predators are generally higher than those of their aphid prey, but in *Hemerobius pacificus* Banks immature stages, the thresholds are remarkably low, being 0.4° for eggs, 4.1° for larvae, and 0.6°C for pupae; the larvae appear to be active during winter in California (Neuenschwander, 1975).

C. Some Relationships of Predators to Prey

1. Searching

Predaceous arthropods respond to a sequence of environmental cues to locate their prey. Most predators, adult stage, unlike most parasitoids, require a minimum number of prey in the egg production and oviposition (Fig. 4). A minimum number of prey must also be eaten in the larval or nymphal stages to provide the required nutrients and energy for maintenance, searching, growth, and development. The number or biomass of prey required for these functions depends in part upon the size of the predator and the extent of its searching and other energy-consuming activities, and, in turn, upon the size and nutritional quality of the prey population. The ease with which the required amount of prey can be found is also dependent upon the predator's searching efficiency and prey

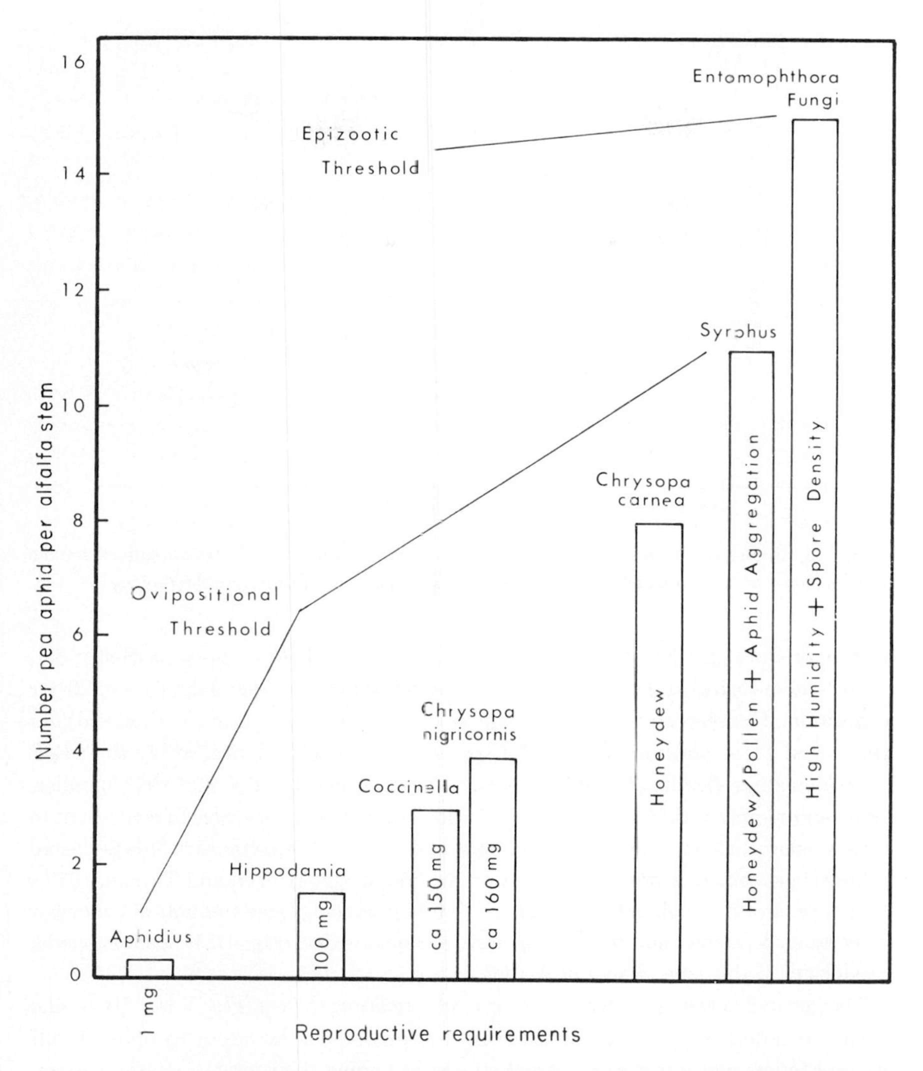

Fig. 4. The bars indicate the relative mean number of pea aphids on alfalfa stems, including leaves, necessary to induce oviposition in various aphid natural enemies. The information within the bars are minimum total biomasses of pea aphids or other requisites for the natural enemies to reproduce and for the fungus to cause an epizootic. One *Hippodamia* female, for example, must search about 50 stems to find enough aphids at 2 per stem to satisfy the 100 mg biomass of aphids needed for egg production and oviposition. (From Hagen, 1976.)

population size and spatial distribution and obstructions in the habitat. Predators usually require higher host population densities than do parasitoids (Chapters 3 and 6).

There are, however, some adult predators that require other foods than their prey in

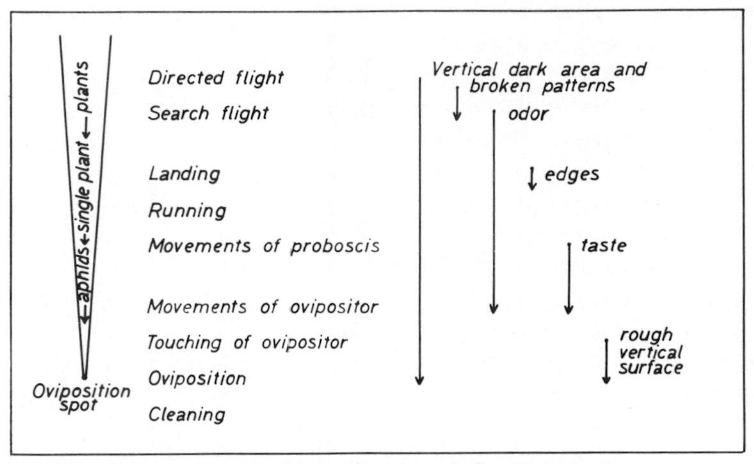

Fig. 5. Sequence of behavioral responses to environmental stimuli which guides female *Syrphus corollae* to an aphid habitat and stimulates oviposition within an aphid colony.

order to produce eggs. For example, syrphid adults require pollen for oogenesis (Schneider, 1969), and once gravid the behavior of the syrphid switches to searching for aphids for oviposition. The searching syrphid female follows a series of physical and chemical cues which lead to oviposition (Fig. 5). All species of *Syrphus* and *Scaeva* lay their eggs in the immediate vicinity of aphids (Bombosh, 1962; Bombosh and Volk, 1966; Chandler, 1968) and require rather high aphid populations to induce oviposition as contrasted to other predator adults which are aphidophagous (Fig. 4). *Chrysopa carnea* is attracted to honeydew, which is a prime food for production of eggs (Hagen and Tassan, 1972). Therefore, a rather high aphid population that is producing large amounts of honeydew is necessary to attract and induce oogenesis and oviposition (Fig. 4). When it is gravid, it oviposits in the general area of feeding.

The general sequence of events leading adult predators to prey (Fig. 5) is quite similar to that of adult parasitoids (Chapter 6), but predaceous larvae or nymphs usually respond to less complex environmental stimuli in finding their prey.

a. Habitat Selection. Finding an appropriate habitat that may harbor potential prey is usually achieved by the adult predator. Adults emerging from immature stages or diapause in perennial plant habitats where they have developed may simply begin searching within the habitat of their origin without dispersing or migrating. However, predators of prey-infesting herbaceous annuals usually must disperse or migrate to find suitable habitats.

The species of plant(s) in the habitat may retain or attract certain predators, but it is the presence of prey that signals attraction to other predators. Some species of *Anthocoris* remain on one type of plant independent of prey density, while others respond to and shift about according to prey densities. *Anthocoris sarothamni* Douglas and Scott is restricted to a species of broom shrub in England where it develops two generations feeding on psyllids. *Anthocoris nemoralis* (L.) produces its first generation on broom and then disperses in the summer to deciduous trees or shrubs to produce its second generation (Dempster, 1968). In British Columbia, *A. nemoralis*, introduced against the pear psylla, remains in the pears while the native *A. melanocerus* Reuter develops on psyllid-infested willows in the spring and migrates to pears in the summer and then to cottonwoods in the fall to feed on aphids. Fields and Beirne (1973) do not view the disappearance of *A. melanocerus* from pear trees as a result of any competitive displacement but simply as a shift to trees supporting higher populations of suitable prey. Dixon and Russel (1972) found that two different species of *Anthocoris* feeding on the sycamore aphid had striking differences in the minimum number of small aphids required for survival and this reflected a difference in the searching abilities of the two species.

A succession of coccinellid species was found to be associated with the age of red pine stands. As trees aged, less aphids were present and the species of coccinellids that were found on older trees had greater searching capacities than the coccinellid species found mainly on the younger trees (Gagne and Martin, 1968).

Iperti (1966) observed in southern France that four common coccinellid species selected different plants even though each plant harbored the same species of aphid, and Banks (1955) found that upon leaving overwintering sites, *Coccinella septempunctata* searched first in any nearby aphid-harboring vegetation.

Hippodamia convergens adults usually migrate long distances between greatly different habitats. From the valleys in the spring where they develop, the hungry young beetles migrate to the mountains where they feed and build up fat, which supports them for 9 months in a dormant state. Upon termination of dormancy in February, the beetles migrate back to California valleys, responding to temperature and light which place them in seasonal winds, bringing them to general areas of appropriate habitats (Hagen, 1962).

Vision and chemoreception (including olfaction) are the two main senses which guide many predators to habitats which are likely to harbor their prey or other foods. Vision has been implicated in *Chrysopa* and syrphid adults seeking flowers, since they are attracted to different colors (Ickert, 1968; Schneider, 1969). Also, *Chrysopa carnea* adults and some coccinellids respond to crops that have attained a certain height (Dickson and Laird, 1962; Hagen *et al.*, 1971; Iperti, 1966). Usinger (1963) suggests that dytiscids and notonectids that are attracted to shiny surfaces probably find their aquatic habitats visually.

Chemoreception of volatile substances from plants or prey from a distance is used by some predators to find suitable habitats. *Chrysopa carnea* adults fly toward honeydew, and artificial honeydew has been used in the field to attract them (Hagen *et al.*, 1971). The kairomone in the honeydew is possibly a tryptophan product, since acidified tryptophan is extremely attractive to this species (Hagen *et al.*, 1976; van Emden and Hagen,

1976). When flying, *C. carnea* seems to respond by anemotaxis by flying upwind toward the artificial honeydew source. They often then land and search for food. If no "honeydew" is located they again fly upwind toward "honeydew" and when they reach the source, aggregation occurs (Hagen *et al.*, 1971).

Predators of bark beetles respond to scolytid aggregation pheromones and to certain volatile tree terpenes. The bark beetle pheromones are used as kairomones by trogostids, clerids, dolichopodids, and perhaps histerids to locate concentrations of bark beetles attacking trees (Borden, 1974; Camors and Payne, 1973).

The coccinellid *Anatis ocellata* (L.) is attracted to aromatic substances associated with pine needles and not to odors coming from aphids (Kesten, 1969).

It will probably be found that many other predaceous insects find their prey habitats through chemoreception.

b. Prey Finding. Once in a suitable habitat under favorable climatic conditions, the adult predators may respond, while searching for prey, to light, gravity, and chemicals. Vision is obviously the main sense used in finding prey in the ambush- and true hunting-type predators, i.e., mantids (Mittlestaedt, 1957; Roeder, 1963; Holling, 1966); Odonata (Corbet, 1962); Heteroptera like *Zelus* (Edwards, 1966); *Podisus* (Morris, 1963), and *Notonecta*, which also responds to vibrations (Ellis and Borden, 1970); many aculeate wasps (Evans and Eberhard, 1970); asilid flies (Lavigne and Holland, 1969; Oldroyd, 1964); cicindelids (Balduf, 1935); dytiscids (Young, 1967); and spiders (Kaestner, 1968; Turnbull, 1973), like thomisids (Haynes and Sisojevic, 1966), lycosids (Anderson, 1974; Kiritani *et al.*, 1972), and salticids (Horner and Starks, 1972).

Some predators, like army ants, are blind (Wilson, 1971), but many predators with well-developed eyes seem not to use them to find their prey. Most of these predators must contact the prey to perceive them. There is a response to light and dark as well as gravity in some of aphidophagous coccinellids, syrphid larvae, and neuropterans. Some species are positively phototactic and geotactic while others have negative or mixed responses to light and gravity (Banks, 1957; Bänsch, 1966; Dixon, 1959; Ewert and Chiang, 1966; Fleschner, 1950; Dixon and Russel, 1972; Murdoch and Marks, 1973). Many of these same authors observed the predators searching "randomly" until contact with the prey. Some coccinellid larvae sense their prey by contacting them with their forelegs (Dixon, 1970), adults by palpi, and *Anthocoris* by their antennae (Dixon and Russel, 1972). *Nabids*, *Geocoris*, and some adult coccinellids seem to respond visually to objects a few millimeters in front of them. After contact or feeding, the searching behavior changes to making more frequent turns. This "tight" turning lasts variable lengths of time. This is an important response because it can concentrate their searching activities in regions of high prey density. Hassell and May (1974) formulated a mathematical model of the random searcher's path and the tight turning behavior, including the speeds involved.

The area covered by predaceous aphidophagous larvae varies with age and size. Even within the first instar the speed and time wasted in areas where aphids are not present and the efficiency of the larvae in capturing aphids differs with age (Dixon, 1970).

There are stimuli from leaf surfaces that direct some predators to search edges and veins of leaves, some areas being searched more than once (Arzet, 1973; Bänsch, 1966; Dixon, 1959, 1970; Dixon and Russel, 1972), but Murdock and Marks (1973) found *Coccinella septempunctata* larvae to make thorough searches, visiting most leaves and the stem while largely avoiding repeated visits to the same area. Leaf surface characteristics of some plants can reduce the effectiveness of certain predators (Arzet, 1973; Scopes, 1969; Wilbert, 1972).

c. Prey Acceptance. Once a prey has been contacted, besides specificity, the age of the predator and the size of the prey can influence the success of the predator attack. Specificity of prey attack varies with the species of the predator concerned, and this has been briefly discussed earlier in this chapter and shown for families in Table 1. However, even within a genus of predators, there can be species which may depart radically from the majority of species in the type of prey they require for complete development.

The composition of the prey cuticle may be important in eliciting the biting or sucking response of those predators that seem to find their prey by bumping into them. Perhaps the composition of cuticular waxes may be involved in prey acceptance. The paraffin coating of artificial diets appears to contain chemicals that stimulate biting and piercing of aphidophagous coccinellids and perhaps *Chrysopa*, although *Chrysopa* larvae seem to probe any projecting object (Hagen and Tassan, 1966). *Nabis* and *Geocoris* also pierce the food droplets coated with paraffin.

The age of the predator and the size of the prey have been found to influence significantly the capture or penetration of the prey of *Chilocorus orbus* Casey. First instar larvae can eat crawlers and newly settled diaspine scales but cannot penetrate the larger scales (Drea, 1956). Dixon (1958) describes the various methods of an aphid avoiding being eaten by *Adalia*. When both the aphid and coccinellid are about the same size the aphid secretes wax from its siphunculi and then attempts to pull away. The waxing of the coccinellid frequently allows the aphid to escape by pulling. If the aphid is larger it often frees itself without waxing. Aphids smaller than the coccinellid are usually quickly overpowered and eaten. As already mentioned, even within the first instar *Adalia*, the percentage efficiency in capturing first instar aphids can vary up to 40%, with the older first instar coccinellid larvae being most successful (Dixon, 1970). The efficiency of four species of coccinellids capturing a wheat aphid varied with the stage of development. First instars were always least efficient, but efficiency improved with each succeeding instar and adults were less efficient than the older larval instars; furthermore predator efficiency also depended on the species of aphid and its stage of development, with older aphids being more difficult to capture (Brown, 1972).

Hunger can influence the behavior of predators in many ways. Holling (1966) found in the mantids that varying the length of time food was deprived influenced the stalking and striking distance, the amount of prey eaten, and the time required to eat the prey. When prey population densities are low many predators turn to cannibalism which increases the chances of survival. Larvae and adults of coccinellids often eat eggs of their own species, which not only permits survival but also increases searching ability when prey abundance is low (Hodek, 1973).

Some predators feed on a variety of prey species, in proportion to their relative abundance in the environment. Other predators, however, may exhibit "switching" behavior, in which the most abundant prey species represents a greater proportion of the predators' diet than it does in the environment as a whole. Murdoch and Marks (1973) found that *Coccinella septempunctata*, offered two aphid species, attacked the species in direct proportion to their relative abundance, and training had no effect on the subsequent diets of the predators, indicating that the coccinellid exhibited no switching. *Notonecta glauca* L., on the other hand, indicated progressive development of switching during 10-day tests by attacking proportionally more of one prey than expected (Lawton *et al.*, 1974). These authors have formulated functional response models for nonswitching and switching predators. A food preference mathematical model dealing with predators associated with one or more prey species has also been formulated by Rapport and Turner (1970) and Manly (1973). The latter author considers the influence of predators on the evolution of prey, depending upon whether the prey species is rare or common.

d. Prey Suitability. The internal composition of prey either illicits continued feeding or rejection. However, as already mentioned, some coccinellids will eat prey which may be poisonous to them and also the prey eaten may influence differentially the reproduction and development of coccinellids (Hodek, 1973) and of *Chrysopa* (Canard, 1970). There are either special nutritional requirements or differing levels of nutrients present in various prey which leads to specificity of predators, because certain artificial diets presented to some predators in paraffin droplets (Hagen and Tassan, 1965) support complete development, while other predators feeding on the same diet do not develop or lay eggs (Hagen, unpublished data).

2. Ovipositional Sites and Predatory Stages

The ovipositional site of the predator in relation to the prey and the stages of the predator that are predaceous vary among predators. Sweetman (1958) outlined these relationships and they are presented here for the important predatory insects.

a. Eggs Deposited in the Immediate Vicinity of Prey. i. Immature stages only are predatory. A few Lepidoptera (Balduf, 1939), the egg predators found in the Hymenoptera (Clausen, 1940), many of the larger aculeate Hymenoptera, e.g., Sphecidae (Evans and Eberhard, 1970; Grandi, 1961; Iwata, 1964; Krombein, 1967) and some Diptera, e.g., Cecidomyiidae, most Syrphids, Chamaeyiidae, and a few Chloropidae, are predatory only in their larval stages. Oviposition usually takes place on or near their prey; thus the adult females themselves behaviorly search for a "prey" situation. The adults may feed upon pollen, nectar, or honeydew; some species do not feed at all as adults. Some of the Hymenoptera that are egg predators may feed on the contents of the eggs "stung" in egg masses, but it is their larvae that destroy the bulk of the eggs. The aphidophagous Syrphidae have received the most attention among the families of this habit group, but not all aphidophagous syrphids oviposit in the immediate vicinity of their prey (Bombosh and Volk, 1966; Chandler, 1968). The larvae of many Cecidomyiidae are predaceous on spider mites and Homoptera, with eighteen genera containing species which attack Coccoidea (Harris, 1967). Many other dipterous families have predaceous species which fall in this category.

ii. Immature and adult stages are predatory on similar types of prey. Many important insect predators in biological control have this type of prey relationship. They deposit their eggs in the immediate vicinity of their prey. In the Neuroptera (Balduf, 1939), the families Chrysopidae (Principi, 1956; Ridgway and Kinzer, 1974), Hemerobiidae (Neuenschwander, 1975), Sympherobiidae, and Coniopterygidae (Badgley *et al.*, 1955) have received the most attention in biological control. About one-half of the species of *Chrysopa* that have been studied are predaceous in the adult stage and oviposit where they find prey, but the other species that feed upon honeydew and pollen as adults oviposit only in the general area of prey (Hagen *et al.*, 1971). In the Thysanoptera, some scattered species in three families are important predators of thrips and spider mites and deposit their eggs where they are feeding upon their prey (Putnam, 1965; Lewis, 1973). In the Coleoptera, most species in the Coccinellidae are predaceous (Balduf, 1935; Hagen, 1962; Hodek, 1973; Sasaji, 1971). The Epilachninae are all phytophagous and the small species in Psylloborini feed on mildews. The Coccinellinae are mainly aphidophagous, but some species also feed upon immature stages of Coleoptera and Lepidoptera. The Coccidulinae feed on coccids; and Chilocorinae feed on both coccids and diaspine scales, while some species are aphidophagous; Sticholotinae eat diaspine scales and white flies; in the Scymninae, the Stethorini are predators of spider mites, while the Scymnini and Hyperaspini feed on mealybugs, coccids, aphids, and psyllids. The biologies of only a few Staphylinidae have been studied (Balduf, 1935), but the arboreal species that feed on spider mites are fairly well known (Badgley and Fleschner, 1956). Some Staphylinidae that are parasitic on soil-inhabiting Diptera have been studied in detail, but the habits of the vast number of staphylinids which live in soil as predators are unknown but undoubtedly are important natural enemies of insects in many soil communities. How close most staphylinids place their eggs to their prey is really unknown. The coleopterous families Cleridae, Histeridae, and Trogostidae (Balduf, 1935; Kline and Rudinsky, 1964) contain predatory species that attack wood-boring beetles, and some Histeridae also prey upon immature Diptera in dung. The species of Pyrophorinae of the Elateridae are mostly predaceous on wood-boring beetles and some genera in the Elaterinae feed upon insects in the soil, although many of those same species will also feed on plant material (Tostowaryk, 1972). How close elaterid eggs are placed near their prey is mostly unknown. Balduf (1935), Clausen (1940), and Sweetman (1958) reviewed the habits of entomophagous Coleoptera. The aculeate Hymenoptera, e.g., vespids that maxillate the prey and feed their larvae with the chewed prey, and the predaceous ants also belong in this category (Evans and Eberhard, 1970; Wilson, 1971).

iii. Immature and adult stages are predatory on different types of prey. Some Anthomyiidae larvae are predaceous on aquatic insects, locust eggs, and dung-inhabiting dipterous larvae, but the adults prey upon different insects (Clausen, 1940).

b. Eggs Deposited Only in the General Environment of the Prey. *i. Immature stages only are predatory.* The greatest number of groups displaying this type of prey relationship are aquatic. Included are the Plecoptera (Vaught and Stewart, 1974), Trichoptera, and the following families of Neuroptera whose larvae are aquatic: Corydalidae, Sialidae, and Sisyridae (Azam and Anderson, 1969; Stewart *et al.*, 1973; Usinger, 1963). The larvae of the neuropterous families Myiodactylidae, Myrmeleontidae, Nemopteridae, Nymphidae, and Osmylidae largely search or build pit traps on the soil surfaces while their adults do not seem to be predaceous (Balduf, 1939). The larvae of a few Lepidoptera (Balduf, 1939) and of *some species* of the dipterous families (Cole and Schlinger, 1969) Culicidae, Tipulidae, Chironomidae, Tabanidae, Rhagionidae, Drosophilidae, Lonchaeidae, and Muscidae are entirely entomophagous, attacking aquatic prey or soil-inhabiting insects. Many Bombyliidae and some Phoridae are predaceous on insect egg masses. The Pallopteridae larvae prey on bark beetle larvae, but the habits of the adult flies are unknown (Morge, 1967).

ii. Immature and adult stages are predatory on different types of prey. The Odonata, the neuropterous family Ascalaphidae, and the dipterous families Asilidae, Dolichopodidae, Empididae, Therevidae, and Ceratopogonidae, as well as a few families of Coleoptera, have predatory representatives that oviposit in the general environment of the prey of their larvae although the adults feed on different prey than their larvae.

c. Eggs Deposited Independent of Prey. *i. Immature stages only are predatory.* In the Coleoptera, the Meloidae exemplify this type, with the larvae of the Epicautini and Mylabrini feeding on eggs of grasshoppers while the adults are phytophagous and may be pests. The adults deposit their eggs in the soil in grasshopper breeding areas and the triungulin larvae find and consume grasshopper eggs (Balduf, 1935; Selander, 1960).

ii. Immature and adult stages are predatory on similar types of prey. Many predaceous Orthoptera, Thysanura, Psocoptera, some Thysanoptera, and a few Hemiptera and Coleoptera fall in this category (Sweetman, 1958).

iii. Immature and adult stages are predatory on different types of prey. The Mantispidae and some Raphidoidea of the Neuroptera belong here. The mantispids rely on their triungulin larvae to find spider egg masses. The Cantharidae of the Coleoptera might also belong here.

iv. Adult stage only is predatory. Species displaying this habit are some Mecoptera, Diptera, and Hymenoptera. Many mecopteran larvae are scavengers while the adults at times prey upon insects (Balduf, 1939). Bornemissza (1957) reports that adult *Harpobittacus* is probably preventing the establishment of the cinnibar moth in Australia to control tansy ragwort. Adult Scatophagidae of the Diptera are voracious predators. In the Hymenoptera many Tenthredinoidea adults are predatory while their larvae are phytophagous. Many species of Chalcidoidea and Ichneumonidae feed upon the hemolymph oozing from wounds made by the ovipositor or they may construct special tubes for this "host feeding" and kill the prey in this way; however, their larvae are generally parasitic on the same species (Clausen, 1940).

IV. PREDATORY ARANEAE AND ACARI

Among the great diversity of species in Acari there are many predators that prey on a great variety of organisms. The diversity of taxa in Acari and some of their biological affinities are treated by Krantz (1970). The review by Turnbull (1973) deals with the ecology of the true spiders.

The role spiders play in the natural biological control of agricultural pests has received limited investigation. Using constant 12-hour observations of bollworm larvae placed on untreated cotton plants in Arkansas, Whitcomb (1967) observed many different kinds of predators feeding on the larvae, but of this complex, the spiders destroyed the greatest number of bollworms. Over 150 species representing 19 families of true spiders have been collected in Arkansas cotton fields (Whitcomb and Bell, 1964). From studies in various agricultural crops, it has been found that a considerable diversity of species is usually present. In field crops, lycosids, erigonids, and tetragnathids dominate (see references in Yeargan and Cothran, 1974), and in orchards salticids, thomisids, and theridiids are most numerous (Dondale, 1958; Hukusima, 1970; and Putnam, 1967).

In agricultural crops, as well as in natural habitats, spiders are considered general predators. Riechert (1974) points out that spiders have not adapted to the fluctuations in numbers of specific pest species. Their success in maintaining themselves through periods of low prey densities, as well as their ability to take advantage of high densities of given acceptable prey species, make them well adapted to most habitats. Since spiders are generally present in rather constant numbers and may shift from prey to prey, they must be considered a stabilizing influence at the invertebrate community level. The stabilizing element of such polyphagous predators stems from their ability to restrain pest outbreaks during the interval between initial pest increase of a given species and a significant numerical response of more prey-specific or host-specific natural enemies (Riechert, 1974).

Kiritani *et al.* (1972) attempted to evaluate quantitatively the predation of spiders on leafhoppers in rice fields. A team of observers walked a line along rice hills and made sight counts of actual feeding by predators twice a week. The rice plots had similar populations of several dominant spider species but the densities of leafhoppers were varied in three plots. During 1969 and 1970, 4.4 to 63.3% nymphs and 8.2 to 28.4%

adults of the green rice leafhopper were preyed on by the complex of spiders present those years. *Lycosa* was mainly responsible during 1969 and *Octothorax* during 1970 for the spider predation observed. Sasaba and Kiritani (1975) formulated a simulation model of the green rice leafhopper—spider system and compared the calculated values with observed generation-to-generation change in egg density of leafhoppers with *Lycosa* abundance and temperature for four generations. The calculated values coincided with observed onces except in the fourth generation.

In biological control, thus far, the phytoseiid mites have received the most attention, most of this effort following the clear demonstration of a controlling effect of two species on the cyclamen mite on strawberries (e.g., see Huffaker and Kennett, 1956) and other spider mite pests (Huffaker *et al.*, 1970).

A. Phytoseiid Mites and Their Role as Predators of Tetranychid Mites

Predatory mites known to feed on the Tetranychidae include species in the families Bdellidae, Trombidiidae, Anystidae, Erythraeidae, Stigmaeidae, and Phytoseiidae (McMurtry *et al.*, 1970). No close association with tetranychid mites has been observed in species of the first four families; probably this kind of prey is utilized as only one kind of food in a general diet. In the Stigmaeidae, the genera *Zetzellia* (= *Mediolata*) and *Agistemus* contain species known to feed on and reproduce when confined to spider mites as prey, but there is little conclusive information relative to their importance as predators of tetranychids.

Phytoseiids have received considerable attention in the last 15 years as predators of tetranychids on agricultural crops. They have been shown to be capable of controlling these pests in such crops as apples, peaches, strawberries, grapes, and glasshouse-grown vegetables (see McMurtry *et al.*, 1970). Because of their relatively small size (scarcely larger than their spider mite prey), they can survive on low populations of the prey and thus have the potential for regulating spider mite populations at low densities. The number of described species exceeds 600 (Muma and Denmark, 1970), most of which were described in the last 15 years. Phytoseiids occur on a wide range of plants and are also common in litter or humus.

1. Life History, Seasonal Biology, and Response Capabilities

The life stages of the Phytoseiidae consist of the egg, the six-legged larva, protonymph, deutonmph, and adult. At warm temperatures (ca. 25 °C), the duration of the egg stage is about 2 days. The larval stage lasts only about 1 day and, in some species, does not feed. The two nymphal stages each last about 1.5 days. Mating occurs as soon as the adult female emerges, and a preoviposition period of about 2 days follows. A generation can thus be completed in about 8 days, somewhat faster than most tetranychid mites under comparable conditions. With one known exception (Kennett, 1958), mating is necessary for oviposition to occur. However, cytological studies of several species indicate that males are haploid, suggesting arrhenotokous reproduction.

Average fecundity is generally in the range of 40 to 60 eggs per female. The rate of oviposition at warm temperatures commonly averages 1–2 eggs per day, but 3–4 for some *Phytoseiulus* species. The maximum rate of oviposition is less than that for the tetranychid mites under comparable conditions. However, the shorter generation time of the phytoseiids is a compensating factor, and their intrinsic rate of increase may be greater (Laing, 1969).

Reproduction can occur the year around in warmer climates, but in temperate climates, studies indicate that only the fertilized adult female survives the winter. A reproductive diapause, induced mainly by photoperiod, has been demonstrated for several species (see McMurtry *et al.*, 1970; Hoy, 1972). The effectiveness of the numerical response of three photoseiid species is described by Laing and Osborn (1974).

2. Food Habits

A wide variety of food habits exists among species in the Phytoseiidae. Tetranychid mites are the preferred food of some, but probably not a majority of species. *Phytoseiulus* species are examples of specialized predators of tetranychids; no other foods have been found acceptable to them. All active stages of tetranychid mites can be preyed upon. The egg stage of many spider mite species is utilized by photoseiids, but that of some species, such as *Panonychus ulmi* (Koch), is not favorable, apparently because the hard chorion cannot be pierced.

The average number of prey consumed during full development is generally fewer than 20. The daily spider mite consumption of ovipositing female predators probably rarely exceeds 5 adult females or 20 larvae or protonymphs, but some oviposition will occur on considerably fewer prey.

Other families of mites known to be fed upon by phytoseiids include the Eriophyidae, Tydeidae, Tenuipalpidae, and Tarsonemidae. Eriophyids have been found to induce a high rate of reproduction in some species. Tydeids and tenuipalpids have not been extensively tested as food, but they have been found to induce reproduction in some phytoseiids.

Other kinds of prey that may induce some reproduction of phytoseiids include scale crawlers, whitefly nymphs, moth eggs, thrips, and nematodes. However, the importance of these foods in the field is unknown.

Considering nonanimal foods, pollens are utilized by a number of species, especially *Amblyseius*. All *Amblyseius* of the "*findlandicus group*" (Chant, 1959) that have been studied are pollen feeders and reproduction on pollen may be more rapid than on mite prey.

Many species of phytoseiids accept a wide range of foods. These general feeders probably maintain their own numbers at more stable population levels than do the specialized feeders on tetranychids.

The presence of alternate foods may be a critical factor in the ability of phytoseiids to control pest mites. Studies in Washington showed that the eriophyid *Aculus schlechtendali* (Nalepa) is beneficial as a food source for *Metaseiulus occidentalis* (Nesbitt) when *Tetranychus mcdanieli* McGregor is scarce. The alternate prey maintains the predator in sufficient numbers and in an adequate distributional pattern to prevent later outbreaks of *T. mcdanieli* (Hoyt, 1969). Similar situations exist on peach and grape in California, where another eriophyid (on peach) and tydeid mites (on grape) maintain *M. occidentalis* populations during periods when spider mites are scarce (Hoyt and Caltagirone, 1971; Flaherty and Hoy, 1971). Experiments on avocado in the greenhouse showed that *A. hibisci* built up rapidly and controlled *Oligonychus punicae* (Hirst) if pollen was added, whereas in the absence of pollen, the predator's numerical response to the increasing prey population was too slow to effect control (McMurtry and Scriven, 1966).

REFERENCES

Adashkevitch, B. P., and Popov, N. A. (1972). The predators of the leguminous aphis (*Acyrtosiphon pisum* Harr.) in Moldavia. *Defense Veg. Plants, Kishinev* **12**, 36–46.

Aldrich, J. M. (1922). A new genus of two-winged flies with mandible-like labella. *Proc. Entomol. Soc. Wash.* **24**, 145–150.

Anderson, J. F. (1974). Responses to starvation in the spiders *Lycosa lenta* Hentz and *Filistata hibernalis* (Hentz). *Ecology* **55**, 576–585.

Anderson, N. H. (1962a). Bionomics of six species of *Anthocoris* (Heteroptera: Anthocoridae) in England. *Trans. Entomol. Soc. London* **114**, 67–95.

Anderson, N. H. (1962b). Anthocoridae of the Pacific Northwest with notes on distributions, life histories and habits (Heteroptera). *Can. Entomol.* **94**, 1325–1334.

Arzet, R. H. (1973). Suchverhalten der larven von *Chrysopa carnea* Steph. (Neuroptera: Chrysopidae) *Z. Angew. Entomol.* **74**, 64–79.

Asquith, D., and Colburn, R. (1971). Integrated pest management in Pennsylvania apple orchards. *Bull. Entomol. Soc. Amer.* **17**, 89–91.

Azam, K. M., and Anderson, N. H. (1969). Life history and habits of *Sialis rotunda* and *S. californica* in western Oregon. *Ann. Entomol. Soc. Amer.* **62**, 549–558.

Badgley, M. E., and Fleschner, C. A. (1956). Biology of *Oligota oviformis* Casey (Coleoptera: Staphylinidae). *Ann. Entomol. Soc. Amer.* **49**, 501–502.

Badgley, M. E., Fleschner, C. A., and Hall, J. C. (1955). The biology of *Spiloconis picticornis* Banks (Neuroptera: Coniopterygidae) introduced into California from Hong Kong. *Psyche* **62**, 75–81.

Balduf, W. V. (1935). "The Bionomics of Entomophagous Coleoptera," 220 pp. John S. Swift Co., Inc., St. Louis, Missouri.

Balduf, W. V. (1939). "The Bionomics of Entomophagous Insects," Part II, 384 pp. John S. Swift, Inc., St. Louis, Missouri.

Banks, C. J. (1955). An ecological study of Coccinellidae associated with *Aphis fabae* Scop. on *Vicia faba*. *Bull. Entomol. Res.* **45**, 561–587.

Banks, C. J. (1957). The behaviour of individual coccinellid larvae on plants. *Brit. J. Anim. Behav.* **5**, 12–24.

Bänsch, R. (1964). Vergleichende Untersuchungen zur Biologie und zum Beutefangverhalten aphidivorer Coccinelliden, Chrysopiden and Syrphiden. *Zool. Jahrb. Syst.* **91**, 271–340.

Bänsch, R. (1966). On prey-seeking behaviour of aphidophagous insects. *In* "Ecology of Aphidophagous Insects" (I. Hodek, ed.), pp. 123–128. Academia, Prague.

Bartlett, B. R., and Lloyd, D. C. (1958). Mealybugs attacking citrus in California—a survey of their natural enemies and the release of new parasites and predators. *J. Econ. Entomol.* **51**, 90–93.

Bay, E. C. (1974). Predator-prey relationships among aquatic insects. *Annu. Rev. Entomol.* **19**, 441–453.

Bennett, F. D., and Hughes, I. W. (1959). Biological control of insect pests in Bermuda. *Bull. Entomol. Res.* **50**, 423–436.

Bequaert, J. (1922). The predaceous enemies of ants. *Bull. Amer. Mus. Natur. Hist.* **45**, 27–331.

Blackman, R. L. (1965). Studies on specificity in Coccinellidae. *Ann. Appl. Biol.* **56**, 336–338.

Blumberg, D., and Swirski, E. (1974). The development and reproduction of Cybocephalid beetles on various foods. *Entomophaga* **19**, 437–443.

Bombosch, S. (1962). Untersuchungen über die Auslösung der Eiablage bei *Syrphus corollae* Fabr. (Diptera: Syrphidae). *Z. Angew. Entomol.* **50**, 81–88.

Bombosch, S. (1963). Untersuchungen zur Vermehrung von *Aphis fabae* Scop. in Samenrübenbestanden unter besonderer Berücksichtigung der Schwebfliegen (Diptera: Syrphidae). *Z. Angew. Entomol.* **52**, 105–141.

Bombosch, S. (1965). Untersuchungen on *Aphis fabae* Scop. und ihren natürlichen Feinden als Grundlage fur ein integriertes Bekampfungsprogramm. *Mitt. Biol. Bundesanst. Land. Forstwirt. Berlin Dahlem* **115**, 13–21.

Bombosch, S., and Volk, S. (1966). Selection of oviposition site by *Syrphus corollae* F. *In* "Ecology of Aphidophagous Insects" (I. Hodek, ed.), pp. 117–119. Academia, Prague.

Borden, J. H. (1974). Aggregation pheromones in the Scolytidae. *In* "Pheromones" (M. Birch, ed.), pp. 133–160. North-Holland Co., Amsterdam; American Elsevier, New York.

Bornemissza, G. F. (1957). Observations on the hunting and mating behavior of two species of scorpion flies (Bittacidae: Mecoptera). *Aust. J. Zool.* **14**, 371–382.

Boving, A. (1917). A generic synopsis of Coccinellid larvae in the United States National Museum, with a description of the larva of *Hyperaspis binotata* Say. *Proc. U.S. Nat. Mus.* **51**, 621–650.

Bradley, G. A. (1973). Effect of *Formica obscuripes* (Hymenoptera: Formicidae) on the predator-prey relationship between *Hyperaspis congressis* (Coleoptera: Coccinellidae) and *Toumeyella numismaticum* (Homoptera: Coccidae). *Can. Entomol.* **105**, 1113–1118.

Brown, H. D. (1972). Predaceous behaviour of four species of Coccinellidae (Coleoptera) associated with the wheat aphid, *Schizaphis graminum* (Rondani), in South Africa. *Trans. Roy. Entomol. Soc. London* **124**, 21–36.

Brues, C. T. (1946). "Insect Dietary," 466 pp. Harvard Univ. Press, Cambridge, Massachusetts.

Camors, F. B., Jr., and Payne, T. L. (1973). Sequence of arrival of entomophagous insects to trees infested with the southern pine beetle. *Environ. Entomol.* **2**, 267–270.

Canard, M. (1970). Incidence de la valeur alimentaire de divers pucerons (Homoptera: Aphididae) sur le potential de multiplication de *Chrysopa perla* (L.) (Neuroptera: Chrysopidae). *Ann. Zool. Ecol. Anim.* **2**, 345–355.

Carayon, J. (1961). Quelques remarques sur les Hemipteres-Heteropteres: Leur importance comme insectes auxiliaires et les possibilites de leur utilisation dans la lutte Biologique. *Entomophaga* **6**, 133–141.

Ceianu, I., Mihalache, G. H., and Balinschi, I. (1965). "Combatera biologica a daunatorilor Forestieri," 225 pp. Editura Agro-Silvica, ed., Bucuresti.

Chandler, A. E. F. (1968). Some factors influencing the occurence and site of oviposition by aphidophagous Syrphydae (Diptera). *Ann. Appl. Biol.* **61**, 435–446.

Chant, D. A. (1959). Phytoseiid mites (Acarina: Phytoseiidae). Part I. Bionomics of seven species in southwestern England. Part II. A taxonomic review of the family Phytoseiidae, with descriptions of thirty-eight new species. *Can. Entomol.* **91**, (Suppl. 12), 166 pp.

Chiang, H. C. (1974). Predation in retrospect and in prospect. *Entomophaga* **7**, 83–88.

Chu, Y-i. (1969). On the bionomics of *Lyctocoris beneficus* (Hiura) and *Xylocoris galactinus* (Fieber) (Anthocoridae, Heteroptera). *J. Fac. Agr. Kyushu Univ.* **15**, 1–136.

Clausen, C. P. (1940). "Entomophagous Insects," 638 pp. McGraw-Hill, New York.

Cochereau, P. (1969). Contròle bilogique d'*Aspidiotus destructor* Signoret (Homoptera: Diaspinae) dans l'ite Vaté (Nouvelles Hébrides) au moyen de *Rhizobus pulchellus* Montrouzier (Coleoptera: Coccinellidae). *Biol. Cah. ORSTOM* **8**, 57–100.

Cole, F. R., and Schlinger, E. I. (1969). "The Flies of Western North America," 693 pp. Univ. Calif. Press, Berkeley, California.

Collyer, E. (1953). Biology of some predatory insects and mites associated with the fruit tree red spider mite [*Metatetronychus ulmi* (Koch)] in south-eastern England. III. Further predators of the mite. *J. Hort. Sci.* **35**, 98–113.

Collyer, E., and Massee, A. M. (1958). Some predators of phytophagous mites, and their occurrence, in southeastern England. *Proc. 10th Int. Congr. Entomol.* **4**, 623–631.

Coppel, H. C., and Jones, P. A. (1962). Bionomics of *Podisus* spp. associated with the introduced pine sawfly, *Diprion similis* (Htg.), in Wisconsin. *Wis. Acad. Sci. Arts Lett.* **51**, 31–56.

Corbet, P. S. (1962). "A Biology of Dragonflies," 247 pp. Quadrangle Books, Chicago, Illinois.

Cullen, M. J. (1969). The biology of giant water bugs (Hemiptera: Belostomatidae) in Trinidad. *Proc. Roy. Entomol. Soc. London* **44**, 123–136.

Dahlsten, D. L., Garcia, R., Prine, J. E., and Hunt, R. (1969). Insect problems in forest recreation areas. *Calif. Agr.* **23**, 4–6.

Davidson, W. M. (1924). Observations and experiments on the dispersion of the convergent lady beetle [*Hippodamia convergens* (Guerin)] in California. *Trans. Amer. Entomol. Soc.* **50**, 163–175.

Davis, C. J., and Chong, M. (1968). Recent introductions for biological control in Hawaii—XIII. *Proc. Hawaii. Entomol. Soc.* **20**, 25–34.

Davis, J. J. (1919). Contributions to a knowledge of the natural enemies of *Phyllophaga. Ill. Natur. Hist. Surv. Bull.* **13**, 53–138.

DeBach, P. (1951). The necessity for an ecological approach to pest control on citrus in California. *J. Econ. Entomol.* **44**, 443–447.

DeBach, P. (ed.). (1964). "Biological Control of Insect Pests and Weeds." 844 pp. Chapman and Hall, London.

DeBach, P. (1974). "Biological Control by Natural Enemies," 323 pp. Cambridge Univ. Press, London.

DeLoach, C. J. (1971). The effect of habitat diversity on predation. *Proc. Tall Timbers Conf. Ecol. Anim. Control Habitat Manage.* **2**, 223–241.

Dempster, J. P. (1968). Intra-specific competition and dispersal: as exemplified by a psyllid and its anthocorid predator. *In* "Insect Abundance" (T.R.E. Southwood, ed.), pp. 8–17. Symp. Roy. Entomol. Soc., London.

Dickson, R. C., and Laird, E. F. Jr. (1962). Green peach aphid populations in desert sugar beets. *J. Econ. Entomol.* **55**, 501–504.

Dixon, A. F. G. (1958). The escape responses shown by certain aphids to the presence of the coccinellid *Adalia decempunctata* (L.). *Trans. Roy. Entomol. Soc. London* **110**, 319–334.

Dixon, A. F. G. (1959). An experimental study of searching behaviour of the predatory coccinellid beetle *Adalia decempunctata* (L.). *J. Anim. Ecol.* **28**, 259–281.

Dixon, A. F. G. (1970). Factors limiting the effectiveness of the coccinellid beetle, *Adalia bipunctata* (L.) as a predator of the sycamore aphid, *Drepanosiphum platanoides* (Schr.). *J. Anim. Ecol.* **39**, 739–751.

Dixon, A. F. G., and Russel, R. J. (1972). The effectiveness of *Anthocoris nemorum* and *A. confusus* (Hemiptera: Anthocoridae) as predators of sycamore aphid, *Drepanosiphum platanoides*. II. Searching behaviour and the incidence of predation in the field. *Entomol. Exp. Appl.* **15**, 35–50.

Dondale, C. D. (1958). Note on population densities of spiders (Aranae) in Nova Scotia apple orchards. *Can. Entomol.* **90**, 111–113.

Doutt, R. L. (1951). Biological control of mealybugs infesting commercial greenhouse gardenias. *J. Econ. Entomol.* **44**, 37–40.

Doutt, R. L. (1959). The biology of parasitic Hymenoptera. *Annu. Rev. Entomol.* **4**, 161–182.

Doutt, R. L., and DeBach, P. (1964). Some biological control concepts and questions. *In* "Biological Control of Insect Pests and Weeds" (P. DeBach, ed.), Chapt. 3, pp. 118–142. Chapman and Hall, London.

Doutt, R. L., and Hagen, K. S. (1950). Biological control measures applied against *Pseudococcus maritimus* on pears. *J. Econ. Entomol.* **43**, 94–96.

Drea, J. J., Jr. (1956). A biological analysis of the California Chilocorini (Coleoptera: Coccinellidae). Ph.D. thesis, Univ. Calif., Berkeley, California.

East, R. (1974). Predation on the soil-dwelling stages of the winter moth at Wytham Woods, Berkshire. J. Anim. Ecol. **43**, 611–626.

Edwards, J. S. (1966). Observations on the life history and predatory behavior of *Zelus exsanguis* (Stal) (Heteroptera: Reuviidae). *Proc. Roy. Entomol. Soc. London* **41**, 21–24.

Ehler, L. E., and van den Bosch, R. (1974). An analysis of the natural biological control of *Trichoplusia ni* (Lepidoptera: Noctuidae) on cotton in California. *Can. Entomol.* **106**, 1067–1073.

Ellis, R. A., and Borden, J. H. (1970). Predation by *Notonecta undulata* (Heteroptera: Notonectidae) on larvae of the yellow-fever mosquito. *Ann. Entomol. Soc. Amer.* **63**, 963–972.

Elson, J. A. (1937). A comparative study of Hemiptera. *Ann. Entomol. Soc. Amer.* **30**, 579–597.

Escherich, K. (1942). Diptera. In "Die Forstinsekten Mitteleuropas" Vol. 5, pp. 505–705. P. Parey, Berlin.

Essig, E. O. (1930). "A History of Entomology," 1029 pp. Macmillan, New York.

Evans, H. E., and Eberhard, M. J. W. (1970). "The Wasps," 265 pp. Univ. Michigan Press, Ann Arbor, Michigan.

Eveleens, K. G., van den Bosch, R., and Ehler, L. E. (1973). Secondary outbreak induction of beet armyworm by experimental insecticide applications in cotton in California. *Environ. Entomol.* **2**, 497–503.

Ewert, M. A., and Chiang, H. C. (1966). Dispersal of three species of coccinellids in corn fields. *Can. Entomol.* **98**, 999–1003.

Ferenz, H. J., and Holters, W. (1975). Corpus allatum hormone induced maturation in males of three carabid species of the genus *Pterostichus* (Coleoptera: Carabidae). *Entomol. Exp. Appl.* **18**, 238–243.

Fields, G. J., and Beirne, B. P. (1973). Ecology of anthocorid (Hemiptera: Anthocoridae) predators of the pear psylla (Homoptera: Psyllidae) in the Okanagan Valley, British Columbia. *J. Entomol. Soc. Brit. Columbia* **70**, 18–19.

Finnegan, R. J. (1974). Ants as predators of forest pests. *Entomophaga Mem. H. S.* **7**, 53–59.

Fisher, T. W. (1963). Mass culture of *Cryptolaemus* and *Leptomastix* natural enemies of citrus mealybug. *Univ. Calif. Agr. Exp. Sta. Bull.* **797**, 38 pp.

Flaherty, D. L., and Hoy, M. A. (1971). Biological control of pacific mites in San Joaquin Valley vineyards. III. Role of tydeid mites. *Res. Population Ecol.* **13**, 80–96.

Fleschner, C. A. (1950). Studies on searching capacity of the larvae of three predators of the citrus red mite. *Hilgardia* **20**, 233–265.

Fleschner, C. A., Hall, J. C., and Ricker, D. W. (1955). Natural balance of mite pests in an avocado grove. *Calif. Avocado Soc. Yearb.* **39**, 155–162.

Franz, J. (1958). Studies on *Larricobius erichsonii* Rosench (Coleoptera: Derodontidae) a predator of chermesids. *Entomophaga* **3**, 109–196.

Franz, J. M. (1961). Biologische Schadlingsbekampfung. *In* "Handbuch der Pflanzenkrankheiten," (P. Sorauer, ed.) Vol. 6. 1–302 pp. Parey, Berlin.

Gagne, W. C., and Martin, J. L. (1968). The insect ecology of red pine plantations in central Ontario. *Can. Entomol.* **100**, 835–846.

Gilbert, N., and Hughes, R. D. (1971). A model of an aphid population—three adventures. *J. Anim. Ecol.* **40**, 525–534.

Gould, H. J. (1970). Preliminary studies of an integrated control programme for cucumber pests and an evaluation of methods of introducing *Phytoseilus persimilis* Athias-Henriot for the control of *Tetranychus urticae* Koch. *Ann. Appl. Biol.* **66**, 503–513.

Grandi, G. (1961). "Studi Di un Entomologo Sugli Imenotteri Superiori," 659 pp. Edizioni Calderini Bologna.

Grasse, P. P. (ed.). (1951). "Traite de Zoologie," Vol. 10, pp. 1–975. Libr. Acad. Med., Paris.

Greathead, D. J. (1963). A review of insect enemies of Acridoidea. *Trans. Roy. Entomol. Soc. London* **114**, 437–523.

Griffiths, K. J., and Holling, C. S. (1969). A competition submodel for parasites and predators. *Can. Entomol.* **101**, 685–796.

Guagliumi, P. (1966). Insetti e Aracnidiedelle plante comuni del Venezuela segnalati nel periodo 1938–1963. *Relaz. Monogr. Agrar. Subtrop. Trop. No.* **86**, 391 pp.

Gurney, B., and Hussey, N. W. (1970). Evaluation of some coccinellid species for the biological control of aphids in protected cropping. *Ann. Appl. Biol.* **65**, 451–458.

Hagen, K. S. (1962). Biology and ecology of predaceous Coccinellidae. *Annu. Rev. Entomol.* **7**, 289–326.

Hagen, K. S. (1974). The significance of predaceous Coccinellidae in biological and integrated control of insects. *Entomophaga Mem. H.S.* **7**, 25–44.

Hagen, K. S. (1976). Role of nutrition in insect management. *Proc. Tall Timbers Conf. Ecol. Anim. Control Habitat Manage.* **6**, 221–261.

Hagen, K. S., and Hale, R. (1974). Increasing natural enemies through use of supplementary feeding and non-target prey. *In* "Proceedings of the Summer Institute of Biological Control of Plant Insects and Diseases" (F. G. Maxwell and R. A. Harris, eds.), pp. 170–181. Univ. Press, Jackson, Mississippi.

Hagen, K. S., and Tassan, R. L. (1965). A method of providing artificial diets to *Chrysopa* larvae. *J. Econ. Entomol.* **58**, 999–1000.

Hagen, K. S., and Tassan, R. L. (1972). Exploring nutritional roles of extracellular symbiotes on the reproduction of honeydew feeding adult chrysopids and tephritids. *In* "Insect and Mite Nutrition" (J. G. Rodriguez, ed.), pp. 323–351. North Holland, Amsterdam.

Hagen, K. S., and van den Bosch, R. (1968). Impact of pathogens, parasites, and predators on aphids. *Annu. Rev. Entomol.* **13**, 329–384.

Hagen, K. S., Tassan, R. L., and Sawall, E. F., Jr. (1970). Some ecophysiological relationships between certain *Chrysopa*, honeydews and yeasts. *Boll. Lab. Entomol. Agr. Portici* **28**, 113–134.

Hagen, K. S. Sawall, E. F., Jr., and Tassan, R. L. (1971). The use of food sprays to increase effectiveness of entomophagous insects. *Proc. Tall Timbers Conf. Ecol. Anim. Control Habitat Manage.* **2**, 59–81.

Hagen, K. S., Greany, P., Sawall, E. F., Jr., and Tassan, R. L. (1976). Tryptophan in artificial honeydews as a source of an attractant for adult *Chrysopa carnea*. *Environ. Entomol.*, in press.

Hämäläinen, M., and Markula, M. (1972). Effect of type of food on fecundity in *Coccinella septempunctata* L. (Coleoptera: Coccinellidae). *Ann. Entomol. Fenn.* **38**, 195–199.

Harris, K. M. (1967). A systematic revision and biological review of the cecidomyiid predators

(Diptera: Cecidomyiidae) on world Coccoidea (Hemiptera-Homoptera). *Trans. Roy. Entomol. Soc. London* **119**, 401–494.

Hassell, M. P. (1966). Evaluation of parasite or predator responses. *J. Anim. Ecol.* **35**, 65–75.

Hassell, M. P., and May, R. M. (1974). Aggregation of predators and insect parasites and its effect on stability. *J. Anim. Ecol.* **43**, 567–587.

Haynes, D. L., Sisojevic, P. (1966). Predatory behavior of *Philodromus rufus* Walckenner (Araneae: Thomisidae). *Can. Entomol.* **98**, 113–133.

Helgesen, R. G., and Tauber, M. J. (1974). Pirimicarb, an aphiade nontoxic to three entomophagous arthropods. *Environ. Entomol.* **3**, 99–101.

Hespenheide, H. A. (1973). Ecological inferences from morphological data. *Annu. Rev. Ecol. Syst.* **4**, 213–229.

Hinton, H. E. (1945). "A Monograph of The Beetles Associated with Stored Products," 443 pp. British Museum, London.

Hodek, I. (ed.). (1966). "Ecology of Aphidophagous Insects," 360 pp. Academia, Prague.

Hodek, I. (1967). Bionomics and ecology of predaceous Coccinellidae. *Annu. Rev. Entomol.* **12**, 79–104.

Hodek, I. (1970). Coccinellids and the modern pest management. *Bioscience* **20**, 543–552.

Hodek, I. (1973). "Biology of Coccinellidae," 260 pp. Dr. W. Junk N. V. Publi, The Hague.

Holling, C. S. (1961). Principles of insect predation. *Annu. Rev. Entomol.* **6**, 163–182.

Holling, C. . (1965). The functional response of predators to prey density and its role in mimicry and population regulation. *Mem. Entomol. Soc. Can.* **45**, 60 pp.

Holling, C. S. (1966). The functional response of invertebrate predators. *Mem. Entomol. Soc. Can.* **48**, 86 pp.

Holling, C. S. (1968). The tactics of a predator. *In* "Insect Abundance" (T. R. E. Southwood, ed.), pp. 47–58. Roy. Entomol. Soc. London.

Horner, N. V., and Starks, K. J. (1972). Bionomics of the jumping spider *Metaphidippus galathea*. *Ann. Entomol. Soc. Amer.* **65**, 602–607.

Hoy, M. (1972). Diapause induction and duration in vineyard-collected *Metaseiulus occidentalis*. *Environ. Entomol.* **4**, 262–264.

Hoyt, S. C. (1969). Population studies of five mite species on apple in Washington. *Proc. 2nd Intern. Congr. Acarol. Sutton-Bonnington, England*, 1967, pp. 117–133. Adad. Kiado, Budapest.

Hoyt, S. C., and Caltagirone, L. E. (1971). The developing programs of integrated control of pests of apples in Washington and peaches in California. *In* "Biological Control" (C. B. Huffaker, ed.), pp. 395–421. Plenum, New York.

Huffaker, C. B. (1958). Experimental studies on predation: dispersion factors and predator-prey oscillations. *Hilgardia* **27**, 343–383.

Huffaker, C. B. (1970). The phenomenon of predation and its roles in nature. *Proc. Advan. Study Inst. Dyn. No. Popul. Oosterbeek*, pp. 327–343.

Huffaker, C. B. (ed.). (1971). "Biological Control," 511 pp. Plenum, New York.

Huffaker, C. B., and Kennett, C. E. (1956). Experimental studies on predation: predation and cyclamen-mite populations on strawberries in California. *Hilgardia* **26**, 191–222.

Huffaker, C. B., van de Vrie, M., and McMurtry, J. A. (1970). Ecology of tetranychid mites and their natural enemies: A review. II. Tetranychid populations and their possible control by predators: an evaluation. *Hilgardia* **40**, 391–458.

Hughes, R. D. (1972). Population dynamics. *In* "Aphid Technology" (H. F. van Emden, ed.), pp. 275–293. Academic Press, New York.

Hughes, R. D., and Gilbert, N. (1968). A model of an aphid population—a general statement. *J. Anim. Ecol.* **37**, 553–563.

Hukusima, S. (1970). Responses of populations of phytophagous and predaceous arthropods on apple trees to some pesticides. *Res. Bull. Fac. Agr. Gifu Univ.* **29**, 33–52.

Hussey, N. W., and Bravenboer, L. (1971). Control of pests in glasshouse culture by the introduction of natural enemies. *In* "Biological Control" (C. B. Huffaker, ed.), pp. 195–216. Plenum, New York.

Hussey, N. W., Read, W. H., and Hesling, J. J. (1969). "The Pests of Protected Cultivation," 404 pp. Edward Arnold,

Ickert, G. (1968). Beitrage zur Biologie einheimischer Chrysopiden. *Entomol. Abh.* **36**, 123–192.

Iperti, G. (1966). Comportement naturel des coccinelles aphidiphages du Sud-Est de la France. Leur type de spécificité leur action predatrice sur *Aphis fabae* L. *Entomophaga* **11**, 203–210.

Iperti, G., Laudeho, Y., Braun, J., and Choppin de Janvry, E. (1970). Les entomophages de *Parlatoria blanchardi* Targ. dans les palmerqies de L'adrar Mauritanien. *Ann. Zool. Ecol. Anim.* **2**, 617–638.

Isely, F. B. (1944). Correlation between mandibular morphology and food specificity in grasshoppers. *Ann. Entomol. Soc. Amer.* **37**, 47–67.

Ito, Y., Miyashita, K., and Sekiguchi, K. (1962). Studies on the predators of the rice crop insect pests using the insecticidal check method. *Jap. J. Ecol.* **12**, 1–11.

Iwata, K. (1964). Bionomics of non-social wasps in Thailand. *Nature and Life S.E. Asia. Fauna and Flora Res. Soc., Kyoto Jap.* **3**, 323–383.

Janzen, D. H. (1966). Coevolution of mutualism between ants and acacias in Central America. *Evolution* **20**, 249–275.

Jenkins, D. W. (1964). Pathogens, parasites and predators of medically important arthropods. Annotated list and bibliography. *Bull. WHO Suppl.* **30**, 150 pp.

Johnson, N. E., and Cameron, S. (1969). Phytophagous ground beetles. *Ann. Entomol. Soc. Amer.* **62**, 909–914.

Kaestner, A. (translated by H. and L. Levie). (1968). "Order Araneae, Spiders in Invertebrate Zoology," Vol. 2, pp. 131–203. Interscience, New York.

Kennett, C. E. (1958). Some predaceous mites of the subfamilies Phytoseiinae and Aceosejinae (Acarina: Phytoseiidae, Aceosejidae) from central California with descriptions of new species. *Ann. Entomol. Soc. Amer.* **51**, 471–479.

Kennett, C. E., and Flaherty, D. L. (1974). "Pest Management of Citrus Red Mite on Citrus in the Southern San Joaquin Valley," 27 pp. San Joaquin Valley Agr. Res. Ext. Center, Parlier, California.

Kesten, U. (1969). Zur Morphologie und Biologie von *Anatis ocellata* (L.) (Coleoptera: Coccinellidae). *Z. Angew. Entomol.* **63**, 412–445.

Kiritani, K. and Dempster, J. P. (1973). Quantitative evaluation of natural enemy effectiveness. *J. Appl. Ecol.* **10**, 323–330.

Kiritani, K., Kawahara, S., Sasaba, T., and Nakasugi, F. (1971). An attempt of rice pest control by integration of pesticides and natural enemies. *Gensei* **22**, 19–23.

Kiritani, K., Kawahara, S., Sasaba, T., and Nakasuji, F. (1972). Quantitative evaluation of predation by species on the green rice leafhopper, *Nephotettix cincticeps* Uhler, by a sight-count method. *Res. Population Ecol.* **13**, 187–200.

Kline, L. N., and Rudinsky, J. A. (1964). Predators and parasites of the Douglas-fir beetle: description and identification of the immature stages. *Agr. Exp. Sta. Oregon St. Univ. Corvallis Tech. Bull.* **79**, 52 pp.

Krantz, G. W. (1970). "A Manual of Acarology," 335 pp. Oregon State Univ. Book Stores, Corvallis.

Krombein, K. V. (1967). "Trap-nesting Wasps and Bees: Life Histories, Nests, and Associates," 570 pp. Smithson Press, Washington, D.C.

Kulman, H. M. (1974). Comparative ecology of North American Carabidae with special reference to biological control. *Entomophaga Mem. H. S.* **7**, 61–70.

Laing, J. E. (1969). Life history and life table of *Metaseiulus occidentalis*. *Ann. Entomol. Soc. Amer.* **62**, 978–982.

Laing, J. E., and Osborn, J. A. L. (1974). The effect of prey density on the functional and numerical responses of three species of predatory mites. *Entomophaga* **19**, 267–277.

Lavigne, R. J., and Holland, F. R. (1969). Comparative behavior of eleven species of Wyoming robber flies. *Univ. Wyo. Agr. Exp. Sta. Sci. Monogr. No.* **18**, 61 pp.

Lawton, J. H., Beddington, J. R., and Bonser, R. (1974). Switching in invertebrate predators. *In* "Ecological Stability" (M. B. Usher and M. H. Williamson, eds.), pp. 141–158. Chapman and Hall, London; Wiley, New York.

Ler, P. A. (1964). On the food and importance of robberflies (Asilidae). *Tr. Nauch. Issled. Zasc. Rast.* (*Kazalch Mm. Agr.*), *SSR* **8**, 213–244.

Lewis, T. (1973). "Thrips: Their Biology, Ecology and Economic Importance," 349 pp. Academic Press, New York.

Liotta, G., and Mineo, G. (1963). Prove di lotto biologica artificiale contro lo *Pseudococcus citri* R. *Boll. Ist. Entomol. Agrar. Osservat. Fitopat Palermo* **5**, 3–16.

Lindern, E. (1949). "Di Fliegen der Palaearktischen Region," 422 pp. Stuttgart.

Lord, F. T. (1971). Laboratory tests to compare the predatory value of six mirid species in each stage of development against the winter eggs of the European red mite, *Panonychus ulmi* (Acari.: Tetranychidae). *Can. Entomol.* **103**, 1663.

Luck, R. F., and Dahlsten, D. L. (1975). Natural decline of a pine needle scale [*Chionaspis pinifoliae* (Fitch)], outbreak at South Lake Tahoe, California following cessation of adult mosquito control with malathion. *Ecology*, **56**, 893–904.

MacLeod, E. G. (1967). Experimental induction and elimination of adult diapause and autumnal coloration in *Chrysopa carnea* (Neuroptera). *Insect Physiol.* **13**, 1343–1349.

McMullen, R. D. (1967). The effects of photoperiod, temperature and food supply on rate of development and diapause in *Coccinella novemnotata*. *Can. Entomol.* **99**, 578–586.

McMurtry, J. A., and Scriven, G. T. (1966). The influence of pollen and prey density on the number of prey consumed by *Amblyseius hibisci* (Acarina: Phytoseiidae). *Ann. Entomol. Soc. Amer.* **58**, 106–149.

McMurtry, J. A., Johnson, H. G., and Scriven, G. T. (1969). Experiments to determine effects of mass release of *Stethorus picipes* on the level of infestation of the avocado brown mite. *J. Econ. Entomol.* **62**, 1216–1221.

McMurtry, J. A., Huffaker, C. B., and van de Vrie, M. (1970). Ecology of tetranychid enemies: their biological characters and the impact of spray practices. *Hilgardia* **40**, 331–390.

Manly, B. F. J. (1973). A linear model of frequency-dependent selection by predators. *Rev. Population Ecol.* **14**, 137–150.

Martel, P., and Sharma, M. L. (1970). Quelques notes sur *Microweisea marginata* (LeConte). (Coleoptera: Coccinellidae), predateur de la cochenille du pin, *Phenacaspis pinifoliae* (Fitch). *Ann. Soc. Entomol. Quebec* **49**, 62–65.

Mathys, G., and Baggiolini, M. (1965). Praktishe Anwendung der integrieten Shadlingbekampfung in Obstanlagen der Westschweiz. *Mitt. Biol. Bundesanst. Land. Forstwirt. Berlin, Dahlem* **115**, 21–30.

Matsumoto, B. M., and Nishida, T. (1966). Predator-prey investigations on the Taro leafhopper and its egg predator. *Hawaii Agr. Exp. Sta. Tech. Bull. No.* **64**, 32 pp.

Miller, N. C. E. (1971). "The Biology of the Heteroptera," 206 pp. E. W. Classy, England.

Minoranskii, V. A. (1967). Über die faktoren, die Massenvermehrung der Rubenblattlaus (*Aphis fabae* Scop.) im Suden der europaischen UdSSR verhindern. *Arch. Pflanzenschutz* **3**, 101–114.

Mitchell, B. (1963). Ecology of two carabid beetles, *Bembidion lampros* (Herbit) and *Trechus quadristnatus* (Schrank). I. Life cycles and feeding behaviour. *J. Anim. Ecol.* **32**, 289–299.

Mittlestaedt, H. (1957). Prey capture in mantids. *Rec. Advan. Invertebr. Physiol. Symp.*, pp. 51–71. Univ. Oregon.

Morge, G. (1967). Eine Beobachtung zur Grundfrage der Abhangigkeit von Wirkungsgrad und Wert naturlicher Feinde gegenuber Schadlingen. *Beitr. Entomol.* **17**, 225–233.

Morris, R. F. (1963). The effect of predator age and prey defense on the functional response of *Podisus maculiventris* Say to the density of *Hyphantria cunea* Drury. *Can. Entomol.* **95**, 1009–2000.

Mukerji, M. K., and LeRoux, E. J. (1969). A study on energetics of *Podisus maculiventris* (Hemiptera: Pentatomidae). *Can. Entomol.* **101**, 449–460.

Muma, M. H. (1955). Factors contributing to the natural control of citrus insects and mites in Florida. *J. Econ. Entomol.* **48**, 432–438.

Muma, M. H., and Denmark, H. A. (1970). Phytoseiidae of Florida. *Fla. Dept. Agr. Consumer Serv., Gainesville.* 1950 pp.

Murdoch, W. W., and Marks, J. R. (1973). Predation by coccinellid beetles: experiments on switching. *Ecology* **54**, 160–167.

Neuenschwander, P. (1975). Influence of temperature and humidity on the immature stages of *Hemerobius pacificus. Environ. Entomol.* **4**, 216–220.

Neuenschwander, P. Hagen, K. S., and Smith, R. F. (1975). Predation on aphids in California's alfalfa fields. *Hilgardia* **43**, 53–78.

Newsom, L. D. (1974). Predator insecticide relationships. *Entomophaga Mem. H. S.* **7**, 13–23.

Niemczyk, E., Olszak, R., Miszczak, M., and Bakowski, G. (1974). "The Effectiveness of Some Predaceous Insects in the Control of Phytophagus Insects in the Control of Phytophagous Mites and Aphids on Apple Trees." 84 pp. Res. Inst. Pomology, Skierniewice, Poland.

Nuorteva, M. (1971). Die Borkenkafer (Coleoptera: Scolytidae) und deren Insektenfeinde im Kirchspiel Kuusamo, Nordfinnland. *Ann. Entomol. Fenn.* **37**, 65–72.

Oatman, E. R. (1970). Studies on integrating *Phytoseiulus persimilis* releases and natural predation for control of the two-spotted spider mite on rhubarb in southern California. *J. Econ. Entomol.* **63**, 1177–1180.

Oatman, E. R., McMurtry, J. A., Shorey, H. H., and Voth, V. (1967). Studies on integrating *Phytoseiulus persimilis* releases, chemical applications, cultural manipulations and natural predation for control of the two-spotted spider mite on strawberry in southern California. *J. Econ. Entomol.* **57**, 1344–1351.

Oatman, E. R., McMurtry, J. A., and Voth, V. (1968). Suppression of the two-spotted spider mite on strawberry with mass releases of *Phytoseiulus persimilis. J. Econ. Entomol.* **61**, 1517–1521.

Okamoto, H. (1966). Three problems of aphidophagous coccinellids. *In* "Ecology of Aphidophagous Insects" (I. Hodek, ed.), pp. 45–46. Academia, Prague.

Oldroyd, H. (1964). "The Natural History of Flies," 320 pp. W. W. Norton, New York.

Perkins, P. V., and Watson, T. F. (1972). Biology of *Nabis alternatus* (Hemiptera: Nabidae). *Ann. Entomol. Soc. Amer.* **65**, 54–57.

Penney, M. M. (1969). Diapause and reproduction in *Nebria brevicollis* (F.) (Coleoptera: Carabidae). *J. Anim. Ecol.* **38**, 219–233.

Pickett, A. D., and Patterson, N. A. (1953). The influence of spray programs on the fauna of apple orchards in Nova Scotia. IV. A review. *Can. Entomol.* **84**, 472–478.

Pollard, E. (1969). The effect of removal of arthropod predators on an infestation of *Brevicoryne brassicae* (Hemiptera: Aphididae) on brussels sprouts. *Entomol. Exp. Appl.* **12**, 118–124.

Principi, M. M. (1956). Contributi allo studio dei Neurotteri Italiani. *Boll. Inst. Entomol. Univ. Bologna* **21**, 319–410.

Putnam, W. L. (1965). The predaceous thrips, *Haplothrips faurei* Hood (Thysanoptera: Phloeothripidae), in Ontario peach orchards. *Can. Entomol.* **97**, 1208–1221.

Putnam, W. L. (1967). Prevalence of spiders and their importance as predators in Ontario peach orchards. *Can. Entomol.* **99**, 160–170.

Quezada, J. R., and DeBach, P. (1973). Bioecological and population studies of the cottony-cushion scale, *Icerya purchasi* Mask., and its natural enemies. *Rodolia cardinalis* Mul. and *Cryptochaetum iceryae* Will., in southern California. *Hilgardia* **41**, 631–688.

Rabb, R. L., Todd, F. A., and Ellis, H. C. (1974). Pest management: an interdisciplinary approach to crop protection. Tobacco Pest Management. *AAAS Symp. Feb., 1974, San Francisco, Calif.*

Rapport, D. J., and Turner, J. E. (1970). Determination of predator food preferences. *J. Theoret. Biol.* **26**, 365–372.

Readio, P. A. (1927). Studies on the biology of Reduviidae of America north of Mexico. *Kans. Univ. Sci. Bull.* **17**, 5–291.

Remaudière, G., and Leclant, F. (1971). Le complexe des ennemis naturels des aphides du pècher dans la Moyenne Vallée du Rhone. *Entomophaga* **16**, 255–267.

Remaudière, G., Iperti, G., Leclant, F., Lyon, J. P., and Michel, M. F. (1973). Biologie et écologie des aphids et de leurs ennemis naturels. Application a la lutte intégréé en vergers. *Entomophaga, Mem. H. S.* **6**, 35 pp.

Ridgway, R. L., and Kinzer, R. E. (1974). Chrysopids as predators of crop pests. *Entomophaga, Mem. H.S.* **7**, 45–51.

Riechert, S. E. (1974). Thoughts on the ecological significance of spiders. *Bioscience* **24**, 352–356.

Rivard, I. (1964). Observations on the breeding periods of some ground beetles (Coleoptera: Carabidae) in eastern Ontario. *Can. J. Zool.* **42**, 1081–1084.

Roeder, K. D. (1963). "Nerve Cells and Insect Behavior," 188 pp. Harvard Univ. Press, Cambridge, Massachusetts.

Rogers, C. E., Jackson, H. B., and Eikenbary, R. D. (1972). Responses of an imported coccinellid, *Propylea 14-punctata*, to aphids associated with small grains in Oklahoma. *Environ. Entomol.* **1**, 198–202.

Sasaba, T., and Kiritani, K. (1975). A systems model and computer simulation of the green rice leafhopper populations in control programmes. *Res. Population Ecol.* **16**, 231–244.

Sasaji, H. (1971). "Fauna Japonica, Coccinellidae," 340 pp. Academic Press of Japan, Keigaku Publ. Co., Tokyo.

Savoiskaya, G. I. (1960). Morphology and taxonomy of coccinellid larvae from southeast Kazakhstan. *Entomol. Rev. (USSR)* **39**, 80–88.

Schneider, F. (1969). Bionomics and physiology of aphidophagous Syrphidae. *Ann. Rev. Entomol.* **14**, 103–124.

Scopes, N. E. A. (1969). The potential of *Chrysopa carnea* as a biological control agent of *Myzus persicae* on glasshouse chrysanthemums. *Ann. Appl. Biol.* **64**, 433–439.

Seguy, E. (1950). "La biologie des Dipteres." Vol. 26, 609 pp. Encyclopedie Entomol. Sér. A. Paris.

Shands, W. A., and Simpson, G. W. (1972). Insect predators for controlling aphids on potatoes. 7. A pilot test of spraying eggs of predators on potatoes in plots separated by bare fallow land. *J. Econ. Entomol.* **65**, 1383–1387.

Shands, W. A., Simpson, G. W., Wave, H. E., and Gordon C. C. (1972). Importance of arthropod predators in controlling aphids on potatoes in northeastern Maine. *Univ. Maine Tech. Bull.* **54**, 49 pp.

Schiefelbein, J. W., and Chiang, H. C. (1966). Effects of spray of sucrose solution in a corn field on the populations of predatory insects and their prey. *Entomophaga* **11**, 333–339.

Selander, R. B. (1960). "Bionomics, Systematics, and Phylogeny of *Lytta*, a Genus of Blister Beetles (Coleoptera, Meloidae)," 295 pp. Univ. Ill. Press, Urbana, Illinois.

Sheldon, J. K., and MacLeod, E. G. (1971). Studies on the biology of the Chrysopidae. II. The feeding behavior of the adult of *Chrysopa carnea* (Neuroptera). *Psyche* **78**, 107–121.

Sheldon, J. K., and MacLeod, E. G. (1974). Studies on the biology of the Chrysopidae. IV. A field and laboratory study of the seasonal cycle of *Chrysopa carnea* Stephens in central Illinois. *Trans. Amer. Entomol. Soc.* **100**, 437–512.

Shurovenkov, B. G. (1962). Field entomophagous predators (Coleoptera, Carabidae and Diptera, Asilidae) and factors determining their efficiency. *Entomol. Obozr.* (*Entomol. Rev.*) **41**, 476–485.

Simmonds, F. J. (1960). Biological control of the coconut scale, *Aspidiotus destructor* Sign., in Principe, Portuguese West Africa. *Bull. Entomol. Res.* **51**, 223–237.

Skuhravy, V. (1959). Prispevek k bionomii polnich strevlikovitych (Col. Carabidae). (Additions to the bionomics of field carabids.) *Rozpr. Cesk. Akad. Ved. Rada Mat.* **99**, 1–64.

Smith, H. S., and Armitage, H. M. (1926). Biological control of mealybugs in California. *Calif. State Dept. Agr. Mon. Bull.* **9**, 104–164.

Smith, S. G. (1965). *Chilocorus similis* Rossi (*Coleoptera: Coccinellidae*); disinterment and case history. *Science* **148**, 1614–16.

Smith, R. F. and Hagen, K. S. (1966). Natural regulation of alfalfa aphids in California. *In* "Ecology of Aphidophagous Insects" (I. Hodek, ed.), pp. 297–315. Academia. Prague.

Solomon, M. E. (1949). The natural control of animal populations. *J. Anim. Ecol.* **18**, 1–35.

Solomon, M. E. (1964). Analysis of processes involved in the natural control of insects. *In* "Advances in Ecological Research" (J. B. Cragg, ed.), Vol. 2, pp. 1–58. Academic Press, New York.

Southwood, T. R. E. (1956). The structure of the eggs of the terrestrial Heteroptera and its relationship to the classification of the group. *Trans. Roy. Entomol. Soc. London* **108**, 163–221.

Southwood, T. R. E. (1966). "Ecological Methods with Particular Reference to the Study of Insect Populations," 391 pp. Methuen, London.

Stark, R. W., and Dahlsten, D. L. (eds.). (1970). Studies on the population dynamics of of the western pine beetle, *Dendroctonus brevicomus* LeConte (Coleoptera: Scolytidae). *Univ. Calif. Div. Agr. Sci.*, 174 pp.

Steiner, H. (1964). Zur Prüfung der Wirkungsbreite von Pflanzenschutz mitte In bei der integrierten Bekämpfung im Obstbau. *Mitt. Biol. Bundesanst. Land. Forstwirt. Berlin Dahlem* **115**, 30–34.

Stern, V. M., Smith, R. F., van den Bosch, R., and Hagen, K. S. (1959). The integration of chemical and biological control of the spotted alfalfa aphid. *Hilgardia* **29**, 81–101.

Stewart, K. W., Friday, G. P., and Rhame, R. E. (1973). Food habits of Hellgrammite larvae, *Corydalus corrutus* (Megaloptera: Corydalidae), in Brazos River, Texas. *Ann. Entomol. Soc. Amer.* **66**, 959–963.

Strouhal, H. (1926). Pilzfressende Coccinelliden. *Z. Wiss. Inst. Biol.* **31**, 131–143.

Sundby, R. A. (1966). A comparative study of the efficiency of three predatory insects *Coccinella septempunctata* L., *Chrysopa carnea* St. and *Syrphus ribesii* L. at two different temperatures. *Entomophaga* **11**, 395–404.

Swan, L. A. (1964). "Beneficial Insects," 429 pp. Harper and Row, New York.

Sweetman, H. L. (1958). "The Principles of Biological Control," 560 pp. Wm. C. Brown Co., Dubuque, Iowa.

Swezey, O. H. (1936). Biological control of the sugarcane leafhopper in Hawaii. *Hawaii Sugar Plant. Ass. Exp. Sta. Entomol. Ser. Bull.* **21**, 57–101.

Takeda, S., Hukusima, S., and Yamada, S. (1964). Seasonal abundance of coccinellid beetles. *Res. Bull. Fac. Agr. Gifu Univ.* **19**, 55–63.

Tamaki, G. (1972). The biology of *Geocoris bullatus* inhabiting orchard floors and its impact on *Myzus persicae* on peaches. *Environ. Entomol.* **1**, 559–565.

Tamaki, G. (1973). Spring populations of the green peach aphid on peach trees and the role of natural enemies in their control. *Environ. Entomol.* **2**, 186–191.

Tamaki, G. and Weeks, R. E. (1972a). Biology and ecology of two predators *Geocoris pallens* Stäl and *G. bullatus* (Say). *U.S.Dept. A. Tech. Bull.* **1446**, 46 pp.

Tamaki, G., and Weeks, R. E. (1972b). Efficiency of three predators, *Geocoris bullatus*, *Nabis americoferus* and *Coccinella transversoguttata*, use alone or in combination against three insect prey species, *Myzus persicae*, *Ceramica picta* and *Mamestra configurata* in a greenhouse study. *Environ. Entomol.* **1**, 258–263.

Tamaki, G., and Weeks, R. E. (1973). The impact of predators on populations of green peach aphids on field-grown sugarbeets. *Environ. Entomol.* **2**, 345–349.

Tamaki, G., Landis, B. J., and Weeks, R. E. (1967). Autumn populations of green peach aphid on peach trees and the role of syrphid flies in their control. *J. Econ. Entomol.* **60**, 433–436.

Tamaki, G., McGuire, J. U., and Turner, J. E. (1974). Predator power and efficacy: a model to evaluate their impact. *Environ. Entomol.* **3**, 625–630.

Tao, C-C., and Chiu, S-C. (1971). Biological control of citrus, vegetables and tobacco aphids. *Taiwan Agr. Res. Inst. Spec. Publ. No.* **10**, 110 pp.

Tauber, M. J., and Tauber, C. A. (1970). Photoperiodic induction and termination of diapause in an insect: response to changing day lengths. *Science* **167**, 170.

Tauber, M. J., and Tauber, C. A. (1972). Larval diapause in *Chrysopa nigricornis*: sensitive stages, critical photoperiod, and termination (Neuroptera: Chrysopidae). *Entomol. Exp. Appl.* **15**, 105–111.

Tauber, M. J., and Tauber, C. A. (1973). Insect phenology: criteria for analyzing dormancy and for forecasting postdiapause development and reproduction in the field. *Search Agr.* **3**, 16 pp.

Taylor, T. H. C. (1935). The campaign against *Aspidiotus destructor* Sign., in Fiji. *Bull. Entomol. Res.* **26**, 1–102.

Thompson, W. R. (1951). The specificity of host relations in predaceous insects. *Can. Entomol.* **83**, 262–269.

Thompson, W. R., and Simmonds, F. J. (1964–1965). A catalogue of the parasites and predators of insect pests. 3. Predator host catalogue, 204 pp. 4. Host predator catalogue, 198 pp. Commonw. Agri. Bur. Commonw. Inst. Biol. Control, Bucks, England.

Thorpe, M. A. (1930). The biology, post-embryonic development and economic importance of *Cryptochaetum iceryae* (Diptera: Agromyzidae) parasite on *Icerya purchasi* (Coccidae: Monophlebini). *Proc. Zool. Soc. London* **60**, 929–971.

Toschi, C. A. (1965). The taxonomy, life histories and mating behavior of the green lacewings of Strawberry Canyon (Neuroptera: Chrysopidae). *Hilgardia* **36**, 391–432.

Tostowaryk, W. (1972). Coleopterous predators of the Swaine jack-pine sawfly, *Neodiprion swainei* Middletown (Hymenoptera: Diprionidae). *Can. J. Zool.* **50**, 1139–1146.

Turnbull, A. L. (1973). Ecology of the true spiders (Araneomorphae). *Annu. Rev. Entomol.* **18**, 305–348.

Usinger, R. (ed.), (1963). "Aquatic insects of California," 344 pp. Univ. Calif. Press, Berkeley, California.

van den Bosch, R., and K. S. Hagen, (1966). Predaceous and parasitic arthropods in California cotton fields. *Univ. Calif. Agr. Exp. Sta. Bull.* **820**, 32 pp.

van den Bosch, R., and Messenger, P. S. (1973). "Biological Control," 180 pp. Intext Educat. Publ., New York.

van den Bosch, R., Leigh, T. F., Falcon, L. A., Stern, V. M., Gonzales, D., and Hagen, K. S.

(1971). The developing program of integrated control of cotton pests in California. *In* "Biological Control" (C. B. Huffaker, ed.), pp. 377–394. Plenum, New York.

van Emden, H. F. (1966). The effectiveness of aphidophagous insects in reducing aphid populations. *In* "Ecology of Aphidophagous Insects" (I. Hodek, ed.), pp. 227–235. Academia, Prague.

van Emden, H. F. (ed.). (1972). "Aphid Technology," 344 pp. Academic Press, New York.

van Emden, H. F., Eastop, V. F., Hughes, R. D., and Way, M. J. (1969). The Ecology of Myzus persicae. *Annu. Rev. Entomol.* **14**, 197–270.

van Emden, H. F. and Hagen, K. S. (1976). Olfactory reactions of the green lacewing, *Chrysopa carnea*, to tryptophan and certain breakdown products, *Environ. Entomol.* **5**, in press.

Varley, G. C. (1971). The effects of natural predators and parasites on winter moth populations in England. *Proc. Tall Timbers Conf. Ecol. Anim. Control Habitat Manage* **2**, 103–116.

Varley, G. C., and Gradwell, G. R. (1968). Population models for the winter moth. *In* "Insect Abundance" (T. R. E. Southwood, ed.), No. 4, pp. 132–142. Symp. Roy. Entomol. Soc., London.

Varley, G. C., Gradwell, G. R., and Hassell, M. P. (1973). "Insect Population Ecology on Analytical Approach," 212 pp. Blackwell, Oxford.

Vaught, G. L., and Stewart, K. W. (1974). The life history and ecology of the stone fly *Neoperla clymene* (Newman) (Plecoptera: Perlidae). *Ann. Entomol. Soc. Amer.* **67**, 167–178.

von Koschel, H. (1971). Zur Kenntnis der Raubwanze *Himacerus apterus* F. (Heteroptera, Nabidae) Teill I. *Z. Angew. Entomol.* **68**, 1–24. Teil II. pp. 113–137.

Way, M. J. (1963). Mutualism between ants and honeydew-producing Homoptera. *Annu. Rev. Entomol.* **8**, 307–344.

Whitcomb, W. D. (1940). Biological control of mealybugs in greenhouses. *Mass. Agr. Exp. Sta. Bull.* **375**, 22 pp.

Whitcomb, W. H. (1967). Field studies on predators of the second-instar bollworm, *Heliothis zea* (Boddie). *J. Ga. Entomol. Soc.* **2**, 113–118.

Whitcomb, W. H. (1974). Natural populations of entomophagous arthropods and their effect on the agroecosystem. *In* "Plant Insects and Diseases" (F. G. Maxwell and F. A. Harris, eds.), pp. 150–169. Proc. Summer Inst. Biol. Control. Univ. Press Mississippi, Jackson.

Whitcomb, W. H., and Bell, K. (1964). Predaceous insects, spiders, and mites of Arkansas cotton fields. *Univ. Ark. Agr. Exp. Sta. Bull.* **690**, 84 pp.

Wilbert, H. (1972). Der Einfluss der beutedichte auf die sterblichkeit der larven von *Aphidoletes aphidmyza* (Rond.) (Cecidomyiidae). *Z. Angew. Entomol.* **70**, 347–352.

Willie, J. E. (1958). El control biologico de los insectos agricolas en el Peru. *Proc. Tenth Internat. Congr. Entomol.* **4**, 519–23.

Wilson, E. O. (1971). "The Insect Societies," 548 pp. The Belknap Press, Harvard Univ. Press, Cambridge, Massachusetts.

Wilson, F. (1974). The use of biological methods in pest control. *In* "Biology in Pest and Disease Control" (D. P. Jones and M. E. Solomon, eds.), 398 pp. Halstead Press, New York.

Yeargan, K. V., and Cothran, W. R. (1974). Population studies of *Pardosa ramulosa* (McCook) and other common spiders in alfalfa. *Environ. Entomol.* **3**, 989–993.

Young, A. M. (1967). Predation in the larvae of *Dytiscus marginalis* L. *Pan-Pac. Entomol.* **43**, 113–117.

Zhavoronkova, T. N. (1969). Certain structural peculiarities of the Carabidae (Coleoptera) in relation to their feeding habits. *Entomol. Rev.* (*USSR*) **48**, 462–471.

6

BIOLOGY AND HOST RELATIONSHIPS OF PARASITOIDS

R. L. Doutt, D. P. Annecke, and E. Tremblay

I. THE PARASITOID IN NATURE

A. Role of the Adult Female Parasitoid

The free-living adult female parasitoid deserves special attention from entomologists engaged in biological control. Most of our intricate knowledge of these valuable insects has come from laboratory investigations, and with this information we construct conceptual models of the adult female in the field. While these extrapolations from laboratory data to the activities of a free-living parasitoid are useful, and in most cases closely parallel field measurements, they do not nevertheless explain some of the puzzles encountered in the course of a biological control project. For instance, we do not understand precisely why some species of parasitoids fail to become established even though all conditions appear to be most favorable at the time of colonization. It is most likely that the explanation will be found in a study of the adult female. A careful consideration of her activities in nature and her ability to react to the changing host situation and to the hazards of the environment deserve a high priority in biological control research programs. This is, however, ecological research of a most difficult order.

Certainly one can easily and logically reason that *successful* establishment of an imported parasitoid species depends mostly upon the attributes of the adult female. It is generally agreed that her searching behavior, either individually or collectively as a population of females, to a large degree, determines the true potentiality of her species in regulating the host density at low, or at least tolerable levels (Chapter 3). This searching capacity is, therefore, of prime importance to biological control. Experience has shown that some imported species are readily cultured in the insectary, and furthermore immediately after field colonization perform well at the initially high host density in which they are placed. Eventually, however, when the host population becomes reduced and scattered, they may disappear and be competitively replaced by other, more effective, parasitoid species. It is these latter species that are so immensely valuable to biological control for their females have the ability to find and attack hosts and to distribute progeny efficiently when hosts are sparce. As a group, the hymenopterous parasitoids are unsurpassed in this ability although there are some significant specific exceptions in the parasitic Diptera, e.g., *Ptychomyia remota* Aldrich (Tothill *et al.*, 1930).

When the hymenopterous female parasitoid reaches the adult stage, she begins her free existence by using her mandibles to cut an exit from the protective cocoon, host integument, leaf mine, gallery, or earthen cell in which she passed the pupal stage. In forcing her body through this small aperture she aids the removal of residual pupal integument from her appendages. If her species is bisexual and gregarious, there is a high probability that a male has matured a few hours earlier and will be in the vicinity. This increases the likelihood of mating and in some species, such as *Melittobia chalybii* Ashmead, this tendency for inbreeding with siblings is the normal pattern. If the species is solitary in development, mating may be comparatively delayed; but we know very little about sex attractants among parasitic Hymenoptera. Mating is, of course, unnecessary in thelyotokous species.

B. The Concurrent Environmental Requirements of the Female Parasitoid and Her Host

Emergence tends to occur during the morning hours and the newly emerged female, irrespective of mating, is generally well equipped for her special role of finding hosts in the environment and distributing her progeny among them. If the parasitoid species is to be effective in biological control, it must necessarily be temporally, spatially, and ecologically coincident with its hosts that are in a stage susceptible to its attack. This intimate relationship is ensured through the operation of complex mechanisms that vary with the species of parasitoid and its host. These large requirements of time, space, and ecological coincidence must first be satisfied before the host/parasitoid system can operate and before the host-searching adaptations of the parasite can function well.

1. Time

The association in time is usually assured by the physiological milieu of the host in which an internal parasitoid develops. Here the synchronization of development is frequently governed by the phenomenon of host-induced diapause (vide infra). Seldom does the female adult parasitoid influence this temporal parameter directly, but Simmonds (1946, 1947, 1948) discovered there was a significant influence of maternal physiology on the incidence of diapause in progeny of *Cryptus inornatus* Pratt and *Spalangia drosophilae* Ashmead. These species are widely separated taxonomically, representing two different superfamilies, so this phenomenon may be more common among the parasitic Hymenoptera than we know. On theoretical grounds the female could influence the host/parasitoid association in time by causing an even-brooded condition through an especially effective attack on a single host stage or by limiting her attack to only those host individuals entering diapause. It is often obvious in biological control projects that the normal host/parasitoid synchrony can be disastrously disrupted by the adverse effect of chemical toxicants on the adult female parasitoids. They may be killed outright or their searching ability may be so impaired that they cannot function properly. The pesticide may also disrupt the delicately balanced synchrony by selectively eliminating the host stage that otherwise would be available and susceptible to attack by searching females.

2. Space

The problem of spatial segregation between host and natural enemy tends to be minimized because the female stage in most parasitoids is winged and highly vagile. There may be temporary isolations of a host population, as is frequently seen in dooryard or urban plantings, but this is usually of short duration, since the effective parasitoids soon locate this host colony which briefly enjoyed spatial protection. The spatial barrier can become a seasonally recurrent problem, however, if the host and parasitoid have very different requisites for surviving the winter, e.g., *Anagrus epos* Girlt. cannot overwinter with its host, *Erythroneura elegantula* Osborn, in vineyards but obligatorily must pass this period on another host, *Dikrella cruentata* (Gillette), on *Rubus*. The commercial planting of vineyards miles distant from *Rubus* has created a spatial barrier that *Anagrus* must span each spring if it is to attack the *Erythroneura* on these grapes. This is an artificial situation created by man since under endemic conditions the *Rubus* and *Vitis* grow together as natural components of a riparian flora.

3. The Habitat and the Host Finding Process

Not only must hosts and parasitoids be in harmony in space and time, but the microhabitats of the two species must coincide. The mechanisms which maintain such ecological coincidence are clearly adaptations that reside in the adult female parasitoid. She initially and fundamentally seeks a particular environment irrespective of the presence or absence of hosts, but it also is the special environmental situation especially favorable to her preferred host species. Accordingly, the female parasite responds to a complex of subtle stimuli including microclimates, odors, and physical aspects of the environment that place her in the exact microhabitat most likely to contain hosts.

It is interesting to note that females of some species during a protracted preoviposition period will seek a habitat entirely different from the host-containing habitat to which they are attracted upon becoming gravid. Thus in the tachinid *Eucarelia rutilla* Vill. the habitat selection differs among young and old females. An initial preference for oak changes gradually about the third week of adult life into one for pine trees where the hosts of *Eucarelia* feed almost exclusively upon the needles (Herrebout, 1969). Essentially the same pattern of shifting habitats in accordance with ovarian development has been reported in the braconid *Opius fletcheri* Silv. by Nishida (1956) and in the ichneumonid *Pimpla ruficollis* Grav. by Thorpe and Caudle (1938).

Although the female is in the habitat of the host, she still must go through the process of actually finding a host individual, and the studies on parasite searching ability and the adaptations for locating host individuals involve fascinating aspects of insect behavior.

There is some evidence for preimaginal conditioning in certain parasitoids, but it does not appear to be a widespread phenomenon or a particularly powerful influence on host selection, and until more work is done on this subject, it will be difficult to evaluate. It deserves attention, however, in connection with the common practice of rearing parasites on factitious insectary hosts and then releasing the females to attack their ancestral hosts in the field. Under such circumstances is there any preimaginal conditioning and, if so, does it affect the females in their host selection? So far the evidence seems negative, but this needs to be monitored. The work of Arthur (1966, 1967) shows that an element of associative learning exists among parasitoids. The ichneumonid *Itoplectis conquisitor* (Say) was conditioned to associate color with the presence of hosts. The females could also be conditioned to attack artificial host shelters of particular lengths, diameters, and mountings. The tachinid parasite *Drino bohemica* Mesnil learned to associate the movement of part of its cage with the presence of host larvae (Monteith, 1963).

It is sometimes difficult to separate host habitat finding from the process of actually locating a host individual in the area. They are both part of what Salt in his masterful studies (1934, 1935, 1937) termed ecological selection. Ullyett (1953) noted that the parasitoid female tends to search in the parts of its environment most likely to contain its host and that this is caused by a combination of preadaptation in habits and a specific attraction exerted by the portion of the environment concerned.

Of the senses used by the parasite in detecting the presence of hosts, the ones most commonly reported are tactile and olfactory, but sight is also involved. Many studies have been made on the reactions of parasitoids to hosts and their habitats, and these have been reviewed by Doutt (1959, 1964).

C. Host Selection by the Discriminating Female Parasitoid

When the host is finally located, the female parasitoid may still not accept it as an oviposition site if her senses perceive that it is in some way unsuitable. Rejections are frequently made if the host has been previously visited by any parasitoid, including herself. This avoids the wastage of progeny through competition for the host individual with larvae of the same parasitoid species (superparasitism) or competition with larvae of a different species (multiple parasitism) (vide infra).

Some adult female parasitoids have well developed powers of discrimination between healthy and parasitized hosts and will tend to avoid oviposition in the latter. She perceives the difference through sensory receptors on her antennae, tarsi, and ovipositor. The female parasitic wasps characteristically possess a multipurpose ovipositor. This marvelous instrument enables them (1) to paralyze and subdue hosts; (2) to tap a rich source of proteinaceous nutrients by preparing the host for adult feeding; (3) to determine with remarkable sensitivity the suitability of the host for oviposition; and (4) to place the egg precisely in the optimum site for development, even in a specific host tissue or morphological structure. In other wasps (true wasps) the ovipositor is not adapted for such purposes but is used solely for injecting venom. In such Hymenoptera the eggs do not pass down this stinger, but are discharged directly from the body.

The sensitivity of the ovipositor is not unique to the parasitoid wasps. Monteith (1956) reports that chemotactile stimuli received by the ovipositor are of importance in egg deposition by the tachinid, *Drino bohemica* Mesn. More recently, Herrebout (1969) suggests an element of discrimination in the tachinid *Eucarcelia rutilla* Vill., where egg deposition can be inhibited by stimuli received at the tip of the ovipositor.

Fischer and Ganesalingam (1970) suggested that changes in the protein composition of parasitized hosts may provide the stimuli by which female *Devorgilla canescens* (Grav.) (a species also reported in the genera *Nemeritis*, *Exidecthis*, or *Venturia*) can distinguish between healthy and parasitized hosts when probing them with her ovipositor.

In parasitized *Anagasta* larvae, phenylalanine, present in healthy host hemolymph, had disappeared by the 5th day and leucine was also absent by the 8th day after parasitization. Further, two unidentified ninhydrin-reacting substances appeared in host hemolymph chromatograms by the 5th day and their spots had greatly enlarged by the 8th day at 25 °C.

It is interesting that Arthur *et al.* (1969) report isolation of a proteinlike component of *Galleria* hemolymph which *induces* oviposition by *Itoplectis conquisitor* (Say). Presumably this physiologically active oviposition-inducing factor is detected by receptors on the ovipositor when it pierces the host. They suggest their findings might be of practical use if, in mass culture programs, parasitoids could be induced to oviposit in synthetic diets containing the active substance.

D. The Economy of Oviposition and Supplementary Feeding

When the female parasitoid enters adult life, she usually carries sufficient residual metabolites from her developmental stages to produce a number of eggs without supplementary feeding. However, among the species that are most valuable to biological control and which maintain hosts at low levels, some have adaptations to withhold egg production during periods of host scarcity or other unsuitable environmental conditions. In these cases the eggs are absorbed, recycled, and produced again when conditions for deposition become favorable. The continual production of eggs during periods of host abundance is dependent upon supplementary feeding. It is known that the distribution of some parasites is influenced by sources of free water and certain plants which they visit for nectar, but for some species perhaps the single most important source of supplementary food is derived from host feeding. This is a regular habit among many chalcids and braconids. This may be simple direct feeding at puncture wounds made by the ovipositor or it may involve the construction of a feeding tube. This feeding tube is formed by secretions that are manipulated with the ovipositor and harden in the air to form a structure much like a pipette or drinking straw. The host fluids rise in the tube by capillary action and the internal pressure of the hemolymph. The parasitoid can use the feeding tube device to feed on a host that is protected in a cocoon, gallery, or leaf mine and that would otherwise be inaccessible. This approaches a tool-making and using ability in these highly evolved parasitoids.

E. The Influence of the Female on the Sex Ratio of Her Progeny

The adult female plays a significant role in determining the sex ratio of her progeny, but to outline this aspect of her behavior, a brief review of the sex-determining mechanisms in the parasitic Hymenoptera is primary. Most insect eggs undergo meiosis and development then stops unless the egg is fertilized. This is not the case, however, in Hymenoptera where parthenogenesis is the rule. It appears in three forms, namely:

1. *Thelyotoky*—The progeny are all females that are uniparental, impaternate, and produced by diploid parthenogenesis.
2. *Deuterotoky*—The progeny are all uniparental, impaternate, and mostly females, but occasionally a male is produced.
3. *Arrhenotoky*—This is a system of haplodiploidy with facultative parthenogenesis. The male progeny are uniparental, impaternate, and produced by haploid parthenogenesis. The females are biparental and are produced zygogenetically.

In thelyotoky a recombination of genes is not possible and so this lack of adaptability is a disadvantage, but there are advantages in having no need for mating and a potential for developing more prolifically than a bisexual species. In some parasitoids there are apparently two "races" of the same species, one bisexual and the other thelyotokous.

Wilson and Woolcock (1960) found that certain temperatures to which the female is exposed either prior to emergence or during her oviposition period after emergence may produce males or gynandromorphs in an otherwise thelyotokous species, *Ooencyrtus submetallicus* Howard. Reproduction, however, remained uniparental. In view of this production of males under special environmental conditions to which a thelyotokous species may be subjected, it is interesting that there are no recorded cases of alternating generations among the parasitoids, not even among the parasitic cynipoids where the pattern of alternating unisexual and bisexual generations is common among the phytophagous species.

Arrhenotoky is exhibited generally through the entire order Hymenoptera. This system has rarely arisen among the Metazoa so the Hymenoptera stand somewhat apart for it could only develop under special circumstances and it must have great antiquity in this order of insects.

In biological control projects the widely variable and often greatly fluctuating sex ratios of arrhenotokous species may cause difficulty both in the insectary production and the colonization of such parasitoids. The sex of the egg is changed from male to female by fertilization, and the stimulation of the spermatheca to discharge spermatozoa into the oviduct is often a response of the female parasitoid to external conditions.

Sex ratios may be determined by the duration of the premating period after emergence, the rate of oviposition, the number of eggs deposited at one insertion of the ovipositor, and many other factors associated with the life of the adult free-living female parasitoid. These have been summarized by Flanders (1946).

II. THE DEVELOPING PARASITOID

A. Embryonic Development

The types of eggs of the parasitoids were extensively reviewed by Hagen (1964). According to him, the large variety of parasitoid eggs fall into the following categories: hymenopteriform (generally occurring among Hymenoptera and some Diptera), membranous (tachinids and sarcophagids), acuminate, stalked, pedicellate (commonly occurring in parasitic Hymenoptera and rarely in Diptera), encyrtiform (encyrtids), macrotypic (tachinids), microtypic (tachinids and trigonalids), and acroceriform (acrocerid Diptera).

Among the internal egg structures, the yolk and the oosome deserve comment. The yolk system undergoes rapid reduction and eventually disappears in connection with endoparasitism in Hymenoptera. The consequences of this involution will be discussed in the section dealing with polyembryony. The oosome or germ-cell determinant occurs frequently in the early embryogeny of Hymenoptera and Diptera in the aspect of a round body. Its presence at the posterior pole of the egg and its dissolution in connection with the arrival of some of the cleavage nuclei are usually correlated to the appearance of the first germ or sexual cells at this pole (pole cells). No direct evidence, however, is available at present in support of this correlation. Alkaline phosphatase activity has been demonstrated at the site of oosome dissolution in *Apanteles glomeratus* (L.) (Tawfik, 1957). In insects of other orders (e.g., chrysomelids), the obliteration of the oosome leads to germless larvae. On the other hand, removing of the pole cells by x-raying the eggs of *Pimpla turionellae* (L.) has no effect on the fertility of the adults (Günther, 1971). In this species the migration of the pole cells into the gonadal rudiment was previously described (Bronskill, 1959).

As described above, parthenogenesis is of usual occurrence in Hymenoptera in its types known as thelyotoky (uniparental species) and arrhenotoky (uniparental males and biparental females), the latter being normal for the whole order of Hymenoptera (Doutt, 1959). Deuterotoky (occasional production of males from normally thelyotokous species) is also frequent in parasitic Hymenoptera. Change over from thelyotoky to arrhenotoky may be produced also, as an effect of environmental conditions (e.g., high temperature) (Flanders, 1965).

In Hymenoptera (Ivanova-Kazas, 1961) the embroyonic development of the ectoparasitoids does not differ substantially from that of the nonparasitic groups (e.g., bees and wasps). The same is true for some primitive types of embryogenies of endoparasites (*Pimpla turionellae* L.) in which the more significant deviation from the general scheme for insects is the reduction of the amnion. Most endoparasitoids, on the contrary, evolved toward the complete reduction of the enveloping membranes (e.g., *Anagrus incarnatus* Hal.) or toward an increased trophic function of the serosal envelope (Aphidiinae) and polyembryony (production of more than one individual from a single egg). Among parasitic Hymenoptera polyembryony has arisen independently in the encyrtids, platygasterids, braconids, and dryinids from the most evolved types of monoembryonic development along with a reduction of the oocyte size, complete loss of yolk material, and total cleavage (Ivanova-Kazas, 1961).

The individuals which derive from a single egg range from a few (in several platygasterids) to more than one thousand (some encyrtids). The real number of the embryos and larvae is sometimes higher; a number of them may die, being unable to complete development in consequence of intrinsic defects (asexual or teratoid larvae: Silvestri, 1905; Doutt, 1947, 1952) or as a result of competitive suppression (*Macrocentrus ancylivorus* Roh., Daniel, 1932). In the polyembryonic type of development the parasitic embryos are separated, enclosed, and nourished by a syncytial and proliferating envelope or trophamnion, which derives from a consistent part of the egg cytoplasm including the polar nuclei (Silvestri, 1905). This peculiar, thick envelope, which is an active site of protein synthesis and of energy supplying materials (Koscielska, 1962), may be also present in some monoembryonic species (e.g., *Platygaster dryomiae* Silv. and some genera of evolved Aphidiinae, in which it is of oligocellular type). In the genus *Trioxys* Hal., in which the conditions conducive to polyembryony are attained, the latter being, however, not compatible with the host—parasite dimensional relationship, two of the first eight nuclei of the embryo associate with the polar bodies and end up in the trophamniotic envelope (Tremblay and Calvert, 1972) (Fig. 1). It seems possible that the normal nuclear cytoplasmatic balance in the early embryo may be thus preserved. This balance has been suggested by Doutt (1947) as stimulating polyembryonic proclivity when changed in favor of the nuclear material. The suggestion is also supported by the fact that sperm entrance into the egg determines the polyembryony in the female sex only (*Platygaster hyemalis* Forb., Leiby, 1926) or highly increases the production of polyembryos in the female broods (Doutt, 1947).

Information on the embryology of parasitoid Diptera is extremely scanty. Detailed data exist on the embryology of the species belonging to the genus *Calliphora* Rob. (see Johannsen and Butt, 1941) to which tachinid and sarcophagid embryologies should show similarities. In the majority of the tachinid species, the eggs undergo partial or complete (microtype eggs) uterine incubation, though in frequent cases the entire embryonic development takes place outside the body of the parent female (*Parasetigena* type of Herting, 1960) as it is the case of the macrotype eggs. Among the species that inject their eggs directly into the host's body only a few deposit them before any appreciable embryonic development has taken place. In these species a partial uterine incubation occurs. The incubation period may range from a few days to several months of interrupted development inside the host. Hatching may be greatly extended over complete incubation in a number of cases (microtype eggs of tachinids and trigonalids).

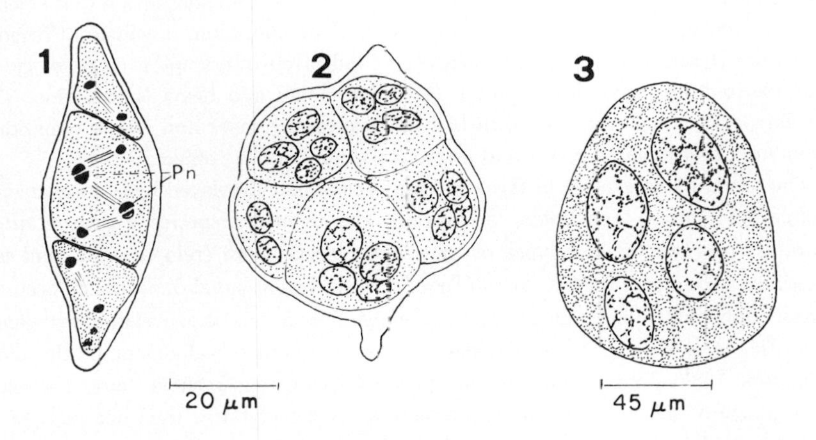

Fig. 1. Origin of the embryonic envelope and of trophocytes in *Trioxys pallidus* Hal. (1) Association of the two first derivatives of the *polar nucleus* (Pn) to the cleavage nuclei during the third division. (2) Two nuclei which derive from the polar nucleus plus two smaller nuclei of embryonic derivation in each of the eight envelope cells which at hatching evolve in trophocytes. (3) One of the trophocytes. Original by E. Tremblay.

B. Larval Development

An extensive general description of the larval stages of the parasitoids is given by Clausen (1940) and Hagen (1964). The first instar larvae exhibit a wide range of types often showing no resemblance to other insect larvae. In these cases they receive names which derive from the original taxonomic group to which they belong (viz., mymariform, encyrtiform, agriotypiform, eucoiliform, and teleaform). In other cases their names are expressions of some external features (mandibulate, caudate, vermiculate, sacciform). Some types derive their denomination from other insects or animals (triungulin, cyclopiform). The mandibulate type is quite common in endophagous parasitoid Hymenoptera, often in combination with other types (e.g., mandibulate-caudate), and the possession of sharp, opposable mandibles seems to be an efficient weapon in inter- and intraspecific competition, being not expressly correlated with feeding (Salt, 1961; Fisher, 1971). The wide range of structures in possession by the primary larvae of parasitoids, fall into the five classes listed by Grandi (1959) in his study on the morphological adaptations in insects that have a specialized diet: (a) involutions, rudimentations, or disappearance of organs or portions of the organs; (b) abnormal (hypertelic) developments of organs or portion of organs; (c) displacement of organs; (d) transformation of organs; (e) development of new parts in preexisting organs and also of new organs.

The "appendages" and processes which often characterize these larvae have no ontogenetic value, being secondary or coenogenetic structures, easily associating by convergences, due to environmental effects, species belonging to phylogenetically unrelated groups. An impressive example of such convergences is the "planidium" larva (intermediate parasitic phase of Malyshev, 1968), which is an adaptation to the necessity of "moving to the host by its own efforts." The planidium is widely diffused in parasitoid Hymenoptera, Diptera, Coleoptera, and Strepsiptera and is morphologically characterized by heavy sclerotization of the segments and heavy spinulation. This peculiar larva is also provided with high resistance to hunger and shows remarkable locomotory and jumping power and attraction to any moving object.

The larval heterogeneity in Hymenoptera is especially displayed by some chalcidoid groups (mymarids, aphelinids, encyrtids). The mymarid species *Patasson nitens* (Girault) has two distinct types of first instar larvae, both endoparasitic, and each correlated with one of the two sexes (Tooke, 1953). Sexual larval dimorphism occurs in certain aphelinids (e.g., the genus *Coccophagus*) and this is correlated with endoparasitic development of the females and ectoparasitic development of the males (Flanders, 1937; Zinna, 1961). In the species *Coccophagus gurneyi* Comp. the sexual dimorphism of this type is further complicated by a facultative dual ontogeny of the male sex itself which possesses two types of larvae correlated to two distinctive sites of deposition of the haploid egg (Zinna, unp. res., *in* Flanders, 1964). The asexual or teratoid larva which is a production of some polyembryonic encyrtids is another interesting example of larval dimorphism. In the species *Copidosoma koehleri* Blanchard the teratoid larva appears only in the progeny of the fertilized females (Doutt, 1952) and is of the ichneumonid type with a definite internal anatomy (Silvestri, 1906). It dies prematurely, being unable to reach pupation.

The instars following the first until maturity are variable in number according to the genera and also to the species (and to the authors!). In the trichogrammatids, for instance, the number of larval instars is said to range from one to five (Hagen, 1964). In the Aphidiinae the authors report from three to five instars. No uncertainty exists about three instars, the primary, the penultimate, and the final stage. There is some controversy about the existence of another stage between the primary and the penultimate (Tremblay, 1964). Interesting cases of first larval ecdysis occurring still inside the chorion are known for the male sex of some aphelinids (Zinna, unpub. res., in Flanders 1964; Gerling, 1966). The first instar larva usually molts into a stage which has good similarities with the last larval stage. Some exceptions are known of which the most striking is the "histriobdellid" second instar larva of some mymarid species of the genus *Anagrus* Hal. The mature larvae of the parasitoid Hymenoptera and Diptera show less heterogeneity than the first instar larvae. The presence of 13 postcephalic segments is usual in the mature larva of Hymenoptera while 11–12 segments are found in the final stage of the Diptera. The occurrence of fleshy tubercles on the body segments is a rather common feature in Hymenoptera. The head structures of Hymenoptera larvae are often of a complicated nature and provide satisfying clues, which are valid as an aid in the classification of the adult parasites (Căpek, 1970; Short, 1970). In the chalcidoids the head structures often undergo simplification and desclerotization. In the Cyclorrhapha (Diptera) an important framework (the buccopharyngeal skeleton) develops in association with a striking reduction of the external cephalic structures.

C. Adaptations for Endoparasitic Life

In the endophagous Hymenoptera the process of assimilation starts during the egg stage with the absorbtion of a relatively large quantity of water, increasing the egg size thousands of times. This type of absorption continues after hatching during the first larval stage in a number of cases, through a hydrophilic permeable cuticle. Among the Diptera, as in the case of *Cryptochaetum*, cutaneous feeding may also occur. Discharge of cytolytic or digestive enzymes from the salivary glands and gut into the host's body fluids is largely quoted in literature (Fisher, 1971) as an explanation for the histolysis of its tissues in connection with the presence of the parasitic larva. Electrophoretic data collected by Mellini and Callegarini (1968) for a chrysomelid parasitized by a tachinid species point out the high histolytic incidence of the digestive juices deriving from the host itself whose gut becomes injured by the parasite. In ectophagous Hymenoptera another alimentary adaptation evolved which consists in the discharge from the first instar larvae of a secretion which paralyzes the host or several hosts (Principi, 1958; Viggiani, 1963; Zinna, 1962). Very striking examples of the trophic adaptations achieved by endoparasitic Hymenoptera are the so-called trophic membranes which may exert their nutritive function not only in favor of the embryo but also during the larval development of the parasite. Their segregation into separate cells at hatching is not followed in some groups (e.g., braconids) by disintegration or degeneration, but by a process of absorption and storage of nutrients from the host body. This trend reaches its extreme in the two most evolved braconid groups, Euphorinae and Aphidiinae (Tremblay, 1966; Sluss, 1968; Sluss and Leutenegger, 1968), and different stages toward this adaptation may be traced in the other less evolved braconid genera (Kitano, 1965; Vinson, 1970). Trophic cells or trophocytes attain large size in the two evolved braconid groups, up to 500 μm in *Perilitus coccinellae* (Schr.), in which their surface is covered by microvillae (Sluss, 1968). The relation existing between the trophic evolution of the embryonic envelope and the mandibular involution of the intermediate larval instars of a number of endophagous braconids was analyzed by Tremblay (1966a).

A digestive feature common to many endophagous parasitoids is the posterior closure of the midgut which serves to avoid excretory contamination of the host. Storage of urates in the fat body and of insoluble salts in the midgut cells and other waste accumulation mechanisms are obligatory until the communication between the midgut and the hindgut is established and the meconium is expelled before or at the end of pupation (Fisher, 1971).

The Malpighian tubules undergo a number of modifications in parasitic Hymenoptera in connection with endoparasitism. They may also be utilized for silk production (Noble, 1938) or involved in secreting the membranous sheath which surrounds the mature larva (Zinna, 1959; Saakjan-Baranova, 1965) for respiratory purposes.

Of special interest are the respiratory adaptations of the endoparasitoids whose problems have been often assumed to be similar to those of aquatic insects. The eggs of the endoparasitic Hymenoptera characterized by a delicate chorion and by an underlying trophamnion of the embryo are supposed to acquire oxygen and nutrition by diffusion (Clausen, 1950). These parasites may acquire atmospheric air through the "aeroscopic plate" of the egg stalk as in the young larvae of many encyrtids (Zinna, 1959). Mature larvae of encyrtids and aphelinids envelop themselves and the tracheae of the host in the membranous sheaths produced by the ileolabial glands. Cutaneous respiration is very common in endophagous first instar larvae of Hymenoptera, and is the usual system during the whole larval life of some large groups of egg parasites (Mymaridae and Trichogrammatidae). Interesting results were obtained by means of oxygen consumption measurements in relation to competition and survival (Fisher, 1961, 1963). Another special type of respiratory adaptation is the anal vesicle (evagination of the proctodeum) of some braconid endoparasitic larvae (e.g., *Apanteles*) during all their preimaginal life. Peculiar respiratory adaptation occurs also in ectoparasitic species of Hymenoptera. This is the case of the floating air-filled ribbons of the agriotypids (Fisher, 1932). In parasitoid Diptera there are frequent cases of cutaneous respiration among young larval instars, but in the last instar these species make connections to respire metapneustically the atmospheric air. In the simple cases, the host tracheae or air sacs, or, in addition, intersegmental membranes, are ruptured for respiration without any intermediary funnel or tube. In the majority of tachinids, nemestrinids, and acrocerids a special respiratory funnel is induced by the parasite in the host by stimulating wound tissue formation and hemocytic reaction at the place where the connection with the external air is made (tegumental or tracheal). The primary type of funnel is induced in the penetration point through the integument. The secondary type is produced in later stages through the integument or tracheae (Mellini, 1965).

D. Pupation

Pupation occurs under a variety of circumstances according to the groups and genera or species. In parasitic Hymenoptera the mid-hindgut connection is established in the prepupal stage for the voiding of the fecal material accumulated during larval development. Most species utilize the host remains or the environment where the host lived for shelter during the pupal stage or pupate externally or in the soil.

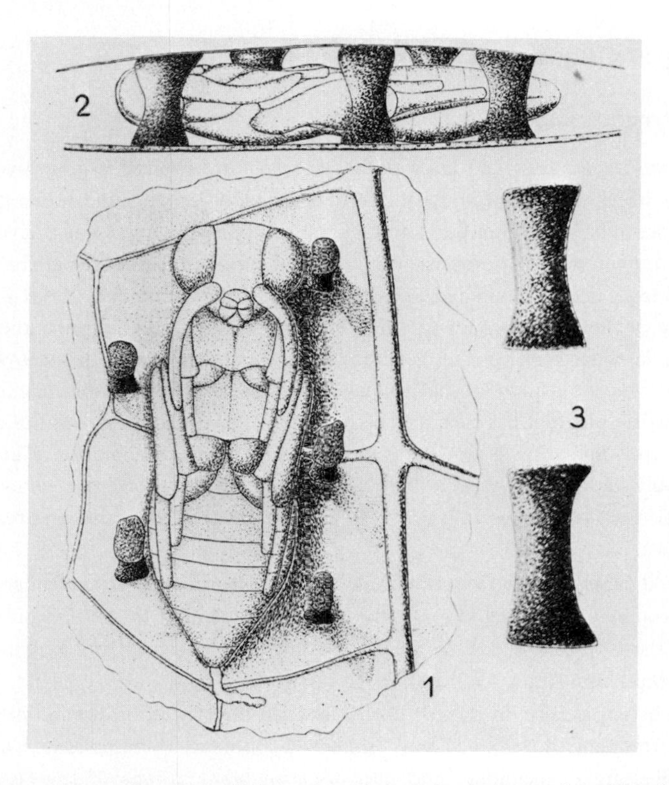

Fig. 2. Supporting pillars of fecal derivation (3) made by the larva of the Eulophid *Chrysocharis gemma* Walk. (2) A pupa of the parasite in lateral view surrounded by the pillars in the mine. (1) The pupa surrounded by the pillars viewed after detachment of one of the mine walls. (From Viggiani, 1964).

In Hymenoptera a cocoon is frequently spun, especially among the Ichneumonoidea, by production of true silk threads (Baccetti, 1958). In chalcidoids silk cocoons are rarely produced although cocoon-forming substances are often secreted from ileac glands (Flanders, 1938). Fecal material which is excreted in pellets in a thin sac (or without it) is sometimes utilized for protection, as in the case of some eulophids that attack leaf miners (Doutt, 1957; Viggiani, 1964). The larvae of these parasites use such meconial pellets in the construction of rows of pillars around them in order to protect the naked pupae from being injured by the two walls of the narrow mine (Fig. 2).

Interesting cases of reciprocal attraction (*plesiotropism*, Goidanich, 1956) induce special ordered arrangements (in rows, in lines, in rays) of the pupae as in the case of the chalcidoids *Euplectrus bicolor* (Swed.) and *Eulophus larvarum* (L.).

Among the tachinids there are species which pupate in the respiratory funnels and species which leave them some time before maturity. Pupation in the soil is common in this group. In rare cases (some cecidomyiids) a sort of cocoon is produced.

III. SPECIAL HOST RELATIONSHIPS

A. Hyperparasitism

The term hyperparasitism is properly reserved for advanced trophic levels of parasitism by which the hyperparasitic larva feeds in or on the immature instars of a primary, secondary, or other parasitoid. The hyperparasitic habit has evolved independently in numerous genera representing especially four superfamilies of the Hymenoptera, namely, Ichneumonoidea, Chalcidoidea, Proctotrupoidea, and Cynipoidea. Especially in the first two superfamilies, obligate secondary parasitoids abound. Their larval development is completed, generally ectoparasitically, at the expense of a primary parasitoid which is usually in the larval, prepupal, or pupal stage. In some families, the hyperparasitic mode of life may tend to be more or less genus-restricted (e.g., Encyrtidae) but in others this is not the case (Ichneumonidae, Eulophidae, Eurytomidae). Occasionally, a secondary species may parasitize a primary congener, as in two *Tetrastichus* species found in certain wax scale insects in the Mediterranean area (Domenichini, 1966).

Reported cases, mainly from the Chalcidoidea, of tertiary and quaternary parasitism are less common, and few if any have been demonstrated to develop obligatorily at either of these trophic levels. In a very few species (Muesebeck and Dohanian, 1927), tertiary status appears to be usual, or preferred.

Direct hyperparasitoids (Smith, 1916) oviposit directly on or in another parasitoid. They are frequent in the chalcidoid families Eulophidae, Pteromalidae, Eurytomidae, Torymidae, and Chalcididae, and also occur in other groups of the Hymenoptera. These hyperparasitoids are active in most insect populations, especially where cocoon-forming primaries occur. Indirect hyperparasitoids are also common: the parent females reach their hosts by first finding the host of suitable primary parasitoid. In this way certain Ichneumonidae, Encyrtidae, and Aphelinidae search for and oviposit in host insects for the sake of primary parasitoids they may contain. Certain indirectly hyperparasitic Perilampidae and Trigonalidae oviposit on plants, the young stages entering plant-feeding hosts, which are or may become parasitized by suitable primary parasitoid hosts.

Many hymenopterous parasitoids are able to develop as either primaries or secondaries, according to the host available, and such behavioral variation is often accompanied, as in many Eupelmidae, by a change from endophagous feeding as primaries to ectophagy when developing in the hyperparasitic role.

In biological control, hyperparasitoids are commonly considered to be injurious. Great care is exercised to exclude them when transshipping insects (Doutt and DeBach, 1964). Yet, despite frequent records of high levels of hyperparasitism, particularly in aphid and psyllid populations (Hagen and van den Bosch, 1968; Clark, 1962; Catling, 1969), their impact in the field has been seldom studied in detail.

The reasoning of Muesebeck and Dohanian (1927) that hyperparasitoids play a role in the balance between insect species is attractive. However, their precise role in this regard remains to be demonstrated, and the mounting record of successful biological pest control by the importation and establishment of primary parasitoids without their native hyperparasitoids certainly supports the policy of exclusion of the latter. (See further Chapter 3.)

B. Adelphoparasitism

A special case of hyperparasitism known as adelphoparasitism (Greek *adelphos*—brother, twin) or autoparasitism that has arisen in the chalcidoid family Aphelinidae was first identified and described by Flanders (1937). In most such species, a highly specialized and remarkable sexual divergence in ontogeny, which is usually accompanied by radical changes in adult female behavior at mating, is superimposed on the normal chalcidoid arrhenotoky. Males develop as hyperparasitoids from unfertilized eggs which are deposited by the parent female behaving as an indirect hyperparasitoid. Females, on the other hand, arise from fertilized eggs which are deposited, and develop to maturity, as primary parasitoids. The adult life of many female parasitoids is characterized by a behavioral climacteric, unique in parasitic animals, occasioned by the act of mating: prior to mating such females lay unfertilized eggs as hyperparasitoids; following this they have the ovipositional instincts of a primary parasitoid. In some adelphoparasitoids, ovipositional behavior remains unaltered by mating.

Important elements of the biologies of a number of species have been reported and reviewed by Flanders (1937, 1959, 1967, 1969), Zinna (1961, 1962), Flanders *et al.* (1961), Broodryk and Doutt (1966), and Gerling (1966). These show that among those species whose ovipositional behavior changes at mating from hyperparasitic to primarily parasitic, there are species which lay male eggs externally on primary parasitoids within their homopterous hosts; such ectophagous hyperparasitic males occur, for example, in *Coccophagoides utilis* Doutt, *Coccophagus lycimnia* (Walker), *Euxanthellus* sp., and *Prospaltella clypealis* Silvestri. Others oviposit internally in the primary parasitoid and the developing larva feeds as an endophagous hyperparasitoid; included among these are *Coccophagus scutellaris* (Dalman), *Physcus fulvus* Compere and Annecke, *Encarsia pergandiella* Howard, and *Aneristus ceroplastae* Howard.

In *Coccophagus basalis* Compere and *C. gurneyi* Compere male-producing eggs are laid in the same way as are female-producing eggs, but whereas the female eggs hatch and develop as primary parasitoids, the male eggs may exhibit "inhibited hatching." The appearance of the first instar male larva is delayed until a suitable host is available upon which its hyperparasitic development can be completed. The hyperparasitic male larva of *C. gurneyi* is apparently facultatively ectophagous or endophagous according to the species of primary parasitoid attacked, while that of *C. basalis* is an obligate ectophage.

Coccophagus basalis is able to deposit male- and female-producing eggs after mating, but in other species such as *Coccophagoides utilis*, apparently only female-producing eggs are laid after mating.

Many adelphoparasitoids, perhaps most, are able to produce males at the expense of females, or even males as in *Coccophagoides utilis*, of their own species (Fig. 3); others

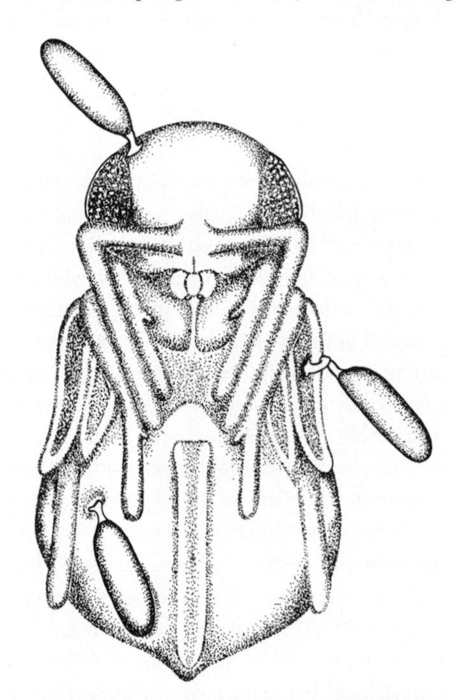

Fig. 3. Female pupa of the adelphoparasitoid, *Coccophagoides utilis* Doutt, removed from the host, *Parlatoria oleae* (Colvée), showing indirectly hyperparasitic male-producing eggs laid by unmated females of the same species. (Reproduced from Broodryk and Doutt, 1966.)

again, such as *Coccophagus basalis*, require for male production hosts which, unlike females of their own species, pupate in a dry environment within their homopterous hosts. A most remarkable departure in male production is seen in certain species of *Prospaltella* and *Encarsia* (Flanders, 1925; Beingolea, 1958) which utilize the eggs of Lepidoptera in which males are produced as primary parasitoids, while the females develop in Homoptera, also as primaries.

Sexual dimorphism among immature instars is of general occurrence among adelphoparasitoids (Cendaña, 1937; Flanders, 1937, 1959; Hagen, 1964); male- and female-producing eggs may differ as in *Coccophagoides utilis*, and first and second instar larvae may also show striking structural differences correlated with their behavior and environment.

Despite the tortuous and seemingly precarious nature of adelphoparasitic biologies, these Aphelinidae are diversified, abundant, and successful in homopterous hosts. Some are credited (see Flanders, 1967) with host regulation at low densities.

C. Synchronization with Hosts

As discussed above, successful and persistent host—parasitoid relationships are characterized by synchronization in time and place of the two partners. They coincide in such a way that stages of the host suitable for parasitism are available to ovipositing parasitoids. Perfectly synchronized parasitoids are those that retain their synchrony throughout the whole, or the greater part, of the favorable zone of climatic conditions under which the host can survive and breed. The degree of synchrony often reflects on the likelihood of stability and permanence of the host—parasitoid relationship and the efficiency of biological control.

*1. Voltinism**

Life cycles of insects from egg to adult vary greatly in duration from a few days to several years. Univoltine species pass through an entire generation each year, while those with shorter life cycles may complete two (bivoltine species) or more (multivoltine species) generations each year.

Overlapping host generations may be important in biological control as exemplified by a parasitoid selectively attacking a single age group of the host. If the parasitoid develops at a rate different from that of its host, and is efficient, nonoverlapping host generations may be produced, with serious consequences (Schoonhoven, 1962).

Many parasitoids accurately match the host, generation for generation, through the seasons. Notable examples are the braconid egg-larval parasites which oviposit during a restricted part, the egg stage, of the host's life cycle. *Chelonus curvimaculatus* Cameron attacks the egg of potato tuber moth, *Phthorimaea operculella* (Zeller), and adults are produced simultaneously with the adult moths which escaped parasitism in the egg

* Derived from the Italian *volta* meaning time in the sense of occasion.

stage (Broodryk, 1969). This parasitoid is able to maintain the synchrony in other lepidopterous hosts which develop at considerably slower rates than tubermoth.

Other species such as *Comperiella bifasciata* Howard attacking Florida red scale, *Chrysomphalus aonidum* (L.), may utilize a wide range of the host's developmental stages for oviposition (Cilliers, 1971). In such cases, synchronization of generations may not be a striking feature of the association, yet the parasitoid is assured of suitable hosts at all seasons.

2. Diapause

The state of arrested morphogenesis known as diapause permits many insects to mentain their populations through unfavorable seasons. Among parasitoids, diapause is important also in synchronizing the life cycles of parasitoid and host (Schoonhoven, 1962). The condition of strict diapause grades almost imperceptibly to quiescence (Way, 1962) and certain states of arrested growth in parasitoids remain difficult to classify. Lees (1955) discusses diapause and quiescence in several hymenopterous and dipterous parasitoids.

In parasitic Hymenoptera, developmental arrests of one sort or another are recorded in every stage of development including eggs within the parental female body, all developmental instars, and also in the adults of a few species (Doutt, 1959).

Parasitoids may enter diapause (Schoonhoven, 1962) autonomously, or the host's physiology may determine the duration of the life cycle of the parasitoids. *Apanteles glomeratus* (L.) is able to attack and oviposit in various larval instars of *Pieris brassicae* L., and may even attack the same individual host repeatedly through several instars. The parasitoid eggs hatch but development of the resulting larvae is arrested for varying periods while the host continues to feed and grow (Fig. 4). Thus the host's physiology, evidently through the composition of its hemolymph (Corbet, 1968), determines the duration of diapause (or perhaps quiescence) of the first instar larva. Shortly before the host's pupal molt, all the parasitoids simultaneously resume development, and kill and leave the host for pupation before the latter reaches the pupal stage. At this point, depending on external factors such as photoperiod and temperature experienced by the larvae while still in the host, the parasitoid pupae may diapause. Though the responses of host and parasitoid to these critical environmental factors differ slightly, it is clear that the system generates synchrony: arrested larval development of the parasitoids produces adults when suitable hosts are available, while factors inducing diapause in the parasitoid pupae also normally do so in unparasitized host pupae, leading to availability of hosts when needed by the following generation of parasitoids.

Working with the ichneumonid, *Diplazon fissorius* Grav., an internal larval-pupal parasitoid of hover flies (Syrphidae), Schneider (1948, 1951) illustrated the plasticity of response of the parasitoid to influences experienced within host syrphids of different species. The parasitoid is able to match its life cycle to univoltine or multivoltine hosts, to hosts which estivate, and to those which exhibit a prolonged larval diapause.

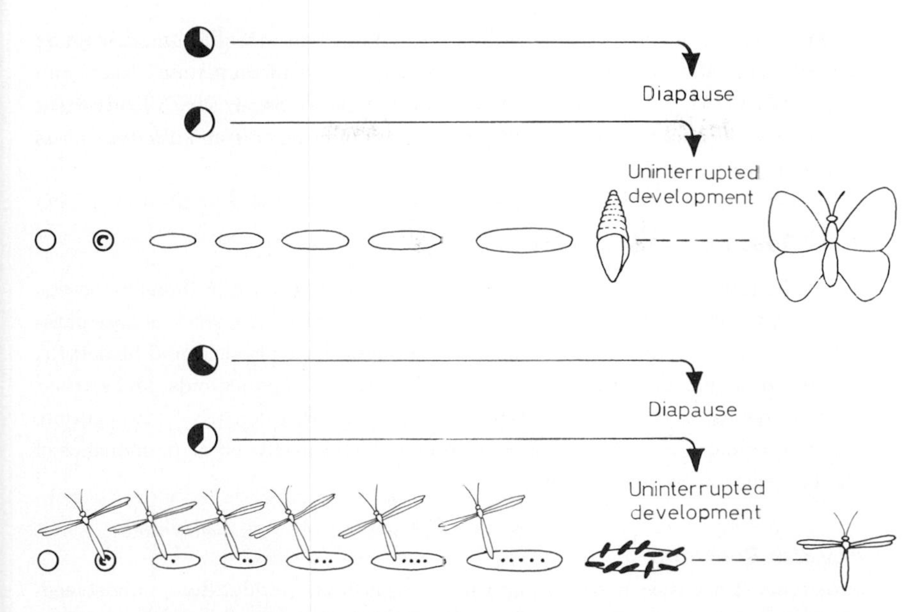

Fig. 4. Schematic representation of the synchronization of *Pieris brassicae* L. and *Apanteles glomeratus* L. by means of the parasitoid's larval arrest, the duration of which is determined within the host caterpillar; the pupal diapause which is controlled by the external environment. (Reproduced from Schoonhoven, 1962.)

3. *Alternate Hosts*

Parasitoids are restricted in the number of host species they exploit in nature. Published lists of the hosts of parasitic Hymenoptera (Peck, 1963; Shenefelt, 1969, 1970; Muesebeck *et al.*, 1951) give frequent cases of a single recorded host, more commonly up to about 10 hosts, and occasionally up to 50—100, and even more, for a single species of parasitoid. Despite apparently perfect attunement to a host species, probably relatively few parasitoids will prove to be rigorously restricted in their natural range of distribution to a single host.

The range of hosts used in natural populations by a parasitoid is generally narrow by comparison with that which will be accepted when the barriers of time and space are removed, as in laboratory cultures. This attribute of parasitoids has been repeatedly exploited in biological pest control operations in which large numbers of parasitoids have been produced under artificial conditions in or on factitious hosts (Doutt, 1959; Finney and Fisher, 1964).

Alternate hosts are not ordinarily of critical importance in maintaining host—parasitoid synchrony in biological control projects that have led to suppression of introduced pests. However, agricultural changes in the environment may, as in the case of the grape leafhopper and its mymarid egg parasitoid in California, disrupt an ancient and critically important synchrony between host and parasitoid (Doutt and Nakata, 1965; Doutt, 1965). This occurred by establishing vineyards with inadvertent disregard for the accessibility of the egg parasitoid's winter refuges and alternate hosts in the native vegetation.

D. Host Reaction to Parasitoids

Bearing in mind the great diversity of host—parasitoid relationships found in insects, it is not surprising that hosts have evolved a variety of reactions which oppose parasitization. These reactions may be conveniently classed as physical and hemocytic. Under certain circumstances they may limit multiplication of parasitoids, and so their relevance to biological control is unquestionable. Moreover, interpreted with caution, they may illuminate the nature, degree of intimacy, and likelihood of permanence of specific host—parasitoid relationships.

1. Physical Reactions

Parasitoids attack their hosts for purposes of oviposition, paralyzation, or host feeding. Many evoke little or no observable reaction in their hosts, such as those attacking eggs and other more or less immobile stages of the host. Hosts exhibiting maternal care are exceptional, such as certain mantids (Faure, 1940), which straddle their egg packages for long periods and repulse approaching egg parasitoids, and certain solitary bees (Nixon, 1954), which attack Chrysidid parasitoids threatening their young.

Many parasitoids of Homoptera induce at most a slight reflex action at the act of penetration of the ovipositor, as in aphidiine Braconidae, which oviposit rapidly without paralyzing the aphid host. Lepidopterous larvae often react more violently, making obvious attempts to elude or repel the parasitoid. Some noctuid caterpillars (*Plusia* species) strike out at attacking parasitoids with their heads, and even attempt to drown them with an excessive frothy secretion at the mouth (Malyshev, 1968). In South Africa *Plusia acuta* (Walker) behaves similarly (Neser, 1973), and also, after oviposition by an *Euplectrus* species (Eulophidae) and after paralysis has worn off, attempts to remove and destroy the externally placed eggs. This defensive host reaction seems to have so challenged the *Euplectrus* that there is a strong tendency for eggs to be laid where they are inaccessible to the host's mouth.

2. Hemocytic Reactions

The hemocytic defense mechanisms of insects to foreign bodies and particles of many kinds, and the means employed by parasitoids to resist them, are reviewed in detail by Salt (1963, 1968, 1970). Salt reserves the term phagocytosis, traditionally too loosely used by entomologists, for the action of a phagocyte which "engulfs particles by drawing them into its cytoplasm, as an amoeba takes in food" (Salt, 1970, p. 2).

a. Encapsulation and Melanization. The related hemocytic process by which hosts normally oppose parasitoid eggs or larvae in the hemocoel is termed encapsulation, where "a number of blood cells surround the object and adhere to its surface" (Salt, 1970). This aggregation of hemocytes leads (Salt, 1963) to the formation of relatively thick-walled cellular capsules of a range of types, or less commonly to thin-walled, tough, sheath capsules. In cellular capsules, melanin may sometimes be produced and deposited on the invasive parasitoid.

Successful encapsulation leads to the death of the parasitoid, or the latter is at least rendered harmless. However, the course and outcome of encapsulation is influenced by "the species, genetic strain, stage, instar, size, health, and physiological state of the host; and by the species, genetic strain, physical and physiological activity, and health of the parasite. Environmental temperature and the presence of other parasites of the same or different species" (Salt, 1963, p. 633) are also influential.

Most insects are capable of encapsulating internal parasitoids, and each species of host generally opposes all invasive parasitoids in this way except a few favored ones which are not so molested. The latter are usually habitually parasitic on the given host. Occasionally, however, parasitoids long and intimately associated with a given host are also faced with encapsulation. In these cases various mechanisms have evolved which nullify the host's inimical hemocytic reaction (Salt, 1968).

REFERENCES

Arthur, A. P. (1966). Associative learning in *Itoplectis conquisitor* (Say) (Hymenoptera: Ichneumonidae). *Can. Entomol.* **98**, 213–223.

Arthur, A. P. (1967). Influence of position and size of host shelter on host-searching by *Itoplectis conquisitor* (Hymenoptera: Ichneumonidae). *Can. Entomol.* **99**, 877–886.

Arthur, A. P., Hegdekar, B. M., and Rollins, L. (1969). Component of the host haemolymph that induces oviposition in a parasitic insect. *Nature (London)* **223**, 966–967.

Baccetti, B. (1958). Ghiandole labiali e fabbricazione del bozzolo negli Imenotteri. Richerche comparative su Ichneumonidi e Braconidi. *Redia* **43**, 215–294.

Beingolea, O. (1958). Dos estudios sobre accion de insecticidas sobre la fauna benefica del algodonero. *Estac. Exp. Agr. "La Molina" Min. Agr. Lima, Peru, Informe* **107**, 1–37.

Bronskill, J. F. (1959). Embryology of *Pimpla turionellae* (L.) (Hymenoptera: Ichneumonidae). *Can. J. Zool.* **37**, 655–688.

Broodryk, S. W. (1969). The biology of *Chelonus* (*Microchelonus*) *curvimaculatus* Cameron (Hymenoptera: Braconidae). *J. Entomol. Soc. S. Afr.* **32**, 169–189.

Broodryk, S. W., and Doutt, R. L. (1966). The biology of *Coccophagoides utilis* Doutt (Hymenoptera: Aphelinidae). *Hilgardia* **37**, 233–254.

Căpek, M. (1970). A new classification of the Braconidae (Hymenoptera) based on the cephalic structures of the final instar larva and biological evidence. *Can. Entomol.* **102**, 846–875.

Catling, H. D. (1969). The bionomics of the South African citrus psylla, *Trioza erytreae* (Del Guercio) (Homoptera: Psyllidae). 2. The influence of parasites and notes on the species involved. *J. Entomol. Soc. S. Afr.* **32**, 209–223.

Cendaña, S. M. (1937). Studies on the biology of *Coccophagus* (Hymenoptera), a genus parasitic on nondiaspine Coccidae. *Univ. Calif. Berkeley Publ. Entomol.* **6**, 337–399.

Cilliers, C. J. (1971). Observations on circular purple scale, *Chrysomphalus aonidum* (Linn.), and two introduced parasites in western Transvaal citrus orchards. *Entomophaga* **16**, 269–284.

Clark, L. R. (1962). The general biology of *Cardiaspina albitextura* (Psyllidae) and its abundance in relation to weather and parasitism. *Aust. J. Zool.* **10**, 537–586.

Clausen, C. P. (1940). "Entomophagous Insects," 688 pp. McGraw-Hill, New York.

Clausen, C. P. (1950). Respiratory adaptations in the immature stages of parasitic insects. *Arthropoda* **1**, 197–224.

Corbet, S. A. (1968). The influence of *Ephestia kuehniella* on the development of its parasite *Nemeritis canescens*. *J. Exp. Biol.* **48**, 291–304.

Daniel, D. M. (1932). *Macrocentrus ancylivorus* Rohwer, a polyembryonic braconid parasite of the oriental fruit moth. *N.Y. Agr. Exp. Sta. Tech. Bull.* **187**, 1–101.

Domenichini, G. (1966). "The Index of Entomophagous Insects. Hym. Eulophidae. Palearctic Tetrastchini," 101 pp. Le Francois, Paris.

Doutt, R. L. (1947). Polyembryony in *Copidosoma koehleri* Blanchard. *Amer. Natur.* **81**, 435–453.

Doutt, R. L. (1952). The teratoid larva of polyembryonic Encyrtidae (Hymenoptera). *Can. Entomol.* **84**, 247–250.

Doutt, R. L. (1957). Biology of *Solenotus begini* (Ashmead). *J. Econ. Entomol.* **50**, 373–374.

Doutt, R. L. (1959). The biology of parasitic Hymenoptera. *Annu. Rev. Entomol.* **4**, 161–182.

Doutt, R. L. (1964). Biological characteristics of entomophagous adults. *In* "Biological Control of Insect Pests and Weeds" (P. DeBach ed.), pp. 141–167. Chapman and Hall, London.

Doutt, R. L. (1965). The proper role of biological control in pest suppression. *In* "Research in Pesticides" (C. O. Chichester, ed.), pp. 257–264. Academic Press, New York.

Doutt, R. L., and DeBach, P. (1964). Some biological control concepts and questions. *In* "Biological Control of Insect Pests and Weeds" (P. DeBach, ed.), pp. 118–142. Chapman and Hall, London.

Doutt, R. L., and Nakata, J. (1965). Parasites for control of grape leafhopper. *Calif. Agr.* **19**, 3.

Faure, J. C. (1940). Maternal care displayed by mantids (Orthoptera). *J. Entomol. Soc. S. Afr.* **3**, 139–150.

Finney, G. L., and Fisher, T. W. (1964). Culture of entomophagous insects and their hosts. *In* "Biological Control of Insect Pests and Weeds" (P. DeBach, ed.), pp. 328–355. Chapman and Hall, London.

Fisher, K. (1932). *Agriotypus armatus* (Walk.) (Hymenoptera) and its relations with its host. *Proc. Zool. Soc. London.* pp. 451–461.

Fisher, R. C. (1961). A study in insect multiparasitism II. The mechanism and control of competition for possession of the host. *J. Exp. Biol.* **38**, 605–628.

Fisher, R. C. (1963). Oxygen requirements and the physiological suppression of supernumerary insect parasitoids. *J. Exp. Biol.* **40**, 531–540.

Fisher, R. C. (1971). Aspects of the physiology of endoparasitic Hymenoptera. *Biol. Rev.* **46**, 243–278.

Fisher, R. C., and Ganesalingam, V. K. (1970). Changes in the composition of host haemolymph after attack by an insect parasitoid. *Nature (London)* **227**, 191–192.

Flanders, S. E. (1925). *Prospaltella* as an egg parasite of the codling moth. *Pan-Pac. Entomol.* **2**, 188–189.

Flanders, S. E. (1937). Ovipositional instincts and developmental sex differences in the genus *Coccophagus*. *Univ. Calif. Berkeley Publ. Entomol.* **6**, 401–422.

Flanders, S. E. (1938). Cocoon formation in endoparasitic chalcidoids. *Ann. Entomol. Soc. Amer.* **31**, 167–180.

Flanders, S. E. (1946) Control of sex and sex-limited polymorphism in the Hymenoptera. *Quart. Rev. Biol.* **21**, 135–143.

Flanders, S. E. (1959). Differential host relations of the sexes in parasitic Hymenoptera. *Entomol. Exp. Appl.* **2**, 125–142.

Flanders, S. E. (1964). Dual ontogeny of the male *Coccophagus gurneyi* Comp. (Hymenoptera: Aphelinidae): A phenotypic phenomenon. *Nature (London)* **204**, 944–946.

Flanders, S. E. (1965). On the sexuality and sex ratios of hymenopterous populations. *Amer. Natur.* **99**, 489–494.

Flanders, S. E. (1967). Deviate-ontogenies in the aphelinid male [Hymenoptera] associated with the ovipositional behavior of the parental female. *Entomophaga* **12**, 415–427.

Flanders, S. E. (1969). Herbert D. Smith's observations on citrus blackfly parasites in India and Mexico and the correlated circumstances. *Can. Entomol.* **101**, 467–480.

Flanders, S. E., Bartlett, B. R., and Fisher, T. W. (1961). *Coccophagus basalis* (Hymenoptera: Aphelinidae). Its introduction into California with studies of its biology. *Ann. Entomol. Soc. Amer.* **54**, 227–236.

Gerling, D. (1966). Studies with whitefly parasites of southern California I. *Encarsia pergandiella* Howard (Hymenoptera: Aphelinidae). *Can. Entomol.* **98**, 707–724.

Goidanich, A. (1956). Sui concetti contrapposti di plesiotropismo e di interattrazione specifica nelle associazioni omogenee di alcuni imenotteri. *Men. Soc. Entomol. Ital. Genova* **35**, 183–224.

Grandi, G.(1959). The problems of "morphological adaptation" in insects. *Smithson. Misc. Collect.* **137**, 203–230.

Günther, J. (1971). Entwicklungfahigkeit, Geschlechtsverhaltnis und Fertilitat von *Pimpla turionellae* L. mach Rontgenbestrahlung oder Abschnurung des Eihinterpols. *Zool. Jabrh. Anat.* **88**, 1–46.

Hagen, K. S. (1964). Developmental stages of parasites. *In* "Biological Control of Insect Pests and Weeds" (P. DeBach, ed.), pp. 168–246. Chapman and Hall, London.

Hagen, K. S., and van den Bosch, R. (1968). Impact of pathogens, parasites, and predators on aphids. *Annu. Rev. Entomol.* **13**, 325–384.

Herrebout, W. M. (1969). Some aspects of host selection in *Eucarcelia rutilla* Vill. (Diptera: Tachinidae). *Neth. J. Zool.* **19**, 1–104.

Herting, B. (1960). Biologie der westpaläarktischen Raupenfliegen Dipt., Tachinidae. *Monogr. Angew. Entomol.* **16**, 1–188.

Ivanova-Kazas, O. M. (1961). "Studies on the Comparative Embryology of the Hymenoptera," 266 pp. Univ. Leningrad Press, Leningrad (in Russ.).

Johannsen, O. A., and Butt, F. H. (1941). "Embryology of Insects and Myriapods," 462 pp. McGraw-Hill, New York.

Kitano, H. (1965). Studies on the origin of giant cells. *Zool. Mag. Tokyo* **74**, 192–197.

Koscielska, M. K. (1962). Investigation upon trophamnion *Ageniaspis fuscicollis* Dalm. (Chalcidoidea: Hymenoptera). *Stud. Soc. Sci. Torun.* (Zool.) **6**, 1–9.

Lees, A. D. (1955). "The Physiology of Diapause in Arthropods," 150 pp. Cambridge Univ. Press, Cambridge.

Leiby, R. W. (1926). The origin of mixed broods in polyembryonic Hymenoptera. *Ann. Entomol. Soc. Amer.* **19**, 290–299.

Malyshev, S. I. (1968). "Genesis of the Hymenoptera and the Phases of their Evolution," 319 pp. Methuen, London (trans. from Russian, O. W. Richards and B. Uvarov, eds.).

Mellini, E. (1965). L'Imbuto respiratorio negli ospiti dei Ditteri Larvevoridi. *Atti Accad. Naz. Ital. Entomol.* **12**, 47–62.

Mellini, E., and Callegarini, C. (1968). Analisi elettroforetica delle emoproteine delle larve di *Chrysomela herbacea* Duft. (Col. Chrysomelidae) parassitizzate da *Meigenia mutabilis* Fall. *Boll. 1st Entomol. Univ. Bologna* **29**, 49–60.

Monteith, L. G. (1956). Influence of host movement on selection of hosts by *Drino bohemica* Mesn. (Diptera: Tachinidae) as determined in an olfactometer. *Can. Entomol.* **88**, 583–586.

Monteith, L. G. (1963). Habituation and associative learning in *Drino bohemica* Mesn. (Diptera: Tachinidae). *Can. Entomol.* **95**, 418–426.

Muesebeck, C. F. W., and Dohanian, S. M. (1927). A study of hyperparasitism, with particular reference to the parasites of *Apanteles melanoscelus* (Ratzeburg). *U.S. Dept. Agr. Bull.*, **1487**, 36 pp.

Muesebeck, C. F. W., Krombein, K. V., and Townes, H. K. (1951). Hymenoptera of America North of Mexico—Synoptic Catalog. *U.S. Dept. Agr. Monogr.* **2**, 1043 pp.

Neser, S. (1973). Biology and behaviour of *Euplectrus* species near *laphygmae* Ferriere (Hymenoptera: Eulophidae). *Entomol. Mem. Dep. Agr. Tech. Serv. Repub. S. Afr.* **32**, 1–33.

Nishida, T. (1956). An experimental study of the ovipositional behavior of *Opius fletcheri* Silvestri (Hymenoptera: Braconidae), a parasite of the melon fly. *Proc. Hawaii. Entomol. Soc.* **16**, 126–134.

Nixon, G. E. J. (1954). "The World of Bees," 214 pp. Hutchinson, London.

Noble, N. S. (1938). *Euplectrus agaristae* Craw., a parasite of the grape vine moth (*Phalaenoides glycine* Lew.). *N. S. W. Dept. Agr. Bull.* **63**, 27 pp.

Peck, O. (1963). A catalogue of the Nearctic Chalcidoidea (Insecta: Hymenoptera). *Can. Entomol. Suppl.* **30**, 1092 pp.

Prinicipi, M. M. (1958). Ricerche di morfologia e di etologia su di un Dittero Cicidomiide galligeno, la "*Putoniella marsupialis*" F. Loew, vivente su piante del gen. "Prunus". *Boll. 1st Entomol. Univ. Bologna* **23**, 35–68.

Saakjan-Baranova, A. A. (1965). Functional peculiarities of the secreting organs of the parasites of the Soft Scale (*Coccus hesperidum* L.). *In* "Ecology of Pest and Entomophagous Insects." *Zool. Inst. Acad. Sci. SSSR* **36**, 96–111 (In Russ.).

Salt, G. (1934). Experimental studies in insect parasitism. II. Superparasitism. *Proc. Roy. Soc. London* **B114**, 455–476.

Salt, G. (1935). Experimental studies in insect parasitism. III. Host selection. *Proc. Roy. Soc. London* **B117**, 413–435.

Salt, G. (1937). The sense used by Trichogramma to distinguish between parasitized and unparasitized hosts. *Proc. Roy. Soc. London* **B122**, 57–75.

Salt, G. (1961). Competition among insect parasitoids. *Symp. Soc. Exp. Biol.* **15**, 96–119.

Salt, G. (1963). The defence reactions of insects to metazoan parasites. *Parasitology* **53**, 527–642.

Salt, G. (1968). The resistance of insect parasitoids to the defence reactions of their hosts. *Biol. Rev.* **43**, 200–232.

Salt, G. (1970). "The Cellular Defence Reactions of Insects," 118 pp. Cambridge Univ. Press, Cambridge.

Schneider, F. (1948). Beitrag zur Kenntnisder generationsverhältnisse und diapause räuberischer schwebfliegen (Syrphidae: Dipt.). *Mitt. Schweiz. Entomol. Ges.* **23**, 155–194.

Schneider, F. (1951). Einige physiologische beziehungen swischen syrphidenlarven und ihren parasiten. *Z. Angew. Entomol.* **33**, 150–162.

Schoonhoven, L. M. (1962). Synchronization of a parasite/host system, with special reference to diapause. *Ann. Appl. Biol.* **50**, 617–621.

Shenefelt, R. D. (1969). Braconidae 2. *In* "Hymenopterorum Catalogus" (C. Ferriere and J. van der Vecht, eds.), pp. 1–175. Dr. W. Junk, Holland.

Shenefelt, R. D. (1970). Braconidae 1. *In* "Hymenopterorum Catalogus" (C. Ferriere and J. van der Vecht, eds.), pp. 177–305. Dr. W. Junk, Holland.

Short, J. R. T. (1970). On the classification of the final instar larvae of the Ichneumonidae (Hymenoptera). *Trans. Roy. Entomol. Soc. London Suppl.* **122**, 185–210.

Silvestri, F. (1905). Un nuovo interessantissimo caso di germinogronia (poliembrionia specifica) in um imenottero parassita endofogo con particulare destino dei globuli polari e dimorfismo larvale. *Regia Accad. Lincei Sci. Fis. Mat. Natur.* **14**, 534–542.

Silvestri, F. (1906). Contribuzioni alla conosenza biologica degli Imenotteri parassiti. I. Biologia del *Litomastix truncatellus* (Dalm.). *Ann. Regia Scu. Super. Agr. Portici* **6**, 3–51.

Simmonds, F. J. (1946). A factor affecting diapause in hymenopterous parasites. *Bull. Entomol. Res.* **37**, 95–97.

Simmonds, F. J. (1947). Some factors influencing diapause. *Can. Entomol.* **79**, 226–232.

Simmonds, F. J. (1948). The influence of maternal physiology on the incidence of diapause. *Phil. Trans. Roy. Soc. London* **B233**, 385–414.

Sluss, R. (1968). Behavioral and anatomical responses of the convergent lady beetle to parasitism by *Perilitus coccinellae* (Schrank) (Hymenoptera: Braconidae). *J. Invertebr. Pathol.* **10**, 9–27.

Sluss, R., and Leutenegger, R. (1968). The fine structure of the trophic cells of *Perilitus coccinellae* (Hymenoptera: Braconidae), *J. Ultrastr. Res.* **25**, 441–451.

Smith, H. S. (1916). An attempt to redefine the host relationships exhibited by entomophagous insects. *J. Econ. Entomol.* **9**, 477–486.

Tawfik, M. F. S. (1957). Alkaline phosphatase in the germ-cell determinant of the egg of *Apanteles*. *J. Insect Physiol.* **1**, 286–291.

Thorpe, W. H., and Caudle, H. B. (1938). A study of the olfactory responses of insect parasites to the food plant of their host. *Parasitology* **30**, 523–528.

Tooke, F. G. C. (1953). The eucalyptus snout-beetle, *Gonipterus scutellatus* Gyll. A study of its biology and control by biological means. *Union S. Afr. Dept. Agr. Entomol. Mem. No.* **3**, 282 pp.

Tothill, J. D., Taylor, T. H. C., and Paine, R. W. (1930). The coconut moth in Fiji, a history of its control by means of parasites. *Publ. Imp. Bur. Entomol. London*, 269 pp.

Tremblay, E. (1964). Ricerche sugli imenotteri parassiti, I. Studio morfobiologico sul *Lysiphlebus fabarum* (Marshall) (Hymenoptera: Braconidae: Aphidiinae). *Bull. Lab. Entomol. Agr. Portici* **22**, 1–122.

Tremblay, E. (1966). Ricerche sugli imenotteri parassiti. II. Osservazioni sull'origine sul destino dell'involucro embrionale degli Afidiini (Hymenoptera: Braconidae: Aphidiinae) e considerazioni sul significato generale delle membrane embrionali. *Bull. Lab. Entomol. Agr. Portici* **24**, 119–166.

Tremblay, E. and Calvert, D. (1972). New cases of polar nuclei utilization in insects. *Ann. Soc. Entomol. Fr.* **8**, 495–498.

Ullyett, G. C. (1953). Biomathematics and insect population problems. *Mem. Entomol. Soc. S. Afr. No.* **2**, 89 pp.

Viggiani, G. (1963). Contributi alla conoscenza degli insetti fitofagi minatori e loro simbionti. III. Reperti etologici sulla *Lithocolletes blancardella* F. in campania e studio morfo—biologico dei suoi entomoparassiti. *Bull. Lab. Entomol. Agr. Portici* **22**, 1–62.

Viggiani, G. (1964). La specializzazione entomoparassitica in *Alcuni eulofide* (Hymenoptera: Chalcidoidea). *Entomophaga* **9**, 111–118.

Vinson, S. B. (1970). Development and possible functions of teratocytes in the host-parasite association. *J. Invertebr. Pathol.* **16**, 93–101.

Way, M. J. (1962). Definition of diapause. *Ann. Appl. Biol.* **50**, 595–596.

Wilson, F., and Woolcock, L. T. (1960). Environmental determination of sex in a parthenogenetic parasite. *Nature (London)* **186**, 99–100.

Zinna, G. (1959). Ricerche sugli insetti entomofagi. I. Specializzazione entomoparassitica negli Encyrtidae: Studio morfologico, etologico e fisiologico del *Leptomastix dactylopii* Howard. *Bull. Lab. Entomol. Agr. Portici* **18**, 1–148.

Zinna, G. (1961). Ricerche sugli insetti entomofagi. II. Specializzazione entomoparassitica negli Aphelinidae: Studio morfologico, etologico e fisiologico del *Coccophagus bivittatus* Compere, nuovo parassita del *Coccus hesperidum* L. per l'Italia. *Bull. Lab. Entomol. Agr. Portici* **19**, 301–357.

Zinna, G. (1962). Ricerche sugli insetti entomofagi. III. Specializzazione entomoparassitica negli Aphelinidae: Interdipendenze biocenotiche tra due specie associate. Studio morfologico, etologico a fisiologico del *Coccophagoides similis* (Masi) e *Azotus matritensis* Mercet. *Bull. Lab. Entomol. Agr. Portici* **20**, 73–184.

7

HOST RELATIONSHIPS AND UTILITY OF PATHOGENS

J. Weiser, G. E. Bucher, and G. O. Poinar, Jr.

I. INTRODUCTION

The role of pathogens in the natural and induced control of insects and mites is well known from the early history of applied entomology in agriculture, forestry, and medicine. There are many field observations of efficient pest control arising from disease outbreaks. However, there are only a few data on the occurrence of pathogens in normal populations and these include examples of diseases which do not necessarily kill the host. There is now an increasing series of reports about the application of pathogens for pest control. There is a striking difference between good natural pathogens and good biological preparations: for example, while *Bacillus thuringiensis* Berliner is one of the most effective pathogens known to man, its occurrence in nature is rather rare, and while entomophthorous fungi cause extensive epizootics in nature, their practical use on cultivated plants does not induce a high incidence of disease. Only in a few instances, such as with viruses of sawflies, which in one case resulted from the innoculation of a minute amount of pathogen into an intact population, do widespread epizootics result from introductions of microbial material. These examples suggest that, undoubtedly, many pathogens occurring in nature can be used to serve man if handled properly.

Several factors can explain the different behavior of arthropod pathogens occurring naturally and those used in artificial dissemination. These factors are the preciseness of conditions necessary for spread of the disease, successful infection, and production of viable material. This chapter on host relations and utility of pathogens in biological control covers only certain general aspects. Since different groups of pathogens act in different ways, they are treated in the various chapters relating to use of pathogens.

II. VIRUS INFECTIONS

Insect viruses are known to be host-specific, most of them infecting only one host species or one insect group. They are considered density-dependent reducing factors although the responsiveness to host density may be also dependent upon concurrence of a number of other environmental and host conditions. They are found in most environments, including both aquatic and above and below ground terrestrial habitats. Viruses come into contact with their hosts in water or on contaminated leaves; but they can also be transmitted on ovipositors of parasites or via the eggs of infected females. Most viruses are able to produce latent infections which, after some stress, may suddenly manifest themselves as epizootics. A large supply of virus particles is produced in infected insects (reproduction rates up to 10^5). The polyhedral viruses are protected as sets of particles surrounded by common protein envelopes. The spread of certain viruses, the polyhedral viruses, for example, is facilitated by the protection provided to viral particles by these protein envelopes, by the responses of infected hosts which produce watery, diarrheic feces containing masses of polyhedra, and by the movement of infected hosts ("Wipfeln") to the tops of plants after which their bodies break open and allow polyhedra-containing hemolymph to drip down onto underlying foliage.

Viruses may not destroy a whole host population since there is often a different virulence for different hosts or subpopulations. The most virulent viruses often disappear after they kill all available hosts. After an outbreak, polyhedroses of the gypsy moth or nun moth as well as gut polyhedroses of sawflies, enter a long period of latency in the insects when they are not observed in host populations. In other viruses (iridescent virus of mosquitoes, spindle-shaped viruses, etc.) the disease is commonly encountered, but at very low infection levels under normal conditions.

Host specificity varies from group to group in the viruses. Nuclear polyhedroses from the gut of sawflies, granuloses of Lepidoptera, and the densonucleosis of the wax moth have been shown to be species-specific. The situation is not clear with the polyhedroses of the fat body of Lepidoptera, where direct feeding of the virus does not produce overt infection in every case and the lack of reaction in some cross-infections is also unclear. In other groups of viruses there is lower specificity; cytoplasmic polyhedroses (CPV) from the gut of Lepidoptera are transmitted to other closely related insects. Neilson (1964) was able to introduce the CPV from *Vanessa cardui* (L.) to 11 additional hosts. Similarly, *Tipula* iridescent virus (TIV) was infectious for 20 host species and, while some of them were Diptera, many were Lepidoptera and Coleoptera.

After administration of an adequate dose, most viruses produce symptoms in 8–10 days. If primary infection does not result, later appearances are quite unlikely or depend on some environmental stress.

With a higher frequency of transmissions, the virulence of the virus for a selected host changes to some degree. Different field populations of *Pieris brassicae* L. varied in susceptibility to a granulosis virus (Rivers, 1958) and Smirnoff (1961) demonstrated that subsequent passages increased the mortality of *Neodiprion swainei* Middl. from a gut polyhedrosis from 39.9 to 88% with the same dosage. The first eight transfers are essential for the adaptation of the virus to a given host.

Further accentuation of virus infections may be caused by stressors. In some cases, polyhedral proteins from other hosts are able to incite acute infections of a latent virus in the host (Tanada, 1959, 1971). A very efficient stressor of latent virus infections is the long storage of diapausing eggs (polyhedrosis of the silkworm, gypsy moth, and nun moth, granulosis of the fall webworm). In double infections with microsporidia, viruses appear first and cause high mortality in a shorter time than in single infection. (Weiser, 1966).

For polyhedral viruses the infective dose may vary from 5000 to 50,000 polyhedra. The lower the dose, the higher the risk that the resulting infection will remain latent. In production experiments (Ignoffo and Hink, 1971), the output from dead larvae was 1 to 150 $\times$ 10^8; in other cases, 4 $\times$ 10^3 or 10^4 times the initial doses. The same results were produced with gut polyhedroses of sawflies. In cytoplasmic gut polyhedroses, outputs are usually much higher than indicated above, i.e., 10^{10} and more in one animal and similar results were obtained with granuloses of *Hyphantria cunea* Drury and *Choristoneura murinana* (Hubn.) (Weiser, 1966).

Several viruses have been used in field trials, mostly with good results. The inoculum for such applications was produced from larvae in insectaries (see Ignoffo and Hink, 1971) or from field-collected material (Steinhaus and Thompson, 1949). Methods using tissue culture for production of polyhedra are studied. Virus inoculum is resistant when stored in the refrigerator as crude cadavers or dry cleaned polyhedral powder. Polyhedroses preparations survive usually for 4 to 5 years, granuloses for 2 to 3 years, and cytoplasmic polyhedroses for 1 year. For application, the inoculum is mixed with water and industrial spreaders and stickers (Angus and Luthy, 1971). Polyhedral suspensions in oil are used as aerosols, and watery suspensions in sprays were most efficient when used against the first instars of susceptible hosts. Treatments can be applied over either wide acreages or limited areas from which develop secondary centers of infection. Introductions of viruses by the release of infected animals and entomophages are also efficient, such as with *Neodiprion sertifer* Geof. and *Gilpinia hercyniae* Hartig in

Canada. Direct sunshine and phytotoxins from some plants can damage the viruses after their application.

III. BACTERIAL INFECTIONS

Bacteria are found in all dead insects, but only in a relatively few instances are these bacteria the primary cause of the mortality. They may cause transient nonlethal infections, but these generally go unrecognized and only a few lethal cases may be noticed. Pathogenic bacteria bring about diseases in healthy insects. With deliberate administration of inoculum usually the percentage of affected insects increases with an increase in the dosage. The resulting data can be interpreted by plotting the effect in probits against the log dose. The resulting linear regressions are characterized by low slopes, usually between 0.5 and 2.0, which contrasts strongly with probit regressions for insects treated with insecticides. The low slopes may indicate that populations of insects are very heterogenous in regards to bacterial infections. In practical terms, this means that if a dose X kills 50% of the test insects, a dose of 20 X to 1000 X may be necessary to kill 99%.

Bacteria, in general, seem unable to penetrate the intact exoskeleton of a susceptible insect host. They enter the insect with its food, remaining concealed in the peritrophic membrane of the gut. When they enter the body (through wounds) they cause a general septicemia and do not concentrate in any tissue. As the cellular and humoral immunity mechanisms of insects are organized on different principles than those of mammals (Stephens, 1963), the main defenses of insects against infection are: (1) to prevent penetration into the hemocoel by an intact chitinous covering, (2) to rapidly eleminate the small number of bacteria that may break this armor, and (3) to provide conditions in the blood unfavorable to the multiplication of bacteria. There is still insufficient evidence of the latter type of defense.

Since the taxonomic position of bacterial pathogens bears little relationship to the way they cause insect disease, they are arranged in groups based on their pathogenic properties.

A. Obligate Pathogens

Obligate pathogens require special conditions for multiplication, so they occur in nature only in specific insects where they cause specific diseases. They have narrow host ranges, are difficult to culture "*in vitro*," and are able to invade susceptible host tissue without the help of extraneous factors such as wounds, or physical or chemical damage (Weiser, 1966). *Bacillus popilliae* Dutky and similar forms cause milky disease in larvae of Scarabaeidae where they multiply and sporulate in the blood. The spores are ingested by the larvae in the soil, but it is not clear how the bacteria cross the gut wall. *Bacillus larvae* White causes American foulbrood in honeybees. When young bee larvae ingest spores of this bacterium before their peritrophic membrane is completed, the spores germinate and grow between the membrane and the gut epithelium, damaging the cells and invading the hemocoel. *Clostridium brevifaciens* Bucher and similar forms cause brachytosis by growing and sporulating in the gut of tent caterpillar larvae, for example; they never invade the hemocoel. *Streptococcus pluton* (White) causes European foulbrood by growing in the gut of larval bees. There is no evidence of a toxin present in this disease.

The obligate pathogens have limited ecological niches closely associated with their hosts. They possess resistant stages (spores and resistant vegetative stages in *S. pluton*) to help them survive in the absence of hosts. The insect gut is anaerobic and frequently alkaline. The obligate pathogens are either strict anaerobes (*Clostridium*) or tolerant of anaerobic conditions and high pH. The obligate pathogens do not totally destroy host populations unless applied in large dosages since they usually occur only at one period in a host's life (Bailey, 1963), and some hosts usually are also resistant.

B. Potential Pathogens

Potential pathogens do not require special conditions or hosts for infection and are ubiquitous in soil, water, foods, and the guts of insects. They do not multiply well in the gut of healthy insects and thus do not produce sufficient enzymes or toxins to damage the gut wall and cross this barrier. They cross the wall when the insect is stressed, for example, by temporary asphyxia, stoppage of peristalsis, intoxication, or damage due to activity of another pathogen. Once they enter the hemocoel, they multiply and produce a septicemia. Death may result from experimental injection of only 10 to 20 bacterial cells, and less than 1000 cells are usually enough (Bucher, 1960). The most common ports of entry are wounds resulting from the bite of predators, of a sibling or other means. Many insects are cannibalistic and will attack each other if crowded or deprived of food or water. Damage to the midgut cells from gregarines, microsporidia, and viruses provide the entry for this group of bacteria in the same way as do abrasive components in the food (Bucher, 1963).

The potential pathogens are taxonomically diverse and include *Pseudomonas aeruginosa* (Schroeter), *Enterobacter aerogenes* (Kruse), *Serratia marcescens* Bizio, *Proteus morganii* (Winslow *et al.*) Rauss. or *Bacillus cereus* Frankland and Frankland. *Achromobacter nematophilus* Poinar and Thomas is carried in the gut of the DD-136 and agriotos strains of the nematode *Neoaplectana carpocapsae* Weiser (Poinar, 1966). The ensheathed nematode larvae enter the body cavity of insects, and release the bacteria that then grow and produce a lethal septicemia. The nemas feed on the bacteria and breakdown products of the host. The bacteria are then transported to new hosts by the third stage nematodes. In a sense, *A. nematophilus* is an obligate pathogen, requiring a very special host—parasite relationship although it can be grown on artificial media. *Steinernema krausei* in *Cephaleia abietis* has a symbiotic Flavobacter transported in the same way.

Potential pathogens produce proteolytic enzymes causing tissue destruction. Specific toxic proteases have been isolated from *Pseudomonas aeruginosa* (Lysenko, 1967) which cause enzymatic damage to cells and tissues resulting in death of the insect.

In contrast to the obligate pathogens, the potential pathogens are aerobic bacteria, more tolerant of acidic pH, and less of alkaline pH. These intolerances may be reasons for their failure to multiply extensively in the normal insect gut.

Some potential pathogens, such as *Serratia marcescens* Bizio, *Streptococcus faecalis* Andrews, and Horder or *Cloaca* spp. are able to produce infections in insects under special conditions of constant association. The first invades the hemolymph when respiration of the host is affected, with decrease of current in running waters, and when the food plant is wilted. *Bacillus thuringiensis* which differs from *B. cereus* in producing a parasporal crystal in its sporangium at sporulation, is a specific case (Rogoff and Yousten, 1969). The crystal toxin is in fact a protoxin activated when ingested by the gut juices of Lepidoptera. Ingested parasporal crystals are toxic only to larvae of Lepidoptera where the pH of the intestine is high. As a family, the noctuids have a lower pH and are more resistant to the toxin. If sporulated cultures of *B. thuringiensis* are fed to insects, one of three principal results follow, depending on the susceptibility of the insect species and on the size of the dose: (1) the insects are poisoned by the toxic crystals, are rapidly paralyzed, show pathological changes in tissues, and may die before any true growth or infection by *B. thuringiensis* develops; (2) the insects show signs of poisoning (e.g., cessation of feeding) and deterioration of the midgut epithelium that allows penetration of bacteria into the blood and a lethal septicemia results with or without prior growth of bacteria in the gut; and (3) the insects are relatively undamaged by the crystal, in which case *B. thuringiensis* behaves like *B. cereus* and acts as a facultative or potential pathogen capable of producing lethal septicemia if the hemocoel is involved.

About 50 strains of *B. thuringiensis* have been isolated from insects and classified into 12 groups by esterase patterns and by serological and biochemical tests (de Barjac and Bonnefoi, 1962). The crystals of different serotypes and strains have different toxicities to various insects. After the host dies, *B. thuringiensis* may multiply in the hemocoel but frequently does not sporulate, in which case the dead insect presents little danger to its siblings. *Bacillus thuringiensis* is readily cultured but little growth occurs anaerobically and sporulation is strictly aerobic. Like *B. cereus*, it produces a variety of enzymes, an antibiotic, and different toxins. Some strains produce an exotoxin of nonspecific toxicity. Another spore-forming bacterium, *Bacillus sphaericus*, was tested recently as a pathogen for mosquito larvae.

C. Occurrence and Role in Nature

All bacteria directly incriminated as insect pathogens have been diagnosed in reared insects, i.e., insects under close scrutiny and monitoring of man. Bacterial epizootics in nature have, however, been reported under similar conditions (stored products, insects in colonies, freshwater populations under special physical or chemical influences). In other circumstances, bacterial epizootics seldom occur or are not recognized by observers. Little is known of the role of bacterial pathogens in the control of insect pests in nature in the absence of man's influences. Obtaining evidence of bacterial infections is sometimes difficult since insects dying from bacterial diseases rot quickly, with little evidence remaining in the host niche.

D. Commercial Usage

At present only two bacteria are commercially produced and registered for use in controlling insects in agriculture. One of these, *Bacillus popilliae* was artificially spread in areas of heavy populations of the Japanese beetle in the eastern United States (Beard, 1945; Dutky 1963). Large dosages of spores were added to soil areas arranged in a grid pattern. The disease spreads slowly under natural conditions and its artificial dissemination by man resulted in suppression of beetle populations throughout most of the insect's range. The bacterium does not sporulate readily in artificial media, and can only be produced in living larvae, which makes the production cost high and largely limits it use to colonizations in new infestations or to the treatment of particularly valuable areas. (The clostridial pathogens of tent caterpillars also fail to sporulate in culture and must be produced in living hosts at high cost.) High costs can be tolerable for introductions in new areas, if such introductions result in the permanent establishment and spread of the disease. It is prohibitive when large amounts of material are used repeatedly as with insecticides.

The second registered bacterium is *B. thuringiensis*, which acts as a specific insect poison and is not established in nature as a constant population-reducing agent. The function of *B. thuringiensis* is mainly to produce toxic crystals and the effort of man is to control that toxicity to his benefit. Different strains produce more or less specific toxins for a selection of insects. Since the crystal toxin is harmless to plants and vertebrates (Fisher and Rosener, 1959), it is safer and less polluting than chemical insecticides and will be more commonly used in the future. Others may be developed as well.

IV. FUNGUS INFECTIONS

The degree of specificity of entomogenous fungi is quite variable. There is the heterogeneous group of fungi adapted to more proteinaceous media and living only on insect hosts, which include the Entomophthoraceae, the Chytrids, Blastocladiales, Lagenidiales or Xylariales, and *Cordyceps*; and the less specific groups, living on insects as well as on other organic substances, which include the Deuteromycetes with the genera *Beauveria*, *Metarrhizium*, *Sorosporella*, *Cephalosporium*, and others. This difference in food requirements has to be respected in growing both groups on artificial media. The completion of the developmental cycle of the fungi requires certain conditions of moisture and temperature. Germination occurs in the soil, in galleries in plants, or in dense plant growth. The fungi enter the host mainly from the exterior after contacting the cuticle. The spores adhere to the host by surface adhesion, and penetration is possible only after the joint action of the enzymes proteinase, lipase, and chitinase, which are produced by the fungal conidia (Samšiňáková *et al.*, 1971). Transmission is mainly by conidia, but resting spores, blastospores, and hyphal bodies are also involved. In the lower fungi, zoospores enter the host with the food or they dissolve their way through the cuticule. The longevity of fungal spores varies from a few hours in zoospores to several months in blastospores and hyphal bodies or several years in resting spores (Müller-Kögler, 1965). Infections in insect populations depend upon the distribution of the spores in the environment and their contact with the target insect. Some general abiotic stressors (moisture, temperature, and food) are able to sensitize large insect populations to infection and massive outbreaks may suddenly occur. Especially common are those caused by Entomophthoraceae on *Operophtera brummata* L., *Panolis flammea* Schiff. on 150,000 ha of forests (Garbowski, 1927), on *Plusia gamma* L., or *Agrotis segetum* (L.). At other times, only local foci of infection occur around infected individuals in optimal niches (*Metarrhizium*, *Sorosporella*, *Cordyceps*).

Host group specificity occurs in *Entomophthora* infections [*E. muscae* (Cohn), *E. grylli* Fres., *E. destruens* Weiser, in *Tarichium* or *Massospora*], in Chytrids (*Coelomycidium* in blackflies and mosquitoes, *Myiophagus* in scale insects), and in *Cordyceps* fungi (Weiser, 1966). Specificity results from similar ecological conditions to that of the hosts (water habitat, identical food, or shelter). In other cases, infected animals transmit the pathogens during mating (*Strongwellsea*, Batko and Weiser, 1964, and *Massospora*), e.g., in *Melanosella mors apis* and in Laboulbeniales. In specific fungi, artificial cultures are most difficult to obtain while in nonspecific forms isolation on ordinary media is quite easy. For efficient transmission, *Coelomomyces* fungi need an intermediary host, such as a copepod or ostracod.

The adaptation to the specific host is mainly dependent on the method of transmission. *Entomophthora tetranychi* Weiser produces resting spores, besides normal conidia, which are spread with air currents and borne on the webbing of mites attached by a special narrow polar cap. The fungus *Massospora* splits open the last segments of the adult cicada and spores are brought to optimal infection counts by the flying adults. Conidia generally germinate when placed on a new host or nutrient medium. However, only 20% of the resting spores of *Entomophthora* germinate immediately under such conditions. The remaining 80% are viable but dormant and germinate only after a second or third stimulus.

Winter shelters are specific foci where fungal infections may spread. *Beauveria* spreads among hibernating lady beetles in dry grass, and the same may happen with *Eurygaster integriceps* Puton when the bugs hibernate under bark of trees or with overwintering codling moth larvae. *Entomophthora destruens* kills up to 80% of the hibernating populations of *Culex* in caves (Weiser, 1966).

There is little information concerning the amount of spores or conidia needed to infect a given insect. In infections with deuteromycetes, the dosages used were between 50,000 and 150,000 conidia. Veen (1968) found that 150,000 conidia of *Metarrhizium anisopliae* (Metch.) Sorokin were required for an LD_{50} dosage for *Schistocera gregaria* Forskäl. In Entomophthoraceae only a small number of conidia (200) are necessary.

Only in rare cases are fungi collected and stored for field application. *Coelomomyces* sporangia were collected in mud and introduced into new localities for experimental infections (Laird, 1971). *Myiophagus* on citrus scales was colonized in new localities by moving branches bearing infected scales. Mites infected with *Entomophthora floridana* Weiser and Muma were spread mixed in water by sprinkling water on uninfected mite populations. *Aschersonia* is widely colonized for control of citrus scales in the sub-caucasian region and other fungi have been used in control experiments with white grubs, the sugar beet curculio, the Colorado potato beetle, wheat bugs, and codling moths.

In most cases, fungi for field applications have been produced on artificial media (Samšiňáková, 1961; Weiser, 1966). One mass-produced preparation of *Beauveria* is Beauverin (U.S.S.R.), and methods are also known for *Metarrhizium*, *Sorosporella*, *Cephalosporium*, *Spicaria*, and others. Insecticides in 1/10 normal dosages are used as synergists in mixtures of Beauverin. Available fungus preparations are less resistant than bacteria to long periods of storage.

V. PROTOZOAN INFECTIONS

Besides being associated with protozoa in a phoretic or symbiotic manner, insects may serve as vectors of pathogenic protozoa or themselves be affected and killed by rather specific protozoan parasites. The protozoa which cause death in insects are typical intracellular parasites of tissues and can be grown outside their hosts only in tissue culture.

These pathogenic protozoa which invade tissues and cause high mortality are mainly schizogregarines, coccidia, and microsporidia. Most damage to the host is caused during the vegetative, schizogonial period of development when they destroy host tissues. Less damage occurs during spore formation (sporogony). The parasites are often discharged from the host in feces, when sporogony occurs in the gut, the Malpighian tubes, or in the silk glands. When localized in the fat body, the muscles, the hypoderm or hemocytes, they leave the host only after its death, in the feces of predatory insects, or, e.g., borne on the surface of the ovipositors of the parasitic Hymenoptera. In cases where the ovary is infected, the microsporidia may be transferred to the new generation in infected eggs. The spores are usually viable for at least 12 months in the dead insect or in moist soil, but they are killed by sunshine or drought within a few days when placed on plants (Weiser, 1966).

All these protozoa enter the host body with contaminated food, and the composition of the gut juices is one of the major factors governing host specificity. A second factor is the susceptibility of tissues to a specific pathogen. An overdose of infective spores may result in death of the host from a bacterial septicemia in 3–5 days (Weiser, 1961). Infections from low doses produce masses of spores at the end of the infection cycle. The severity of infection depends upon its localization in the tissues. In the fat body, the spores fill the whole tissue and death occurs at the time of pupation. Infections in the gut, Malpighian tubules, muscles, or general infections are in most cases acute and quickly kill the host. Infections of later instars result in spore formation in pupae and adults. High temperatures may reduce the development of the protozoa and favor the host. A decrease in metabolism during hibernation or diapause reduces the development of the pathogen and the disease continues only after the host resumes its normal metabolic activity.

Spores of the pathogen may be brought into contact with new hosts through infected feces and eggs (e.g., microsporidia and schizogregarines of bark beetles). Spores from infected silk glands are distributed on leaves. Examples of ovarial transmission are that of *Nosema bombycis* Nägeli in the silkworm and of *N. otiorrhynchi* Weiser or *N. mesnili* (Paillot) and other cases. Ovarial infection reduces the number of viable eggs and thus the following generation size, as shown for *N. stegomyiae* Fox and Weiser (1959). In the gut parasite *N. mesnili*, the muscles, fat body, and hypodermis of *Pieris brassicae* L., *P. rapae* (L.), *P. napi* (L.), *Aporia crataegi* (L.), and several other hosts, the ovaries are also infected, and the infection is transmitted in the eggs.

In some cases the ovarial transmission is very efficient and may be the sole method of transmission. The microsporidian *Parathelohania legeri* Hesse was found in one study to be present in about 2% of female mosquitoes and the eggs of these infected females produced progeny with both latent and overt infection. All larvae containing spores die before pupation. These larvae are male larvae and only female adults emerge. The microsporidian develops during pupation and the initial imaginal development. The eggs of infected individuals were infected with the microsporidian (Hazard and Weiser, 1968). Infected isolated colonies died out because no males were produced.

Parasitic Hymenoptera may serve as vectors (as well as hosts) of protozoan pathogens. *Mattesia dispora* Naville in the flour moth is transmitted by *Habrobracon* wasps. *Apanteles glomeratus* (L.), *Hyposoter eveninus* Gravenhorst, *Pimpla instigator* (Fab.), and several other Hymenoptera transmit spores of *N. mesnili* on their ovipositors, but are also subject to the infection themselves, thus creating secondary centers of spread for the disease (Weiser, 1966; Hostounský, 1970) by infected eggs.

Microsporidia, Coccidia, and Schizogregarina are most efficient in dense host populations where they cause considerable mortality, e.g., *Nosema lymantriae* Weiser in the gypsy moth, *N. bombycis* in the silkworm, and *N. otiorrhynchi* in snout beetles. *Nosema whitei* Weiser, *Farinocystis tribolii* Weiser, *Mattesia dispora*, and *Adelina tribolii* Bhatia cause reductions of pests of stored products.

Protozoan infections may render insect hosts more susceptible to insecticides (Weiser, 1961). The natural spread of protozoan infections in insect populations is sometimes dramatic. Introduction and application of Microsporidia, Schizogregarina, and Coccidia has been attempted in the field in several cases. In treatments of the cotton bollworm with a mixture of protozoans (McLaughlin, 1971), positive results were obtained.

A drawback to use of protozoa is that spore production is possible only in living hosts. However, after use of low dosages, spore production may reach 150,000 times the initial dosage for *Nosema plodiae* Kellen and Lindgren in just 18 days (Weiser and Hostounský, 1972). The first spores are usually formed 8 days after infection. Some Microsporidia are pathogenic in useful insects such as honeybees and silkworms. This could prejudice their use as microbial insecticides, or their introduction as new pathogenic agents in a country. However, this prejudice could be unwarranted as the different species are quite host-specific in nature and are promising for experimental use and for introduction.

VI. NEMATODE INFECTIONS

In dealing with entomogenous nematodes, one is presented with a diverse assemblage of nematode groups showing various relationships with insects. Thus the polyphyletic origin of insect parasitism by nematodes becomes quite clear. The facultative parasites can infect healthy insects, yet still retain their ability to reproduce and develop exogenously in the host's environment (Poinar, 1972). Welch (1963) and Poinar (1971) have recently summarized the results dealing with the host—parasite relationship and use of nematodes as biological control agents of insects in general.

Representatives of the Mermithidae, Steinernematidae, and Tylenchida are discussed here. Members of the Mermithidae obtain their nourishment within the body cavity of insects. Adult stages are found in the host environment and the infective juveniles or eggs initiate the parasitic cycle. The host usually dies after the nematode leaves it and this fact makes mermithids attractive as control agents. *Mermis subnigrescens* Cobb. and *Agamermis decaudata* C.S.Ch. attack grasshoppers in the northeastern United States (Christie, 1937) and *Filipjevimermis leipsandra* Poinar and Welch attacks *Diabrotica balteata* LeC. in the southern United States (Cuthburt, 1968). In Japan, 76.6% of a population ($n = 372$) of *Chilo suppressalis* Wlk. was parasitized by mermithids and this was considered an important factor in controlling the pest. Cutworm larvae in India were heavily attacked (92.8%) by mermithids, which held this insect in check during pest outbreaks (Khan and Hussain, 1964). Mermithids also attack aquatic insects and recent studies show their influence on populations of mosquitoes, midges, and blackflies. Of special interest is *Reesimermis nielseni* Tsai and Grundmann which is capable of infecting a large number of culicine species. This mermithid has been mass reared and the infective stage juveniles introduced into aquatic environments in California, Taiwan, and Thailand against target hosts (H. Chapman, personal communication). In some localities in Russia and the United States mermithids have been considered responsible for temporarily eliminating blackfly larvae.

Mermithid infections are generally spotty in occurrence. Thus, they would have to be artificially distributed throughout a pest area for effectiveness. This would require an efficient method of mass production and release. This has been one of the major drawbacks regarding the practical use of mermithids. Natural spread occurs by the movement of parasitized adult insects which are infected during the pupal or late larval stage.

Mermithids were originally thought to be host-specific, but many forms have been shown to infect insects belonging to two or more families. Potential insect hosts are often able to encapsulate and melanize juvenile mermithids. Some mermithids escape this fate by directly entering one of the nerve ganglia of the host (*Filipjevimermis leipsandra*, for instance) during invasion. In this manner, they escape the blood cells and later, after they grow out of the ganglia, they do not attract host hemocytes (Poinar, 1968).

The Steinernematidae containing the genus *Neoaplectana* constitute an important group for use in biological control. They are generally nonspecific in regards to host selection and very quickly bring about insect mortality. The bionomics of many species needs clarification. *Neoaplectana glaseri* Steiner was the first described and extensively studied species, in connection with control projects on the Japanese beetle (Steinhaus, 1949). Although little was known of its natural distribution in beetle populations, it was cultivated (Glaser, 1932) on artificial media and established in noninfected plots (Martignoni, 1964). One culture was grown on veal medium and applied to an extensive area, resulting in a 40% reduction of the grub population. Eventually, the nematodes were distributed at 2.5-mile intervals over the state of New Jersey in an attempt to widely distribute the parasite (Glaser *et al.*, 1940). Unfortunately, field populations of this nematode have not been recovered from the Japanese beetle for the past 20 years. Cultures of *N. glaseri* were also shipped to the southern U.S.A. for use against white-fringed beetle, *Pantomorus peregrinus* Buch., and to New Zealand for control of the grass-grub *Costelytra zealandica* (White).

Some information is known about the natural distribution of other *Neoaplectana*. Recently, Poinar and Lindhardt (1971) reisolated *Neoaplectana bibionis* Bovien (Bovien, 1937) from three localities in Denmark, indicating a rather constant association between bibionid host and parasite. *Neoaplectana bibionis* undoubtedly reduces fly numbers by a significant amount and may account for the only occasional occurrence of serious outbreaks of bibionid larvae in young barley fields. Weiser and Koehler (see Weiser, 1966) reported finding 33% of the sawfly *Acantholyda nemoralis* L. attacked by *Neoaplectana* in Poland, and Kirjanova and Puchkova (1955) recovered *N. bothynoderi* Kirj. and Puch. from 13 to 67% of individuals of a beet weevil population in Russia.

One of the most popular species used for field trials is *N. carpocapsae* Weiser. This nematode is composed of several geographical strains, all capable of interbreeding and all practically identical morphologically. Weiser described the species from codling moth in Czechoslovakia and Dutky and Hough (1955) isolated the DD-136 strain from the same host in America. Recently, Poinar and Veremtchuk (1970) reported another strain in larvae of the beetle *Agriotes lineatus* L. collected near Leningrad. The biology of all strains appear similar and it was shown that the associated bacterium in DD-136 (U.S.A.) and *Agriotes* (U.S.S.R.) were nearly identical and belonged to the species *Achromobacter nematophilus* (Poinar *et al.*, 1971). Thus, the bacterium is carried in the intestinal lumen of infective stage juveniles and introduced into the host after penetration. Without its presence, nematode multiplication in the insect is reduced or impossible. Field trials with the DD-136 strain were reported by Welch and Briand (1961) and summarized by Poinar (1971). Moore (1970) demonstrated the ability of the infective stages of *N. carpocapsae* to enter bark beetle tunnels and kill larvae of *Dendroctonus frontalis* Zimm. To facilitate field trials, this nematode can be reared in large numbers under laboratory conditions using living insects or artificial media. Another member of the group, *Steinernema krausei*, has the same system of giant females and minute subsequent generations. It is specific for sawflies and less suitable for laboratory rearings.

Members of the Tylenchida include both true and facultative insect parasites. Although they generally do not directly kill their host, they often cause a partial or complete sterilization, producing a more subtle yet effective result on insect populations. At least three genera have been used as biological control agents. *Tripius sciarae* (Bovien) parasitizes sciarid flies and one of its hosts is *Bradysia paupera* Tuom., a glasshouse pest in England (Poinar, 1965). The infective stage female enters the body cavity of the larval host by direct penetration of the cuticle and produces eggs and larvae. Heavily infected larvae are killed, while lightly infected hosts carry the parasites through their metamorphosis into the adult stage. Infected female flies oviposit packets of nematodes instead of eggs over the soil. This nematode was introduced into the Rothamsted glasshouses in England and became so efficient that in about 1 month it was no longer necessary to spray for sciarid flies. *Diplogaster entomophaga* and further related species interfere with *Neoaplectana* in soil insects. They are able to survive on organic media without living hosts, but they infect and kill insects when ingested.

The neotylenchids, *Deladenus wilsoni* Bedding and *D. siricidicola* Bedding occur in the body cavity of *Sirex* wood wasps. These nematodes also have a free-living generation that feeds on the fungus *Amylostereum chailletii* (Pers.), symbiotic with *Sirex*. They can be cultured on the fungus and released for field trials. Producing sterility in both sexes of *Sirex*, they are potentially good biological control agents, and have been released in Tasmania and Southern Australia (Bedding, 1968; Chapter 12).

Much attention has been given to a *Heterotylenchus* infecting the face fly, *Musca autumnalis* DeG., in the U.S.A. and a similar species found parasitizing the bushfly, *Musca vetustissima* Walk. in Australia (Hughes and Nicholas, 1969). Poinar recently received specimens of *Musca sorbens* Wied. from Senegal, West Africa infected with a *Heterotylenchus*; thus, these parasites appear to be widespread in muscoid flies. *Musca autumnalis* sterilizes the female flies. It is possible to increase the natural occurrence of these nematodes by artificially applying the infective stages to the fly breeding sites or by releasing parasitized flies.

VII. FUTURE PROSPECTS

Resistance of insect pests to insecticides and modern concern with environmental protection should increase the use of pathogens to control insect pests. Insect diseases are relatively specific and present less threat than chemicals to the environment or to non-target organisms. There are two ways in which pathogens can be manipulated to control insects.

One is the classic method of biological control wherein a pathogen is introduced and spread against an exotic pest in its new home; it includes the use of pathogens from related pests and even the colonization or redistribution of pathogens found on the native insect itself or on its relatives. The key action is colonization and the goal is to establish a disease that will reduce the pest population and be self-sustaining. The cost is expected to be surpassed by long-lasting economic benefits, as continuous manipulation is unnecessary. Successful examples include sawflies in Canada and the Japanese beetle in the United States; we can expect other successes in the future on selected insects.

The other method is analogous to insecticide application and involves treating the crop with doses of pathogens sufficiently large to give economic protection. The goal is the protection of a specific crop or field and it implies treatment whenever justified by cost-benefit analysis without regard for self-sustaining population regulation. In the past pathogens have not competed well with insecticides for several reasons: low virulence; high cost of production; problems of maintaining viability during formulation, packaging, storage, and distribution; rapid loss of viability in spray deposits; and slow production of mortality in the target insect. Even the high specificity of pathogens works against their use, for industry is loath to invest heavily in a product that has limited sale potential because the number of target species is small. In addition, governmental agencies have hesitated to register for use products containing living organisms that might conceivably mutate and attack non-target forms. As a result, commercial preparations of pathogens are rare and are best represented by those containing *Bacillus thuringiensis*, which can be propagated inexpensively and has a relatively large number of target pests. The increased use of pathogens with this method will depend on the solution of the aforestated problems by research.

REFERENCES

Angus, T. A., and Luthy, P. (1971). Formulation of microbial insecticides. *In* "Microbial Control of Insects and Mites" (H. D. Burges and N. W. Hussey, eds.), pp. 623–638. Academic Press, New York.

Bailey, L. (1963). Infectious diseases of the honey bee. Land, *London* 176 p.

Batko, A., and Weiser, J. (1964). On the taxonomic position of the fungus discovered by Strong, Wells, and Apple. *Strongwellsea castrans* gen.et sp. nov. Phycomyc., Entomophthor. *J. Invertebr. Pathol.* **7**, 455–463.

Beard, R. (1945). Studies on the milky disease of Japanese beetle larvae. *Conn. Agr. Exp. Sta. Bull.* **491**, 505–583.

Bedding, R. A. (1968). *Deladenus wilsoni* n. sp. and *D. siricidicola* n. sp. (Neotylenchidae) entomophagous-mycetophagous nematodes parasitic in siricid woodwasps. *Nematologica* **14**, 515–525.

Bovien, P. (1937). Some types of association between nematodes and insects. *Bidensk. Medd. Dansk Naturhist. Foren.* **101**, 1–114.

Bucher, G. E. (1960). Potential bacterial pathogens of insects and their characteristics. *J. Insect Pathol.* **2**, 172–195.

Bucher, G. E. (1963). Nonsporulating bacterial pathogens. *In* "Insect Pathology: An Advanced Treatise" (E. A. Steinhaus, ed.), Vol. 2, pp. 117–147. Academic Press, New York.

Christie, J. R. (1937). *Mermis subnigrescens*, a nematode parasite of grasshoppers. *J. Agr. Res.* **55**, 353–364.

Cuthburt, F. P., Jr. (1968). Bionomics of a mermithid (Nematoda) parasite of soil inhabiting larvae of certain chrysomelids (Coleoptera), *J. Invertebr. Pathol.* **12**, 283–287.

de Barjac, H., and Bonnefoi, A. (1962). Essai de classification biochimique et sérologique de 24 souches de *Bacillus* du type *B. thuringiensis*. *Entomophaga* **7**, 5–31.

Dutky, S. R. (1963). The milky diseases. *In* "Insect Pathology: An Advanced Treatise" (E. A. Steinhaus, ed.), Vol. 2, pp. 75–115. Academic Press, New York.

Dutky, S. R., and Hough, W. S. (1955). Note on a parasitic nematode from codling moth larvae *Carpocapsa pomonella* (Lepid., Olethreut.). *Proc. Entomol. Soc. Wash.* **57**, 244.

Fisher, R., and Rosener, L. (1959). Toxicology of the microbial insecticide, Thuricide. *Agr. Food Chem.* **7**, 686–688.

Fox, R. W., and Weiser, J. (1959). A microsporidian parasite of *Anopheles gambiae* in Liberia. *J. Parasitol.* **45**, 21–30.

Garbowski, L. (1927). Sur les Entomophthorees. *Empusa aulicae* Reich en relation avec les ravages causes dans les forets de Posnanie et Pomeranie en 1923/24 par *Panolis flammea* Sch. *Prace Wydz. Chorob Rosl. Panstw. Inst. Nauk Roln. Bydgoszcz* **4**, 1–24.

Glaser, R. W. (1932). Studies on *Neoaplectana glaseri*, a nematode parasite of the Japanese beetle (*Popillia japonica*). *N.J. Dep. Agr. Circ.* **211**, 3–34.

Glaser, R. W., McCoy, E. G., and Girth, H. B. (1940). The biology and economic importance of a nematode parasite in insects. *J. Parasitol.* **26**, 479–495.

Hazard, E. I., and Weiser, J. (1968). Spores of *Thelohania* in adult female *Anopheles*: Development and transovarial transmission and redescriptions of *T. legeri* Hesse and *T. obesa* Kudo. *J. Protozool.* **15**, 817–823.

Heimpel, A. M., and Angus, T. A. (1963). Diseases caused by certain sporeforming bacteria. *In* "Insect Pathology: An Advanced Treatise" (E. A. Steinhaus, ed.), Vol. 2, pp. 21–73. Academic Press, New York.

Hostounský, Z. (1970). *Nosema mesnili* (Paillot), a microsporidian of the cabbageworm, *Pieris brassicae* (L.) in the parasites *Apanteles glomeratus* (L.), *Hyposoter ebeninus* (Grav.) and *Pimpla instigator* (F.). *Acta Entomol. Bohemoslov.* **67**, 1–5.

Hostounský, Z., and Weiser, J. (1972). Production of spores of *Nosema plodiae* Kellen et Lindegren in *Mamestra brassicae* L. after different infective dosage. I. *Věstn. Česk. Spolecnosti Zool.* **36**, 97–100.

Hughes, R. D., and Nicholas, W. L. (1969). *Heterotylenchus* spp. parasitizing the Australian bush fly. *J. Econ. Entomol.* **62**, 520–521.

Ignoffo, C. M., and Hink, F. W. (1971). Propagation of arthropod pathogens in living systems. *In* "Microbial Control of Insects and Mites" (H. D. Burges and N. W. Hussey, eds.), pp. 541–580. Academic Press, New York.

Khan, M. Q., and Hussain, M. (1964). Natural control of *Laphygma exigua* by a *Mermis* sp. (Nematoda). *Indian J. Entomol.* **26**, 124–125.

Kirjanova, E. S., and Puchkova, L. B. (1955). A new parasite of the beet weevil *Neoaplectana bothynoderi* n. sp (Nematoda). *Tr. Zool. Inst. Akad. Nauk SSSR* **18**, 53–62.

Laird, M. (1971). Microbial control of arthropods of medical importance. *In* "Microbial Control of Insects and Mites" (H. D. Burges and N. W. Hussey, eds.), pp. 387–406. Academic Press, New York.

Lysenko, O. (1967). Bacterial toxins. *In* "Insect Pathology and Microbial Control" (P. A. van der Laan, ed.), pp. 219–237. North Holland, Amsterdam.

McLaughlin, R. E. (1971). Use of protozoans for microbial control of insects. *In* "Microbial Control of Insects and Mites" (H. D. Burges and N. W. Hussey, eds.), pp. 151–172. Academic Press, New York.

Martignoni, M. E. (1964). Mass production of insect pathogens. *In* "Biological Control of Insect Pests and Weeds" (P. DeBach, ed.), pp. 579–609. Reinhold, New York.

Moore, G. E. (1970). *Dendroctonus frontalis* infection by the DD-136 strain of *Neoaplectana carpocapsae* and its bacterial complex. *J. Nematol.* **2**, 341–344.

Müller-Kögler, E. (1965). "Pilzkrankheiten bei Insekten," 444 pp. Parey, Berlin.

Neilson, M. M. (1964). A cytoplasmic polyhedrosis virus pathogenic for a number of lepidopterous hosts. *J. Insect Pathol.* **6**, 41–52.

Poinar, G. O., Jr. (1965). The bionomics and parasitic development of *Tripius sciarae* (Bovien) (Sphaerulariidae: Aphelenchoidea) a nematode parasite of sciarid flies (Sciaridae: Diptera). *Parasitology* **55**, 559–569.

Poinar, G. O., Jr. (1966). The presence of *Achromobacter nematophilus* Poinar and Thomas in the infective stage of a *Neoaplectana* sp. (Steinernematidae: Nematoda). *Nematologica* **12**, 105–108.

Poinar, G. O., Jr. (1968). Parasitic development of *Filipjevimermis leipsandra* Poinar and Welch (Mermithidae) in *Diabrotica u. undecimpunctata* (Chrysomelidae). *Proc. Helm. Soc. Wash.* **35**, 161–169.

Poinar, G. O., Jr. (1971). Use of nematodes for microbial control of insects. *In* "Microbial Control of Insects and Mites" (H. D. Burges and N. W. Hussey, eds.), pp. 181–203. Academic Press, New York.

Poinar, G. O., Jr. (1972). Nematodes as facultative parasites of insects. *Annu. Rev. Entomol.* **17**, 103–122.

Poinar, G. O., Jr., and Lindhardt, K. (1971). The re-isolation of *Neoaplectana bibionis* Bovien (Nematodea) from Danish bibionids (Diptera) and their possible use as biological control agents. *Entomol. Scand.* **2**, 301–303.

Poinar, G. O., Jr., and Veremtchuk, G. V. (1970). A new strain of entomopathogenic nematode and the geographical distribution of *Neoaplectana carpocapsae* Weiser (Rhabditida, Steinernematidae), *Zool. Zh.* **49**, 966–969.

Poinar, G. O., Jr., Thomas, G. M., Veremtschuk, G. V., and Pinnock, D. E. (1971). Further characterization of *Achromobacter nematophilus* from American and Soviet populations of the nematode *Neoaplectana carpocapsae* Weiser. *Int. J. Syst. Bacteriol.* **21**, 78–82.

Rivers, C. (1958). Virus resistance in larvae of *Pieris brassicae*. *Trans. 1st Int. Conf. Insect. Pathol. Biol. Control*, pp. 205–210.

Rogoff, M. H., and Yousten, A. A. (1969). *Bacillus thuringiensis*: Microbial considerations. *Annu. Rev. Microbiol.* **23**, 357–386.

Samšiňáková, A. (1961). L'utilisation d'une préparation de champignon dans la lutte contre le Doryphore. *Agron. Glas.*, pp. 563–565.

Samšiňáková, A., Mišíková, S., and Leopold, J. (1971). Action of enzymatic system of *Beauveria bassiana* on the cuticle of the greater wax moth larvae (*Galleria melonella*). *J. Invertebr. Pathol.* **18**, 322–330.

Smirnoff, W. A. (1961). A virus disease of *Neodiprion swainei* Middleton. *J. Insect Pathol.* **3**, 29–46.

Steinhaus, E. A. (1949). "Principles of Insect Pathology," 757 pp. McGraw-Hill, New York.

Steinhaus, E. A., and Thompson, C. G. (1949). Preliminary field tests using a polyhedrosis virus in the control of the alfalfa caterpillar. *J. Econ. Entomol.* **42**, 301–305.

Stephens, J. M. (1963). Immunity in insects. *In* "Insect Pathology: An Advanced Treatise" (E. A. Steinhaus, ed.), Vol. 1, pp. 273–297. Academic Press, New York.

Tanada, Y. (1959). Synergism between two viruses of the armyworm *Pseudaletia unipuncta* (Haworth) (Lepidopt., Noct.). *J. Insect Pathol.* **1**, 215–231.

Tanada, Y. (1971). Interactions of insect viruses with special emphasis on interference. *In* "The Cytoplasmic Polyhedrosis Virus of the Silkworm" (H. Aruga and Y. Tanada, eds.), pp. 185–200. Univ. Tokyo, Press, Tokyo.

Veen, K. H. (1968). Recherches sur la maladie due a *Metarrhizium anisopliae* chez le criquet pelerin. *Medelel. Landbouwhogesch. Wageningen* **68**, 5.

Weiser, J. (1961). Mikrosporidien als Parasiten der Insekten. *Z. Angew. Entomol.* **17**, 149 pp.

Weiser, J. (1966). "Nemoci hmyzu" (Diseases of Insects), 554 pp. Academia, Praha.

Weiser, J. (1968). *Triplosporium tetranychi* sp. n. (Phycomycetes, Entomophthoraceae), a fungus infecting the red mite, *Tetranychus althaeae* Hanst. *Folia Parasitol.* **15**, 115–122.

Weiser, J., and Hostounský, Z. (1972). Production of spores of *Nosema plodia* Kellen et Lindegren in *Mamestra brassicae* L. after different dosage. *Věst. Česk. Spolecnosti Zool.* **36**, 97–100.

Welch, H. E. (1963). Nematode infections. *In* "Insect Pathology: An Advanced Treatise" (E. A. Steinhaus, ed.), Vol. 2, pp. 363–392. Academic Press, New York.

Welch, H., and Briand, J. L. (1961). Field experiments on the use of a nematode for the control of vegetable crops insects. *Proc. Entomol. Soc. Ont.* **91**, 197–202.

SECTION III

METHODOLOGY

8

FOREIGN EXPLORATION AND IMPORTATION OF NATURAL ENEMIES

H. Zwölfer, M. A. Ghani, and V. P. Rao

I. INTRODUCTION

The objective of foreign exploration is to detect, select, and export natural enemies showing promise as biological control agents and to provide information facilitating their establishment in the country of introduction. The selected agents should be imported in healthy condition, in suitable stages, and in sufficient quantities. If possible, the foreign explorer should also collect data for later comparative evaluations of the natural enemies in their original and adopted ecosystems, thus contributing to an improved understanding of the principles of natural or applied biological control.

The organization of a foreign exploration program is dictated by the objectives of the project; therefore, it is necessary that these objectives initially be well defined. The type of target organism (pest animal or weed), its noxious status worldwide, the results of previous attempts at its biological control, the quantity and quality of information available on its natural enemies, the level of financial, collaborative, and logistical support, among other factors, greatly influence the scope and depth of the program.

The various phases of a foreign exploration program of the pioneer type are dia-

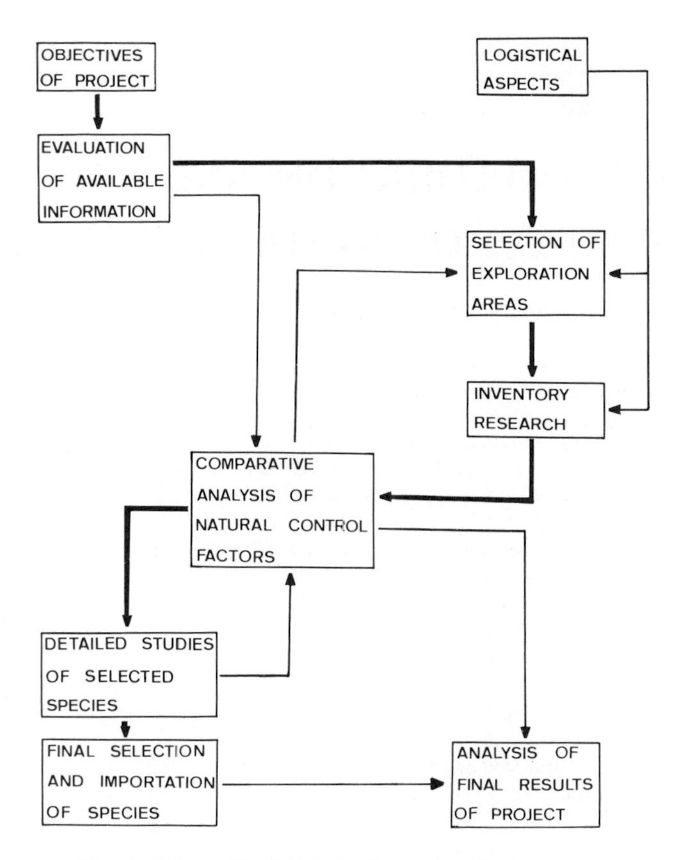

Fig. 1. Information flow in foreign exploration programs.

grammed in Figure 1. The flow chart illustrates how information gained during one phase of the work interacts with previous and subsequent phases. Under actual conditions several phases usually proceed concurrently. Bartlett and van den Bosch (1964) reviewed and discussed foreign exploration for biological control in considerable detail.

II. PLANNING AND PREPARATION OF THE PROGRAM

A. Accumulation and Evaluation of Available Information

Ideally, the following information on the pest species should be obtained before search for its natural enemies begins: (a) taxonomic position, life history, and economic importance in the country of origin, (b) native geographical distribution, (c) total present distribution, (d) host plant distribution, (e) probable center of origin of the organism and its close relatives, (f) coextensive occurrence of related species, (g) occurrence and distribution of related and ecologically similar species in regions where the target organism does not occur, but where exploration for an enemy agent seems desirable because of climatic similarities to infested areas of the country of introduction, and (h) available records of natural enemies and other mortality factors. These records should be augmented by faunistic surveys of the pest in the country of introduction.

It is fundamental to obtain at the outset authoritative identifications of specimens of the target organism obtained from both foreign and domestic sources. Incorrect identifications or nomenclatural inconsistencies can greatly impede progress. Whenever identification of the target organism proves questionable, the foreign explorer should carry with him preserved specimens of the target organism. (See also Chapter 4).

Handbooks and review journals such as the "Review of Applied Entomology" or "Weed Abstracts" may provide useful surveys of the biology, ecology, and distribution of a target organism. Detailed information can be obtained from specialists, museum collections, or catalogs. Research institutes, university agricultural departments, quarantine stations, and organizations involved in biological control are sometimes important sources of information. As access to this information and its extraction is usually laborious and time-consuming, the need for regional or global information storage and retrieval systems is urgently felt.

B. Selection of Target Organisms and Exploration Areas

1. Logistical Aspects

Logistical problems may considerably modify exploration schemes based solely on biological or ecological criteria. Inman (1970a) listed a number of the logistical problems encountered: (a) inaccessibility of a search area because of national or international political considerations, unavailability of transportation, or inclement weather, (b) lack of laboratory facilities within the search area, (c) restrictive customs procedures barring importation of specialized equipment into the search areas. (d) maintenance of cultures or collections during extended research periods, and (e) quarantine restrictions on material imported for testing purposes.

At present, certain potential areas for exploration for natural enemies of Palearctic origin, such as important zoogeographical centers in central and eastern USSR and in China, frequently are inaccessible logistically.

2. Target Organisms

Most foreign explorations concern searches for natural enemies of the target species itself. If, however, no promising control agents of the target species have been found after reasonable efforts, or if control of native pests is sought, species closely related to the target organism can be used as sources of natural enemies. This approach is justified by examples of successful biological control of pests by agents originally associated with related host species. Some such agents, which did not coevolve with the target host, may even be more effective for biological control purposes because of the tendency of homeostasis to evolve between some hosts and certain types of their natural enemies, e.g., true parasites and their hosts. The devastating effect of the fungus *Endothia parasitica* (Mürr) on American chestnut, in contrast to its mild effect on Asian chestnut, is an example. (See further Chapters 3 and 21.)

3. Exploration Areas

The following criteria are considered when selecting exploration areas:

1. Searches initially should be undertaken in areas where the target species is native as this greatly increases the chances of finding a rich, diversified complex of natural enemies, some of which may be highly host-specific, and, in general, are better prospects. (See also Chapter 3.) However, this principle does not preclude the possibility that suitable control agents exist outside the native areas of the host, either because they have spread with the host or because species in the new ecosystem have adopted the target host, a process which apparently has occurred frequently in tropical environments (Greathead, 1971).

2. The climate and other environmental conditions should be reasonably similar to those of the country where introductions are intended. This will increase chances of a successful establishment and a rapid adaptation of the natural enemy to the target organism in the new environment.

3. At least a part of the search area should include or be as near as possible to the center of origin of the target species and its close relatives (if climatically comparable to the target area) because the diversity of complexes of specialized natural enemies tends to be a function of the duration of association of an organism with a given ecosystem, and hence there is a better chance of obtaining one or more highly specialized and effective natural enemies. Goeden (1971) illustrated this for the weed *Solanum elaeagnifolium* Cav.

4. Faunal and floristic history, vegetation, and habitat structures should present a maximal diversity in the area selected for exploration (Greathead, 1971). This criterion is particularly useful in the case of species where the native range (1) and the center of origin (3) cannot be established, or where the range in conditions in the target area is great.

5. The exploration areas should be as wide and diversified as possible to maximize the chances of finding numerous and ecologically differentiated natural enemies of broad genetic variability, and of acquiring information on the ecology of individual species under different conditions (Section III,B). The exploration of large areas is, of course, expensive and time-consuming, but it will increase the chances of success.

III. INVENTORY RESEARCH AND INVESTIGATIONS ON SELECTED SPECIES

A. Inventory Research

The objectives of inventory research are to provide (a) a survey of the available natural enemies, and (b) criteria for the selection of suitable control agents. Hence, a reasonably broad inventory of natural enemies, including pertinent ecological details (e.g., hyperparasitic habit, host range), should be completed before detailed studies begin. When possible, species related to the target host should be included in the survey, as this provides preliminary information on the host specificities and host preferences of the natural enemies.

Often the explorer will be tempted to concentrate his search at places of high host density, where collections are easier and more enemies might be found. However, search must also be carried out at localities of low host density, since, if the low density is due to natural enemies, ones having good searching powers are encountered (Chapter 3). The densities of field populations of the host can be manipulated to investigate the searching capacities of natural enemies (Section III,C,3).

If natural host densities are very low, eggs, larvae, or pupae of the host can be artifically concentrated in the field. This "exposure method" has been successfully applied in many exploration programs, e.g., recently, during studies on several sawfly pests carried out by the Commonwealth Institute of Biological Control (CIBC). Another application of the "exposure method" is the experimental grouping of plant species which do not occur together in nature. Thus, a "thistle garden" at the European Station of the CIBC that currently contains about 30 different thistle and knapweed species has provided valuable information on the host preferences of insects associated with these weeds.

The routine work of inventory research may involve regular sampling at selected localities, sampling along transects through ecologically or climatically distinct areas, field observations, laboratory rearings, dissection of host material, and preparation and identification of specimens. As the equipment needed for the work depends on the nature of the target organism and various logistical considerations, no detailed descriptions are given here. A useful tool, not often mentioned, is a reference collection of host and natural enemy species which should include developmental stages, if possible. As the identity of many natural enemies is not known, as a rule, during early phases of exploration, symbols are frequently used to designate individual species.

Standardized data sheets for recording details of field observations, samples, dissections, or rearings facilitate evaluation of inventory results and help prevent the inadvertent omission of data. Often exploration programs yield so much data that registers, sets of card files, or punch card systems are used for information storage and retrieval. Thus, the CIBC survey program on European thistle insects employs a punch card system (EKAHA-Sichtloch-kartei, 05000 SL) with a capacity for up to 5000 samples of individual host plant demes. This system allows a quick evaluation of such parameters as distribution, phenology, host range, constancy of association with a given host, and co-occurrence of natural enemies.

Apart from data on natural enemies, inventory research also provides information on the status of the target organism in different parts of the exploration area (e.g., population densities, economic damage, and various factors influencing the populations).

B. Comparative Analysis of Natural Control Factors

Comparative analysis of accumulated data has a central role in any foreign exploration program (see Fig. 1), and results in the selection of species for detailed studies as candidate biological control agents.

It may provide additional data and feedback to the selection of exploration areas, suggesting that searches be undertaken in new regions. Information provided by comparative analysis of survey data ideally should also be relevant in evaluating the final results of the biological control project.

1. Effect of Natural Enemies

As the general principles describing the role of enemies in the population dynamics of their hosts are discussed elsewhere, especially in Chapter 3, only those phenomena of major concern to foreign exploration are treated here. Development of population ecology has produced elaborate methods of analyzing enemy action, e.g., the life table approach (reviewed by Varley and Gradwell, 1970, 1971) and other techniques as well to assess the responses of natural enemies (Huffaker *et al.*, 1968, 1971; and Chapter 11). Employment of such techniques during foreign exploration programs would be scientifically rewarding and provide a more satisfactory basis for a comparative analysis of natural enemy activities and host responses. However, the dilemma of the foreign explorer is that—apart from restrictions in time and labor—he has to deal with a broad range of populations over wide regions, while the methods of population dynamics require intensive investigation of particular localized populations. For certain purposes, studies of the population dynamics of a host pest and its natural enemies requires that some 8—10 or more generations of the insects be considered in order to provide a reasonable basis for conclusions. Also, homeostatic processes (e.g., by hyperparasitism) and the extreme complexity of interactions characterizing many native enemy—host systems may mask the control potential which natural enemies might display in the simplified ecosystem where an introduction is intended. Extensive comparative studies of the conventional type, combined with analyses of interactions of natural enemies, may sometimes better indicate the potential over a wide area than the intensive investigations of the "key factors" operating for a single host population. This is illustrated by the successful biological control of winter moth, *Operophtera brumata* (L.), in Canada by two introduced European parasites, *Cyzenis albicans* (Fall.) and *Agrypon flaveolatum*

Grav. (Embree, 1971). (See also Chapter 13.) Comparative analysis of samples collected in six European countries indicated that the potential of both parasites (but particularly of *Cyzenis*) was considerably greater than the actual mortality they caused (Sechser, 1970). A key factor analysis covering 17 years of population changes of the winter moth at a single locality in England (Varley and Gradwell, 1968, 1971) did not indicate this potential, since both parasites are ineffective in that study area.

Future explorations will undoubtedly make increasing use of modern approaches of population dynamics (i.e., of intensive multifactor studies of selected populations), but those approaches will not replace conventional methods which primarily compare the structure of enemy—host complexes and attempt to assess effects of enemy—host interactions over wide areas. These conventional methods attempt to assess the following parameters for parasites and predators:

1. *Host range*. This is the range of acceptable host species, estimated by the number of supraspecific systematic categories (e.g., genera, families) attacked.

2. *Host preference*. This parameter indicates the acceptance of the target host relative to other acceptable host species.

3. *Constancy*. This statistic represents the percentage occurrence of a natural enemy in all samples taken in the exploration area. It is a measure of the frequency of association between enemy and host.

4. *Abundance*. This term designates the numerical occurrence of one species within individual samples. Abundance is usually expressed as "apparent percentage parasitism."

5. *Intrinsic competitive capacity*. This indicates the outcome of intrinsic competition (direct competition within or on an individual host) with other natural enemies.

"Host range" and "host preference" are deduced from comparative field studies, laboratory experiments, or other sources of information. The values for "constancy" and "abundance" are calculated from available samples and for maximum usefulness should be segregated, in their determination, for high and low host population densities. Competitive behavior is usually studied by sequential dissections of samples of host stages or by competition experiments. The comparative analyses of the native parasite complexes of the European fir budworm, *Choristoneura murinana* (Hb.) (Zwölfer, 1961), the European skipper, *Thymelicus lineola* (Ochs.) (Carl, 1968), and the winter moth (Sechser, 1970) provide examples of the assessment of such parameters.

Comparative evaluation of the effects of phytophagous insect enemies of weeds is more difficult, because these agents may, and often do, reduce the competitive capacity of the host rather than kill it outright. Phytophages attacking the reproductive system of a plant may provide such comparative quantitative data as the average numbers of flowers, fruits, or seeds destroyed by an individual. If the phytophagous organisms exploit the vegetative stage of a weed, a comparative analysis will consider the mode of attack (e.g., leaf feeding, girdling, stem- or root-mining, and production of galls). Some of the problems involved in assessing the control potential of organisms attacking weeds have been discussed by Wapshere (1970). Harris (1973a) suggested that the effect on the carbohydrate cycle of the host be used to compare the effectiveness of phytophagous insects. He (Harris, 1973b) proposed a scoring system to select the most effective agents. In general, the quantitative assessment of the effect of phytophagous insect enemies needs greater attention than for entomophagous enemies in foreign exploration programs.

2. Interactions among Natural Enemies

Common among the stabilizing mechanisms that enable concurrent exploitation of a given host population by several enemy species is the type of interaction where well-adapted enemies with good searching abilities coexist with less specialized but intrinsically superior species, often ones capable of both primary and secondary parasitism. Here the extrinsic superiority of species A may be compensated by the intrinsic superiority of species B or it is countered by the hyperparasitism of species B. Examples of such systems of "balanced competition" among natural enemies associated with the same host population are discussed by Pschorn-Walcher *et al.* (1969) and Zwölfer (1971). Cleptoparasitism forms a particular type of interaction among parasites. In the European parasite complex of *Rhyacionia buoliana* (Schiff.) the extrinsically superior species, *Orgilus obscurator* Nees, is exploited by four intrinsically superior cleptoparasites, *Temelucha interruptor* Grav., *Eremochila ruficollis* Grav., *Campoplex mutabilis* Hlmgr., and *Pristomerus* sp. (Schröder, 1974). If a parasite species shows a marked tendency to act as a cleptoparasite it should be classified as an ecological hyperparasite which should not be introduced.

If the analysis of interactions within a multispecies system reveals "balanced competition" or apparent competition involving any well-adapted enemy species, the foreign explorer should recommend that the species concerned be introduced in a predetermined sequence, not simultaneously, starting with the intrinsically inferior enemies (Franz, 1961).

3. Hyperparasites and Other Secondary Enemies

The potential rate of increase of the natural enemies of a target host can be greatly restricted by their own natural enemies, organisms that occupy a superior trophic level, e.g., hyperparasites, parasites of predators, and predators and parasites of phytophagous agents. Therefore, comparative analysis of an enemy complex should consider the nature of and extent to which the secondary enemies are associated with the primary enemies studied. Determining whether these secondary enemies are biologically or ecologically highly specialized (e.g., hyperparasites belonging to the ichneumonid genus *Mesochorus*) or rather unspecific (e.g., ants) is equally important. The detection of highly specialized secondary enemies of a primary agent in an exploration area suggests that the primary enemy may have an enhanced biological control potential when transferred to a new ecosystem. Vulnerability to attack by general predators is an unfavorable characteristic in an enemy agent, as its populations are more liable to be rendered ineffective by similar components of the new ecosystem. Simmonds (1949) predicted that ants might prevent *Physonota alutacea* (Boh.) from becoming an effective agent for control of the weed *Cordia macrostachya* (Jacq.) in Mauritius, since he had found that this chrysomelid was restricted to areas in its native Trinidad where the predaceous ant *Solenopsis geminata* (F.) was absent.

4. Stabilizing Behavior Patterns

Recent investigations (e.g., Birch, 1971; Gruys, 1971; Hassell, 1971) produce ever-increasing evidence that inbuilt behavior patterns can help to stabilize populations of insects. Examples relevant to biological control of weeds are the territory behavior of adults of certain species of tephritids which reduces the probability of overexploitation of flower heads. The strongly clumped dispersion of adult *Altica carduorum* (Guer.) on *Cirsium* at high beetle densities results in extreme intraspecific competition between adults and first instar larvae, to the detriment of the latter on the individual host plants, but the majority of hosts thus remain unattacked. Imperfect synchronization of their life cycles protects certain European predator species from overexploiting their host, *Dreyfusia piceae* (Ratz.) (Pschorn-Walcher and Zwölfer, 1956). In general, such behavior patterns are undesirable attributes for biological control agents.

5. Factors other than Natural Enemies Affecting Target Organisms

Comparative analysis may reveal that factors other than enemies are more important in the natural control of the target organism in the exploration area. Competition with surrounding vegetation, for example, can be a decisive factor determining weed densities. The physiological state of a crop plant or the timing and pattern of its culture may adversely influence its pest insects. The key factor in the population dynamics of *Dreyfusia piceae* on *Abies alba* (L.) in Europe is a density-dependent defense mechanism of the host tree (temporary formation of necrotic tissue). Native predators are biologically adapted to exploit surplus *D. piceae*, but their inverse density-dependence renders them unable to control mass outbreaks of the pest insect (Eichhorn, 1969). These relationships may account for the fact that the four predators established in Canada against *D. piceae* on *Abies balsamea* (L.), a tree lacking such a defense reaction, are ineffective (Clark *et al.*, 1971). These considerations resulted in selecting new exploration areas and allied host species as sources of natural enemies (Eichhorn, 1969).

C. Detailed Studies on Selected Species

Although the foreign exploration effort cannot determine all the factors regulating a species in its native environment, it can establish important parameters by field and laboratory investigations. Admittedly, such investigations cannot predict whether a natural enemy species will become established or effective in a new environment, but they can discover a clear unsuitability of an agent for particular areas and provide criteria for the priority selection of candidate species.

1. Reproductive Capacity and Impact on Host

Preintroduction studies can determine the fecundity of natural enemies as well as the number of hosts killed by a single individual. The effective rate of reproduction and the number of hosts killed may be much less than the oviposition capacity of a parasite because of super- and multiple parasitism or host encapsulation of parasite eggs. On the other hand, host paralyzation without oviposition, feeding of adult parasites on hosts, or interference between parasite adults and hosts [e.g., annoyance of *Acyrthosiphon pisum* Harris by *Aphidius smithi* Rao and Sharma (Ghani, 1969)] can cause considerable additional host losses.

2. Adaptation to Different Climates

Preintroduction studies should elucidate the climatic requirements of natural enemies. The various species of *Chrysolina* attacking *Hypericum* in Europe prefer different moisture zones. This ecological specialization was the decisive factor responsible for the successes and failures to establish *C. hyperici* (Forst.), *C. quadrigemina* Suff., and *C. varians* (Schall.) in different parts of Canada (Harris *et al.*, 1969). Comparing *Apanteles sesamiae* Cam. and *A. flavipes* (Cam.), Mohyuddin (1971) found the later species more tolerant of dry conditions. A study of two East African pupal parasites of graminaceous borers indicated that *Pediobius furvus* Gah. is more suitable for arid conditions than *Dentichasmias busseolae* Hein. (Mohyuddin, 1971). Such findings can help to plan releases in target countries.

3. Searching Ability

The assessment of searching ability and searching behavior of natural enemies is discussed in Chapters 3, 5, and 6 and elsewhere in this treatise. A simple but effective method to compare the relative searching capacities of parasites was used in European studies on the sawfly *Neodiprion sertifer* (Geoffr.). Field experiments with prearranged patterns of host colonies demonstrated that certain species of parasites were able to locate isolated hosts while other species mainly attacked foci of high densities (Pschorn-Walcher, 1973).

4. Host Selection

Host specificities of natural enemies can be empirically defined by field observations and laboratory tests. More satisfactorily they are assessed through causal analysis of such processes as habitat finding, host plant finding, host finding, host acceptance, and/or host suitability. Tests to establish the degree of host specificity and safety of candidate phytophages represent the most important phase of preintroduction studies for biological control of weeds (Chapter 19). If a natural enemy is to be used against a target organism which is different from the native host, an understanding of the host selection processes is particularly necessary. If it is not inherently highly host-specific, it is necessary that an enemy species possess or be able to develop a high preference for the target organism in the particular new ecosystem. Actually, most successful biological control agents are not "*strictly*" host-specific (Simmonds, 1969), but all have or developed a high preference for the target pest. Our present understanding of genetic and behavioral mechanisms (e.g., associative learning, conditioning) is still insufficient to be able to predict the adaptability of host preferences of nonspecific natural enemies transferred to new environments. However, pioneer work by Arthur (1966), Huettel and Bush (1972), and others suggests that such research may have applications. (See also Chapter 9).

5. Synchronization

Synchronization of natural enemies with their host is often a complex process involving numerous abiotic factors (e.g., circadian rhythms, light, temperature, and humidity) and biotic factors (e.g., host and parasite behavior and influence of plant species on which the host is feeding; see Cheng, 1970). If environmental conditions in the country of introduction differ much from the donor country, the aspect of synchronization has to be investigated.

6. Genetics

Genetic studies may find application in solving taxonomic problems as discussed in Chapter 9. Preintroduction investigation of genetic variability may aid in selection of enemy strains suitable for target areas. Genetic differences in host resistance and enemy agent virulence can create problems when phytopathgenic fungi (e.g., *Puccinia* spp.) are considered for use against noxious weeds (Inman, 1970b; Hasan, 1970). Similarly, the resistance of populations of the sawfly *Pristiphora erichsonii* (Htg.) to certain strains of the ichneumonid *Mesoleius tenthredinis* Morl. required genetic studies of both insect host and parasite. These studies resulted in the recent successful introduction of the virulent Bavarian strain of *M. tenthredinis* into Canada (Turnock and Muldrew, 1971) (see Chapter 12).

IV. IMPORTATION OF NATURAL ENEMIES

A. Selection of Agents for Introduction

The selection of natural enemies for introduction is perhaps the most crucial phase in any biological control program. Although this phase remains an empirical process to a considerable extent, the following criteria are of importance: (a) The candidate species must not develop harmful effects in the country of introduction. This criterion is of top priority in biological weed control, for such an agent must not attack desired plants. (b) It should possess a good searching capacity and be able to operate effectively, preferably at both low and high host population densities. (c) It should be well adapted to the climatic conditions, host habitats, and other ecological factors in the country where biological control is intended. (d) A relatively high host specificity or the capability of developing a high host preference in the situation is highly desirable (but see Section III,C,4). (e) Its life cycle should be well synchronized with its host, thus enabling full exploitation of the latter. This is related to (c) and (d). (f) High reproductive capacity, preferably combined with short generation time and a high "effective rate of oviposition" (Smith, 1939), are valuable assets. (g) In some instances it may prove useful that the agent is able to integrate itself into the system of mortality factors already existing in the target country. The biological control of *Promecotheca coeruleipennis* (Blanch.) in Fiji by the introduced parasite *Pleurotropis parvulus* Ferr. (Taylor, 1937) provides a good example of such integration and of the high dividends which may result from careful preselection of a natural enemy.

In recent years attention has been given to the population genetics of biological control agents. Introductions should incorporate broad-spectrum genetic variability on which selection can act to favor the genotype optimal for the new environment. Hence, Remington (1968) warns against introducing material from laboratory rearings started from a few individuals although some such introductions have been highly successful. He suggests that a large sample from a large, central, wild source population in an environment most similar to that of the region of intended establishment should be introduced. He further suggests that a closely spaced series of introductions of samples from many wild source populations in moderately similar environments would even better maximize genetic variability. Lucas (1969) criticized Remington's model but similarly concluded that the most successful introductions should arise from ecologically diverse central populations. Ratcliffe (1966) mentions the desirability of introducing several local strains of an enemy, the cross-breeding of which may provide a broader gene pool and facilitate its adaptation to the new environment. According to this author, the success of the parasite, *Trissolcus basalis* (Woll.), against *Nezara viridula* (L.) in inland Australia was possibly due to the infusion of genes from a strain of *T. basalis* introduced from Pakistan into a locally established parasite population originating from another area.

In most biological control programs, "multiple introductions" (i.e., introductions of more than one enemy species) are undertaken to cover a broader ecological range and to provide better compensation against disruption of control. If the analysis of interactions among natural enemies (Section III,B,2) indicates that well-adapted enemies suffer from intrinsic competition with less specialized enemies, introductions should not be made simultaneously, but in a sequence starting with the better adapted species (extrinsically superior ones).

B. Preparation of the Material

1. Laboratory Propagation

If promising natural enemies are found in the exploration area it may be desirable to establish a semipermanent laboratory facility for the propagation of field-collected material to provide stock for shipments to the country of introduction. Moreover, with such a facility it may be possible to prescreen and ensure the shipment of disease-free individuals that can be released directly in the field if adequate rearing facilities are not available in the country of introduction. Laboratory colonies preferably should be started with a large number of individuals to ensure the genetic heterogeneity of the population established in the new country.

Methods of laboratory propagation, problems encountered with inducing mating, the production of suitable sex ratios, and using alternate (factitious) hosts, as well as the different types of rearing cages, etc., are described in detail by Fisher and Finney (1964).

2. Timing the Shipments

The phenology of the pest may differ in the donor country and country of introduction. This necessitates manipulations to ensure the availability of the control agent when the vulnerable stage of the target organism is present in adequate numbers in areas where releases are planned. This problem particularly arises when natural enemies are transferred from the northern to the southern hemisphere or vice versa. Thus, one of the major difficulties of introducing European parasites of the clover casebearer, *Coleophora alcyonipennella* Koll., into New Zealand was the need to terminate artificially the diapause of the parasites.

3. Screening for Parasites and Diseases

When exporting parasites or predators to another country, great caution is exerted to prevent hyperparasites or parasites of predators from gaining access to the new region, as these organisms can reduce whatever benefits may arise from the introduction (Section III,B,3). Accordingly, laboratory-reared material is safer to introduce than field-collected material. Particular caution is advisable if there are no quarantine facilities in the country where releases are intended. As a rule, shipping only the adult stage of an enemy species reduces the risk of including hyperparasites in the shipment, but with many predators (e.g., coccinellids) this precaution alone is not sufficient.

Animal or plant pathogens may find their way into the country of introduction if diseased natural enemies or plant material are included in shipments. The use of sterile techniques may be necessary to ensure that no pathogens are transported in the shipping containers. This aspect is particularly important if weed control agents are being introduced. The problem created by latent diseases of phytophages used in biological control of weeds is discussed by Bucher and Harris (1961).

C. Shipping Methods

Satisfactory transporting of natural enemies today is considerably facilitated by air freight facilites, but it still requires considerable planning and ingenuity. The optimal conditions of humidity and temperature vary with different natural enemies; hence, pilot tests should be made to determine the best shipping method for the species concerned.

Shipping containers should be well insulated, sturdy, light, and easy to handle. Some commonly used types of containers are shown in Figures 2–5. They usually are made of wood, cardboard, plastic, or light-weight metal to withstand rough handling

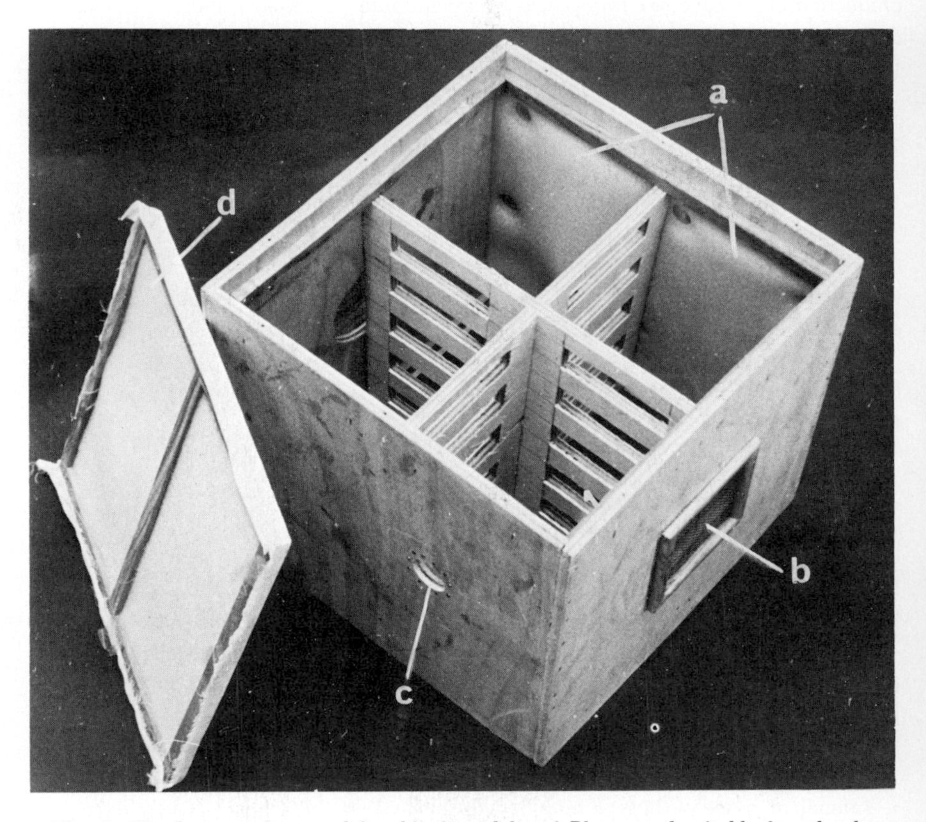

Fig. 2. Wooden container used for shipping adults of *Rhyssa* and suitable for other large Hymenoptera. The synthetic sponge (a) is moistened; the wire screen (b) provides aeration; a hole (c) on one side is used for releasing the parasites into the box; the cloth screen (d) is placed over the box and covered with moistened moss for additional humidity.

Fig. 3. Wooden container used for shipping tachinid puparia. Puparia are placed in the sliding compartment to the right containing moistened moss (a); on emerging the flies enter through the holes (b) into the larger compartment (c) in which food (diluted honey) is provided on a piece of synthetic sponge or cotton swab (d). The larger compartment is screened with a transparent plastic sheet to facilitate customs inspection. At the destination, adult flies are collected by inserting a vial in the hole provided (e).

in transit. A transparent plastic sheet at one side of the container facilitates inspection by customs officials and minimizes their need to open the container.

Most natural enemies require a relative humidity of 50% or more. This can be provided by placing moist materials such as sterilized moss or *Sphagnum*, synthetic sponges, or cotton pads in the container in such a way that the insects do not come into direct contact with the wet surfaces. Low temperature can be maintained by using

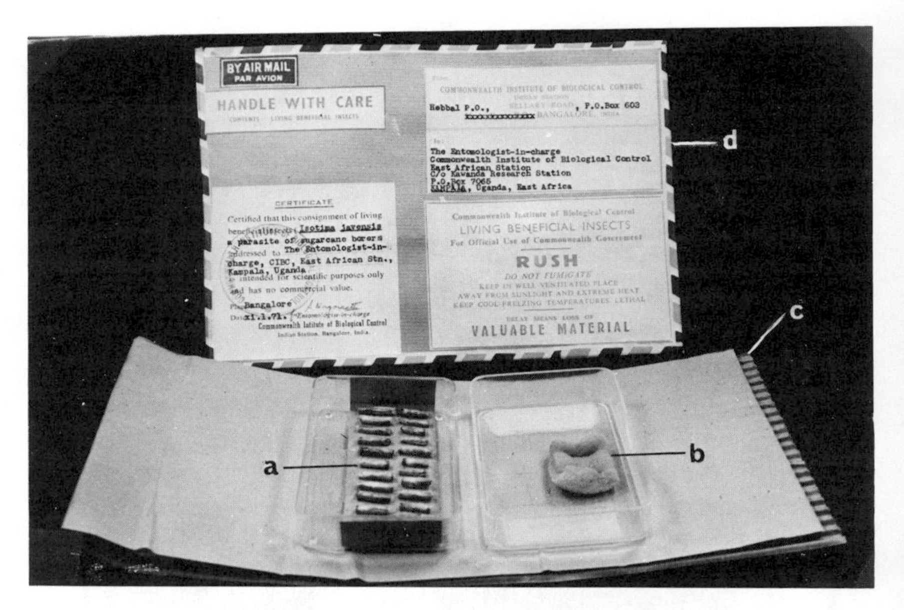

Fig. 4. Plastic container prepared for shipment of tough cocoons of the ichneumonid, *Isotima javensis*. The cocoons (a) are stuck on a strip of gummed paper which is in turn affixed to a plastic container; dilute honey is provided on a piece of synthetic sponge (b). The container is then wrapped in a double layer of corrugated cardboard (c) and inserted in the envelope shown (d) for mailing. Note the labels affixed.

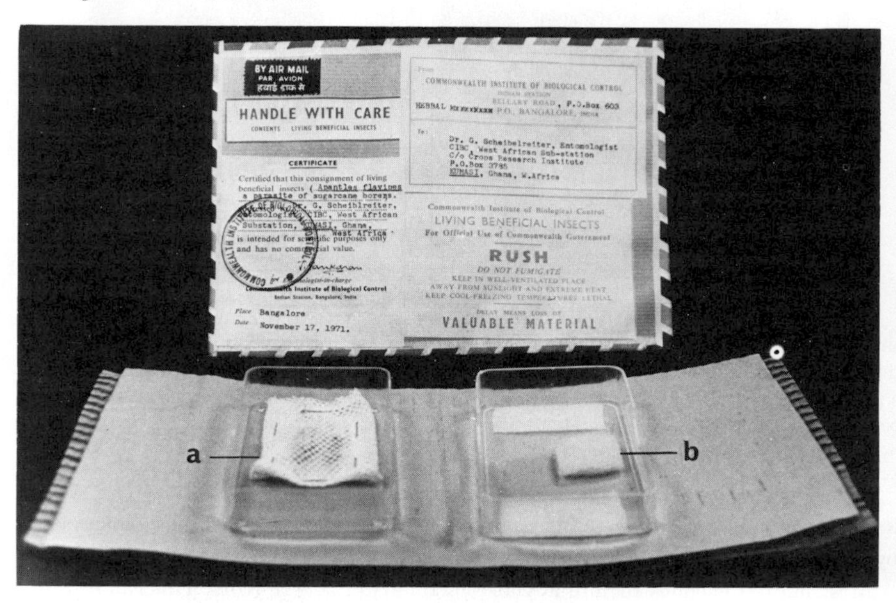

Fig. 5. Similar plastic container used for shipping cocoons of small braconids like *Apanteles* spp. The cocoons are placed in a net bag (a) which is stapled to the container. Food (honey) is provided on a synthetic sponge piece (b).

coolant materials in the packing. Unfortunately, these cooling elements often make the shipment bulky and thus increase mailing costs. Whether food is provided or not depends on the type and stage of organism shipped. Honey, sugar solution, or raisins are usually suitable foods for adult parasites. For adult predators (e.g., coccinellids) it may be preferable to incorporate semisolid droplets of a mixture of honey and agar.

The shipment packages should bear instructions for the postal authorities. A request to the postmaster to telephone the addressee on receipt of the parcel may avoid delays. When sending shipments by air freight, affixing instructions concerning special handling, transfers, etc., may facilitate quick transportation. As most countries today have adopted strict quarantine regulations, certified importation permits usually must be obtained and attached to every package.

REFERENCES

Arthur, A. P. (1966). Associative learning in *Itoplectis conquisitor* (Say) (Hymenoptera: Ichneumonidae). *Can. Entomol.* **98**, 213–223.

Bartlett, B. R., and van den Bosch, R. (1964). Foreign exploration for beneficial organisms. *In* "Biological Control of Insect Pests and Weeds" (P. DeBach, ed.), pp. 283–304. Chapman and Hall, London.

Birch, L. C. (1971). The role of environmental heterogeneity in determining distribution and abundance. *In* "Dynamics of Populations" (P. J. den Boer and G. R. Gradwell, eds.), pp. 109–128. Center Agr. Publ. Doc., Wageningen.

Bucher, G. E., and Harris, P. (1961). Food-plant spectrum and elimination of disease of Cinnabar moth larvae, *Hypocrita jacobaeae* L. (Lepidoptera: Arctiidae). *Can. Entomol.* **93**, 931–936.

Carl, K. P. (1968). *Thymelicus lineola* (Lepidoptera: Hesperidae) and its parasites in Europe. *Can. Entomol.* **100**, 785–801.

Cheng, L. (1970). Timing of attack by *Lypha dubia* Fall. (Diptera: Tachinidae) on the winter moth *Operophtera brumata* (L.) (Lepidoptera: Geometridae) as a factor affecting parasite success. *J. Anim. Ecol.* **39**, 313–320.

Clark, R. C., Greenbank, D. O., Bryant, D. G., and Harris, J. W. E. (1971). *Adelges piceae* (Ratz.), Balsam Woolly Aphid (Homoptera: Adelgidae). *Tech. Commun. Commonw. Inst. Biol. Control* **4**, pp. 113–127.

Eichhorn, O. (1969). Natürliche Verbreitungsareale und Einschleppungsgebiete der Weisstannen-Wolläuse (Gattung *Dreyfusia*) und die Möglichkeiten ihrer biologischen Bekämpfung. *Z. Angew. Entomol.* **63**, 113–131.

Embree, D. G. (1971). The biological control of the winter moth in eastern Canada by introduced parasites. *In* "Biological Control" (C. B. Huffaker, ed.), pp. 217–226. Plenum, New York.

Fisher, T. W., and Finney, G. L. (1964). Insectary facilities and equipment. *In* "Biological Control of Insect Pests and Weeds" (P. DeBach, ed.), pp. 381–401. Chapman and Hall, London.

Franz, J. M. (1961). Biologische Schädlingsbekämpfung. *In* "Handbuch der Pflanzenkrankheiten" (H. Richter, ed.), 2nd ed., Vol. 6, No. 3, pp. 1–302. Parey, Berlin.

Ghani, M. A. (1969). Natural enemies of forage and grain legume aphids in Pakistan. *Annu. Rep. Commonw. Inst. Biol. Control, Pakistan Sta.*, unpublished.

Goeden, R. D. (1971). Insect ecology of Silverleaf Nightshade. *Weed. Sci.* **19**, 45–51.

Greathead, D. J. (1971). "A Review of Biological Control in the Ethiopian Region". *Tech. Commun. Commonw. Inst. Biol. Control* **5**, 162 pp.

Gruys, P. (1971). Mutual interference in *Bupalus pinarius*. *In* "Dynamics of Populations" (P. J. den Boer and G. R. Gradwell, eds.), pp. 199–207. Center Agr. Publ. Doc., Wageningen.

Harris, P. (1973a). Selection of effective agents for the biological control of weeds. *Proc. 2nd Int. Symp. Biol. Control of Weeds*, pp. 29–34. (*Misc. Publ. Commonw. Inst. Biol. Control* **6**.)

Harris, P. (1973b). The selection of effective agents for the biological control of weeds. *Can. Entomol.* **105**, 1495–1503.

Harris, P., Peschken, D., and Milroy, J. (1969). The status of biological control of the weed *Hypericum perforatum* in British Columbia. *Can. Entomol.* **101**, 1–15.

Hasan, S. (1970). The possible control of Skeleton Weed, *Chondrilla juncea* L., using *Puccinia chondrillina* Bubak & Syd. *Proc. 1st Int. Symp. Biol. Control of Weeds*, pp. 11–14. (*Misc. Publ. Commonw. Inst. Biol. Control* **1**.)

Hassell, M. P. (1971). Parasite behaviour as a factor contributing to the stability of insect host-parasite interactions. *In* "Dynamics of Populations" (P. J. den Boer and G. R. Gradwell, eds.), pp. 366–379. Center Agr. Publ. Doc., Wageningen.

Huettel, M. D., and Bush, G. L. (1972). The genetics of host selection and its bearing on sympatric speciation in Procecidochares (Dipt.: Tephritidae). *Entomol. Exp. Appl.* **15**, 465–480.

Huffaker, C. B., Kennett, C. E., Matsumoto, B., and White, E. G. (1968). Some parameters in the role of enemies in the natural control of insect abundance. *In* "Insect Abundance" (T. R. E. Southwood, ed.), pp. 59–75. Blackwell, Oxford.

Huffaker, C. B., Messenger, P. S., and DeBach, P. (1971). The natural enemy component in natural control and the theory of biological control. *In* "Biological Control" (C. B. Huffaker, ed.), pp. 16–67. Plenum, New York.

Inman, R. E. (1970a). Problems in searching for and collecting control organisms. *Proc. 1st Int. Symp. Biol. Control Weeds*, pp. 105–108. (*Misc. Publ. Commonw. Inst. Biol. Control* **1**.)

Inman, R. E. (1970b). Host resistance and biological weed control. *Proc. 1st Int. Symp. Biol. Control Weeds*, pp. 41–45. (*Misc. Publ. Commonw. Inst. Biol. Control* **1**.)

Lucas, A. M. (1969). The effect of population structure on the success of insect introductions. *Heredity* **24**, 151–157.

Mohyuddin, A. I. (1971). Comparative biology and ecology of *Apanteles flavipes* and *A. sesamiae* as parasites of graminaceous borers. *Bull. Entomol. Res.* **61**, 33–39.

Pschorn-Walcher, H. (1973). Die Parasiten der gesellig lebenden Kiefern-Buschhornblattwespen (Fam. Diprionidae) als Beispiel für Koexistenz und Konkurrenz in multiplen Parasit-Wirt-Komplexen. *Verh. Deut. Zool. Ges.* (*Jahresversammlung*) **66**, 136–145.

Pschorn-Walcher, H., and Zwölfer, H. (1956). The predator complex of the white-fir woolly aphids (Gen. *Dreyfusia*, Adelgidae). *Z. Angew. Entomol.* **39**, 63–75.

Pschorn-Walcher, H., Schröder, D., and Eichhorn, O. (1969). Recent attempts at biological control of some Canadian forest insect pests. *Tech. Bull. Commonw. Inst. Biol. Control* **11**, 1–18.

Ratcliffe, F. N. (1966). Biological control. *Aust. J. Sci.* **28**, 237–240.

Remington, C. L. (1968). The population genetics of insect introduction. *Annu. Rev. Entomol.* **13**, 415–426.

Schröder, D. (1974). A study of the interactions between the internal larval parasites of *Rhyacionia buoliana* (Lep. Olethreutidae). Entomophaga **19**, 145–171.

Sechser, B. (1970). Der Parasitenkomplex des Kleinen Frostspanners (*Operophthera brumata* L.) (Lep., Geometr.) unter besonderer Brücksichtigung der Kokonparasiten. Teil I und II. *Z. Angew. Entomol.* **66**, 1–35, 144–160.

Simmonds, F. J. (1949). Insects attacking *Cordia macrostachya* (Jacq.) Roem & Schult in the West Indies. 1. *Physonota alutacea* Boh. (Col., Cassididae). *Can. Entomol.* **81**, 185–199.

Simmonds, F. J. (1969). "Commonwealth Institute of Biological Control. Brief Resume of Activities and Recent Successes Achieved," 16 pp. Commonw. Agr. Bureaux Publ., Ferozsons Ltd., Rawalpindí.

Smith, H. S. (1939). Insect populations in relation to biological control. *Ecol. Monogr.* **9**, 311–320.

Taylor, T. H. C. (1937). "The Biological Control of an Insect in Fiji. An Account of the Coconut Leaf-Mining Beetle and its Parasite Complex," 239 pp. Imp. Inst. Entomol., London.

Turnock, W. J., and Muldrew, J. A. (1971). *Pristiphora erichsonii* (Hartig), Larch sawfly (Hymenoptera: Tenthredinidae). *In* "Biological Control Programs Against Insects and Weeds in Canada 1959–1968," pp. 175–194. (*Tech. Commun. Commonw. Inst. Biol. Control* **4**.)

Varley, G. C., and Gradwell, G. R. (1968). Population models for the winter moth. *In* "Insect Abundance" (T. R. E. Southwood, ed.), pp. 132–142. Blackwell, London.

Varley, G. C., and Gradwell, G. R. (1970). Recent advances in insect population dynamics. *Annu. Rev. Entomol.* **15**, 1–24.

Varley, G. C., and Gradwell, G. R. (1971). The use of models and life tables in assessing the role of natural enemies. *In* "Biological Control" (C. B. Huffaker, ed.), pp. 93–112. Plenum, New York.

Wapshere, A. J. (1970). The assessment of the biological control potential of organisms for controlling weeds. *Proc. 1st Int. Symp. Biol. Control Weeds*, pp. 79–89. (*Misc. Publ. Commonw. Inst. Biol. Control* **1**.)

Zwölfer, H. (1961). A comparative analysis of the parasite complexes of the European Fir Budworm, *Choristoneura murinana* (Hb.), and the North American Spruce Budworm, *C. fumiferana* (Clem.). *Tech. Bull. Commonw. Inst. Biol. Control* **1**, 1–162.

Zwölfer, H. (1971). The structure and effect of parasite complexes attacking phytophagous host insects. *In* "Dynamics of Populations" (P. J. den Boer and G. R. Gradwell, ed.), pp. 405–418. Center Agr. Publ. Doc., Wageningen.

9

VARIATION, FITNESS, AND ADAPTABILITY OF NATURAL ENEMIES

P. S. Messenger, F. Wilson, and M. J. Whitten

I. INTRODUCTION

For the last 10 to 15 years increasing interest has been displayed by biological control specialists in the variation, fitness, and adaptability of the organisms which are the natural enemies of insect pests and weeds. This interest derives from attempts to better understand the factors involved in two aspects of the successful use of introduced natural enemies, their ability to colonize a new environment, and their efficacy in suppressing or controlling the target pest populations. A concomitant interest in the fitness and adaptability of natural enemies is the strong desire to increase the proportion of successful importations (Clausen, 1956; Wilson, 1960a; Turnbull, 1967; van den Bosch, 1968).

Variation, fitness, and adaptability of organisms involve aspects of genetics. Biological control specialists therefore seek answers to such questions as: What is the relation of genetic diversity to fitness and adaptability? How does genetic diversity of imported enemies influence colonization and control efficacy? In what ways can fitness and adaptability of imported and colonized natural enemy populations be increased? How can genetic diversity be measured? How should source populations of natural enemies be sampled? How many individuals should be imported? How do importation procedures influence the genetic diversity that does occur? These questions involve the genetics of colonizing species, species intentionally colonized by man, but species which are expected not only to colonize well, but also to function as efficient pest population control agents. It is stimulating that during the past decade these topics have attracted the attention of geneticists, specialists who study such aspects of population genetics as population gene structure, natural selection and adaptation, fitness, colonizing species, and the ways in which fitness and adaptability can be influenced.

Such geneticists have been actively working on questions of colonizing species (Baker and Stebbins, 1965). Biological control specialists have been more concerned with facilitating improvements in the control by natural enemies (Simmonds, 1963; DeBach, 1964, 1965; Wilson, 1965; Messenger and van den Bosch 1971). In this chapter all these aspects shall be considered: the variability of natural enemies as encountered in biological control work, the fitness of individuals and populations, the ways fitness is being or can be influenced by importation procedures, and the ability of imported natural enemies to adapt to the new environments.

II. VARIABILITY IN NATURAL ENEMIES

A. How Natural Enemies Vary in Nature

The natural enemies of an insect or plant pest are numerous in species and diverse in characteristics. The more thorough the research undertaken, the greater the diversity that will be found in the natural enemies. For example, studies of the enemies of the California red scale, *Aonidiella aurantii* (Mask.), have extended over some 75 years and the complex of associated predators and parasites is still far from completely known (see Chapter 15). Also, it is not uncommon to find that what was earlier considered a single taxonomic entity actually comprises several distinct species or sibling species, subspecies, or races. An illustration is the complex of parasites associated with *Eurygaster* and related genera (Delucchi, 1961) (see also Chapter 15). Such related forms may or may not be morphologically distinguishable, but they are often first recognized by biological differences and referred to as biological, physiological, or geographical races, or strains. The biological differences discovered are often important in biological control (Sabrosky, 1955). It is this variability at the species level or below that is of great interest in biological control research.

The natural enemy complexes of a pest in different parts of its natural range usually differ greatly in composition and relative abundance, as Thompson and Parker (1928) noted for *Ostrinia nubilalis* (Hübn.) in Europe. This is a reflection of the high degree of ecological specialization that natural enemies commonly exhibit. This may affect any characteristic, whether morphological, physiological, ecological, behavioral, or otherwise, and may involve climatic tolerances, searching capacities, host relationships or other important characteristics. Such differences occur not only between species, but also between subspecies or races. These high degrees of diversity and specialization are a consequence of long-term inter- and intraspecific competition, which have produced more specialized natural enemies better able to survive the competition, though at the same time often becoming more restricted as to habitat.

The following are examples of the kinds of variability that occur in nature, and that are being found with increasing frequency during biological control research.

Stocks of *Tiphia popilliavora* Roh. from Korea and Japan differ in reproductive capacity and time of emergence (Clausen, 1936). The egg parasite, *Trichogramma minutum* Riley, has races that differ in developmental rates, mortalities at different humidities, length of life cycle, host range, distribution, and habitats (Flanders, 1931; Lund, 1934; Quednau, 1960). Quednau considers that this genus contains a number of very complex species with populations that can adapt in respect to host relationships and habitats occupied. *Aphytis maculicornis* (Masi), a parasite of olive scale, has three reproductively isolated strains that differ in duration of developmental stages, sex ratio, and fecundity (Hafez and Doutt, 1954). In the United States there are three host-specific races of *Trichopoda pennipes* (F.) limited as to distribution; a race in the eastern States that attacks *Anasa tristis* (Deg.) and other coreids; a race in the southeastern States that parasitizes *Nezara viridula* (L.) and other pentatomids; and a California race that is parasitic on the pyrrhocorid, *Euryophthalmus cinctus californicus* Van D. (Clausen, 1956; Dietrick and van den Bosch, 1957).

Simmonds (1963) assembled some examples of racial differences in entomophages. *Metagonistylum minense* Tnsd., a tachinid parasite of the sugarcane borer, has two races that differ in adaptation to the wetness of the habitat. The stem borer parasite *Paratheresia claripalpis* Wulp has several races in Central America and the West Indies that differ in the hosts attacked, puparial period, and fecundity. Among parasites of citrus scale insects in California, three species each have two races that parasitize different hosts: a race of *Comperiella bifasciata* How. from Japan utilizes the yellow scale, *Aonidiella citrina* (Coq.), and a Chinese race attacks California red scale, *A. aurantii*; two races of *Prospaltella perniciosi* Tow. parasitize respectively the San Jose scale, *Quadraspidiotus perniciosus* (Comst.), and California red scale; and in *Aspidiotiphagus citrinus* (Craw.), one race attacks the oleander scale, *Aspidiotus nerii* Bouche, and the yellow scale, while the other race parasitizes California red scale (DeBach, 1958). There may also be differences in the stages of the host attacked.

Metaphycus luteolus (Timb.) in California reproduces on the earlier stages of *Saissetia oleae* (Oliv) as a solitary parasite, and on all stages of *Coccus hesperidum* L., being gregarious in the later stages; in Mexico, on the other hand, this parasite attacks all stages of *S. oleae*, being gregarious in the later stages, but it was not reared at all from *C. hesperidum* (DeBach, 1958).

Another kind of variability is exemplified by *Chrysolina quadrigemina* (Suffr.), used in several countries for control of the weed *Hypericum perforatum* L. This insect is polymorphic, the adults occurring in various color morphs, which are found in different proportions in different areas. This polymorphism may involve physiological characteristics, and may provide this univoltine insect with a mechanism for adjusting populations to the differing environmental conditions to which the species is exposed in different areas and seasons (Wilson, 1965). In this connection, it is of interest to note that in California *C. quadrigemina* is the dominant and most effective control agent against *H. perforatum*, while the related species, *C. hyperici*, also introduced and once common to abundant, is less well adapted and has now been displaced by the former (Huffaker and Kennett, 1959, 1969). However, in British Columbia where again both species were colonized, it was *C. hyperici* which was the more abundant, while *C. quadrigemina* originally was rare and presumably, in this environment, less fit (Smith, 1958).

From these examples it is evident that entomophagous or phytophagous species, or races or strains of species, may differ greatly in respect of climatic tolerances, host relations, or any of the innumerable properties of natural enemies that are of significcance for their ability to regulate the numbers of a pest in a required area (Chapter 3).

The habitat specialization of such species or races often imposes restrictions of their usefulness. The coccinellid *Cryptolaemus montrouzieri* Muls. is a tree-frequenting predator that cannot operate effectively against its mealybug hosts in vineyards or other low-growing cultures. Similarly, *Rodolia cardinalis* (Muls.) is unable to exploit the cottony-cushion scale, *Icerya purchasi* Mask. on some plants, such as Scotch broom and maple, while *Chrysolina quadrigemina* is markedly less effective in the control of *Hypericum perforatum* where shaded by other vegetation. *Aphelinus mali* (Hald.) is effective against the woolly aphid, *Eriosoma lanigerum* (Hausm.), when the aphid's colonies are on the aerial parts of apple trees, but not when on the roots. Parasites may also be unable to develop on their hosts when these are growing on certain host plants, e.g., *Habrolepis rouxi* Comp. develops satisfactorily on California red scale on citrus, but not on the same insect on *Cycas revoluta* Thunb. (Flanders, 1940).

Considerable genetic variability also exists, of course, at the level of the individuals of a species or race, and this has great importance for the ability of an introduced population to adapt to a new habitat for, as Allen (1958) commented, genetic differences exist that fit segments of a general population of a species to particular ecological environments. Evidence of this is seen in those cases in which it has been possible by selection to modify the characteristics of laboratory strains. For example, Wilkes (1942, 1947) produced in four generations a strain of *Dahlbominus fuscipennis* (Zett.) that had lower temperature preferences, greater fecundity and longevity, and a higher proportion of female progeny; Urquijo (1951) increased by fourfold the host-finding ability of *Trichogramma minutum* in 3 years, and Pielou and Glasser (1952) increased by several times the DDT tolerance of *Macrocentrus ancylivorus* Rohw.

B. Variation at the Molecular Level

Thus far variation in the whole individual has been considered, realizing, of course, that phenotypical variation has its basis in the genes. It is also worthwhile to consider intraspecific variation at the molecular level.

Analysis of genetic variation in natural populations has been restricted in the past to that fraction of the genome which manifests itself at the morphological level, such as, variation in organ size, shape, pigmentation, and bristle number, and to a lesser extent for physiological and behavioral traits. Wild populations of some of the better studied species, e.g., *Drosophila* spp., have been characterized by a lack of variation at the more obvious morphological level and this had led to the "classical" school of thought typified by the works of H. J. Muller and J. Crow, that most loci in the genome carry only wild-type alleles. Some loci might display two or more alleles but these were regarded as exceptional. The converse viewpoint held by the "balanced" school of thought, exemplified by T. Dobzhansky and B. Wallace, is now regarded as closer to the truth. Estimates of the number of polymorphic loci in a population, based on electrophoresis which detects net electrical charge differences in the actual gene products, the enzymes, are of the order of 30%, with some 12% of the genome heterozygous per individual. Thus each individual has two alleles segregating at many loci. These levels seem to hold for diverse organisms, e.g., flies, frogs, mice . . . and man (Selander *et al.*, 1970), and plants (Allard and Kahler, 1971). Since some 75% of mutations are not expected to alter electrophoretic mobility, these estimates are probably too low. Thus at the level of the gene product there might appear to be an unlimited amount of raw material in natural enemies for natural selection to act upon. Such an inference has to be treated with caution until the significance of this wealth of heterogeneity is better understood. It is not known how much is adaptive or even whether much of it affects enzyme activity. The knowledge is of little advantage until the electrophoretic variation can be related to some developmental process, whether it be physiological, morphological or behavioral, which has some ultimate ecological value.

This relationship has been tentatively established in some instances, e.g., organophosphate resistance in *Lucilia cuprina* (Wied.) (Schuntner and Roulston, 1968) and *Boophilus annulatus* (Say) (Schuntner *et al.*, 1968) and alcohol metabolism in *D. melanogaster* Meig. (Gibson, 1970; for a review see Manwell and Baker, 1970). However, it would be optimistic to suppose that in general one could determine the effects of particular variants and thereby select the superior natural enemy without reference to performance of the whole organism. For the present, the exposure of this vast array of genetic variation seems only of direct relevance to the geneticist, but as is discussed in Section III,B, it could be used for purposes related to biological control in the following ways: to clarify the taxonomic status of strains of a natural enemy (cf. Johnson and Selander, 1971) to provide a basis of whether samples of natural enemies should be pooled or held and released separately; to identify ecological centers and margins of the distribution of a natural enemy; to monitor strains subsequent to their release, determining which particular release was most effective; to measure the extent that postcolonization adaptation alters the genetic structure of the released material. The remarkable similarity at the molecular level between island and continental populations (Ayala *et al.*, 1971) suggests that postcolonization changes may not be as substantial as is usually supposed. Rapid advances in the analysis of electrophoretic variants (cf. Brewer, 1970, for review of techniques) should permit the extension of this tool to a great number of natural enemies.

III. FITNESS OF NATURAL ENEMIES

A. The Meaning of Fitness

The success of a natural enemy following its release into a new area depends largely on its genetic fitness and this, in turn, is determined by the genetic composition of the source material. Some attention should thus be given to the genetic structure of this source material. A major concern, then, is to determine whether there are any genetic considerations which would influence the tactics employed by entomologists above and beyond those already dictated by common sense or imposed by technical limitations, such as inaccessibility of source material, its uncertain taxonomic status, shipping, and quarantine restrictions. It has to be assumed that detailed information on the population genetics of a particular natural enemy will generally be lacking. Indeed, sometimes even the taxonomic status of the natural enemy in question is uncertain. In such circumstances are there general guidelines whose adoption would, on the average, maximize the chances of introducing a strain which will adapt successfully to its new environment, or is each introduction sufficiently unique that even rough guidelines have little merit?

The first consideration is what is usually understood by the terms "fitness" and "adaptability" and whether natural selection will always act on the available genetic variability to maximize both of these attributes. Following this, the problem of what tactics might gainfully be employed to ensure that a fit and adaptable strain of natural enemy is released must be considered.

Biological control workers tend to measure fitness of a natural enemy by its ability to colonize successfully its target area and by its ability to effectively regulate its host. These two abilities are quite distinct (Messenger and van den Bosch, 1971), and, in fact, may conflict in some instances as evidenced by many natural enemies of some dipterans whose establishment is readily accomplished and whose survival seems rather secure but whose regulatory effects on the numbers of their host(s) are minimal. Conversely, it is conceivable that a natural enemy might be so effective that its own survival could be in jeopardy, though no evidence has been produced that demonstrates this.

B. The Measurement of Fitness

For biological control purposes, then, fitness can be inferred as the capability of the natural enemy to survive in the new environment and control (effectively) its host pest. Natural selection acts to maximize short-term fitness (i.e., ability to survive) but unfortunately little can be said a priori about how it will impinge upon effectiveness.

However, in respect to effectiveness, natural selection can be expected to favor those natural enemy individuals (females) which are successful in finding hosts when the latter are scarce, that is, whose searching capacity at low host densities is high. It is in this area of behavior that survival and effectiveness (as a biological control agent) become identical. This latter aspect of fitness cannot be adequately measured except by observing an actual release. Various features of survival ability can be measured before releases are made but there is little or nothing that population genetics can suggest to the entomologist in this area. A complete genotypical description of all the individuals in the population would not permit an adequate measurement of fitness. Even attempts at fitness estimates in well specified situations, e.g., between alleles in cage populations of *D. melanogaster*, are troubled by technical and methodological problems (see Prout, 1971). Relative fitness of genotypes is not independent of the environment and just two among a host of unspecifiable facets of this environment are density (Sokal, 1967) and the frequency of other genotypes (Harding *et al.*, 1966). Thus in the complex situation of using a natural enemy to colonize a new territory, there may be little point in detailing a description of the material for release, either morphologically or even electrophoretically. Rather than measure the fitness of the natural enemy with a view to selecting and releasing a preadapted strain, emphasis should be given to adequate sampling of existing populations, and rearing and releasing them in a manner which maximizes the prospects of evolving an effective natural enemy in the new environs. If this point of view is accepted, the ability to adapt rather than fitness is the important criterion.

While most geneticists endorse the viewpoint that the broadest genetic base should permit maximum evolutionary advance, there is difference of opinion about how this broad base should be used (Lewontin, 1965b). It must be borne in mind that conclusions must relate to individual situations where the collector may have no genetic expertise and, indeed, may have little or no information about the ecology, population structure, or taxonomy of the material he is collecting. Unfortunately, these limitations tend to restrict direct applications of much of the possibilities presented in the following sections.

A better basis for understanding this problem will be in hand after consideration, in the next section, of the ways in which current practices in biological control affect potential fitness.

C. Fitness and Current Importation Practices

Importation and colonization of exotic biological control agents involve a sequence of processes which often reduce the genetic diversity of the imported sample (Bartlett and van den Bosch, 1964; Messenger and van den Bosch, 1971). This sequence includes the search for and selection of one or more samples from one or more source populations, their shipment in closed containers to the receiving laboratory, the receipt, identification, isolation, and microculture of the new import in a quarantine facility, usually the mass culture of the colony under artificial conditions after release from quarantine, and the colonization of greater or lesser numbers of the species at various places in the target environment.

The procurement of samples of natural enemies from source populations has usually involved very few individuals relative to the numbers present in the source environment. Natural enemies, by their very nature, are, paradoxically, numerically scarce where most effective. Furthermore, a particular natural enemy species may be scattered over entire continents, it may when discovered be new to science, or it may appear as but one species in a complex of species in association with the target pest. It is therefore not surprising that initial samples of the source population of the sought after natural enemy are small.

Good foreign explorers are cognizant of this problem. Indeed, for decades biological control explorers have been counseled to select as large a sample of the natural enemy as possible in order to increase the genetic diversity of the colonizing inoculum and thereby maximize its adaptive potential (Clausen, 1936; Smith, 1941; Allen, 1958; Simmonds, 1963; Doutt and DeBach, 1964; Wilson, 1965; Force, 1967). However, practicalities, some of which have already been mentioned, often interfere. Furthermore, initial sampling is most often made at one or two points in the area of occupancy of the species. Choice of geographical sites for collecting candidate material is often guided by similarities in climatic conditions with those of the habitats into which the agents are to be introduced. Little attempt is made to determine the extent of the source population distribution, let alone its population structure (i.e., whether there is one central population or a central population plus several peripheral, partially isolated populations) (Remington, 1968; Lucas 1969; Whitten, 1970). Once a sample of a natural enemy is collected, it is either shipped promptly to the home reception facility or is held "in camp" for additional development to an appropriate stage for shipping (Bartlett and van den Bosch, 1964). At present, such a shipment is received 5–10 days later in the quarantine laboratory where it is held for identification, life history study, and transfer to local host cultures. All of these procedures and processes take time, usually involve close crowding under highly artificial conditions, and thus are potentially or actually deleterious to the survival of the introduced individuals (Fisher, 1964; Messenger and van den Bosch, 1971).

A further restriction or selective "sieve" on the genetic diversity of the imported sample occurs in the processes of mass culture which usually precede intensive colonization efforts (Mackauer, 1972). Usually several to many generations are propagated to build up numbers from a few dozen to thousands or hundreds of thousands. Each such generation undergoes some artificial selection under the insectary culture conditions prevailing [high densities in cages, artificial illumination, usually continuously favorable, warm, moist atmosphere, sometimes entirely foreign or artificial food supplies (factitious live hosts, chemical diets)]. Oftentimes the adults produced in culture are held in cold storage in small, crowded containers in preparation for field release. This also reduces survival.

It is thus clear that in an introduction effort, even when the goal of maximized genetic diversity is recognized, technical procedures tend to interfere with such an objective. There are other practical issues involved. One is whether or not to merge several samples of a natural enemy species collected in a source environment (at time of shipment, quarantine processing, or mass culture). For reasons of space, labor, and budget, hardly ever are such samples maintained or colonized separately. Advice from geneticists is not clear as to whether or not pooling affects diversity or potential adaptability (Lewontin, 1965a; Force, 1967; Remington, 1968; Whitten, 1970; Mackauer, 1972). This will presumably be dependent partly on the degree of artificial selection that occurs during the importation procedures.

In conventional colonization, usually considerable numbers are released, from several hundreds to several thousands at one place, under presumed favorable conditions. The rationale is that if a few individuals might not establish, a larger number might. Where only small numbers are released it is mainly because of problems which restrict mass culture. Clearly, however, the greater the number of individuals released the greater is the likelihood that an increased proportion of the potential genetic diversity will be established.

Finally, where large releases are made at one locality, there is the matter of the impact of repeating releases periodically for an entire season or for several years in a sequence. It is quite possible in such cases for first releases to lead to incipient adaptation to the local habitat, which adapted stock might then become swamped by repeated releases of less adapted insectary stock. Hence the question is whether the best strategy is one of one or two large releases or many smaller, repeated releases.

IV. DESIRABLE CHARACTERISTICS IN NATURAL ENEMIES

Though the ability of a natural enemy to control a pest in a new habitat is readily disclosed by its introduction, it is seldom possible to forecast with confidence the result of an introduction, and never possible to describe precisely all the characteristics necessary for effectiveness in any particular instance (but see Chapters 3, 8). Such prescience is at present precluded by the rather extreme complexity of the interactions between the pest, the introduced natural enemies, and the physical and biotic environment.

It is essential that introduced natural enemies have marked degrees of ecological specialization and fixity of habits, for these characteristics ensure against their attacking other than the target organisms. This is obviously essential, for example, in the biological control of weeds. Fortunately, as was stated above, such specialization is a common characteristic of entomophagous and phytophagous insects; it is also an essential element in the high level of adaptation to their hosts that makes it possible for many natural enemies to regulate their host's numbers.

That some properties of such natural enemies are, to a substantial degree, fixed is evident from experience in biological control work, for their characteristics often prevent them from exploiting their hosts fully throughout the range of the latter. Experience shows that such deficiencies are not necessarily mitigated by subsequent adaptation. For example, the Egyptian race of *Asolcus basalis* (Woll.) remainded for decades incapable of controlling *Nezara viridula* (L.) in areas of southeast Australia having a cold winter climate (Wilson, 1960b). The French strain of *Trioxys pallidus* (Hal.), effective in southern California against the walnut aphid, was unable to establish in host-infested areas of central and northern California (Messenger, 1970). On the other hand, in some instances (DeBach, 1965), slow improvement in the fitness of entomophages may have occurred, but the limited evidence does not suggest that any long-term improvement in ability to control should be expected. This is not to deny that genetic variability in an introduced population of a natural enemy may well allow natural selection of subpopulations that have greater local survival value and control effectiveness.

An introduced population of a natural enemy should be preadapted to the pest and the new environment, and should have no important needs that cannot be satisfied in this environment. Some of the necessary preadaptations to the host and habitat can be indicated. The natural enemy must be able to tolerate the range of temperatures, humidities, and other physical conditions that occur in the different seasons, including those that occur during extremes of weather. The natural enemy must be so adapted that these physical conditions permit it a fecundity, or, more precisely, a power of increase, that is ample to overtake the host numerically during a part of the year, especially at those times that are particularly favorable for host increase. Developmental rate and voltinism are of importance in assessing capacity to overtake host numbers. The enemy ought also to have a coevolutionary power comparable to that of its host, i.e., it ought to be able to overcome any host resistance which may tend to arise in time (Pimentel, 1961, 1968).

It is also desirable that an entomophage should be able to attack the host on all of its important host plant species. This requires that the natural enemy should frequent all such species and be able to develop on hosts that use any of them for food.

The range of hosts attacked by the natural enemy should be rather narrow or even extremely so (especially, of course, in phytophagous insects used to control weeds), but, in some cases polyphagous natural enemies are effective when several of their host species are present together in a given habitat (Ehler and van den Bosch, 1974). The biology of the natural enemy must be well synchronized with that of the host, so that they are in phase and susceptible stages of the host are available at the required time. This synchronization must extend through quiescent stages of the host's annual cycle (as in diapause) or there must be alternative hosts that can carry the natural enemy through such periods. For example, it is considered that the failure to establish *Triaspis thoracicus* Curt. against *Bruchus pisorum* (L.) in Western Australia was attributable to the absence of alternative hosts. Lack of a similar requirement has been judged to be a reason why the parasite *Macrocentrus ancylivorus* has not become established in many regions of North America where its host, the oriental fruit moth, occurs. An alternate host has also been found to be important for the grape leafhopper egg parasite, *Anagrus epos* (Girault) (Doutt and Nakata, 1965).

A natural enemy must have a high level of ability for discovering the host; for an entomophage, this involves discovering its host's host plants and finding individual hosts on these plants. These activities necessitate appropriate high powers of dispersal, search, and persistence. These qualities must be present at a level that allows the discovery of hosts even when they are at low density. It is commonly found that well-adapted entomophages are able to discriminate between healthy and parasitized hosts, and to exercise restraint in oviposition except in or on healthy hosts.

It is sometimes thought of entomophages that parasites are more likely than predators to be effective controlling agents. This view is supported by experience which shows that four times as many parasites as predators have provided control of pests (DeBach, 1965), and on the general finding that predatory insects are less host-specific. However, predators should not be neglected. There is reason to believe that some are highly specific both in host range and in habitats selected for host search (Thompson, 1951). Evidence is accumulating of the importance of rather general predators in the control of a complex of pests in frequently disturbed situations (i.e., annual crops) (Ehler and van den Bosch, 1974). In any event, it is clear that some predators provide an extremely high level of control as *R. cardinalis* does with the cottony-cushion scale (see also Chapter 5).

There exist many examples of the disturbance of the balance between natural enemies and their hosts following the release of insecticides into the ecosystem. Host species often evolve resistance to the insecticide before any associated natural enemy shows increased tolerance. This low ability on the part of parasites and predators to evolve resistance may be the combined effect of two factors. The natural enemies have not experienced during their evolutionary past the same procession of noxious substances produced by plants as part of the latter's arsenal of defense mechanisms. These substances have been detoxified in part or wholly by an enemy's host, which consequently has had a history of this type of experience. Thus the host may be better preadapted to cope with yet other noxious substances, the synthetic pesticides. This is certainly not the whole story as many parasites, e.g., myiasis flies, whose evolutionary experiences were presumably similar to those of predators and parasites, readily evolve resistance.

Perhaps a more general explanation of the natural enemy's tardiness to evolve resistance is simply that it lacks the same level of opportunity as the host species. Until the host has become resistant its numbers are often too low to support the predator which may in fact become extinct in many local areas where insecticide is regularly applied (Huffaker, 1971). Once the host has become resistant to one insecticide an alternative substance is substituted. However, occasionally when the main target species is another pest, e.g., in the control of codling moth with organophosphates, the insecticide is still used despite development of resistance in species of tetranychid mite pests, thereby allowing predatory phytoseiid mites an opportunity also to evolve resistance (Croft and Barnes, 1971).

Should a natural enemy exhibit increased resistance to an insecticide, occasional screening with the insecticide might be advisable if a long sojourn in the laboratory is anticipated. In the absence of insecticide treatment the ability to detoxify them is sometimes quickly lost. A gene for OP-resistance in population cages of *Lucilia cuprina*, with initial frequencies around 0.5, was virtually lost within 20 generations due to adverse pleiotrophic effects of the resistance gene in the absence of the insecticide (M. J. Whitten, unpublished data).

Where several natural enemies influence the abundance of a single host it is possible that their relative importance might have changed significantly following insecticide usage. This could explain why phytoseiid mites play only a minor role to the predatory beetles of the genus *Stethorus* (Coccinellidae) in Australia in the control of tetranychid mites (J. L. Readshaw, unpublished data) yet are important natural enemies of such mites in Europe and the United States (Huffaker *et al.*, 1970). It would be unfortunate if resistant phytoseiid mites were not imported from the United States and at least tested against tetranychid mites simply because their pest control role, as susceptible individuals, has been considered minor. This has now been done and the results seem promising.

It is not to be expected that all the desirable characteristics listed above will be found to be present in perfect form in a natural enemy; or is it necessary, for even in highly successful examples of biological control the controlling species often have deficiences in some of these qualities.

V. IMPROVING THE ADAPTABILITY OF IMPORTED SPECIES

A. Sampling Procedures

The restrictive effects of conventional exploration and sampling procedures on the genetic diversity of imported natural enemies have been previously mentioned (Section III,C). To the extent that broad genetic diversity increases the adaptability of a natural enemy once established in a new environment, it appears possible that adaptability can be enhanced by modification of sampling plans. Two aspects may be considered here: where to sample, and how many individuals to collect. On the latter point, unfortunately, no population geneticist has analyzed and estimated the *number* of organisms of an interbreeding population which should be collected in order to provide a given percentage (80, 90, and 95%) of the existing genetic diversity. Instead, as has already been stated, entomological advice is at hand which simply says to collect as high a number of the source population as can be conveniently done (Clausen, 1936; Smith, 1941; Allen, 1958; Wilson, 1960b; Simmonds, 1963). However, it may be possible with electrophoresis techniques to measure the increase in phenotypical diversity contained in differently sized samples of a given population, and thereby gain some idea of the relation sought (Whitten, 1970).

Advice from population geneticists is more abundant, but conflicting regarding where to sample. In most cases it is necessary to have a fair idea of the population structure of the species of natural enemy under consideration; that is, whether made up of one panmictic population or of a group of partially isolated subpopulations, whether composed of "central" and "marginal" populations, and so on. Remington (1968, p. 421) has proposed that in those cases where the target natural enemy species is composed of both large, central populations and small, marginal populations, the better sampling strategy for both initial establishment and subsequent adaptive improvement is to:

> Introduce a large, wild sample from a large, central source population which has an environment most similar to that of intended establishment. Even better would be to introduce a closely spaced succession of wild samples from several source populations from various environments moderately like the area of intended colonization; this maximizes the relevant genetic variability on which selection can then act to produce an optimal genotype in the new environment.

Remington (1968) goes on to warn against introducing a few individuals from a large source population or releasing from laboratory (insectary) mass cultures initiated from only a few founders.

Suffice it to say here that current quarantine restrictions will not, except under unusual circumstances, allow the direct release of imported individuals without first culturing in the laboratory for at least one generation. This is to minimize the possibility of introducing a noxious organism. Further, given the usually large size of the environment to be colonized, coupled with the generally few individuals the foreign explorer will collect, the caveat against laboratory culture prior to release assumes added importance.

Although Lucas (1969) has criticized certain of Remington's (1968) deductions as being inconsistent with his model, he accepts the model proposed, and notes that Remington's conclusions regarding beneficial insect importation seem to be correct interpretations of the model. Lucas (1969, p. 154) then restates the predictions concerning establishment and adaptive improvement as follows:

> The best chance of establishment of a population would be when founders come from an ecologically central population. The best chance for the evolution of a new type in the new environment would occur when the founders come from a population that was ecologically central.

Whitten (1970), while agreeing with both Remington and Lucas that provision of the maximum genetic variability is the best approach for purposeful colonizations, concludes that a better way to attain this is to introduce many separate samples drawn from all sorts of source environments, marginal as well as central. By this approach the chances of colonizing a more highly preadapted sample are increased. This would enhance the probability of establishment, and, if the sample is large enough, provide for post-colonization adaptive improvement. Whitten (1970) suggests that the amount of genetic variability contained in a species population could be estimated by electrophoretic techniques, "genetic fingerprinting," which in turn can provide information about proper sample size, effects of pooling samples, whether samples are derived from marginal or central populations, and what happens (genetically) to the imported population after colonization. It might be pointed out that such a procedure (genetic fingerprinting) may help disclose the effects of intercontinental shipping, quarantine handling, and insectary mass-culture procedures on the genetic diversity of the sampled materials.

Levins (1969) has considered the genetic aspects of colonizations for biological control purposes on quite a different basis. He focused on the population structure of the host or pest species, and from this developed several possible control strategies which, in turn, affect sampling plans. First of all he considers the pest environment as one which varies spatially and temporally (seasonally), and proposed a pest population model composed of numerous subpopulations, some of which will become locally extinct while others will arise anew by migration. There is balance between these two events such that the whole population is in a state of equilibrium. From this model, Levins concluded that a parasitoid can be used to increase the extinction rate of subpopulations.

There are several arguments against such a model and the suggested uses of natural enemies that derive from it. First, not many applied entomologists would recognize the applicability of such a population substructure, with localized extinctions balanced by localized reinvasions, to the insect pest species with which they commonly deal. Most crops in any one area are year after year infested with the same major pest species. Extinctions and reinvasions are rarely, if ever, observed.

Second, experience with natural enemy population dynamics suggests that effective ones never bring their host populations to extinction except perhaps in extremely localized situations, i.e., on one leaf or plant. The property of density dependence, brought about by mutual interference or other forms of competition at high natural enemy densities, or ultimate limitation of host-finding ability at very low host densities, serves to reduce to a very low value the probability of host extinction. Both host and natural enemy vagility also serve to reduce the likelihood of host extinction.

B. Alteration of Introduction and Propagation Procedures

1. Shipping Procedures

A certain amount of mortality usually occurs when natural enemies are shipped from one region of the world to another. Although modern air transportation enables most such parcels to be received in from 5 to 10 days, nevertheless this degree of confinement, sometimes coupled with exposure to high temperatures during handling at airports, occasionally causes loss of stock. These problems are recognized, and considerable ingenuity is used to reduce the losses (Bartlett and van de Bosch, 1964; see also Chapter 8).

When the shipment parcels are prepared the sample isolates should be put in separate containers. If the foreign explorer has obtained several samples of the natural enemy from separate localities, these samples should be kept separate during and after shipment (Lewontin, 1965b; Force, 1967; Whitten, 1970).

2. *Quarantine Handling*

Quarantine reception and processing requires that introduced stocks be reared to the adult stage, identified, isolated into pure culture, and propagated for at least one generation on locally derived host stocks. Frustrations arise so far as maintenance of maximum genetic diversity is concerned when the natural enemy proves difficult to propagate. Often such problems arise as a consequence of complications in the natural enemy life cycle as, for example, the self-parasitic habit in some of the aphelinid Hymenoptera (Flanders, 1959). Sometimes the required host is unavailable at the location of the quarantine laboratory, as in the case of the western cherry fruit fly, *Rhagoletis indifferens* Curran, whose distribution is confined by plant quarantine restriction to its wild habitat in western North America (Messenger, 1970).

In any event, processing of natural enemies in quarantine requires the shipments and sample stocks to be restricted as to numbers (Fisher, 1964). Propagation in quarantine may result in unwitting artificial selection of the imported stock by the culture conditions pertaining. Care must be exercised to enable the maximum number of individuals to reproduce, outbreeding should be encouraged and inbreeding reduced, crowding minimized, and superparasitism which destroys potential progeny must be minimized by use of large host to enemy exposure ratios or short host to enemy exposure periods. Fortunately, past experience suggests that quarantine culture need not extend for more than one or two generations, by which time the necessary quarantine procedures have been completed and the stock is released to the mass culture insectary.

3. *Insectary Propagation*

The genetically restrictive effects of mass culture on the potential adaptability of imported natural enemies has already been mentioned (Section III,B, above; see also Boller, 1972; and Mackauer, 1972). When we consider that among the characteristics desired in a natural enemy (see Section IV) are ability to survive for long periods at low host densities (involving aspects of semistarvation for predators, or host deprivation for female parasites), possession of a high mobility and searching capacity for hosts when the latter are scarce, ability to distinguish between healthy and already parasitized hosts and to avoid superparasitizing them, ability to find food, mates, and other scattered resources or needs when the enemy itself is scarce, and ability to withstand extremes of climate and weather, it is clear that normal mass culture procedures provide wholly contrary conditions. That is, hosts are abundant and proximate, hence placing no requirement on enemy searching ability; enemy exposure periods to hosts are short and occur as soon as enemy sexual maturity is attained, so that little is needed in the way of high survival power or exceptional longevity; ovipositing parasite females are held in crowded conditions with hosts such that superparasitism may be prevalent; temperature, humidity, and light conditions are constantly favorable for rapid development, high fecundity, good longevity, and minimal occurrence of diapause (Finney and Fisher, 1964).

On the contrary, procedures ought to be incorporated in mass-culture programs which will challenge the culture, either once each generation or intermittently, in just those properties deemed important in (a) a colonizing species (Mayr, 1965), and (b) an effective enemy control agent (Doutt and DeBach, 1964; Wilson, 1965). Intermittent selection for longevity and survival in the face of unfavorable conditions of food, hosts, mates, and reduced temperature and/or humidity should be allowed for. High adult mobility and searching power for hosts should be occasionally required. Periodically, brief exposures to extremes of temperature and humidity should be imposed. Most importantly, the capability for diapause, when it can be seen to be a required or desired property for the natural enemy in the target environments, as would be particularly indicated in those situations where the host itself undergoes a seasonal diapause or quiescent period, must be maintained by appropriate environmental stimuli. This becomes particularly important when considering the ease with which a normal diapause capability in a species can be eliminated by laboratory selection for continuous development. (Ironically, this selection for diapause-free, continuously developing cultures is one of the first things intentionally done in a mass-culture insectary where species are involved which normally pass into a diapause state each generation or sequence of seasonal generations.)

In the final analysis, to avoid a substantial artificial selection in the insectary away from those properties considered desirable it would appear that the best strategy would be to maintain conditions, host ratios, and densities of natural enemies under culture that approach values expected to prevail in nature when control of the host is accomplished. As has been indicated above, in most mass-culture cases almost the exactly opposite conditions, host ratios, and densities are maintained. It will take a considerable revision of conventional mass-culture procedures to adapt these suggestions. As Mackauer (1972) has stated, in particular, the policy of maximum numerical production at minimum cost will need reconsideration. Perhaps a compromise between maximizing production and maintaining "natural" habitat conditions will be the most practical strategy.

4. Colonization Procedures

In general, colonization releases are made at times and places seemingly most favorable for establishment (see Section III,C). If different samples of the natural enemy were collected from source environments and maintained separate from one another until time of release, then a plan must be developed for release of these separate cultures. It seems intuitively logical that a proper plan would be to release each "subspecific" culture in a different and somewhat isolated locality. Such isolation need not be complete, but only of sufficient temporal duration to prevent immediate hybridization to enable several generations of the given culture to establish themselves and begin to adapt to the local environment. After initial adaptation, subsequent dispersal and crossing with other partially adapted cultures released in nearby localities ought to reinforce segregation and selection of fitter phenotypes.

One final caveat about conventional procedures in colonizing natural enemies concerns the continued release week after week in the same locality of individuals from a uniform insectary culture. This is probably not a good idea, for any tendency for local adaptation (through natural selection) by some of the progeny of the first releases will probably be swamped by crossing with individuals from later releases of the, by then, less fit insectary stock. This is not to argue against release of large numbers in one locality, for this is one way to assure an adequate genetic diversity. Rather it suggests that such releases be made only over a relatively short period of time (within one generation of the natural enemy).

C. Artificial Selection

Some of the limitations of direct selection for increasing the fitness of natural enemies have been indicated earlier in this chapter. These problems are now discussed in more detail. Where it is possible to attribute failure to a particular shortcoming of the natural enemy, such as inability to tolerate high or low temperatures, it is preferable to exhaust sources of natural variation for remedying the defect before embarking on a program of artificial selection. This was, in fact, the procedure followed in the case of the recently successful biological control campaign against the walnut aphid in California (Messenger, 1970; Messenger and van den Bosch, 1971). Although White *et al.* (1970) were able to produce increased tolerance to higher temperatures in *Aphytis lingnanensis* by direct selection for use against the California red scale, the work lost some of its significance with the successful release of the naturally better adapted *A. melinus* from Pakistan.

In the absence of precise information about the optimal relationship between seasonal adaptation of the host and its natural enemies, source material must be collected which embodies sufficient variation to permit natural selection in the new habitat to mold an effective predator or parasite, and, unfortunately, there is little guarantee that natural selection, in maximizing the survival prospects of the natural enemy, will simultaneously maximize its effectiveness as a biological control agent. Only when it is established that suitable strains are unavailable and that there are no alternative natural enemies, would collections from dissimilar regions or artificial selection for a desired trait seem warranted. Thus considerable ad hoc experimentation is indicated before more detailed studies are initiated.

Nevertheless, the need for direct selection for desired traits may be sufficient to justify an outline of the major steps to be taken to enhance response to selection. These are: (1) to determine what characteristics are desirable, (2) to provide a broad base of genetic variability, (3) to measure the heritability of the character and determine a suitable selection regime, and (4) to minimize undesirable effects of selection such as inbreeding depression or genetic correlation with adverse characteristics.

Step (1) involves a decision where the entomologist has to draw from experience and intuition. Many of the desired characteristics, described in a purely qualitative way, have been listed (Section II,C, and Chapter 3). Major difficulties lie in determining which character is limiting in any given case, how this character is to be measured quantitatively, and how much improvement is required. A large measure of empiricism will be involved in this step.

Regarding step (2), there has been a considerable amount of discussion among plant breeders on how to obtain a broad genetic base on which to impose selection. Frankel and Bennett (1970) stress the importance of exploiting variation among natural populations, while Brock *et al.* (1972) have indicated a role for induced mutation to boost this variation. Muller (1927) first demonstrated that irradiation can induce mutations, but chemical mutagens such as methyl ethyl sulfonate (Krieg, 1963) are now preferred as they tend to cause less chromosome damage. Induced mutation is probably best considered as a last resort, particularly in biological control, to be employed only after unsuccessful exploitation of natural variation.

Apart from the sampling strategies described previously, several specific techniques for making better use of natural variation have been proposed. Hybridization between strains or races has been suggested as a mechanism for providing a species with new sources of genetic material. Birch and Andrewartha (1966) suggested that the Queensland fruit fly, *Dacus tryoni* (Frog.), extended its range into southeastern Australia by recruiting genes from *D. neohumeralis* (Perkins). This hypothesis was discarded by Birch and Vogt (1970) and the current extension of *D. tryoni*'s range must be explained in terms of exploitation of its own gene pools. There is no good evidence that paucity within a gene pool combined with natural mutation rate has been the limiting factor in a species' ability to extend its range or colonize a new territory.

Step (3), measurement of heritability and choice of selection regime, lies in the domain of the quantitative geneticist and cannot be adequately discussed here. [For a review of this subject, see Falconer (1960).] A great deal is dependent upon the particular natural enemy species involved, as well as the genetic basis of the character being selected.

Step (4), undesirable side effects can be minimized if the program is carried out under simulated or actual field conditions. This may be possible in a few instances, e.g., selection for insecticide resistance, but generally field conditions would prevent adequate supervision of the selection regime and generation time. Where the character selected for has a simple genetic basis, e.g., insecticide resistance, it could be recruited from a laboratory strain by the otherwise suitable field strain and here adverse effects of selection might not be serious. The predatory mite *Typhlodromus occidentalis* Nesbitt in southern California acquired OP-resistance in such a manner (Croft and Barnes, 1971).

VI. CONCLUSIONS

From the foregoing analysis it can be seen that more questions have been asked than answered about the genetic aspects of natural enemy importation. Much advice is available based on theoretical considerations or logical grounds, but with little or no concrete evidence relative to population structures, sampling schemes, colonizing procedures, and techniques for improving fitness by artificial selection or for facilitating postcolonization adaptability.

Indeed it is doubtful whether, important as it is for sampling strategies, the population structure of any natural enemy, exotic or native, or the effect of size of sample on successful establishment or subsequent adaptability have ever been investigated. Perhaps the answers to such questions can come best from experimental or observational evidence. In each case, it seems that the best place for such studies is in the actual biological control project itself. For it is here that the appropriate kind of organism will be investigated, that its endemic populations will be at least visited and spot-sampled, that actual importations with all attendant procedures will be carried out, and where intentional colonizations (releases) will be made and evaluated.

One immediate suggestion that comes to mind is the monitoring of the genetic diversity of initial samples of natural enemies from the time they are first collected until they are released in the target areas and subsequently recovered. It has already been proposed that this might be done by the electrophoretic technique of "genetic fingerprinting." Such a technique might also be of value in determining the proportion of the natural population's genetic diversity that is attainable in samples of different sizes and from different places, and any effects, positive or negative, of sample pooling, etc. And, most important, insight may be gained as to how such genetic features vary from species to species, predator versus parasite, or localized versus widely distributed enemies.

One thing is clear, that in order for the planning and execution of any such programs as suggested above to be the most fruitful there will be required the close collaboration of biological control specialists and geneticists.

REFERENCES

Allard, R. W., and Kahler, A. L. (1971). Allozyme polymorphism in plant populations. *In* "Stadler Symposium No. 3" (L. J. Stadler, ed.). University of Missouri, Columbia, Missouri.

Allen, H. W. (1958). Evidence for adaptive races among oriental fruit moth parasites. *Proc. 10th Int. Congr. Entomol., Montreal, 1956*, **4**, 743–749.

Ayala, F. J., Powell, J. R., and Dobzhansky, T. (1971). Polymorphism in continental and island populations of *Drosophila willistoni*. *Proc. Natl. Acad. Sci. U.S.* **68**, 2480–2483.

Baker, H. G., and Stebbins, G. L. (eds.). (1965). "The Genetics of Colonizing Species," 588 pp. Academic Press, New York.

Bartlett, B. R., and van den Bosch, R. (1964). Foreign exploration for beneficial organisms. *In* "Biological Control of Insect Pests and Weeds" (P. DeBach, ed.), pp. 283–304. Reinhold, New York.

Birch, L. C., and Andrewartha, H. G. (1966). Queensland Fruit Fly: a study in evolution and control. *New Sci.*, pp. 204–207.

Birch, L. C., and Vogt, W. G. (1970). Plasticity of taxonomic characters of the Queensland fruit flies *Dacus tryoni* and *Dacus neohumeralis* (Tephritidae). *Evolution* **24**, 320–343.

Boller, E. (1972). Behavioral aspects of mass rearing of insects. *Entomophaga* **17**, 9–26.

Brewer, G. J. (1970). "An Introduction to Isozyme Techniques," 186 pp. Academic Press, New York.

Brock, R. D., Shaw, H. F., and Callen, D. F. (1972). Induced variation in quantitatively inherited characters. *In* "Induced Mutations and Plant Improvement," pp. 317–322. Int. Atom. Energy Agency, Vienna.

Clausen, C. P. (1936). Insect parasitism and biological control. *Ann. Entomol. Soc. Amer.* **29**, 201–223.

Clausen, C. P. (1956). Biological control of insect pests in the continental United States. *U.S. Dep. Agr. Tech. Bull.* **1139**, 1–151.

Croft, B. A., and Barnes, M. M. (1971). Comparative studies on four strains of *Typhlodromus occidentalis*. III. Evaluation of releases of insecticide resistant strains into an apple orchard ecosystem. *J. Econ. Entomol.* **64**, 845–850.

DeBach, P. (1958). Selective breeding to improve adaptations in parasitic insects. *Proc. 10th Int. Congr. Entomol., Montreal, 1956*, **4**, 759–768.

DeBach, P. (ed.). (1964). "Biological Control of Insect Pests and Weeds," 844 pp. Reinhold, New York.

DeBach, P. (1965). Some biological and ecological phenomena associated with colonizing entomophagous insects. *In* "Genetics of Colonizing Species" (H. G. Baker and G. L. Stebbins, eds.), pp. 287–306. Academic Press, New York.

Delucchi, V. L. (1961). Le complexe des *Asolcus* Nakagawa (*Microphanurus* Kieffer) (Hymenoptera: Proctotrupoidea) parasites oophages des punaises des cereales au Maroc et au Moyen Orient. *Cah. Rech. Agron. (Rabat)* **14**, 41–67.

Dietrick, E. J., and van den Bosch, R. (1957). Insectary propagation of the squash bug and its parasite *Trichopoda pennipes* Fabr. *J. Econ. Entomol.* **50**, 627–629.

Doutt, R. L., and DeBach, P. (1964). Some biological control concepts and questions. *In* "Biological Control of Insect Pests and Weeds" (P. DeBach, ed.), pp. 118–144. Reinhold, New York.

Doutt, R. L., and Nakata, J. (1965). Overwintering refuge of *Anagrus epos* (Hymenoptera: Mymaridae). *J. Econ. Entomol.* **58**, 586.

Ehler, L. E., and van den Bosch, R. (1974). An analysis of the natural biological control of *Trichoplusia ni* (Lepidoptera: Noctuidae) on cotton in California. *Can. Entomol.* **106**, 1067–1073.

Falconer, D. S. (1960). "Introduction to Quantitative Genetics," 365 pp. Ronald Press, New York.

Finney, G. L., and Fisher, T. W. (1964). Culture of entomophagous insects and their hosts. *In* "Biological Control of Insect Pests and Weeds" (P. DeBach, ed.), pp. 328–355. Reinhold, New York.

Fisher, T. W. (1964). Quarantine handling of entomophagous insects. *In* "Biological Control of Insect Pests and Weeds" (P. DeBach, ed.), pp. 305–327. Reinhold, New York.

Flanders, S. E. (1931). The temperature relationships of *Trichogramma minutum* as a basis for racial segregation. *Hilgardia* **5**, 395–406.

Flanders, S. E. (1940). Environmental resistance to the establishment of parasitic Hymenoptera. *Ann. Entomol. Soc. Amer.* **33**, 245–253.

Flanders, S. E. (1959). Differential host relations of the sexes in parasitic Hymenoptera. *Entomol. Exp. Appl.* **2**, 125–142.

Force, D. C. (1967). Genetics in the colonization of natural enemies for biological control. *Ann. Entomol. Soc. Amer.* **60**, 722–729.

Frankel, O. H., and Bennett, E. (1970). "Genetic Resources in Plants: Their Exploration and Conservation," 554 pp. Blackwell, Oxford.

Gibson, J. (1970). Enzyme flexibility in *Drosophila melanogaster*. *Nature (London)* **227**, 959–960.

Hafez, M., and Doutt, R. L. (1954). Biological evidence of sibling species in *Aphytis maculicornis* (Masi) (Hymenoptera: Aphelinidae). *Can. Entomol.* **86**, 90–96.

Harding, J., Allard, R. W., and Smeltzer, D. G. (1966). Population studies in predominantly self-pollinating species. IX. Frequency-dependent selection in *Phaseolus lunatus*. *Proc. Natl. Acad. Sci. U.S.* **56**, 99–104.

Huffaker, C. B. (1971). The ecology of pesticide interference with insect populations. *In* "Agricultural Chemicals—Harmony or Discord for Food, People and the Environment" (J. E. Swift, ed.), pp. 92–104. Univ. Calif., Div. Agr. Sci., Berkeley, California.

Huffaker, C. B., and Kennett, C. E. (1959). A ten-year study of vegetational changes associated with biological control of Klamath weed. *J. Range Manage.* **12**, 69–82.

Huffaker, C. B., and Kennett, C. E. (1969). Some aspects of assessing efficiency of natural enemies. *Can. Entomol.* **101**, 425–447.

Huffaker, C. B., van die Vrie, M., and McMurtry, J. A. (1970). Ecology of tetranychid mites and their natural enemies, a review. II. Tetranychid populations and their possible control by predators, an evaluation. *Hilgardia* **40**, 391–458.

Johnson, W. E., and Selander, R. K. (1971). Protein variation and systematics in Kangaroo rats (Genus *Dipostomys*). *Syst. Zool.* **20**, 377–405.

Krieg, D. R. (1963). Ethyl methanesulfonate-induced reversion of bacteriophage T4 r II mutants. *Genetics* **48**, 561–580.

Levins, R. (1969). Some demographic and genetic consequences of environmental heterogeniety for biological control. *Bull. Entomol. Soc. Amer.* **15**, 237–240.

Lewontin, R. C. (1965a). Discussion of P. DeBach's paper "Colonizing Entomophagous Insects." *In* "Genetics of Colonizing Species" (H. G. Baker and G. L. Stebbins, eds.), pp. 304–306. Academic Press, New York.

Lewontin, R. C. (1965b). The genetics of colonizing ability. *In* "Genetics of Colonizing Species" (H. G. Baker and G. L. Stebbins, eds.), pp. 79–94. Academic Press, New York.

Lucas, A. M. (1969). The effect of population structure on the success of insect introductions. *Heredity* **24**, 151–157.

Lund, H. O. (1934). Some temperature and humidity relations of two races of *Trichogramma minutum* Riley (Hymenoptera: Chalcididae). *Ann. Entomol. Soc. Amer.* **27**, 324–340.

Mackauer, M. (1972). Genetic aspects of insect production. *Entomophaga* **17**, 27–48.

Manwell, C., and Baker, C. M. A. (1970). "Molecular Biology and the Origin of Species," 394 pp. Sedgwick and Jackson, London.

Mayr, E. (1965). Summary. *In* "Genetics of Colonizing Species" (H. G. Baker and G. L. Stebbins, eds.), pp. 553–562. Academic Press, New York.

Messenger, P. S. (1970). Bioclimatic inputs to biological control and pest management programs. *In* "Concepts of Pest Management" (R. L. Rabb and F. E. Guthrie, eds.), pp. 84–102. North Carolina State University, Raleigh, North Carolina.

Messenger, P. S., and van den Bosch, R. (1971). The adaptability of introduced biological control agents. *In* "Biological Control" (C. B. Huffaker, ed.), pp. 68–92. Plenum, New York.

Muller, H. J. (1927). Artificial transmutation of the gene. *Science* **66**, 84–87.

Pielou, D. P., and Glasser, R. F. (1952). Selection for DDT resistance in a beneficial insect parasite. *Science* **115**, 117–118.

Pimentel, D. (1961). Animal population regulation by the genetic feedback mechanism. *Amer. Natur.* **95**, 65–79.

Pimentel, D. (1968). Population regulation and genetic feedback. *Science* **159**, 1432–1437

Prout, T. (1971). The relation between fitness components and population prediction in *Drosophila*. I. The estimation of fitness components. *Genetics* **68**, 127–149.

Quednau, W. (1960). Uber die Identitat der *Trichogramma*-Arten und einiger Okotypen (Hymenoptera: Chalcidoidea: Trichogrammatidae). *Mitt. Biol. Bundesanst. Land. Forstwirt. Berlin Dahlem* **100**, 11–50.

Remington, C. L. (1968). The population genetics of insect introduction. *Annu. Rev. Entomol.* **13**, 415–426.

Sabrosky, C. W. (1955). The interrelations of biological control and taxonomy. *J. Econ. Entomol.* **48**, 710–714.

Schuntner, C. A., and Roulston, W. J. (1968). A resistance mechanism in organophosphorous-resistant strains of sheep blowfly. *Aust. J. Biol. Sci.* **21**, 173–176.

Schuntner, C. A., Roulston, W. J., and Schnitzerling, H. J. (1968). A mechanism of resistance to organophosphorous acaracides in a strain of the cattle tick *Boophilus microplus*. *Aust. J. Biol. Sci.* **21**, 97–109.

Selander, R. K., Yang, S. Y., Lewontin, R. C., and Johnson, W. E. (1970). Genetic variation in the horseshoe crab (*Limulus polyphemus*), a phylogenetic "relic." *Evolution* **24**, 402–414.

Simmonds, F. J. (1963). Genetics and biological control. *Can. Entomol.* **95**, 561–567.

Smith, H. S. (1941). Racial segregation in insect populations and its significance in applied entomology. *J. Econ. Entomol.* **34**, 1–13.

Smith, J. M. (1958). Biological control of Klamath weed, *Hypericum perforatum* L., in British Columbia. *Proc. 10th Int. Congr. Entomol., Montreal, 1956*, **4**, 561–565.

Sokal, R. R. (1967). A comparison of fitness characters and their responses to density in stock and selected cultures of *wild type* and *black Tribolium castaneum*. *Tribolium Inf. Bull.* **10**, 142–147.

Thompson, W. R. (1951). The specificity of host relations in predaceous insects. *Can. Entomol.* **83**, 262–269.

Thompson, W. R., and Parker, H. L. (1928). The European corn borer and its controlling factors in Europe. *U.S. Dep. Agr. Tech. Bull.* **59**.

Turnbull, A. L. (1967). Population dynamics of exotic insects. *Bull. Entomol. Soc. Amer.* **13**, 333–337.

Urquijo, P. (1951). Aplicacion de la genetica al aumento de la eficacia del *Trichogramma minutum* en la lucha biologica. *Bol. Patol. Veg. Entomol. Agr.* **18**, 1–12.

van den Bosch, R. (1968). Comments on population dynamics of exotic insects. *Bull. Entomol. Soc. Amer.* **14**, 112–115.

White, E. B., DeBach, P., and Garber, M. J. (1970). Artificial selection for genetic adaptation to temperature extremes in *Aphytis lingnanensis* Compere (Hymenoptera: Aphelinidae). *Hilgardia* **40**, 161–192.

Whitten, M. J. (1970). Genetics of pests in their management. *In* "Concepts of Pest Management" (R. L. Rabb and F. E. Guthrie, eds.), pp. 119–137. North Carolina State University, Raleigh, North Carolina.

Wilkes, A. (1942). The influence of selection on the preferendum of a chalcid (*Microplectron fuscipennis* Zett.) and its significance for biological control of an insect pest. *Proc. Roy. Entomol. Soc. (London) Ser.* **B130**, 400–415.

Wilkes, A. (1947). The effects of selective breeding on the laboratory propagation of insect parasites. *Proc. Roy. Entomol. Soc. (London), Ser.* **B134**, 227–245.

Wilson, F. (1960a). The future of biological control. *Rep. 7th Commonw. Entomol. Conf., London, 1960*, pp. 72–79.

Wilson, F. (1960b). A review of the biological control of insects and weeds in Australia and Australian New Guinea. *Commonw. Inst. Biol. Control Tech. Commun.* **1**, 1–102.

Wilson, F. (1965). Biological control and the genetics of colonizing species. *In* "Genetics of Colonizing Species" (H. G. Baker and G. L. Stebbins, eds.), pp. 307–329. Academic Press, New York.

10

CONSERVATION AND AUGMENTATION OF NATURAL ENEMIES

R. L. Rabb, R. E. Stinner, and Robert van den Bosch

I. INTRODUCTION

Concepts, principles, techniques, and examples of the conservation and augmentation of natural enemies have been effectively presented by several authors (DeBach and Hagen, 1964; van den Bosch and Telford, 1964; Wilson, 1966; National Academy of Science, 1969; Huffaker, 1971; and some aspects are treated in Chapter 24 in this treatise). In this chapter, we focus attention on developments in conservation and augmentation procedures during the past 10 years, emphasizing natural enemy—pest relationships in agroecosystems.

Conservation, as used here, means premediated actions for protecting and maintaining natural enemies. Augmentation refers to actions taken to increase the populations or beneficial effects of natural enemies. While conservation and augmentation can be contrasted theoretically, they cannot be separated on a practical basis because the techniques involved usually produce effects relating to both. Therefore, we have not used the differences in the two concepts for organizing this presentation. We group the techniques into two categories: (1) manipulations of biotic agents (DeBach and Hagen, 1964) and (2) manipulations of environmental elements (van den Bosch and Telford, 1964).

II. CONCEPTUAL BASIS FOR CONSERVATION AND AUGMENTATION OF NATURAL ENEMIES

A. Cost-Benefit Considerations

Appraisal of cost-benefit ramifications should be an initial consideration before launching a project to develop or implement any of the techniques discussed in this chapter. Long-term as well as short-term gains must be estimated as accurately as possible. A technique or combination of techniques which may seem superfluous or unnecessarily demanding, when viewed as a short-term option, may prove invaluable in achieving and maintaining economically and environmentally viable pest management from a long-range perspective.

B. A Realistic Ecosystem Perspective

In developing strategies and tactics for manipulating ecosystem elements to manage pests and their natural enemies, it is instructive to contrast man-dominated (i.e., agroecosystems) and undisturbed segments of an ecosystem (Southwood and Way, 1970; Solomon, 1973). In brief summary: (1) Natural ecosystems are self-perpetuating, whereas, agroecosystems are not. (2) Vegetation in an undisturbed area is the product of natural selection, but, in crop land, man selects the species and varieties to be planted and reduces interspecific competition. (3) Being increasingly selected by modern man principally for yield and quality, crop plants are often more susceptible to climatic variations, insects, nematodes, and disease pathogens than are the types which evolved naturally in undisturbed areas. (4) Man's crop breeding program, coupled with his planting schedules, imposes a high degree of age uniformity on his crops in which flowering and other phenological events are highly synchronized. (5) Fertilization and irrigation practices result in higher nutritional levels, faster growth, and more succulent tissues in crops than generally found in naturally occurring plants. As a consequence, outbreaks of pests (disease organisms and invertebrate and vertebrate herbivores) independent of the influence of natural enemies, are much more frequent in crop lands than in undisturbed areas (Smith, 1972).

The central element in an agroecosystem is the crop, with the use of an enemy as but one of many energy inputs interacting with other essential practices. Some of these interactions may be positive and some negative with respect to optimal production. Care must be taken to minimize the negatives and maximize the positives by choosing techniques which can be used within the economic and ecological constraints imposed in the long term by the crop production system and society.

Contrasting views with respect to the amount of basic ecological knowledge required for successful use of natural enemies have been presented (Turnbull, 1967; van den Bosch, 1968; and Chapter 3). For some pest problems, a rough appraisal of the ecological dimensions has focused attention on a factor which was then successfully manipulated to improve natural enemy action and pest control. The use of natural enemies in the solution of some of our more persistent and complex problems, however, may require prior in-depth ecological studies for successful manipulations.

One important objective of the ecological analysis should be distinguishing "real" and induced pests (see Chapter 27), because enhancement procedures for enemies of induced pests a priori must involve modifications of control actions aimed at the "real" or "key" pests.

Natural enemies with potential for manipulation will be identified through in-depth studies of pest ecology. However, additional studies focused specifically on the life systems of the enemy may be required to discover this potential. Such investigations should result in evaluations of the role of alternate hosts, secondary enemies, intrinsic control mechanisms, and interspecific competition in influencing natural enemy populations relative to their synchrony with and impact on the pest.

C. Criteria for Successful Natural Enemy Manipulation

A single natural enemy technique may solve a pest problem, but this is rare in our more complex pest situations. Techniques should be viewed as potential inputs into a management system for a complex pest situation in which each input serves a unique control function. One manipulation may serve primarily to lower the general level of pest abundance; another may temporarily suppress pest population peaks which reach injury thresholds; still another may preserve and augment the density-dependent action of an enemy complex already present. In some cases, no one of these manipulations alone will solve the problem, but together they may make continuing economic management possible.

Pests which must be held to very low levels because of extremely low injury thresholds are less likely to be controlled by the techniques discussed here than are ones with higher economic injury thresholds. Nevertheless, a technique which can be used to regulate a pest population at a lower level may make its control by other techniques more feasible and may, for example, make it possible to use more environmentally acceptable pesticides or to reduce the rate and frequency of pesticide applications.

III. NATURAL ENEMY RELEASES

We include here only those procedures in which an entomophagous species is directly manipulated through releases. Four steps are necessary: (1) candidate selection, (2) mass rearing, (3) release, and (4) evaluation. Since evaluation is discussed in Chapter 11, it is not dealt with here. The use of insect pathogens is discussed principally in Chapter 7.

A. Candidate Selection

The combination of attributes optimum for a specific enemy (Huffaker and Kennett, 1969; and Chapter 3) depends upon the strategy involved. In this respect, one can separate the release possibilities into three overlapping strategies: (1) inoculative releases made with the expectation that the species will survive permanently in the system and regulate the pest species at a new and lower density, (2) inoculative releases made with the expectation that the species will survive and reproduce only for a limited number of generations and prevent the pest density from rising above the economic threshold, and (3) periodic, inundative releases for *immediate* control of a pest population, with an expectation of immediate host mortality, but not long-term regulation (i.e., the use of "biological insecticides" with thresholds which may differ from those established for chemical usage).

Essential agronomic practices, including insecticidal and cultural control for other pests of the same crop or for pests of adjacent crops, should be considered in selecting a strategy for release. This is particularly true when the pest species of concern is an induced pest. For example, control strategies for *Heliothis* spp. on cotton in the southeast United States are severely limited by the procedures (predominately insecticidal) used to control the boll weevil. Insecticidal drift from adjacent fields often inhibits natural enemy control, such as has been observed with releases of *Trichogramma* spp. against *Heliothis* spp. (U.S. Dep. Agr., 1972).

Various authors have recognized the importance of the survival of the natural enemy in the geographical regions of interest, but a critical review has not been given to the requirements for survival as dictated by the intended usage (strategy).

For classical biological control, the species must be suited to the environment into which it is introduced. In many pest management programs, however, the release of entomophagous insects for short-term control (McMurtry *et al.*, 1969; Ridgway, 1969) might provide suppression of localized infestations of those pests for which the classical approach has not yet been effective (see also Beirne, 1966; Simmonds, 1966).

A natural enemy may be used in inoculative releases, as was reported by Huffaker and Kennett (1956) for control of the cyclamen mite, *Steneotarsonemus pallidus* (Banks), on strawberries; early season colonization compensated for the low mobility of the predaceous mite. For this strategy then, the classical requirements of high mobility and long-term survival are relaxed. Additional examples of this technique include: (1) control of the sugarcane borer, *Diatraea saccharalis* (F.), in many Central and South American countries with the parasites *Lixophaga diatraea* (Townsend), *Paratheresia claripalpis* (Wulp), *Metagonistylum minense* Townsend, and *Trichogramma* spp. (see review by Bennett, 1971); (2) experimental control of the two-spotted spider mite, *Tetranychus urticae* (Koch), with early-season releases of *Phytoseiulus persimilis* Athias-Henriot (Oatman *et al.*, 1968; Hussey and Bravenboer, 1971); (3) experimental control of brown soft scale, *Coccus hesperidum* L., with *Microterys flavus* (Howard) (Hart, 1972); and (4) release of fish for temporary or permanent mosquito management (Bay, 1967; Sholdt *et al.*, 1972; and Chapter 18).

A further relaxation of classical criteria (Huffaker and Kennett, 1969) may be justified if the enemy is to be used for periodic, inundative releases for short-term control of localized infestations. Here the emphasis must be on the overall ability of the natural enemy to cause a rapid, high mortality of the pest under the conditions of a given system (i.e., insecticide use patterns and drift from adjacent crops). Little importance may be placed on either mobility or survival for more than several days, or on host specificity.

Control by *Trichogramma* spp., largely experimental in the U.S.A., has been reported for *Heliothis* spp. (Ridgway, 1969), the cabbage looper, *Trichoplusia ni* (Hbn.) (Oatman and Platner, 1971), and many other lepidopterous species. A review of the European work has been given by Kot (1964), and for China by the National Academy of Sciences (C. B. Huffaker, pers. commun.).

Experimental releases of *Chrysopa* have been effective against lepidopterous pests on cotton (Ridgway and Jones, 1969) and aphids on potatoes (Shands *et al.*, 1972).

Aside from effective natural control often afforded by coccinellids (Hagen, 1962; Hodek, 1970), experimental releases also have shown promise against aphids on potatoes by *Coccinella septempunctata* L. (Shands *et al.*, 1972) and against the avocado brown mite, *Oligonychus punicae* (Hirst), by *Stethorus picipes* Casey (McMurtry *et al.*, 1969).

For inoculative and inundative releases, consideration must be given to the economics of production and release, as well as to the control potential per individual. The cost of rearing is such a major factor (Knipling, 1966) that a species which is only half as effective as another may be used if it can be reared for less than half the cost.

It is also of more importance to have consistent production than to maximize production at the risk of wide variation in quantity and quality of the insects, particularly in periodic mass releases where a short time lag exists between the recognition of the infestation and the need for treatment. The ability to stockpile a given natural enemy over weeks or months (Schread and Garman, 1934; Stinner *et al.*, 1974) for sudden massive releases may more than counterbalance a lower searching capacity or rate of increase.

Use of insecticides and problems of insecticidal drift may warrant selection of an inherently less efficient species if it is significantly resistant. Resistance in certain predators has been demonstrated (Hoyt and Caltagirone, 1971; Lingren and Ridgway, 1967; van den Bosch and Hagen, 1966), and laboratory selection for resistance in certain species could prove valuable. Selection for strains with certain other desirable characteristics, such as high or low temperature tolerance, has some potential (see review by Mackauer, 1972; and Chapter 9), but, to our knowledge, has not yet been successfully utilized in any applied program.

Other biological parameters which must be of concern in candidate selection include: (1) stage of host attacked in relation to damaging stage, (2) host plant preferences, and (3) behavioral patterns which may affect efficiency for inundative releases. Finally, consideration must be given to the potential of the species when the release strategy is coupled with indirect manipulations as discussed below.

B. Mass Rearing

Many authors have discussed mass rearing of selected entomophagous species (Shorey and Hale, 1965; Smith, 1966), but little attention has been paid to technical problems of a general nature. Often the goal of such programs becomes maximum numbers rather than optimum production of quality individuals.

Sanitation (e.g., disease prevention) has often been a problem and is, at times, the limiting factor (Dysart, 1973; Raun, 1966; Steinhaus, 1958). With adequate sanitation and proper controls for early recognition of disease, this pitfall can be avoided.

Another major problem involves subtle biological changes induced by the rearing itself. Genetic deterioration and loss in field efficiency is one of the greatest challenges to such a program and as yet has received only minimum attention. A great deal more effort must be expended on obtaining a broad genetic base for initial cultures (Harley and Kassulke, 1971), and a system for adequate renewal of variability must be included. Excellent reviews of the problems associated with laboratory selection have been presented by Boller (1972) and Mackauer (1972), and the remaining discussion is essentially a synopsis of their views (see also Chapter 9): (1) Physical conditions of rearing—the use of constant temperatures and humidities has been reported to greatly affect the gene pool (Dysart, 1973; White *et al.*, 1970). The use of cycling temperatures could well eliminate this problem. (2) Diet or host—although excellent diets are available for many phytophagous insects, artificial media for entomophagous insects are lacking, with few notable exceptions (Hagen and Tassen, 1966; Simmonds, 1966; Yazgan and House, 1969). While use of facticious hosts often makes mass rearing of certain natural enemies possible, this use can lead to measurable differences in efficiency of the natural enemy (Simmonds, 1966). (3) Handling and rearing procedures—inadequate confinement can allow the escape of the most vagile or aggressive individuals, resulting in a deterioration of the colony's "vigor."

Production estimation and quality control procedures are a necessity, yet most rearing programs obtain only highly variable estimates of numbers produced and even less information on quality. The cost of developing and maintaining procedures for accurately determining the quality and quantity of individuals seems a small price to pay for consistent and satisfactory performance in the field. The procedures necessary will vary with the entomophagous species and the intended usage.

C. Actual Releases

The optimum stage of a natural enemy for release, as well as the timing of releases, varies with each natural enemy—pest complex. In choosing the stage for release, biological efficiency must be weighed against cost; e.g., releasing pupae may be less expensive than releasing adults, but mortality prior to adult emergence may more than offset the additional cost.

In some cases, the time element is critical and may preclude release of any but the most efficient stage (Butler and Hungerford, 1971). In inoculative releases, however, timing may not be so exacting (see Bennett, 1971; Parker *et al.*, 1971, for contrasting situations).

Prerelease conditioning and release procedures also vary with the species and stage. Acclimatization to outside conditions is often helpful. Prerelease feeding, release only late in the evenings, and release after egg maturation may be profitable. Provision of supplemental food in the release site may be critical and affect the method or pattern of release (Parker *et al.*, 1971).

IV. ENVIRONMENTAL MANIPULATIONS

A. General Considerations

Environmental factors influencing survival and activity of an individual natural enemy may be broadly categorized as (1) weather, (2) food, (3) resources other than food (i.e., a particular place to live), (4) other organisms of the same species, and (5) other organisms of a different species, including man (Solomon, 1949).

Wide-area patterns of land use impose constraints in variable degrees on natural enemies. The ratio of cultivated to noncultivated land is of particular significance to those species which periodically must move from one habitat to another. Certain crop mixes do not provide resources and conditions adequate for effective action of natural enemies in certain crops. The effects of manipulating diversity are complex and cannot be predicted without detailed understanding of the cause and effect pathways involved. Southwood and Way (1970) state "the aims of pest management should be to determine what elements of environmental diversity need be retained or added and what need to be eliminated to enhance management of the pest's population" and that "any management practice affecting diversity must be assessed separately in respect to its effects on the pest and those on its natural enemies." Although a number of authors have given theoretical consideration to, and presented some experimental results concerning, optimal cropping patterns (e.g., Pimentel, 1970; DeLoach, 1970), major decisions relative to wide-area land use are seldom made on the basis of pest problems or natural enemy efficiency.

The potential of each natural enemy may be further limited by management practices within each land-use unit (field, woodland, marsh, etc.), but the effects of these practices on pest—natural enemy relationships are imperfectly understood. However, where pest problems do not exist or are being managed satisfactorily by existing techniques, natural enemies are an essential input.

Most of the techniques discussed here are of potential rather than realized value in pest management. For some techniques, more research is needed to add practicality to promise; in other cases, growers have not adopted practical techniques because insecticidal control seems to them more reliable and simpler.

Van den Bosch and Telford (1964) discuss both positive and negative effects of changing agricultural practices on pests and natural enemies. The advent of irrigation in California desert valleys created favorable conditions for *Labidoura riparia* (Pallas), a predaceous earwig (Schlinger *et al.*, 1959). On the other hand, the spread of wheat farming in North America was accompanied at times by severe infestations of the western stem sawfly, *Cephus cinctus* Nort., apparently because the local parasites of the sawfly did not adapt to the newly created environment in wheat (Criddle, 1922). Van den Bosch and Telford cite other cases where cultural practices have disrupted the synchrony between pest and natural enemy populations or otherwise adversely affected the enemies. Synchronization can be improved and enemy effectiveness enhanced in the following ways: (1) reducing direct mortality inflicted by cultural practices on natural enemies, (2) providing supplemental resources, (3) controlling secondary enemies, and (4) manipulating host plant attributes which indirectly influence natural enemies.

B. Reducing Direct Mortality

A practice such as cultivation, harvesting, or use of a pesticide may cause direct mortality of natural enemies as well as pests. The significance of such mortality depends upon the importance of the enemy and upon the differential mortality of pest and natural enemy. A practice resulting in local eradication of pest and enemy may trigger a pest outbreak if reinvasion of the pest sufficiently precedes that of the enemy. The effect may be accentuated if the reproductive potential of the enemy is low in comparison to that of the pest.

Microclimates unfavorable to natural enemies may be improved by management of water, soil, and ground cover and provision of windbreaks. Hall and Dunn (1958) devised irrigation schedules to provide humidity conditions favorable for the action of entomophthorous fungi on populations of the spotted alfalfa aphid, *Therioaphis trifolii*

(Monell). Dry conditions, coupled with cultivation and elimination of ground cover, result in dust, which inhibits effective parasitism or predation, particularly of certain scales in citrus (DeBach, 1958) and spider mites in vineyards (Flaherty and Huffaker, 1970). Dust may be reduced by irrigation, use of ground covers, and "dust-proofing" road surfaces. While the role of wind has not been studied adequately, it seems to influence the searching behavior of some adult parasites and may be modified by windbreaks. Reed *et al.* (1970) found lower populations of the brown soft scale, *Coccus hesperidum* L., as well as higher parasitization of it by *Aphycus stanleyi* (Compere), adjacent to windbreaks in citrus in Texas.

Soil management practices often inflict heavy mortality on both pest and natural enemy populations; however, possible differential effects have been little studied. In perennial crops such as orchards and vineyards, the substitution of ground cover for clean cultivation has been widely suggested to improve natural enemy—pest ratios. Annual crop culture imposes more rigid constraints. While plowing and cultivation generally inflict mortality on both pests and natural enemies, the net effect usually depends upon which species sustains the heaviest mortality. Van den Bosch and Telford (1964) report a case where a change from mouldboard plowing decreased the mortality inflicted on *Caenocrepis bothynoderes* Grom., a pteromalid parasite of the beet pest *Bothynoderes punctiventris* Germ. in the U.S.S.R. The new shallow plowing resulted in increased survival of the parasite and a corresponding decrease in the pest population. The recent introduction of no-tillage techniques and expanded use of herbicides will alter many pest—natural enemy relationships which will need intensive study.

Pruning, harvesting, and crop residue disposal practices also have effects on pests and natural enemies. Wilson (1966) notes the mortality inflicted on *Aphelinus mali* (Haldeman) by pruning of apple wood infested with the woolly aphid, *Eriosoma lanigerum* (Hausm.), in Australia, and reports that some orchardists store the prunings and return them to the orchards in the spring to boost the parasite population. Van den Bosch *et al.* (1967) found that solid-harvesting of alfalfa was more devastating to the parasite *Aphidius smithii* Sharma and Rao than to its aphid host, *Acyrthosiphon pisum* (Harris). Strip-harvesting resulted in an improved microclimate for the parasite, better host—parasite synchrony, and increased parasite efficacy. The effects of removing and/or burning harvest residues on pests and natural enemies are exceedingly complex and difficult to evaluate. Burning of sugarcane trash has been a particularly controversial subject relative to its effects on parasites of the sugarcane borer, *Diatraea saccharalis*, and other natural enemies. Recommendations vary regionally (van den Bosch and Telford, 1964).

Chemical pest control has become a standard crop production practice for many crops, frequently in response to pest outbreaks induced by disruptive effects of other cultural practices on naturally occurring biological control. The use of many such biocides has often created complex and sometimes serious problems by immediate and time-lag effects on natural enemies and other beneficial organisms. Pesticidal resistance, toxic residues, and off-target environmental effects of these chemicals are high priority problems. Consequently, there has been more progress in the selective use of chemical control than in the other techniques discussed in this chapter (see Chapter 23).

C. Providing Supplementary Resources

Essential resources for natural enemies differ among species but may be broadly categorized as host, food other than hosts, water, special sites for reproduction, and protective refuges. Resources for some enemies may be found in sufficiency within a given crop, whereas, other enemies may have more complex requirements (hedgerows, adjacent fields, or woodlands) for some requisite. Researchers have demonstrated the potentiality of manipulating these resources to improve natural enemy action, but practical implementation remains disappointingly low (see Chapter 22).

Alternate hosts or prey may be of critical importance to effective enemy action. The importance of subtle, as well as apparent, predator—prey relationships is illustrated by Flaherty and Huffaker (1970) in their study of the predatory activity of *Metaseiulus occidentalis* (Nesbitt) in the control of the Pacific mite, *Tetranychus pacificus* McGregor, and the Willamette mite, *Eotetranychus willamettei* Ewing, in California vineyards. The Pacific mite is by far the more serious of the two pests and more readily attacked by the predator. However, the Willamette mite is an important additional food resource for the predator, particularly at the low Pacific mite densities which normally occur early and late in the season. Willamette mites (plus tydeid mites) can provide the necessary food for the production of goodly numbers of overwintering *M. occidentalis*. Presence of pollen also influences the abundance of the tydeids and, in turn, the *M. occidentalis*.

Thus environmental manipulations, especially pesticidal applications, should in many cases be devised to provide or protect a complex food resource as well as the natural enemy itself. Subsidiary hosts or prey are commonly provided indirectly by planting or conserving appropriate host plants within the crop or in other important habitats. The potential for using selected species of plants to provide alternate hosts or prey is more restricted for annual than perennial crops. Dempster (1969) found the survival of *Pieris rapae* (L.) larvae to be lower and populations of the predatory *Harpalus rufipes* Degeer to be higher in weedy than in hoed plots of brussel sprouts; however, the advantages were outweighed by the harmful effect of the weeds on the crop.

Plants external to a crop may provide alternate hosts or prey essential to effective biological control of a pest. Doutt and Nakata (1965) found that the egg parasite, *Anagrus epos* Girault, is effective in controlling the grape leafhopper, *Erythroneura elegantula* Osborn, in California vineyards adjacent to wild blackberries (*Rubus* spp). The blackberries harbor a noneconomic leafhopper, *Dikrella cruentata* Gillette, whose eggs serve as the only overwintering resource for the parasite. Although encouraged to plant small patches of blackberries near their vineyards, growers have not yet adopted this manipulation.

Interplanting or rotation with selected plant species may also spatially and temporally synchronize enemies with their hosts or prey. In Peruvian valleys, corn is particularly favorable to production of some entomophagous species and its planting adjacent to crops such as cotton is encouraged as an enemy reservoir (Beingolea, 1957). In Oklahoma, Burleigh *et al.* (1973) also found natural enemy activity in cotton to be enhanced by adjacent plantings of sorghum. Rotation, including a succession of crops or ground cover, may be especially significant in tropical and subtropical areas where enemy activity and reproduction are more continuous. Pest outbreaks in Peruvian valleys have been reduced in this way (Wille, 1951; Smith and Reynolds, 1972).

Strip-cutting, in addition to providing a more favorable microclimate, can be used to increase the continuity of hosts or prey. Stern *et al.* (1967) found that *Apanteles medicaginis* Mues. is less mobile than its host, *Colias eurytheme* Boisduval. Where alfalfa fields are solid-cut, the butterflies move from maturing to early-growth stands, which are often rather widely separated and cause economic damage before the parasite can "catch up." Strip-cutting preserves the spatial synchrony of host and parasite and reduces the frequency of "outbreaks."

In addition to supplying hosts or prey indirectly by manipulating host plants, there is potential in directly adding hosts or prey to the crop. When cultural practices have temporarily eliminated pests, early establishment of pest—natural enemy populations by artificial infestations may lead to effective control and regulation. The scientific feasibility of this approach was demonstrated by inoculating newly planted strawberries with both phytophagous and predatory mites, but the practice has not been adopted commercially (Huffaker and Kennett, 1956, 1969).

Research by Parker and Pinnell (1972) exemplifies the practical potential of direct manipulation of host and enemy populations in a total pest management system. They harmonized deliberate inoculations of cole crops with *Pieris rapae* and two parasites, *Trichogramma evanescens* Westwood and *Apanteles rubecula* Marshall, to correct disruptions caused by normal cultural practices and time lags between natural host and parasite populations. Parker (personal communication) summarized this work as follows:

> The natural population lag between host and parasites was overcome in the preliminary tests by releases of both the pest and its parasites in a one-acre field of cabbage. The continuous release of fertile butterflies increased the pest population nearly tenfold above normal spring populations and enabled the egg parasites to increase early and maintain themselves at an effective level throughout the experiment. Over 95% of the pest eggs were killed by *Trichogramma*. During the 90-day experiment only 3 eggs/plant hatched, but 85% of these were killed by releases of the "back-up" parasite, *Apanteles*. The important result of these experiments was that a marketable crop of cabbage was produced—96% of the plants produced U.S. No. 1 heads. At the check site where natural pest populations increased because of the ineffective natural parasite and predator populations, none of the plants produced No. 1 heads.

While Parker and associates added fertile hosts to the crop itself he points to the potential of adding noncrop or sterilized hosts. Eggs of *Sitotroga cerealella* (Olivier) and *Galleria mellonella* (L.), neither of which attacks crop plants, are suitable hosts for trichogrammatids and can be produced cheaply and in large quantities. Their eggs could be applied to crops to increase *Trichogramma* populations. Adults of pest species, reared and sterilized in the laboratory, also might be released in target crops. Eggs laid would provide food for parasites and predators but those missed would not hatch.

Foods such as pollen, nectar, and honeydew are essential to many natural enemies, which may be dependent upon noncrop plants for these requisites. Leius (1967) found that parasitization was 18-fold greater in tent caterpillar pupae, 4-fold greater in tent caterpillar eggs, and 5-fold greater in codling moth larvae in apple orchards with a rich undergrowth of wild flowers as contrasted to orchards with little or no undergrowth. Leuis also cited numerous papers illustrating the influence of nonhost foods on the reproduction and longevity of parasites and called attention to increasing interest in wild flowers as a food source. Since, however, many pests also feed on nectar, pollen, and honeydew, the potential for manipulating such food can be evaluated adequately only by thorough studies of the population ecology of both the pest and enemies (see Chapter 22).

Various mixtures of nutrient ingredients devised to simulate natural foods such as honeydew have been applied to crops to increase the numbers of entomophagous insects (Schiefelbein and Chiang, 1966). To date, mixtures of protein hydrolyzate, sugar, and water have given the best results. Hagen *et al.* (1970) applied such a mixture to alfalfa and cotton and found that adults of *Chrysopa carnea* Stephens, were attracted to treated plots where they deposited over threefold the number of eggs found in control plots. Syrphid adults were also attracted but not stimulated to oviposit in the absence of aphids. Coccinellids and malachiid beetles (*Collops* spp.) became concentrated in the food-sprayed areas apparently because of retarded dispersal. Conceivably, a supplemental food could retard dispersal, attract from nontreated surroundings, and/or increase reproduction. In spite of the increased interest in food sprays, this technique remains experimental rather than an accepted practice.

Essential resources other than food vary greatly among natural enemy species. Some predatory wasps and birds have requirements for nesting which can be partly satisfied by providing the proper sites in hedgerows and areas adjacent to crops. Artificial structures for nesting of *Polistes annularis* (L.) have been used to increase predation on the cotton leafworm, *Alabama argillaceae* (Hbn.), in St. Vincent (Ballou, 1915). The same approach was used experimentally to reduce populations of *Manduca sexta* (L.) in North Carolina tobacco fields (Lawson *et al.*, 1961) but has not been adopted by tobacco growers. Kirkton (1970) tried the same technique to induce greater predation on lepidopterous pests of cotton in Arkansas but was unable to develop a practical application. Provision of supplementary nesting sites has been suggested for a predaceous ant (Gosswald, 1951), a complex of Hymenoptera (Janvier, 1956), and a parasitic nemestrinid (Spencer, 1958).

Populations of insectivorous birds and small mammals, in some instances, can be augmented by provision of nesting sites and other resources. While the role of these vertebrate predators is poorly understood, there is considerable support for the concept that their consistent impact is critical in preventing outbreaks of certain forest pests. Providing nesting boxes for insectivorous birds has been particularly successful in German forests (Bruns, 1959), but has had minimal attention elsewhere (Coppel and Sloan, 1970). One exception is the experimental work of Dahlsten and Herman (1965) in California with the chickadee *Penthestes gambeli gambeli* (Ridgway) which preys upon the lodgepole pine needle miner, *Coleotechnites milleri* (Busck).

Overwintering refuges present another potential for manipulation. Consideration should be given to conserving and augmenting the hibernation and aestivation sites of major natural enemies which move from cultivated areas to adjacent overwintering sites (Hagen, 1962; Hodek, 1967). For some natural enemies, refuges may be provided in hedgerows and nearby woodlands by inclusion of specific kinds of vegetation or artificial microhabitats. Banding trees in orchards has been widely suggested but not widely practiced. Tamaki and Halfhill (1968) evaluated use of bands on peach trees and found that more than 90% of the arthropods overwintering successfully under the bands were entomophagous. They found a favorable ratio of predaceous to phytophagous mites and suggested that bands might be used to augment biological control of mites on the banded trees and for transfer of predatory mites to predator-poor trees or orchards.

Removing resources of an enemy also has potential in directing its activity to target areas. Rabb and Lawson (1957) found that foraging sites and the prey of *Polistes* species were greatly influenced by mowing, and suggested that cutting of soybeans might be timed to channel foraging wasps to tobacco hornworms. Watson and Wilde (1963) found the green lacewing, *Chrysopa oculata* Say, in abundance on cover crops in and adjacent to pear orchards in British Columbia. Immediately after mowing, there was a fourfold increase in lacewings on the pear foliage where they were important predators of pear psylla, *Psylla pyricola* Foerster.

Environmental manipulations of food and other resources of enemies and pests has occurred fortuitously on a grand scale as land-use patterns have evolved in different geographical area. Van Emden (1965) presents a penetrating review of the role of uncultivated land in the biology of crop pests and their natural enemies and concludes that "almost every advantage offered to beneficial insects (natural enemies) by uncultivated land is at least to some extent offset by a similar advantage to pests." He suggests, however, that the disadvantageous aspects of uncultivated land have had more impact on land-use (e.g., bordering weeds are considered a source of pests and destroyed) practices than have the beneficial aspects and that the potential for readjusting the ratio and quality of hedgerows and verges for more efficient conservation of beneficial species should be carefully examined.

D. Controlling Secondary Enemies

In less disturbed parts of an ecosystem, carnivores of higher trophic levels may contribute to ecosystem stability and hence be judged beneficial. Secondary parasites may function to prevent primary parasites from heavily exploiting (perhaps overexploiting) their hosts. Heavy exploitation of herbivorous hosts, however, is in accord with the aims of crop production. Therefore, secondaries are excluded in quarantine when introducing natural enemies (Chapter 3). The control of well-established secondaries in the field, however, has received little attention except for the control of honeydew feeding ants, which have been shown to interfere with the activities of parasites and predators (e.g., van den Bosch and Telford, 1964; DeBach and Huffaker, 1971). Selective chemical treatments of ant colonies in such situations are known to improve biological control. Way *et al.* (1969) showed that fall plowing destroyed most parasites (primary and secondary) in old brussel sprouts fields. Their data strengthens Sedlag's (1964) suggestion that the efficiency of *Diaeretiella rapae* (M'Intosh) against *Brevicoryne brassicae* L. could be increased by collecting mummified aphids in the fall before plowing, allowing the primary parasites to emerge close to new crops, and destroying the mummies containing secondaries after peak emergence of the primaries.

There are references to problems of establishing introduced enemies because of secondaries. Mackauer (1971) suggests that hyperparasites may have destroyed small founder colonies of *Aphidius smithii* introduced into Nova Scotia for control of the pea aphid, *Acyrthosiphon pisum*. Ballou (1934) also reported that attempts to establish *Polistes annularis* on several islands of the West Indies met with little success due to a small moth, *Calcoela* (= *Dicymolomia*) *pegasalis* (Wlk.), whose larvae killed the wasp larvae.

E. Manipulating Host Plant Attributes

Plant breeders and entomologists are increasingly using genetic qualities in plants which impart resistance against insects and diseases. Although such resistance is usually evaluated in terms of antibiosis, tolerance, or factors influencing preference, differences in plant susceptibility in some cases can also be related to plant attributes influencing natural enemy action on the pests (Bartlett and van den Bosch, 1964). Although many of these attributes theoretically could be altered advantageously, breeders have not yet consciously done so. Chemical factors in a host plant may affect parasites indirectly through modification of the physiology of the herbivorous host (Gilmore, 1938). Morphological characteristics and exudates of plants also may be deleterious to certain enemies. Rabb and Bradley (1968) found parasitization of *Manduca sexta* eggs by *Trichogramma minutum* Riley and *Telenomus sphingis* (Ashmead) to be inhibited by sticky exudates of tobacco leaves, suggesting the breeding of less sticky tobacco varieties. Corn husk characteristics (Collins and Kempton, 1917) and various characteristics of floral bracts of cotton (Leigh *et al.*, 1972; Shepard *et al.*, 1972) appear to have noticeable effects on enemy efficacy and could be manipulated by breeding.

The general vigor and growth characteristics of plants, which can be influenced greatly by fertilization, irrigation, and other cultural practices, have subtle effects on natural enemies. However, the potential for manipulating plant vigor and growth to enhance enemy action has been little studied.

V. MANIPULATIONS IN GLASSHOUSES

Greenhouse culture represents the extreme in artificial ecosystems. Here, physical factors and the spatial and temporal dispersion of a relatively few selected plant species are closely controlled. In spite of attempted exclusion, however, insect pests and plant pathogens invade glasshouses. Pesticidal control is the dominant technique used, but resistance, phytotoxicity, and application problems may speed further development of integrated control, which has already proved feasible in glasshouses.

We shall not discuss in detail the control of glasshouse pests with natural enemies, but wish to draw attention to this specialized application of many of the principles discussed above. Promising results have included parasites of the genus *Encarsia* for control of the greenhouse whitefly, *Trialeurodes vaporariorum* (Westwood) (McClanahan, 1971). Hussey and Bravenboer (1971) reviewed the use of biological control in glasshouses and summarized their own research on the use of enemies to control the two-spotted mite (*Tetranychus urticae*), greenhouse whitefly, cotton aphid (*Aphis gossypii* Glover), green peach aphid (*Myzus persicae* (Sulz)), and a leaf miner (*Phytomyza syngenesiae* Hardy). They also presented integrated programs for controlling pest complexes on cucumbers and chrysanthemums. While there has been extensive commercial application of these programs in the U.S.S.R. and Finland, for example (see Fig. 1), widespread adoption elsewhere has not occurred. One deterrent is the lack

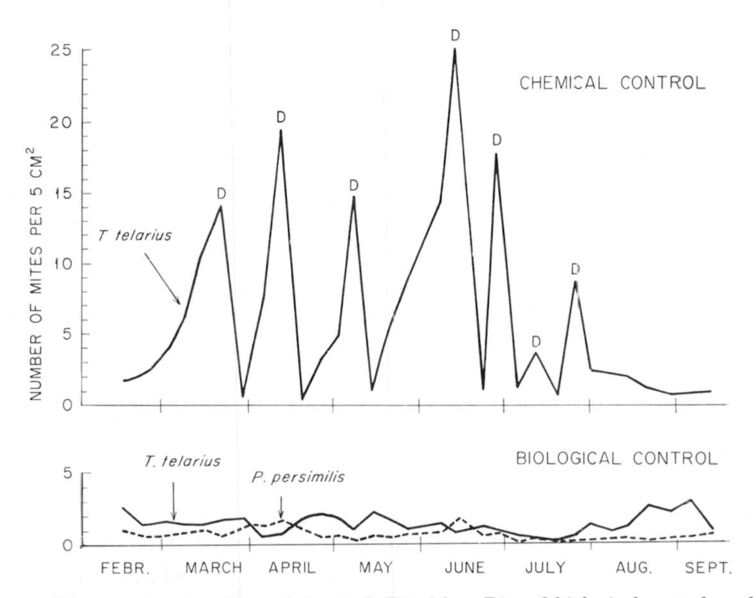

Fig. 1. The comparative effect of chemical (Dicofol = D) and biological control on the number of *Tetranychus telarius* in a commercial greenhouse in Finland. (Dicofol was applied in the chemical control greenhouse when the pest reached a damaging level, and *Phytoseiulus persimilis* was employed in the other.) (After Markkula *et al.*, 1972.)

of efficient production systems for the enemies. However, Hussey and Bravenboer (1971) considered that, "It seems unlikely that any large-scale scheme will be implemented unless the difficulties caused by resistance to pesticides become almost insurmountable."

VI. FUTURE PROSPECTS

The future of the conservation and augmentation techniques discussed in this chapter, and those which potentially could be developed, is closely linked with the nature of man's response to the energy crisis. Continuing human population growth will increase demand for greater food production and will tend to increase demands for pest control. On the other hand, as fossil fuels diminish, priorities as to their use will become more limiting and alternate sources of energy subsidies for pest control will be sought. Since many of the techniques discussed in this chapter require less expenditure of energy than does chemical control, they likely will receive more favorable consideration (Corbet, 1970 and Chapter 27).

We can expect continued selective use of insecticides, cultural manipulations such as strip-harvesting and interplanting of trap crops, mass, strategic, or inoculative colonization of natural enemies, use of artificial sheltering and nesting places, etc. However, we will also surely see the development of innovative techniques such as the use of artificial nutrients in the "assembly line" insectary production of natural enemies as well as in the attraction and augmentation of predator and parasite populations in the field. We will certainly witness the use of behavior-influencing chemicals such as pheromones and kairomones to enhance the performance of natural enemies.

Perhaps the greatest advances in conservation and augmentation will come from our increased understanding of the biology, ecology, ethology, phenology, and physiology of natural enemies that will directly result from the expanded investigations of integrated control. In California, a complex of predators in untreated cotton was found to have sufficient impact on the eggs and larvae of the bollworm, *Heliothis zea* (Boddie), beet armyworm, *Spodoptera exigua* (Hbn.), and cabbage looper, *Trichoplusia ni*, to prevent damaging outbreaks of these pests. However, prior to these intensified integrated control studies, these predator populations were often destroyed by ill-advised chemical treatments for lygus bug (*Lygus* spp.) control, and serious outbreaks of lepidopterous larvae frequently resulted (van den Bosch *et al.*, 1971). Where integrated control has been practiced, natural enemies suffer minimum disruption, and caterpillar outbreaks are virtually unknown.

The increasing concern over diminishing resources and deteriorating environmental quality is focusing sharper attention on land planning and procedures for more effective resource management. The degree to which such procedures are developed and implemented will be largely a function of changing economic, sociological, and political guidelines and institutions. Wide-area population management has proved ecologically and economically sound for many pest problems, as has been so convincingly illustrated by the examples of classical biological control, the use of the sterile male technique to control large area populations of screwworms (Baumhover, 1966), and some of the more recent pest management programs in which multiple techniques have been used to lower the mean level of pest abundance. The potential for wide-area pest management programs in terms of scientific feasibility has hardly been tapped, largely because such programs require modifications of existing philosophies and systems. However, consensus and economic systems continue to evolve and there seems to be a distinctly discernible trend toward the recognition of the essentiality of managing critical resources, including agroecosystems and forest lands, from a more holistic viewpoint. If this trend continues, much existing knowledge, as well as that to be generated with respect to conservation and augmentation of natural enemies, will be used more effectively in pest population management systems. Indeed, even wide-area cropping patterns may be altered in a premeditated manner to better utilize ecosystem diversity in alleviating pest problems.

Although there will be a tendency to increase the use of simplistic, unilateral, and ecologically destructive pest control procedures, the feedback from these actions will create greater problems, involving insect resistance, pesticide residues, and undesirable off-target effects. Consequently, control of the more persistent, complex, and economically significant pest species will, hopefully, evolve into integrated pest management systems based on in-depth ecological understanding. The genetic potential of host plants for resisting pests will be more fully utilized, and an increased understanding of host selection within the context of population dynamics will lead to modifications of cropping patterns, planting and harvesting dates, and rotations. Within this changing agroecosystem structure, microbial agents, parasites, and predators will have to be more effectively manipulated.

In summary, on the basis of concern for his future, it seems inevitable that man will soon place much greater emphasis on conservation and augmentation of natural enemies of pests, basically because the methods involved are "in tune" with "natural laws" which man cannot ignore as flagrantly as he has in the past without jeopardizing his future.

REFERENCES

Ballou, H. A. (1915). West Indian wasps. *Agr. News* **14**, 298.

Ballou, H. A. (1934). Notes on some insect pests in the Lesser Antilles. *Trop. Agr.* **11**, 210–212.

Bartlett, B. R., and van den Bosch, R. (1964). Foreign exploration for beneficial organisms. *In* "Biological Control of Insect Pests and Weeds" (P. DeBach, ed.), pp. 283–304. Reinhold, New York.

Baumhover, A. H. (1966). Eradication of the screwworm fly—an agent of myiasis. *J. Amer.* Med. Ass. **16**, 240–248.

Bay, E. C. (1967). Potential for naturalistic control of mosquitoes. *Proc. Calif. Mosq. Control Ass.* **35**, 34–37.

Beingolea, O. (1957). El sembrio del maiz y la fauna benefica del algodonero. *Estac. Exp. Agr. La Molina, Lima. Informe No.* **104**, 19 pp.

Beirne, B. P. (1966). "Pest Management," 123 pp. Loenard Hill, London.

Bennett, F. D. (1971). Current status of biological control of the small moth borers of sugarcane *Diatraea* spp. (Lep. Pyralididae). *Entomophaga* **16**, 111–124.

Boller, E. (1972). Behavioral aspects of mass rearing of insects. *Entomophaga* **17**, 9–25.

Bruns, H. (1959). The economic importance of birds in forests. *Bird Study* **7**, 192–208.

Burleigh, J. G., Young, J. H., and Morrisson, R. D. (1973). Strip-cropping's effect on beneficial insects and spiders associated with cotton in Oklahoma. *Environ. Entomol.* **2**, 281–285.

Butler, G. D., Jr., and Hungerford, C. M. (1971). Timing field releases of eggs and larvae of *Chrysopa carnea* to insure survival. *J. Econ. Entomol.* **64**, 311–312.

Collins, G. N., and Kempton, J. H. (1917). Breeding sweet corn resistant to the corn earworm. *J. Agr. Res.* **11**, 549–572.

Coppel, H. C., and Sloan, N. F. (1970). Avian predation, an important adjunct in the suppression of larch casebearer and introduced pine sawfly populations in Wisconsin forests. *Proc. Tall Timbers Conf. Ecol. Anim. Control Habitat Manage.* **2**, 259–272.

Corbet, P. S. (1970). Pest management: objectives and prospects on a global scale. *In* "Concepts of Pest Management" (R. L. Rabb and F. E. Guthrie, eds.), pp. 191–208. North Carolina State University, Raleigh, North Carolina.

Criddle, N. (1922). The western wheat stem sawfly and its control. *Can. Dep. Agr. Pamphlet* **6**, [N.S.], 8 pp.

Dahlsten, D. L., and Herman, S. G. (1965). Birds as predators of destructive forest insects. *Calif. Agr.* **19**, 8–10.

DeBach, P. (1958). Application of ecological information to control of citrus pests in California. *Proc. 10th Int. Cong. Entomol.* **3**, 187–194.

DeBach, P., and Hagen, K. S. (1964). Manipulation of entomophagous species. *In* "Biological Control of Insect Pests and Weeds" (P. DeBach, ed.), pp. 429–458. Reinhold, New York.

DeBach, P., and Huffaker, C. B. (1971). Experimental techniques for evaluation of the effectiveness of natural enemies. *In* "Biological Control" (C. B. Huffaker, ed.), pp. 113–140. Plenum, New York.

DeLoach, C. J. (1970). The effect of habitat diversity on predation. *Proc. Tall Timbers Conf. Ecol. Anim. Control Habitat Manage.* **2**, 223–241.

Dempster, J. P. (1969). Some effects of weed control on the numbers of the small cabbage white *Pieris rapae* (L.) on brussel sprouts. *J. Appl. Ecol.* **6**, 339–345.

Doutt, R. L., and Nakata, J. (1965). Parasites for the control of grape leafhopper. *Calif. Agr.* **19**, 3.

Dysart, R. J. (1973). The use of *Trichogramma* in the U.S.S.R. *Proc. Tall Timbers Conf. Ecol. Anim. Control Habitat Manage.* **4**, 165–173.

Flaherty, D. L., and Huffaker, C. B. (1970). Biological control of Pacific mites and Willamette mites in San Joaquin Valley vineyards. I. Role of *Metaseiulus occidentalis*. II. Influence of dispersion patterns of *Metaseiulus occidentalis*. *Hilgardia* **40**, 267–330.

Gilmore, J. U. (1938). Notes on *Apanteles congregatus* (Say) as a parasite of tobacco hornworms. *J. Econ. Entomol.* **31**, 712–715.

Gosswald, K. (1951). "Die rote Waldameise in dienste der Waldhygiene," 160 pp. Metta Kinau Verlag, Wolfu. Tauber, Luneburg, Germany.

Hagen, K. S. (1962). Biology and ecology of predaceous coccinellidae. *Annu. Rev. Entomol.* **7**, 289–326.

Hagen, K. S., and Tassen, R. L. (1966). Artificial diet for *Chrysopa carnea* Stephens. *In* "Ecology of Aphidophagous Insects," (I. Hodek, ed.), pp. 83–87. Proceedings of a Symposium held in Liblice near Prague. Academia, Prague.

Hagen, K. S., Sawall, E. F., Jr., and Tassen, R. L. (1970). The use of food sprays to increase effectiveness of entomophagous insects. *Proc. Tall Timbers Conf. Ecol. Anim. Control Habitat Manage.* **2**, 59–81.

Hall, I. M., and Dunn, P. H. (1958). Artificial dissemination of entomophthorous fungi pathogenic to the spotted alfalfa aphid in California. *J. Econ. Entomol.* **51**, 341–344.

Harley, K. L. S., and Kassulke, R. C. (1971). Tingidae for biological control of *Latana carnara* (Verbenaceae). *Entomophaga* **16**, 389–410.

Hart, W. G. (1972). Compensatory releases of *Microterys flavus* as a biological control agent against brown soft scale. *Environ. Entomol.* **1**, 414–419.

Hodek, I. (1967). Bionomics and ecology of predaceous Coccinellidae. *Annu. Rev. Entomol.* **12**, 79–1004.

Hodek, I. (1970). Coccinellids and the modern pest management. *Bioscience* **20**, 543–552.

Hoyt, S. C., and Caltagirone, L. E. (1971). The developing programs of integrated control of pests of apples in Washington and peaches in California. *In* "Biological Control" (C. B. Huffaker, ed.), pp. 395–421. Plenum, New York.

Huffaker, C. B., (ed.). (1971). "Biological Control," 511 pp. Plenum, New York.

Huffaker, C. B., and Kennett, C. E. (1956). Experimental studies on predation: predation and cyclamen-mite populations in California. *Hilgardia* **26**, 191–222.

Huffaker, C. B., and Kennett, C. E. (1969). Some aspects of assessing efficiency of natural enemies. *Can. Entomol.* **101**, 425–447.

Hussey, N. W., and Bravenboer, L. (1971). Control of pests in glasshouse culture by the introduction of natural enemies. *In* "Biological Control" (C. B. Huffaker, ed.), pp. 195–216. Plenum, New York.

Janvier, H. (1956). Hymenopterous predators as biological control agents. *J. Econ. Entomol.* **49**, 202–205.

Kirkton, R. M. (1970). Habitat management and its effects on population of *Polistes* and *Iridomyrmex*. *Proc. Tall Timbers Conf. Ecol. Anim. Control Habitat Manage.* **2**, 243–246.

Knipling, E. F. (1966). Introduction. *In* "Insect Colonization and Mass Production" (C. N. Smith, ed.), pp. 1–12. Academic Press, New York.

Kot, J. (1964). Experiments in the biology and ecology of species of the genus *Trichogramma* Westwood and their use in plant protection. *Ekol. Pol. Ser.* **A12**, 243–303.

Lawson, F. R., Rabb, R. L., Guthrie, F. E., and Bowery, T. G. (1961). Studies of an integrated control system for hornworms on tobacco. *J. Econ. Entomol.* **54**, 93–97.

Leigh, T. F., Hyer, A. H., and Rice, R. E. (1972). Frego bract condition of cotton in relation to insect populations. *Environ. Entomol.* **1**, 390–391.

Leius, K. (1967). Influence of wild flowers on parasitism of tent caterpillar and codling moth. *Can. Entomol.* **99**, 444–446.

Lingren, P. D., and Ridgway, R. L. (1967). Toxicity of five insecticides to several insect predators. *J. Econ. Entomol.* **60**, 1639–1641.

McClanahan, R. J. (1971). *Trialeurodes vaporariorum* (Westwood), green house whitefly (Homoptera: Aleyrodidae). *Commonw. Inst. of Biol. Control (Trinidad) Tech. Commun.* **4**, 51–52.

Mackauer, M. (1971). *Acythosiphon pisum* (Harris) pea aphid (Homoptera: Aphididae). *Commonw. Inst. Biol. Control (Trinidad) Tech. Commun* **4**, 3–10.

Mackauer, M. (1972). Genetic aspects of insect production. *Entomophaga* **17**, 27–48.

McMurtry, J. A., Johnson, H. G., and Scriven, G. T. (1969). Experiments to determine effects of mass releases of *Stethorus picipes* on the level of infestation of the avocado brown mite. *J. Econ. Entomol.* **62**, 1216–1221.

Markkula, M., Tiittanen, K., and Nieminen, M. (1972). Experiences of cucumber growers on control of the two-spotted spider mite, *Tetranychus telarius* (L.), with the phytoseiid mite *Phytoseiulus persimilis* A. H. *Ann. Agr. Fenn.* **11**, 74—78.

National Academy of Sciences. (1969). "Insect Pest Management and Control" (Volume 3 of Principles of Plant and Animal Pest Control), 508 pp. Nat. Acad. Sci. Publ. 1695, Washington, D.C.

Oatman, E. R., and Platner, G. R. (1971). Biological control of the tomato fruitworm, cabbage looper, and hornworms on processing tomatoes in Southern California, using mass releases of *Trichogramma pretiosum*. *J. Econ. Entomol.* **64**, 501—506.

Oatman, E. R., McMurtry, J. A., and Voth, V. (1968). Suppression of the two-spotted spider mite on strawberry with mass releases of *Phytoseiulus persimilis*. *J. Econ. Entomol.* **61**, 1517—1521.

Parker, F. D., and Pinnell, R. E. (1972). Further studies of the biological control of *Pieris rapae* using supplemental host and parasite releases. *Environ. Entomol.* **1**, 150—157.

Parker, F. D., Lawson, F. R., and Pinnell, R. E. (1971). Suppression of *Pieris rapae* using a new control system: mass releases of both the pest and its parasites. *J. Econ. Entomol.* **64**, 721—735.

Pimentel, D. (1970). Population control in crop systems: monocultures and plant spatial patterns. *Proc. Tall Timbers Conf. Ecol. Anim. Control Habitat Manage.* **2**, 209—221.

Rabb, R. L., and Bradley, J. R., Jr. (1968). The influence of host plants on parasitism of eggs of the tobacco hornworm. *J. Econ. Entomol.* **61**, 1249—1252.

Rabb, R. L., and Lawson, F. R. (1957). Some factors influencing predation of *Polistes* wasps on the tobacco hornworm. *J. Econ. Entomol* **50**, 778—784.

Raun, E. S. (1966). European corn borer. *In* "Insect Colonization and Mass Production" (C. N. Smith, ed.), pp. 332—338. Academic Press, New York.

Reed, D. K., Hart, W. G., and Ingle, S. J. (1970). Influence of windbreaks on distribution and abundance of brown spot scale in citrus groves. *Ann. Entomol. Soc. Amer.* **63**, 792—794.

Ridgway, R. L. (1969). Control of the bollworm and tobacco budworm through conservation and augmentation of predaceous insects. *Proc. Tall Timbers Conf. Ecol. Anim. Control Habitat Manage.* **1**, 127—144.

Ridgway, R. L., and Jones, S. L. (1969). Inundative releases of *Chrysopa carnea* for control of *Heliothis* on cotton. *J. Econ. Entomol.* **62**, 177—180.

Schiefelbein, J. W., and Chiang, H. C. (1966). Effects of spray of sucrose solution in corn fields on the populations of predatory insects and their prey. *Entomophaga* **12**, 475—479.

Schlinger, E. I., van den Bosch, R., and Dietrick, E. J. (1959). Biological notes on the predaceous earwing *Labidura riparia* (Pallas) a recent immigrant to California (Dermaptera: Labiduridae). *J. Econ. Entomol.* **52**, 247—249.

Schread, J. C., and Garman, P. (1934). Some effects of refrigeration on the biology of *Trichogramma* in artificial breeding. *J. N. Y. Entomol. Soc.* **42**, 263—269.

Sedlag, W. (1964). Zur Biologie und Bedeutung von *Diaeretiella rapae* (McIntosh) als Parasit der Kohlblattlaus (*Brevicoryne brassicae* (L.)). *Nachrichtenbl. Deut. Pflanzenschutzdienst (Berlin)* **18**, 31—86.

Shands, W. A., Simpson, G. W., and Storch, R. H. (1972). Insect predators for controlling aphids on potatoes. 3. In small plots separated by aluminum flashing strip-coated with a chemical barrier and in small fields. *J. Econ. Entomol.* **65**, 799—805.

Shepard, M., Sterling, W., and Walker, J. K., Jr. (1972). Abundance of beneficial arthropods on cotton genotypes. *Environ. Entomol.* **1**, 117—121.

Sholdt, L. L., Ehrhardt, D. A., and Michael, A. G. (1972). A guide to the use of the mosquito fish, *Gambusia affinis*, for mosquito control. *Navy Environ. Prev. Med. Unit No. 2, Publ. EPMU2 PUB6250*, pp. 1—18.

Shorey, H. H., and Hale, R. L. (1965). Mass-rearing of the larvae of nine noctuid species on a simple artificial diet. *J. Econ. Entomol.* **58**, 522–524.

Simmonds, F. J. (1966). Insect parasites and predators. *In* "Insect Colonization and Mass Production" (C. N. Smith, ed.), pp. 489–499. Academic Press, New York.

Smith, C. N. (ed.). (1966). "Insect Colonization and Mass Production," 618 pp. Academic Press, New York.

Smith, R. F. (1972). The impact of the green revolution on plant protection in tropical and subtropical areas. *Bull. Entomol. Soc. Amer.* **18**, 7–14.

Smith, R. F., and Reynolds, H. T. (1972). Effects of manipulation of cotton agroecosystems on insect pest populations. *In* "The Careless Technology, Ecology and International Development" (M. T. Farvar and J. P. Milton, eds.), pp. 373–406. The Natural History Press, Garden City, New York.

Solomon, M. E. (1949). The natural control of animal populations. *J. Anim. Ecol.* **18**, 1–35.

Solomon, M. E. (1973). Ecology in relation to the management of insects. *In* "Insects: Studies in Population Management" (Geier *et al.*, eds.), pp. 154–167. Ecol. Soc. Aust. (Memoirs 1), Canberra.

Southwood, T. R. E., and Way, M. J. (1970). Ecological background to pest management. *In* "Concepts of Pest Management" (R. L. Rabb and F. E. Guthrie, eds.), pp. 6–29. North Carolina State University, Raleigh, North Carolina.

Spencer, G. J. (1958). On the Nemestrinidae of British Columbia dry range lands. *Proc. 10th Int. Congr. Entomol.* **4**, 503–509.

Steinhaus, E. A. (1958). Crowding as a possible stress factor in insect disease. *Ecology* **39**, 503–514.

Stern, V. M., van den Bosch, R., Leigh, T. F., McCutcheon, O. D., Sallee, W. R., Houston, C. E., and Garber, M. J. (1967). Lygus control by strip-cutting alfalfa. *Univ. Calif. Agr. Ext. Serv.* **AXT-241**, 1–13.

Stinner, R. E., Ridgway, R. L., and Kinzer, R. E. (1974). Storage, manipulation of emergence, and estimation of numbers of *Trichogramma pretiosum*. *Environ. Entomol.* **3**, 305–307.

Tamaki, G., and Halfhill, J. E. (1968). Bands on peach trees as shelters for predators of the green peach aphid. *J. Econ. Entomol.* **61**, 707–711.

Turnbull, A. L. (1967). Population dynamics of exotic insects. *Bull. Entomol. Soc. Amer.* **13**, 333–337.

U.S. Dept. Agr. (1972). Wasps that guard cotton. *Agr. Res.* **20**, 3–4.

van den Bosch, R. (1968). Comments on population dynamics of exotic insects. *Bull. Entomol. Soc. Amer.* **14**, 112–115.

van den Bosch, R., and Hagen, K. S. (1966). Predaceous and parasitic arthropods in California cotton fields. *Univ. Calif. Agr. Exp. Sta. Bull.* **820**, 32 pp.

van den Bosch, R., and Telford, A. D. (1964). Environmental modification and biological control. *In* "Biological Control of Insect Pests and Weeds" (P. DeBach, ed.), pp. 459–488. Reinhold, New York.

van den Bosch, R., Lagace, C. F., and Stern, V. M. (1967). The interrelationship of the aphid, *Acyrthosiphon*, and its parasite *Aphidius smithii*, in a stable environment. *Ecology* **48**, 993–1000.

van den Bosch, R., Leigh, T. F., Falcon, L. A., Stern, V. M., Gonzales, D., and Hagen, K. S. (1971). The developing program of integrated control of cotton pests in California. *In* "Biological Control" (C. B. Huffaker, ed.), pp. 377–394. Plenum, New York.

van Emden, H. F. (1965). The role of uncultivated land in the biology of crop pests and beneficial insects. *Sci. Hort.* **17**, 121–136.

Watson, T. K., and Wilde, W. H. A. (1963). Laboratory and field observations on two predators of pear psylla in British Columbia. *Can. Entomol.* **95**, 435–438.

Way, M. J., Murdie, G., and Galley, D. J. (1969). Experiments on integration of chemical and biological control of aphids on brussel sprouts. *Ann. Appl. Biol.* **63**, 459–475.

White, E. B., DeBach, P., and Garber, M. J. (1970). Artificial selection for genetic adaptation to temperature extremes in *Aphytis lingnanensis* Compere (Hymenoptera: Aphelinidae). *Hilgardia* **40**, 161–191.

Wille, J. E. (1951). Biological control of certain cotton insects and the application of new organic insecticides in Peru. *J. Econ. Entomol.* **44**, 13–18.

Wilson, F. (1966). The conservation and augmentation of natural enemies. *Proc. FAO Symp. Integrated Pest Control* **3**, 21–26.

Yazgan, S., and House, H. L. (1969). A hymenopterous insect, the parasitoid *Itoplectis conquisitor*, reared axenically on a chemically-defined synthetic diet. *Can. Entomol.* **102**, 1304–1306.

11

EVALUATION OF THE IMPACT OF NATURAL ENEMIES

Paul DeBach, C. B. Huffaker, and A. W. MacPhee

I. INTRODUCTION

The importance of natural enemies in the control of their host (or prey) populations has long been a subject of contention. While today natural enemies are viewed by an increasingly wide scope of ecologists as highly significant natural control factors, there is still disagreement both as to their role and the proper methods of their evaluation. Suitable methods of evaluation are essential for three reasons: (1) to show the value and shortcomings of existing natural enemies, the need for introducing new ones, as well as the need to manipulate the environment, or the natural enemies, so as to make resident species more effective; (2) to provide insights into the principles of population ecology relating to the interplay of biotic and abiotic factors; (3) to demonstrate the effectiveness of natural enemies and thus ensure continued support for biological control research and development.

DeBach and Huffaker (1971) dealt with this subject recently. They wrote:

> So much disagreement has taken place among ecologists concerning the academic and semantic complexities of the question (of how population regulation occurs) that it has led some of the more theoretical workers to conclude that true regulation by natural enemies either does not occur or cannot be measured. It is emphasized that a variety of comparative experimental techniques may be employed to accurately evaluate and demonstrate the precise contribution of natural enemies in prey (host) population regulation in any particular habitat. Such techniques can be devised to exclude the possible contributing or compensatory effects of other environmental parameters such as weather, competitors, or genetic variation, etc., so as to evaluate the actual total regulatory effect brought about essentially by the action of natural enemies.
>
> By *population regulation* we mean what is commonly referred to as natural balance—the maintenance of an organism's population density over an extended period of time between characteristic upper and lower limits. A regulatory factor is one which is wholly or partially responsible for the observed regulation under the given environmental regime and whose removal or adverse change in efficiency or degree will result in an increase in the average pest population density. Note that this definition is amenable to experimental testing. The ecological term, biological control, connotes prey population regulation by natural enemies. Note that economic qualifications purposely are excluded. The degree of economic achievement must be defined for each case; biological control can occur at very low or very high average densities. Temporary suppression or control by some enemies at times would not satisfy the connotation of "regulation," but in practice, reliable long-term *control* by an effective natural enemy is largely indistinguishable from *regulation* by that enemy.

These authors also noted that much more has been written about the mechanisms involved in population regulation by natural enemies than about the quantitative or experimental demonstration of the fact itself. Varley and Gradwell (1970, 1971), Huffaker *et al.* (1971), Huffaker and Stinner (1971), Hassell and Varley (1969), DeBach and Sundby (1963), and the authors herein (Chapter 3) have dealt with certain aspects of the mechanisms of host regulation by entomophagous parasites. Much less has been written about the mechanisms involving invertebrate predators (Varley, 1971), although the work of Holling (1959, 1966), Huffaker *et al.* (1963), Laing and Huffaker (1969), Haynes and Sisojevic (1966), and Sandness and McMurtry (1970) among others, go part way in explaining the mechanisms relative to some arthropod predators. This chapter deals mainly with methods of establishing the *fact* of natural enemy regulation, and only to a limited degree, or as incidentally derived, with methods leading to insights as to *mechanisms*. Moreover, the emphasis will be on use of experimental methods rather than on the traditional collection and analysis of quantitative population dynamics data which at best provide only *indications*.

The emphasis herein on experimental means of evaluation is not meant to imply that there are no other useful methods. However, the authors do not consider that other methods provide the required proof of control or regulation by enemies. Periodic census and life table data provide much valuable information, but such methods, including regression and modeling techniques, have many weaknesses for rating the regulating or controlling power of natural enemies (DeBach and Bartlett, 1964; Huffaker and Kennett, 1969; Hassell and Huffaker, 1969).

Legner (1969) states, "The ultimate and probably only reliable method for judging a parasite's effectiveness is the reduction in host equilibrium position following liberation" (i.e., an evaluation of their effectiveness by using *before* introduction and *after* introduction density comparisons). Hassell and Varley (1969), also stated, "Laboratory studies can go some way to predict which species (among several introduced) may be successful, but an introduction provides the only real test." The important point is that neither precise laboratory studies nor field population studies that do not include experimental check methods involving paired comparative plots, having natural enemies present in one series and natural enemies absent in another series, will furnish adequate means of evaluating the effectiveness of a natural enemy.

While strongly advocating the use of experimental methods (many of which are extremely simple to apply), it is recognized that some cases will be difficult especially with hosts (or prey) or parasite (or predator) species that disperse widely. For such species some adequate system of dealing with the movements is essential, including use of very extensive experimental or comparison areas or mark and recapture techniques, as well as detailed and prolonged population census data.

II. SELECTION OF STUDY AREAS AND THE DURATION OF STUDIES

The areas for study by any method of evaluation should be carefully chosen to minimize the possibility that cryptic factors will mask the potential for measuring the effect of a natural enemy or lessen its actual effectiveness. Plots disturbed by use of adverse chemicals, either in the plots themselves or in adjacent surrounding areas are, in general, suspect. If, for example, only a small plot is left untreated in the middle of a 100- to 1000-acre cotton, grape, or plum planting, and toxic chemicals are periodically applied all around it, the ability of natural enemies to control a host (or prey) population cannot be assessed. Obviously the enemies will be decimated by drift of the chemicals and by dispersal into the treated area. The results from such plots are sometimes used by the undiscerning as evidence that natural enemies are not capable of controlling the respective host species.

The size of the study area must be large enough that normal movements into or out of the area of either pest or natural enemy species would not significantly affect the potential for control. Of equal importance, the size and location of the area must be such that drift into the area, not only of conventional pesticides but of certain "nontoxic" industrial or agricultural materials (e.g., ash or dust), would be minimal (DeBach and Huffaker, 1971).

It is necessary to make arrangements with the grower to assure that only agreed upon materials or practices are used. Some herbicides and fungicides, and, of course, insecticides may cause an increase in a pest's fecundity in addition to causing severe adverse effects on its natural enemies. Certain plant nutrient formulations to correct minor element deficiencies in citrus drastically affect natural enemies. It may be necessary that the grower practice good ant control since ants are known to affect dramatically the efficiency of a number of natural enemies of several important honeydew-producing insects as well as others, such as, armored scales and mites. Clean cultivation or use of a cover crop in orchards or row crops can present very different potentials for effective natural enemy action (DeBach and Huffaker, 1971). Hedgerows and weedy fence areas may have either a beneficial or adverse effect and need to be evaluated with respect to the particular ecosystem being studied (Southwood and Way, 1970).

The time factor is important in any study. Genetic changes can occur in the host and in the natural enemy over a period of time. While for introduced entomophagous insects these changes appear to have been predominantly toward improved adaptation and efficiency of control on the part of the natural enemy, a few examples have occurred in which the host has attained an improved capacity to resist successful attack (Messenger and van den Bosch, 1971, and Chapters 9 and 12). Plot studies must be continued long enough following cessation of adverse chemical treatments for natural balance to be reattained. This may involve adequate reestablishment of some essential nonpest alternate prey or host species of the natural enemy, as well as of the natural enemy itself (DeBach, 1969; Flaherty and Huffaker, 1970). In some cases 3 or 4 years have been shown to be required. Thus, the mere fact that a field or orchard has not been sprayed for 1 or 2 years does not mean that the maximal potential degree of biological control has been reattained. Also, the role of natural enemies in population regulation as partially assessed by use of life table data requires study over a long series of host generations (e.g., Varley and Gradwell, 1970; Klomp, 1966). The relationship of natural enemy efficiency to host density and to the variability in abiotic factors, among others, cannot be ascertained by a short-term study. With use of experimental methods, however, certain early indications of results can oftentimes be obtained (see Section IV).

III. POPULATION SAMPLING

Sampling arthropod populations often involves the estimation of numbers of small, active animals which develop rapidly and change their appearance, redistribute themselves as their environment changes, and which may be so numerous that only a tiny fraction of their numbers may be counted in a practical time interval. The precision required of the estimate varies with the kind of information needed. For example, lower precision may be tolerated in the measurement of the effect of a chemical spray on a pest population than in the measurement of the effect of some biological component such as a predator or a parasite. The problem is most difficult where the objective is to analyze and understand a complex dynamic natural ecosystem *in situ*. This fact is thoroughly demonstrated in the monograph edited by Morris (1963) on spruce budworm, *Choristoneura fumiferana* (Clemens), populations in New Brunswick forests where the problem is complicated by the number of species and stages, by the duration of sampling, and by the need to determine fecundity and the causes of mortality.

In evaluating the impact of natural enemies, we must measure population levels of both prey and natural enemies over a number of generations and on some common basis. Information is needed on the functional and numerical response of the enemies, on the magnitude of population fluctuations, on the economically acceptable level of the pest, on the capacity of the natural enemies to contain the pest population, and on the impact of many other mortality factors.

Although there are many concepts and detailed techniques reported in the literature, work is still needed on sampling problems. A consensus suggests we should examine the best mechanics of sampling the species, that is, how to count, or measure it in its plant, soil or other habitat. Consideration should be given to stratification of the study universe for the purpose of efficiency. The size of the sample unit should be considered since much population interaction information can be missed by taking too large a sample on a lumped basis, or time may be wasted by taking too small a sample. Taking of a sequence of samples to be kept separate for each relatively limited habitat unit (a tree or a square meter) is essential where phases of predator–prey interaction may be different from tree to tree, e.g., in the preliminary study or planning stage the selection of basic or absolute units of measurement should be appraised since in the final analysis there will be a need to consider two or more species on a common basis. Finally, both the accuracy and precision of the data must meet the requirements of the final objective and both should be considered in the planning stage (see Southwood, 1966).

A. Techniques

The techniques of sampling depend so much on the species concerned and on its life history that it is difficult to generalize. The most commonly used technique for large specimens is to search samples visually, such as a part of the plant, an area of soil, or aliquot of grain, for the stage and species concerned. If natural enemies are involved it may also be useful to search or dissect the pest for parasitism or for signs of predation. Another common approach is the determination of relative density from such evidence as tissue damage, feeding signs, or frass. The labor of searching and counting may be reduced through mechanical removal of the target individuals from a unit of the habitat, as for example the removal of phytophagous and predaceous mites from apple foliage by a mechanical brushing machine ("Henderson and McBurnie") or a motorized vacuum suction apparatus to collect and transfer arthropods from foliage into a cloth cage.

Greater demands are placed on the sampling techniques when a number of species of prey, predators, and parasites are to be sampled in their common ecosystem. Lord (1968) sampled apple clusters using a "shaker" for predators, and using a branch as the sample unit. He then also used a portion of the leaf clusters from the same branch for visual counts of European red mite. The total populations of predators on apple trees were calculated by Muir (1958) by releasing a known number of marked individuals, then later sampling the species on a part of the tree and observing the proportion of marked to unmarked individuals. Sonleitner and Bateman (1963) used a similar method for sampling populations of the Queensland fruit fly, *Dacus tryoni* (Frog.). A serological method of some promise in detecting predation, but not predator abundance or total numbers of prey killed, was tested by MacLellan (1954) on suspected predators using a precipitin test to detect the prey protein in the gut of a suspected predator. Dempster (1960) used a similar method on predators of brown beetle, *Phytodecta olivacea* (Forster). The tagging of prey with radioactive tracers and measuring the activity of suspected predators has had some success (Southwood, 1966).

B. Stratification

Sampling error may be reduced by classifying the universe into strata in which the population levels or habitats differ appreciably. When this is done, only variation within the stratum contributes to sampling error (Yates, 1960). If strata are not established by means of some preliminary study, we can proceed by choosing random samples in the universe and then keeping a record of the location or special habitat of each sample until the sampling universe can be classified. In some orchard studies, the separating of samples from different elevations and quadrants has proved helpful (LeRoux and Reimer, 1959). Morris (1955) discussed criteria for stratification, particularly for sampling the spruce budworm. It is not always expedient to stratify, particularly where many species are being counted from the same samples (Lord, 1972); stratification suitable to one species might well not apply to others. Allowances must also be made for changes in the proportion of the population found in a given stratum with change in time and in population density. These two factors are both relevant in brown mite, *Bryobia rubriculus* (Sheuten), populations (Herbert, 1965).

C. Sample Unit, Size, and Number

Morris (1955) suggests that the size of the sampling unit should be chosen to minimize both variance and cost, and that samples should be sufficiently small so that enough can be taken to provide an adequate estimate of sampling variance. Southwood (1966) suggests that too much stress should not be placed on a precise determination of the optimum size of a sampling unit. In at least the preliminary stage of a study, it is important to use small sample units so that variation and distribution can be assessed and optimum sample size determined.

The total number of units required, of any given sample size, depends on the degree of precision desired. Morris (1955) points out that the amount of sampling can be determined to satisfy a specified degree of precision or to an optimum allocation of sampling resources and that the latter may be the more useful since the required level of precision is rarely known. Morris (1955), LeRoux and Reimer (1959), Harcourt (1964), and others have recommended a procedure for the optimum allocation of sampling resources which may be useful in comparing sampling designs. This may be particularly helpful where there is a choice of selecting one or many samples from each of a large number of plants in the universe. In this procedure variance of samples and cost of sampling are minimized.

The attainment of an error as low as 10% of the mean is often accepted as at least a preliminary rule of thumb. Although it may be impossible to evaluate nonsampling errors, the same 10% level based on the best information available may be a useful guideline (Morris, 1955).

D. Absolute Units

Many authors call attention to the need for absolute sample units in which the sample is a specified part of a universe, such as ground area, and cannot change.

In many agricultural crops where a known number of plants are grown per acre, the single plant can be the basic unit to which numbers are related. By sampling parts of a fruit tree such as clusters, for example, and counting total clusters on the tree, the absolute population can be estimated. Frazer (1971) used this method in estimating total populations of walnut aphids and of its natural enemies in walnut trees in California. In studies in an experimental apple orchard in Nova Scotia, clusters are chosen on a systematic basis from trees with known numbers of clusters (MacLellan, 1963). In a study on orchard fauna, Lord (1972) used this method, recording the counts from each cluster and also the type of cluster (from a classification of six types); the data can then be classified into strata for study and analysis for some of the species. In sampling for winter moth, Embree (1965) used a method relating branch diameter to the number of leaf clusters on oaks to arrive at an estimate of the absolute number. In many situations it may be desirable, or essential, to relate both pest and natural enemy populations to one absolute unit so that mortality from the various causes can be related to the number of pests and natural enemies present in the universe (e.g., Frazer, 1971).

E. Conclusion

Sampling procedures have been improved over the years to meet the needs of the study of population dynamics, but in some ways have fallen short of the hopes of the applied ecologist. Where the purpose is to measure the impact of natural enemies in a dynamic ecosystem, a high precision is necessary due to the species complexity and to the large number of sampling occasions required to resolve causes of population change. Further improvements could be made by increased initial planning and the involvement of the statistician with the biologist.

IV. EXPERIMENTAL OR COMPARISON METHODS OF EVALUATION

In assessing natural enemy effectiveness we commonly consider the problems associated with (1) newly introduced, exotic natural enemies, and (2) indigenous or already established, imported natural enemies. DeBach and Bartlett (1964) discuss in detail various methods of evaluation of natural enemy effectiveness and conclude that use of experimental comparisons is the best. These methods can be classified as (1) addition, (2) exclusion (or subtraction), and (3) interference (DeBach and Huffaker, 1971). These authors give a documentary summary of studies exemplifying these three techniques and the subject is dealt with here in similar fashion.

A. The Addition Method

The addition technique has been discussed by DeBach and Huffaker. They note the following:

> The addition method applies primarily to measurement of the results of importation (including subsequent transfer or relocation) of new natural enemies. The procedure involves "before and after" type comparisons involving initially comparable plots, some having natural enemies colonized, and others not receiving any. Thus, for example, an imported parasite can be colonized in 10 plots and its population increase as well as the host-population decrease measured. Host population trends in 10 plots not receiving parasites would be measured simultaneously. Any differences between the two series can be ascribed to the parasites. Such tests, of course, must be planned well in advance of colonization, because dispersal of some parasites or predators is so rapid as to eliminate any enemy-free plots within a short time. Obviously, plots should be separated initially by a suitable distance. Replication of this type of test can be achieved by conducting the same sort of experiments following transfer of the natural enemy to a new area or country. Results from 10 olive groves where *Aphytis maculicornis* (Masi) alone was present and 10 groves where a second parasite, *Coccophagoides utilis* Doutt, additionally was present, were used by Huffaker and Kennett (1966) to evaluate the net effect of the addition of the second species in the control of olive scale, *Parlatoria oleae* (Colvée) in California. (See further DeBach *et al.*, 1971; Huffaker *et al.*, 1971.)

In many biological control projects, results are so striking that "before and after" photographs are more spectacular than are population census data. Preferably, both types of measurement should be obtained. While there are many examples of the "addition" method, and other experimental techniques reported in the literature, the authors will rely mostly on their own studies for illustration here.

The addition method is illustrated by some well-documented examples of "before and after" evaluation of pests brought under control by introduced natural enemies. For quite a few, excellent photographs which supplement census data are available. Unfortunately, only in a relatively few cases have a good series of "before and after" population density measurements been obtained. For some, however, detailed "before and after" census records over a wide range of conditions are available which furnish a good basis of evaluation. For illustrative purposes, summary accounts have been selected for the winter moth and European spruce sawfly in Canadian forests, olive parlatoria scale in California, prickly pear in Australia, and St. Johnswort in the western United States. (See also Chapters 12, 13, 15, and 19.)

1. Winter Moth, Operophtera brumata *L., in Eastern Canada*

Embree (1971) summarizes the results from introductions of six species of parasites from Europe into the Eastern Provinces of Canada from 1955 to 1960 for control of winter moth. Only two species, *Cyzenis albicans* Fall., a tachinid, and *Agrypon flaveolatum* (Grav.), an ichneumonid, became established. Before the parasites were established, the moth population fluctuated at high densities, causing enormous losses in oak forests, a resource not then nearly as valuable as it is today and will be in the future. Embree estimated that in Nova Scotia alone some $12,000,000 of loss was prevented by the introductions. Figure 1 from Embree (1966) illustrates the role of the parasites. For the first 5 years of the study, previous to achievement of significant parasitization, the moth densities remained high, but with the achievement of substantial parasitization,

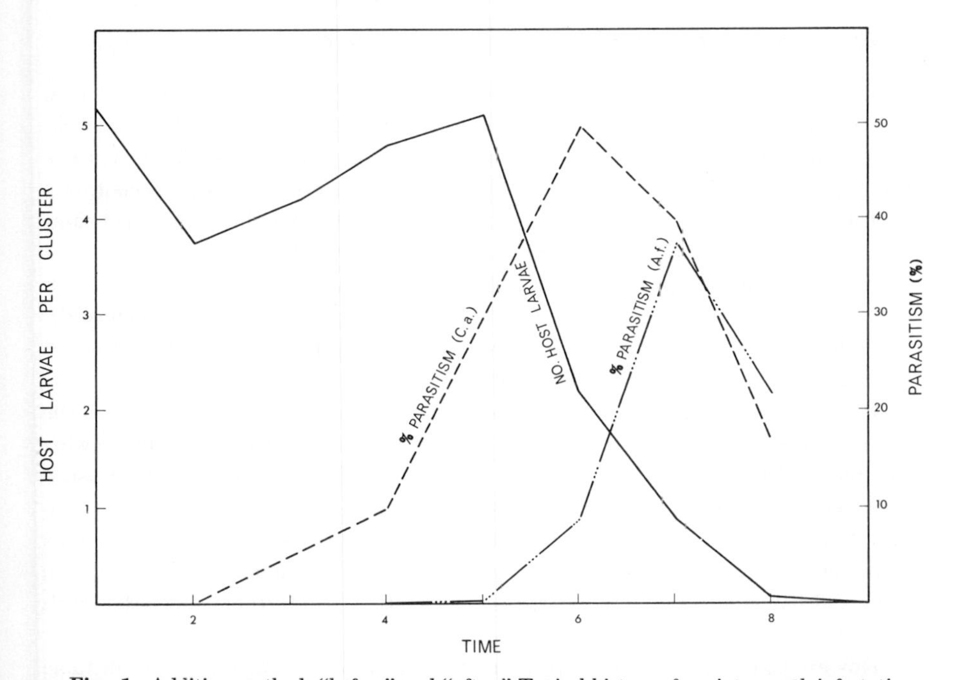

Fig. 1. Addition method: "before" and "after." Typical history of a winter moth infestation and parasitism following introduction of *Cyzenis albicans* (Fall.) (C.a. in figure) and *Agrypon flaveolatum* (Grav.) (A.f. in figure). Time refers to the number of years the outbreak persisted. (After Embree, 1966.)

the population crashed and persists now only at very low densities (see Chapter 12). (See further documentation based on life table data below.)

This example illustrates the case where natural enemies may possibly be more effective in an area where both they and the host are introduced exotics than in the indigenous area. It may also illustrate the case where introduction of a large complex of parasites would have been a mistake.

The winter moth is not considered to be under good biological control on oaks in its native home area of Europe where it is attacked, not only by *C. albicans* and *A. flaveolatum*, but by some 15 or more competing parasites (and additional predators): three or four hyperparasites also attack *C. albicans*. Pschorn-Walcher *et al.* (1969) consider it plausible that the lesser effectiveness of *C. albicans* and *A. flaveolatum* in Europe, where they are native, than in eastern Canada, where they are introduced, is due to the inhibiting effects of hyperparasites and competition by intrinsically superior but extrinsically inferior parasites. Sechser (1970) found that in multiparasitism, *C. albicans* and *A. flaveolatum* are inferior (intrinsic competition) to all the other parasites. However, they presumably possess superior extrinsic competitive abilities (high searching capacities) since they coexist among the complex in Europe.

2. European Spruce Sawfly, *Diprion hercyniae* (Hartig)

In the 1930's the European spruce sawfly became a major threat to spruce forests in eastern Canada, causing widespread outbreaks (Bird and Elgee, 1957; Neilson *et al.*, 1971). Subsequent to 1945 it was not a pest of economic importance. The change has been generally ascribed to the effects of biological control introductions, especially of a nuclear polyhedrosis virus which was inadvertently introduced and two species of parasites, *Drino bohemica* (Mesn.) and *Exenterus vellicatus* Cush.

Reeks and Cameron (1971) estimated that the sawfly may have directly killed one-half of the salable spruce in the Gaspé Peninsula up to 1946, or about 8.5 million cords. They also estimated that the biological control program, which cost $300,000, saved $6,000,000 by checking further destruction.

The best documented record for this biological control effort is that for the Maritime Provinces. Neilson and Morris (1964) stated that for some 20 years following the initial crash in the outbreak in this region, brought about by the disease, parasites appeared to be responsible for the very high degree of control at the low densities then prevailing. However, Reeks and Cameron (1971) concluded that, while either the virus acting alone or the parasites alone can control the sawfly population at low levels, both virus and parasites together reduce the pest population even further. This latter result might be due to a spreading of the pathogen by the parasites.

This example illustrates the contention (Huffaker and Kennett, 1969) that life table and correlation techniques as used thus far (although clarifying some of the relationships in natural enemy performance, and other factors as well) do not furnish a measure of the "power of control" of the prey (or host) population. That is, they do not answer the very practical question of just how high densities would go in the absence of the natural enemy (or enemies). Only comparisons of densities before and after introductions, or use of check methods comparing natural-enemy-present (and uninhibited) areas with natural-enemy-absent (or inhibited) areas will do this. For this case, Neilson and Morris did indeed point to the considerable value of life table and correlation analysis in detecting certain properties of the natural enemies and in relating their action to *changes* in host density. However, it is only because we have census data of the sawfly populations both before and after the impact of the introduced agents, that we see the actual control power of these natural enemies. The original populations show how high densities would presumably still go (to tree killing levels) if the natural enemies were not present.

Both Neilson and Morris (1964) and Bird and Elgee (1957) present data from study plots wherein sawfly larval densities prior to the effects of the disease and the parasites were of the order of 10 to 30 per tree sample, but subsequent to the establishment of the natural enemies and achievement of their full effects, endemic levels of a fraction of a larva per tree sample have been characteristic.

Bird and Elgee also noted that, prior to the introductions, sawfly populations in the outbreak area continued to rise "until limited by the supply of foliage." They suggested that the parasites might have been able to bring the outbreak under control without the action of the disease, but noted that the disease caused a crash before the potential value of the parasites could be realized.

Neilson *et al.* (1971) further note that DDT treatments in 1960—1962 so reduced sawfly numbers in study areas in New Brunswick that the virus and parasites were virtually eliminated. Within five generations following cessation of treatments, the sawfly populations had approached the densities of the peak outbreak years. This, in effect, is an experimental check. They add, "Now, 24 years after the collapse of the outbreak, it can be stated with confidence that *D. hercyniae* is presently not a problem in Canada, and furthermore, barring major disruption such as widespread forest spraying, it is not likely to become one in the forseeable future."

3. *Olive Parlatoria Scale in California*

While the check-method data secured on the biological control of olive scale well illustrates the "interference" method (Huffaker *et al.*, 1962; DeBach and Huffaker, 1971), other data clearly show the process of the simple "before and after" comparison of the addition method. Extensive census data on the incidence of scaly olives during the early years of the establishment of the first introduced parasite, *Aphytis maculicornis* (Masi), in areas untreated with insecticides show the uniform seriousness of the scale on a very general basis in the San Joaquin Valley (Tables 1 and 2).

Moreover, Fig. 2 shows (A) a typical untreated olive tree during the years prior to establishment of either *A. maculicornis* or the supplementing species, *Coccophagoides utilis* Doutt, introduced later; and (B) a typical olive tree under otherwise comparable

TABLE 1

Representative Commercial Fruit Cullage in Various Olive Groves prior to Fully Effective Establishment of Two Parasites of *Parlatoria oleae*[a]

Location	Percentage cullage
Fluetch, Merced	30.0 (1956)
E. 21st St. Merced	24.0 (1956)
Canal Farm Inn, Los Banos	34.0 (1956)
523 Washington St., Los Banos	46.0 (1956)
Cunha, Gustine	16.0 (1956)
Hwy. 33 and West Ave., Gustine	96.0 (1956)
Murray, Gustine	12.0 (1956)
Cunningham Ranch, Bakersfield	100.0 (1956)
Whitney, Exeter	85.0 (1956)
Clovis Ave. #1, Clovis	35.0 (1956)
Clovis Ave. #2, Clovis	16.0 (1956)
Clovis Ave. #3, Clovis	14.0 (1956)
Cypress St., Madera	84.0 (1956)
Austin St., Madera	30.0 (1956)
Olive St., Fresno	40.0 (1956)
Clinton St., Fresno	16.0 (1956)
Bell, Hills Valley	23.7 (1955)
Oberti, Madera	60.2 (1958)
Colusa St., Willows	49.1 (1957)

[a]Considerable effect of *Aphytis maculicornis* already exerted in places.

TABLE 2

Representative Commercial Fruit Cullage in Various Olive Groves following Fully Effective Establishment of Two Parasites of *Parlatoria oleae*

Location	1966	1967
Duncan, Herndon	0.0	0.0
River Ranch, Herndon	0.3	0.8
Oberti, Madera	1.0	0.2
Sheeler, Madera	0.0	0.5
Daggett, Madera	0.3	0.3
Martinelli, Madera	0.0	0.0
Ransom, Madera	0.5	0.5
Snelling, Merced	0.8	1.0
Bell, Hills Valley	0.9	0.5
Bell, Clovis	0.3	0.0
Greer, Lemon Cove	0.3	0.0
Kirkpatrick, Lindsay	0.0	0.0
Kennedy, Lindsay	1.0	0.8
Lucca, Lindsay	0.0	0.0
Houghton, Lindsay	0.0	0.0
Hatakeda, Seville	0.0	0.0
Shimaji, Ivanhoe	0.0	0.0
McKellar, Ivanhoe	0.0	0.0
Baird, Springville	0.0	0.0
Brun, Orland	0.0	0.0
Delano, Richgrove	1.0	0.0

conditions after establishment of these two parasites. Later check-method comparisons of adjacent trees, ones treated with DDT to interfere with the then well-established parasites and others left undisturbed, verify the conclusion from such a "before and after" situation shown here (Huffaker *et al.*, 1962; DeBach and Huffaker, 1971).

Table 3 shows the additional effect of adding a second natural enemy to an ecosystem already including a more effective one, the two species being able to coexist. Far from reducing the effectiveness of the previously well-established and superior species, *A. maculicornis*, whose effectiveness has been demonstrated by three different methods (Huffaker and Kennett, 1966), the addition of the second species (*C. utilis*) greatly added to the reliability and degree of biological control. It is clear that *A. maculicornis* greatly reduced host populations in general but that in some situations *A. maculicornis* alone (prior to the addition of *C. utilis*) gave only erratic, marginal, or entirely unsatisfactory commercial control. After *C. utilis* also became well established in these groves, full commercial control was achieved every year. This has been true generally in the state. Extensive life table data were acquired during these studies but were not suitable to evaluate the role of the two natural enemies as were the comparative methods.

Fig. 2. Addition method: "before" and "after." (A) Typical untreated olive tree infested with *Parlatoria oleae* during years prior to introduction of parasites for biological control of the scale. (B) Typical untreated tree with *P. oleae* under complete biological control by introduced parasites, *Aphytis maculicornis* and *Coccophagoides utilis*. Note differences in foliage and vigor. (See further Fig. 9.) Photos by F. E. Skinner and J. Nakata.

TABLE 3

Percentage of *Parlatoria* Scale-Marked Olives (Culls) in Relation to Chronological Establishment of Introduced Parasites

	Duncan grove	Oberti grove
1955	3.8[a]	—[c]
1956	5.4[a]	—[c]
1957	4.4[a]	22.2[d]
1958	27.4[a]	60.2[d]
1959	7.4[b]	4.4[b]
1960[e]	0.4	1.0
1961	0.5	0.5
1962	0.0	0.6
1963	0.0	0.4
1964	0.0	0.3
1965	0.5	0.7
1966	0.0	1.0
1967	0.0	0.2

[a] *Aphytis* alone established.
[b] *Aphytis* fully, *Coccophagoides* intermediately established.
[c] Neither species established.
[d] *Aphytis* alone and not fully established.
[e] From 1960 to 1967, both *Aphytis* and *Coccophagoides* well established.

4. Prickly Pear in Australia and Klamath Weed in Western United States

Two of the world's outstanding examples of biological control of widespread devastating weeds also illustrate the addition method. The pioneer example is that of control of the prickly pear, *Opuntia stricta* Haw., in Australia and Klamath weed, or St. Johnswort, *Hypericum perforatum* L., in the western United States.

The prickly pear virtually threatened economic disaster in much of Queensland and New South Wales, being economically uncontrollable by conventional weed control measures. Some 50 species of exotic insects were imported for use against it, but with the establishment of an Argentine moth, *Cactoblastis cactorum* (Berg), in the late

Fig. 3. Addition method: "before" and "after." Destruction by the introduced moth, *Cactoblastis cactorum*, of dense prickly pear in belar scrub country, Chinchilla, Queensland, Australia (after Dodd, 1940). (A) Taken in October, 1926, the prickly pear in its virgin state. (B) Three years later, in October, 1929, showing the characteristic destruction resulting from the feeding on the fleshy pads of pear by larvae of *Cactoblastis*. (C) Taken in December, 1931, after trees had been cut and burnt off and the land put back into use, showing a prolific growth of rhodesgrass.

C18
A

C18
B

C

1920's, the picture was changed drastically within a few years (Dodd, 1940). Some 50,000,000 acres of land was reclaimed in Queensland alone. Figure 3 presents a picture of the process of reclamation that occurred over vast tracts of land. This was replicated time and again as the moth spread or was moved to new areas, so, in effect, a large series of experiments was conducted. Now, 45 years later, this weed is still under highly effective biological control.

Klamath weed control by the "addition" of a natural enemy has been documented by extensive plot studies and, again, the use of photographs to show the "before and after" infestations (Holloway and Huffaker, 1952; Huffaker and Kennett, 1959; Huffaker, 1967). A good series of photographs are particularly useful in such cases as this where the extent of the infestation is so clearly revealed (Fig. 4). The process of control following the addition of the imported insects (*Chrysolina quadrigemina* Suffrian being most effective) was repeated in hundreds of places all over California and somewhat less quickly in other western states.

B. The Exclusion Method

The "exclusion or subtraction" method involves the elimination and subsequent exclusion of resident natural enemies from a number of plots which can then be compared with a like number of otherwise comparable plots where the natural enemies are not disturbed. Differential "before and after" pest population densities showing different equilibrium levels for the two groups of plots serve as a direct measure of the control and regulating effectiveness of the natural enemies. The experiments must be designed so as to be biologically realistic, considering, for example, significant behavioral and ecological characteristics, so that aside from excluding enemies, the technique does not modify other parameters which may exert an appreciable influence on results. This method is most feasible with insects having low powers of dispersal.

DeBach and Huffaker (1971) stated:

> Elimination of natural enemies can be accomplished in various ways, both chemical and mechanical (Smith and DeBach, 1942; DeBach *et al.*, 1949; Huffaker and Kennett, 1956). The pest may or may not be eliminated simultaneously. If so, it must be subsequently reintroduced into the plot. Following true elimination, the enemies are generally excluded by mechanical means, usually cages, but barriers of various sorts [or spatial separation] would suffice for non-flying enemies.
>
> Whatever exclusion technique is used, utmost care is required that the comparisons involve only the one major variable, natural enemies. Thus, if cages are used for exclusion, similar cages are used in the non-exclusion, enemy-present plots, with the difference being that openings are present to permit ingress and egress of the enemies. If the method of exclusion is by use of a chemical that might have a stimulative effect on the power of increase of the pest species, this possibility should be checked independently.

DeBach and Huffaker (1971) present a number of illustrations of the results of evaluation using the "exclusion or elimination" method. The proof of the effective-

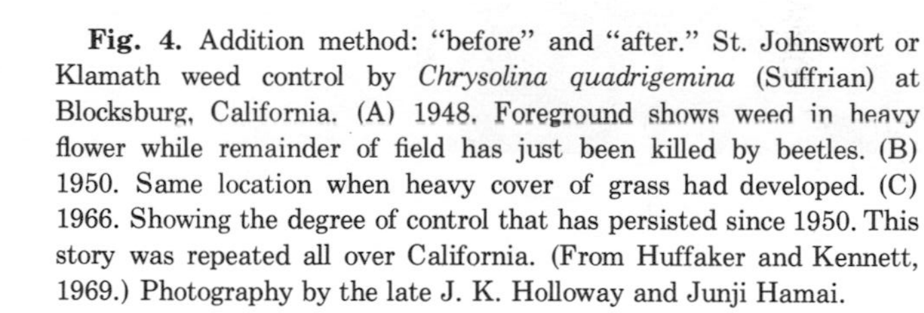

Fig. 4. Addition method: "before" and "after." St. Johnswort or Klamath weed control by *Chrysolina quadrigemina* (Suffrian) at Blocksburg, California. (A) 1948. Foreground shows weed in heavy flower while remainder of field has just been killed by beetles. (B) 1950. Same location when heavy cover of grass had developed. (C) 1966. Showing the degree of control that has persisted since 1950. This story was repeated all over California. (From Huffaker and Kennett, 1969.) Photography by the late J. K. Holloway and Junji Hamai.

Fig. 5. Exclusion method, paired sleeve cages. (Top) Open and closed cages in position. (Center) Open cage (sleeve removed), showing undamaged branch exposed to action of natural enemies. (Bottom) Closed cage, sleeve open to show heavy damage by California red scale, *Aonidiella aurantii*, where parasites were excluded. (After DeBach and Huffaker, 1971.)

ness of natural enemies can be seen either by use of comparative photographs or by census data. Use of both methods is advisable. Figure 5 shows the results when paired sleeve cages (one open to permit ingress and egress of parasites) were used to study the influence of natural enemies of California red scale, *Aonidiella aurantii* (Mask.), on citrus in southern California. Paired branch cages or cages used over whole trees can also be used. Paired leaf cages present other possibilities as shown in Figure 6. In this instance, paired plastic cells (one having apertures for ingress and egress of parasites and one lacking them) were used to study the effectiveness of parasites of California red scale infesting English ivy.

Census data can be used to express the patterns of changes in density of the infestation and the host's mortality over a period of several generations of the host in such paired cage experiments. DeBach and Huffaker (1971) reported that the mortality of California red scale in cells (as in Fig. 7) remained low, well below the equilibrium mortality for this species, whereas in the open cells, the mortality over a period of several months was consistently much higher, and toward the later months of the experiment, exceeded the equilibrium mortality at 85 to 90%.

C. The Interference Method

Again, DeBach and Huffaker (1971) deal in detail with the "interference or neutralization" method which involves greatly reducing the efficiency of natural enemies in one group of plots, as contrasted to another group having natural enemies undisturbed. Comparative differences in pest density over a period of time are measured as in the "exclusion" method. Any increase in density in the interference plots, relative to the normal biological control plots, demonstrates the effectiveness of the natural enemies. Such comparisons reveal only a part of the total extent of host population control, because the natural enemies are not entirely removed and may be producing some limiting effect.

DeBach and Huffaker stated:

> In other words, an increased average host density following interference demonstrates that enemies were responsible for the original lower level, but the maximum level the host would attain in their *complete absence* remains unknown, because enemies whose efficiency is reduced by interference may still remain the chief regulating agency at some considerably higher average pest density level, or contribute to the collective process.

Percentage parasitization may commonly be as great in the interference plots as in the normal activity plots. The explanation for this has been furnished by Smith (1955). (See also Huffaker and Messenger, 1964, pp. 82–83.) It is based on the theory that the density of the pest population at equilibrium is determined by the rate at which premature mortality of a pest increases with the density. If the rate of parasitization increases rapidly as the host population starts to increase, the host equilibrium density will be low; if it increases slowly, the density will be high. This is because the equilibrium

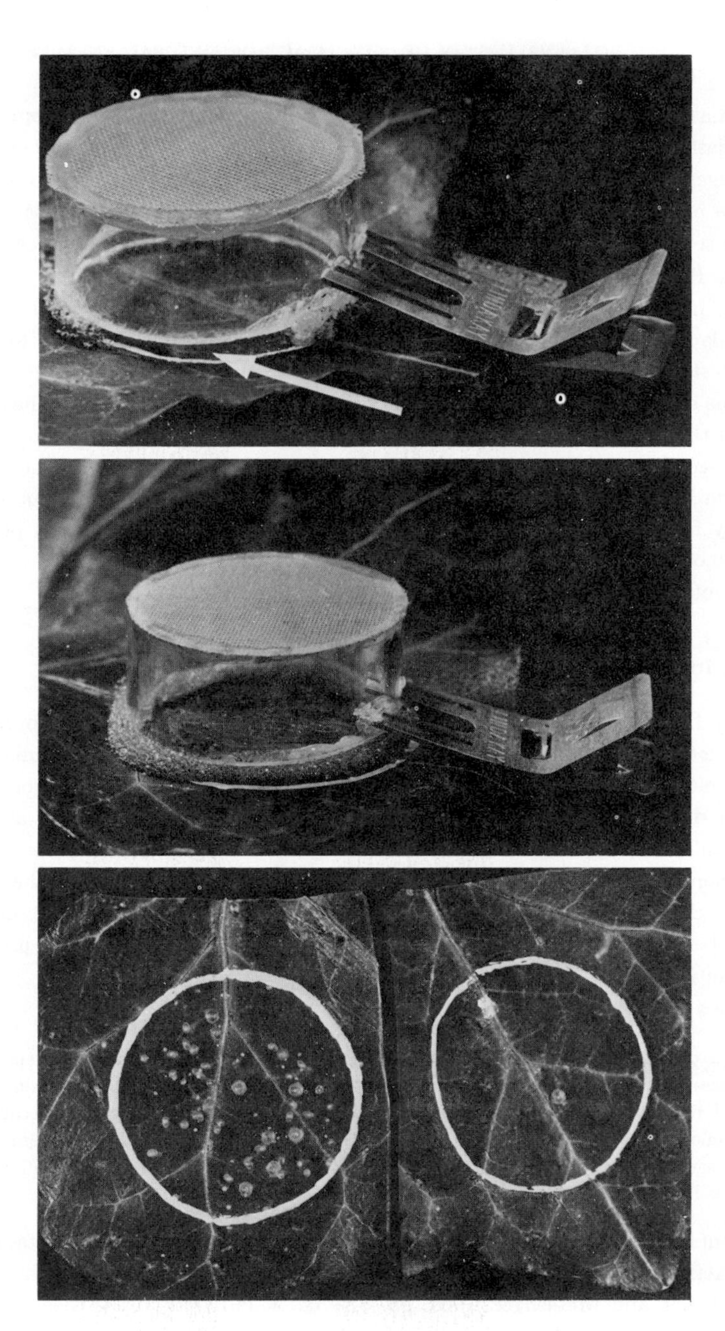

Fig. 6. Exclusion method, paired leaf cages. (Top) Open cage (cell with small apertures permitting ingress and egress of parasites—see arrow) in position on California red scale [*Aonidiella aurantii* (Mask.)] infested English ivy (*Hedera helix* L.) leaves. (Center) Closed cage in position excluding parasites. (Bottom) Cages removed to show relative scale density in closed cage (left, parasites excluded) and open cage (right, parasites present). (After DeBach and Huffaker, 1971.)

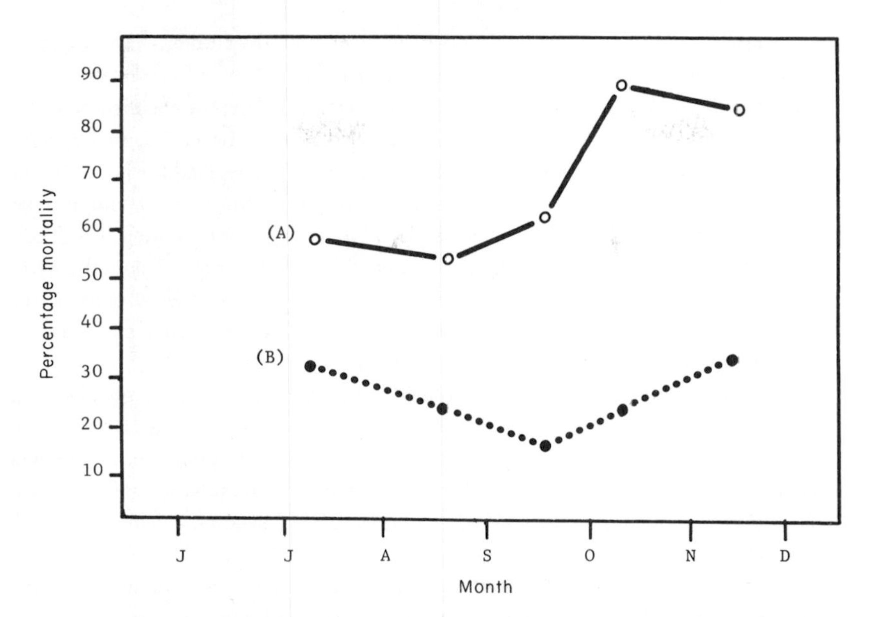

Fig. 7. Exclusion method. Average California red scale [*Aonideilla aurantii* (Mask.)] mortality on English ivy (*Hedera helix* L.). (A) in open cells permitting entry of parasites and (B) in closed cells excluding parasites. (After DeBach and Huffaker, 1971.)

mortality level (where the mortality equals the natality) is reached quickly and at low levels in the first instance but not in the latter. This illustrates the difference between an effective and an ineffective natural enemy. It also explains why pesticides—an interference measure—cause pest upsets when effective natural enemies are inhibited and *why parasites may be even more abundant in a chemically treated habitat* and yet unable to control the pest, or maintain it at low density.

It also suggests why the percentage parasitization, taken alone, does not indicate the efficiency of a parasite. In the graph presented by Huffaker and Messenger (1964), illustrating the two types of habitats (A, producing a high rate of density-dependent response, and B, a low rate) any given percentage of mortality (e.g., parasitization) occurs at a lower density in habitat A than in B. This is an important reason why use of experimental methods is stressed. There are many built-in compensations and other forms of interaction tending to mask discovery of the role of a natural enemy when using life table, regression, and modeling approaches, whereas the *net result* of the natural enemy is, in any event, measured by the experimental method.

Interference techniques include the "insecticidal check method," the "biological check method," "hand removal method," and the "trap method" (DeBach and Bartlett, 1964). If applied stringently enough, these might serve as exclusion methods. As used, they either kill or reduce the efficiency of the natural enemies, resulting in an increase to a higher density of the host (or prey) species, and thus demonstrate that the natural enemies were the responsible regulating agency at the former, lower density (within the existing climatic regime and other conditions).

The insecticide or chemical check method has been used widely; the chemical selected must exhibit a marked differential in its adverse effect upon the host (pest) species as contrasted to the natural enemy. DDT has been most widely used for a number of reasons (DeBach and Huffaker, 1971), but any material (even "inert" dust) might be so used in appropriate situations. One of a pair of plots is, of course, treated, the other left untreated, and there should be adequate replication. Huffaker and Kennett (1966) were able to use DDT as a "check method" to selectively inhibit two species of parasites of olive parlatoria scale, by precise timing of the treatments, to assess their comparative roles in regulation of olive parlatoria scale densities.

Any possible stimulating effect on the host's (or prey's) fecundity by the chemical should receive attention, and where possible more than a single material or "check method" should be used concurrently. Huffaker and Kennett (1956) used hand removal as a "check" on the chemical "check method" they employed in cyclamen mite studies. DeBach (1955) and Huffaker *et al.* (1962) found that DDT used as a "check method" had no effect on the fecundity of the pest species they studied.

Three examples of the "interference" method will be given, one involving biological or ant interference with natural enemies and two involving insecticide interference.

1. Interference by Ants: Biological Check Method

Ant interference with natural enemies which attack honeydew-producing insects is widely documented (DeBach and Bartlett, 1964). This interference has also been used as a means of partially assessing the role of natural enemies in the population regulation and control of honeydew-producing phytophagous species. By use of a suitable chemical insecticide, appropriately applied, or by means of barriers, ants can be removed or excluded from one group of test trees and left fully active in another group. DeBach *et al.* (1951) demonstrated the utility of this biological "check method" of evaluation. Figure 8 illustrates results obtained when (a) the Argentine ant, *Iridomyrmex humilis* Mayr, was fully active (top photo) versus (b) where it was eliminated (bottom photo) in citrus infested with California red scale. The relative visibility of the observer in each photo accents the degree of defoliation in the ant-present tree versus that in the ant-free tree. This method has also been used to evaluate the role of various natural enemies of such pests as citrus mealy bug, *Planococcus citri* (Risso), citrus red mite, *Panonychus citri* (McG.), black scale *Saissetia aleae* (Oliv.), brown soft scale, *Coccus hesperidum* L., and citrus aphids on citrus in southern California (DeBach and Huffaker, 1971).

2. Interference by Insecticide: Insecticide Check Method

Tests conducted on olive parlatoria scale on olive trees in the San Joaquin Valley of California subsequent to the full establishment, first of *Aphytis maculicornis*, and then of

Fig. 8. Interference method: biological check method. Tree damage from California red scale, *Aonidiella aurantii* (Mask.), resulting from ants interfering with the performance of natural enemies (top) as compared to a check tree where ants were eliminated (bottom). The relative visibility of the researcher in the two photos is an indication of differences in infestation as represented by pest-caused defoliation and thus represents the degree of biological control. (After DeBach and Huffaker, 1971.) Photo by K. Middleham.

Coccophagoides utilis, further substantiate the utility of the interference: insecticidal check method of evaluation. Thus, even if they had not had the benefits of the addition method available which, as already discussed, contrasts degree of pest infestation "before and after" establishment of parasites, the use of the insecticidal check method would have proved the value of the parasites. Huffaker *et al.* (1962) treated two trees with DDT in each of six olive groves under biological control by *A. maculicornis* and measured the olive scale population densities and fruit losses over a period of four host generations. Treatments to inhibit parasites were repeated as required. Host densities multiplied from 75- to nearly 1000-fold in the *Aphytis*-inhibited trees with the result of virtually a total loss of the crop. However densities remained very low with no loss of the crop in the untreated biological control trees.

Figure 9A shows a typical DDT-treated *Aphytis*-inhibited tree (left) beside a row of adjacent biological control trees (right) after about 2 years. Note the extreme loss of vigor, lack of new growth, and defoliation in the *Aphytis*-inhibited tree versus the block of biological control trees. Figure 9B and C presents a close-up of branches from each type of tree showing the contrasting scale densities and state of the foliage.

The scale *Lecanium coryli* (L.) in Nova Scotia orchards presents another example. It has been present in Nova Scotia orchards for a long time, but first became a serious problem in the 1950's and then only in a small number of orchards. A survey of possible causes for the outbreak indicated that scale increase was associated with the use of DDT sprays. The factors involved in its control or in influencing its increases were not known. An experiment using an insecticidal check method was initiated in 1963, after initial studies had been conducted on the scale and its biological enemies. The experiment was carried out over four scale generations from 1963 to 1967 in an orchard with an initial small scale population then being attacked by two parasite species and by predators (MacPhee and MacLellan, 1971). These natural enemies were suppressed by the use of DDT sprays (one application in June, 1963, and one in June, 1964) during the following years when no insecticidal sprays were used. Predators were low in 1963 and 1964, the parasite *Blastothrix sericea* (Dalm.) was reduced to a low level, and the parasite *Coccophagus* sp. remained at about its normal attack level, which is usually less than 30%. The reduction of natural enemies allowed a fivefold increase of the host scale by the 1964—1965 generation. However, damage was not extensive until the summer of 1965, when the scale had markedly increased and also the survival of nymphs was high during the summer months. The subsequent increase of predators and *B. sericea* caused high scale mortality by the spring of 1966 and the scales which did hatch in July,

---

Fig. 9. Interference method: insecticide check method. (A) Tree to left treated with DDT to inhibit action of *Aphytis maculicornis*, and showing lack of vigor, defoliation, and dieback resulting from heavy *P. oleae* densities. Trees to right had no treatments with DDT, and *Aphytis* kept the scale under good control, with tree growth and condition excellent; (B) twig from representative DDT-treated, largely parasite-free tree showing nature of damage by *P. oleae*; (C) twig from representative untreated parasite-present tree, showing clean condition. Photography by F. E. Skinner. (After Huffaker *et al.*, 1962.)

▶

A

B

C

1966 were quickly reduced in density. In 1967 the scale was again under good control at levels below that causing significant injury. The control of the pest had been brought about the second year following the last application, by the action of the two species of parasites and predators killing nymphs, and by hemipteron predators feeding on the maturing adults before their completion of egg laying.

For additional examples of the interference method, including the hand removal method, the reader is referred to DeBach and Bartlett (1964) and DeBach and Huffaker (1971).

V. CENSUS, LIFE TABLE, AND CORRELATION METHODS OF EVALUATION OF NATURAL ENEMIES

In Chapter 8, use of general census data is discussed in appraising the potential of natural enemies as candidates for introduction into other areas where their host (or prey) has become established without them. The characteristics of a good natural enemy are also discussed in Chapter 3. Some significant aspects of evaluation of natural enemies are treated in those chapters and they will not be directly dealt with here, although some of the material presented in this chapter is related to concepts discussed in Chapters 3 and 8. Moreover, we will also not attempt to treat adequately the use of life table data in analysis of population dynamics and natural control. Life table data and quantitative correlation techniques of evaluation of natality and mortality phenomena have value well beyond their limited utility in assessing the regulative power of natural enemies relative to a given host (or prey) population. Also, it is commonly important to have a measure of the total generation mortality in order to appraise the importance of a given level of natural enemy caused mortality. A full life table study makes this possible in some instances (Varley, 1970). Even so, many pitfalls lie along the way in attempting to appraise the role of natural enemies when no complementary approaches along experimental lines are used as discussed here. The difficulties arise from a number of sources. Correlation can be spurious. Interaction, including compensatory reactions, can mask a real cause and effect relationship. In the absence of other evidence, correlation itself may not suggest what is the cause and what the effect. The lack of a clearly independent variable in many such assessments is a problem. The lack of distinct or separate generations is another problem. Nevertheless, progress is being made and such approaches promise to be used more widely, particularly as a means of furnishing better insights into natural enemy performance and inadequacies. Such knowledge should encourage further biological control effort along indicated lines rather than the common negative reaction of abandonment as fruitless of an effort which is not immediately successful.

In several specific areas, quantitative techniques have been developed for appraising the behavior and characteristics of natural enemies in the field and these are discussed below.

It should be clearly understood that the "key" factor in the analysis of Morris (1963) and the K-factor of Varley and Gradwell (1968) do *not* carry the connotation of being the principal natural control factor or main density-regulating factor. The terms have been used to refer only to the factor accounting, not for the *mean* population density, but for the greatest *variation* in population size from generation to generation. A *main density-regulating* factor would be one most responsive to changes in host (or prey) population size (density-dependent), therefore most instrumental in regulating population *size*. The method of Morris (1963) is faulty on a number of counts, not only as an appraisal of the regulative power of a natural enemy, and as a means of even *detecting* the delayed density-dependent action such agents display, but with reference also to other considerations (Ito, 1967; Hassell and Huffaker, 1969; Varley and Gradwell, 1970; Southwood, 1967). For example, Luck (1971) showed that while the Morris technique would *not*, the K-factor analysis of Varley and Gradwell *would* identify the four basic kinds of mortality factors: density-independent, direct density-dependent, delayed density-dependent, and inverse density-dependent factors. With respect to the action of natural enemies, Varley and Gradwell's technique, as such, will detect their delayed density-dependent action but will not directly assess their power of host population regulation. Their techniques, however, can be used to assess partially this power in that the *functional* or behavioral aspect of response is picked up as a direct density-dependent factor and the slope representing this response is a partial measure of the regulation potential of the natural enemy. Moreover, if one computes a regression between progeny production per parent and host density, the degree of correlation, if any, and the slope are both measures of the numerical response capability of the parasite, and thus of the density-dependence of the latter. Such measures are, therefore, indicators of the delayed density-dependent responsiveness of the parasite to host population size, that is to say, the power of regulation of this enemy.

In Fig. 10, parasite A, having the steepest slope, would be a better regulating natural enemy than parasite B. It would control its host at a lower density.

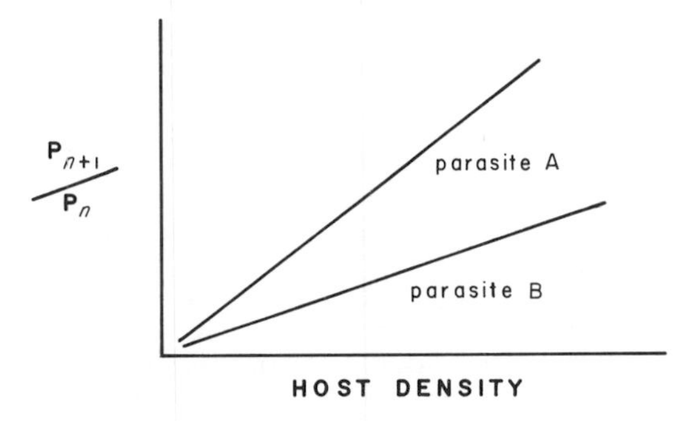

Fig. 10. Hypothetical graph illustrating the difference in numerical response to host density, and therefore control capabilities, of two parasites, A and B. P_n, parasites in generation 1; P_{n+1}, parasites in generation 2.

Both the direct density-dependence (functional response) and the delayed density-dependence (reproductive numerical response) are essential in assessing the full regulating potential of a given enemy. Embree (1966) (Chapter 13) demonstrated a functional response component in *Cyzenis albicans* (Fall.), a tachinid parasite of winter moth in the Maritime Provinces (Canada), and Huffaker and Stinner (1971) illustrated an overall density-dependent performance in *Aphytis maculicornis* on olive parlatoria scale in California. In the latter instance, a marked numerical response in the parasite occurs within a single host generation. They used Varley and Gradwell's statistical test of significance of the relationship that was suggested when the K-values were plotted against log host densities. This test is necessary because of the lack of an independent variable.

Nevertheless, in the field there are so many other contemporaneously interacting factors, including dispersal and movements of any kind, that use of these life table and regression techniques gives relatively little assurance that the true role of a natural enemy will necessarily be discovered by their use. We reiterate that the best, the clearest, and only reliable means of evaluation is through use of experimental or comparison methods. We are convinced that they warrant far greater use, despite the fact that substantial problems also sometimes present themselves using these methods (e.g., with highly mobile organisms).

REFERENCES

Bird, F. T., and Elgee, D. E. (1957). A virus disease and introduced parasites as factors controlling the European spruce sawfly, *Diprion hercyniae* (Htg.) in central New Brunswick. *Can. Entomol* **89,** 371–378.

DeBach, P. (1955). Validity of the insecticidal check method as a measure of the effectiveness of natural enemies of diaspine scale insects. *J. Econ. Entomol.* **44**, 373–383.

DeBach, P. (1969). Biological control of diaspine scale insects on citrus in California. *Proc. 1st Int. Citrus Symp. Riverside, Calif.* **2**, 801–815.

DeBach, P., and Bartlett, B. R. (1964). Methods of colonization, recovery and evaluation. *In* "Biological Control of Insect Pests and Weeds" (P. DeBach, ed.), pp. 402–426. Reinhold, New York.

DeBach, P., and Huffaker, C. B. (1971). Experimental techniques for evaluation of the effectiveness of natural enemies. *In* "Biological Control" (C. B. Huffaker, ed.), pp. 113–140. Plenum, New York.

DeBach, P., and Sundby, R. A. (1963). Competitive displacement between ecological homologues. *Hilgardia* **34**, 105–166.

DeBach, P., Dietrick, E. J., and Fleschner, C. A. (1949). A new technique for evaluating the efficiency of entomophagous insects in the field. *J. Econ. Entomol.* **42**, 546.

DeBach, P., Fleschner, C. A., and Dietrick, E. J. (1951). A biological check method for evaluating the effectiveness of entomophagous insects. *J. Econ. Entomol.* **44**, 763–766.

DeBach, P., Rosen, D., and Kennett, C. E. (1971). Biological control of coccids by introduced natural enemies. *In* "Biological Control" (C. B. Huffaker, ed.), pp. 165–193. Plenum, New York.

Dempster, J. P. (1960). A quantitative study of the predators on the eggs and larvae of the broom beetle, *Phytodecta olviacea* Forster, using the precipitin test. *J. Anim. Ecol.* **29**, 149–167.

Dodd, A. P. (1940). "The Biological Campaign against Prickly Pear," 177 pp. Commonw. Prickly Pear Bd., Brisbane.

Embree, D. G. (1965). The population dynamics of the winter moth in Nova Scotia 1954–1962. *Mem. Entomol. Soc. Can.* **46**, 1–57.

Embree, D. G. (1966). The role of introduced parasites in the control of the winter moth in Nova Scotia. *Can. Entomol.* **98**, 1159–1168.

Embree, D. G. (1971). The biological control of the winter moth in eastern Canada by introduced parasites. *In* "Biological Control" (C. B. Huffaker, ed.), pp. 217–226. Plenum, New York.

Flaherty, D. L., and Huffaker, C. B. (1970). Biological control of Pacific mites and Willamette mites in San Joaquin Valley vineyards. *Hilgardia* **40**, 267–330.

Frazer, B. D. (1971). Biological control and population dynamics of the walnut aphid, *Chromaphis juglandicola* (Kalt.). Ph. D. Thesis, Univ. Calif., Berkeley, California.

Harcourt, D. G. (1964). Population dynamics of *Leptinotarsa decemlineata* (Say) in eastern Ontario. II. Population and mortality estimation during six age intervals. *Can. Entomol.* **96**, 1190–1198.

Hassell, M. P., and Huffaker, C. B. (1969). The appraisal of delayed and direct density dependence. *Can. Entomol.* **101**, 353–361.

Hassell, M. P., and Varley, G. C. (1969). New inductive population model for insect parasites and its bearing on biological control. *Nature (London)* **223**, 1133–1137.

Haynes, D. L., and Sisojevic, P. (1966). Predatory behavior of *Philodromus rufus* Walchenaer (Araneae: Thomisidae). *Can. Entomol.* **98**, 113–133.

Herbert, H. J. (1965). The brown mite, *Bryobia arborea* Morgan and Anderson (Acarina: Tetranychidae), on apple in Nova Scotia. I. Influence of intratree distribution on the selection of sampling units. II. Differences in habits and seasonal trends in orchards with bivoltine and trivoltine populations. III. Diapause eggs. IV. Hatching of diapause eggs. *Can. Entomol.* **97**, 1303–1318.

Holling, C. S. (1959). The components of predation as revealed by a study of small mammal predation of the European pine sawfly. *Can. Entomol.* **91**, 293–320.

Holling, C. S. (1966). The functional response of invertebrate predators to prey density. *Mem. Entomol. Soc. Can.* **48**, 1–86.

Holloway, J. K., and Huffaker, C. B. (1952). Insects to control a weed. *In* "Insects, Yearbook of Agriculture for 1952" (A. Stafferud, ed.). 730 pp. U.S. Dept. Agri., Wash. D.C.

Huffaker, C. B. (1967). A comparison of the status of biological control of St. Johnswort in California and Australia. *Mushi* **39**, 51–73.

Huffaker, C. B., and Kennett, C. E. (1956). Experimental studies on predation: I. Predation and cyclamen mite populations on strawberries in California. *Hilgardia* **26**, 191–222.

Huffaker, C. B., and Kennett, C. E. (1959). A ten-year study of vegetational changes associated with biological control of Klamath weed. *J. Range Manage.* **12**, 69–82.

Huffaker, C. B., and Kennett, C. E. (1966). Studies of two parasites of olive scale, *Parlatoria oleae* (Colvée). IV. Biological control of *Parlatoria oleae* (Colvée) through the compensatory action of two introduced parasites. *Hilgardia* **37**, 283–335.

Huffaker, C. B., and Kennett, C. E. (1969). Some aspects of assessing efficiency of natural enemies. *Can. Entomol.* **10**, 425–447.

Huffaker, C. B., and Messenger, P. S. (1964). The ecological basis of biological control. *In* "Biological Control of Insect Pests and Weeds" (P. DeBach, ed.), pp. 45–73. Reinhold, New York.

Huffaker, C. B., and Stinner, R. E. (1971). The role of natural enemies in pest control programs.

In "Entomological Essays to Commemorate the Retirement of Professor K. Yasumatsu" (Editorial Committee S. Ashahina, J. L. Gressitt, Z. Hikada, T. Nichida, K. Nomura), pp. 333–350. Hokuryukan Publ., Tokyo.

Huffaker, C. B., Kennett, C. E., and Finney, G. L. (1962). Biological control of olive scale, *Parlatoria oleae* (Colvée) in California by imported *Aphytis maculicornis* (Masi) (Hymenoptera: Aphelinidae). *Hilgardia* **32**, 541–636.

Huffaker, C. B., Shea, K. P., and Herman, S. G. (1963). Experimental studies on predation. III. Complex dispersion and levels of food in an acarine predator-prey interaction. *Hilgardia* **34**, 305–330.

Huffaker, C. B., Messenger, P. S., and DeBach, P. (1971). The natural enemy component in natural control and the theory of biological control. *In* "Biological Control" (C. B. Huffaker, ed.), pp. 16–62. Plenum, New York.

Ito, Y. (1967). Life tables in the population ecology with special reference to insects. *Seibutsu-Kagaku* **18**, 127–134, 167–175.

Klomp, H. (1966). The dynamics of a field population of the pine looper *Bupclus pinicrius* L. *Advan. Ecol. Res.* **3**, 207–304.

Laing, J. E., and Huffaker, C. B. (1969). Comparative studies of predation by *Phytoseiulus persimilis* Athias-Henriot and *Metaseiulus occidentalis* (Nesbitt) (Acarine: Phytoseiidae) on populations of *Tetranychus urticae* Koch (Acarine: Tetranychidae). *Res. Population Ecol.* **11**, 105–126.

Legner, E. F. (1969). Distribution pattern of host and parasitization by *Spalangia drosophilae* (Hymenoptera: Pteromalidae). *Can. Entomol.* **101**, 551–557.

LeRoux, E. J., and Reimer, C. (1959). Variation between samples of immature stages, and of mortalities from some factors, of the eye-spotted bud moth, *Spilonota ocellana* (D. & S.) (Lepidoptera: Olethreutidae) and the pistol casebearer, *Coleophora serratella* (L.) (Lepidoptera: Coleophoridae), on apple in Quebec. *Can. Entomol.* **91**, 428–449.

Lord, F. T. (1968). An appraisal of methods of sampling apple trees and results of some tests using a sampling unit common to insect predators and their prey. *Can. Entomol.* **100**, 23–33.

Lord, F. T. (1972). Comparisons of the abundance of the species composing the foliage inhabiting fauna of apple trees. *Can. Entomol.* **104**, 731–749.

Luck, R. F. (1971). An appraisal of two methods of analyzing insect life tables. *Can. Entomol.* **103**, 1261–1271.

MacLellan, C. R. (1954). The use of the precipitin test in evaluating codling moth predators. M.A. Thesis, Queen's University, Kingston, Ontario.

MacLellan, C. R. (1963). Predator population and predation on the codling moth in an integrated control orchard—1961. *Mem. Entomol. Soc. Can.* **32**, 41–54.

MacPhee, A. W., and MacLellan, C. R. (1971). Cases of naturally-occurring biological control in Canada. *In* "Biological Control" (C. B. Huffaker, ed.), pp. 312–328. Plenum, New York.

Messenger, P. S., and van den Bosch, R. (1971). The adaptability of introduced biological control agents. *In* "Biological Control" (C. B. Huffaker, ed.), pp. 68–89. Plenum, New York.

Morris, R. F. (1955). The development of sampling techniques for forest insect defoliators, with particular reference to the spruce budworm. *Can. J. Zool.* **33**, 225–294.

Morris, R. F. (1963). The dynamics of epidemic spruce budworm populations. *Mem. Entomol. Soc. Can.* **31**, 1–32.

Muir, R. C. (1958). On the application of the capture-recapture method to an orchard population of *Blepharidopterus angulatus* (Fall.) (Hemiptera-Heteroptera, Miridae). *Rep. E. Malling Res. Sta. 1959*, pp. 140–147.

Neilson, M. M., and Morris, R. F. (1964). The regulation of European spruce sawfly numbers in the Maritime Provinces of Canada from 1937 to 1963. *Can. Entomol.* **96**, 773–784.

Neilson, M. M., Martineau, R., and Rose, A. H. (1971). *Diprion hercyniae* (Hartig), European spruce sawfiy (Hymenoptera: Diprionidae). *Commonw. Inst. Biol. Control Tech. Commun.* **4**, 136–143.

Pschorn-Walcher, H., Schroder, D., and Eichhorn, D. (1969). Recent attempts at biological control of some Canadian forest insect pests. *Commonw. Inst. Biol. Control Tech. Bull.* **11**, 1–100.

Reeks, W. A., and Cameron, J. M. (1971). Current approach to biological control of forest insects. *Commonw. Inst. Biol. Control Tech. Commun.* **4**, 105–113.

Sandness, J. N., and McMurtry, J. A. (1970). Functional response of three species of Phytosiidae (Acarina) to prey density. *Can. Entomol.* **102**, 692–704.

Sechser, B. (1970). Der Parasitenkomplex des kleinen Frostspanners (*Operophtera brumata* L.) (Lep., Geometridae) unter besonderer Berucksichtigung der Kokon-Parasiten. II. Teil. *Z. Angew. Entomol.* **66**, 144–160.

Smith, H. S. (1955). Ecological aspects of insect population dynamics. Unpubl. manuscript. Presented as a symposium paper before the Entomol. Soc. Amer., Cincinnati, Ohio, 1955.

Smith, H. S., and DeBach, P. (1942). The measurement of the effect of entomophagous insects on population densities of their hosts. *J. Econ. Entomol.* **35**, 845–849.

Sonleitner, F. J., and Bateman, M. A. (1963). Mark-recapture analysis of a population of Queensland fruit-fly, *Dacus tryoni* (Frogg.), in an orchard. *J. Anim. Ecol.* **32**, 259–269.

Southwood, T. R. E. (1966). "Ecological Methods with Particular Reference to the Study of Insect Populations," 391 pp. Methuen, London.

Southwood, T. R. E. (1967). The interpretation of population change. *J. Anim. Ecol.* **36**, 519–529.

Southwood, T. R. E., and Way, M. J. (1970). Ecological background to pest management. *In* "Concepts of Pest Management" (R. L. Rabb and F. E. Guthrie, eds.), pp. 6–29. N. Carolina State Univ., Raleigh, North Carolina.

Varley, G. C. (1970). The need for life tables for parasites and predators. *In* "Concepts of Pest Management" (R. L. Rabb and F. E. Guthrie, eds.), pp. 59–70. N. Carolina State Univ. Raleigh, North Carolina.

Varley, G. C. (1971). The effects of natural predators and parasites on winter moth populations in England. *Proc. Tall Timbers Conf. Ecol. Anim. Control Habitat Manage.* **2**, 103–116.

Varley, G. C., and Gradwell, G. R. (1968). Population models for winter moth. *In* "Insect Abundance" (T. R. E. Southwood, ed.), pp. 132–142. Blackwell, Oxford.

Varley, G. C., and Gradwell, G. R. (1970). Recent advances in insect population dynamics. *Annu. Rev. Entomol.* **15**, 1–24.

Varley, G. C., and Gradwell, G. R. (1971). The use of models and life tables in assessing the role of natural enemies. *In* "Biological Control" (C. B. Huffaker, ed.), pp. 93–110. Plenum, New York.

Yates, F. (1960). "Sampling Methods for Censuses and Surveys." Hafner, New York.

SECTION IV

BIOLOGICAL CONTROL IN SPECIFIC PROBLEM AREAS

12

BIOLOGICAL CONTROL OF PESTS OF CONIFEROUS FORESTS

W. J. Turnock, K. L. Taylor, D. Schröder, and D. L. Dahlsten

I. INTRODUCTION

Coniferous forests occur extensively, particularly in the northern hemisphere, in both monospecific and mixed stands. The demand for coniferous wood, and the rapid growth of some species, has also led to their widespread use in plantations. In many areas, native coniferous forests are relatively simple in structure, with one or a few species dominating large areas. These forests seem particularly prone to attack by "opportunistic pests" (Watt, 1968) which may be able to subsist for some time at low densities and then reach very high densities in response to improved environmental conditions. These high densities profoundly alter the physical and biotic environment of the forest, destroy host trees, and thereby may play an important role in modifying or maintaining the composition and structure of the forest. The western pine beetle, *Dendroctonus brevicomis* Le Conte, tends to eliminate ponderosa pine, *Pinus ponderosa* Laws., from parts of the mixed forest on the west side of the Sierra complex in California (Stark and Dahlsten, 1970) while the spruce budworm, *Choristoneura fumiferana* (Clemens), through preferential feeding on balsam fir, tends to increase the proportion of spruce in mixed spruce-fir forests (Turner, 1952). Even-aged stands of lodgepole pine in Yosemite National Park, California, are perpetuated by outbreaks of the lodgepole needle miner, *Coleotechnites milleri* (Busck), and the mountain pine beetle, *Dendroctonus ponderosae* Hopkins, which kill older trees and release the young growth (Patterson, 1921; Telford, 1961). Similarly, outbreaks of the spruce beetle *Dendroctonus rufipennis* (Kirby), maintain two-storied, two-aged stands of Engle-mann spruce in the Rocky Mountains by removing mature and overmature trees (Wygant and Nelson, 1949; Miller, 1970).

Anthropogenic pests, which reach damaging levels due to changes in the environment resulting from man's activities (Turnock and Muldrew, 1971a), are particularly severe in plantations and "natural" forests under intensive management, e.g., *Hylobius pales* (Hbst.), *H. radicis* Buch., *Neodiprion lecontei* (Fitch) in North America (Davidson and Prentice, 1967). Since biotic agents continue to be effective in adjacent undisturbed areas, human actions seem to have created an environment that limits the effectiveness of these agents.

Exotic pests, introduced from other regions, are serious pests in both native coniferous forests and plantations. These pests, such as *Diprion hercyniae* (Hartig), *Coleophora laricella* (Hübner), *Neodiprion sertifer* (Geoff.), *Rhyacionia buoliana* (Schiff.), and *Sirex noctilio* (F.), are usually considered prime targets for biological control because they lack their native complement of natural enemies.

Low-density pests, which are not subject to wide variations in numbers but which cause economic damage at low population levels (Turnock and Muldrew, 1971a) are not of major concern in native coniferous forests. Such pests of seeds, shoots, and high-value stands with a low threshold for damage (e.g., Christmas tree plantings) can be expected to become more important under intensive management systems.

Exotic conifers may be attacked by pests of related native species, and be much more severely damaged. For example, the North American root collar weevil, *Hylobius radicis*, prefers exotics, particularly Scots pine, *Pinus sylvestris* L., to native pines (Davidson and Prentice, 1967). In Manitoba, the depredations of this pest led to the abandonment of Scots pine for plantations. Native insect species may also switch from native hosts to exotic conifers that are not closely related. In South Africa, Australia, and New Zealand, several species of native insects have become major pests on pine plantations, even though *Pinus* spp. are not represented in the natural flora. The appearance of hitherto unimportant species as major pests of plantations suggests that some phytophagous species are more adaptable to simplified forests under intensive management than are their natural enemies.

It has been suggested that high-intensity management will reduce the severity of insect problems in the coniferous forest because many of the major pests, which affect mature or suppressed trees in forests that are unmanaged or have low-intensity management, would be absent. Although intensive management may avoid tree mortality from specific pests, e.g. thinning dramatically reduces mortality caused by *Dendroctonus pseudotsugae* Hopkins in Douglas fir (Williamson and Price, 1971) and by the mountain pine beetle, *D. ponderosae*, in ponderosa pine (Sartwell, 1971), it often creates new problems. It was once thought that the bark beetle problems affecting the major timber species of the western United States would subside once the virgin timber was cut. However, the mountain pine beetle has become a major tree killer in dense second-growth stands. Such stands, particularly on poor growing sites, provide ideal conditions for bark beetle outbreaks (Sartwell, 1969). It was also thought that harvesting, by breaking up large even-aged stands and removing mature trees, would reduce the susceptibility of spruce-fir forest to the spruce budworm (Greenbank, 1963). However, harvesting also encourages the development of pure stands of fir, which are more susceptible to attack than mixed spruce-fir stands.

Natural control of insects in coniferous forests is poorly known (see reviews by Hagen *et al.* and MacPhee and MacLellan, in Huffaker, 1971) because most insects have been studied only at high densities. The role of natural control agents may be revealed when some disturbance affects the system. For example, the pine needle scale, *Phenaecaspis pinifoliae* (Fitch), normally occurs in low numbers on lodgepole and Jeffrey pines in California but massive infestations were recorded in 1968 in areas where a Malathion fogging program was conducted against adult mosquitoes. After the fogging program was reduced in 1969 and stopped in 1970 this scale outbreak collapsed due to the actions of two coccinelids, *Chilocorus orbus* Casey and *Cryptoweisea atronitens* (Casey), an aphelinid, *Prospaltella bella* Gahan, and an eulophid, *Achrysochorella* sp. (Luck and Dahlsten, 1975).

Similarly, the application of DDT for 3 consecutive years against the spruce budworm in New Brunswick, Canada, reduced population densities of the European spruce sawfly, *Diprion hercyniae*, to very low levels and thus virtually eliminated its virus disease and parasites from the area. After spraying ceased, the sawfly populations increased to damaging levels, and receded to prespray equilibrium only after the disease and parasites became reestablished (Neilson *et al.*, 1971). In Texas, new infestations of the southern pine beetle, *Dendroctonus frontalis* Zimmerman, became more numerous and the outbreak area became larger in forest blocks where chemical control was applied promptly and efficiently than in blocks where control operations were less efficient (Vité, 1971).

The tendency of coniferous forests to grow in simple communities, which are relatively unstable and prone to wide fluctuations in insect abundance, makes biological control more difficult. The situation may become worse with increases in the area of planted conifers and the stated objective of many forest managers, particularly in North America, that the next generation of forests, the so-called "Third Forest," must be designed to allow clear-cutting with mechanical harvesters in even-aged, monospecific stands with a minimum of genetic variability. Although such a design could minimize harvesting costs, the costs will be high for seed orchards, planting programs, and fertilizers. Protection costs, due to the attacks of known and currently unknown insect species can be expected to increase. The induced necessity for the use of chemical insecticides may fail to solve the problems or may create new insect problems because the chemicals themselves may act as disrupters and simplifiers (Dahlsten *et al.*, 1969). For these reasons, and because such stands are ill-suited for recreation and other multiple purpose uses, the Third Forest concept should be challenged. The problems of forest protection through biological and other control approaches must be given serious consideration in the management plans for coniferous forests. It seems likely that protection costs could be reduced by encouraging stand diversity, thus maximizing the effectiveness of natural and biological control agents and minimizing the use of chemical insecticides.

II. EVALUATION OF PAST BIOLOGICAL CONTROL ATTEMPTS

Biological control has been rather widely attempted in coniferous forests and these efforts encompass a broad range of approaches. Parasites and predators have been involved in the following approaches: colonization, periodic inoculation, inundation, augmentation, and conservation. The use of microbial agents in colonization (here enzootic establishment), periodic inoculation, and as microbial insecticides, will be considered separately.

Examination of the record of past attempts reveals considerable variation in the thoroughness with which they were planned, executed, and reported. The first step in evaluation must therefore be an analysis of the integrity of each reported attempt: the analysis of programs that were ill-conceived, poorly executed, or inadequately reported can be of little help in improving future efforts. Therefore, where appropriate, an attempt will be made to identify serious or "valid" attempts and only these will be used as a basis for suggesting future improvements.

A. Colonization

Colonization, the establishment of relatively small numbers of a new (usually exotic) parasite or predator species in a region where it is expected to become a permanent part of the natural control complex (Turnock and Muldrew, 1971a), has been the most common approach. The following criteria were used to divide the attempts against the 37 target species into three groups:

1. *Major Attempts*—programs which included adequate planning, operation, and evaluation of the establishment of introduced biological control agents and their subsequent impact on the target species (Table 1).

2. *Secondary Attempts*—opportunistic programs, often utilizing surplus natural enemies from major programs against other target species. These attempts were usually inadequately evaluated (Table 2).

3. *Futile Attempts*—programs doomed by inadequate selection of natural enemies and poor handling and release techniques (Table 3).

This grouping of programs provides the basis for separating the effects of organizational failures in planning, operation, and assessment from the scientific problems associated with the colonization of natural enemies and from possible inherent inadequacies of these agents. Information on the releases of each species of natural enemy was examined to see if "adequate" provisions had been made to ensure establishment. Many of the species released (termed "doomed" in Tables 1—3) had very little chance of establishing themselves because: (a) the number of individuals released was too small (less than 100 individuals were arbitrarily taken to be inadequate, although smaller numbers of healthy, well-adapted individuals have become established in several programs, e.g., against *Sirex noctilio*); (b) the natural enemy was unsuitable because of taxonomic problems arising from misidentification of the host or the natural enemy; (c) the natural enemies were released against target species that were widely dissimilar, taxonomically or ecologically (habitat, phenology, behavior, etc.), from their hosts or prey in the donor area—surplus natural enemies from major biological control programs were often used in this way; and (d) the enemy was already present in the target area.

The major programs (Table 1), usually directed against pests causing serious damage, have benefited from a relatively high level of support in personnel and finances. Nearly all of these programs took place in North America, the exceptions being against *Sirex noctilio* in New Zealand and Australia and against *Rhyacionia buoliana* in Argentina. Although these programs were directed against serious pests of coniferous forests, and were generally well supported, they vary widely in their success in colonizing the introduced agents. For all major programs, 29% of the species released were "doomed" and only 30% resulted in establishment. The percentage established rises to 43% for the "adequate" releases and to about 75% for programs where both the agents and the target species had been carefully studied (e.g., the *S. noctilio* program and recent releases against *R. buoliana*). Future adequately planned and financed attempts to colonize natural enemies should expect to achieve approximately 75% success in establishment.

The major programs also must be judged in terms of success of colonized agents in terminating the pest outbreaks that existed at the time of colonization and then in maintaining the pest population below the economic damage threshold. Both of these objectives have been attained in three of the ten major programs:

1. *Coleophora laricella* in eastern North America was controlled by the action of two specific larval parasites colonized from Europe (Webb and Quedneau, 1971).

2. The *Diprion hercyniae* outbreak was terminated largely through the action of the fortuitously introduced *Borrelina* virus. This virus, in conjunction with two species of parasites, has maintained *D. hercyniae* at endemic levels since 1945 (Neilson *et al.*, 1971).

3. *Diprion similis* (Hartig) near Montreal was controlled by the relocation of *Monodontomerus dentipes* (Dalm.). This parasite, accidentally introduced to the United States, appears to regulate its host populations at low levels throughout eastern North America (Dowden, 1962; McGugan and Coppel, 1962).

TABLE 1

Major Biological Control Attempts Using Colonization of Parasites and Predators

Target species	Target area	Date	Agents released		Agents established		Assessment[b]	Reference
			Doomed	Adequate	Following releases	Accidentally		
Adelges piceae (Ratz.)	Eastern North America	1933–1960	3	15	5		F-P	McGugan and Coppel, 1962
		1961–1969	7	13	7		F-P[c]	Dowden, 1962
		1933–1969	8	24	8			Clark *et al.*, 1971
	Western North America	1957–1968	12	11	5		F[c]	Mitchell and Wright, 1967
								Clark *et al.*, 1971
Choristoneura fumiferana (Clem.)	Eastern North America	1946–1955	9	9			F	Miller and Angus, 1971
								Dowden, 1962
Coleophora laricella (Hübner)	Eastern North America	1934–1939	1	4	2		S	Webb and Quedneau, 1971
	Western North America	1960–1969		1	1		F[c]	Webb and Denton, 1967
Diprion hercyniae (Hartig)	Eastern North America	1933–1951	4	31	7[a]		S	McGugan and Coppel, 1962
								Dowden, 1962
Diprion similis (Hartig)	United States	?				1	?	Dowden, 1962
	Canada	1935		1	1		S	McGugan and Coppel, 1962
Neodiprion sertifer (Geoff.)	Eastern North America	1940–1951	6	5	3	2	P	Griffiths, *et al.*, 1971
		1959–1962	1	2	1			
Pristiphora erichsonii (Hartig)	Canada	1910–1913		1 ?	1	1	P	McGugan and Coppel, 1962
	North America	1935–1953		3	1	1	F	Dowden, 1962
	Canada	1961–1964		6	2		P-S[c]	Turnock and Muldrew, 1971b

Rhyacionia buoliana (Schiff.)	Eastern North America	1932–1958	4	14	3		F	Dowden, 1962
	Canada	1959–1968		3	2		F[c]	Syme, 1971
	Argentina	1970–1971		1	1		P[c]	Brewer and Varas, 1971
Rhyacionia frustrana bushnelli (Busck)	United States	1925	8	3	1		S	Dowden, 1962
Sirex noctilio (F.)	New Zealand	1928–1931		1	1		P	Miller and Clark, 1938
		1954–1958		1	1	1	P	Zondag, 1959
		1963–1970	1	2	2		P	Zondag, personal communication
	Australia	1957–1969	2	4	4		P[c]	Taylor, 1967
		1970–1971	2	4	4 ?		[c]	Taylor, unpublished

[a] Only four of these species have been recovered since 1958 and only two of them contribute to the control of host populations (Neilson *et al.*, 1971).

[b] S, success; P, partial success; F, failure; ?, not assessed.

[c] Assessment incomplete.

TABLE 2

Secondary Biological Control Attempts Using Colonization of Parasites and Predators

Target species	Target area	Date	Agents released		Agents established		Assessment[a]	Reference
			Doomed	Adequate	Following releases	Accidentally		
Dendroctonus frontalis Zimmerman	United States	1894		1	?		F ?	Dowden, 1962
Diprion frutetorum (F.)	Canada	1935–1942	1	5		1[b]	F ?	McGugan and Coppel, 1962
Eriococcus araucariae (Maskell)	Australia	1900 ?		1		1	?	Wilson, 1960
Lambdina fiscellaria (Guen.)	Canada	1949, 1951		1			F	McGugan and Coppel, 1962
Megastigmus spermotrophus (Wachtl.)	New Zealand	1955, 1957		1			F	R. Zondag, personal communication
Neodiprion abietis (Harr.)	Eastern North America	1941, 1947		2	1	1[b]	F	McGugan and Coppel, 1962 Dowden, 1962
N. fulviceps (Cresson)	United States	1941		1			?	Dowden, 1962
N. lecontei (Fitch)	Canada	1937–1949	1	8			F	Bird, 1971
N. nanulus (Schedl.)	Canada	1943–1945		4	1	1	?	McGugan and Coppel, 1962
N. pratti banksianae (Roh.)	Canada	1943–1951		5	2 ?	1	?	McGugan and Coppel, 1962

N. swainei (Midd.)	Eastern North America	1940–1953	2		4[b]	P	McLeod and Smirnoff, 1971; Price, 1970; Dowden, 1962
N. tsugae (Midd.)	Canada	1948–1951	1	1		?	McGugan and Coppel, 1962
N. virginianus complex	Canada	1940	1		2[b]	?	McGugan and Coppel, 1962
Orgyia leucostigma J. E. Smith	United States	1910–1929	3		2	?	Dowden, 1962
O. pseudotsugata McD.	United States	1930–1937	3			?	Dowden, 1962
Pikonema alaskensis (Roh.)	Canada	1948–1951	1	1	1[b]	?	McGugan and Coppel, 1962

[a]P, partial success; F, failure; ?, not assessed.
[b]Established from massive releases against *Neodiprion sertifer* and *Diprion hercyniae*.

TABLE 3

Futile Biological Control Attempts Using Colonization of Parasites and Predators

Target species	Target area	Date	Agents released		Agents established		Reference
			Doomed	Adequate	Following releases	Accidentally	
Acleris variana Fern.	Canada	1950–1951	1				McGugan and Coppel, 1962
Carulaspis visci (Schrank)	Canada	1947	1				McGugan and Coppel, 1962
Dendroctonus piceaperda Hopk.	Canada	1933–1934	2				McGugan and Coppel, 1962
Euproctis terminalis Walker	S. Africa	1941–1944		1	1[a]		Greathead, 1971
Exotelia pinifoliella (Chamb.)	Canada	1950	2	1			McGugan and Coppel, 1962
Hylastes ater (Paykull)	New Zealand	1934?	2	1			Miller and Clark, 1934
Laspeyresia youngana Kearf.	Canada	1947–1948	2				McGugan and Coppel, 1962
Pineus boerneri (Ann.)	Australia New Zealand	1932–1939 1934?	5				Wilson, 1960 Miller and Clark, 1934
Pineus strobi (Hartig)	Canada	1939–1940	1	1			McGugan and Coppel, 1962
Pissodes strobi (Peck)	Canada	1947	1			1	McGugan and Coppel, 1962
Recurvaria starki (Free.)	Canada	1950	1				McGugan and Coppel, 1962

[a]The released parasite, *Compsilura concinnata* (Meig), was already present in the area.

Partial control, consisting of the termination of outbreaks and a reduction in their frequency and duration, has been achieved in several cases through the action of a single species of introduced parasite. *Mesoleius tenthredinis* Morely has been a major controlling factor against *Pristophora erichsonii* (Hartig) in British Columbia, Idaho and Montana since 1935 (Turnock, 1972) and the recent colonization of *Parasierola nigrifemur* (Ash.) against *R. buoliana* in Argentina (Brewer and Varas, 1971) and *Olesicampe benefactor* Hinz against *P. erichsonii* in central and eastern Canada (Turnock and Muldrew, 1971b) show similar promise. In two cases, partial control by a single colonized parasite species has been negated by changes in the host population: *Campoplex frustranae* Cushman was successful against *Rhyacionia frustrana bushnelli* (Busck) in Nebraska until the host was replaced by a subspecies of *R. frustrana* which is not parasitized by *C. frustranae* (Dowden, 1962); and the controlling influence of *M. tenthredinis* in central and eastern North America (McGugan and Coppel, 1962) was negated by the appearance and spread of the "resistant" strain of *P. erichsonii* which is capable of encapsulating and killing the parasite eggs (Muldrew, 1953, 1964). This specific defense mechanism against *M. tenthredinis* is being overcome by the colonization of a Bavarian strain of *M. tenthredinis*. Experiments showed that this strain was capable of overcoming the encapsulation reaction of North American *P. erichsonii*, and that this ability is inherited by the progeny of crosses of Bavarian and Canadian strains of *M. tenthredinis* (Turnock and Muldrew, 1971b). The establishment of a second parasite species, *O. benefactor*, in addition to the Bavarian strain should reduce the possibility of the host redeveloping the defense mechanism.

The colonization of *Agathis pumila* (Ratz.) against *Coleophora laricella* in Montana and Idaho was designed to test the hypothesis that either of the two parasite species that were responsible for the control of *C. laricella* in eastern North America were capable of reaching high levels of effectiveness (Webb and Quedneau, 1971). *Agathis pumila* was successfully colonized in Montana in 1960 but its spread and impact have not been satisfactory. Lack of documentation, apparently due to inadequate support for the program, prevents evaluation of the experiment, but the results indicate that the single species of parasite is less effective than two species.

The excellent programs against *Sirex noctilio* illustrate the improvement of the level of control as additional biological control agents are colonized. In New Zealand, the control attained by the colonization of *Rhyssa persuasoria* L. between 1929 and 1932 was not satisfactory and was subsequently improved by the colonization of *Ibalia leucospoides* (Hochenw.) between 1954 and 1958 (Zondag, 1959). The fortuitous introduction of a nematode parasite of the siricid (Zondag, 1969) had led to the reduction of siricid populations to very low levels in at least some areas. In Australia, the well-planned biological control program has attempted to deliberately colonize parasites, e.g., *Ibalia drewsini* (Borries), which are suited to different climatic zones and different developmental stages and patterns of the siricid (Spradbery, 1970). *Rhyssa persuasoria* and *I. leucospoides* were colonized from New Zealand but the latter was found to be badly synchronized with *S. noctilio* in Tasmania. Colonization of a Mediterranean strain of *I. leucospoides*, together with *I. ensiger* Norton, brought about a much more satisfactory synchronization (Taylor, 1967). New parasites and strains have been colonized which are more efficient or attack parts of the siricid population that have previously escaped attack, and parasitism is increasing rapidly (Fig. 1). These parasites plus the parasitic nematode are expected to further improve the level of control and the program seems certain to be successful.

The programs against *Neodiprion sertifer* and *Rhyacionia buoliana* in North America have succeeded in colonizing several species of predators or parasites without significantly altering the level of damage caused by these pests. With *N. sertifer*, the successfully colonized cocoon parasites have not reduced pest populations to a tolerable level but the careful selection and recent successful colonization of the larval parasite *Lophyroplectus luteator* (Thunberg) in Ontario, plus the switching of indi-

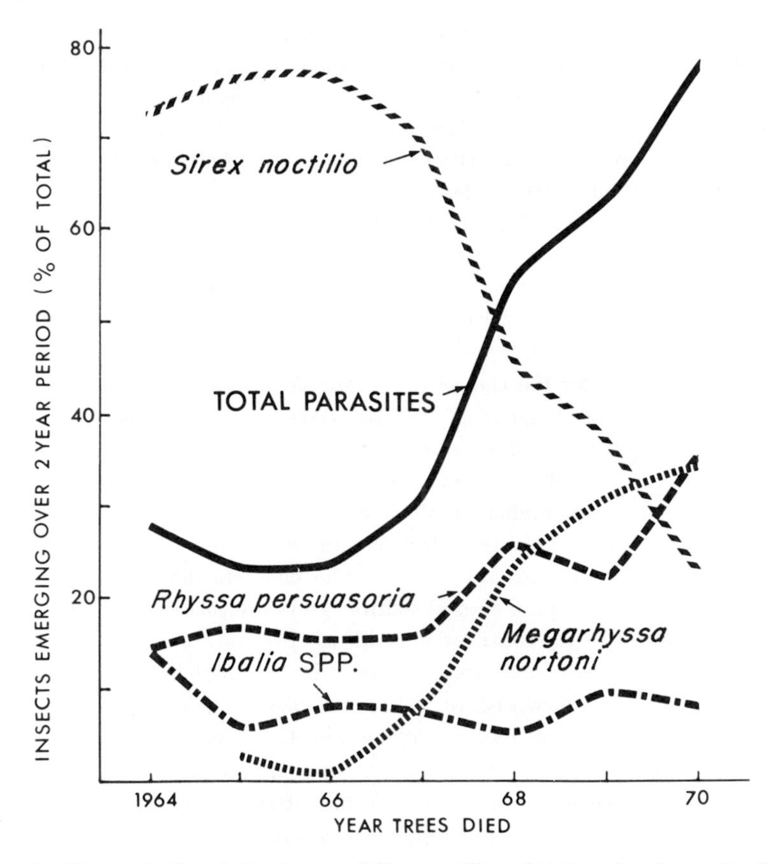

Fig. 1. Changes in the relative density of *Sirex noctilio* and its introduced parasites in Pittwater Forest, Tasmania, 1964—1970, based on the emergence of adults in cages on about 60 trees each year.

genous larval parasites to *N. sertifer*, may improve the level of control (Griffiths, *et al.*, 1971). With *R. buoliana*, the release of a broad spectrum of parasites from Europe, without knowing their potential interaction with other introduced and native parasites, led to the colonization of three cleptoparasitic species which compete to the detriment of the potentially effective *Orgilus obscurator* (Nees). More recently, knowledge of parasite interactions has been the basis for the selection and release of *Lympha dubia* (Fallén) (Schröder, 1969; Syme, 1971).

The failure of attempts to colonize parasites of the spruce budworm, *Choristoneura fumiferana*, was apparently related to their inability to adapt to the environmental conditions associated with the very high host population levels in the release areas. The objectives of the biological control program have recently been changed and efforts are being made to obtain parasite species that would be more efficient at endemic levels and thus reduce the chances that the budworm populations would be "released" by favorable food and weather conditions (Miller and Angus, 1971).

This program and the one against *Rhyacionia buoliana* are examples of how past errors have led to better identification of the type of biological control agent needed and the application of stricter criteria for selection before release.

The program against *Adelges piceae* (Ratz.), involving the introduction of about 40 species of aphid predators, is the only example of the colonization of invertebrate predators against a coniferous forest pest. Some of the established predators contribute to the reduction of heavy aphid populations on the stems of host trees in the continental climate regions of eastern North America (Clarke *et al.*, 1971) but an effective predator against these aphids in twig infestations has not been colonized. Failures in establishing species and in attaining satisfactory control from colonized species has been attributed to the inability of the predators to adapt to *A. piceae* and other new hosts, and to a lack of cold-hardiness and suitable overwintering conditions (Clarke *et al.*, 1971; Amman and Speers, 1971; Mitchell and Wright, 1967). In addition, the discovery that predators of *A. piceae* in Europe are ineffective at high prey densities (Eichhorn, 1969) makes it unlikely that they could significantly contribute to the suppression of outbreaks in North America.

A unique aspect of the Canadian program against the larch sawfly was the colonization of the masked shrew, *Sorex cinereus* Kerr, in Newfoundland (Turnock and Muldrew, 1971b). Since this island had no insectivores and few small fossorial mammals, *S. cinereus* has occupied a broad spectrum of habitats. The predator is rapidly spreading across the island, and although the evaluation is incomplete, it appears to have increased the level of predation and average mortality on larch sawfly cocoons. *Sorex cinereus* can be expected to have an impact on a variety of invertebrate species; for example, slug populations have apparently declined in one area occupied by the shrews.

Secondary programs (Table 2) which share the quality of opportunism, were initiated because a source of agents was available rather than because the target species was a major problem. In most cases, this source was a major colonization program, but sometimes the fortuitous discovery of a source of agents led to a one-shot colonization attempt, e.g., abundant clerids observed feeding on bark beetles in Germany were sent to the United States for release against *Dendroctonus frontalis* (Dowden, 1962). The target species for the natural enemy used in secondary programs were selected with some care and the release operations were adequately conducted but evaluation tended to be minimal. In general, no improvement in the level of control of such target species has been recorded for coniferous forest programs. The exception, the improved control of *Neodiprion swainei* (Midd.), is the result of the fortuitous establishment of beneficial parasites released against *Diprion hercyniae* rather than against *N. swainei* itself (Price, 1970).

The "futile" programs (Table 3) also tend to be opportunistic in nature but are characterized by the release of enemy species without adequate attention to the selection of suitable ones, or care in handling and releasing them. Most of the colonized species had very little chance of becoming established ("doomed," Table 3) and, in fact, none of them was established as a result of the releases.

The failure of these "futile" and secondary programs to decrease the level of damage caused by the target pests must be attributed largely to the lack of critical scientific judgment associated with the selection and release of the introduced species. In some cases, particularly those using surplus material from other biological control programs, the low chances of success were balanced by the low cost of the program. However, low cost alone does not justify a poorly managed attempt because it may discourage subsequent, more serious, attempts. To avoid this hazard, it is highly desirable that all programs should include provision for documenting the handling and release techniques and the subsequent assessment of establishment and impact of the introduced species.

B. Periodic Inoculation

Periodic inoculative releases are designed to establish a natural enemy in local areas from which it is temporarily absent due to adverse environmental conditions (Turnock and Muldrew, 1971a). As in the colonization approach, the success of the effort depends on the establishment and reproduction of the agent. In Europe, inoculative releases have been made with relatively small numbers of *Trichogramma* spp. against *Rhyacionia buoliana* in Germany (Fankhänel, 1963), Poland (Koehler, 1970b), and the U.S.S.R. (Krushev, 1960). *Trichogramma embryophagum* (Hartig) and *Telenomus verticillatus* (Kieffer) have similarly been used against *Dendrolimus pini* L. and *Diprion pini* L. in the U.S.S.R. (Ryvkin, 1955). The release of *Parasierola nigrifemur* against *R. buoliana* in Argentina (Brewer and Varas, 1971) could be considered inoculative. Inoculation with diseases is discussed in a later section.

The relocation (usually with added protection) of natural enemies to augment their numbers and effectiveness is also a form of inoculation. In Europe, red wood ants (*Formica* spp.) have been used against defoliators of conifers in Germany (Otto, 1967), Italy (Ortisi, 1965), Poland (Koehler, 1965), and the U.S.S.R. (Marikovskij, 1963;

Malyseva, 1968). Reviews of these programs (Wellenstein, 1954; Adlung, 1966; Otto, 1967) agree that positive results were mainly observed in pine forests, and that the ants were most effective against dipterous and lepidopterous larvae, less effective against sawflies, and ineffective against beetles. Otto (1967) states that effective protection of coniferous forests has been achieved using ants against: *Coleophora laricella* (Germany, Switzerland, and Italy), *Bupalus piniarius* L. (Germany and U.S.S.R.), *Panolis flammea* Schiff. (Germany and U.S.S.R.), *Lymantria monacha* L. (Germany, C.S.S.R., and U.S.S.R.), *Thaumatopoea pytiocampa* Schiff. (Italy), *Acantholyda nemoralis* Thoms. (Germany and Poland), *Pristiphora abietina* (Christ) (Germany), and *Diprion pini* L. (Germany and U.S.S.R.).

The inoculative approach seems particularly well suited to situations where pest foci occur in isolated areas, e.g., on plantations of exotic conifers, and should be given greater attention in the development of pest management systems.

C. Inundation

Inundative releases differ from the periodic inoculative approach in that the objective is to control the pest through the release of large numbers of natural enemies. Control is expected through the action of the released predators or the immediate progeny of released parasites. In Europe, the control of localized outbreaks of *Diprion pini*, *Pannolis flammea*, and *Theocodiplosis brachyntera* Schwaegrichen, has been achieved by: (a) mass collection of hosts in outbreak areas and release of the emerging parasites in areas with increasing host densities (Ceballos and Zarco, 1952; Postner, 1962; Cankov, 1964; Fankhänel and Zeletzki, 1964; Rossi, 1966) and (b) mass breeding of local parasites and their release during the phase of population increase of the pest, when parasites are generally scarce (Ryvkin, 1955; Szmidt, 1959; Schwenke, 1964; Shchepetilnikova, 1970). This method has not received much attention in forestry, but it might be used to advantage in relatively small areas where chemical control is undesirable. (See also this volume, Chapter 13.)

D. Enhancement

The enhancement of the effectiveness of existing biotic agents through environmental manipulation offers a wide scope for biological control. The abundance of avian predators, especially in monocultures of pine and spruce, may be limited by a shortage of nesting sites (Berlepsch, 1929). Artificial nesting sites have increased bird populations and thus have given long-term protection to pine stands in Germany from damaging defoliation by *Bupalus piniarius*, *Pannolis flammea*, *Dendrolimus pini*, and *Diprion frutetorum* (F.) (Herberg, 1960; Bösenberg, 1970). The effectiveness of local parasites and predators may also be enhanced by improving the physical environment through the encouragement of specific plants for shelter and food, protection from natural enemies, provision of shelter for overwintering parasites and predators, etc. (Buckner, 1971;

Sailer, 1971; Turnock and Muldrew, 1971b). In Poland, a large-scale program was initiated in 1958 (Burzynski, 1970; Koehler, 1970a) to provide protection in permanent outbreak areas by increasing the effectiveness of natural enemies through environmental improvement (application of fertilizers, planting of deciduous trees and shrubs, and provision of nectar plants for parasites and predators). In California, the availability of nesting sites is a limiting factor for the mountain chickadee and studies are continuing to evaluate the role of these birds in the population dynamics of several pests (D. L. Dahlsten and W. A. Copper, unpublished). In general, the enhancement of avian predators can be expected to aid in preventing rather than suppressing insect outbreaks.

III. MICROBIAL AGENTS

The use of microorganisms in the control of insect pests has recently enjoyed increased emphasis due to the need to find safer, more specific alternatives to chemical insecticides (Burges and Hussey, 1971; Katagiri, 1969). Microorganisms or microbial agents may be used either in attempts to colonize microorganisms (enzootic establishment) or to inundate as microbial insecticides. In coniferous forests, the accidental colonization in the 1930's of a nuclear polyhedrosis into New Brunswick, Canada, quickly terminated a major outbreak and is still a major factor in the control of the European spruce sawfly, *Diprion hercyniae* (Neilson *et al.*, 1971). The successful introduction of this pathogen to Newfoundland should have stimulated other attempts to colonize pathogens of exotic pests, but very few additional attempts have been made. The colonization of *Bacillus thuringiensis* (Berl.) var. *dendrolimus* against *Dendrolimus sibericus* Tschtv. in the U.S.S.R. has given long-term control (Talalajev, 1961; Falcon, 1971) but although an exotic nuclear polyhedrosis of *Neodiprion sertifer* has been used successfully as a microbial insecticide in Canada, its colonization has not led to continued suppression of the pest outside the treated areas (Griffiths, *et al.*, 1971). The possibility of inoculating pathogens into areas of increasing pest populations has been discussed and some tests have been made (Martignoni and Auer, 1957; Niklas and Franz, 1957), but this approach for conifers does not seem to be receiving much attention (but see Chapters 7, 13, 17, and 27).

Microbial insecticides have received much more attention because of their potential to replace chemical insecticides. Because of its relatively broad toxicity, *Bacillus thuringiensis* was tested against several lepidopterous pests of coniferous forests but the results have been variable (Falcon, 1971). More virulent strains and better techniques of application will be necessary before the commercially prepared *B. thuringiensis* can be used in major control operations. Development of operational techniques is often slow: for example, promising field tests using *B. thuringiensis* against *Thaumatopoea pytiocampa* in Europe were done in 1954 but subsequent laboratory work and field tests have not yet led to operational use (Grison, 1970). Toxins derived from this and other microorganisms may also prove to be valuable insecticides but their use is still in its infancy (Lysenko and Kucera, 1971). Another broad-spectrum disease organism, the fungus *Isaria farinosa* (Dicks.) Fr., was successfully tested against *Dendrolimus spectabilis* Butler in Japan but further use was discontinued because the fungus was potentially dangerous to the silkworm industry (Katagiri, 1969). Ying (1970) reports effective control of *Dendrolimus punctatus* Walker on Taiwan with applications of *Isaria* sp., a cytoplasmic virus, or *B. thuringiensis*, the microbial agent being selected for application to the appropriate stage of the pest insects.

Host-specific microorganisms for use in inundative control attempts have been obtained by selecting virulent strains of nuclear and cytoplasmic polyhedroses or granuloses from native populations. The application of these, as sprays in field experiments and in small plantings of high value, has been quite successful (Bird, 1971; Lukjancikov, 1964; Rivers, 1964; Schönherr, 1965, 1969; Stairs, 1971). However, they are not in general use because commercial development has been inhibited by the lack of accepted protocols for licensing microbial preparations and the noncommercial production is impractical. In addition, methods of measuring dosages are crude, application techniques need refinement, and little is known of the factors affecting the dosage/mortality relationship. Such problems may lead to unusual results in the field, as when a cytoplasmic polyhedrosis reduced survival among larvae of *Dendrolimus spectabilis* at low and intermediate densities but showed no effect at high densities (Katagiri, 1969). The most successful large-scale operations involve the use of the European nuclear polyhedrosis against *Neodiprion sertifer* (Griffiths, *et al.* 1971) and a selected native strain against *N. lecontei* (Bird, 1971) in Canadian pine plantations.

Promising evidence of synergism of microorganisms and chemical insecticides is described by Benz (1971) but little has been done against coniferous forest pests. A mixture of cytoplasmic and nuclear polyhedroses has shown promise against *Lymantria fumida* Butler in Japan (Katagiri, 1969) while mixtures of *B. thuringiensis* and insecticides have been tested in Europe.

The use of microbial agents against coniferous forest pests has great potential, although most of the successes have been of technological rather than of operational nature (Munroe, 1971). Realization of this potential will require persistence in the efforts to solve the problems of virulence, high cost, safety, and licensing. In too many cases, the experimenters have been satisfied by the demonstration of experimental and technological successes without attempting to solve the problems of formulation and application that prevent operational use. Every effort should be made to encourage the development of operational techniques.

IV. PROSPECTS FOR BIOLOGICAL CONTROL

The various approaches to biological control have not received equal attention in coniferous forests. Colonization has been the most widely used method and the recorded successes suggest that further attempts would be worthwhile. Initial failures or subeconomic successes should not be taken as the final result, because persistence and careful study of the characteristics of potential colonizing species have led to increases in the number of successfully colonized species and the level of control, e.g., the programs against *Sirex noctillo* in Australia and New Zealand.

Colonization of exotic natural enemies against "native" pests of conifers has not been adequately tested, but a number of exotic parasites have been accidentally established on native phytophages of conifers. In Canada, seven of the parasite species released against *Diprion hercyniae* and *Neodiprion sertifer* have been recovered from eight native species of conifer-feeding sawflies; and the polyphagous parasites, *Meteorus versicolor* (Wesm.) and *Compsilura concinnata* Meigen, released against *Stilpnotia salicis* L., were recovered from 19 native phytophages on conifers (from McGugan and Coppel, 1962). These data suggest that well-planned attempts against native pests, using selected parasites and predators of taxonomically or ecologically related hosts from foreign areas, would have a good chance of establishing new natural enemies. However, the chances of success appear to be greatly reduced if the introductions are made at the peak of outbreaks of opportunistic pests when the forest stand is highly disturbed. Examination of the attempts to introduce parasites against spruce budworm in outbreak years led Miller and Angus (1971) to suggest that releases designed to increase the natural control complex during the endemic period would be more promising.

Much greater attention should be given to the colonization of microorganisms, with particular attention to mechanisms for intra- and intergeneration spread. In some cases, the success of the colonization of a microorganism could require the previous colonization of a parasite vector.

Different combinations of inoculation, inundation, and environmental enhancement have been suggested for use against opportunistic and anthropogenic pests by Turnock and Muldrew (1971a). The selected strategy must be developed as part of a broader pest management plan (Stark, 1971) and should be based on comparisons of the structure and characteristics of the natural control complexes (Zwölfer, 1971; Price, 1970) in different host populations. Even relatively crude data on host population patterns and life systems can be used in such comparisons (Turnock, 1972). Programs to introduce or enhance the effectiveness of natural enemies must be compatible with forest management plans. In general, attempts to enhance effectiveness will require more stand diversity than is currently deemed desirable by the managers of coniferous forests. Flowering plants, as a source of adult food to increase the longevity and fecundity of parasites of *Rhyacionia buoliana* (Syme, 1971), may require openings in the forest stand while oak woods are necessary for females of *Eucarcelia rutilla* Vill. during their preoviposition period although their host, *Bupalus piniarius*, lives in pine woods (Herrebout, 1967).

Biological control in coniferous forests should be approached from the basis of a sound knowledge of the ecosystem. Too often, biological control, like other control methods, is applied as a result of an emotional response to insect damage rather than any justified knowledge of the effects of this damage on the final crop (Stark, 1971). Development of biological control techniques as part of an integrated control or pest management program is necessary to avoid costly failures and, perhaps, costly unnecessary successes. Thus, a biological control program that reduces the "damage" caused by a species that only helps to thin overstocked stands could hardly be called an "economic" success. The successful application of biological control methods in coniferous forests faces a major obstacle in the current trend toward monospecific, even-aged forests adapted to mechanized harvesting. Much more must be known about the population ecology of pests and their natural enemies before we can develop pest management systems that minimize crop damage and the costs of production and maximize the economic returns.

ACKNOWLEDGMENT

We thank the many foresters, entomologists, and biological control workers who aided us by supplying references and technical information, and particularly Dr. B. V. Ryvkin, Belorussian Science Research, Gomel, U.S.S.R. and Mr. R. Zondag, Forest Research Institute, Rotorua, New Zealand.

REFERENCES

Adlung, K. G. (1966). A critical evaluation of the European research on the use of red wood ants (*Formica rufa*-group) for the protection of forests against harmful insects. *Z. Angew. Entomol.* **57**, 167–189 (In German).

Amman, G. D., and Speers, C. F. (1971). Introduction and evaluation of predators from India and Pakistan for control of the balsam woolly aphid (Homoptera: Adelgidae) in North Carolina. *Can. Entomol.* **103**, 528–533.

Benz, G. (1971). Synergism of micro-organisms and chemical insecticides. *In* "Microbial Control of Insects and Mites" (H. D. Burges and N. W. Hussey, eds.), pp. 327–355. Academic Press, New York.

Berlepsch, H. V. (1929). "Der gesamte Vogelschutz," 12th ed. Neumann, Neudamm.

Bird, F. T. (1971). *Neodiprion lecontei* (Fitch), red-headed pine sawfly (Hymenoptera: Diprionidae). *Commonw. Inst. Biol. Control Tech. Commun.* **4**, 148–150.

Bösenberg, K. (1970). The importance of birds in biological control, especially in the forest. *Tagungsber. Deut. Akad. Landwirtschaftwiss.* Berlin **110**, 71–82. (In German.)

Brewer, M., and Varas, D. (1971). Cría masiva de *Parasiola nigrifemur* (Ash.), (Hymenoptera: Bethylidae). *Rev. Peruana Entomol.* **14**, 352–361.

Buckner, C. H. (1971). Vertebrate predators. *U.S. Dep. Agr. Forest. Serv. Res. Paper NE*-**194**, 21–31.

Burges, H. D., and Hussey, N. W. (eds.). (1971). "Microbial Control of Insects and Mites," 861 pp. Academic Press, New York.

Burzynski, J. (1970). Biologische Bekampfungs—methoden von Forstschadlingen. *Tagungsber. Deut. Akad. Landwirtschaftwiss. Berlin* **10**, 37–42. (In German.)

Cankov, G. (1964). The use of certain entomophagous insects in the biological control of the European pine shoot moth. *Gorskostop. Nauka* **4**, 61–70.

Ceballos, G., and Zarko, E. (1952). The biological control of an outbreak of *Diprion pini* (L.) on *Pinus silvestris* in the Sierra de Albarracin. *Madrid Inst. Esp. Entomol.*, pp. 1—38. (In Spanish.)

Clark, R. C., Greenbank, D. O., Bryant, D. G., and Harris, J. W. E. (1971). *Adelges piceae* (Ratz.), balsam woolly aphid (Homoptera: Adelgidae). *Commonw. Inst. Biol. Control Tech. Commun.* **4**, 113—127.

Dahlsten, D. L., Garcia, R., Prine, J. E., and Hunt, R. (1969). Insect problems in forest recreation areas. *Calif. Agr.* **23** (7), 4—6.

Davidson, A. G., and Prentice, R. M. (eds.), (1967). Important Forest Insects and Diseases of Mutual Concern to Canada, the United States and Mexico. Canada Dep. Forestry and Rural Development, Ottawa.

Dowden, P. B. (1962). Parasites and predators of forest insects liberated in the United States through 1960. *U.S. Dep. Agr. Handb.* **226**, 70 pp.

Eichhorn, O. (1969). Naturliche Verbreitungsareale und Einschleppungsgebiete der Weissenannen-Wollause (Gattung *Drefusia*) und die Moglichkeiten ihrer biologischen Bekampfung. *Z. Angew. Entomol.* **63**, 113—131.

Falcon, L. A. (1971). Use of bacteria for microbial control of insects. *In* "Microbial Control of Insects and Mites" (H. D. Burges and N. W. Hussey, eds.), pp. 67—95. Academic Press, New York.

Fankhänel, H. (1963). The use of egg parasites of the genus *Trichogramma* against the pine shoot moth, *Rhyacionia buoliana* Schiff., during 1960—62. *Beitr. Entomol.* **13**, 643—653. (In German.)

Fankhänel, H., and Zeletzki, C. (1964). The development of the endoparasite *Misocyclops pini* Kieffer (Proctotrupoidea: Scelionidae) and its use against *Thecodiplosis brachyntera* Schwaegrichen (Diptera: Cecidomyiidae). *Beitr. Entomol.* **14**, 707—730. (In German.)

Greenbank, D. O. (1963). Host species and the spruce budworm. *Mem. Entomol. Soc. Can.* **31**, 219—223.

Greathead, D. J. (1971). A review of biological control in the Ethiopian region. *Commonw. Inst. Biol. Control Tech. Commun.* **5**, 162.

Griffiths, K. J., Rose, A. H., and Bird, F. T. (1971). *Neodiprion sertifer* (Geoff.), European pine sawfly. *Commonw. Inst. Biol. Control Tech. Commun.* **4**, 150—162.

Grison, P. (1970). Lucha microbiologica contra la "procesionaria del pino." *Bol. Serv. Plagas For.* **13**, 133—143.

Herberg, H. (1960). Three decades of bird protection and the control of forest insect pests by colonizing birds in artificial nesting boxes. *Arch. Forstwesen.* **9**, 1015—1048. (In Ger.)

Herrebout, W. M. (1967). Habitat selection in *Eucarcelia rutilla* Vill. (Diptera: Tachinidae). I. Observations on the occurrence during the season. *Z. Angew. Entomol.* **60**, 219—229.

Huffaker, C. B. (ed.). (1971). "Biological Control," 511 pp. Plenum, New York.

Katagiri, K. (1969). Review on microbial control of insect pests in forests in Japan. *Entomophaga* **14**, 203—214.

Koehler, W. (1965). The effect of the red wood ant (*Formica rufa* L.) on the population dynamics of *Acantholyda nemoralis* Thoms. *Collana Verde* **16**, 219—230. (In Ital.).

Koehler, W. (1970a). The theoretical background of the "Complex Method." *Tagungsber. Deut. Akad. Landwirtschaftwiss. Berlin* **110**, 31—35. (In Ger.)

Koehler, W. (1970b). The importance of *Trichogramma* species in the reduction of *Rhyacionia buoliana* Schiff. populations. *Tagungsber. Deut. Akad. Landwirtschaftwiss Berlin* **110**, 177—183 (In Ger.)

Krushev, L. I. (1960). The use of *Trichogramma* in the biological control of the European pine shoot moth. *Sb. Nauch. Rab. Inst. Les. Kchozj. 13*, 198—204. (In Russ.)

Luck, R. F. and Dahlsten, D. L. (1975). Natural decline of a pine needle scale (*Chionaspis pini-*

foliae [Fitch]), outbreak at South Lake Tahoe, California following cessation of adult mosquito control with malathion. *Ecology* **56**, 893–904.

Lukjancikov, V. P. (1964). The use of a granulose to control the Siberian pine moth. *Lesn. Khoz.* **16**, 49. (In Russ.)

Lysenko, O., and Kucera, M. (1971). Micro-organisms as sources of new insecticidal chemicals: Toxins. *In* "Microbial Control of Insects and Mites" (H. D. Burges and N. W. Hussey, eds.), pp. 205–227. Academic Press, New York.

McGugan, B. M., and Coppel, H. C. (1962). A review of the biological control attempts against insects and weeds in Canada. II. Biological control of forest insects, 1910–1958. *Commonw. Inst. Biol. Control Tech. Commun.* **2**, 35–216.

McLeod, J. M., and Smirnoff, W. A. (1971). *Neodiprion swainei* Midd., Swaine jack-pine sawfly (Hymenoptera: Diprionidae). *Commonw. Inst. Biol. Control Tech. Commun.* **4**, 162–176.

Malyseva, M. S. (1968). The method of the colonization of the red wood ant *Formica polycytena* Foerst. in pine forests, with a discussion of the results. *Tr. Vses. Nauch. Issled Zasc. Rast.* **31**, 244–255.

Marikovskij, P. I. (1963). *Experimental colonization of Formica rufa* L. to protect forests against pest insects. *Sb. Zasc. Les. Sibir Nasekomych-vredet. Akad. Nauk S.S.S.R.*, pp. 85–89. (In Russ.)

Martignoni, M. E., and Auer, C. (1957). The experimental use of a granulosis to control *Eucosma griseana* (Hubner) (Lepidoptera: Tortricidae). *Mitt. Schweiz. Anst. Forstl. Versuchsw.* **33**, 73–93. (In Ger.)

Miller, C. A., and Angus, T. A. (1971). *Choristoneura fumigerana* (Clemens), spruce budworm (Lepidoptera: Tortricidae). *Commonw. Inst. Biol. Control Tech. Commun.* **4**, 127–130.

Miller, D., and Clark, A. F. (1934). Distribution of parasites in New Zealand. *N. Z. J. Sci. Technol.* **15**, 301–307.

Miller, D., and Clark, A. F. (1938). The establishment of *Rhyssa persuasoria* in New Zealand. *N. Z. J. Sci. Technol.* **19**, 63–64.

Miller, P. C. (1970). Age distributions of spruce and fir in beetle-killed forests on the White River Plateau, Colorado. *Amer. Midl. Natur.* **83**, 206–212.

Mitchell, R. G., and Wright, K. H. (1967). Foreign predator introductions for control of the balsam woolly aphid in the Pacific Northwest. *J. Econ. Entomol.* **60**, 140–147.

Muldrew, J. A. (1953). The natural immunity of the larch sawfly [*Pristiphora erichsonii* (HTG.)] to the introduced parasite *Mesoleius tenthredinis* Morley, in Manitoba and Saskatchewan. *Can. J. Zool.* **31**, 313–332.

Muldrew, J. A. (1964). The biological control programme against the larch sawfly. *Proc. Entomol. Soc. Manitoba* **20**, 63.

Munroe, E. G. (1971). Status and potential of biological control in Canada. *Commonw. Inst. Biol. Control Tech. Commun.* **4**, 213–255.

Neilson, M. M., Martineau, R., and Rose, A. H. (1971). *Diprion hercyniae* (Hartig), European spruce sawfly (Hymenoptera: Diprionidae). *Commonw. Inst. Biol. Control Tech. Commun.* **4**, 136–143.

Niklas, O. F., and Franz, J. M. (1957). Factors limiting a gradation of the European pine sawfly [*Neodiprion sertifer* (Geoffr.)] in southwestern Germany between 1953–1956. *Mitt. Biol. Bundesanst. Land. Forstwirt. Berlin-Dahlem* **89**, 1–39. (In Ger.)

Ortisi, A. (1965). Populations of ants of the *Formica rufa*-group of the Province of Bergamo as a source for ant colonizations in Italy. *Collana Verde* **16**, 236–249. (In Ital.)

Otto, D. (1967). The importance of *Formica*-colonies in the reduction of important pest insects. A literature review. *Waldhygiene* **7**, 65–90. (In German.)

Patterson, J. E. (1921). Life history of *Recurvaria milleri*, the lodgepole needle-miner, in the Yosemite National Park, California. *J. Agr. Res.* **21**, 127–142.

Postner, M. (1962). Biological control of *Agevillea abietis* Hubault (Diptera: Cecidomyiidae). *Verh. 11th Int. Congr. Entomol. Wien, 1960* **2**, 711–713. (In Ger.)

Price, P. W. (1970). Characteristics permitting coexistence among parasitoids of a sawfly in Quebec. *Ecology* **51**, 445–454.

Rivers, C. F. (1964). The use of a polyhedral virus disease in the control of the pine sawfly *Neodiprion sertifer* Geoffr. in northwest Scotland. Colloq. Int. Pathol. Insectes, Lutte Microbiol (Paris, 1962). *Entomophaga Mem. No.* **2**, 477–480.

Rossi, D. (1966). Biological control of *Evetria buoliana* Schiff. *Note Appunti Sper. Entomol. Agr. Perugia* **11**, 1–14. (In Ital.)

Ryvkin, B. V. (1955). Some new methods to control forest insect pests. *Les. Khoz.* **8**, 58–60. (In Russ.)

Sailer, R. I. (1971). Invertebrate predators. *U.S. Dep. Agr. Forest. Serv. Res. Paper NE-***194**, 32–44.

Sartwell, C. (1969). Role of mountain pine beetle in the population ecology of ponderosa pine. *Bull. Oreg. Entomol. Soc.* **35**, 255.

Sartwell, C. (1971). Thinning ponderosa pine to prevent outbreaks of the mountain pine beetle. *In* "Precommercial Thinning of Coastal and Intermountain Forests in the Pacific Northwest" (D. M. Baumgartner, ed.), pp. 41–52. Wash. State Univ., Pullman, Washington.

Shchepetilnikova, V. A. (1970). The knowledge about egg parasites of the genus *Trichogramma* and their use against insect pest in agriculture and forestry. *Tagungsber. Deut. Akad. Landwirtschaftwiss. Berlin* **110**, 117–136. (In Ger.)

Schönherr, J. (1965). The use of a virus disease in the biological control of the European pine sawfly, *Neodiprion sertifer* (Geoffr.), and the spread of the disease after application. *Z. Pflanzenkr. Pflanzepathal. Pflanzenshutz.* **72**, 466–477. (In Ger.)

Schönherr, J. (1969). Field experiments using a granulose for the biological control of the European fir budworm, *Choristoneura murinana* (Hubner). *Entomphaga* **14**, 251–260. (In Ger.)

Schröder, D. (1969). *Lypha dubia* (Fall.) (Diptera: Tachinidae) as a parasite of the European pine shoot moth, *Rhyacionia buoliana* (Schiff.) (Lepidoptera: Eucosmidae) in Europe. *Commonw. Inst. Biol. Control Tech. Bull.* **12**, 43–60.

Schwenke, W. (1964). Successful experiments using Chalcidids (Hymenoptera) for biological control of *Diprion pini* L. (Hymenoptera: Tenthredinidae) and *Pannolis flammea* Schiff. (Lepidoptera: Noctuidae). *Z. Angew. Entomol.* **53**, 179–186. (In Ger.)

Spradbery, J. P. (1970). The biology of *Ibalia drewseni* Borries (Hymenoptera: Ibaliidae), a parasite of siricid woodwasps. *Proc. Roy. Entomol. Soc. London* **A 45**, 104–113.

Stairs, G. R. (1971). Use of viruses for microbial control of insects. *In* "Microbial Control of Insects amd Mites" (H. D. Burges and N. W. Hussey, eds.), pp. 97–124. Academic Press, New York.

Stark, R. W. (1971). Integrated control, pest management or protective population management? *U.S. Dep. Agr. Forest. Serv. Res. Pap. NE-***194**, 111–129.

Stark, R. W., and Dahlsten, D. L. (eds.). (1970). "Studies on the Population Dynamics of the Western Pine Beetle, *Dendroctonus brevicomis* LeConte (Coleoptera: Scolytidae)," 174 pp. Univ. Calif. Div. Agr. Sci., Berkeley, California.

Syme, P. D. (1971). *Rhyacionia buoliana* (Sciff.), European pine shoot moth (Lepidoptera: Olethreutidae). *Commonw. Inst. Biol. Control Tech. Commun.* **4**, 113–127.

Szmidt, A. (1959). The use of *Dahlbominus fuscipennis* Zett. (Hymenoptera: Chalcididae) against sawflies (Hymenoptera: Diprionidae). *Poznan. Tow. Przyj. Nauk*, **1**, 1–57. (In Polish.)

Talalajev, E. V. (1961). Bacterial control of *Dendrolimus sibiricus*. *Zastch. Rast.* (*Moscow*) **6**, 20–22. (In Russ.)

Taylor, K. L. (1967). The introduction, culture, liberation and recovery of parasites of *Sirex noctilio* in Tasmania, 1962–1967. *Commonw. Sci. Ind. Res. Organ. Div. Entomol. Tech. Pap.* **8**, 19 pp. and 16 photos.

Telford, A. D. (1961). Lodgepole needle miner parasites: biological control and insecticides. *J. Econ. Entomol.* **54**, 347–355.

Turner, K. B. (1952). The relation of mortality of balsam fir caused by the spruce budworm to forest composition in the Algoma forest of Ontario. *Can. Dep. Agr. Publ.* **875**.

Turnock, W. J. (1972). Geographical and historical variability in population patterns and life systems of the larch sawfly. *Can. Entomol.* **104**, 1883–1900.

Turnock, W. J., and Muldrew, J. A. (1971a). Parasites. *U.S. Dep. Agr. Forest. Serv. Res. Pap. NE-194*, pp. 59–87.

Turnock, W. J., and Muldrew, J. A. (1971b). *Pristiphora enrichsonii* (Hartig), larch sawfly. *Commonw. Inst. Biol. Control Tech. Commun.* **4**, 113–127.

Vité, J. P. (1971). Silviculture and the management of bark beetle pests. *Proc. Tall Timbers Conf. Ecol. Anim. Control Habitat Manage.* **3**, 155–168.

Watt, K. E. F. (1968). "Ecology and Resource Management," 450 pp. McGraw-Hill, New York.

Webb, F. E., and Denton, R. E. (1967). Larch casebearer, *Coleophora laricella* (Hbn.). *In* "Important Forest Insects and Diseases of Mutual Concern to Canada, the United States and Mexico" (A. G. Davidson and R. M. Prentice, eds.), pp. 85–88. Can. Dep. Forest. and Rural Develop., Ottawa.

Webb, F. E., and Quedneau, F. W. (1971). *Coleophora laricella* (Hubner), larch casebearer (Lepidoptera: Coleophoridae). *Commonw. Inst. Biol. Control Tech. Commun.* **4**, 131–136.

Wellenstein, G. (1954). What can we expect from the red wood ant in forest protection. *Beitr. Entomol.* **4**, 117–138. (In German.)

Williamson, R. L., and Prince, F. E. (1971). Initial thinning effects in 70 to 150 year old Douglas-fir in western Oregon and Washington. *U.S. Dep. Agr. Forest. Serv. Res. Pap. PNW*-**117**. 15.

Wilson, F. (1960). A review of the biological control of insects and weeds in Australia and Australian New Guinea. *Commonw. Inst. Biol. Control Tech. Commun.* **1**, 102 pp.

Wygant, N. D., and Nelson, A. L. (1949). Four billion feet of beetle-killed spruce. *In* "Trees," (A. Stefferud, ed.) pp. 417–422. (U.S. Dep. Agr.). Yearb. Agri., Washington D.C.

Ying, Sze-ling (1970). Application of *Isaria* sp., cytoplasmic polyhedrosis and *Bacillus thuringensis* against the pine caterpillar *Dendrolimus punctatus* Walker (Lepidoptera: Lasciocampidae). *Quart. J. Chinese Forest.* **4**, 51–67.

Zondag, R. (1959). Progress report on the establishment in New Zealand of *Ibalia leucospoides* (Hochenw.) a parasite of *Sirex noctilio* (F.). *N. Z. Forest. Res. Notes* **20**.

Zondag, R. (1969). A nematode infection of *Sirex noctilio* (F.) in New Zealand. *N. Z. J. Sci.* **12**, 732–747.

Zwölfer, H. (1971). The structure and effect of parasite complexes attacking phytophagous host insects. *In* "Dynamics of Populations" (P. J. den Boer and G. R. Gradwell, eds.), pp. 405–418. Centre Agr. Publ. Doc., Wageningen.

13

BIOLOGICAL CONTROL OF PESTS OF BROAD-LEAVED FORESTS AND WOODLANDS

W. E. Waters, A. T. Drooz, and H. Pschorn-Walcher

I. INTRODUCTION

The forests of the world comprise its largest plant ecosystem, covering approximately 30% of the total land area. Of this, two-thirds is in broad-leaved forest. This proportion varies, increasing generally from boreal to tropical regions.

Timber-related activities and products account for a large part of the economic value derived from forest areas, but additional values and benefits derive from watershed protection and prevention of wind erosion, the food and cover afforded wildlife and livestock, and increasing use of forest areas for recreational purposes. Aesthetic values of the forest—as a place for physical and spiritual release and enjoyment—are less tangible but nonetheless vital to society today. Broad-leaved trees in restricted woodland areas not formally classified as forest (e.g., park and shade trees in urban and suburban areas and shelterbelts) also serve important uses difficult to assess in monetary terms. Protection of forests and woodlands against destructive insects and diseases is an integral part of their management and preservation, particularly in North America, Europe, and the U.S.S.R. where intensive use is made of them and the values at stake are greater. In the tropics, increasing utilization of the forests undoubtedly will require greater attention to pest control.

Insects and diseases attack all parts of the living tree and all stages of growth from seed to harvest (or natural demise). They kill trees outright and, in less direct ways, further reduce and limit the productivity, value, and usefulness of forest stands and woodlands. Their depredations interact (often synergistically) with fire, drought, wind, snow, ice, animals, and man-directed activities. More broadly, they create profound and long lasting disturbances in the ecosystems in which they operate. In so doing, they disrupt the planned use and management of the resources involved.

Forests provide a unique opportunity to probe into the principles of pest management on a long-term basis, and to develop and apply truly efficient strategies for biological (and integrated) control. As ecological systems, forests are variable and dynamic, but their continuity in space and time permits comprehensive measurement and analysis of key components, such as pest and natural enemy populations, to a degree not possible with short-term crop ecosystems. Continual feedback from and adjustment of applied biological control measures allow rapid development of improved strategies for specific pest situations.

There are constraints, however, in the development and application of biological control in forest ecoystems. Certain limiting factors applicable to forests in general, and broad-leaved forests and woodlands in particular, have made it difficult to develop rational and efficient protection strategies. The pests themselves are rated lower in importance than those affecting human health and food crops, and the benefits to be derived from the control of forest pests are less obvious. The diversity of conditions under which the trees grow, the range of uses and values placed on them, and variation in the kind and degree of damage inflicted preclude any simple or generally applicable formula for calculating the economic threshold for a specific pest or pest complex. Similarly, determination of the ecological threshold or critical density of a pest—the point at which a specified control action would be most efficient from the biological standpoint—is dependent on the population dynamics of the pest, the objective (or kind) of control strategy contemplated, and the relative effectiveness of the agent(s) or method(s) to be employed.

Since adequate quantitative data on the requisite socioeconomic and ecological elements have been lacking, generally, biological control efforts against pests in broad-leaved forests and woodlands have been minimal and have had only sporadic success. Nonetheless, much valuable information and experience have been gained. Moreover, present evidence of the hazards of chemical pesticides and the vital need for effective protection of forest resources in the face of the human population boom emphasize the importance of biological, nonchemical alternatives.

The examples of biological control efforts and successes that follow give evidence of the tremendous potential of this approach to protection of the resource values in broad-leaved forests and woodlands.

II. BIOLOGICAL CONTROL OF SPECIFIC PEST INSECTS

The rationale and achievements of biological control efforts against insect and disease pests of broad-leaved forests and woodlands are best considered in an ecological, rather than a taxonomic, context. This chapter is therefore structured according to the techniques and strategies of biological control and presents specific examples.

A. Introduction and Augmentation of Exotic Parasites and Predators against Exotic Pests

The number of destructive pests accidentally established in broad-leaved forests from foreign lands is quite small. This is largely because relatively few broad-leaved tree species have been transported (even as seed) from one country to another for forestry purposes. Also, potentially dangerous pests from deciduous ornamental stock generally have been successfully intercepted through quarantine activities. The examples given here, therefore, are few.

1. Gypsy Moth

Native to temperate Europe, Asia, and North Africa, the gypsy moth, *Porthetria* (= *Lymantria*) *dispar* (L.), was first introduced into the United States in 1868 or 1869 at Medford, Massachusetts. More time, effort, and money have been expended to control this species in the United States by use of imported natural enemies than any other exotic forest insect. Parasites from its native environs have received most attention, but several insect predators have been involved in the importation program as well. The efforts fall into two distinct phases: (1) early large-scale importations from Europe and Japan in the periods 1905—1914 and 1922—1933, released in the New England States; and (2) recent large-scale rearing and releases in New Jersey (1963 to present) and Pennsylvania (1969 to present), and small importations of new candidate species from India. The early work is interestingly described by Howard and Fiske (1911) and Burgess and Crossman (1929). A complete summary of introductions in this period, plus sporadic releases made subsequently, up to 1960, is given by Dowden (1962). The recent work is summarized only in unpublished reports of Dowden, Metterhouse, and Rhoads.*

Up to 1934, over 40 species of gypsy moth parasites and predators were imported and released in southern Maine, New Hampshire, and Vermont, throughout Massachusetts and Rhode Island, and in eastern Connecticut. The exact number is not certain, due to some misidentifications in the earliest collections and subsequent

*P. B. Dowden, unpublished reports 1963—1971 on gypsy moth rearing, on file at Gypsy Moth Methods Improvement Laboratory, U.S. Dept. Agr., Otis AFB, Falmouth, Massachusetts; W. W. Metterhouse, annual reports 1963—1971, Plant Laboratory, New Jersey Dept. Agr., Trenton, New Jersey; L. D. Rhoads, annual reports of gypsy moth parasite releases in Pennsylvania 1969—1971, Bur. Plant Industry, Pa. Dept. Agr., Harrisburg, Pennsylvania.

changes in the names of a few species. Over 93 million parasites were released from 1905 to 1933. About 91 million of these were two species of egg parasites (see below). These totals include some foreign stock imported and released directly, ones reared from foreign and established stock at the Gypsy Moth Laboratory in Melrose Highlands, Massachusetts, and ones collected and redistributed from locations in New England.

Nine parasite and two predator species were definitely established in the early program (Dowden, 1962). Another species, *Brachymeria intermedia* (Nees), which had been introduced but not recovered, was found in Connecticut in 1965 (Leonard, 1966) and subsequently in Maine (Leonard, 1971).* The introduced species now established in the United States are shown in the following tabulation:

Egg parasites	*Anastatus disparis Ruschka*	Eupelmidae
	Ooencyrtus kuwanai (How.)	Encyrtidae
Larval parasites	*Apanteles melanoscelus* (Ratz.)	Braconidae
	Phobocampe (=*Hyposoter*) *disparis* (Vier.)	Ichneumonidae
	Blepharipa (*Sturmia*) *scutellata* (R.-D.)	Tachinidae
	Compsilura concinnata (Meig.)	Tachinidae
	Exorista larvarum (L.)	Tachinidae
	Parasetigena silvestris (=*agilis*) (R.-D.)	Tachinidae
Pupal parasites	*Brachymeria intermedia* (Nees)	Chalcididae
	Monodontomerus aereus Walk.	Torymidae
Larval—pupal predators	*Calosoma sycophanta* (L.)	Carabidae
	Carabus auratus L.	Carabidae

Evaluation of the effectiveness or capabilities of the foregoing species is clouded by the influence of the insecticidal eradication and control efforts and the regulatory activities that were carried out concurrently almost from the beginning. Though the parasites and predators presently established have persisted, even in areas receiving intensive treatment with chemical insecticides (Dowden, 1961), other potentially effective species may have been eliminated by such treatments. Moreover, the abrupt changes in abundance of the gypsy moth (and alternate hosts, perhaps) effected by these treatments undoubtedly have modified the long-term effectiveness of the introduced natural enemies. Evaluation is hampered, too, by the lack of quantitative data on gypsy moth numbers and parasite and predator incidence. Detailed data were obtained at times for special studies (Bess, 1961; Campbell, 1967), but these data

*One other parasite, *Apanteles liparidis* (Bouchè), originally introduced into the United States from Europe and Japan, was recovered from a collection of gypsy moth larvae at Waterbury, Connecticut, in 1965. It probably, however, was introduced with recent parasite shipments from Spain (R. I. Sailer, personal communication, 1971).

are not sufficient to draw generalized conclusions. Clearly, however, the parasites and predators successfully introduced early into the New England states are now part of the ecosystem(s) in which the gypsy moth operates. More specifically, they are a significant part of the complex of mortality-causing agents that regulates its numbers.

The second phase of activity with parasites and predators against the gypsy moth in the United States began in 1963. This was a joint effort of state agencies and the U.S. Department of Agriculture (USDA), and had two main thrusts: (1) laboratory studies to develop improved means of rearing specific parasites in large numbers on a schedule, and (2) mass rearing and release, supplemented by field collecting and transfer, of selected species in New Jersey and Pennsylvania. This activity is continuing as part of a comprehensive program aimed at integrated control of this pest in northeastern United States.

The parasite species studied in most detail were *Apanteles melanoscelus* (Ratz.), *A. porthetriae* Mues., and *Exorista segregata* (Rond.). The latter two species were among those imported and released earlier, but not established. Additional rearing studies were made of six parasites of *Lymantria obfuscata* Walk. in India that were considered to be potentially useful against the gypsy moth in its southward extension in the United States.* These species were *Exorista rossica* Mesn. (Tachinidae), *Drino discreta* Wulp (Tachinidae), *D. inconspicuoides* Bar. (Tachinidae), *Rogas indiscretus* Reardon (Braconidae), *Apanteles solitarius* (Ratz.) (Braconidae), and *Palexorista* sp. (Tachinidae). Limited field releases were made of these new parasites for colonization in southern New England, along with small numbers of *Apanteles porthetriae* and *Exorista segregata*.

Beginning with small lots received from the USDA Gypsy Moth Methods Improvement Laboratory at Falmouth, Massachusetts, supplemented with importations from other sources, approximately 80 million gypsy moth parasites were reared and released in New Jersey by State personnel from 1963 to 1971. Seven parasites and one predaceous beetle were established: *Ooencyrtus kuwanai*, *Apanteles melanoscelus*, *Phobocampe disparis*, *Blepharipa scutellata*, *Compsilura concinnata*, *Parasetigena silvestris*, *Brachymeria intermedia*, and *Calosoma sycophanta*. With the possible exception of *O. kuwanai*, the numbers released each year have been of an inoculative, rather than inundative, scale. However, Metterhouse reports that *O. kuwanai*, *B. intermedia*, and *P. silvestris* appear to have had a significant role in the reduction of high populations and maintenance of stable, tolerable populations in certain areas in 1971.†

In Pennsylvania, parasite releases against the gypsy moth were initiated in 1969. Approximately 16.5 million *O. kuwanai* and 9600 *B. intermedia* were released that

*Obtained through a Forest Service-sponsored Special Foreign Currency Research (P.L. 480) Grant to the Indian Station, Commonwealth Institute of Biological Control, at Bangalore. Reports on file in the Division of Forest Insect and Disease Research, U.S. Forest Service, Washington, D.C.

†W. W. Metterhouse, annual Rpt. 1970–1971, Plant Laboratory, N. J. Dept. Agr., Trenton, New Jersey.

year and in 1970. Many of the releases were made at locations where only male moth catches occurred. However, *O. kuwanai* has become established wherever the gypsy moth is present in significant numbers. *Brachymeria intermedia* was recovered from one site in 1970 (L. D. Rhoads, personal communication, 1971).

No biological control attempts have been instituted against the gypsy moth in Canada. The earliest infestations, which occurred at two locations in southern Quebec in 1924, were eliminated by mechanical and chemical means. A subsequent infestation discovered in 1936 at St. Stephen, New Brunswick, was similarly treated with apparent success. Current infestations are limited to areas near the New York and Vermont borders. Cooperative quarantines with the U.S. Department of Agriculture, intensive surveys for male moths and egg masses, and spraying with insecticides are relied upon to hold the insect in check (Brown, 1968).

2. Brown-Tail Moth

This insect, *Nygmia phaeorrhoea* (Donov.), was first discovered in the United States in 1897 at Somerville, Massachusetts, in the same general area where the gypsy moth first became established.

Large-scale importations of parasites and predators were carried out from 1905 to 1911, concurrently with those for control of the gypsy moth (Burgess and Crossman, 1929). Approximately 20 species were introduced from Europe in this period (Clausen, 1956). Of these, eight parasites and one predator were definitely established (Dowden, 1962): (see following tabulation):

Larval parasites	*Apanteles lacteicolor* Vier.	Braconidae
	Meteorus versicolor (Wesm.)	Braconidae
	Eupteromalus nidulans (Thoms.)	Pteromalidae
	Alsomyia nidicola (Tns.)	Tachinidae
	Carcelia laxifrons Vill.	Tachinidae
	Compsilura concinnata (Meig.)	Tachinidae
	Exorista larvarum (L.)	Tachinidae
Pupal parasite	*Monodontomerus aereus* Walk.	Torymidae
Larval–pupal predator	*Carabus auratus* L.	Carabidae

According to Dowden (1962), *A. lacteicolor*, *A. nidicola*, and *M. versicolor* have been most effective against the brown-tail moth. Two others, *C. concinnata* and *E. larvarum*, have been abundant at times, but both have other lepidopterous hosts, including the gypsy moth, and their importance as natural enemies of the brown-tail moth is uncertain. *Carabus auratus*, along with *Calosoma sycophanta* and native predators, may have had a role in limiting spread and build-up of brown-tail moth populations. However, since the insect was subjected early to intensive suppressive and quarantine measures, and climatic factors such as low winter temperatures appear to have a stringent effect on its survival, it is difficult to evaluate the effectiveness of these biological control agents. Nevertheless, the brown-tail moth has been extremely scarce since the late 1920's. It is now limited to isolated infestations in coastal areas of Massachusetts, New Hampshire, and southern Maine.

The brown-tail moth spread early into the Maritime Provinces of Canada. It was first discovered at St. John, New Brunswick, in 1902 (McIntosh, 1903) and subsequently in Nova Scotia (Pickett and Payne, 1938). Three species of parasites and one predator were released in infested areas of New Brunswick and Nova Scotia in the period 1912–1916 (McGugan and Coppel, 1962). The parasites involved were *Apanteles lacteicolor*, *Compsilura concinnata*, and *Meteorus versicolor*. The predator was *Calosoma sycophanta*. Only *A. lacteicolor* became well established. Host collection records indicate that relatively high levels of parasitism by this species occurred in the years immediately following the release program, when brown-tail moth populations also were high. Whatever the cause(s), the brown-tail moth disappeared from New Brunswick and Nova Scotia by 1927 (Pickett and Payne, 1938).

3. Satin Moth

The satin moth, *Stilpnotia salicis* (L.), was first found in North America in 1920 at two widely separate locations, near Boston, Massachusetts and at New Westminister, British Columbia (Burgess, 1921; Glendenning, 1932). Several of the parasites introduced from Europe to help control the gypsy moth and brown-tail moth were already at work on the satin moth when it was found in Massachusetts. The carabid *Calosoma sycophanta* and the tachinid *Compsilura concinnata* were found the first year (Burgess, 1921). Later, the tachinid *Blepharipa scutellata* and the pteromalid *Eupteromalus nidulans* were reported (Burgess and Crossman, 1927).

Between 1927 and 1934, seven species of parasites were introduced from Europe specifically to control the satin moth. Two species became established (McLeod, 1951; Dowden, 1962). The most important of these is the braconid *Apanteles solitarius* (Ratz.), in both eastern and western regions (Jones *et al.*, 1938; Reeks and Smith, 1956). The other, *Meteorus versicolor* (Wesm.), occurs in the satin moth only in the West. *Compsilura concinnata* and *Eupteromalus nidulans* take a high toll of this host at times in the East (Dowden, 1962; Forbes and Ross, 1971).

Although these introduced parasites are not completely successful in maintaining the satin moth at tolerable levels, native parasites have contributed little in this regard, and the introduced species undoubtedly are an important factor in the general dampening of satin moth populations and in terminating local outbreaks before significant damage occurs.

4. Winter Moth

The winter moth, *Operophtera brumata* (L.), project is one of the best documented examples among the introduced forest insects in North America. Its biology and population dynamics have been carefully studied by Canadian entomologists concurrent with the importation of parasites from Europe.

The winter moth is native to Sweden south to North Africa. It was first found in North America along the south shore of Nova Scotia. It was determined from specimens collected late in 1949, although its larvae had apparently been mistaken for spring cankerworm, *Paleacrita vernata* Peck, for about 20 years (Hawboldt and Cuming, 1950). It is now generally distributed throughout Nova Scotia and in isolated locations in Prince Edward Island and eastern New Brunswick (Embree, 1971).

Some 63 species of parasites are reported from the winter moth in Europe (Wylie, 1960). None of these came to Nova Scotia with the host insect and attacks by native parasites were rare (Smith, 1950). Consequently, a program was initiated in 1954 for the introduction and study of selected European parasites. Between 1954 and 1958, six species of parasites were liberated in infested areas of Nova Scotia (Graham, 1958). Only the tachinid *Cyzenis albicans* (Fall.) and the ichneumonid *Agrypon flaveolatum* (Grav.) became established. *Cyzenis albicans* was introduced during 1954—1957; it was first recovered in 1956. *Agrypon flaveolatum* was released during 1956—1957; it was recovered in significant numbers from mature larvae of the winter moth in 1957. When success with these parasites was indicated, further introductions were halted to minimize the possible chances of adverse competition (Embree, 1966).

The status of the winter moth in the Maritime Region has changed significantly since 1958 because of the success of *C. albicans* and *A. flaveolatum* (Embree, 1971). Defoliation and consequent effects on red oak in forest areas is now minimal, and the insect is one of the less common pests of hardwoods. Noticeable populations persist in commercial apple orchards; these are controlled through regular spray programs. Occasional localized outbreaks occur in abandoned orchards, but are of short duration. Both *C. albicans* and *A. flaveolatum* respond efficiently to changes in host density, with the latter more effective at low densities and the former at higher densities (Embree, 1966). Analysis of life table data developed before these parasites were established showed that the key factor affecting winter moth population fluctuations, generation to generation, was the degree of phenological synchrony between the winter moth and its host plant(s). Following establishment of *C. albicans* and *A. flaveolatum*, parasitism became the principal factor maintaining and regulating populations below levels of economic damage (Embree, 1966).

A nuclear polyhedral virus was also detected occasionally in 1961, and is now generally present over the Maritimes range of the winter moth. It has contributed to collapse of at least two infestations (Embree, 1971). The nature and degree of parasite-virus interaction has not as yet been evaluated.

5. Birch Leaf-Mining Sawfly

Widely distributed in Europe, the birch leaf-mining sawfly, *Heterarthrus nemoratus* (Fall.), was first noticed in Nova Scotia in 1905 (MacGillivray, 1909). It now occurs throughout northeastern United States and southeastern Canada (Baker, 1972).

This sawfly was especially abundant in New England and the Adirondack region of New York from 1927 to 1933. In the period 1931–1935, five species of parasites were imported from central Europe and released in Maine, New Hampshire, Massachusetts, and Vermont (Dowden, 1941). Two of these, *Chrysocharis laricinellae* (Ratz.) (Eulophidae) and *Phanomeris phyllotomae* Mues. (Braconidae), became established.

Chrysocharis laricinellae was also introduced into Canada and the United States from Europe in the late 1920's and early 1930's against the larch casebearer, *Coleophora laricella* (Hbn.). However, Dowden (1941) states that its establishment in sawfly populations resulted from the liberations directed against it. He reports also that predatory feeding by the adult parasite females kill large numbers of sawfly larvae.

Little is known of the effectiveness of *P. phyllotomae* in controlling the sawfly. A general decrease in sawfly populations followed release and establishment of this species and *C. laricinellae* but the role of these parasites in this is not certain. Other natural control factors, such as competition from other birch leaf-miners and foliage feeders, are known to affect survival of the sawfly.

6. Oriental Moth

A native of Asia, the oriental moth, *Cnidocampa flavescens* (Walk.), was first found in the United States in 1906 in Boston, Massachusetts. Though periodically abundant, it has not spread from eastern Massachusetts in all these years (Baker, 1972). In its present environs, it feeds principally on certain species of maple (*Acer*), buckthorn (*Rhamnus*), and a variety of fruit and ornamental trees.

The first introduction of a natural enemy against this pest was made in 1917–1918, with the importation of *Chrysis shanghaiensis* Smith from China. This parasite was only recovered the year subsequent to its release (Fernald, 1920).

In 1929–1930 a tachinid, *Chaetexorista javana* B. & B., was introduced from Japan and immediately became established. Within 3 years, parasitization reached 60% at the liberation sites and oriental moth populations were reduced to noninjurious levels (Dowden, 1946). Since then, host and parasite numbers have fluctuated, with the lower winter temperatures causing proportionately higher mortality of the parasite and allowing upsurges of the host (Dowden, 1962). Rapid increase of the parasite associated with milder winters quickly reduces host numbers again.

Though definitive data are lacking, parasitism by *C. javana* may be considered a critical factor in maintaining the numbers of oriental moth at innocuous levels. Considering its wide geographical range and diversity of host plants in Asia, this insect otherwise would appear capable of becoming a serious forest pest in North America.

7. Elm Leaf Beetle

A native of Europe, the elm leaf beetle, *Pyrrhalta* (=*Galerucella*) *luteola* (Mull.), was first found in the United States at Baltimore, Maryland, in 1838 (Berry, 1938).

Two parasites have been found useful in suppressing elm leaf beetle populations in California. One of these, *Erynniopsis rondanii* Tns. (Tachinidae), a larval parasite, was first imported from Italy in 1909 and 1911, and additionally from France in 1924–1925 and 1932–1935, for release at a number of points in eastern United States (Clausen, 1956). These were not successful. However, in 1939 stock of *E. rondanii* from southern France was released near Stockton and Manteca, California, and later transferred elsewhere in the State. It rapidly became established and parasitism reached 20% by late 1940 (Flanders, 1940) and much higher (e.g., 70 to over 90%) by 1947 (C. B. Huffaker, unpublished data). Transfers to Pennsylvania, Arkansas, Idaho, Maryland, Nevada, and Virginia have not been successful (Dowden, 1962).

The other parasite, *Tetrastichus brevistigma* Gahan (Eulophidae), is a native to eastern North America (Berry, 1938). It was introduced into California in 1934 and quickly became established. It is now abundant in infested areas throughout the State and in Oregon and Washington (Muesebeck *et al.*, 1951).

According to Clausen (1956), the activity of *E. rondanii* and *T. brevistigma* has resulted in a substantial reduction of injury caused by the elm leaf beetle in California. Spraying has, however, continued, and because of spring movement of unparasitized adults into areas of high (e.g., 98%) previous fall parasitization by *E. rondanii*, defoliation may occur unless treatments are made (C. B. Huffaker, personal communication, 1971). The potential role of the parasites is clouded by these treatments.

8. Fall Webworm

The fall webworm, *Hyphantria cunea* (Drury), is a native of North America. It is the rare case of an insect pest of broad-leaved forests and woodlands that has been introduced inadvertently from the New to the Old World. First reported from a single location near Budapest, Hungary, in 1940, it now has become established in parts of central and in southeastern Europe, Japan, and Korea.

In Japan, the webworm is not considered a forest pest, as it propagates only on ornamentals and roadside trees (Y. Ito, personal communication, 1971). Therefore, biological control attempts against it in that country are not discussed here.

Warren and Tadić (1970) list an array of 65 insect parasites and 28 predators that attack *Hyphantria cunea* in the United States and Canada. According to Tadić (personal communication, 1971), a total of seven parasite species and three predator species have been introduced into central and eastern Europe. Over 18,000 individuals of selected species have been released in Yugoslavia, those in greatest numbers being *Mericia ampelus* Walk. (Tachinidae), *Apanteles hyphantriae* Riley (Braconidae), and *Coleomegilla maculata* DeG. (Coccinellidae). Some species have been introduced into Czechoslovakia, the Moldavian region of U.S.S.R., and other countries; these include M. *ampelus*, *A. hypantriae*, *Campoplex validus* Cress., and *Phobocampe* (= *Hyposoter*) sp. Most of these releases were made by 1955, but releases of the coccinellid *C. maculata* and two species of pentatomids, *Podisus maculiventris* Say and *P. placidus* Uhl., were not made until 1968.

At present only one species, the braconid *A. hyphantriae*, is known to be established in Yugoslavia and Czechoslovakia (M. Tadić, personal communication, 1971). The fate of the other introductions is uncertain. However, a large number of European parasites and predators have accepted the fall webworm as host or prey. As synchronization and selectivity increase over time a greater impact on webworm numbers may be expected.

9. Eucalyptus Snout Beetle

With increased planting of *Eucalyptus* spp. in many African countries, serious problems arose with the alien eucalyptus snout beetle, *Gonipterus scutellatus* Gyll. (Tooke, 1955). This weevil arrived in South Africa from Australia on planting stock some time before 1916. It subsequently spread into Rhodesia and Malawi, and in the 1940's and later it was found in Kenya, Mauritius, Madagascar, and St. Helena.

Biological control efforts began in Australia in 1926 with the immediate discovery of an egg parasite and its shipment to Cape Town and Pretoria, South Africa (Tooke, 1955). This parasite, *Patasson nitens* (Gir.), was reared at Pretoria and released that year at Johannesburg. Up to 1931, a total of 620,000 *P. nitens* were released in all major infested areas in South Africa. By 1935, damage fell below economic levels in all areas except the High Veld. Climatic conditions in the latter region apparently hampered the buildup of *P. nitens* to consistently effective numbers.

The parasite spread naturally into Rhodesia and Malawi and increased rapidly so that the weevil no longer is an injurious pest in those countries.

Transfer of the parasite to East Africa was begun in 1945, with equal success. It was then introduced into Mauritius, and in 1948 into Madagascar where egg parasitism ranged as high as 67% within 6 months. In 1958, it was introduced into St. Helena where a rapid increase in parasitism resulted in significant protection to the eucalyptus plantations from 1959 onward.

Overall, this project was a resounding success (Greathead, 1971). It proves how rapidly an efficient exotic parasite can effect control of an introduced pest in an intensive timber culture system.

Control of this weevil has also been achieved by *P. nitens* in New Zealand. Introductions from Australia in 1927–1929 resulted in establishment in both the North and South Islands. Subsequent records are scant, but high levels of parasitism have been observed at times and it is considered an important factor in the regulation of weevil populations throughout the range of the pest in New Zealand (E. W. Valentine, personal communication, 1971).

10. Oak Leaf Miner

This European insect, *Lithocolletis messaniella* Zell., was first found in New Zealand in 1951 (Wise, 1953). Its spread was rapid, and in its first years of abundance it extended its host range from exotic oaks (*Quercus*) and chestnut (*Castanea*) to

European beech (*Fagus sylvatica* L.), birches (*Betula*), and other species, and also to native beech, *Nothofagus* sp., thus threatening the indigenous forests of New Zealand.

Chemical control was considered impracticable and, following a study of its parasites in Europe, two species were selected for introduction into New Zealand (Delucchi, 1958): *Apanteles circumscriptus* Nees, a winter parasite in Europe, and *Enaysma splendens* Delucchi, active during summer months. *Apanteles circumscriptus* was received from Italy in 1957, 1958, and 1959 and releases were made each year. *Enaysma splendens* was received from Switzerland in 1957 and again in 1959–1960.

The results were at first uncertain (Given, 1959), but both species did become established and a noticeable reduction of leaf miner infestations occurred in the 1960's (E. W. Valentine, personal communication, 1971). No outbreaks have since occurred and leaf miner populations are now generally at a tolerable level.

B. Introduction and Augmentation of Exotic Parasites and Predators against Native Pests

Few attempts have been made to use exotic parasites or predators to control indigenous insects in broad-leaved forests or woodlands. Native forest insects generally host a complex of natural enemies and the probability of reducing the average density or significantly modifying the fluctuations of the pest insect by the addition of an exotic parasite or predator is generally low. There is also some possibility of hyperparasitism, interference, or competitive displacements by the exotic species that might result in increased densities and/or greater instability in the pest population system. In certain, perhaps fortuitous, instances the introduction of an exotic parasite or predator has had no measurable effect either way on the target native insect—e.g., releases of *Compsilura concinnata* and *Calosoma sycophanta* against the forest tent caterpillar in Minnesota and the Great Basin tent caterpillar in New Mexico (Dowden, 1962) and of *C. sycophanta* against the oak looper on Vancouver Island (McGugan and Coppel, 1962).

There have been several specific cases, however, where additions to the indigenous natural control complex of native forest pests appeared needed, and introductions of exotic parasites or predators were made that proved helpful.

1. Gypsy Moth

Since egg parasites of the gypsy moth were absent in Czechoslovakia, attempts were made recently to introduce potentially useful species from the Mediterranean region (Capek, 1971). Introduced were *Anastatus disparis* from Bulgaria and Spain, and *Ooencyrtus kuwanai* from Montenegro (Yugoslavia) and Spain. *Ooencyrtus kuwanai* did not become established. Releases of *A. disparis* in 1960 to 1965 resulted in establishment in all areas. This species has just one generation a year and is unable to reach all eggs in a gypsy moth egg mass in oviposition. However, parasitism by *A. disparis* has consistently ranged from 10 to 15% in recent years and this assists in holding gypsy moth numbers in Czechoslovakia at tolerable levels (Capek, 1971).

2. Pustule Scale

The pustule scale, *Asterolecanium pustulans* (Ckll.), injures a wide variety of shrubs and broad-leaved trees throughout the tropical and subtropical world. It was first noted in Puerto Rico in 1900 (Wolcott, 1953). Tree species decimated by the scale included the introduced Australian silk oak, *Grevillea robusta* Cunn., and cassia de Siam, *Cassia siamea* (Lam.) Britt., the native abeyuelo, *Colubrina arborescens* (Mill.) Sarg., and the valuable maga, *Montezuma speciosissima* Sessé & Moc.

In 1938, the coccinellid *Chilocorus cacti* (L.), was introduced into Puerto Rico from Cuba (Wolcott, 1953) and was rapidly established. Its predation was so effective that the pustule scale virtually disappeared in 3 years. The scale reappeared in 1955, and no new evidence of the predator was reported up to 1960 (Wolcott, 1960). If significant damage to maga occurs again, this highly efficient predator is the likely choice for a renewed biological control effort.

3. Teak Defoliators

Teak, *Tectona grandis* L., is fed upon by numerous lepidopterous defoliators, including *Hapalia machaeralis* Wlk., *Maruca testulalis* Geyer, and *Hyblaea puera* Cram. The braconid *Cedria paradoxa* Wlkn. was the first parasite introduced to enhance natural control of these defoliators in India and Burma (Beeson, 1941). Native to northern India and China, this parasite was reared on a number of hosts at the Forest Research Institute, Dehra Dun. Many thousands were sent to south India and Burma from 1937 to 1940 and subsequently for control of *H. machaeralis*. It has since been recovered from this host and from *M. testulalis* and *H. puera* at a number of locations.

No real assessment of the work with *C. paradoxa* has been made beyond the confirmation of establishment and its frequent recovery (Sankaran, 1972). However, its frequent recovery from these teak defoliators indicates that it may have some regulating influence. Further study of this and other potentially useful exotic parasites in India and neighboring countries is greatly needed.

C. Augmentation of Native Parasites and Predators against Native Pests

As with any strategy of biological control, effective augmentation of indigenous natural enemies by indirect or direct means requires a sound and thorough knowledge of the biologies of the pest insects and the natural enemies. More than any other habitat management strategy, perhaps, it requires detailed quantitative information on the numerical and functional relations between the enemies and their pest hosts. This approach has a logical appeal for long-term regulation or control of destructive native forest insects. It implies less uncertainty in the selection of potentially effective agents and less risk of interference or competitive displacement* than is inherent in the introduction of exotic species. However, insufficient information on key indigenous parasites and predators and/or inadequate technology for truly augmenting the numbers of those considered as such have limited the practical application of this approach in biological control of broad-leaved forest and woodland pest insects. A few examples can be cited below.

1. Gypsy Moth

A variety of tactics have been attempted to augment the natural enemies of the gypsy moth, especially in Europe. An interesting approach, involving maintenance of host numbers at a level supportive to the parasite complex, was tested recently in Yugoslavia (Maksimovič *et al.*, 1970). For a period of 4 years, 1964—1967, 14 to 55 kg of gypsy moth egg masses were collected annually in outbreak areas and transferred to an area where the insect was in an incipient outbreak phase. Detailed data were taken on the parasitism of eggs and larvae and on defoliation in the "treated" area and in a nearby "untreated" area where a similarly anticipated population increase was in progress. By 1967, and confirmed by observations in 1969, the upward trend in gypsy moth numbers in the "treated" area was halted and the population stabilized at a relatively low density. Defoliation was very light. In contrast, the gypsy moth population in the "untreated" area escalated, and complete defoliation ensued. Changes in egg parasitism by *Anastatus disparis* and *Ooencyrtus kuwanai* in the host-seeded area were slight. Parasitism of the young larvae by *Apanteles porthetriae*, however, appeared to be an important control factor. The ratio of *A. porthetriae* to host larvae in the seeded area became 8.5 times greater than that in the unseeded area. Sapiro and Malyseva (1970) also report that transfer of *Apanteles porthetriae* from one area to another in the U.S.S.R. resulted in a fivefold reduction of gypsy moths in the seeded area.

In the U.S.S.R. the timing factor has been utilized in insecticide treatments against the gypsy moth and the tent caterpillar, *Malacosoma neustria* L., to enhance the ratio of natural enemies to these pests (Sapiro and Malyseva, 1970). The optimal period for application of the DDT dust used was in the spring when both insects were in larval instars I and II and most of the critical parasites and predators were still in hibernation or otherwise not exposed to the insecticide. Of the major parasites recorded, only *Apanteles solitarius* appeared early enough to be vulnerable to the treatment. *Apanteles liparidis*, *Parasetigena silvestris*, *Blepharipa scutellata*, and several other parasites of lesser importance appeared later and were little affected by the treatment. The most active period of *Calosoma sycophanta* was also such that it was not affected adversely.

*The editors do not view competitive displacement as normally a "risk."

Mortality of the target pest insects from the insecticide was 93–98%. The ratio of host insects to parasites and predators was shifted dramatically in favor of the latter, resulting in further reduction in pest numbers.

2. Miscellaneous Defoliators—Europe

Fankhänel (1963) reported success with transfers of the egg parasites *Teleas* (= *Telenomus*) *laeviusculus* Ratz. and *Telenomus ovulorum* Bouché for control of *Malacosoma neustria* in the U.S.S.R. Parasitism was increased up to 95% in release areas, with consequent reductions in densities and damage by *M. neustria*.

Artificial propagation and colonization of wood ants (*Formica* spp.) is standard practice in many forest areas in Europe, chiefly in conifer stands but also in some broad-leaved forests (Gösswald, 1951; Otto, 1970; and Pavan, 1955). *Formica polyctena* Först. is the most commonly used species. It is restricted mainly to forest stands at lower elevations, as is *F. rufa* L. which also has been propagated and colonized frequently. At higher elevations, e.g., in the European Alps above 1000 meters, two other species, *F. lugubris* Zett. and *F. aquilonia* Yarr., are predominant but here artificial augmentations are not generally needed (Eichhorn, 1964).

From the European experience with predatory wood ants, it is clear that they must be used discriminately. The ant colonies must be established in effective numbers and pattern, and they must be protected from destructive agents for a substantial time. Generally, use of *Formica polyctena* and related species can be an effective complement in an integrated control program, especially for pests such as the green oak leaf roller, *Tortrix viridana* L., having regularly recurring and long-lasting periods of abundance.

Augmentation of insectivorous bird populations by provision of artificial nesting sites is of older standing in Europe, but its practicality and effectiveness are questionable. As stated by Bösenberg (1970), "Insect-eating birds have a prophylactic importance at least in pine forests, while only occasional or perhaps sometimes illusory results have been obtained against certain pests of deciduous forests, even with very high bird densities induced by an anomalously ample supply of nesting boxes." This conclusion was supported by a thorough study of Altenkirch (1968) in which the occurrence and periodic outbreaks of *Tortrix viridana* were influenced little by increasing bird densities by addition of nesting boxes.

D. Microbial Control

Disease is an important mortality component in the population dynamics of many major insect pests of broad-leaved forests and woodlands. The importance of disease as a natural control factor and the possibility of using certain pathogens for applied control were recognized years ago (e.g., Glaser and Chapman, 1913). However, only since 1950 has research in basic and applied aspects of insect pathology been oriented to developing practical, effective, and safe means of microbial control. This field is well covered in a recent book edited by Burges and Hussey (1971). (See also Chapter 7).

All forms of microorganisms investigated thus far for microbial control—bacteria, viruses, fungi, and protozoa—may have some potential use in forest insect control. However, the bacteria and viruses have definite theoretical and practical advantages over the other forms, and the work on them is further advanced.

1. Bacteria

A wide variety of bacterial pathogens have been recovered from insects infesting broad-leaved trees. For example, in a study of bacteria associated with the gypsy moth, Podgwaite and Cosenza (1966) isolated types representing eight major families. Reviews of the use of bacteria for control of forest and agricultural crop insects are given by Heimpel and Angus (1960) and Falcon (1971). Only *Bacillus thuringiensis* Berliner (*Bt*) has been developed and used on a scale sufficient to warrant discussion here.

Isolates of spore-forming, crystalliferous bacteria classifiable as *Bt* have been recovered from several forest Lepidoptera in natural environments, e.g., the gypsy moth (Cosenza and Lewis, 1966). However, investigations aimed at operational use of *Bt* have concentrated on commercially available preparations, which vary considerably in their biologically active components, adjuvants, and formulations. Bioassays have demonstrated the variability in potency of the different products against particular broad-leaved forest insects (e.g., Morris, 1969).

Formulations of *Bt* have been applied successfully in the United States against the gypsy moth (Lewis and Connola, 1966; Doane, 1966; Yendol *et al*, 1973); the linden looper, *Erannis tiliaria* (Harr.) (Duda, 1962); the fall cankerworm, *Alsophila pometaria* (Harr.) (Thompson, 1962; Quinton and Doane, 1962); the Great Basin tent caterpillar, *Malacosoma fragile* (Stretch) (Stelzer, 1965); and the California oakworm, *Phryganidia californica* Pack. (Pinnock and Milstead, 1971). The particular *Bt* formulations for which data on efficacy and safety were presented in evidence are currently registered for operational use against these insects. The spring cankerworm, *Paleacrita vernata* (Peck), is included in the registration involving the fall cankerworm. Jaques (1961) reported successful results with several *Bt* formulations against the winter moth in Nova Scotia. At least one of these carries a United States registration for use against this pest.

In Europe, satisfactory control of the gypsy moth was obtained with *Bt* dust formulations (Rupérez, 1967) and water suspensions (Adroić, 1967). *Bt* was also reported successful against the green oak leaf roller (Franz *et al.*, 1967), the winter moth, and the ermine moth, *Hyponomeuta padella* L. (Küthe, 1965). Franz and Krieg (1967) report other examples of use of *Bt* in Europe and the U.S.S.R.

The effectiveness of *Bt* can be enhanced by combination with chemicals or other microbial agents. Examples of synergism of *Bt* by DDT and other insecticides are summarized by Benz (1971). The combination of *Bt* with a nuclear-polyhedrosis virus was shown to be more effective than either agent alone against the Great Basin tent caterpillar (Stelzer, 1965, 1967). Further study is needed concerning combination of *Bt* and nuclear-polyhedrosis viruses or chemical toxicants-stressors in forest insect control.

2. Viruses

Viruses are a primary cause of disease in many forest insects, particularly leaf-feeding Lepidoptera and Hymenoptera. The high degree of specificity and virulence of many insect viruses make them ideal candidates for microbial control. Many are transmitted from one host generation to the next and thus may be used to initiate epizootics and produce long-term population regulation. Among the four major types, the granuloses (GV) are most specific, followed in order by the nuclear polyhedroses (NPV), cytoplasmic polyhedroses (CPV), and noninclusion viruses (NIV) (Ignoffo, 1968). All types occur in the pest insects of broad-leaved forests and woodlands.

Recently, Stairs (1971, 1972) and Franz (1970) summarized the status and potential of insect viruses for short-term control and long-term regulation of major forest insect populations. Studies of the NPV of gypsy moth and forest tent caterpillar are well advanced in this regard.

The "wilt" disease of gypsy moth, caused by an NPV is an important component in the disease complex of this pest (Campbell and Podgwaite, 1971) and plays a critical role in its population dynamics (Campbell, 1963, 1967). A small-scale field application of the virus was made in Connecticut in 1963 with a concentration of 2.7×10^8 polyhedra/ml (approximately one trillion per U.S. gallon) (Rollinson *et al.*, 1965). Applied at four gallons per acre mainly against instars II and III, significant reduction in the population and noticeable foliage protection were obtained. Field trials in Spain (Rupérez, 1964) have been promising also, and in Sardinia concentrations of 10^8 to 10^9 polyhedra/ml applied against mainly instar II larvae caused 90–100% mortality (Magnoler, 1967). Magnoler (1968) also indicated that nonpurified suspensions of this NPV may be more effective than purified preparations.

Definitive large-scale experiments are needed to determine the best treatment combinations of dosage, formulation, timing, and method of application for a practical range of host insect and environmental conditions. Adequate safety tests are needed, too, to meet the requirements for United States registration of this NPV as a pesticide.

Tent caterpillars, *Malacosoma* spp., are distributed worldwide. All species feed on a variety of deciduous trees and recur in outbreak numbers more or less regularly. All major species are susceptible to NPV epizootics, which are generally a prime factor in reducing high host populations to low levels. The NPV in North American species of *Malacosoma* is freely transmissible among them and, possibly, among all species in the genus (Stairs, 1972).

Field trials have shown that infection and mortality varies with dosage and age of the larvae when treated (Stairs, 1964, 1965). These studies indicate that applications of NPV at a dosage of 10^{11} polyhedra/A or more against newly hatched larvae may effectively suppress damaging populations of the forest tent caterpillar, and lower dosages may be used to initiate epizootics for longer term control (Stairs, 1971). In laboratory studies, Raheja and Brooks (1971) demonstrated that two applications of the NPV were more effective than a single dose and that no increased resistance resulted from the initial dose. This has important implications for field use of the virus.

3. Fungi

Fungal pathogens have been found in nearly all major pest insects of broad-leaved trees. *Entomophthora*, *Beauveria*, and *Aspergillus* are most frequently reported. High host mortality has been observed at times, but the data on disease occurrence have mostly been qualitative and fragmentary. Laboratory and field studies of specific fungal pathogens have tended to emphasize their limitations for applied microbial control. Research and experience to date are well summarized by Roberts and Yendol (1971).

One of the earliest applications of a fungus for control of a forest insect was conducted by Speare and Colley (1912) against the brown-tail moth, *Nygmia phaeorrhoea*, using *Entomophthora aulicae* Reich. Colonization was achieved in a large number of infestations in Massachusetts. One may speculate on the role of this fungus in the observed dramatic decline of brown-tail moth numbers in subsequent years.

4. Protozoa

Protozoans pathogenic to insects are diverse in form and host relations. They may infect and persist in a wide range of hosts. They are slow acting, as compared with *Bt* or the viruses. With few exceptions, they reproduce only in living hosts. Their primary effect on host insects in nature is impairment of function and reduced fecundity, rather than outright mortality. Their use in microbial control, therefore, is aimed at developing and maintaining a stable pathogen—host balance at a tolerable host density. Weiser (1963) and McLaughlin (1971) comprehensively review the protozoans as insect pathogens and as potential microbial pesticides.

Very little experimentation has been done with protozoans of insect pests of broad-leaved trees. Weiser and Veber (1957) sprayed small infestations of the fall webworm with spore suspensions of a microsporidian, *Thelohania hyphantriae*, with resultant high mortality of the webworm and reduced tree injury. Similar treatments in an area where the webworm was very abundant gave negligible control and no establishment of the disease in the general population.

III. BIOLOGICAL CONTROL OF TREE DISEASES

Though the basic principles and advantages of biological control are applicable to forest tree diseases, little research has been directed to developing specific biotic agents or techniques for control of tree pathogens. No practicable means of preventing or ameliorating pathogen invasion, establishment, or growth in a broad-leaved tree species by use of other microorganisms has been developed to date, but Krstić (1956), Hansbrough (1965), and Franz (1971) have reviewed the prospects for biological control of forest tree diseases and emphasized the complexities, difficulties, and limitations of this approach. (See also Chapter 21.)

However, the occurrence of parasitic, predaceous, and particularly antagonistic microorganisms, arthropods, and other biotic and abiotic agents or factors affecting the success of tree pathogens is well documented. Detailed studies of organism interactions in living trees, such as those of Shigo (1967), provide clues as to how the natural sequence of pathogen invasion might be interrupted biologically. Comprehensive studies of the pathology, etiology, and epidemiology of specific diseases such as those of True *et al.* (1960) on oak wilt [caused by the fungus *Ceratocystis fagacearum* (Bretz) Hunt] provide further information on antagonistic organisms and the ecological requirements for their augmentation that might be exploited. Feasibility of this approach will depend on the nature of the pathogen-host interactions involved and the reduction in disease incidence expected. It must be considered to have a real potential in forest disease control, however, at least as part of an integrated control strategy. (See further Chapter 21.)

IV. CRITIQUE

Biological control clearly has a place in the management of pest populations in broad-leaved forests and woodlands. The examples cited here represent efforts that have been at least partially "successful" or, as with the microbial agents, have reached the stage in development where they appear to be operationally feasible for at least one target pest. They demonstrate, moreover, the unique compatibility of the biological control approach with the ecological and socioeconomic features of forest ecosystems.

Some serious limitations in concept and execution are apparent, however. Definitive criteria for selection and evaluation of techniques and strategies have been lacking. Thus, decisions on when, where, and how to attempt biological control have been arbitrary and, in some cases, capricious. Attention has been focused on certain exotic pests of potential concern and on a very few indigenous pests causing severe damage to trees of commercial importance. Nearly all attempts have been fragmentary—in time or place. In most cases, a strictly unilateral approach was taken by using parasites, *or* predators, *or* a microbial agent. Few clues have been yielded on what would be an optimal biological control strategy for a given pest. Information is scant on how biological control techniques should be integrated with chemical or silvicultural methods for long-term management of a particular pest or pest complex. (For other situations, see Chapters 10, 14–17, 21–24, 27.)

Experience to date and recent population dynamics research on the gypsy moth, winter moth, and other major pests strongly emphasize the need for comprehensive quantitative data on the life systems of target pests. The evaluation of techniques under study must be carried out over a sufficient span of time, and in a truly representative array of ecological situations, to determine not only whether *some* effect on the average density or trend in pest population numbers was accomplished, but how maximal effectiveness might be achieved. Experimental flexibility and continuity, coupled with thorough data analysis and interpretation, are needed to obtain definitive results and to adequately document the results obtained for extension to applications.

A corollary to the research needs is that a simplistic, unilateral approach to biological control operations against pest insects and diseases of broad-leaved trees is not likely to succeed. The probabilities of success with one, or a very few, biological control agents may be higher with pests in plantations or monocultures. In general, however, the diversity and complexity of ecological and socioeconomic conditions in broad-leaved forest ecosystems will require that biological control of specific pests be conceived and applied as part of a broad, *flexible* management system, utilizing a variety of tactics.

In summary, dynamic improvements are needed in the concept, organization, and support of biological control research and operations against pest insects and diseases in broad-leaved forests and woodlands. As emphasized by Munroe (1971), a much expanded effort is needed to achieve planning and delivery systems that will provide the best knowledge and the most suitable methods and material when and where they are required.

REFERENCES

Adroić, M. (1967). Application of a new strain of *Bacillus thuringiensis* Berliner in the control of the gypsy moth (*Lymantria dispar* L.). *Zast. Bilja* **18**, 343–348.

Altenkirch, W. (1968). Vogelschutz und Eichenwickler (*Tortrix viridana* L.) Synökologische Untersuchungen der Kronenfaune eines Eichen-Hainbuchenwaldes. *Z. Angew. Zool.* **55**, 1–69.

Baker, W. L. (1972). Eastern forest insects. *U.S. Dep. Agr. Forest. Serv. Misc. Publ.* **1175**, 642 pp.

Beeson, C. F. C. (1941). "The Ecology and Control of the Forest Insects of India and the Neighbouring Countries," 767 pp. Forest Res. Inst. and Coll., Dehra Dun (reprinted in 1961).

Benz, G. (1971). Synergism of micro-organisms and chemical insecticides. *In* "Microbial Control of Insects and Mites" (H. D. Burges and N. W. Hussey, eds.), pp. 325–355. Academic Press, New York.

Berry, P. A. (1938). *Tetrastichus brevistigma* Gahan, a pupal parasite of the elm leaf beetle. *U.S. Dep. Agr. Circ.* **485**, 11 pp.

Bess, H. A. (1961). Population ecology of the gypsy moth *Porthetria dispar* L. (Lepidoptera: Lymantridae). *Conn. Agr. Exp. Sta. Bull.* **646**, 43 pp.

Bösenberg, K. (1970). Zur Bedeutung der Vogelwelt in Rahmen der biologischen Schädlingsbekämpfung, besonders in Wald. *Tagungsber. Biol. Bek. Meth. von Forstschädl.* **110**, 71–82.

Brown, G. S. (1968). The gypsy moth, *Porthetria dispar* L., a threat to Ontario horticulture and forestry. *Proc. Entomol. Soc. Ont.* **98**, 12–15.

Burges, H. D., and Hussey, N. W. (1971). "Microbial Control of Insects and Mites," 861 pp. Academic Press, New York.

Burgess, A. F. (1921). The satin moth: an introduced enemy of poplars and willows. *U.S. Dep. Agr. Bur. Entomol. Circ.* **167**, 16 pp.

Burgess, A. F., and Crossman, S. S. (1927). The satin moth a recently introduced pest. *U.S. Dep. Agr. Bull.* **1469**, 23 pp.

Burgess, A. F., and Crossman, S. S. (1929). Imported insect enemies of the gipsy moth and the brown-tail moth. *U.S. Dep. Agr. Tech. Bull.* **86**, 147 pp.

Campbell, R. W. (1963). The role of disease and desiccation in the population dynamics of the gypsy moth *Porthetria dispar* (L.) (Lepidoptera : Lymantriidae). *Can. Entomol.* **95**, 426–434.

Campbell, R. W. (1967). The analysis of numerical change in gypsy moth populations. *Forest. Sci. Monogr.* **15**, 33 pp.

Campbell, R. W., and Podgwaite, J. D. (1971). The disease complex of the gypsy moth. I. Major components. *J. Invertebr. Pathol.* **18**, 101–107.

Capek, M. (1971). Results of experiments with the introduction of egg parasites of *Lymantria dispar* into Slovakia. *Les. Casop.* **17**, 127–137.

Clausen, C. P. (1956). Biological control of insect pests in the continental United States. *U.S. Dep. Agr. Tech. Bull.* **1139**, 151 pp.

Cosenza, B. J., and Lewis, F. B. (1966). Taxonomic considerations of four "wild"-type crystalliferous bacilli and their toxicity to larvae of the gypsy moth, *Porthetria dispar*. *J. Invertebr. Pathol.* **8**, 520–525.

Delucchi, V. L. (1958). *Lithocolletis messaniella* Zeller (Lep. Gracilariidae): analysis of some mortality factors with particular reference to its parasite complex. *Entomophaga* **3**, 203–270.

Doane, C. C. (1966). Field tests with newer materials against the gypsy moth. *J. Econ. Entomol.* **59**, 618–620.

Dowden, P. B. (1941). Parasites of the birch leaf-mining sawfly (*Phyllotoma nemorata*). *U.S. Dep. Agr. Tech. Bull.* **757**, 56 pp.

Dowden, P. B. (1946). Parasitization of the oriental moth [*Cnidocampa flavescens* (Walk.)] by *Chaetexorista javana* B. & B. *Ann. Entomol. Soc. Amer.* **39**, 225–241.

Dowden, P. B. (1961). The persistence of gypsy moth parasites in heavily sprayed areas on Cape Cod, Massachusetts. *J. Econ. Entomol.* **54**, 873–875.

Dowden, P. B. (1962). Parasites and predators of forest insects liberated in the United States through 1960. *U.S. Dep. Agr, Agr. Handb.* **226**, 70 pp.

Duda, E. J. (1962). An observation on the behavior of linden looper larvae following treatment with *Bacillus thuringiensis* Berliner. *Sci. Tree Top.* **2**, 14.

Eichhorn, O. (1964). Die höhen- und waldtypenmässige Verbreitung der nützlichen Waldameisen in den Ostalpen. *Waldhygiene* **5**, 129–135.

Embree, D. G. (1966). The role of introduced parasites in the control of the winter moth in Nova Scotia. *Can. Entomol.* **98**, 1159–1168.

Embree, D. G. (1971). *Operophtera brumata* (L.), winter moth (Lepidoptera: Geometridae). *Commonw. Inst. Biol. Control Tech. Commun.* **4**, 167–175.

Falcon, L. A. (1971). Use of bacteria in microbial control. *In* "Microbial Control of Insects and Mites" (H. D. Burges and N. W. Hussey, eds.), pp. 67–95. Academic Press, New York.

Fankhänel, H. (1963). Zu Fragen der biologischen Methode der Schädlingsbekämpfung in der Sowjetunion. *Beitr. Entomol.* **13**, 72–78.

Fernald, H. T. (1920). Ten years of the oriental moth. *J. Econ. Entomol.* **13**, 210–212.

Flanders, S. E. (1940). Observations on the biology of the elm leaf beetle parasite *Erynnia nitida* R-Desv. *J. Econ. Entomol.* **33**, 947–948.

Forbes, R. S., and Ross, D. A. (1971). *Stilpnotia salicis* (L.), satin moth (Lepidoptera: Liparidae). *Commonw. Inst. Biol. Control Tech. Commun.* **4**, 205–212.

Franz, J. M. (1970). Part I. Biological and integrated control of pest organisms in forestry. *Unasylva* **24**, 37–46.

Franz, J. M. (1971). Biological and integrated control of pest organisms in forestry. Part II. *Unasylva* **25**, 45–56.

Franz, J. M., and Krieg, A. (1967). *Bacillus thuringiensis*—Präparate gegen Forstschädlinge—Erfahrungen in der Alten Welt. *Gesundh. Pflanz.* **19**, 175–176, 178–180, 182.

Franz, J. M., Krieg, A., and Reisch, J. (1967). Freilandversuche zur Bekämpfung des Eichenwicklers (*Tortrix viridana* L.) (Lep., Tortricidae) mit *Bacillus thuringiensis* im Forstamt Hanau. *Nachrichtsbl. Deut. Pflanschutzdienst. Braunscheweig* **19**, 36–44.

Given, B. B. (1959). Biological control factors influencing populations of oak leaf-miner, *Lithocolletis messaniella* Zeller, in New Zealand including the introduction of parasites. *N.Z. J. Agr. Res.* **2**, 124–133.

Glaser, R. W., and Chapman, J. W. (1913). The wilt disease of gypsy moth caterpillars. *J. Econ. Entomol.* **6**, 479–488.

Glendenning, R. (1932). The satin moth in British Columbia. *Can. Dep. Agr. Entomol. Br. Pam.* **50**, [N.S.], 15 pp.

Gösswald, K. (1951). "Die rote Waldameise in Dienste der Waldhygiene," 160 pp. M. Kinau Verlag, Luneberg (West Germany).

Graham, A. R. (1958). Recoveries of introduced species of parasites of the winter moth, *Operophtera brumata* (L.) (Lepidoptera: Geometridae), in Nova Scotia. *Can. Entomol.* **90**, 595–596.

Greathead, D. J. (1971). Eucalyptus snout beetle—*Gonipterus scutellatus* Gyll. *Commonw. Inst. Biol. Control Tech. Commun.* **5**, 49–50.

Hansbrough, J. R. (1965). Biological control of forest tree diseases. *J. Wash. Acad. Sci.* **55**, 41–44.

Hawboldt, L. S., Cuming, F. G. (1950). Cankerworms and the European winter moth in Nova Scotia. *Can. Dep. Agr. Sci. Serv. Forest Insect Invest. Bimonth. Progr. Rep.* **6**(1), 1–2.

Heimpel, A. M., and Angus, T. A. (1960). Bacterial insecticides. *Bacteriol. Rev.* **24**, 266–288.

Howard, L. O., and Fiske, W. F. (1911). The importation into the United States of the parasites of the gypsy moth and the brown-tail moth. *U.S. Dep. Agr. Bur. Entomol. Bull.* **91**, 344 pp.

Ignoffo, C. M. (1968). Specificity of insect viruses. *Bull. Entomol. Soc. Amer.* **14**, 265–276.

Jaques, R. P. (1961). Control of some lepidopterous pests of apple with commercial preparations of *Bacillus thuringiensis* Berliner. *J. Insect Pathol.* **3**, 167–182.

Jones, T. H., Webber, R. T., and Dowden, P. B. (1938). Effectiveness of imported insect enemies of the satin moth. *U.S. Dep. Agr. Circ.* **459**, 24 pp.

Krstić, M. (1956). Prospects of application of biological control in forest pathology. *Botan. Rev.* **22**, 38–44.

Küthe, K. (1965). Frostspanner (*Cheimatobia brumata* L.) und Gespinstmotten (*Hyponomeuta padella* L.) Bekämpfung mit *Bacillus thuringiensis* Berliner. *Mitt. Biol. Bundesanst. Land. Forstwirt. Berlin Dahlem* **115**, 55–59.

Leonard, D. E. (1966). *Brachymeria intermedia* (Nees) (Hymenoptera: Chalcididae) established in North America. *Entomol. News* **77**, 25–27.

Leonard, D. E. (1971). *Brachymeria intermedia* (Hymenoptera: Chalcididae) parasitizing gypsy moth in Maine. *Can. Entomol.* **103**, 654–656.

Lewis, F. B., and Connola, D. P. (1966). Field and laboratory investigations of *Bacillus thuringiensis* as a control agent for gypsy moth, *Porthetria dispar* (L.) *U.S. Dep. Agr., Forest Serv. Res. Pap. NE-50*, 38 pp.

MacGillivray, A. D. (1909). A new genus and some new species of Tenthredinidae. *Can. Entomol.* **41**, 345–346.

McGugan, B. M., and Coppel, H. C. (1962). Biological control of forest insects, 1910—1958. *Commonw. Inst. Biol. Control Tech. Commun.* **2**, 35—216.

McIntosh, W. (1903). The brown-tail moth (*Euproctis chrysorrhoea*). Note on first recovery in Canada. *33rd Rep. Entomol. Soc. Ont.*, p. 93.

McLaughlin, R. E. (1971). Use of protozoans for microbial control of insects. *In* "Microbial Control of Insects and Mites" (H. D. Burges and N. W. Hussey, eds.), pp. 151—172. Academic Press, New York.

Magnoler, A. (1967). L'applicazione di un virus poliedrico nucleare nella lotta contro larve di *Lymantria dispar* L. *Entomophaga* **12**, 199—207.

Magnoler, A. (1968). The differing effectiveness of purified and nonpurified suspensions of the nuclear-polyhedrosis virus of *Porthetria dispar*. *J. Invertebr. Pathol.* **11**, 326—328.

Maksimović, M., Bjegović, P., and Vasiljević, Lj. (1970). Maintaining the density of the gypsy moth enemies as a method of biological control. *Zast. Bilja* **21** (107), 1—15 (in Serbo-Croat., with English figure and table captions and summary.)

McLeod, J. H. (1951). Biological control investigations in British Columbia. *Entomol. Soc. Brit. Columbia Proc.* **47**, 27—36.

Morris, O. N. (1969). Subsceptibility of several forest insects of British Columbia to commercially produced *Bacillus thuringiensis*. II. Laboratory and field pathogenicity tests. *J. Invertebr. Pathol.* **13**, 285—295.

Muesebeck, C. F. W. *et al.* (1951). Hymenoptera of America north of Mexico—synoptic catalog. *U.S. Dep. Agr. Monogr.* **2**, 1420 pp.

Munroe, E. G. (1971). Status and potential of biological control in Canada. *Commonw. Inst. Biol. Control Tech. Commun.* **4**, 213—255.

Otto, D. (1970). Einige grundsätzliche Feststellungen zur Einsatzmöglichkeit von *Formica polyctena* Först. im Forstschutz. *Biol. Bek. Meth. von Forstschädl. Tagungsber. No.* **110**, 87—108.

Pavan, M. (1955). La lotta biologica con *Formica rufa* L. contro gli insetti dannosi alle foreste. *Min. Agr. For. Italy Collana Verda* **3**, 1—75.

Pickett, A. D., and Payne, H. G. (1938). The history of entomology in Nova Scotia particularly in respect to the activities of provincial authorities. *69th Rept. Entomol. Soc. Ont.*, pp. 11—15.

Pinnock, D. E., and Milstead, J. E. (1971). Control of the California oakworm with *Bacillus thuringiensis* preparations. *J. Econ. Entomol.* **64**, 510—513.

Podgwaite, J. D., and Cosenza, B. J. (1966). Bacteria of living and dead larvae of *Porthetria dispar* (L.). *U.S. Dep. Agr. Forest Serv. Res. Note NE-50*, 7 pp.

Quinton, R. J., and Doane, C. C. (1962). *Bacillus thuringiensis* against the fall cankerworm, *Alsophila pometaria*. *J. Econ. Entomol.* **55**, 567—568.

Raheja, A. K., and Brooks, M. A. (1971). Inability of the forest tent caterpillar, *Malacosoma disstria*, to acquire resistance to viral infection. *J. Invertebr. Pathol.* **17**, 136—137.

Reeks, W. A., and Smith, C. C. (1956). The satin moth, *Stilpnotia salicis* (L.), in the Maritime Provinces and observations on its control by parasites and spraying. *Can. Entomol.* **88**, 565—579.

Roberts, D. W., and Yendol, W. G. (1971). Use of fungi for microbial control of insects. *In* "Microbial Control of Insects and Mites" (H. D. Burges and N. W. Hussey, eds.), pp. 125—149. Academic Press, New York.

Rollinson, W. D., Lewis, F. B., and Waters, W. E. (1965). The successful use of a nuclear-polyhedrosis virus against the gypsy moth. *J. Invertebr. Pathol.* **7**, 515—517.

Rupérez, A. (1964). Epoque optima pour la proliferation de la polyhedrose dans les chenilles de *Lymantria dispar* (L.) (Lep., Lymantriidae). *Entomophaga Mem. Hors. Ser.* **2**, 514—520.

Rupérez, A. (1967). Utilisation de *Bacillus thuringiensis* dans la lutte contre *Lymantria dispar* en Espagne. *Proc. Int. Colloq. Insect Pathol. Microbial Control, Wageningen, 1966*, pp. 266–274.

Sankaran, T. (1972). *Commonw. Inst. Biol. Control Tech. Commun.* **6**.

Sapiro, V. A., and Malyseva, M. S. (1970). Zur Begründung von methoden integrierter Forstschutzmassnahmen gegen laub- und nadelfressende Schädlinge. *Biol. Bek. Meth. von Forstschädl.* 51–58.

Shigo, A. L. (1967). Successions of organisms in discoloration and decay of wood. *Int. Rev. Forest. Res.* **2**, 237–299.

Smith, C. C. (1950). Notes on the European winter moth in Nova Scotia. *Can. Dep. Agr. Sci. Serv. Forest Insect Invest. Bimonth. Progr. Rept.* **6**(2), 1.

Speare, A. T., and Colley, R. H. (1912). "The Artificial Use of the Brown-tail Fungus in Massachusetts," 31 pp. Wright & Potter, Boston, Massachusetts.

Stairs, G. R. (1964). Dissemination of nuclear polyhedrosis virus against the forest tent caterpillar, *Malacosoma disstria* (Hübner) (Lepidoptera: Lasiocampidae). *Can. Entomol.* **96**, 1017–1020.

Stairs, G. R. (1965). Artificial initiation of virus epizootics in forest tent caterpillar populations. *Can. Entomol.* **97**, 1059–1062.

Stairs, G. R. (1971). Use of viruses for microbial control of insects. *In* "Microbial Control of Insects and Mites" (H. D. Burges and N. W. Hussey, eds.), pp. 97–124. Academic Press, New York.

Stairs, G. R. (1972). Pathogenic microorganisms in the regulation of forest insect populations. *Annu. Rev. Entomol.* **17**, 355–372.

Stelzer, M. J. (1965). Susceptibility of the Great Basin tent caterpillar, *Malacosoma fragile* (Stretch), to a nuclear-polyhedrosis virus and *Bacillus thuringiensis* Berliner. *J. Invertebr. Pathol.* **7**, 122–125.

Stelzer, M. J. (1967). Control of a tent caterpillar, *Malacosoma fragile incurva*, with an aerial application of a nuclear-polyhedrosis virus and *Bacillus thuringiensis*. *J. Econ. Entomol.* **60**, 38–41.

Thompson, H. E. (1962). Effect of delayed spraying on cankerworm control. *J. Econ. Entomol.* **55**, 558–559.

Tooke, F. G. C. (1955). The eucalyptus snout beetle, *Gonipterus scutellatus* Gyll., a study of its ecology and control by biological means. *Entomol. Mem. Dep. Agr. S. Afr.* **3**, 282 pp.

True, R. P., Barnett, H. L., Dorsey, C. K., and Leach, J. G. (1960). Oak wilt in West Virginia. *W. Va. Univ. Agr. Exp. Sta. Bull.* **448T**, 119 pp.

Warren, L. O., and Tadić, M. (1970). The fall webworm, *Hyphantria cunea* (Drury). *Arkansas Agr. Exp. Sta. Bull.* **759**, 106 pp.

Weiser, J. (1963). Sporozoan infections. *In* "Insect Pathology—An Advanced Treatise" (E. A. Steinhaus, ed.), Vol. 2, pp. 291–334. Academic Press, New York.

Weiser, J., and Veber, J. (1957). Die Mikrosporidie *Thelohania hyphantriae* Weiser des weissen Bärenspinners und anderer Mitglieder seiner Biozönose. *Z. Angew. Entomol.* **40**, 55–70.

Wise, K. A. J. (1953). Host plants of *Lithocolletis messaniella* Zeller (Lepidoptera: Gracillariidae) in New Zealand. *N.Z. J. Sci. Technol.* **A35**, 172–174.

Wolcott, G. N. (1953). Biological control of the pustule scale in Puerto Rico. *Univ. Puerto Rico J. Agr.* **37**, 228–233.

Wolcott, G. N. (1960). Efficiency of lady beetles (Coccinellidae: Coleoptera) in insect control. *Univ. Puerto Rico J. Agr.* **44**, 166–172.

Wylie, H. G. (1960). Insect parasites of the winter moth, *Operophtera brumata* (L.) (Lepidoptera: Geometridae), in western Europe. *Entomophaga* **5**, 111–129.

Yendol, W. G., Hamlen, R. A., and Lewis, F. B. (1973). Evaluation of *Bacillus thuringiensis* for gypsy moth suppression. *J. Econ. Entomol.* **66**, 183–186.

14

BIOLOGICAL CONTROL OF PESTS OF TEMPERATE FRUITS AND NUTS

A. W. MacPhee, L. E. Caltagirone, M. van de Vrie, and Elsie Collyer

I. INTRODUCTION

The world production of pome fruits, stone fruits, and tree nuts is very large. The apple crop alone is over 10 million metric tons annually. Fruit and nut crops are all relatively intensively cultivated with the aid of many man-imposed factors such as fertilizers, sprays, associated crops, plant spacing, pruning, and harvesting. Accordingly, the pest population dynamics and the studies of them are in an environment greatly modified from a natural one in which the species and their interrelationships evolved. The examples of biological control, which include control of pests by both native and introduced parasites, predators, and pathogens come from studies of natural processes in this modified environment.

In order to evaluate the impact of biological controls on arthropod pests of temperate fruits and nuts, it is of value to consider pests as falling into three categories: (1) those species which occur for long periods at endemic levels and cause no significant damage; (2) those which cause damage and whose dynamics on preliminary examination indicate they are affected by natural enemies; (3) those which cause damage and for which preliminary studies reveal little evidence of natural enemies.

There has been little practical need for ecological research on the first group. The third group is fortunately small and the failure to find any encouraging indications in exploratory studies has often discouraged further work. The second group, where exploratory work or general observation reveals promising avenues of approach for useful biological control, has received intensive study. It is the results of these studies which are summarized in this chapter.

The difficulty of interpreting the significance of these studies is due in large part to the extremely applied approach which has motivated the investigations. When the practical benefits of further study are not indicated, a study is usually discontinued whether it is a success or a failure. For example, an introduced parasite once established will either succeed or fail. Further study may not be seen to directly affect the value of the results and the extent of the benefits have often not been evaluated unless they have been extremely obvious.

In this chapter only certain pests and certain of the more important fruits and nuts of temperate areas are treated. Space does not permit a more complete account. The examples are not restricted entirely to those where biological control has been successful. The pest species discussed and the natural enemies represent a wide range taxonomically, biologically, and ecologically, and in the nature of damage caused by the pests. A number of major pests which have received some biological control attention, some of which were partial successes, are not discussed in detail. (See Appendix.)

The San Jose scale, *Aspidiotus perniciosus* (Comstock), of Asian origin threatened the deciduous fruit industry with ruin when it arrived in California and, as well, has become a serious pest in many parts of the world. The introduction of *Prospaltella perniciosi* Tower from Georgia, together with action of a complex of other resident parasites and predators, considerably alleviated the situation. For some 20 years, i.e., until the broad-spectrum organosynthetic insecticides began to be widely used, San Jose scale was of only minor occurrence, but since has again become more serious. The vine scale, *Eulecanium persicae* (F.), in western Australia (Wilson, 1960), the apple leafhopper, *Typhlocyba frogatti* Baker, in Tasmania (Wilson, 1960), the apple leaf curling midge, *Dasyneura mali* Kieffer, and pear leaf midge, *D. pyri* (Bouché), in New Zealand (Todd, 1959; Miller *et al.*, 1936), the fig scale, *Lepidosaphes ficus* (Sign.), in California (Clausen, 1956), the Japanese beetle, *Popillia japonica* Newm., in the eastern United States (Clausen, 1956; Hall, 1964), and the winter moth, *Operophtera brumata* L., in eastern Canada (Embree, 1971; MacPhee, 1967) are examples for which partial to complete biological control has been claimed.

There have also been some attempts to achieve biological control of many other pests of crops in the United States alone. These include the pecan nut casebearer, *Acrobasis caryae* Grote (Clausen, 1956), the peach twig borer, *Anarsia lineatella* Zell. (Clausen, 1956), several species of orangeworms (Clausen, 1956; Caltagirone *et al.*, 1964), cherry fruit flies, *Rhagoletis* spp. (Clausen, 1956; Messenger, 1970), the walnut husk fly, *R. completa* Cress. (Clausen, 1956), Kuno scale, *Lecanium kunoensis* (Kuw.) (Clausen, 1956), and white peach scale, *Pseudaulacaspis pentagona* (Targ.). These examples have not shown any significant success but efforts for some of them have been minimal or are in an early stage of development.

II. APPLE PESTS

A. Tetranychid Mites

The most important species of tetranychid mites on apple are *Panonychus ulmi* (Koch), *Tetranychus urticae* Koch, *T. mcdanieli* McGregor, *T. viennensis* Zacher, and *Bryobia rubrioculus* (Sheuten). These pests, in common with other phytophagous mites, are almost universally subject to predation, and less so to disease epidemics. They are not known to be attacked by insect parasites (parasitoids). Unfortunately, predation has been difficult to evaluate as a regulating factor in the control of pests and thus it is necessary to examine and interpret large amounts of rather circumstantial evidence. Nevertheless, the consensus of workers in widely scattered areas (e.g., Japan, New Zealand, Canada, United States, and Europe) is that except for the adverse effects of modern pesticides on their natural enemies, spider mites would not be major pests of apples (and of many other crops).

1. European Red Mite

The long-term and relatively detailed studies on biological control of *Panonychus ulmi* in Nova Scotia apple orchards have provided an opportunity for evaluation of the nature of the control in that area. Gilliatt (1935b) made many observations on the predatory activity of a number of species on *P. ulmi*, and Lord (1949) reported on extensive studies of the effect of predators and of a number of spray programs on the predators. Throughout these studies and up to the present time no one has reported outbreaks of *P. ulmi* on neglected trees. Outbreaks occurred most often in orchards using sulfur fungicides for apple scab control in the 1930's. Where DDT was used for codling moth control in the late 1940's and 1950's, predators decreased and mites increased to outbreak numbers. Lord and Stewart (1961) found that the influence of natural enemies dominated mite regulation and that tree nutrition in their study did not change the eventual mite density.

A 13-year study in an integrated control orchard (Patterson, 1966) showed that *P. ulmi* densities decreased rapidly and remained low, to the end of the study, in the presence of modest densities of a number of predaceous species. Sanford and Lord (1962) found that if all predators except the mirid *Hyaliodes harti* Knight were removed from apple trees, *P. ulmi* would still be controlled but larger population fluctuations, with longer periods, would be experienced.

The species complex of predators varies both in time and place in Nova Scotia orchards. Although biological control can be confidently predicted, under prescribed conditions, the kinds of predaceous species involved cannot be predicted with similar confidence. The predaceous complex will, however, include phytoseiid mites, thrips, and mirids (Lord, 1949; Sanford and Herbert, 1970).

The similar situation of biotic control of *P. ulmi* seems to occur in much of North America. Parent (1967) reported *P. ulmi* as a major pest of apple orchards in the Province of Quebec but where insecticides are not applied it does not attain a high density. In orchards withdrawn from insecticide programs, predators increased and European red mites decreased. Cutright (1963) reported that in unsprayed apple orchards in Ohio the low numbers of mites are due to biological controls which cannot operate under the conditions imposed by the use of modern sprays for insects and diseases. Clancy and Pollard (1952) found that in abandoned orchards in Virginia there apparently was a favorable balance between phytophagous mites and predators. The coccinellid *Stethorus* sp., predaceous thrips, and phytoseiids, were the most important predators. In Pennsylvania an integrated control program for apple pests has been developed in which *P. ulmi* is effectively controlled by *Stethorus punctum* (LeConte) (Asquith and Colburn, 1971).

Panonychus ulmi is the most important pest mite of apple in New Zealand, Tasmania, and Japan. In New Zealand, phytoseiid mites, of which *Typhlodromus pyri* (Sheuten) is the most abundant species, are an important group as they are able to maintain *P. ulmi* populations at a low level. *Stethorus bifidus* Kapur, *Sejanus albisignata* (Knight), and *Agistemus longisetus* González are also of some importance (Collyer, 1964a). In Tasmania, the coccinellid *Stethorus vagans* (Blackburn) is considered more important than phytoseiid mites. In Japan (Mori, 1967), the anthocorids *Orius* sp. and *Anthocorus* sp., the rove beetle *Oligota* sp., and a number of Neuroptera are among the useful predators of this mite.

In Europe a number of workers (Collyer, 1958, 1964b, c; Dosse, 1960, 1962; van de Vrie and Kropczynska, 1965; van de Vrie and Boersma, 1970; Gratwick, 1965) have provided strong evidence that predators can keep *P. ulmi* in check. They also provide evidence that agricultural pesticides have a marked effect on predators. Again, apparently this mite would not be a problem on apple in the absence of harmful sprays. In England a mirid, *Blepharidopterus angulatus* (Fallen), and a phytoseiid mite, *T. pyri*, are considered particularly important (Collyer, 1952, 1960). Having but one generation a year, *B. angulatus* moderately controlled *P. ulmi* but allowed wide density fluctuations. Collyer and Kirby (1959) presented evidence that *T. pyri*, with its low food requirements and its many generations per year, may be a better regulator of such a prey species.

2. Two-Spotted Spider Mite

The two-spotted spider mite, *Tetranychus urticae* Koch, is in some ways potentially more destructive than *P. ulmi*, having a wider range of hosts. However, it appears to be more subject to biological regulation. In Nova Scotia this mite has rarely been a problem in apple. A few isolated outbreaks occurred in the 1950's where DDT was used for codling moth control, followed by a parathion application for European red mite or aphid control. Predator pressure appears to be particularly effective on this species and biological control is reestablished rapidly. In Quebec province, Parent (1967) found *T. urticae* to be more easily suppressed by predators than *P. ulmi*. In California, *T. urticae* on peaches is controlled by *Metaseiulus occidentalis* (Nesbitt) when insecticides destructive to this predaceous mite are kept to a minimum and if a continuous alternate food supply, such as an eriophyid mite, is available (Hoyt and Caltagirone, 1971). *Tetranychus urticae* as well as *T. mcdanieli* and *P. ulmi* were all considered to be secondary pests in western North America and of little or no importance prior to the widespread use of chlorinated hydrocarbons and organophosphorus and carbamate compounds to control primary pests.

Tetranychus urticae is the principal mite pest of apple in the mainland states of Australia. It is preyed on by *Stethorus vagans* which appears effective even at low densities. In New Zealand, *T. urticae* is an occasional pest of apple and is preyed on by *Stethorus bifidus*, *Anthrocnodax* sp., and phytoseiid mites (Collyer, 1964a). It is also a minor pest on apple in Japan and predators there include *Amblyseius longispinosus* (Evans) and *A. tsugawai* Ehara (Mori, 1967).

It therefore appears that *T. urticae*, on apple, is controlled by predation over all or most of the world except where predators are interfered with by harmful sprays, the level of control being modified by the presence or absence of an alternative food supply for its predatory enemies.

3. McDaniel Spider Mite

Tetranychus mcdanieli is a serious pest of apple in western North America, particularly Washington, Oregon, and British Columbia. It is considered to be a secondary pest and was of little or no economic importance before the use of broad-spectrum pesticides. It is adequately controlled by *M. occidentalis* when a suitable pesticide management program is employed that allows this predator to function naturally (Hoyt, 1969; Hoyt and Caltagirone, 1971).

B. Apple Mealybug

The apple mealybug, *Phenacoccus aceris* (Signoret), is native to Asia and Europe where it is a pest on apple, pear, plum, cherry, small fruits, and many other food plants (Gilliatt, 1935a). It has one generation per year but because of its high fecundity can increase quickly and cause economic injury. Heavy populations cause fruit discoloration, reduced tree vigor, and roughen the fruit surface.

Rau (1942) stated that apple mealybug was probably present in the New England States as early as 1910. This species was first recorded in Canada in British Columbia in 1927 and later in Nova Scotia in 1932 (Gilliatt, 1935a). Two outbreaks occurred in Nova Scotia, one in 1934–1938 and the other in 1964–1965, but they quickly subsided. These widely spaced, large fluctuations suggest that effective biological control agents were causative factors in maintaining endemic levels during the intervening periods.

Gilliatt (1939) investigated *P. aceris* in Nova Scotia and found the parasite *Allotropa utilis* Muesebeck attacking it in large numbers. His observations suggest that this parasite was the main controlling agent of mealybug and was responsible for suppressing the epidemic (Marshall and Pickett, 1944). This conclusion was supported by Chachoria (1967) who found that 60% mortality was caused by *A. utilis* in mealybug populations entering hibernation in two orchards. There was also high mortality in earlier stages due to dispersion and predation. Outbreaks of *P. aceris* appear to be due to the failure of the parasite *A. utilis*, but the cause of the failure is unknown.

The early success of *A. utilis* in Nova Scotia led to its introduction into British Columbia in the 1930's where it successfully controlled the pest by the middle 1940's (Wishart, 1946).

No relationship has been determined in the Canadian studies between biological control of *P. aceris* and the type of spray programs followed, probably because the parasite overwinters as a mature larva in the mummy of the host situated in the crevices and cracks on the tree trunk. It does not emerge until after the spray season in July and is thus probably protected.

Panis (1969) found a complex of four species of natural enemies attacking *P. aceris* in Europe. Two parasites, *Aphycus apicalis* (Dalman) and *Leptomastidea bifasciata* (Mayr), were considered the most important.

C. Woolly Apple Aphid

The woolly apple aphid, *Eriosoma lanigerum* Hausmann, is found in nearly all apple growing areas of the world. deFluiter (1931) described its biology and ecology in the Netherlands. Massee (1954) in England, Georgala (1953) in South Africa, Dumbleton and Jeffrey (1938) in New Zealand, and Hukusima (1960) in Japan discussed its bionomics and control. It has been a potentially serious pest in Nova Scotia during, at least, most of this century, with local temporary outbreaks occurring at damaging levels. It is not considered an important pest in Europe as opposed to the situation in America, South Africa, and Australia.

Eriosoma lanigerum spends the winter on the roots, trunk, and branches in wounds, cankers, and other shelters. All stages are present in early winter but with the exception of the younger stages most die before spring. Activity begins in March to May in the northern hemisphere at which time the production of "wool" is very conspicuous. After the fourth molt the aphids become adult and start producing living young; many generations may develop during the summer.

This aphid is considered by several authors as native to North America, although Thomas (1879, cited by LeRoux, 1971) believed it native to Europe from where it reached eastern North America. Eventually it reached the Pacific Coast of North America but without its specific parasite, thus becoming a major pest of apple. Petit in 1897 reported it in Michigan and recommended an attempt at eradication, noting that it had long been known in eastern North America as being destructive. Because it attacks the roots as well as aerial parts of the plant, the use of woolly apple aphid-resistant apple rootstocks has given some control in many countries, particularly New Zealand and Australia.

This aphid is attacked by an apparently monophagous parasite, *Aphelinus mali* Haldeman (Aphelinidae), which has been introduced into apple growing areas of the world. This parasite was found originally in areas of the eastern United States and Canada. *Aphelinus mali* was introduced into Europe from North America in the 1920's, into South Africa in 1920, into other parts of North America between 1921 and 1939, and into Uruguay, Chile, and Perú in 1921–1922 (Clausen, 1936; Gonzalez and Rojas, 1966; Graf and Cortes, 1939; Isla, 1959; Wille, 1958), into Mexico in 1953 (Coronado, 1955), into New Zealand in 1921 and 1922 (Tillyard, 1921, 1922), into Western Australia and Queensland in 1923, and subsequently into Tasmania, Japan, and India (Hukusima, 1960). By 1960 it had been introduced into 25 countries (DeBach, 1964).

In all cases, *A. mali* was successfully established and either provided complete economic control or appreciably reduced the need for treatments. For example, in Western Australia, the parasite was effective within four years and in Queensland pesticide applications for this pest were reduced from 6 or 7 to 1 or 2. In many cases the parasite also eliminated the need for use of resistant rootstocks. In Tasmania, the aphid was reduced from a major to a minor pest. In Europe, *A. mali*, although established in most regions, has not been a complete success. In the Netherlands, Evenhuis (1958) studied the ecology of this parasite and its host in detail, as did Bonnemaison (1965) in France. Evenhuis compared the capacity of increase of the woolly apple aphid and *A. mali* and concluded that the aphid has greater reproductive capacity. The parasite also has difficulty finding some host individuals in more inaccessible sites. In some areas, winter mortality of the aphid is very high, while more of the parasites survive. Percentage parasitism may be very high in the first generation, then the aphid may, in some situations, outdistance the parasite and reach fairly high levels. In midsummer, aphid fecundity may decrease and the parasite may overtake it again.

These observations appear to explain the occasional fluctuation above economic levels even in situations where no artificial deterrence to parasite development, such as sprays, are present. In Nova Scotia orchards under an integrated spray program this pest is rarely serious and it is unusual for chemical controls to be required.

Predators do not seem to have an important role in the control of *E. lanigerum*. *Chrysopa septempunctata* Wesmael (Withycombe, 1923, in England; Marchal, 1929, in France), *Exochomus quadripustulatus* (L.) (Evenhuis, 1958, in the Netherlands; Clausen, 1956, in the United States), and various Syrphidae are among the frequently mentioned predators. *Coccinella undecimpunctata* L., *Leis conformis* Bois, and *Adalia bipunctata* L. appear to have minor suppressing effects in New Zealand.

The widespread use of broad-spectrum chemicals, such as DDT against codling moth and other pests, has interfered with the development of parasites, particularly early in the season (Newcomer *et al.*, 1946). In South Africa, pesticide applications reduce the effectiveness of *A. mali* and have caused aphid increases to damaging densities requiring additional chemical measures (Georgala, 1953). Similar situations prevail in England (Moreton, 1969), Germany (Kotte, 1958), Italy (Castellari, 1967), France (Chaboussou, 1961), Bulgaria (Balevski and Vasey, 1962), and Japan (Hukusima, 1960). Integrated control programs have allowed *A. mali* to recover. Some spray programs used generally in apple pest control, in addition to killing *A. mali*, also control *E. lanigerum* so that the natural control disturbance may not become an economic factor (Holdsworth, 1970).

In Tadzhikestan, U.S.S.R., *A. mali* is capable of suppressing the woolly apple aphid in orchards in the Gisar Valley when chemical treatment is not used and where nectar-producing plants are grown as a ground cover (Boldyreva, 1970). The nectar provides food for the parasite. Where chemical control is used, *A mali* is ineffective but a reservoir is maintained on underground parts of the trees.

The extent of woolly apple aphid as a worldwide pest of apple and the effectiveness of *A. mali* as a control agent make this one of the most successful biological control programs of a tree fruit pest. It is also one of the least difficult to manage since *A. mali* establishment has proved to be relatively easy. The absence of significant references to hyperparasites of *A. mali* seems to indicate the importance of this factor to the quick response and persistent effectiveness of the parasite.

D. Oyster Shell (Mussel) Scale

The oyster shell scale, *Lepidosaphes ulmi* (L.), has been a serious pest of tree crops in many temperate areas of the world. The scale has one generation per year in most of its range, but two have been reported in the warmest latitudes. There are many references to the destructive nature of *L. ulmi*, the earliest in North America probably being by Enoch Perley who is quoted by Griswold (1925) as describing the problem with this pest in 1794. Low scale densities are relatively harmless but high densities in which a large percentage of the bark surface is covered will suppress tree growth and eventually kill the lower branches.

The results of biological control of *L. ulmi* have been outstanding and are well documented. The earliest records of parasitism of *L. ulmi* were by Fitch, in 1856, and Le Baron, in 1870, who described the parasite *Aphytis* (*Chalcis*) *mytilaspidis* (LeBaron) (Griswold, 1925). The predatory activity of the mite *Hemisarcoptes malus* Shimer was mentioned by Shimer in 1868 (Samarasinghe and LeRoux, 1966). Its importance as a control agent was documented by Tothill (1919) for the year 1916 for 37 locations across Canada and 1 in New York State. His report provided circumstantial evidence that *H. malus* was causing high mortality and possibly controlling the scale in eastern Canada. He did not find it in British Columbia, and in 1917 (Lord, 1971) *H. malus* was introduced there from New Brunswick. It became established and was eventually transferred to other locations where it has been effective in suppressing the scale (LeRoux, 1971).

A sporadic apple pest in Nova Scotia prior to 1930, *L. ulmi* usually occurred on a few isolated trees. In the 1930's a serious outbreak occurred. Lord (1947) discovered and documented the fact that sulfur fungicides in use at that time were largely responsible for the outbreak. Flotation sulfur treatments killed *H. malus* and suppressed *A. mytilaspidis*, rendering both species ineffective. Lime sulfur acted similarly on these scale enemies but its direct toxicity gradually controlled the scale without the help of the natural enemies. In an experimental orchard, two ferbam spray programs, one started in 1943, the other in 1944, were found innocuous to both the predatory mite and the parasite and *L. ulmi* was then completely controlled by these agents by 1946. Similar results were reported by Pickett (1965) who measured the mortality of scales caused by *A. mytilaspidis* and *H. malus*.

In the years following these findings, scale-infested orchards in Nova Scotia were changed to spray programs using the fungicide ferbam to replace the sulfur fungicides and within 1 to 2 years the scale ceased to be a problem. In subsequent years other fungicides and pesticides, also harmless to these beneficial species, have been used in the program. No further need for chemical control or damage from this scale has been experienced.

In New Brunswick the failure of *A. mytilaspidis* to survive the low temperatures (Lord and MacPhee, 1953) resulted in greater fluctuations in scale density in some areas because *H. malus* alone then provided the control.

In Quebec, Samarasinghe and LeRoux (1966) quantitatively confirmed the effectiveness of *H. malus* and *A. mytilaspidis* as regulating factors for *L. ulmi* at low and medium densities on apple. Other parasites do attack *L. ulmi*; however, they cause very little mortality. There is some feeding by insect predators. *Hemisarcoptes malus* and *A.*

mytilaspidis appear to be the only effective regulating agents which contain the pest below economic levels in eastern Canada.

E. Lecanium Scale

Lecanium coryli is one of a complex of species of European and North American *Lecanium*, and its specific identity is uncertain. A lecanium scale, native to Europe, reached British Columbia on nursery stock in 1903. It was free of at least some of its natural enemies (Graham and Prebble, 1953). By 1920 it was causing damage and the parasite *Blastothrix sericea* (Dalman) was introduced from England. The parasite quickly became established and its spread was assisted through further distributions. The scale populations remained low for the next 10 years in the lower Fraser River area. Scale infestations were reported on Vancouver Island in 1941 by Graham and Prebble (1953). From 1941 to 1945 they studied the parasitism by *B. sericea* on scales on many species of host trees in British Columbia. The scale caused some damage during this time and there was high parasitism but often equal mortality resulted from winter conditions and underdevelopment. Scales parasitized in the fall by *B. sericea* are killed in early spring when the parasite matures. However, the next generation of parasites attack the nearly mature larvae and do not usually kill the scales until a large complement of the eggs are laid. The eggs survive and hatch normally.

The first reported damage from *L. coryli* in Nova Scotia in the 1930's led to the introduction of *B. sericea* from England and from British Columbia in 1934 (McLeod *et al.*, 1962).

Lecanium coryli occurs on many species of trees in Nova Scotia and on rare occasions is destructive only to isolated trees, notably ash and oak. It occurs in low numbers almost universally on apple trees and has never been found destructive in unsprayed orchards. In certain situations it may become a harmful pest capable of destroying fruit, foliage, and eventually main branches of apple. Such outbreaks have occurred in a small number of commercial orchards over the past 20 years and it gradually became clear that they were associated with certain spray practices, specifically ones with a history of some years of use of a broad-spectrum pesticide. Examination of orchards usually revealed the presence of the parasites *B. sericea* and a *Coccophagus* sp. Experimental work in Nova Scotia (MacPhee and MacLellan, 1971) showed that DDT sprays could precipitate an increase of this scale to outbreak numbers over a period of years. An experimental orchard initially lightly infested was studied for 5 years. An application of DDT was made in the spring in each of the first 2 years. There was no increase in scales the first year but a fivefold increase occurred the second year and moderately destructive levels were attained the third year.

In the two following years, *B. sericea* greatly increased and a number of species of predators increased in number, killing scales and reducing fecundity. *Coccophagus* sp. continued to kill about 20% of the scales. The pest then became light in the fourth year and very scarce in the fifth. These events have been observed in a number of commercial orchards where scale densities remain regulated well below economic levels.

F. Codling Moth

The codling moth, *Laspeyresia pomonella* (L.), was known in central Europe in the early 17th century. The date of its arrival in North America is not known but it was present in the early 19th century and probably earlier in South America. It was reported in Australia and New Zealand in the latter half of the 19th century and in South Africa about 1885. It was known in California in 1874, but not until about 1912 was it established in British Columbia. It is still one of the more important pests of apple and pear in most of the pome fruit areas of the world.

The control of codling moth has been dominated for the past 25 years by broad-spectrum insecticides and the role of natural enemies has received very little attention. Chemical control soon led to large fluctuations in various apple pest populations, to minor or innocuous species becoming major pests, to the development of resistance in strains of pests, and to cumulative contamination of the environment. These problems have led to recent interest in pest management or integrated control, and the resulting more stable and economic pest control possibilities. Integrated control commonly refers to the use of chemicals in association with natural enemies and other compatible measures so that the best of each measure may be utilized. Natural enemies, i.e., predators, parasites, and pathogens, if left undisturbed by noxious chemicals, generally increase with increasing host or prey numbers but they seldom exert adequate control of codling moth by themselves. The use of selective insecticides, or selective low dosages of broad spectrum pesticides, is the most reliable means at present of achieving good codling moth control in Nova Scotia. The codling moth population is thereby suppressed by the chemical and its natural enemies are then more capable of containing it, while at the same time natural enemies of other species are not made ineffective.

Although the codling moth has a large number of natural enemies which attack all stages in its life cycle, attempts at controlling it using these natural enemies have been regarded as failures (Clausen, 1956; LeRoux, 1971; MacLellan, 1971b; Turnbull and Chant, 1961; Wille, 1958). However, serious biological control work against this moth has been scanty and surprisingly few natural enemies have been introduced from one country to another. A few parasites and predators that were imported against other pests have adopted the codling moth as another host or prey. Attempts to discover effective natural enemies in any part of the world, and especially in Central Asia, apparently the native home of codling moth, and import them into other countries, have been very few.

In California, up to 1974 only two parasites had been purposely imported to control codling moth. They were the ichneumonid *Liotryphon caudatus* (Ratz.) (= *Apistephialtes caudata*), imported from Spain 1904—1905, and again from France in 1935—1936, and the bethylid *Parasierola* (as *Perisierola*) *emigrata* (Roh.) from Hawaii in 1947.

Neither species became established (Clausen, 1956). Parasites from Central Asia are now being introduced.

In Western Canada the ichneumonids *Liotryphon caudatus* and *Cryptus sexmaculatus* (Grav.) were introduced from France in 1939–1940 (LeRoux, 1971). They failed to become established.

In Peru the braconid *Ascogaster quadridentata* Wesm. (= *A. carpocapsae* Vier.) was introduced in 1937 (Wille, 1958). It has never been recovered.

Since economic control by natural enemies alone has seldom occurred, chemical control has been emphasized for many years. Investigations on natural enemies have concentrated chiefly on parasites, particularly in Europe where egg parasitism by *Trichogramma* spp. and larval parasitism by a number of species of ichneumonids, braconids, chalcids, and tachinids contribute to a general reduction of codling moth populations. Also the fungus *Beauveria bassiana* (Balsamo) Vuillemin caused a decrease in the number of wintering larvae in Austria (Russ, 1964).

Geier (1961) reported a dearth of natural enemies in Australia and Wood (1965) observed 14% egg parasitism by *Trichogramma* sp. and small amounts of predation on eggs and young larvae on unsprayed trees in New Zealand. In pre-DDT days, Nel (1942) in South Africa reported that *Trichogramma* sp. increased in percentage parasitism as the season advanced and parasitized eggs in the 80% range. Larval and pupal parasites included *Ascogaster quadridentata* Wesmael, *Cremastus* sp., *Pimpla heliophila* Cam., *Cryptus* sp., and a few ichneumonids of uncertain status. Only two predators were observed, the ant, *Iridomyrmex humilis* Mayr, and the reduviid, *Coranus papillosus* Thnb. In the second and third years of one experiment in pear orchards in which the effectiveness of natural enemies was measured, infestation by codling moth was no greater than when spraying was still practiced.

Natural enemies have not been very effective in the pome fruit areas of North America except in Quebec and Nova Scotia where they are supplemented when necessary by selective chemical controls. In Quebec, LeRoux (1960) reported that arthropod predators and parasite species numbered 83 and 39 species, respectively, and there were 23 species of insectivorous birds, many of which fed upon codling moth. All contributed to the success of an integrated control program which was as effective as a chemical control program in a 5-year comparison.

The most success with natural enemies of the codling moth has occurred in Nova Scotia where the insect has a single generation and where the botanical insecticide ryania has provided the necessary selectivity while reducing high populations of the pest. MacLellan (1960, 1962, 1963, 1971a) reported a sequence of mortality agents attacking all stages except the adult. The eggs are attacked by *Trichogramma* sp. and the egg-larval parasite *A. quadridentata*. Egg-larval predators include the predaceous mites *Anystis agilis* Banks and *Atomus* sp., the thrips *Haplothrips faurei* Hood and *Leptothrips mali* (Fitch), the anthocorid *Orius insidiosus* (Say), the mirids *Blepharidopterous angulatus* (Fallen), *Deraeocoris fasciolus* Knight, *Deraeocoris nebulosus* (Uhler), *Diaphnocoris* spp., *Hyaliodes harti* Knight, *Phytocoris* spp., *Pilophorus perplexus* Douglas and Scott, *Plagiognathus obscurus* Uhler, and several species of coccinellids, pentatomids, nabids, clerids, and chrysopids. Wintering larvae are attacked by two species of woodpecker, *Dendrocopus pubescens medianus* (Swainson) and *Dendrocopus villosus villosus* (L.), an ostomatid beetle, *Tenebroides corticalis* Melsh, and at least six species of fungi of which *Beauveria* is the most common (Jaques and MacLellan, 1965), and are killed as well by winter temperatures of −27°C or below (MacPhee, 1964).

In Nova Scotia there are two hazardous periods for the codling moth, from egg deposition until the larva enters the fruit, and during the overwintering phase. A decrease in normal effectiveness of natural enemies in either period will cause a significant increase in codling moth within one generation. Thus, although natural enemies are capable of containing the pest, their failure in one season may require the use of a selective chemical the following year. Integrated control of codling moth has worked very well in Nova Scotia.

III. PEAR PESTS

A. Pear Psylla

The pear psylla, *Psylla pyricola* (Foerster), was accidentally introduced into eastern United States from Europe about 1832 (Slingerland, 1892). It is now found in nearly all areas where pears are grown in North America. It has from two to four or more generations per year. Damage is caused by a black fungus growing on psylla honeydew, by the feeding of the psylla on foliage and fruit, and by transmission of the viruses which cause pear decline and pear leaf curl (Griggs *et al.*, 1968; Jensen *et al.*, 1964).

In North America a number of parasites and predators attack pear psylla (Jensen, 1957; Madsen *et al.*, 1963; McMullen, 1966, 1971; McMullen and Jong, 1967; Nickel *et al.*, 1965; Westigard *et al.*, 1968; Rasmy and MacPhee, 1970). Their control efficiency varies according to circumstances. Apparently no known natural enemy suppresses the psylla population to densities low enough to render the pest unimportant as a virus vector but natural enemies are able to reduce psylla populations below economic injury levels in trees that are not susceptible to viruses.

In California, the predator *Anthocoris antevolens* White is the most abundant and effective in the coastal region, while the parasites *Trechnites insidiosus* (Crawford) and *Psyllaephagus* sp. seem to be most common in inland situations.

In 1963 the anthocorid *Anthocoris nemoralis* (Fabricius) was imported into British Columbia from Switzerland and became established. Its effect on psylla populations is not yet known, but it may be of significance as it has shown the capability of displacing native species of *Anthocoris*, and is better adapted than *A. antevolens*, an important native predator, to withstand cold temperatures (McMullen, 1971).

Conditions prevailing in orchards under heavy insecticide treatments applied to control the codling moth are certainly not conductive to biological control of pear psylla. However, there are good possibilities that integrated control programs will be developed for pests of pears if there is minimal use of insecticides to control codling moth. According to Nickel *et al.* (1965) reduced dosages of insecticides adequate to control codling moth do not affect too drastically the predators of pear psylla.

In Nova Scotia, where codling moth is a minor pest of pears, this psyllid commonly remains at low levels but occasionally reaches densities requiring chemical control. There has been a recent (1969 and 1970) instance of some damage from this pest associated with an increase of the second generation. Normally, however, the second generation is decreased in numbers. Rasmy and MacPhee (1970) found a 95–98% reduction from the first to the second generation in an experimental orchard apparently due mainly to predation by coccinellids, anthocorids, and mirids.

General observation and limited measurements therefore indicate that good natural control of pear psylla is likely in at least part of its range wherever natural enemies are not suppressed and vector role is not significant. However, the reason for occasional increases to economically harmful levels even where no harmful materials are used is not known. It would seem that this may be a natural outcome of interplay of relationships among a complexity of species since *P. pyricola* enemies appear to attack and buildup on many prey species on a variety of hosts.

B. Grape Mealybug

The grape mealybug, *Pseudococcus maritimus* complex, is probably native to North America. Its host plants include grape, citrus, walnut, apple, pear, and others. It is effectively controlled by its natural enemies. When DDT was used to control codling moth, natural enemies were decimated and it became an important pest of pears. Hagen *et al.* (1971) reported that pear orchards treated with DDT had up to 75 times more mealybugs than unsprayed control trees, and by harvest, 70% of the fruits were infested, contrasted to 3.3% in unsprayed trees. The green lacewing, *Chrysopa carnea* Stephens, was found to be its most important natural enemy (Doutt and Hagen, 1950). It was controlled in experimental commercial orchards by periodic colonization of *C. carnea* eggs at the rate of 250 eggs per tree. This was possible because the eggs and larvae of *C. carnea* were resistant to the DDT used by the industry.

IV. PEACH PEST

Oriental Fruit Moth

The oriental fruit moth, *Grapholitha molesta* (Busck), an important pest of peaches, is believed to have been introduced into the United States from Japan in 1913 and from the United States to Canada in 1925 (McLeod *et al.*, 1962). The difficulty in controlling this pest with chemicals before the advent of DDT led to extensive efforts to develop an alternative control. The most important parasite of oriental fruit moth is the larval parasite *Macrocentrus ancylivorus* Rohwer, which was present in eastern North America, apparently a native parasite of the strawberry leaf roller, *Ancylis comptana fragariae* (Walsh and Riley), before *G. molesta* arrived.

The discovery of the oriental fruit moth in California in 1942 led to mass production of *M. ancylivorus*. It began in 1943 and continued through 1946; a total of 58,165,000 were colonized (Finney *et al.*, 1947). In spite of the massive effort it was almost impossible to evaluate the effect of the liberated parasites due to the low populations of the moth. There was evidence that eradication may have been accomplished in a number of areas in California. However, in 1954, and since, this moth has been a major pest of peaches in the state.

Macrocentrus ancylivorus proved to be a valuable natural control agent in eastern North America where it provided a measure of control and prevented serious peach losses during a period when no suitable insecticide was available (McLeod *et al.*, 1962; Allen, 1958) and has continued to provide useful supplementary control.

Attempts to control this moth with natural enemies in Argentina have been made since 1936. That year, *M. ancylivorus*, *Ascogaster quadridentata* Wesmael, *Glypta rufiscutellaris* Cresson, and *Bassus diversus* Muesebeck were imported. A subsequent importation of *M. ancylivorus* was made in 1946 (deCrouzel, 1963). *Macrocentrus delicatus* Cresson was, apparently, inadvertently introduced with shipments of *M. ancylivorus*; it is established in the Parana Delta area where it has displaced *M. ancylivorus* (deCrouzel, 1963).

It should be noted that *G. molesta* is not native to either Europe or Japan, yet its natural enemies have been collected in these areas for importation into the United States and other areas in the New World. No exploration for its natural enemies has been made in continental Asia where presumably the oriental fruit moth is indigenous. This is unfortunate. These "made over" natural enemies have not regulated the oriental fruit moth populations to levels below economic injury. The moth is still a major pest of peaches in Japan, Europe, the United States, Canada, Brazil, and Argentina. Perhaps it is time that we look closely in continental Asia.

V. WALNUT PEST

Walnut Aphid

The walnut aphid, *Chromaphis juglandicola* (Kaltenbach), native to the Old World, was accidentally introduced to California prior to 1900 and by 1911 was considered a pest. Native predators (coccinellids and green lacewings) were not able to reduce the density of the aphid (van den Bosch *et al.*, 1970). In 1959 the aphidiid *Trioxys pallidus* (Halliday) was imported from France. It established itself and dispersed rapidly in southern California, but it failed to do so in the rest of the state (van den Bosch *et al.*, 1962; Sluss, 1967).

In 1968, *T. pallidus* from a more comparable environment, Iran, was colonized. This time the parasite became established in northern California and dispersed rapidly. In some of the release sites it quickly and effectively controlled the aphid (van den Bosch *et al.*, 1970), and since has spread throughout most of the infested areas (Chapter 9).

VI. OLIVE PEST

Olive Scale

The olive scale, *Parlatoria oleae* (Colvée) can be a most severe pest of olives in California. It attacks limbs, twigs, and leaves and infests fruits, rendering them unacceptable commercially. The host plants include some 200 species. Among them are olive, many ornamental shrubs, and deciduous fruits of the family Rosaceae. *Parlatoria oleae* has two generations a year. In spring and fall the crawlers settle on limbs, twigs, leaves, and on fruits, when present, where in the latter case even comparatively few individuals can cause considerable economic damage.

This scale became a pest in California in 1934. In 1949 the ectoparasitic aphelinid *Aphytis masculicornis* (Masi) was introduced from Egypt. Subsequently, several natural enemies of olive scale were imported from India, Pakistan, Iran, Iraq, Syria, Lebanon, Israel, Cyprus, Egypt, Greece, and Spain.

Among the imported species there were four morphologically identical but biologically different strains of *A. maculicornis*. Of these, the Persian strain which was imported from Iran and Iraq was the most promising strain, but it could not reliably control the scale in all infested situations by itself. The Persian *A. maculicornis* reaches large numbers in the spring, often parasitizing 90% or more of its host. However, this parasite does not tolerate the hot, dry summer of the Central Valley, so a large portion of the second generation of the scales escapes parasitization, and this can result in economic damage, but does not always do so.

In 1957—1958 the endoparasitic aphelinid *Coccophagoides utilis* Doutt was imported from Pakistan and colonized in the Central Valley. This parasite proved to be capable of improving the control of the olive scale, acting in a complementary fashion to parasitization by the superior *A. maculicornis*. Destruction of the scale by *C. utilis* is not high in either of the two scale generations, reaching only some 30—60% of the population, but since it fills the gap left by the unsatisfactory performance of *A. maculicornis* in summer, parasitization by *C. utilis* is enough that the two species together consistently keep the scale population well below the economic injury level.

Detailed accounts of this successful biological control project have been published by Huffaker *et al.* (1962), Huffaker and Kennett (1966), Kennett *et al.* (1966), and DeBach *et al.* (1971).

VII. SUMMARY

The natural enemies discussed in this chapter include many parasites and predators; pathogens are notably absent as reliable regulating agents of tree fruit pests, although there is some evidence they can cause at least temporary control; more often they are reported as catastrophic and disturbing in their effects.

Phytophagous mites appear to be under excellent biological control over most of the world due to the action of predators; exceptions occur, particularly where the predators are removed or suppressed by spray programs.

Grape mealybug, walnut aphid, and olive scale are controlled by biological agents; the latter two providing excellent examples of the value of introduced parasites.

Apple mealybug and the woolly apple aphid both have numerous enemies but each is controlled in most situations by a specific parasite. These controls can fail for any one of a number of reasons and the populations then rise above economic levels. However, chemical control is required only in a small percentage of cases.

The oyster shell scale and lecanium scale are controlled by natural enemies over most of their ranges whenever the balances are not disturbed. For the former, either a specific parasite or a predator (or both) gives good control; in the latter, a complex of species appears to operate in a complementary way.

The lepidopterous pests, codling moth and oriental fruit moth, have many natural enemies but the natural enemy controls are less certain, more population variation occurs, and economically acceptable levels are often exceeded over a large part of the range of these species.

The biological control of pear psylla and black scale (see Chapter 15) has varied from good to poor and more information on control mechanisms is needed and will probably improve management programs.

Many pest species which attack temperate fruit and nut crops, including the successful biological control of grape leaf skeletonizer on grape in California, have not been reviewed here, some no doubt having equally interesting and informative biological control histories. We cannot claim the review is in any sense exhaustive and it is probably slanted by the prior knowledge and interests of the authors.

REFERENCES

Allen, H. W. (1958). Orchard studies on the effects of organic insecticides on parasitism of the oriental fruit moth. *J. Econ. Entomol.* **51**, 82–87.

Asquith, D., and Colburn, R. (1971). Integrated pest management in Pennsylvania apple orchards. *Bull. Entomol. Soc. Amer.* **17**, 89–91.

Balevski, A., and Vasey, A. (1962). Contribution to the clarification of the causes of the decrease in effectiveness of the parasite of the woolly aphid *Aphelinus mali* Hald. *Izv. Tsent. Nauchnoizsled. Inst. Zash. Rast.* **2**, 143–173. (In Bulga.)

Boldyreva, T. P. (1970). The ecology of *Aphelinus mali* Hal. (Hymenoptera: Aphelinidae) a parasite of the woolly apple aphid in Tadzhibestan. *Entomol. Rev.* **4**, 457–459.

Bonnemaison, L. (1965). Observations ecologiques sur *Aphelinus mali* Hald. parasite du puceron lanigere (*Eriosoma lanigerum* Hausmann). *Ann. Soc. Entomol. Fr.* **1**, [N. S.] 143–176. (In French, Ger. summary.)

Caltagirone, L. E., Shea, K. P., and Finney, G. L. (1964). Parasites to aid control of naval orangeworm. *Calif. Agr.* **18**, 10–12.

Castellari, P. L. (1967). Research on the bionomics and ecology of *Eriosoma lanigerum* Hausm. and its parasite *Aphelinus mali* in Emilia, with particular regard to the secondary effects of chemical control. *Bol. Inst. Entomil. Univ. Bologna* **28**, 177–231. (In Ital., Engl. summary.)

Chaboussou, F. (1961). Action de divers insecticides et notamment de certains produits endotherapiques vis-a-vis d'*Aphelinus mali* Hald. evoluant a l'interieur du puceron lanigere du pommier: *Eriosoma lanigerum* Hausm. *Rev. Pathol. Veg. Entomol. Agr. Fr.* **40**, 17–29. (In French, Engl. summary.)

Chachoria, H. S. (1967). Mortality in apple mealybug, *Phenacoccus aceris* (Homoptera: Coccidae) populations in Nova Scotia. *Can. Entomol.* **99**, 728–730.

Clancy, D. W., and Pollard, H. N. (1952). The effect of DDT on mite and predator populations in apple orchards. *J. Econ. Entomol.* **45**, 108–114.

Clausen, C. P. (1963). Insect parasitism and biological control. *Ann. Entomol. Soc. Amer.* **29**, 201–223.

Clausen, C. P. (1956). Biological control of insects in the continental United States. *U.S. Dep. Agr. Tech. Bull.* **1139**, 151.

Collyer, E. (1952). The biology of some predatory insects and mites associated with the fruit tree red spider mite (*Metatetranychus ulmi* Koch) in southeastern England. I. The biology of *Blepharidopterus angulatus* (Fall.). *J. Hort. Sci.* **27**, 117–129.

Collyer, E. (1958). Some insectary experiments with predaceous mites to determine their effect on the development of *Metatetranychus ulmi* (Koch) populations. *Entomol. Exp. Appl.* **1**, 138–146.

Collyer, E. (1960). The control of phytophagous mites on fruit. *Sci. Rev.* (*Melbourne*) **20**, 33–35.

Collyer, E. (1964a). Phytophagous mites and their predators in New Zealand orchards. *N.Z. J. Agr. Res.* **7**, 551–568.

Collyer, E. (1964b). A summary of expcriments to demonstrate the role of *Typhlodromus pyri* Scheut. in the control of *Panonychus ulmi* (Koch) in England. *Acarologia* **9**, 363–371.

Collyer, E. (1964c). The effect of alternate food supply on the relationship between two *Typhlodromus* species and *Panonychus ulmi* (Koch). *Entomol. Exp. Appl.* **7**, 120–124.

Collyer, E., and Kirby, A. H. M. (1959). Further studies on the influence of fungicide sprays on the balance of phytophagous and predacious mites on apple in southeast England. *J. Hort. Sci.* **34**, 39–50.

Coronado, R. (1955). Control quimico y biologico del pulgon lanigero del manzano. *Agr. Tec. Mex.* **1**, 10–47.

Cutright, C. R. (1963). The European red mite in Ohio. *Ohio Agr. Exp. Sta. Res. Bull.* **953**.

DeBach, P. (1964). Successes, trends and future possibilities. *In* "Biological Control of Insect Pests and Weeds" (P. DeBach, ed.), pp. 673–713. Chapman and Hall, London.

DeBach, P., Rosen, D., and Kennett, C. E. (1971). Biological control of coccids by introduced natural enemies. *In* "Biological Control" (C. B. Huffaker, ed.), pp. 165–194. Plenum, New York.

deCrouzel, I. S. (1963). Sobre el control biologico de *Grapholitha molesta* (Busck) en la Republica Argentina. *Rev. Soc. Entomol. Argent.* **26**, 129–131.

deFluiter, H. J. (1931). De bloedluis *Eriosoma lanigerum* (Hausmann) in Nederland. *Diss. Rijksuniv. Leiden*, 126 pp.

Dosse, G. (1960). Über den einfluss der raubmilbe *Typhlodromus tiliae* Oud. auf die obstbaumspinnmilbe *Metatetranychus ulmi* Koch. *Pflanzenschutzberichte* **24**, 113–137.

Dosse, G. (1962). Die feinde der raubmilben als reduzierender faktor der spinnmilben. *Entomophaga* **7**, 227–236.

Doutt, R. L., and Hagen, K. S. (1950). Biological control measures applied against *Pseudococcus maritimus* on pears. *J. Econ. Entomol.* **43**, 94–96.

Dumbleton, L. J., and Jeffreys, F. J. (1938). The control of the woolly aphids by *Aphelinus mali*. *N.Z. J. Sci. Technol.* **20**, 183–190.

Embree, D. G. (1971). The biological control of the winter moth in eastern Canada by introduced parasites. *In* "Biological Control" (C. B. Huffaker, ed.), pp. 217–226. Plenum, New York.

Evenhuis, H. A. (1958). Ecological investigations on the woolly aphid, *Eriosoma lanigerum* (Hausm.) and its parasite *Aphelinus mali* (Hald.) in the Netherlands. *Diss. Rijksuniv. Groningen*, 103 pp. Also: *Tijdschr. Plantenzieten* **64**, 1–103. [In Dutch, Engl. summary.]

Finney, G. L., Flanders, S. E., and Smith, H. S. (1947). Mass culture of *Macrocentrus ancylivorus* and its host, the potato tuber moth. *Hilgardia* **17**, 437–483.

Geier, P. W. (1961). Numerical regulation of populations of the codling moth, *Cydia pomonella* (L.). *Nature (London)* **190**, 561–562.

Georgala, M. B. (1953). The woolly aphis of apple and its control. *Farming S. Afr.* **28**, 21–27.

Gilliatt, F. C. (1935a). A mealy bug, *Phenacoccus aceris* Signoret, a new apple pest in Nova Scotia. *Can. Entomol.* **67**, 161–164.

Gilliatt, F. C. (1935b). Some predators of the European red mite, *Paratetranychus pilosus* C. & F., in Nova Scotia. *Can. J. Res.* **D13**, 19–38.

Gilliatt, F. C. (1939). The life history of *Allotropa utilis* Mues., a hymenopterous parasite of the orchard mealy bug in Nova Scotia. *Can. Entomol.* **71**, 160–163.

Gonzalez, R. H., and Rojas, S. (1966). Estudio analitico del control biologico de plagas agricolas en Chile. *Agr. Tec. (Santiago de Chile)* **26**, 133–147.

Graf, A., and Cortes, R. (1939). Introduction de hyperparasitos en Chile: resumen de las importaciones hechas y des sus resultados. *Proc. 6th Pac. Sci. Congr.* **4**, 351–357.

Graham, K., and Prebble, M. L. (1953). Studies of lecanium scale, *Eulecanium coryli* (L.) and its parasite, *Blastothrix sericea* (Dalm.), in British Columbia. *Can. Entomol.* **83**, 153–181.

Gratwick, M. (1965). Laboratory studies of the relative toxicities of orchard insecticides to predatory insects. *Rep. E. Malling Res. Sta. 1964*, pp. 171–176.

Griggs, W. H., Jensen, D. D., and Iwakiri, B. T. (1968). Development of young pear trees with different root stocks in relation to psylla infestation, pear decline and leaf curl. *Hilgardia* **39**, 153–204.

Griswold, G. H. (1925). A study of the oyster-shell scale, *Lepidosaphes ulmi* (L.), and one of its parasites, *Aphelinus mytilaspidis* LeB. *Mem. Cornell Univ. Agr. Exp. Sta.* **Mem. 93**, 3–67.

Hagen, K. S., van den Bosch, R., and Dahlsten, D. L. (1971). The importance of naturally occurring biological control in the western United States. *In* "Biological Control" (C. B. Huffaker, ed.), pp. 253–293. Plenum, New York.

Hall, I. M. (1964). Use of micro-organisms in biological control. *In* "Biological Control of Insect Pest and Weeds" (P. DeBach, ed.) pp. 610–628. Reinhold, New York.

Holdsworth, R. P., Jr. (1970). Aphids and aphid enemies: effect of integrated control in an Ohio apple orchard. *J. Econ. Entomol.* **63**, 530–535.

Hoyt, S. C. (1969). Integrated control of insects and biological control of mites on apple in Washington. *J. Econ. Entomol.* **62**, 74–86.

Hoyt, S. C., and Caltagirone, L. E. (1971). The developing programs of integrated control of pests of apples in Washington and peaches in California. *In* "Biological Control" (C. B. Huffaker, ed.), pp. 395–421. Plenum, New York.

Huffaker, C. B., and Kennett, C. E. (1966). Biological control of *Parlatoria oleae* (Colvee) through the compensatory action of two introduced parasites. *Hilgardia* **37**, 283–335.

Huffaker, C. B., Kennett, C. E., and Finney, G. L. (1962). Biological control of the olive scale, *Parlatoria oleae* (Colvee) in California by imported *Aphytis maculicornis* (Masi) (Hymenoptera: Aphelinidae). *Hilgardia* **32**, 541–636.

Hukusima, S. (1960). The interrelationship between the woolly apple aphid, *Eriosoma lanigerum* Hausmann and its parasite. *Jap. J. Ecol.* **10**, 15–22.

Isla, R. (1959). Nota sobre la lucha biologica contra las plagas agricolas en Chile. *Bol. Fitossanit.* **8**, 3–7.

Jaques, R. P., and MacLellan, C. R. (1965). Fungal mortality of overwintering larvae of the codling moth in apple orchards in Nova Scotia. *J. Invertebr. Pathol.* **7**, 291–296.

Jensen, D. D. (1957). Parasites of the Psyllidae. *Hilgardia* **27**, 71–99.

Jensen, D. D., Griggs, W. H., Gonzalez, C. Q., and Schneider, H. (1964). Pear decline virus transmission by pear psylla. *Phytopathology* **54**, 1346–1351.

Kennett, C. E., Huffaker, C. B., and Finney, G. L. (1966). The role of an auto-parasitic aphelinid, *Coccophagoides utilis* Doutt, in the control of *Parlatoria oleae* (Colvee). *Hilgardia* **37**, 255–282.

Kotte, W. (1958). "Krankheiten und Schadlinge im Obstbau und ihre Bekampfung," 519 pp. Parey, Berlin, Hamburg.

LeRoux, E. J. (1960). Effects of "modified" and "commercial" spray programs on the fauna of apple orchards—Quebec. *Ann. ent. Soc. Queb.* **6**, 87–121.

LeRoux, E. J. (1971). Biological control attempts on pome fruits (apple and pear) in North America, 1860–1970. *Can. Entomol.* **103**, 963–974.

Lord, F. T. (1947). The influence of spray programs on the fauna of apple orchards in Nova Scotia. II. Oystershell scale, *Lepidosaphes ulmi* (L.). *Can. Entomol.* **79**, 196–209.

Lord, F. T. (1949). The influence of spray programs on the fauna of apple orchards in Nova Scotia. III. Mites and their predators. *Can. Entomol.* **81**, 202–230.

Lord, F. T. (1971). Biological control of agricultural insects in Canada, 1959–1968. *Commonw. Inst. Biol. Control Tech. Commun.* **4**, 24–25.

Lord, F. T., and MacPhee, A. W. (1953). The influence of spray programs on the fauna of apple orchards in Nova Scotia. VI. Low temperatures and the natural control of the oystershell scale, *Lepidosaphes ulmi* (L.) (Homoptera: Coccidae). *Can. Entomol.* **85**, 282–291.

Lord, F. T., and Stewart, D. K. R. (1961). Effects of increasing the nitrogen level of apple leaves on mite and predator populations *Can. Entomol.* **93**, 924–927.

MacLellan, C. R. (1960). Cocooning behaviour of overwintering codling moth larvae. *Can. Entomol.* **92**, 469–479.

MacLellan, C. R. (1962). Mortality of codling moth eggs and young larvae in an integrated control orchard. *Can. Entomol.* **94**, 655–666.

MacLellan, C. R. (1963). Predator populations and predation on the codling moth in an integrated control orchard—1961. *Mem. Entomol. Soc. Can.* **32**, 45–54.

MacLellan, C. R. (1971a). Woodpecker ecology in the apple orchard environment. *Proc. Tall Timbers Conf. Ecol. Anim. Contr. Habitat Manage.* **2**, 273–284.

MacLellan, C. R., Proverbs, M. D., and Jaques, R. P. (1971b). *Carpocapsa pomonella* (L.), codling moth (Lepidoptera: Olethreutidae). *Commonw. Inst. Biol. Control Tech. Commun.* **4**, 12–15.

McLeod, J. H., McGugan, B. M., and Coppel, H. C. (1962). A review of the biological control attempts against insects and weeds in Canada. *Commonw. Inst. Biol. Control Tech. Commun.* **2**, 1–127.

McMullen, R. D. (1966). New records of chalcidoid parasites and hyperparasites of *Psylla pyricola* Foerster in British Columbia. *Can. Entomol.* **98**, 236–239.

McMullen, R. D. (1971). *Psylla pyricola* Foerster, pear psylla (Hemiptera: Psyllidae). *Commonw. Inst. Biol. Control Tech. Commun.* **4**, 33–38.

McMullen, R. D., and Jong, C. (1967). New records and discussion of predators of the pear psylla, *Psylla pyricola* Foerster, in British Columbia. *J. Entomol. Soc. Brit. Columbia* **64**, 35–40.

MacPhee, A. W. (1964). Cold-hardiness, habitat and winter survival of some orchard arthropods in Nova Scotia. *Can. Entomol.* **96**, 617–625.

MacPhee, A. W. (1967). The winter moth, *Operophtera brumata* (Lepidoptera: Geometridae), a new pest attacking apple orchards in Nova Scotia, and its cold hardiness. *Can. Entomol.* **99**, 829–834.

MacPhee, A. W., and MacLellan, C. R. (1971). Cases of naturally-occurring biological control in Canada. *In* "Biological Control" (C. B. Huffaker, ed.), pp. 312–328. Plenum, New York.

Madsen, H. F., Westigard, P. H., and Sisson, R. L. (1963). Observations on the natural control of the pear psylla, *Psylla pyricola* Foerster, in California. *Can. Entomol.* **95**, 837–844.

Marchal, P. (1929). Les ennemis du puceron lanigere, conditions biologiques et cosmiques de sa multiplication. *Ann. Epiphyt.* **15**, 125–181.

Marshall, J. V., and Pickett, A. D. (1944). The present status of the apple mealy bug, *Phenacoccus acertis* (Sig.) in British Columbia and Nova Scotia. *Can. Entomol.* **76**, 19.

Massee, A. M. (1954). "The Pests of Fruits and Hops," 325 pp. Crosby Lockwood & Son, London.

Messenger, P. S. (1970). Bioclimatic inputs to biological control and pest management programs. *In* "Concepts of Pest Management" (R. L. Rabb and F. E. Guthrie, eds.), pp. 84–102. North Carolina State Univ., Raleigh, North Carolina.

Miller, D., Clark, A. F., and Dumbleton, L. J. (1936). Biological control of noxious insects and weeds in New Zealand. N.Z. J. Sci. Technol. **18**, 579–593.

Moreton, B. D. (1969). Beneficial insects and mites. *Min. Agr. Fish. Food Bull.* **20**, 97–100.

Mori, H. (1967). A review of biology on spider mites and their predators in Japan. *Mushi* **40**, 47–65.

Nel, R. I. (1942). Biological control of the codling moth in South Africa. *J. Entomol. Soc. S. Afr.* **5**, 118–137.

Newcomer, E. J., Dean, F. P., and Carlson, F. W. (1946). Effect of DDT, Xanthone and Nicotine Bentonite on the woolly apple aphid. *J. Econ. Entomol.* **39**, 674–676.

Nickel, J. L., Shimizu, J. T., and Wong, T. Y. (1965). Studies on natural control of pear psylla California. *J. Econ. Entomol.* **58**, 970–976.

Panis, A. (1969). *Aphycus apicalis* (Dalman) and *Leptomastidea bifasciata* (Mayr) (Hymenoptera: Encyrtidae), parasites of *Phencoccus aceris* (Signoret) (Homoptera: Pseudococcidae) in the Maritime Alps. *Entomophaga* **14**, 383–391.

Parent, B. (1967). Population studies of phytophagous mites and predators on apple in southwestern Quebec. *Can. Entomol.* **99**, 771–778.

Patterson, N. A. (1966). The influence of spray programs on the fauna of apple orchards in Nova Scotia. XVI. The long-term effect of mild pesticides on pests and their predators. *J. Econ. Entomol.* **59**, 1430–1435.

Petit, R. H. (1897). Some insects of the year 1897. *Mich. Exp. Sta. Bull.* **160**, 408.

Pickett, A. D. (1965). The influence of spray programs on the fauna of apple orchards in Nova Scotia. XIV. Supplement to II. Oystershell scale, *Lepidosaphes ulmi* (L.). *Can. Entomol.* **97**, 816–821.

Rasmy, A. H., and MacPhee, A. W. (1970). Studies on pear psylla in Nova Scotia. *Can. Entomol.* **102**, 586–591.

Rau, G. J. (1942). The Canadian apple mealy bug, *Phenacoccus aceris* (Signoret) and its allies in northeastern America. *Can. Entomol.* **74**, 118–125.

Russ, K. (1964). Uber ein bemerkenswertes Auftreten von *Beauveria bassiana* (Bals.) Vuill. an *Carpocapsa pomonella* (L.). *Pflanzenschutzberichte* **31**, 105–108.

Samarasinghe, S., and LeRoux, E. J. (1966). The biology and dynamics of oyster shell scale, *Lepidosaphes ulmi* (L.) on apple in Quebec. *Ann. ent. Soc. Queb.* **11**, 204–292.

Sanford, K. H., and Herbert, H. J. (1970). The influence of spray programs on the fauna of apple orchards in Nova Scotia. XX. Trends after altering levels of phytophagous mites or predators. *Can. Entomol.* **102**, 592–601.

Sanford, K. H., and Lord, F. T. (1962). The influence of spray programs on the fauna of apple orchards in Nova Scotia. XII. Effects of perthane on predators. *Can. Entomol.* **84**, 928–934.

Slingerland, M. V. (1892). The pear tree psylla. *Bull. Cornell Univ. Agr. Exp. Sta.* **44**, 161–168.

Sluss, R. R. (1967). Population dynamics of the walnut aphid, *Chromaphis juglandicola* (Kalt.) in northern California. *Ecology* **48**, 41–58.

Tillyard, R. J. (1921). The introduction into New Zealand of *Aphelinus mali*, a valuable parasite of the woolly aphis. *N.Z. J. Agr.* **23**, 7–19.

Tillyard, R. J. (1922). Parasitizing the woolly-aphis. Progress of the work of breeding and distribution of *Aphelinus mali* in New Zealand. *N.Z. J. Agr.* **25**, 31–34.

Todd, D. H. (1959). The apple leaf-curling midge, *Dasyneura mali* Kieffer, seasonal history, varietal susceptibility and parasitism, 1955–58. *N.Z. J. Agr. Res.* **2**, 859–869.

Tothill, J. D. (1919). Some notes on natural control of the oyster-shell scale [*Lepidosaphes ulmi* (L.)]. *Bull. Entomol. Res.* **9**, 183–196.

Turnbull, A. L., and Chant, D. A. (1961). The practice and theory of biological control of insects in Canada. *Can. J. Zool.* **39**, 697–753.

van den Bosch, R., Schlinger, E. I., and Hagen, K. S. (1962). Initial field observations in California on *Trioxys pallidus* (Halliday), a recently introduced parasite of the walnut aphid. *J. Econ. Entomol.* **55**, 857–862.

van den Bosch, R., Fraser, R. D., Davis, C. S., Messenger, P. S., and Hom, R. (1970). *Trioxys pallidus*—An effective new walnut aphid parasite from Iran. *Calif. Agr.* **24**, 8–10.

van de Vrie, M., and Boersma, A. (1970). The influence of the predaceous mite *Typhlodromus potentillae* (Garman) on the development of *Panonychus ulmi* (Koch) on apple grown under various nitrogen conditions. *Entomophaga* **15**, 291–304.

van de Vrie, M., and Kropczynska, D. (1965). The influence of predatory mites on the population development of *Panonychus ulmi* (Koch) on apple. *Boll. Zool. Agr. Bachicolt. Ser. II* **7**, 119–130.

Westigard, P. H., Gentner, L. G., and Berry, D. W. (1968). Present status of biological control of the pear psylla in southern Oregon. *J. Econ. Entomol.* **61**, 740–743.

Wille, J. E. (1958). El control biologico de los insectos agricolas en el Peru. *Proc. 10th Int. Congr. Entomol.* **4**, 519–523.

Wilson, F. (1960). A review of the biological control of insects and weeds in Australia and Australian New Guinea. *Commonw. Inst. Biol. Control, Tech. Commun. I. Ottawa, Canada*, 102 pp.

Wishart, G. (1946). Important reduction of three introduced pests in British Columbia by introduced parasites. *77th Annu. Rep. Entomol. Soc. Ontario*, pp. 35–37.

Withycombe, C. L. (1923). Notes on the biology of some British Neuroptera (Planipennia). *Trans. Roy. Entomol. Soc. London*, pp. 501–594.

Wood, T. G. (1965). Field observations on flight and oviposition of codling moth [*Carpocapsa pomonella* (L.)] and mortality of eggs and first-instar larvae in an integrated control orchard. *N.Z. J. Agr. Res.* **8**, 1043–1059.

15

BIOLOGICAL CONTROL OF PESTS OF TROPICAL FRUITS AND NUTS

F. D. Bennett, David Rosen, P. Cochereau, and B. J. Wood

I. INTRODUCTION

Most tropical fruit and nut crops are perennial plants which persist for several to many years without abrupt, major changes other than seasonal leaf formation, flowering, and fruit development. They also grow in the absence of rigorous climatic changes such as are experienced by temperate crops. Thus, such crops provide a relatively stable environment, offering good opportunities for biological control and effective pest management programs. The many examples of successful biological control of pests of such crops, starting with *Icerya purchasi* Maskell by *Rodolia cardinalis* (Mulsant) in California in 1888, bear witness to the value of this approach (see also Chapters 1–3). Similarly, important work on pests of oil palm and cocoa, reviewed below, provides fundamental knowledge to the concept of conservation of natural enemies and to their use in integrated control programs (see also Chapters 10, 22, 24, and 27).

Space restrictions prevent detailing all the attempts at biological control of pests of these crops, but some outstanding successes as well as some less successful but otherwise illustrative examples are described. The examples are grouped by crops, except for fruit flies and fruit-piercing moths which, because of their wide range of hosts, are treated separately.

II. PESTS OF CITRUS

Citrus is a leading cash and export crop in most subtropical areas. It has numerous pests, the most important commercially usually being coccids, mites, and fruit flies. Projects aimed at control of citrus pests have played a major role in the development of biological control as a science. Ever since the first outstanding success of classical biological control—achieved against cottony-cushion scale on citrus in California—more such efforts have been made on citrus than on any other crop. In general, the number of successes in biological control has been proportional to the amount of effort (DeBach, 1964b). The high proportion of successes on citrus is also partly because most of its serious pests are introduced ones, notably coccids which, for various reasons, are perhaps more amenable to biological control than other groups of organisms (DeBach *et al.*, 1971). Work on citrus has consequently contributed heavily to development of basic concepts and methods of biological control. The crucial importance of sound systematics of both the pests and their natural enemies to success of such efforts (Chapter 4), cryptic effects of host plants on survival of natural enemies (Chapter 22), the hypothesis that competitive displacement of one ecologically comparable species by another results in improved biological control (Chapter 3), the general utility of multiple importations (Chapter 3), the general superiority of importation to various techniques of augmentation (Chapter 3), the harmful effects of modern pesticides on natural enemies (Chapters 10, 11, 14, 17, 22, and herein), and various check methods for evaluating the effectiveness of natural enemies (Chapter 11)—all these and others have been demonstrated in various projects on citrus.

Some of the more outstanding projects on biological control of citrus pests are discussed below. (See also fruit flies, this chapter.)

1. Cottony-Cushion Scale, *Icerya purchasi* Maskell

Successful biological control of cottony-cushion scale on citrus in California in the late 1880's is recognized as the crucial impetus in the early advance of biological control. This fascinating story, which established biological control as a valid method of pest control, was recounted by Doutt (1964). *Icerya purchasi*, a polyphagous tropical–subtropical species, accidentally introduced into California around 1868 on *Acacia*, rapidly assumed alarming proportions; by 1887 it threatened the citrus industry with extinction. Since Australia was assumed to be its native home, a search for its natural enemies was made there. The vedalia lady beetle, *Rodolia cardinalis*, and a parasitic fly, *Cryptochaetum iceryae* (Williston), were introduced from Australia in 1888–1889. Both species were readily established. Although the fly was at first regarded the more promising, its performance in northern California noncommercial citrus was soon overshadowed by that of vedalia on commercial citrus in southern California. The vedalia was distributed in all infested areas, increased at a fantastic rate, and brought the scale under complete control by the end of 1889, and has held it so ever since except for localized upsets caused by use of pesticides (DeBach and Bartlett, 1951; Clausen, 1956a). Recent studies in southern California have shown that both *R. cardinalis* and *C. iceryae* are still established and effective in maintaining this high level of control. *Rodolia* is dominant in desert areas and *Cryptochaetum* in coastal areas; they appear to coexist in intermediate interior areas, where fierce competition between them does not disrupt the excellent biological control (Quezada and DeBach, 1973; DeBach *et al.*, 1971).

Following this spectacular success in California, the vedalia was transferred with similar results to numerous other world areas where *I. purchasi* was causing losses. Complete control resulted in 25 countries, including major citrus producers such as Florida, Israel, Italy, Spain, North Africa, South Africa, and various South American countries; substantial control has been recorded in four additional countries (see DeBach, 1964b and Appendix herein). *Icerya purchasi* continues to be accidentally introduced into new areas and to be successfully controlled by introduction of *R. cardinalis*. A spectacular buildup of this pest in St. Kitts, West Indies, in 1966 collapsed following the introduction of less than 500 *Rodolia* (Bennett, 1971).

2. California Red Scale, *Aonidiella aurantii* (Maskell)

The long, involved history of biological control of California red scale on citrus in California and elsewhere is of special interest, since that intensive project, spanning eight decades, has elucidated some basic principles of biological control (see DeBach, 1969; DeBach *et al.*, 1971).

Aonidiella aurantii is a highly polyphagous species of Oriental origin, a major pest of citrus in California, Mexico, parts of South America, Australia, South Africa, North

Africa, and the eastern Mediterranean Basin. Apparently invading California between 1868 and 1875, it became a serious pest in most citrus growing areas. Intensive biological control efforts have been made against it ever since the 1889 success with cottony-cushion scale. Numerous natural enemies were repeatedly introduced from several continents, but most failed to become established or proved ineffective. Substantial success was achieved only during the last 20 years or so (DeBach *et al.*, 1971; Rosen and DeBach, 1976).

Most of the early efforts concentrated on predators and endoparasites of red scale. Ectoparasites of the genus *Aphytis* were largely ignored because they were commonly misidentified as *A. chrysomphali* (Mercet), which had been present in California since its accidental introduction in the early 1900's. Several predators of red scale were included among some 40 coccinellids imported from Australia and certain Pacific islands during 1889—1892; only *Lindorus lophanthae* (Blaisdell) and *Orcus chalybeus* (Boisduval) became established. *Chilocorus similis* Rossi, imported from China in 1924—1925, and *Coccidophilus citricola* Brèthes, from South America in 1934—1935, also became established. However, these general predators proved to be inadequate and, following the introduction of more effective, host-specific natural enemies, have become very rare or extinct.

Repeated attempts to introduce endoparasites failed for various reasons. Certain Oriental species of *Aonidiella* were commonly mistaken for California red scale, and their parasites were repeatedly introduced against red scale but failed because they were not adapted to it. On several occasions, *Cycas* palms infested with California red scale were taken to the Orient and exposed there, but no parasites were obtained; only later was it found that certain endoparasites cannot complete development in red scale on *Cycas* whereas the same scale on citrus is suitable. Attempts to introduce the endoparasite *Pteroptrix chinensis* (Howard) from China failed due to lack of essential information on its biology or its possible need of an alternate host. Numerous other predators and endoparasites failed to become established for unknown reasons. In the mid-1930's, all hopes for discovery of effective natural enemies were abandoned, and the project was generally considered a complete failure.

Habrolepis rouxi Compere, an endoparasite introduced in 1937 from South Africa, became established in a limited area in California but also proved ineffective.

After some of the main reasons for failure were understood, the search for natural enemies was resumed in the Orient. Two endoparasites, *Comperiella bifasciata* Howard in 1941, and *Prospaltella perniciosi* Tower in 1947, were introduced from China and established. Both have become important factors in the control of *A. aurantii*.

Discovery of new species of *Aphytis* attacking California red scale in the Orient opened a new era. *Aphytis lingnanensis* Compere, introduced from southern China, was readily established in 1948 and became the dominant parasite, gradually displacing *A. chrysomphali* (DeBach and Sundby, 1963). It was not adequate in the interior citrus areas, and efforts were made to augment its populations by periodic releases (DeBach *et al.*, 1950) and by selection of strains resistant to temperature extremes (White *et al.*, 1970).

Aphytis melinus DeBach, introduced from India and Pakistan in 1956–1957, is better adapted to the climatic conditions of interior areas, and has displaced *A. lingnanensis* from most of the citrus areas (DeBach *et al.*, 1971). It is currently the most effective natural enemy of this scale in California.

The introduced parasites have demonstrated their ability to maintain satisfactory control in numerous untreated citrus plots. *Aphytis melinus* is the dominant natural enemy in interior areas, with some aid from *C. bifasciata*, while *A. lingnanensis* is dominant in coastal areas, where it acts in combination with *P. perniciosi*. A general decline of red scale was evident in most of California by 1962 (DeBach *et al.*, 1971). It remains, however, a major pest in the lower San Joaquin Valley where its parasites are still inadequate.

Following these successful efforts in California, several parasites have been transferred to other countries. *Aphytis melinus*, introduced into Greece in 1962, appears to have largely displaced *A. chrysomphali* from red scale populations in the Peloponnesus. *Aphytis lingnanensis*, *P. perniciosi*, and *C. bifasciata* failed to become established. Although still considered serious, the scale problem has been considerably alleviated since the establishment of *A. melinus* (DeBach and Argyriou, 1967; Argyriou, 1974). Partial biological control has occurred in western Turkey, where *A. melinus* was apparently established by ecesis from Greece (DeBach, 1971).

Aphytis melinus was introduced into Israel in 1961. Although establishment and dispersal were initially slow, the parasite increased rapidly in the late 1960's and improved control. *Comperiella bifasciata*, released in 1961, acts as a complementary agent in certain areas.

Aphytis melinus was established in Cyprus and Sicily (Rosen and DeBach, 1976). In Morocco both *A. melinus* and *A. lingnanensis* apparently became established (Benassy, 1969). *Aphytis lingnanensis* and *L. lophanthae* were established in Chile in 1966 (Rosen and DeBach, 1976) and *A. melinus* in Argentina (de Crouzel, 1971).

In South Africa, *L. lophanthae* was established as early as 1900. More recently, several species of *Aphytis* were introduced from California in 1962–63, and *C. bifasciata* from Australia in 1966. Both *C. bifasciata* and *A. melinus* are firmly established (Bedford, 1968, 1973; Annecke, 1969) and *A. aurantii* in some areas is under satisfactory control.

In Australia, early importations against California red scale comprised mainly predators, but *Aphytis* sp. from China was established in Western Australia as early as 1905. *Comperiella bifasciata*, introduced from California in 1942–1947, became established in South Australia and Victoria (Wilson, 1960). More recently, *A. chrysomphali* was introduced into Victoria from California and then to New South Wales in 1954, followed by *A. melinus* from California in 1961. Both became established, and with *C. bifasciata*, produce satisfactory control in certain areas of Victoria (I.W. McLaren, personal communication).

3. Yellow Scale, *Aonidiella citrina* (Coquillett)

Yellow scale, a polyphagous Oriental species and a serious pest of citrus in certain areas, is closely related to the California red scale and was long considered a strain of the latter. Its successful biological control may be regarded a by-product of the campaign against red scale in California.

In California, a race of *Comperiella bifasciata*, obtained from *Aonidiella taxus* Leonardi and *Chrysomphalus bifasciculatus* Ferris in Japan, introduced in 1924–1925, failed to develop in red scale, but became established on *C. bifasciculatus* on ornamentals, and in 1931 was found to develop freely in yellow scale. This so-called "Japanese race" of *C. bifasciata* was readily established on yellow scale and proved particularly effective in residential areas. An extensive dissemination program on commercial citrus proved effective, and yellow scale was gradually eliminated as an economic pest of citrus in southern California, but not in the San Joaquin Valley (Rosen and DeBach, 1976). Other red scale parasites were also tried on yellow scale, but only *Aphytis melinus*, obtained from both of these scales in India and Pakistan, became established.

Aphytis melinus was subsequently transferred from California, mainly against red scale, to various other countries (above). Complete biological control of yellow scale has resulted in Victoria, Australia, where *A. melinus* acts in combination with the yellow scale race of *C. bifasciata*, introduced from California in 1946 (I. W. McLaren, personal communication).

4. Dictyospermum Scale, *Chrysomphalus dictyospermi* (Morgan)

Successful biological control of dictyospermum scale has been another significant side benefit from the campaign against California red scale. A widespread, polyphagous species, *C. dictyospermi* is a major pest of citrus in the western Mediterranean Basin. As *Aphytis melinus* controlled dictyospermum scale on citrus in California (DeBach, 1969), it was colonized against both red and dictyospermum scales in Greece (DeBach and Argyriou, 1967), and has achieved complete control of the latter pest (Argyriou, 1974). Following establishment of *A. melinus* in Turkey by ecesis from Greece, a reduction of dictyospermum scale on citrus was observed (Tunçyürek and Oncuer, 1974). Likewise, introduction of *A. melinus* into Sicily in 1964 resulted in substantial control of *C. dictyospermi* (Rosen and DeBach, 1976).

5. Florida Red Scale, *Chrysomphalus aonidum* (L.)

Florida red scale, a polyphagous Oriental species, is a serious pest of citrus in Florida, Texas, Mexico, Brazil, the eastern Mediterranean Basin, South Africa, and Australia. It invaded Israel around 1910 and by the mid-1950's had become the most important citrus pest. *Aphytis holoxanthus* DeBach, misidentified as *A. lingnanensis*, was introduced against it from Hong Kong in 1956–1957. Complete biological control resulted throughout the coastal plain of Israel. Although sporadic infestations persist in a few groves in the hot Jordan Valley, in spite of periodic releases of *A. holoxanthus*, this scale has been virtually eliminated as an economic problem (DeBach *et al.*, 1971; Rosen and Debach, 1976). This success was the key to development of a vigorous program of integrated control of citrus pests in Israel (Harpaz and Rosen, 1971). *Pteroptrix smithi* (Compere), introduced at the same time, persists but at low levels.

Following its remarkable success in Israel, *A. holoxanthus* was introduced, with similar results, into Florida (Muma, 1969), Mexico (Maltby *et al.*, 1968), South Africa (Annecke, 1969), and Brazil (Rosen and DeBach, 1976).

6. Purple Scale, *Lepidosaphes beckii* (Newman)

Purple scale is an important, rather specific pest of citrus in both semiarid and humid areas of the world. Of Oriental origin, it was apparently accidentally introduced into California from Florida in 1889. Efforts to introduce its natural enemies into California began in the early 1890's. The first introductions comprised general scale predators: *Orcus chalybeus* and *Lindorus lophanthae* from Australia in 1891–1892, *Cybocephalus* sp. from China in 1932–1933, and *Coccidophilus citricola* from Brazil in 1934–1935. Although all became established on purple scale (Clausen, 1956a), they are now rather rare and ineffective (Rosen and DeBach, 1976).

Aphytis lepidosaphes Compere, a host-specific ectoparasite, introduced into California from China in 1948, was mass released and established in subsequent years. While not completely effective, due mainly to high mortaility of its immature stages during winter and spring, it has greatly retarded the rate of increase of scale populations so that integrated control by means of alternate years of strip treatment has become possible (DeBach and Landi, 1961; Rosen and DeBach, 1976).

Following its success in California, *A. lepidosaphes* has been released in Texas, Mexico, Chile, Peru, Brazil, Cyprus, Greece, and Crete and spread by ecesis or unintentional introduction into Louisiana, Florida, Puerto Rico, Jamaica, Guadeloupe, El Salvador, Argentina, Hawaii, Australia, Fiji, New Caledonia, Turkey, and Israel. It appears to be responsible for substantial to complete biological control in nearly all of those countries (DeBach, 1971).

7. Mediterranean Black Scale, *Saissetia oleae* (Olivier)

Until recently, all records of "black scale" in the world were referred to *Saissetia oleae* (Olivier), which is indeed a highly polyphagous, nearly cosmopolitan species of South African origin and a serious pest of citrus, olive, and various ornamentals in many subtropical regions (see De Lotto, 1971a). However, since Bartlett (1960) recorded existence of so-called biological races of black scale, having specific parasites, a complex of several discrete species has been found masquerading under the general designation of "black scale" (see De Lotto, 1971b). Apparently, except in California, *S. oleae* (Olivier) is not the most abundant species of black scale in North and Central America. Hence the presence of cryptic species, easily confused with Mediterranean black scale, may partly explain certain biological peculiarities reported in the life history and host preferences of "black scale" and provide clues to certain puzzling failures of parasite introductions.

Saissetia oleae apparently invaded California between 1868 and 1875. Importation of its natural enemies has again been one of the most intensive campaigns in the history of biological control, spanning seven decades. Some 50 species of parasites and predators, introduced from all subtropical and tropical regions of the world, have been colonized in California. Numerous other species were reared from black scales in various countries. The project was greatly complicated because many species of *Coccophagus* exhibiting differential host relations of the sexes were encountered (Flanders, 1959). Altogether, 17 parasites are established in California and while considerable success has been attained, the project is still actively pursued (Clausen, 1956a; R. L. Doutt, personal communication).

Of interest, *Scutellista cyanea* Motschulsky, an egg predator and facultative ectoparasite, was established in California from four females obtained from Italy in 1901. Initially, it appeared very successful but later it declined and proved rather ineffective. This was partly ascribed at the time to heavy attack by a predaceous mite, *Pyemotes ventricosus* Newport, as well as to the fact that part of the progeny of "parasitized" scales escapes predation by the larvae of *S. cyanea*. *Scutellista cyanea* may have largely displaced another pteromalid egg predator, *Moranila californica* (Howard), without contributing greatly to the control of black scale (Flanders, 1958).

Continuous efforts to introduce natural enemies from Africa, the Orient, and Australia led to establishment of *Metaphycus lounsburyi* (Howard), an endoparasite of "rubber stage" and mature black scales, from Australia in 1916. For some years it effected satisfactory control in coastal areas where black scale populations exhibit "uneven hatch" development, but was ineffective in interior, "even hatch" areas. However, it soon declined drastically, due to heavy hyperparasitism by the introduced *Quaylea whittieri* Girault and presumably to the "evening off" of scale populations in coastal areas, brought about by its own activity (Clausen, 1956a). Various other species introduced in that period became established but show very limited distributions and are of no economic significance (Clausen, 1956a).

Importations from Africa were renewed in 1936–1937. Altogether, 28 species were received alive in California. Of the 10 species colonized, 4 from South Africa were established [*Coccophagus cowperi* Girault, *C. rusti* Compere, *Metaphycus helvolus* (Compere), and *M. stanleyi* Compere]. *Metaphycus helvolus*, a solitary endoparasite of young nymphs, proved to be the most effective and rapidly brought about complete control of the pest in coastal areas. It has been less effective in interior southern California and much less so on the olives of the San Joaquin Valley, where cold winters, combined with an "even-hatch" condition of scale populations, limit its efficiency (Clausen, 1956a; C. B. Huffaker, personal communication). Efforts to increase its effectiveness in such areas by periodic colonization (DeBach and Hagen, 1964) or environmental manipulations (van den Bosch and Telford, 1964) have not been commercially successful. Nevertheless, this pest has been reduced in California to an estimated one-fifth of its former population level (van den Bosch *et al.*, 1955). Savings accrued have been estimated at more than $1.6 million annually in California alone (DeBach, 1964a). (See also Chapter 28.)

Additional search in 1953 in areas of northern and eastern Africa climatically similar to inland California resulted in establishment in California of *Diversinervus elegans* Silvestri and *Metaphycus citrinus* Compere from Eritrea, and *Coccophagus eleaphilus* Silvestri from both Eritrea and Morocco. *Diversinervus elegans* now plays a significant role in the biological control of this scale (Bartlett and Medved, 1966). More recently, a "race" of *Microterys flavus* (Howard) that attacks black scale was introduced in 1957 from Pakistan and established (Bartlett and Lagace, 1961). Still more recently, *Coccophagus capensis* Compere, *C. scutellaris* (Dalman), and *S. cyanea* have been introduced from South Africa in 1972 for use against black scale on olive. Results are as yet unknown (R. L. Doutt, personal communication).

Metaphycus lounsburyi was introduced into Australia from South Africa in 1902, and *S. cyanea* from California in 1904. Both species are effective natural enemies in some regions. *Metaphycus helvolus*, introduced from California in 1942, has given complete biological control of black scale on citrus at Mildura, Victoria (Wilson, 1960; I. W. McLaren, personal communication).

Saissetia oleae has been a major pest of olive in Chile since the early 1800's (Duran and Cortes, 1941). In 1933, *M. lounsburyi*, *Coccophagus ochraceus* Howard, and *S. cyanea* were introduced from California, and all became established (Isla, 1959). *Metaphycus lounsburyi* and *S. cyanea* became very abundant, but did not provide satisfactory control. In 1951, *M. helvolus* was imported from California, became quickly established, and reduced the scale populations in coastal regions. Again, in the interior it has failed (Gonzalez and Rojas, 1966; Isla, 1959).

Prior to establishment of *M. helvolus* in Chile, a closely related species, *M. flavus* (Howard), was a very common parasite of *S. oleae* (Caltagirone, 1957). It is not known when *M. flavus* was introduced in Chile. As *M. helvolus* spread from the release sites and increased in density, *M. flavus* became scarce; in some areas it disappeared in about 5 years, presumably by competitive displacement (L. Caltagirone, personal communication).

In Peru the parasite *M. lounsburyi* and the egg predators *S. cyanea* and *Lecaniobius utilis* Compere were imported in 1936 and established in the Yauca Valley (Wille, 1941) where they brought the scale under satisfactory control (Wille, 1958). In 1961, *M. helvolus* was imported from California and colonized in infested olives. It became well established and provided excellent control (DeBach, 1964b; Salazar, 1966; Beingolea, 1969). *Metaphycus helvolus* was also introduced from California into Greece in 1962 (Argyriou and DeBach, 1968), into Florida in 1967–1968 (McCoy and Selhime, 1971), into France in 1969, and into Israel in 1972.

An interesting by-product of the campaign against the black scale has been the complete suppression of the nigra scale, *Saissetia nigra* (Nietner), a pest of ornamentals, by *M. helvolus* in California. *Metaphycus helvolus* is also credited with complete elimination of *Eucalymnatus tesselatus* (Signoret) before it became a serious pest in California (Bartlett, 1969). Similarly, *M. helvolus* has effected complete biological control of hemispherical scale, *Saissetia coffeae* (Walker), in Peru (Beingolea, 1969).

8. Citricola Scale, *Coccus pseudomagnoliarum* (Kuwana)

Citricola scale, a serious citrus pest in California, occurs also in Japan but is presumably native to arid mainland Asia. It is attacked in California by *Coccophagus caridei* (Brèthes), *Metaphycus stanleyi*, and *M. helvolus*, introduced against black scale, and by *M. luteolus* (Timberlake), a parasite of brown soft scale, *Coccus hesperidum L. Metaphycus helvolus* and *M. luteolus* are credited with partial, often satisfactory, control (Bartlett, 1969). Five additional parasites introduced in 1951 from Japan were propagated and colonized in California; *Coccophagus japonicus* Compere, *C. yoshidae* Nakayama, and *C. hawaiiensis* Timberlake have been recovered from this scale (Gressitt *et al.*, 1954). Active efforts are continuing.

9. Red Wax-Scale, *Ceroplastes rubens* Maskell

Ceroplastes rubens has been a serious pest of citrus, persimmon, tea, and various other plants in Japan since its introduction from China in 1897. Attempts to introduce its parasites via California and Hawaii during 1932—1938 failed. However, in 1946 an effective parasite, *Anicetus beneficus* Ishii and Yasumatsu, was discovered on Kyushu Island. By transfer, the species was then established on the islands of Honshu and Shikoku, where it spread rapidly. Commercial control was usually effected within 3 to 4 years after release (Yasumatsu, 1958, 1969).

10. Brown Soft Scale, *Coccus hesperidum* L.

Coccus hesperidum, a polyphagous, cosmopolitan species, is usually kept under satisfactory control in most areas by a complex of natural enemies. In the U.S.S.R., where *C. hesperidum* has been a serious pest of citrus and other plants, its indigenous natural enemies were considered inadequate. *Metaphycus luteolus* and *Microterys flavus*, introduced from California in 1959, were established in greenhouses and out of doors and are reported providing effective control (Saakian-Baranova, 1966).

11. Citrus Mealybug, *Planococcus citri* (Risso)

Planococcus citri is a serious citrus and greenhouse pest in various parts of the world. Biological control efforts against it in California started with introduction of *Cryptolaemus montrouzieri* Mulsant from Australia in 1891—1892. It readily cleaned up heavy infestations, but subsequently proved to be generally slow in response, and was unable to survive winter in large enough numbers, eventually persisting only along the immediate coast (DeBach and Hagen, 1964). *Leptomastidea abnormis* (Girault), introduced from Sicily in 1914, effected partial control (Clausen, 1956a).

Development of efficient techniques for mass rearing of citrus mealybug on potato sprouts in 1917 has permitted mass rearing and periodic colonization of natural enemies on a commercial scale. Insectaries were established in the 1920's, and *C. montrouzieri* was released against various mealybugs. Ten beetles per tree during summer were adequate for control of most infestations of this mealyby. In 1928, over 42 million beetles were released on citrus in California. While most of the *Cryptolaemus* work was discontinued following successful control of citrophilus mealybug by introduced parasites (below), a few insectaries have continued production against sporadic outbreaks (DeBach and Hagen, 1964).

Several parasitic encyrtids have also been successfully propagated and colonized periodically against citrus mealybug in California. Of these, *Leptomastix dactylopii* Howard, introduced from Brazil in 1934, and *Hungariella peregrina* (Compere), introduced from China in 1949, have become established. Importations in the 1950's resulted in establishment of *Allotropa citri* Muesebeck from China and *Anagyrus pseudococci* (Girault) from Brazil (Bartlett and Lloyd, 1958).

Partial control has been effected in Hawaii by *L. abnormis*, introduced from California in 1915, and in Chile by *L. abnormis* and *C. montrouzieri*, introduced from California in 1931 (DeBach, 1964b).

Following establishment of *C. montrouzieri* on *Dactylopius* spp., this predator has effected complete control of the citrus mealybug in South Africa (Greathead, 1971). It was also transferred to and propagated commercially in other countries, e.g., France and Italy (Smith, 1948). It failed in Israel (then Palestine) where a native hemerobiid predator, *Sympherobius sanctus* Tjeder, was also mass released, but gave unsatisfactory control (Bodenheimer, 1951). *Leptomstidea abnormis* and *L. dactylopii* were imported from the United States into the U.S.S.R. in 1960 and established on citrus mealybug on grape in the Caucasus (Trjapitzin, 1963).

12. Citrophilus Mealybug, Pseudococcus fragilis *Brain*

Pseudococcus fragilis, found in California in 1913, soon became a serious pest of citrus, deciduous fruits, and ornamentals. Periodic colonization of *C. montrouzieri* against citrus mealybug (above) was fairly effective against citrophilus mealybug. Also, *Scymnus binaevatus* (Mulsant), introduced from South Africa in 1921, was established in California but was ineffective (Clausen, 1956a). Two endoparasites, *Coccophagus gurneyi* Compere and *Hungariella pretiosa* (Timberlake), were introduced from Australia in 1928. Both species were highly effective, and complete control was attained in 2 years (Clausen, 1956a). Savings accrued since 1930 are estimated at $2 million annually (DeBach, 1964a).

Coccophagus gurneyi was introduced from California into South Africa in 1934 and became established (Greathead, 1971) and into Chile in 1936 where it is credited with substantial control of this mealybug (DeBach, 1964b). Both *H. pretiosa* and *C. gurneyi* were introduced from the United States into the U.S.S.R in 1962–1963; *C. gurneyi* was established and effected such excellent control that the mealybug has ceased to be an economic pest (Simmonds, 1969).

13. Green's Mealybug, Pseudococcus citriculus Green

Pseudococcus citriculus was first noticed in Israel in 1937 where it became a major pest along the coast. *Clausenia purpurea* Ishii, a parasite of *P. comstocki* (Kuwana), introduced from Japan in 1940, quickly displaced various native natural enemies and effected complete biological control within a few seasons (Rivnay, 1968).

14. Citrus Blackfly, Aleurocanthus woglumi Ashby

Native to Asia, the citrus blackfly was first discovered in the New World in Jamaica in 1913; by 1920 it had spread to Cuba, the Bahamas, Panama, and Costa Rica. Best known as a pest of citrus, it has a wide host range, and is a severe pest of coffee in the New World. Close study by Clausen and Berry (1932) indicated that its relative unimportance in the Far East was due to its natural enemies. The aphelinid *Eretmocerus serius* Silvestri was introduced into Cuba in 1930 and rapidly cleaned up heavily infested citrus groves. Transferred to Haiti, Panama, Jamaica, and Hawaii, it produced generally satisfactory results. However, in Mexico, its introduction on several occasions, while eventually resulting in some control in humid regions, did not prove satisfactory, especially in dry areas. Intensive studies in semiarid regions of Asia in 1948–1950 resulted in importations into Mexico of seven parasites and two predators. Four species, including *Prospaltella smithi* Silvestri, became established, three of which, *Amitus hesperidum* Silvestri, *P. clypealis* Silvestri, and *P. opulenta* Silvestri, have become dominant in controlling this pest following the largest distribution project in the history of biological control. *Prospaltella opulenta* is better adapted to a wide climatic range than the other two. According to Smith *et al.* (1964), "the blackfly is satisfactorily controlled in the entire biological control zone of Mexico."

Eretmocerus serius and *P. opulenta* were introduced in Kenya many years after the citrus blackfly was reported present in 1913, with successful results. The former was released in the humid coastal region in 1959 and the latter in a drier inland area in 1966 (Greathead, 1971).

Aleurocanthus woglumi invaded Barbados, West Indies, only in 1963. Both *E. serius* (from Jamaica) and *P. opulenta* (from Mexico) were introduced in 1964. The latter, predominantly, provided excellent control (Pschorn-Walcher and Bennett, 1967). *Prospaltella opulenta* was released the same year in Jamaica where, despite earlier claims that *E. serius* gave successful control, sporadic outbreaks of *A. woglumi* still occurred (van Whervin, 1968). *Prospaltella opulenta* has now almost completely displaced *E. serius*, providing much more efficient control. This example supports the position presented in Chapter 3 that additional species should be introduced if the problem persists.

Aleurocanthus woglumi was discovered in 1959 in South Africa. *Eretmocerus serius* was then introduced from Jamaica and also Kenya, providing satisfactory control (Bedford and Thomas, 1965). *Eretmocerus serius* (from Jamaica in 1955) has also provided effective control in the Seychelles Islands (Greathead, 1971).

Recently invading El Salvador, heavy infestations of this pest at the release site were eliminated in 4 months by *Prospaltella opulenta* imported from Mexico. Distribution to other areas is continuing (J. R. Quezada, personal communication).

15. Spiny Blackfly, *Aleurocanthus spiniferus* (Quaintance)

Native to Malaya, spiny blackfly was discovered in Japan in 1922 and caused severe damage. *Prospaltella smithi*, introduced from China, has provided complete biological control (Watanabe, 1958).

This pest was recorded in Guam in 1951 and, in 1952, *P. smithi, E. serius*, and *Amitus hesperidum* Silvestri were released in good numbers, along with some *P. opulenta* and *P. clypealis*. Only *P. smithi* and *A. hesperidum* are firmly established. Complete control was quickly effected on citrus, with *P. smithi* being dominant. Results on rose and grape were far less spectacular (Peterson, 1955).

16. Wooly Whitefly, *Aleurothrixus floccosus* Quaintance

Woolly whitefly, apparently native to the Americas, invaded California in the mid-1960's. Several parasites were introduced from Mexico in 1967, including *Amitus spiniferus* Brèthes and *Eretmocerus paulistus* Hempel. *Amitus spiniferus* was readily established and showed great promise. *Eretmocerus paulistus*, again introduced, along with other parasites, in 1968 from Mexico and in 1969 from Florida, also became established and prospects for successful control appeared quite good (DeBach and Warner, 1969). However, in 1970, when an intensive (and eventually unsuccessful) chemical eradication campaign was mounted in California, the emphasis of the biological control program was moved to adjacent areas of Baja California, Mexico, where *A. spiniferus* and *E. paulistus* were established in 1969. Additional parasites, including *Cales noacki* Howard, were also introduced from Central and South America and woolly whitefly infestations have been markedly reduced (Anon., 1971; P. DeBach, personal communication).

Aleurothrixus floccosus was recently introduced into Spain and France. *Cales noacki*, introduced from Chile and Peru, has become established and shows great promise (P. DeBach, personal communication).

III. PESTS OF COFFEE

Coffee is attacked by several hundred insects. Numerous records suggest that many of these are under good control by natural enemies, and in a few instances introductions of natural enemies or manipulations of ones already present have provided satisfactory control (Le Pelley, 1968; Greathead, 1971). Perhaps the most significant feature arising from investigations on coffee pests, apart from the spectacular biological control of *Planococcus kenyae* (Le Pelley), is the general acceptance that natural enemies do keep in check many of the minor or potential pests of coffee and that this should be considered in planning insecticide applications or integrated control programs. A difficult problem is posed by *Antestiops* spp. in Africa. While parasites take a heavy toll, the very low economic threshold, often as low as one insect per tree, means that such palliative measures as insecticide applications are necessary. These measures, particularly since the advent of chlorinated hydrocarbons and organophosphorus compounds, have caused outbreaks of previously unknown insects, e.g., a geometrid, *Ascotis* sp., and increased attacks by known pests, e.g., *Leucoptera* spp. and *Planococcus* spp. (Le Pelley, 1968).

1. Coffee Mealybug, Planococcus kenyae *(Le Pelley)*

Planococcus kenyae became a serious pest in Kenya about 1923. Although later shown to be native to Uganda, it was initially misidentified as *P. citri* and *P. lilacinus* (Cockerell), and natural enemies of these species were introduced but none became effectively established as parasites of *P. kenyae*. In 1937, investigations in Uganda revealed a large complex of primary and secondary parasites; 5 of the former were mass reared and liberated in Kenya. *Anagyrus* nr. *kivuensis* Compere proved to be highly efficient, according to Le Pelley (1968). Another introduced parasite, *Anagyrus beneficans* Compere, also became established and may be more important on indigenous shrubs whereas *A.* nr. *kivuensis* is the dominant parasite in coffee plantations. Within 3 years, *P. kenyae* was reduced from a major to a minor pest. Although redistribution of parasites was at times necessary and some attention had to be paid to ant control, the control was excellent until the use of chlorinated hydrocarbons for other pests destroyed its parasites and resulted in new outbreaks of *P. kenyae*. A switch to nonpersistent organophosphorus insecticides permitted effective biological control to be reestablished. Estimates of losses caused by *P. kenyae* prior to this biological control were given by Kirkpatrick (1927) at £100,000 in the first 6 months of 1927. Mellville (1959) estimated that £10,000,000 had been saved by the biological control of this mealybug and that the total cost could not have exceeded £30,000 (see also Greathead, 1971 and Chapter 3).

2. *Coffee Leaf Miners,* Leucoptera *spp.*

Leucoptera coffeela Guérin-Méneville attacks coffee in the neotropics and three other species mine coffee leaves in Africa. Bess (1964) and Tapley (1961), following field studies, concluded that persistent pesticides destroyed parasites thereby triggering prolonged attacks of *L. meyricki* Ghesquière in Africa. Attempts to introduce *Mirax insularis* Muesebeck from the West Indies into Kenya were unsuccessful (Greathead, 1971).

3. *Coffee Berry Borer,* Hypothenemus hampei *(Ferrari)*

Limited efforts to combat this introduced pest in Java, Brazil, and Ceylon have not met with success (Le Pelley, 1968). Releases of parasites have not been followed up (Greathead, 1971).

4. *Citrus blackfly,* Aleurocanthus woglumi

The citrus blackfly became a serious pest of coffee after its introduction into the New World. (For details of its biological control see under citrus.)

IV. PESTS OF COCOA

Attempts to control pests of cocoa by classical methods of biological control are not numerous nor have they been very successful, although investigations on the indigenous natural enemies of several species indicate that they play an important role in regulating populations, a fact which has been further emphasized by the occurrence of outbreaks as a result of the effects of applications of residual contact insecticides on parasites and predators (Conway and Wood, 1964).

1. *Cocoa Mealybugs*

The control of the swollen shoot virus of cocoa by the attempted biological control of its mealybug vectors *Planococcoides njalensis* (Laing), *Planococcus citri*, and *Dysmicoccus brevipes* (Cockerell) was unsuccessful despite the importation of more than 25 species of parasites and predators during 1948–1955. Attempts to use a fungus, *Cephalosporium* sp., were also unsuccessful (Greathead, 1971).

2. *Cocoa Thrips*

The control of *Selenothrips rubrocinctus* (Giard), a pest of cocoa in the West Indies, by the introduction of *Goetheana parvipennis* (Gahan) from Ghana was attempted in

1935. Establishment in Trinidad was reported by Callan (1943), but Greathead (1971) suggested that it may already have been present in the Neotropics. In Trinidad it plays at most an insignificant role in controlling thrips. While it is more common in Jamaica it does not provide adequate control.

3. Cocoa Mirids

Investigations in Ghana on the role of the ants *Oecophylla longinoda* Latreille and *Macromischoides aculeatus* (Mayr) as possible predators of cocoa mirids, including *Sahlbergella singularis* Haglund, *Distantiella theobroma* (Distant), and *Bryocoropsis laticollis* Shaun, failed to demonstrate effective predation (Williams, 1954). However, recent studies suggest that the ants *O. longinoda* and *Crematogaster* spp. do aid in reducing mirid populations, particularly those of *Distantiella* (Greathead, 1971). While parasites and other predators were investigated in Ghana, the Belgian Congo, and Sierra Leone during 1946–1947, the results were not encouraging (Greathead, 1971).

In Cameroun, where the New World ant *Wasmannia auropunctata* Roger is now naturalized, it has been transferred into cocoa plantations for the control of mirids. The ants drove away the mirids but encouraged a buildup of *Saissetia* spp.; further studies were recommended (Greathead, 1971).

V. PESTS OF GUAVAS

In the tropics, guavas are favorite hosts for many fruit flies (below) and are attacked by many coccids and whiteflies. While introductions of natural enemies have seldom been made specifically against scales and whiteflies, releases against these pests on other crops have resulted in establishment and activity on guavas.

Guava Scale, *Chloropulvinaria psidii* (Maskell)

Chloropulvinaria psidii attacks a wide range of hosts, including guava. In Bermuda, outbreaks, mainly on *Ficus* spp. and oleander, occurred in 1953. *Microterys kotinskyi* (Fullaway) and *Aneristus ceroplastae* Howard were introduced from Hawaii in 1953 and the former also from California in 1954 and 1955. *Microterys kotinskyi* was established and is considered an important factor in the control of *C. psidii*. Of several other natural enemies released, only *Azya luteipes* (Mulsant) and *Cryptolaemus montrouzieri* became established and aided in reducing this scale to a noneconomic level (Bennett and Hughes, 1959).

VI. PESTS OF PASSION FRUIT

Passion fruit, *Passiflora edulis* Sims, is grown as a commercial crop in Hawaii where it is attacked mainly by coccids and mites.

Barnacle Scale, *Ceroplastes cirripediformis* Comstock

In 1967 *Coccidoxenus mexicanus* Girault, reared from *Ceroplastes floridensis* Comstock, was sent from Trinidad to Hawaii for release against *C. cirripediformis* which had killed many linear feet of passion fruit on Maui (Davis and Chong, 1968). By June, 1971, parasitism of 75 to 95%, and an associated decline in populations of the scale, were reported (C. J. Davis, personal communication.)

VII. PESTS OF PINEAPPLES

Pineapple Mealybug, *Dysmicoccus brevipes* (Cockerell)

In Hawaii, pineapple mealybug has been implicated in the dissemination of a virus disease. The encyrtid parasites *Hambletonia pseudococcina* Compere and *Anagyrus ananatis* Gahan were introduced from South America (Carter, 1937); *A. ananatis* became established but is seldom encountered (Beardsley, 1969). Both species were established in Puerto Rico during 1936–1937 (Bartlett, 1939); Wolcott (1948) reported *H. pseudococcina* to be most effective. In Florida, where *H. pseudococcina* was released, Clausen (1956a) considered that it played an unimportant role because of the wide use of organic insecticides on the crop.

Cryptolaemus montrouzieri was introduced into South Africa against *D. brevipes* in 1900. Later it became established on other mealybugs on other crops but it was not effective against *D. brevipes* (Greathead, 1971). It was also colonized in Guinea and Mauritius but it did not become established on *D. brevipes* (Greathead, 1971).

In the Philippines, *C. montrouzieri* was introduced against pineapple mealybugs from the United States in 1928–1929, but establishment was reported only at one location. Several other natural enemies were introduced from Hawaii in 1934 and 1956 but their status has not been assessed (Rao *et al.*, 1971).

VIII. PESTS OF BANANAS

1. Banana Weevil Borer, *Cosmopolites sordidus* Germar

Cosmopolites sordidus is a serious pest of bananas in the tropics. The predatory histerid beetle *Plaesius javanus* Erichson was introduced into Fiji in 1913–1914 and in

1918 from Java (Jepson, 1914; Simmonds, 1935). Simmonds (1935) reported that damage by *C. sordidus* had been markedly reduced but Pemberton (1953) considered that *P. javanus* offered only partial control. From Fiji the histerid was sent to Jamaica in 1937–1938; establishment occurred and until the advent of modern insecticides it was common practice to release adults in new banana plantations. In 1942, it was established in Trinidad, from which it, as well as a native histerid, *Hololepta (Leionota) quadridentata* (Fabricius), was sent to other West Indian Islands. Establishment had not been reported (Simmonds, 1958) until 1972, when *H. quadridentata* was recovered in St. Vincent (M. Yaseen, personal communication). Both species were supplied to Cameroun (1952), Mauritius (1959), and the Seychelles (1950–1954) (Greathead, 1971). *Plaesivs javanus* was also shipped to Uganda from Java in 1934–1935. None of these introductions into Africa resulted in establishment (Greathead, 1971). *Plaesius javanus* from Java or Fiji was released in Queensland and other areas of Australia several times but did not become established (Wilson, 1960).

2. Banana Scab Moth, *Nacoleia octasema* (Meyrick)

Banana scab moth damages bananas in Keravet, New Britain, and also blemishes other fruits in Fiji.

Cremastus sp. was introduced into Fiji from Java in 1933–1934 but has not been recovered in recent times. *Argyrophlax fransseni* Bar. and *Goniozus triangulifer* Kieffer from New Britain were released in 1953 and 1956 (O'Connor, 1960; Paine, 1964). *Pentalitomastix nacoleiae* Eady was mass bred; despite the release of a million adults, permanent establishment failed, possibly because this parasite does not frequent the banana plant (O'Connor, 1960). *Chelonus striatigenas* Cameron, introduced from Indonesia in 1960, provides parasitism up to 70% (O'Connor, 1964). This species, also established in Samoa, has little effect (Cochereau, 1972). Attempts to obtain control in New Britain by the introduction of *P. nacoleiae* in 1957 were unsuccessful (Wilson, 1960).

As pointed out by Cochereau (1972) this pest remains one of the unsolved biological control problems of the Pacific, although it should be amenable to this solution.

3. Florida Red Scale, *Chrysomphalus aonidum*

Severe outbreaks of Florida red scale, a serious citrus pest (above), were recorded in the late 1960's in bananas in the Jordan Valley, Israel, where its effective parasite *Aphytis holoxanthus* was repeatedly decimated by the hot summers. Great damage was caused, with entire crops frequently being lost. Periodic releases of laboratory-reared *A. holoxanthus* during fall and winter have practically solved the problem and have become routine. The cultural practice of removing old infested leaves at the end of the growing season, adopted by many growers, also helps to keep this pest in bananas at subeconomic levels (S. Kamburov, personal communication).

IX. PESTS OF AVOCADOS

In California most arthropod pests of avocado remain at innocuous levels due mainly to the action of biotic agents (Fleschner, 1954). Accordingly, to avoid the possibility of biological upsets the release of natural enemies against one pest, even if costly, could be preferable to utilizing chemical pesticides which might cause repercussions by destroying natural enemies of other pests. Mass releases of *Stethorus picipes* Casey assisted in suppressing populations of the mite *Oligonychus punicae* (Hirst) in experimental plots. This method may be of value if releases can be critically timed, even if production costs are not yet economical (McMurtry *et al.*, 1969). Clausen (1956a) reported that in Southern California the biological control of long-tailed mealybug, *Pseudococcus longispinus* (Targioni-Tozzetti), on avocado (as well as citrus) was accomplished by introduction of three encyrtids, *Anarhopus sydneyensis* Timberlake from Australia in 1933, *Hungariella peregrina* from Brazil in 1934, and *Anagyrus fusciventris* Girault from Hawaii in 1936, the first two being the more important.

In the Caroline Islands, Trust Territory, the Egyptian "mealybug" *Icerya aegyptiaca* (Douglas), a serious pest of avocado, banana, breadfruit, citrus, etc., was brought under control in 1947 by *Rodolia pumila* (Wesmael), introduced from the Marianas (Gardner, 1958). In Vaté, New Hebrides, the same pest was recently controlled by the introduction of *R. cardinalis* (P. Cochereau, unpubl. data).

In Israel, avocado groves were virtually free of economic pest injury until the late 1960's, when heavy outbreaks of *P. longispinus* occurred due to destruction of natural enemies by drift of nonselective pesticides from neighboring fields. Severe damage was compounded by various moths, especially *Cryptoblabes gnidiella* Millière, attracted to mealybug honeydew and, in Western Galilee, often caused the culling of entire crops. Since chemical control was not feasible, an integrated approach was adopted. Aerial spraying of insecticides was prohibited within 200 meters of avocado groves. *Hungariella peregrina*, introduced in 1952 from the United States (Rosen, 1967), has been mass reared and recolonized in infested groves. *Anagyrus fusciventris* was introduced from Australia and established in 1972. Efforts have been made to introduce *Anarhopus sydneyensis* from Australia and California, but that species has not yet been established. By the end of 1972, the situation in avocados had been remarkably alleviated, and economic control was anticipated (E. Swirski and D. Rosen, unpubl. data).

X. PESTS OF MACADAMIA NUTS

Southern Green Stink Bug, Nezara viridula (L.)

Severe losses to macadamia nuts, mangoes, and several other fruits and some vegetables resulted following establishment of *N. viridula* in Hawaii in 1961 (Davis, 1964). Several natural enemies were released and the establishment of *Trissolcus basalis* (Wollaston) from Australia and *Trichopoda pennipes* (Fabricius) from Montserrat, West Indies, resulted in considerable control of this pentatomid; although damage to macadamia was reported for 1965 (Davis and Krauss, 1966), no serious outbreaks have occurred since that time (Davis, 1967).

XI. PESTS OF COCONUTS

Coconuts are often subsistence crops; hence biological control, requiring limited material input or grower effort, is very desirable. Coconuts have provided some of the classic examples of the biological control approach, mostly concerning problems with alien pests in new areas, especially islands, and the importation of their enemies. These cases prompted some rather specious conclusions about biological control—that it only worked on islands, only with introduced pests, only with parasites (or predators), and so on. However, it is now evident that the method may work in more subtle ways and the potential benefits are not limited by the constraints implied. Lever (1969) reviewed the pests of coconuts, including their biological control. Rhinoceros beetles, pests of both coconuts and oil palms, are considered under the latter crop.

1. Zygaenid Caterpillars

In the early 1900's, in Viti Levu, Fiji, coconut production virtually ceased when *Levuana iridescens* Bethune-Baker was accidentally introduced about 1877. It caused total defoliation, with only brief periods of respite. A major investigation was initiated following its appearance in other islands (Tothill *et al.*, 1930). Several natural enemies of related insects on various plants in the Pacific were imported. The tachinid *Bessa remota* (Aldrich), a larval parasite of *Brachartona catoxantha* (Hampson) on coconuts, proved decisive. Its establishment throughout the range of *L. iridescens* occurred in 6 months, and this moth was reduced to insignificance in a year. The pest has since been scarce. No other importation played any important role.

Brachartona catoxantha occurs in the Malay archipelago and the Philippines, but is usually scarce due to natural enemies, primarily *B. remota* in Malaya and *Apanteles artonae* Wilkinson in Java (Lever, 1969). Others include the predatory clerid *Callimerus arcufer* Chapin and the fungus *Beauveria bassiana* (Balsamo Vuillemin). Why they sometimes fail to control this pest is not clear. Various investigators have noted that dry weather is a contributory cause, and outbreaks are more common in the drier half of the year (Lever, 1953). Gater (1925) suggested that fires under palms may play a part. Severe defoliation may occur before the pest is restored to its usual insignificance. Lever (1953) recommended avoidance of broad-spectrum residual-contact insecticides, which can cause biological disruption, while Ho *et al.* (1972) showed how selective insecticides can aid in reestablishing natural balance.

2. Hispine Beetles

Serious outbreaks of *Promecotheca coeruleipennis* Blanchard, a leaf miner, occurred in certain climatic zones in Fiji in the 1920's due, paradoxically, to the accidental introduction of the predatory mite, *Pyemotes ventricosus* (Taylor, 1937). Indigenous parasites included *Oligosita utilis* Kowalski on eggs, and *Elasmus hispidarum* Ferrière on larvae (but not pupae); they had been effective before the mite's arrival. The story illustrates not only the importance of natural enemies, but also of the one-stage condition, in outbreaks of pests in the tropics.

In dry periods, when the pest was by chance at a peak in the usual fluctuating balance between it and the indigenous enemies (Taylor's "multiple stage" outbreaks), the mite increased rapidly, destroying all the immature stages of the hispid and then disappeared, allowing normal development from eggs of surviving *P. coeruleipennis* adults. This led to a small one-stage generation and since the native parasites were poorly adapted to cope with this situation outbreaks occurred.

Taylor postulated that in this situation a successful parasite would attack a wide range of the host's stages, have a long adult life, be multivoltine with respect to the host, and be internal so that the host's exoskeleton would protect it from attack by the mite. Such attributes were found in *Pediobius parvulus* (Ferrière), a parasite of the related *Promecotheca cumingi* Baly in Java. Colonization of this eulophid resulted in complete success, ending one widespread outbreak within a year. Two other importations, *Sympiesis javanica* (Ferrière) and *Pediobius painei* (Ferrière), failed although *S. javanica*, not *P. parvulus*, is the most important parasite of *P. cumingi* in Java. *Promecotheca cumingi* itself is normally well controlled in the Malay archipelago and Philippines by these parasites and others, including *Achrysocharis promecothecae* Ferrière on eggs, although occasionally, a damaging flare-up occurs, e.g., in Singapore (Lever, 1951) and Indonesia (Kalshoven, 1950–1951). In 1970, infestations spread devastatingly within a 20- to 25-km radius of Colombo (Sri Lanka), presumably from a new introduction. Native parasites were ineffective, and several importations included *P. parvulus* and *S. javanica*. Only the latter spread rapidly, but already substantial control occurred by December 1972 (Fernando, 1972). Spectacular control has since resulted and benefits estimated at $48,000.00 have been quoted (F. J. Simmonds, personal communication, 1974).

Promecotheca papuana Csiki in New Britain and nearby Pacific Islands is also affected by *Pyemotes ventricosus* in an as yet undefined way. Indigenous enemies include predatory ants, particularly *Oecophylla smaragdina* Fabricius. *Pediobius parvulus* was established in 1938, but localized one-stage outbreaks still occur (Gressitt, 1958). *Promecotheca opacicollis* Gestro was serious in the New Hebrides until it was controlled in the 1930's with *P. parvulus*, obtained from Fiji (Lods and Dupertuis, 1939). Cochereau (1970a) confirmed the good control, although noting that a heavy outbreak occurred when the pest spread to a new island, Tikopia.

Larvae and adults of *Brontispa* spp. feed within the unopened coconut fronds and may kill several in succession. The genus supports a complex of parasites rather different from that on *Promecotheca*. *Brontispa longissima* (Gestro) in Indonesia and several Pacific islands has egg parasites including *Haeckeliana brontispae* Ferrière, *Trichogrammatoidea nana* (Zehntner), and *Ooencyrtus* sp., and the eulophid *Tetrastichus brontispae* (Ferrière) on larvae and pupae. The latter, from *B. longissima* var. *javana* Weise in Java, was imported to Celebes in 1929 and brought very severe infestations of *B. longissima* var. *celebensis* Gestro under control (Awibowo, 1934), although Lever (1936) did not obtain success in the Solomons. The species remains serious in Java perhaps because there are two strains, one susceptible and one not susceptible to parasitization by *T. brontispae* (Tjoa, 1965). The same parasite was brought to New Caledonia (against var. *froggatti* Sharp) where newly imported palm varieties were much more heavily attacked than local ones (Cochereau, 1969). It became established and by 1972 reduced *B. longissima* to a minor pest status. Stapley (1971) imported it to the Russell Islands where it controlled the pest, eliminating the previous need for routine insecticide applications every 2 months on some coconut varieties. The same species in 1948 (from Java on *B. longissima* and from Malaya on *Plesispa nipae* Maulik) rapidly controlled *B. mariana* Spaeth, previously a severe pest in the Mariana and Caroline Islands (Doutt, 1950; Lange, 1950). *Tetrastichus brontispae* is also being released in Madagascar for control of the hispines *Gestronella centrolineata* Fairm and *Xiphispa lugubris* Fairm on coconut (J. Appert, personal communication).

3. Scale Insects

Aspidiotus destructor Signoret was a severe problem in Fiji in the early 1900's on bananas as well as coconuts (Taylor and Paine, 1935). Two aphelinid parasites, *Aphytis chrysomphali* and *Aspidiotiphagus citrinus* (Craw), were introduced from Tahiti in 1920, the latter probably a reintroduction. *Aphytis chrysomphali* became common, but did not give control. Several enemies of *A. destructor* and related scales imported from Java in 1926 did not establish or do well. In 1927–1928, five predators were imported from Trinidad, *Cryptognatha nodiceps* Marshall, *C. simillima* Sicard, *Azya trinitatis* Marshall, *Pentilia insidiosa* Mulsant, and *Scymnus aeneipennis* Sicard. All except *P. insidiosa* were released; *C. nodiceps* became of primary importance in controlling the scale within 9 months. The others did not become permanently established, perhaps because of the host's severe decline (competitive displacement). Taylor and Paine (1935) attribute the success of *C. nodiceps* mainly to its voracity, high reproductive potential (due to its short life cycle), good dispersal and searching power, and survival on other hosts, such as *Pseudaulacaspis pentagona* (Targioni-Tozzetti). Similar success with this predator occurred in Principe (West Africa) in the mid-1950's where Simmonds (1967) discussed the economics of the program (see also Simmonds, 1960; Rosen and DeBach, 1976).

Aspidiotus destructor threatened coconuts in Mauritius in the 1920's and 1930's. Here, *Chilocorus politus* Mulsant from Java (1937) and *C. nigritus* (Fabricius) from

Ceylon (1939) proved effective (Moutia and Mamet, 1946). *Aspidiotus destructor* appeared in Vaté Island in New Hebrides in 1962, but *C. nodiceps* and *Azya trinitatis* from Trinidad and *Pseudoscymnus* sp. from Caroline Islands failed to give control. *Rhizobius pulchellus* Montrouzier (originally misidentified as *Lindorus lophanthae*) from New Caledonia has now effectively reduced the pest to minor status.

Vesey-FitzGerald (1953) reported that other scales, which occur only scarcely elsewhere, were pests in the Seychelles prior to the late 1930's. *Ischnaspis longirostris* (Signoret) was so severe as to literally blacken palms. *Chilocorus distigma* (Klug) from East Africa in 1936 controlled it, but a buildup of *Pinnaspis buxi* (Bouché) then caused even worse damage. This was reduced to negligible proportions by another coccinellid, *Telsimia nitida* Chapin, brought from Guam in 1936 (Pemberton, 1953). *Chilocorus nigritus* from India also controlled *Chrysomphalus aonidum*, which had been locally severe.

4. Coreid Bugs

Immature (or premature) fall of coconuts was attributed to physiological or pathological disorders or various insects until it was traced to coreid bugs, viz *Amblypelta cocophaga* China in the Solomon Islands and *Pseudotheraptus wayi* Brown in East Africa and Zanzibar. Damage is considerable even at very low pest numbers. A single *Amblypelta* on a caged spadix caused 94% premature nutfall (Phillips, 1940), with production in areas having only 0.85—1.92 bugs per palm, reduced from 12 cwt/acre to 4—5. Similar effects occur from *Pseudotheraptus* (Way, 1953). *Amblypelta cocophaga* has three native egg parasites, a reduviid predator of both nymphs and adults, *Euagoras dorycus* Boisduval, and a fungal disease (Phillips, 1940). Importations included egg parasites and two tachinids, *Pentatomophaga bicincta* de Meijere, which attacks *A. lutescens* (Distant) in Queensland, and *Trichopoda pennipes* from Florida (Phillips, 1956). Only some egg parasites became established but little affected the control which was already holding the pest at low numbers.

In both cases, these bugs are completely eliminated and nutfall is controlled by *Oecophylla* ants, a genus which as Way (1954) remarks, includes remarkably efficient predators, used, for example, in China in the 12th century to control citrus pests.

Phillips (1940) related the distribution of *Amblypelta* to that of four predominant ant species. It does not occur with *O. smaragdina* and is found with *Pheidole megacephala* (Fabricius) and *Iridomyrmex myrmecodiae* Emery. *Anoplolepis longipes* (Jerdon) suppresses the pest, a point confirmed by Brown (1959a). Increased nutfall was evidently due to replacement of *Oecophylla* by *Pheidole* and Phillips suggested investigating what conditions might favor *Oecophylla*, but World War II intervened. Afterward, *Oecophylla* was found to have extensively replaced *Pheidole*. O'Connor (1950) suggested that the untended ground vegetation might be responsible, but the evidence was conflicting. Another possible explanation (O'Connor, 1950) was that insecticides used for malaria control had differentially favored *Oecophylla*. Brown (1959b) found that the change continued without these influences and, if anything, *Oecophylla* was favored by clear conditions. Greenslade (1971) suggested an 8-year fluctuation in *Oecophylla* populations, connected with the life span of a colony. Stapley (1972) showed that in areas where both ants were present, herbicide used at the palm bases, followed by three applications of dieldrin on the lower trunks, led to overall replacement of *Pheidole* by *Oecophylla* in about 4 months. With use of herbicides only, there was a similar but slower change. *Oecophylla* was still predominant 18 months later. An intermediate host tree, *Annona* sp., aids the establishment of *Oecophylla* where it is initially scarce (J. H. Stapley, personal communication).

A similar problem occurs with *A. lutescens papuensis* Brown in Southern Papua, with a similar association with ants (Smee, 1965). In the New Hebrides, *Axiagastus cambelli* Distant may cause premature nutfall, apparently due to low rainfall which adversely affects the predatory ants. Two scelionids, *Microphanurus painei* Ferrière from the Solomon Islands and *Trissolcus basalis* from New Caledonia, have been established but it is too early to comment on their efficacy (Cochereau, 1972).

5. Rhinoceros Beetles

Over 40 *Oryctes* spp. (rhinoceros beetles) are major palm pests in the tropics. *Oryctes rhinoceros* (L.), native to southeast Asia and spreading through the Pacific Islands, is rated the world's most serious coconut pest; it also attacks oil palms. Attack is reduced by dense ground cover vegetation and close planting and can give economic control in oil palm plantations (Owen, 1959; Wood, 1969). The principal parasites of *Oryctes* spp. are scoliids and nematodes; predators include carabid, elaterid, and histerid beetles. Several *Scolia* spp. and predatory beetles have been introduced but the results have been disappointing (Lever, 1969). Various natural enemies have been released in Mauritius but are not known to be established (Greathead, 1971). Although the predator, *Platymeris laevicollis* Distant, was introduced into several Pacific Islands and Malaysia and establishment reported, effective control has not occurred (see Hoyt and Catley, 1966; Rao *et al.*, 1971 for other introductions).

The green muscardine fungus *Metarrhizium anisopliae* (Metchnikoff) Sorokin has been recorded from most *Oryctes* spp. Results of microbial applications have been erratic and unpromising (Young, 1971). Attempts to infest *O. rhinoceros* with various pathogenic bacteria have failed (Zelazny, 1971; Surany, 1960).

The only confirmed biological control success is with "Malaya disease" caused by *Rhabdionvirus oryctes* Huger, discovered in the West Malayan Peninsula (Huger, 1966). It affects both larvae which soon die, and adults (Zelazny, 1972, 1973). Transmission is by ingestion or by contact. Infected adults stop feeding and egg-laying, but can still spread the virus.

Infection and establishment is easy (Bedford, 1971). The beetle has apparently reached many Pacific Islands without the virus, and when introduced in Samoa, the disease led to marked reduction in damage (Marschall, 1970). Introductions to other islands have reduced attack from very high levels (Young, 1974; Bedford, 1971) and incidence now resembles the more tolerable situation of territories where the virus has been long endemic (B. Zelazny, personal communication). Controlled introduction to other Pacific Islands is being undertaken, and releases in Mauritius are promising (J. Monty, personal communication).

6. Other Coconut Pests

A complex of tettigoniids, mainly *Sexava* spp., has caused considerable coconut defoliation in New Guinea and other Pacific areas (Smee, 1965). Two egg parasites, *Leefmansia bicolor* Waterston and *Doirania leefmansi* Waterston from the Moluccas, Indonesia, have proved effective (Lever, 1969). A strepsipteran, *Stichotrema dallatorreana* Hofeneder, was associated with good control in parts of New Guinea in 1954 (O'Connor, 1959). Attempts were made to establish it in New Britain (Dun, 1955).

Success has been reported in mass-releasing indigenous parasites against sporadic outbreaks of the caterpillar *Nephantis serinopa* Meyrick in Ceylon and south India. Millions are released annually, principally *Trichospilus pupivora* Ferrière, *Tetrastichus israeli* Mani and Kurian, *Elasmus nephantidis* Rohwer, *Perisierola nephantidis* Muesebeck, and *Bracon brevicornis* Wesmael (Rao *et al.*, 1971).

XII. PESTS OF OIL PALMS

The West African oil palm was introduced to southeast Asia early this century and later to South America. Its pests, specifically in Malaysia, were described by Wood (1968). Little biological control work in the strict sense of importation, augmentation, and manipulation of enemies has been attempted but the wrong use of insecticides and more obscure environmental factors have triggered rapid explosions of some pests, stimulating studies of their natural regulation. The overriding importance of natural enemies has been well demonstrated and clearly must be taken into account in any rational control program.

1. Bagworms in West Malaysia

Major bagworm outbreaks on oil palms were first recorded in West Malaysia in the 1950's, although the crop had been grown since the 1920's. Most of its insect pests had their origin on wild flora or coconuts, but had remained at low numbers on oil palms despite high reproductive potentials. Wood (1971) deduced that the outbreaks, which had no common locale or time factor, were caused by broad-spectrum residual-contact insecticides such as DDT and dieldrin applied against other minor pests.

The most serious bagworm, *Metisa plana* Walker, supports several hymenopterous parasites, both primary and secondary, which are associated with declines in populations and reestablishment and maintenance of low levels when the use of disruptive chemicals is stopped, sometimes aided by a single application of a selective insecticide, such as trichlorphon (Wood, 1971). The tendency to continued buildup after the immediate adverse effect of the insecticides is probably due to a desynchronization of host and enemy life cycles. In outbreaks, the pest occurs in a single-phase condition, probably largely because the chemicals differentially spare one stage of its life cycle, and parasites with short life cycles are thus at a continuing disadvantage. The dispersal powers of these bagworms are low and they concentrate in one area, but some individuals do disperse and may initiate similar trains of events in new areas.

2. Caterpillars in Sumatra

Similar recurrent severe caterpillar outbreaks on oil palms occurred in Sumatra where DDT was applied, suggesting a parallel to the West Malaysian bagworm situation. Efforts are now being made to restore the position by using integrative chemicals (Hutauruk and Situmorang, 1971).

3. Caterpillars in Sabah, East Malaysia

A few oil palm estates in East Malaysia have existed up to 12 years, and although insects have not posed a general problem, outbreaks of another bagworm, *Mahasena corbetti* Tams, occurred in the eastern part of the country. Parasites were scarce in the first outbreak in 1965, but selective chemicals reduced the population, and parasites soon appeared and exercised control of the population surviving the spray (Wood and Nesbit, 1969). Sankaran (1970) subsequently reported up to 55% parasitism by *Eozenillia equatorialis* Townsend. Wood and Nesbit (1969) speculated that the Sabah outbreaks were triggered by chance factors, with continuing buildup due to a high reproductive potential (up to 4000 eggs per female) and initial scarcity of parasites, together with desynchronization of life cycles. (See also Sankaran and Syed, 1972.)

Nettle caterpillars, particularly *Setoria nitens* Walker and *Darna trima* Moore, are also common. Outbreaks have occurred often, particularly in sprayed groves (even with rather selective chemicals). The reason why natural control breaks down needs to be investigated further. The numerous parasites of *S. nitens* include *Chaetexorista javana* Brauer and Bergenstamm and *Spinaria spinator* Guerin. Similar general predators to those in West Malaysia also occur (Syed and Pang, 1972). Few parasites of *D. trima* are recorded but importations are being considered. A virus often eliminates an outbreak and microbial applications show early promise (Syed, 1971).

XIII. FRUIT-PIERCING MOTHS

Adults of several noctuids commonly pierce, feed on, and damage fruits of citrus, mangoes, guavas, etc.; the feeding puncture also serves as a feeding site or entry for other insects and various pathogens. The larvae are invariably defoliators of other plants, cultivated or wild, often occurring some distance from the fruit growing areas, and mass immigrations from the breeding areas may occur. The biology of some species, e.g., *Othreis fullonia* Clerck, which occurs widely in the Pacific Islands, Africa, Asia, and Australia, is well known and widespread fluctuations occur from year to year. In the Americas the major fruit-piercing moths belong to the genus *Gonodonta* (Todd, 1959), whereas species of *Othreis* and *Eumaenas* are widespread in the Old World. In all, about 20 species are known as important pests of tropical fruits.

The possibilities of biological control have often been considered and records of natural enemies are frequent (e.g., Cochereau, 1976), but actual attempts to exploit these have been limited.

In India, attempts have been made to obtain biological control of the castor semi-looper, *Achaea janata* (L.), a pest of castor in its larval stages (the adults pierce citrus fruits). The effort has been aimed at reducing larval damage to castor. *Trichogramma australicum* Girault was released on a large scale in 1942–1949; in 1964 *Telenomus* sp. was imported from New Guinea (Phalak and Raodeo, 1967). In field cages, 60–80%

Fig. 1. The tachinid *Winthemia caledoniae* Mesnil in the act of ovipositing on the head of the caterpillar of *Othreis fullonia* Clerck, a fruit-piercing moth in New Caledonia. Note the position of the ovipositor, doubled back between the parasite's legs, and note the eggs already deposited. Photograph by P. Cochereau.

parasitism occured (Rao *et al.*, 1971). Also in India, field application of *Bacillus thuringiensis* Berliner on castor resulted in 80% larval mortality (Kulshreshtha *et al.*, 1965).

Cochereau (1976) has shown that in New Caledonia damage by *O. fullonia* is normally light due to natural enemies, particularly *Ooencyrtus* sp. Outbreaks only occur after a drought or exceptional dry period at the beginning of summer which initially adversely affects populations of both *O. fullonia* and *Ooencyrtus* sp.; the former recovers more rapidly and the spread of adults from the consequent pest buildup results in heavy damage. Since many fruit-piercing moths have restricted larval host ranges, destruction of host plants (*Erythrina* spp. in the case of *O. fullonia*) may offer a solution (Cochereau, 1976; Fennah, 1942). Cochereau (1976) concluded that detailed studies of the population dynamics, including adult migration of the fruit-piercing moths, are essential to a solution of the problem.

XIV. TROPICAL FRUIT FLIES

While certain major fruit fly problems result from accidental introductions, many important problems are also caused by indigenous species, the latter often constituting serious problems with exotic fruits.

Biological control of fruit flies first received major attention in Hawaii with the invasion of the Mediterranean fruit fly, *Ceratitis capitata* (Wiedemann), in 1910. During 1912–1914, parasites were obtained from West and South Africa and Australia. *Opius humilis* Silvestri from South Africa, *O. tryoni* Cameron from Australia, and *O. fullawayi* Silvestri, *Dirhinus giffardii* Silvestri, and *Tetrastichus giffardianus* Silvestri from West Africa were established by 1914 (Clausen, 1956b). Since the results were not satisfactory, investigations were reopened in 1935–1936, with exploration in West and East Africa, Brazil, Malaysia and India; however, no additional species were established against this fruit fly.

In 1915–1916, *O. fletcheri* Silvestri was also established in Hawaii against the melon fly.

Appearance of the oriental fruit fly, *Dacus dorsalis* Hendel, in Hawaii in 1946 and its serious status on many crops resulted in further importations. Several species were introduced in 1947–1949 (Clausen *et al.*, 1965). While 11 species and varieties were established, interest centered on three species of *Opius*: *O. longicaudatus* (Ashmead) which increased rapidly in 1948; *O. vandenboschi* Fullaway which largely displaced it in 1949; and *O. oophilus* Fullaway which became dominant in 1950 and has remained so ever since. Factors accounting for this "succession" of dominance were as follows: *O. longicaudatus* attacks only large larvae, often less vulnerable to attack; *O. vandenboschi* attacks the more readily vulnerable small larvae; while *O. oophilus* oviposits in the eggs and is intrinsically superior to the other species (van den Bosch and Haramoto, 1953) (see further Chapter 3). Newell and Haramoto (1968) demonstrated that when attacked by *O. oophilus* high percentages, up to 92%, of the eggs died either as a result of traumatic injury or pathogens transmitted by the parasite. After thorough study they concluded that the decline of *D. dorsalis* can be fully accounted for by the introduced parasites. While satisfactory control in some hosts, e.g., guava, has not been achieved, reductions in populations have been sufficient even on these hosts, to permit effective use of cultural and chemical control measures which otherwise would have been quite ineffective. Haramoto and Bess (1970) note that *O. oophilus* is now also the dominant parasite of *C. capitata*.

In Fiji, where *Dacus passiflorae* Froggatt and *D. xanthodes* Brown are important pests, several parasites introduced from Hawaii during 1935–1938 did not provide control. In 1951–1954 *Opius vandenboschi*, *O. oophilus*, *O. longicaudatus*, and *O. incisi* Silvestri were then introduced. *Opius oophilus* and *O.* longicaudatus became established in Vitilevu and fruit fly infestations were considerably reduced mainly due to the activities of *O. oophilus* (O'Connor, 1960; Rao *et al.*, 1971).

Several other efforts at biological control of a variety of fruit flies have been made throughout the world (see Bartlett, 1941; Berry, 1955; Jimenez, 1969; Turica, 1968; Clausen, 1956b; Baranowski and Swanson, 1971, for work in the Neotropics, Etienne, 1969, for the Indian Ocean Region, and Wilson, 1960; Snowball and Lukins, 1964; and Cochereau, 1970b, for elsewhere in the Indo-Australo-Pacific Region). The results have in general been disappointing or are still inconclusive and space limitations do not permit a review of the details.

It is of interest to compare the conditions of Hawaii, where a considerable level of success has been achieved, with Fiji where parasites are well established and have provided some but less effective control, and with efforts in Australia (Snowball and Lukins, 1964; Wilson, 1960) where, due to an array of unfavorable conditions, no appreciable benefits have derived. The subject fruit flies of Hawaii are all imported species and, where they have been controlled, introduced parasites have been involved. In Australia and Fiji some of the important fruit flies are natives, which have adapted to introduced fruits, whereas their native parasites (even if effective on native plant hosts) are not effective on the introduced fruits. Alternatively, the introduced fruit flies have attacked native fruits to which both the native and introduced parasites are not adapted effectively. Also, *O. oophilus* became established only in areas in Queensland where fruit flies occur the year round (Snowball and Lukins, 1964). South of Brisbane *O. oophilus* failed due to its inability to diapause during lower winter temperatures and the seasonal absence of host fruits. However, it would seem possible that other species of *Opius* could be established in Australia and Fiji. The displacement, partial or total, of one species by another can be due to direct competition for hosts or to differences in their respective tolerances to unfavorable ecological conditions.

REFERENCES

Annecke, D. P. (1969). Recent developments in biological and integrated control of citrus pests in South Africa. *Proc. 1st Int. Citrus Symp. Riverside, Calif.* **2**, 849–854.

Anonymous. (1971). A big bet on biological control. *Citrograph* **56**, 315–316, 327.

Argyriou, L. C. (1974). Data on the biological control of citrus scales in Greece. *Bull. SROP* 1974/3: 89–94.

Argyriou, L. C., and DeBach, P. (1968). The establishment of *Metaphycus helvolus* (Compere) (Hymenoptera: Encyrtidae) on *Saissetia oleae* (Bern.) (Homoptera: Coccidae) in olive groves in Greece. *Entomophaga* **13**, 223–228.

Awibowo, R. (1934). Der Klapperbladkever, *Brontispa froggatti* var. *celebensis*, en zijn biologische bestrijding op Celebes. (Voorloopige mededeeling). *Landbouw. Buitenz.* **10**, 76–92.

Baranowski, R. M., and Swanson, R. W. (1971). The utilization of *Parachasma cereum* (Hymenoptera: Braconidae) as a means of suppressing *Anastrepha suspensa* (Diptera: Tephritidae) populations. *Proc. Tall Timbers Conf. Ecol. Anim. Contr. Habitat Manage.* **3**, 249–252.

Bartlett, B. R. (1960). Biological races of the black scale, *Saissetia oleae*, and their specific parasites. *Ann. Entomol. Soc. Amer.* **53**, 383–385.

Bartlett, B. R. (1969). The biological control campaigns against soft scales and mealybugs on citrus in California. *Proc. 1st Int. Citrus Symp., Riverside, Calif.* **2**, 875–878.

Bartlett, B. R., and Lagace, C. F. (1961). A new biological race of *Microterys flavus* introduced into California for the control of lecaniine coccids, with an analysis of its behavior in host selection. *Ann. Entomol. Soc. Amer.* **54**, 222–227.

Bartlett, B. R., and Lloyd, D. C. (1958). Mealybugs attacking citrus in California—a survey of their natural enemies and the release of new parasites and predators. *J. Econ. Entomol.* **51**, 90–93.

Bartlett, B. R., and Medved, R. A. (1966). The biology and effectiveness of *Diversinervus elegans* (Hymenoptera: Encyrtidae), an imported parasite of lecaniine scale insects in California. *Ann. Entomol. Soc. Amer.* **59**, 974–976.

Bartlett, K. A. (1939). Introduction and colonization of two parasites of the pineapple mealybug. *Univ. Puerto Rico J. Agr.* **23**, 67–72.

Bartlett, K. A. (1941). The introduction and colonization in Puerto Rico of beneficial insects parasitic on West Indian fruit flies. *Univ. Puerto Rico J. Agr.* **25**, 25–31.

Beardsley, J. W., Jr. (1969). The Anagyrina of the Hawaiian Islands (Hymenoptera: Encyrtidae) with descriptions of two new species. *Proc. Hawaii. Entomol. Soc.* **20**, 287–310.

Bedford, E. C. G. (1968). The biological control of red scale, *Aonidiella aurantii* (Mask.), on citrus in South Africa. *J. Entomol. Soc. S. Afr.* **31**, 1–15.

Bedford, E. C. G. (1973). Citrus scale insects biological control proves successful. *Citrus SubTrop. Fruit J., February, 1973*, pp. 5–17.

Bedford, E. C. G., and Thomas, E. D. (1965). Biological control of the citrus blackfly, *Aleurocanthus woglumi* (Ashby) (Homoptera: Aleyrodidae) in South Africa. *J. Entomol. Soc. S. Afr.* **28**, 117–132.

Bedford, G. O. (1971). "Virus Release Programme in Fiji. Report of the Project Manager for the Period June 1970 – May 1971." pp. 234–239. [UNDP(SF)/SPC project for research on the control of the coconut palm rhinoceros beetle.] South Pacific Commission, Noumea, New Caledonia. (Multigr.)

Beingolea, G. O. (1969). Biological control of citrus pests in Peru. *Proc. 1st Int. Citrus Symp. Riverside, Calif.* **2**, 827–838.

Benassy, C. (1969). The biological control against coccids of citrus in the country of the French zone. *Proc. 1st Int. Citrus Symp. Riverside, Calif.* **2**, 793–799.

Bennett, F. D. (1971). Some recent successes in the field of biological control in the West Indies. *Rev. Peruv. Entomol.* **14**, 369–373.

Bennett, F. D., and Hughes, I. W. (1959). Biological control of insect pests in Bermuda. *Bull. Entomol. Res.* **50**, 423–436.

Berry, N. O. (1955). Mexican fruit fly and citrus blackfly control in Mexico. *J. Econ. Entomol.* **48**, 414–416.

Bess, H. A. (1964). Populations of the leaf-miner *Leucoptera meyricki* and its parasites in sprayed and unsprayed coffee in Kenya. *Bull. Entomol. Res.* **55**, 59–82.

Bodenheimer, F. S. (1951). "Citrus Entomology in the Middle East," 663 pp. Dr. W. Junk, Publ., The Hague.

Brown, E. S. (1959a). Immature nutfall of coconuts in the Solomon Islands. I. Distribution of nutfall in relation to that of *Amblypelta* and of certain species of ants. *Bull. Entomol. Res.* **50**, 97–133.

Brown, E. S. (1959b). Immature nutfall of coconuts in the Solomon Islands. II. Changes in ant populations, and their relation to vegetation. *Bull. Entomol. Res.* **50**, 523–558.

Callan, E. McC. (1943). Natural enemies of the cacao thrips. *Bull. Entomol. Res.* **34**, 313–321.

Caltagirone, L. (1957). Insectos entomofagos y sus huespedes anotados para Chile. *Agr. Tec. (Santiago de Chile)* **17**, 16–48.

Carter, W. (1937). Importation and laboratory breeding of two chalcid parasites of *Pseudococcus brevipes* (Ckll.). *J. Econ. Entomol.* **30**, 370–372.

Clausen, C. P. (1956a). Biological control of insect pests in the continental United States. *U.S. Dep. Agr. Tech. Bull.* **1139**.

Clausen, C. P. (1956b). Biological control of fruit flies. *J. Econ. Entomol.* **49**, 176–178.

Clausen, C. P., and Berry, P. A. (1932). The citrus blackfly in Asia, and the importation of its natural enemies into tropical America. *U.S. Dep. Agr. Bull.* **320**.

Clausen, C. P., Clancy, D. W., and Chock, Q. C. (1965). Biological control of the Oriental fruit fly (*Dacus dorsalis* Hendel) and other fruit flies in Hawaii. *U.S. Dep. Agr. Tech. Bull.* **1322**.

Cochereau, P. (1969). Installation de *Tetrastichus brontispae* Ferr. (Hymenoptera: Eulophidae), parasite de *Brontispa longissima* Gestro, var. *froggatti* Sharp (Coleoptera: Chrysomelidae: Hispinae) dans la presqu'Ile de Noumea. *Cah. ORSTOM Ser. Biol.* **7**, 139–141.

Cochereau, P. (1970a). "Recherches aux Nouvelles-Hébrides de *Pleurotropis parvulus* Ferr. (Hymenoptera: Eulophidae), parasite de *Promecotheca pulchellus* Gestro (Coleoptera: Hispinae), et expéditions sur la Côte d'Ivoire," 14 pp. Nouméa, N. Caledonia, Off. Rech. Sci. Tech. Outre Mer. (Multigr.)

Cochereau, P. (1970b). Les mouches des fruits et leurs parasites dans la zone indo-australo-pacifique et particulierement en Nouvelle-Caledonie. *Cah. ORSTOM Ser. Biol.* **12**, 15–50.

Cochereau, P. (1972). La lutte biologique dans le Pacifique. *Cah. ORSTOM Ser. Biol.* **16**, 89–104.

Cochereau, P. (1976). "Biologie et dynamique des populations, en Nouvelle-Calédonie, d'un papillon piqueur de fruits: *Othreis fullonia* Clerck (Lepidoptera: Noctuidae, Catocalinae)," 258 pp. Mem. ORSTOM, Paris.

Conway, G. R., and Wood, B. J. (1964). Pesticide chemicals—help or hindrance in Malaysian agriculture. *Malaysian Nature J.* **18**, 111–119.

Davis, C. J. (1964). The introduction, propagation, liberation and establishment of parasites to control *Nezara viridula* var. *smaragdula* (Fabricius) in Hawaii. *Proc. Hawaii. Entomol. Soc.* **18**, 369–375.

Davis, C. J. (1967). Progress in the biological control of the southern stink bug *Nezara viridula* var. *smaragdula* (Fabricius) in Hawaii (Heteroptera: Pentatomidae). *Mushi Suppl.* **39**, 9–16.

Davis, C. J., and Chong, M. (1968). Recent introductions for biological control in Hawaii. XIII. *Proc. Hawaii. Entomol. Soc.* **20**, 25–34.

Davis, C. J., and Krauss, N. L. H. (1966). Recent introductions for biological control in Hawaii. XI. *Proc. Hawaii. Entomol. Soc.* **19**, 201–207.

DeBach, P. (1964a). The scope of biological control. *In* "Biological Control of Insect Pests and Weeds" (P. DeBach, ed.), pp. 3–20. Chapman & Hall, London.

DeBach, P. (1964b). Successes, trends, and future possibilities. *In* "Biological Control of Insect Pests and Weeds" (P. DeBach, ed.), pp. 673–713. Chapman & Hall, London.

DeBach, P. (1969). Biological control of diaspine scale insects on citrus in California. *Proc. 1st Int. Citrus Symp. Riverside, Calif.* **2**, 801–815.

DeBach, P. (1971) Fortuitous biological control from ecesis of natural enemies. "Entomological Essays to Commemorate the Retirement of Professor K. Yasumatsu," pp. 293–301. Hokuryukan, Tokyo.

DeBach, P., and Argyriou, L. C. (1967). The colonization and success in Greece of some imported *Aphytis* spp. (Hymenoptera: Aphelinidae) parasitic on citrus scale insects (Homoptera: Diaspididae). *Entomophaga* **12**, 325–342.

DeBach, P., and Bartlett, B. R. (1951). Effects of insecticides on biological control of insect pests of citrus. *J. Econ. Entomol.* **44**, 372–383.

DeBach, P., and Hagen, K. S. (1964). Manipulation of entomophagous species. *In* "Biological Control of Insect Pests and Weeds" (P. DeBach, ed.), pp. 429–458. Chapman & Hall, London.

DeBach, P., and Landi, J. (1961). The introduced purple scale parasite, *Aphytis lepidosaphes* Compere, and a method of integrating chemical with biological control. *Hilgardia* **31**, 459–497.

DeBach, P., and Sundby, R. A. (1963). Competitive displacement between ecological homologues. *Hilgardia* **34**, 105–166.

DeBach, P., and Warner, S. C. (1969). Research on biological control of whiteflies. *Citrograph* **54**, 301–303.

DeBach, P., Dietrick, E. J., Fleschner, C. A., and Fisher, T. W. (1950). Periodic colonization of *Aphytis* for control of the California red scale. Preliminary tests, 1949. *J. Econ. Entomol.* **43**, 783–802.

DeBach, P., Rosen, D., and Kennett, C. E. (1971). Biological control of coccids by introduced natural enemies. *In* "Biological Control" (C. B. Huffaker, ed.), pp. 165–194. Plenum, New York.

de Crouzel, I. S. (1971). Studies on the biological control of diaspidid scales on citrus in Argentina. Proc. 12th Pac. Sci. Congr. *Canberra* **1**, 200.

De Lotto, G. (1971a). The authorship of the Mediterranean black scale (Homoptera: Coccidae). *J. Entomol.* **B40**, 149–150.

De Lotto, G. (1971b). A preliminary note on the black scales (Homoptera: Coccidae) of North and Central America. *Bull. Entomol. Res.* **61**, 325–326.

Doutt, R. L. (1950). Field studies on the parasites of *Brontispa mariana* Spaeth. *Proc. Hawaii. Entomol. Soc.* **14**, 55–58.

Doutt, R. L. (1964). The historical development of biological control. *In* "Biological Control of Insect Pests and Weeds" (P. DeBach, ed.), pp. 21–42. Chapman & Hall, London.

Dun, G. S. (1955). Economic entomology in Papua and New Guinea 1948–1954. *Papua New Guinea Agr. J.* **9(3)**, 109–119.

Duran, L., and Cortes, R. (1941). La conchuela negra del olivo, *Saissetia oleae* Bern., en Chile. *Boll. Dep. Sanid. Veg.* **1** (2), 37–47.

Etienne, J. (1969). Lutte contre les mouches des fruits. *Rapp. Inst. Rech. Agron. Trop. Reunion*, pp. 251–259.

Fennah, R. G. (1942). The citrus pests investigation in the Windward and Leeward Islands, British West Indies, 1937–1942. *Imp. Coll. Trop. Agr. Trinidad*, 66 pp.

Fernando, H. E. (1972). "Concluding Report on Control of *P. cumingi*, The Introduced Pest of Coconuts, 14 pp. Central Agr. Res. Inst., Dep. Agr. Peradeniya, Sri Lanka (Mimeo.)

Flanders, S. E. (1958). *Moranila californica* as a usurped parasite of *Saissetia oleae*. *J. Econ. Entomol.* **51**, 247–248.

Flanders, S. E. (1959). Differential host relations of the sexes in parasitic Hymenoptera. *Entomol. Exp. Appl.* **3**, 125–142.

Fleschner, C. A. (1954). Biological control of avocado pests. *Calif. Avocado Soc. Yearb.* **38**, 125–129.

Gardner, T. R. (1958). Biological control of insect and plant pests in the Trust Territory and Guam. *Proc. 10th Int. Congr. Entomol. Montreal* **4**, 465–469.

Gater, B. A. R. (1925). Some observations on the Malaysian coconut zygaenid (*Artona catoxantha* Hamps.). *Malaysian Agr. J.* **13**, 92–115.

Gonzalez, R. H., and Rojas, S. (1966). Estudio analitico del control biologico de plagas agricolas en Chile. *Agr. Tec.* (*Santiago de Chile*) **26**, 133–147.

Greathead, D. J. (1971). A review of biological control in the Ethiopian region. *Commonw. Inst. Biol. Control Tech. Commun.* **5**.

Greenslade, P. J. M. (1971). Interspecific competition and frequency changes among ants in Solomon Islands coconut plantations. *J. Appl. Ecol.* **8**, 323–349.

Gressitt, J. L. (1958). Ecology of *Promecotheca papuana* Csiki, a coconut beetle. *Proc. 10th Int. Congr. Entomol. Montreal* **2**, 747–753.

Gressitt, J. L., Flanders, S. E., and Bartlett, B. R. (1954). Parasites of citricola scale in Japan, and their introduction into California. *Pan-Pac. Entomol.* **30**, 5–9.

Haramoto, F. H., and Bess, H. A. (1970). Recent studies on the abundance of the Oriental and Mediterranean fruit flies and the status of their parasites. *Proc. Hawaii. Entomol. Soc.* **20**, 551–566.

Harpaz, I., and Rosen, D. (1971). Development of integrated control programs for crop pests in Israel. *In* "Biological Control" (C. B. Huffaker, ed.), pp. 458–468. Plenum, New York.

Ho Thian Hua, Ahmad bin Baha, and Yusof bin Din (1972). Aerial spraying against *Artona catoxantha* Hamps. on coconut palms. In "Cocoa and Coconuts in Malaysia" (R. L. Wastie & D. A. Earp, eds.), pp. 329–344. Inc. Soc. of Planters, Kuala Lumpur.

Hoyt, C. P., and Catley, A. (1966). Current research on the biological control of *Oryctes* (Coleoptera: Scarabaeidae: Dynastinae). *Mushi* **39**, 3–8.

Huger, A. M. (1966). A virus disease of the Indian rhinoceros beetle *Oryctes rhinoceros* (L.) caused by a new type of insect virus, *Rhabdionvirus oryctes* gen.n., sp.n. *J. Invertebr. Pathol.* **8**, 38–51.

Hutauruk, C., and Situmorang, H. S. (1971). Some notes on the control of the bagworm, *Metisa plana* Wlk., in North Sumatra. *In* "Crop Protection in Malaysia" (R. L. Wastie and B. J. Wood, eds.), pp. 166–172. Inc. Soc. of Planters, Kuala Lumpur.

Isla, R. (1959). Nota sobre la lucha biologica contra las plagas agricolas en Chile. *Bol. Fitossanit.* **8**(3), 3–7.

Jepson, F. P. (1914). A mission to Java in quest of natural enemies for a coleopterous pest of bananas. *Bull. Dep. Agr. Fiji* **7**.

Jimenez-Jimenez, E. (1969). Control biologico de algunas plagas de los citricos en Mexico. *Proc. 1st Int. Citrus Symp. Riverside, Calif.* **2**, 781–784.

Kalshoven, L. G. E. (1950–1951). "De plagen van de cultuurgewassen in Indonesia," Two vol., 1065 pp. W. van Hoeve, The Hague.

Kirkpatrick, T. W. (1927). The common coffee mealybug in Kenya Colony. *Bull. Dep. Agr. Kenya* **18**.

Kulshreshtha, J. P., Sanghi, P. K., and Ravindranath, V. (1965). Microbial control of castor semilooper, *Achaea janata*. *Indian J. Entomol.* **27**, 353–354.

Lange, W. H., Jr. (1950). The biology of the Mariana coconut beetle, *Brontispa mariana* Spaeth, on Saipan, and the introduction of parasites from Malaya and Java for its control. *Proc. Hawaii. Entomol. Soc.* **14**, 143–162.

Le Pelley, R. H. (1968). "Pests of Coffee," 590 pp. Longmans Green, London.

Lever, R. J. A. W. (1936). Control of *Brontispa* in Celebes by *Tetrastichodes*. *Brit Solomon Islands Agr. Gaz. Suppl.* **3**(4), 6.

Lever, R. J. A. W., (1951). A new coconut pest in Singapore. *Malayan Agr. J.* **34**, 78–82.

Lever, R. J. A. W. (1953). Notes on outbreaks, the parasites and habits of the coconut moth *Artona catoxantha* Hamps. *Malayasian Agr. J.* **36**, 20–27.

Lever, R. J. A. W. (1969). "Pests of the Coconut Palm," 190 pp. FAO, Rome.

Lods, G., and Dupertuis, C. B. (1939). "Note sur la lutte contre la 'mouche du cocotier' aux Nouvelles-Hébrides; introduction du *Pleurotropis*, son parasite," 18 pp. Service de l'Agriculture des Nouvelles-Hébrides, Port Vila. (Multigr.)

McCoy, C. W., and Selhime, A. G. (1971). Influence of some natural enemies on black scale, *Saissetia* spp., in Florida. *J. Econ. Entomol.* **64**, 213–217.

McMurtry, J. A., Johnson, H. G., and Scroda, G. T. (1969). Experiments to determine effects of mass releases of *Stethorus picipes* on the level of infestation of the avocado brown mite. *J. Econ. Entomol.* **63**, 1216–1221.

Maltby, H. L., Jimenez-Jimenez, E., and DeBach, P. (1968). Biological control of armored scale insects in Mexico. *J. Econ. Entomol.* **61**, 1086–1088.

Marschall, K. J. (1970). Introduction of a new virus disease of the coconut rhinoceros beetle in Western Samoa. *Nature (London)* **225**, 288–289.

Mellville, A. R. (1959). The place of biological control in the modern science of entomology. *Kenya Coffee* **24**, 81–85.

Moutia, L. A., and Mamet, R. (1946). A review of twenty-five years of economic entomology in the island of Mauritius. *Bull. Entomol. Res.* **36**, 439–472.

Muma, M. H. (1969). Biological control of various insects and mites on Florida citrus. *Proc. 1st Int. Citrus Symp. Riverside, Calif.* **2**, 863–870.

Newell, I. M., and Haramoto, E. H. (1968). Biotic factors influencing populations of *Dacus dorsalis* in Hawaii. *Proc. Hawaii. Entomol. Soc.* **20**, 81–139.

O'Connor, B. A. (1950). Premature nutfall of coconuts in the British Solomon Islands Protectorate. Appendix by R. Leach. *Agr. J. Fiji* **21**, 21–42.

O'Connor, B. A. (1953). Biological control of insects and plants in Fiji. *Proc. 7th Sci. Congr. Aukland Christchurch* **4**, 278–293.

O'Connor, B. A. (1959). The coconut tree-hopper *Sexava* and its parasites. *Papua New Guinea Agr. Gaz.* **11**, 121–125.

O'Connor, B. A. (1960). A decade of biological control work in Fiji. *Agr. J. Fiji* **30**, 1–11.

O'Connor, B. A. (1964). Report on Fiji in Pacific Entomology Report of the Standing Committee Chairman, J. J. H. Szent-Ivany. *Proc. 10th Pac. Sci. Congr.*, pp. 92–93.

Owen, R. P. (1959). "Proposals for Vegetative Barrier Experiments" pp. 1–3. South Pacific Commission, Trust Territory of the Pacific Islands, Koror. Caroline Islands (Multigr.)

Paine, R. W. (1964). The banana scab moth *Nacoleia octasema* Meyrick: its distribution, ecology and control. *S. Pac. Comm. Tech. Pap.* **145**.

Pemberton, C. E. (1953). The biological control of insects in Hawaii. *Proc. 7th Pac. Sci. Congr.* **4**, 220–223.

Peterson, G. D. (1955). Biological control of the orange spiny whitefly in Guam. *J. Econ. Entomol.* **48**, 681–683.

Phalak, V. R., and Raodeo, A. K. (1967). Possibilities of controlling the castor semi looper *Achaea janata* (L.) (Lepidoptera: Noctuidae) using the egg parasite *Telenomus* sp. (Hymenoptera: Scelionidae) introduced from New Guinea. *Commonw. Inst. Biol. Control Tech. Bull.* **9**, 81–92.

Phillips, J. S. (1940). Immature nutfall of coconuts in the Solomon Islands. *Bull. Entomol. Res.* **31**, 295–316.

Phillips, J. S. (1956). Immature nutfall of coconuts in the British Solomon Islands Protectorate. *Bull. Entomol. Res.* **47**, 575–595.

Pschorn-Walcher, H., and Bennett, F. D. (1967). The successful biological control of citrus blackfly [(*Aleurocanthus woglumi* (Ashby)] in Barbados, West Indies. *PANS* **A13**, 375–384.

Quezada, J. R., and DeBach, P. (1973). Bioecological and population studies of the cottony-cushion scale, *Icerya purchasi* Mask., and its natural enemies, *Rodolia cardinalis* Mul. and *Cryptochaetum iceryae* Will., in southern California. *Hilgarida* **41**, 631–688.

Rao, V. P., Ghani, M. A., Sankaran, T., and Mathur, K. C. (1971). A review of the biological control of insects and other pests in south-east Asia and the Pacific Region. *Commonw. Inst. Biol. Control Tech. Commun.* **6**.

Rivnay, E. (1968). Biological control of pests in Israel (a review 1905–1965). *Isr. J. Entomol.* **3**, 1–156.

Rosen, D. (1967). Biological and integrated control of citrus pests in Israel. *J. Econ. Entomol.* **60**, 1422–1427.

Rosen, D., and DeBach, P. (1976). Diaspididae. *In* "Introduced Parasites and Predators of Arthropod Pests and Weeds: A World Review" (C. P. Clausen, ed.), pp. 78–128. U.S. Dept. Agr., Agr. Handbook 480.

Saakian-Baranova, A. A. (1966). The life cycle of *Metaphycus luteolus* Timb. (Hymenoptera: Encyrtidae), parasite of *Coccus hesperidum* (Homoptera: Coccidae), and the attempt of its introduction into the U.S.S.R. *Entomol. Rev.* **45**, 733–751. (In Russ.)

Salazar, J. (1966). Avances en el control biologico de queresas Lecaniinae *Saissetia oleae* Bern. y *Saissetia hemispherica* Targ. *Rev. Peru. Entomol.* **7**, 8–12.

Sankaran, T. (1970). The oil palm bagworms of Sabah and the possibilities of their biological control. *PANS* **16**, 43–55.

Sankaran, T., and Syed, R. A. (1972). The natural enemies of bagworms on oil palms in Sabah, East Malaysia. *Pac. Insects* **14**, 57–71.

Simmonds, F. J. (1958). Recent work on biological control in the British West Indies. *Proc. 10th Int. Congr. Entomol. Montreal* **4**, 475–478.

Simmonds, F. J. (1960). Biological control of coconut scale (*Aspidiotus destructor*) in Principe, Portuguese West Africa. *Bull. Entomol. Res.* **51**, 223–237.

Simmonds, F. J. (1967). The economics of biological control. *J. Roy. Soc. Arts* **115**, 880–898.

Simmonds, F. J. (1969). The work of the Commonwealth Institute of Biological Control relating to citrus insects. *Proc. 1st Int. Citrus Symp. Riverside, Calif.* **2**, 765–767.

Simmonds, H. W. (1935). Annual Report of the Government entomologist for the year 1934. *Annu. Bull. Div. Rep. Dep. Agr. Fiji*, pp. 12–16.

Smee, L. (1965). Insect pests of *Cocos nucifera* in the Territory of Papua and New Guinea: Their habits and control. *Papua New Guinea Agr. J.* **17**(2), 51–64.

Smith, H. D., Maltby, H. L., and Jimenez-Jimenez, E. (1964). Biological control of the citrus blackfly in Mexico. *U.S. Dep. Agr. Tech. Bull.* **1311**.

Smith, H. S. (1948). Biological control of insect pests. *In* "The Citrus Industry" (L. D. Batchelor and H. J. Webber, eds.), Vol. 2, pp. 597–625. Univ. Calif. Press, Berkeley, California.

Snowball, G. J., and Lukins, R. C. (1964). Status of introduced parasites of Queensland fruit fly (*Strumeta tryoni*), 1960–1962. *Aust. J. Agr. Res.* **15**, 586–608.

Stapley, J. H. (1971). The introduction and establishment of the *Brontispa* parasite in the Solomon Islands. pp. 2–6 In *Noumea, S. Pac. Comm. Inf. Circ.* No. 30.

Stapley, J. H. (1972). Field studies on the ant complex in relation to premature nutfall of coconuts in the Solomon Islands. In "Cocoa and Coconuts in Malaysia" (R. L. Wastie & D. A. Earp, eds.), pp. 345–356. Inc. Soc. of Planters, Kuala Lumpur.

Surany, P. (1960). Diseases and biological control of rhinoceros beetle. *S. Pac. Comm., Noumea, New Caledonia Tech. Rep.* **128**.

Syed, R. A. (1971). Biological control possibilities of some insect and weed pests in Sabah. *In* "Crop Protection in Malaysia" (R. L. Wastie and B. J. Wood, eds.), pp. 124–132. Inc. Soc. of Planters, Kuala Lumpur.

Syed, R. A., and Pang, T. C. (1972). Status, history and control of *Setora nitens* (Wlk.) in Sabah. In "Cocoa and Coconuts in Malaysia" (R. L. Wastie & D. A. Earp, eds.), pp. 322–328 Inc. Soc. of Planters, Kuala Lumpur.

Tapley, R. A. (1961). Coffee leaf miner epidemics in relation to the use of persistent insecticides. *Res. Rep. Coffee Res. Sta. Lyamangu, Tanganyika* **10**, 43–55.

Taylor, T. H. C. (1937). "The Bilogical Control of an Insect in Fiji. An Account of the Coconut Leaf-Mining Beetle and Its Parasite Complex," 239 pp. Imp. Inst. Entomol., London.

Taylor, T. H. C., and Paine, R. W. (1935). The campaign against *Aspidiotus destructor* Sign. in Fiji. *Bull. Entomol. Res.* **26**, 1–102.

Tjoa Tjien-Mo. (1965). The occurrence of two strains of *Brontispa longissima* (Gestro) (Coleoptera: Hispidae) based on resistance or non-resistance to the parasite *Tetrastichus brontispae* (Ferr.) (Hymenoptera: Eulophidae) in Java. *Bull. Entomol. Res.* **55**, 609–614.

Todd, E. L. (1959). The fruit piercing moths of the genus *Gonodonta* Hubner (Lepidoptera: Noctuidae). *U.S. Dep. Agr. Tech. Bull.* **1201**.

Tothill, J. D., Taylor, T. H. C., and Paine, R. W. (1930). "The Coconut Moth in Fiji. A History of Its Control by Means of Parasites," 296 pp. Imp. Bur. Entomol., London.

Trjapitzin, V. A. (1963). Survey of acclimatization of Encyrtidae for control of agricultural pests and suggestions for their acclimatization in the U.S.S.R. *In* "Acclimatization of Animals in the U.S.S.R." (A. I. Yanushevich, ed.), pp. 226–229. Isr. Program for Sci. Transl., Jerusalem, 1966. (Engl. Transl.)

Tunçyürek, M., and Oncuer, C. (1974). Studies on aphelinid parasites and their hosts, citrus diaspine scale insects, in citrus orchards of the Aegean region. *Bull. SROP* 1974/3: 95–108.

Turica, A. (1968). Lucha biologica como medio de control de las moscas de los frutos. *IDIA* **241**, 29–38.

van den Bosch, R., and Haramoto, F. H. (1953). Competition among parasites of the Oriental fruit fly. *Proc. Hawaii. Entomol. Soc.* **15**, 201–206.

van den Bosch, R., and Telford, A. D. (1964). Environmental modification and biological control. *In* "Biological Control of Insect Pests and Weeds" (P. DeBach, ed.), pp. 459–488. Chapman & Hall, London.

van den Bosch, R., Bartlett, B. R., and Flanders, S. E. (1955). A search for natural enemies of lecaniine scale insects in northern Africa for introduction into California. *J. Econ. Entomol.* **48**, 53–55.

van Whervin, L. W. (1968). The introduction of *Prospaltella opulenta* Silvestri into Jamaica and its competitive displacement of *Eretmocerus serius* Silvestri. *PANS* **A14**, 456–464.

Vesey-FitzGerald, D. (1953). Review of the biological control of coccids on coconut palm in the Seychelles. *Bull. Entomol. Res.* **44**, 405–413.

Watanabe, C. (1958). Review of biological control of insect pests in Japan. *Proc. 10th Int. Congr. Entomol. Montreal* **4**, 515–517.

Way, M. J. (1953). The relationship between certain ant species with particular reference to biological control of the Coreid, *Theraptus* sp. *Bull. Entomol. Res.* **44**, 669–691.

Way, M. J. (1954). Studies of the life history of the ant *Oecophylla longinoda* Latreille. *Bull. Entomol. Res.* **45**, 93–112.

White, E. B., DeBach, P., and Garber, M. J. (1970). Artificial selection for genetic adaptation to temperature extremes in *Aphytis lingnanensis* Compere (Hymenoptera: Aphelinidae). *Hilgardia* **40**, 161–192.

Wille, J. E. (1941). Resumen de las diferentes labores ejecutadas en el Peru para combatir insectos dañinos por le "método biológico." *Proc. 6th Pac. Congr.*, **4**, 369–371.

Wille, J. E. (1958). El control biológico de los insectos agricolas en el Peru. *Proc. 10th Int. Congr. Entomol.* **4**, 519–523.

Williams, G. (1954). Field studies on the cocoa mirids *Sahlbergella singularis* Hagl. and *Distantiella theobroma* (Dist.) in the Gold Coast. III. Population fluctuations. *Bull. Entomol. Res.* **41**, 725–748.

Wilson, F. (1960). A review of the biological control of insects and weeds in Australia and Australian New Guinea. *Commonw. Inst. Biol. Control Tech. Commun.* **1**.

Wolcott, G. M. (1948). The insects of Puerto Rico. *Univ. Puerto Rico J. Agr.* **32**, 975 pp.

Wood, B. J. (1968). "Pests of Oil Palms in Malaysia and Their Control," 204 pp. Inc. Soc. Planters, Kuala Lumpur.

Wood, B. J. (1969). Studies on the effect of ground vegetation on infestations of *Oryctes rhinoceros* (L.) (Coleoptera: Dynastidae) in young oil palm replantings in Malaysia. *Bull. Entomol. Res.* **59**, 85–96.

Wood, B. J. (1971). Development of integrated control programs for pests of tropical perennial crops in Malaysia. *In* "Biological Control" (C. B. Huffaker, ed.), pp. 422–457. Plenum, New York.

Wood, B. J., and Nesbit, D. P. (1969). Caterpillar outbreaks on oil palms in eastern Sabah. *Planter Kuala Lumpur* **45** (5), 285–299.

Yasumatsu, K. (1958). An interesting case of biological control of *Ceroplastes rubens* Maskell in Japan. *Proc. 10th Int. Congr. Entomol. Montreal* **4**, 771–775.

Yasumatsu, K. (1969). Biological control of citrus pests in Japan. *Proc. 1st Int. Citrus Symp. Riverside, Calif.* **2**, 773–780.

Young, E. C. (1971). "Report on the introduction of *M. anisopliae* into Tongatapu. Report of the project manager for the period June 1970–May 1971," pp. 250–253. [UNDP (SF)/SPC project for research on the control of the coconut palm rhinoceros beetle.] South Pacific Commission, Noumea, New Caledonia. (Multigr.)

Young, E. C., (1974). The epizootiology of two pathogens of the coconut palm rhinoceros beetle. *J. Invert. Path.*, **24**, 82–92.

Zelazny, B. (1971). "Report of the project manager for the period June 1970–May 1971," pp. 42–131. [UNDP (SF)/SPC project for research on the control of the coconut palm rhinoceros beetle.] South Pacific Commission, Noumea, New Caledonia. (Multigr.)

Zelazny, B., (1972). Studies on *Rhabdionvirus oryctes*. I. Effect on larvae of *Oryctes rhinoceros* and inactivation of the virus. *J. Invert. Pathology* **20**, 235–241.

Zelazny, B., (1973). Studies on *Rhabdionvirus oryctes*. II. Effect on adults of *Oryctes rhinoceros*. *J. Invert. Pathology* **22**, 122–126.

16

BIOLOGICAL CONTROL OF PESTS OF RANGE, FORAGE, AND GRAIN CROPS

K. S. Hagen, G. A. Viktorov, Keizo Yasumatsu,
and Michael F. Schuster

I. INTRODUCTION

The grouping of crops in this chapter is rather unnatural. Involved are annual and perennial crops, ones grazed standing and ones cut for hay, and others harvested as grain for human and animal consumption. Moreover, they are grown under widely different geographic, climatic and, edaphic conditions, from dry semidesert grasslands and arctic tundra to tropical irrigated rice paddies.

Of all the agricultural crops, the forage, grains, and range, directly and indirectly, furnish the basic food supply of the world's human population, and there is great pressure today to increase production of these crops. As acreages and monocultures have increased and higher yields have been intensively sought, the greater the insect problems have become. Fortunately natural control, cultural control, resistant plant varieties, and in recent years pesticides and applied biological controls have permitted man to gain the harvest of most of these crops for himself.

Yet, even today the use of insecticides on wheat, barley, and range lands is not great, and in remote areas of the Orient pesticides are seldom used on rice. For centuries rice was grown in the Old World without the use of insecticides. Alfalfa is not regularly treated with insecticides in most countries. Only in the last 40 years have insecticides been used on this crop, and this has come about mainly because of newly introduced pests, and more intense alfalfa hay and seed production.

Not much has been attempted in biological control of pests of grasses used for grazing. Somewhat more, but still a very limited effort has been made to ascertain the scope of possibilities and to conduct biological control campaigns for pests of cereal grains, including rice, whereas a greater effort has been made regarding pests of forage and hay crops, notably alfalfa.

For all the six crops the areas of classical biological control and periodic colonizations are brought together here.

A. Classical Biological Control

The importation of exotic natural enemies to control insect pests in grain, forage, and range has been attempted in remarkably few instances, although for some of these importations varying degrees of control have been achieved. In wheat, an introduced ichneumonid from Europe became widely established and caused heavy parasitism of the European wheat stem sawfly in Canada and eastern United States (Streams and Coles, 1965). Beginning in 1966, three European parasites of the cereal leaf beetle have been imported, released, and recovered in eastern United States. The mymarid egg parasite, *Anaphes flavipes* (Foerst.), was established in Michigan in 1967 and by 1970 it was found over an area of 8000 square miles. Also, two larval parasites, *Tetrastichus julis* (Walker) and *Diaparsis carinifer* (Thomson), have become established at several places in Michigan, and after 10 years it appears that they may become a useful factor in integrated control of the populations (Haynes *et al.*, 1974), or according to Sailer (1972) may have already, and alone, reduced the problems. Recently, control of skeleton weed in wheat and other grain crops has been achieved in Australia by introduction of a species of rust, *Puccinia chondrillina* Bubak and Syd. (See Chapter 19.)

In sorghum, *Aphelinus asychis* (Walker) from France has recently become established against the greenbug, *Schizaphis graminum* (Rondani), in Oklahoma and has also been recovered from *Aphis helianthi* Monel on sun flowers (Jackson *et al.*, 1971). *Aphelinus varipes* (Forester), also imported from France, is widely established in Oklahoma on the corn leaf aphid, *Rhopalosiphum maidis* (Fitch), attacking sorghum and thus supplementing the parasitization of *R. maidis* by the native parasite *Aphelinus nigritus* Howard (Archer *et al.*, 1974).

In rice, two imported egg parasites and three braconid larval parasites were established in the Hawaiian Islands against the rice stem borer, *Chilo suppressalis* (Walker), and the braconid *Spathius helle* Nixon became established against the same stem borer in Japan. In the Philippines, three species of *Trichogramma* were established against rice stem borers. A tachinid from Nigeria was established in the Belgian Congo as a parasite of *Eldana saccharina* Walker where it is now heavily parasitized by native secondary parasites. However, Yasumatsu and Torii (1968) considered that none of these introduced parasites appeared to give successful control. More recent introductions against rice pests have included the tachinid *Sturmiopsis inferens* Tns., introduced from India to the Philippines against rice stem borers, the ichneumonid, *Cremastus* (*Trathala*) *flavoorbitalis* (Gam.), from Hawaii to Fiji against the rice leaf roller *Susumia exigua* (Butler), and *Apanteles marginiventris* (Cresson) from Hawaii to Fiji against the noctuid *Mythimma separata* (Walker), which in spite of being attacked by secondary parasites has prevented outbreaks of the noctuid in rice (Rao *et al.*, 1971).

The range, with its more stable ecosystem and its economic limitations with respect to wide-scale use of pesticides, is more conducive to classical biological control than are the cultivated ephemeral grain and forage crops. Yet, biological control attempts against range and pasture pests have been few. The most outstanding successes have been the use of insects to control range weeds, e.g., *Opuntia* spp. in Australia and many other areas; St. Johnswort in Australia and the western United States and Canada; ragwort in California; Koster's curse in Fiji; and lantana in Hawaii (Huffaker, 1969; Andres and Goeden, 1971; and Chapter 19). An outstanding example of biological control of an insect pest of range is the control of rhodesgrass scale in Texas by *Neodusmetia sangwani* (Rao) introduced from India. This example and a number of other attempts are discussed in the section on range grasslands.

The few attempts to introduce natural enemies against grasshoppers and locusts have essentially failed. Historically, the first purposeful international importation of a natural enemy against an insect pest was the introduction in 1762 of the Indian mynah bird, *Acridotheres tristis* L., to the island of Mauritius to combat the red locust, *Nomadacris septemfasciata* (Serv.). This introduction is credited with reducing the locust outbreaks (Moutia and Mamet, 1946). *Scelio pembertoni* Timberlake, introduced into the Hawaiian Islands from Malaysia, according to Pemberton (1948), has controlled the Chinese grasshopper, *Oxyo chinensis* (Thunb.). Attempts to utilize dipterous parasites to control locusts in Canada failed (McLeod, 1962).

Cultivation of wild lands has increased pest problems, but it has also solved some former severe problems. In the 1870's the Rocky Mountain grasshopper, *Melanoplus spretus* (Walsh), destroyed grain over thousands of square miles in midwestern North America from Canada to Texas, but no specimens have been taken for more than 50 years, and this plague grasshopper is thought to be extinct (Helfer, 1953). However, some researchers (e.g., Uvarov, 1928) believe *M. spretus* may be the swarming phase of *Melanoplus mexicanus atlantis* Riley. One of the first federal (United States) entomological commissions was formed to study and combat this pest. Many natural enemies associated with *M. spretus* were described (Riley *et al.*, 1878). These pioneer entomologists concluded (erroneously) that a period of pest excess is sure to be followed by a period of scarcity, due largely to the increase of its natural enemies. In this case, on the contrary, the apparent extinction of *M. spretus* can hardly be ascribed to natural enemies, but was probably the result of greatly reducing the natural egg deposition areas by the advent of cultivation (e.g., Uvarov, 1928).

Greathead's (1963) recent world review of the natural enemies of grasshoppers and locusts lists 223 species of Acridoidea and 534 species of parasites and predators, belonging to 43 insect families, which attack them. Among this great complex, one would expect some examples of very effective natural control. Many of the grasshopper species rarely manifest outbreaks compared to the migratory locusts. Perhaps natural enemies are more effective against the former, while the latter escape by their migrating habits.

Of the crops discussed in this chapter, alfalfa has received perhaps the most attention by biological control workers. Most attempts to introduce natural enemies against alfalfa pests have been in North America, and recently in South America, for it is in the New World that the serious pests have invaded without their native parasites.

In the United States, two species of Aphidiidae introduced against the pea aphid have become established; two species of the same family and also an aphelinid have been imported and established as parasites of the spotted alfalfa aphid; at least six species of parasitic Hymenoptera have been imported and established for control of the alfalfa weevil, *Hypera postica* (Gyll.); three hymenopterous parasites have recently been introduced and established (California) against the Egyptian alfalfa weevil, *H. brunneipennis* (Roh.), a comparatively recent invader; an ichneumonid recently imported from Italy against the clover leaf weevil, *H. punctata* (F.), is now widely established; and *Hypera nigrirostris* (F.) is also now parasitized in the United States by two Hymenoptera introduced against *H. postica* (see Section V).

In other areas, the sweet clover weevil has been the subject of biological control attempts mainly in Canda, and a braconid from Sweden has been recovered from this weevil, and as well, from a native related *Sitona* sp. In Australia and South Africa, a predatory mite introduced from the Mediterranean region has become established against the lucerne "flea." It is too early yet to draw conclusions as to the benefits to be derived from many of these introductions.

B. Periodic Colonizations

The ephemeral grain crop is not the ideal ecosystem in which to attempt classical biological control, and the vastness of these annual grain monocultures nearly precludes the use of periodic releases of natural enemies. However, extensive releases of *Trichogamma* have been made in the U.S.S.R. against noctuid pests of wheat, and scelionids have been periodically released against migratory grain-infesting pentatomids in the U.S.S.R. and Iran.

One of the early attempts to control aphids by mass releases of parasites in grain was in Kansas (United States). This occurred during a severe outbreak of the greenbug, *Schizaphis graminum*. Hunter (1909) observed parasitism of the introduced aphid only in the extreme southern part of the state. Since the aphid became abundant in Texas and Oklahoma earlier than in Kansas, Hunter transported aphidiid parasitized greenbugs from those states and released them in Kansas early in the season to offset the normal lag in parasite buildup. Several hundred million *Lysiphlebus testaceipes* (Cresson) were transported north to Kansas. The results are not clear. Many farmers claimed it to be a success, but in numerous cases the claims were made within 10 days of the parasite releases, which would appear hardly time enough for the wasps to have had significant impact. However, parasitization does stop the aphid's production of young within 3 days (Hight *et al.*, 1972). Glenn (1909) found that the greenbug could develop and reproduce at temperatures which retarded *Lysiphlebus* oviposition and development. This was considered to be the key factor limiting the effectiveness of *Lysiphlebus* in the critical early spring period (Hunter, 1909). This native parasite, however, is often credited, along with predators, with preventing or terminating outbreaks of the greenbug (Archer *et al.*, 1974). In the High Plains area of Texas, *L. testaceipes* parasitizes 100% of the 6th generation of Biotype C greenbug on grain sorghum, but the use of insecticides is necessary on the 4th and 5th generations of the greenbug (Walker *et al.*, 1973).

Attempts to control this aphid in Uruguay by transporting *Lysaphidus platensis* (Brèthes) into infested oat fields were not successful. When such transfers were made only 5 aphids per 100,000 were estimated to be parasitized in host populations approximating 360 million per hectare (Silveira-Guido and Conde-Jahn, 1945).

Releases of 100 million to 300 million cage-reared *Aphidius smithi* Sharma and Subba Rao were made per year over a 3-year period against the pea aphid in alfalfa in Washington (United States) in order to prevent the migration of aphids from their overwintering sites in alfalfa to about 100,000 acres of peas. The pea aphid populations did not reach economic levels in any of the alfalfa fields where parasites were released, whereas in most control fields or fields where insecticides were used the aphid populations attained economic levels and more alate aphids were generated. However, the parasite release method used appears impractical at present because of the production and distribution requirements involved (Halfhill and Featherston, 1973).

Transporting the coccinellid *Hippodamia convergens* (Guerin-Meneville) from aestivo-hibernating aggregations in the Sierra Nevada of California to lowland valleys to control aphids was begun in the early 1900's, but these releases failed to control the aphids in the grain fields where released because they dispersed far and wide before preying on aphids (Davidson, 1924). Today, farmers and gardeners still purchase these collected beetles in spite of their ineffectiveness (Cooke, 1963; Hagen, 1962; Kieckhefer and Olson, 1974).

II. RICE

Rice cultivation has had a very long history. During the extensive periods prior to use of processed insecticides and fertilizers, farmers had been selecting rice best suited to their needs from various standpoints, such as quality and quantity of yield, adaptation to soil and climatic conditions, pest problems, availability of labor, and water. In Thailand, for example, farmers unknowingly have selected rice varieties somewhat resistant to insect borers or which have a detrimental effect on their development because of rapid decomposition of stubble after harvest (Nishida and Wongsiri, 1972). Thus, in every developing country in Asia, mainly traditional varieties of rice are grown. This is only disturbed at some risk.

In rice paddies, natural enemies, too, play an important role in suppressing the populations of rice pests below economic levels. As a result of experiments in Burma, Lazerevic (FAO, 1970) concluded: "Most of the pests were found in rather low populations and below the level which could affect yield. This was confirmed by spraying experiments in Burma. The plots of paddy sprayed regularly with different insecticides gave about the same yields as the unsprayed controls." Such situations exist in various districts even at the present time (e.g., Lim, 1970; Li, 1970).

On the other hand, effort has recently focused on the improvement of rice production through breeding and introduction of high-yielding varieties. Current utilization of high-yielding rice varieties, chemical fertilizers, and pesticides has considerably changed the complex biotic environment of rice paddies, which was itself formerly conducive to great stability in the populations of rice pests and their natural enemies by providing a varied and complex system of checks and balances. In modern rice culture, some artificial pest control seems unavoidable in obtaining high yields of quality rice; use of natural enemies alone is not sufficient. Application of insecticides is required, and this poses potential disturbance of natural controls and pollution of the environment. Yet the necessary suppression of the pests can be accomplished by the combined tactics of integrated control, employing or conserving natural enemies to a maximum, and using insecticides minimally, only as really needed to protect the crop, and supplemented by cultural tactics.

A. The Natural Enemy Fauna of Rice Paddies

General faunistic study of rice fields was initiated only in 1963, i.e., by one of us (Yasumatsu) and associates in south and east Asia. The natural enemy fauna of rice stem borers is now comparatively well known. By 1966, over 100 species of their natural enemies (parasites, predators, and pathogens) had been recorded in Asia alone, e.g., approximately 56 species for *Chilo suppressalis* (Walker), 40 species for *Tryporyza incertulas* (Walker), and 16 species for *Sesamia inferens*. Also, Mohyuddin and Greathead (1970) published an annotated list of the parasites of all graminaceous stem borers in eastern Africa. Some important species were referred to in several reviews (e.g., Yasumatsu and Torii, 1968).

However, information on the natural enemies of rice pests other than of rice stem borers remains quite fragmentary. The taxonomy of sciomyzids serving as alternate hosts of *Trichogramma* spp. in south and east Asia (paddies and elsewhere) was studied by Yano (e.g., 1970). The taxonomy of Oriental pipunculids, including parasites of leafhoppers and planthoppers in rice, was partly published by Hardy (1972). Recently, study of the spiders in rice in Japan has been initiated, and their importance as predators of leafhoppers and planthoppers has been noted (Kobayashi, 1961). Similar research has been conducted in Thailand (Okuma, 1968) and Taiwan (Chu and Okuma, 1970). Also in south and east Asia paddies, 16 genera and 33 species of coccinellids are reported as predators of rice aphids and other pests, and of egg masses of rice stem borers (Sasaji, 1968).

1. Egg Parasites and Egg Predators

a. Rice Gall Midge, Pachydiplosis oryzae (Wood-Mason). The egg—larval parasite, *Platygaster oryzae* Cameron, is an important factor in the mortality of the rice gall midge. According to Fernando (1972), parasitization is low early in the season, rises to 40—50% in the tillering phase, and reaches 80—95% by the end of the season, which is a representative trend for this species everywhere that a single crop is grown annually.

b. Rice Stem Borers. In many unsprayed paddies, egg parasites of pyralid stem borers are the most important suppressing factor (Bess, 1967, 1972; Li, 1971; review by Yasumatsu and Torii, 1968). Bess (1967) wrote: "It appears that parasites of (borer) eggs and predators of newly hatched larvae are the most effective and promising natural

TABLE 1
Major Rice Stem Borers and Their Egg Parasites

Rice stem borer	Egg Parasite
Chilo suppressalis (Walker)	*Trichogramma australicum* Girault
	Trichogramma chilonis Ishii
	Trichogramma japonicum Ashmead
	Telenomus dignus Gahan
Chilo polychrysus (Meyrick)	*Trichogramma japonicum* Ashmead
	Telenomus dignus Gahan
Tryporyza incertulas (Walker)	*Trichogramma australicum* Girault
	Trichogramma japonicum Ashmead
	Telenomus rowani Gahan
	Telenomus dignoides Nixon
	Tetrastichus schoenobii Ferriere.
Tryporyza innotata (Walker)	*Trichogramma australicum* Girault
	Trichogramma japonicum Ashmead
	Telenomus rowani Gahan
	Aprostocetus sp.

enemies to use in biological control programs." Moreover, Rothschild (1970) reported that the heaviest mortality occurs during the egg and early larval stages, and that parasites, together with predatory gryllids (*Anaxipha* spp.) and tettigoniids (*Conocephalus* spp.), can account for over 90% of these losses in Sarawak. Bess (1972) considered that egg predation is apparently primarily by ants. Coccinellids were also observed feeding on eggs. The bionomics of about ten species of egg parasites were published by Rothschild (1970). The egg parasites of the major rice stem borers are listed in Table 1, and interrelations are shown in Fig. 1.

In Japan, prior to the introduction of organosynthetic insecticides, about 50 to 90% of the eggs of the first emerging stem borer moths were parasitized by *Trichogramma japonicum* Ashmead and *Telenomus dignus* Gahan. Eggs of the next generation were only lightly parasitized, because, apparently, the rice plants are older and borer egg masses are usually laid on the sheathes among dense foliage (Yasumatsu, 1967). In Thailand, the percentage of stem borer eggs parasitized ranged from 48 to 94%. Egg parasites appeared to be active soon after planting and up to the tillering stage (Nishida and Wongsiri, 1972). In tropical countries, the overlapping of rice stem borer generations offers ample opportunity for the host to produce many eggs for attack by the egg parasites; during the off season, also, the parasites oviposit in borer eggs on volunteer or ratoon rice plants (Rothschild, 1970).

c. Leafhoppers and Plant Hoppers. Egg parasites of leafhoppers and plant hoppers may effectively suppress populations of their hosts. An interesting example is given in papers by Kiritani and collaborators (e.g., Sasaba and Kiritani, 1972) on the factors causing mortality of *Nephotettix cincticeps* Uhler: "The parasitism of eggs has been studied for 4 years in an experimental paddy field, in Ino, about 700 m^2 in area, where no insecticides have been used for a successive period of five years since 1967. The

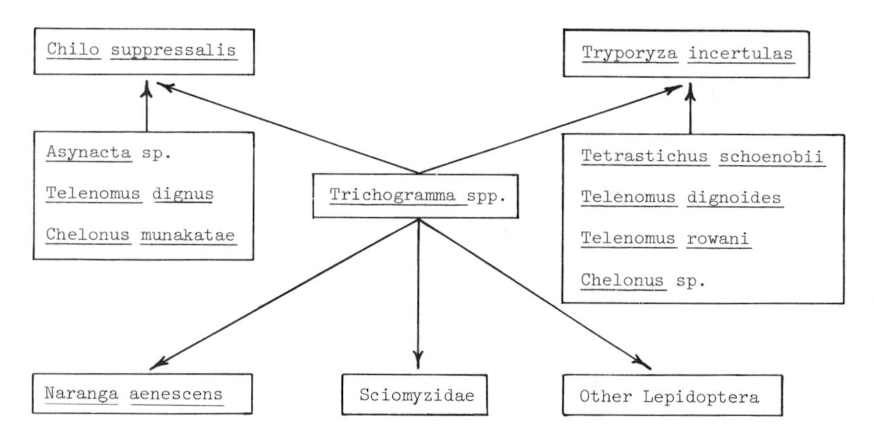

Fig. 1. Relationship between hymenopterous parasites and rice stem borers in the egg stage (After Yasumatsu, 1967).

percentage of eggs parasitized [by *Paracentrobia andoi* (Ishii)] increased year by year . . . the maximum number of eggs per hill throughout the season decreased year after year: 909.3, 673.3, 581.4, and 447.4 in 1968, 1969, 1970, and 1971, respectively. . . . These facts suggest that the population density of leafhoppers was reduced annually more and more by biological control factors. . . . It is conceivable (therefore), that the egg parasites would play an important role in the integrated control program (for) the leafhoppers." Among the egg parasites, species of the genera *Alaptus*, *Anagrus*, *Lymaenon*, *Ooctonus*, and *Oligosita* are important. As egg predators, *Cyrtorrhinus* spp. are not negligible in effect.

d. Southern Green Stink Bug. *Nezara viridula* L. attacks various crops, including rice. Among its egg parasites, two species are most important in reducing its populations, at least in some crops. Although mainly a pest of other crops there, Davis (1967) wrote that introduced parasites, *Trissolcus basalis* (Wollasten) and *Trichopoda pennipes* Fabricius, have exerted considerable control of *Nezara* populations on Oahu and Kauai (Hawaii) and that there had been no serious outbreaks since 1963, a statement that holds to this time (personal communication). In South Pacific areas, including Australia and New Zealand, *Trissolcus basalis* has also been utilized successfully to control this bug on several crops. In the Philippines, this species has never been known to break out in epidemic numbers perhaps because of its egg parasite, *Ooencyrtus* sp.

2. Parasites and Predators of Rice Stem Borer Larvae and Pupae (Fig. 2)

Parasites of larval and pupal stages of rice stem borers are not generally as effective as egg parasites; parasitism seldom exceeds 2–3% (Rothschild, 1970; Nishida and Wongsiri, 1972; Bess, 1972), notwithstanding the abundance of species that attack the larvae and pupae (Rao, 1972; Yasumatsu, 1967; Mohyuddin and Greathead, 1970). There are, however, many exceptions (Yasumatsu, 1967; Yasumatsu and Torii, 1968).

In some areas of Japan, *Temelucha biguttula* Munakata is sufficiently abundant and effective that insecticides are not needed against *Chilo suppressalis*. *Apanteles chilonis* Munakata is also effective against overwintering larvae of *C. suppressalis* in rice stems. It completes several generations during the spring and early summer (Kajita and Drake, 1969), parasitizing larvae to an increasing extent. Active also is the fungus *Spicaria farinosa* (Fries) and two species of predatory anthocorids. Nearly 100% of overwintering larvae in rice stems held for pulp factory use in Japan were attacked by *Lyctocoris beneficus* (Hinra) and *Xylocoris galactinus* (Fieber). Parenthetically, in a study of *Sesamia inferens* Walker attacking sugarcane in southern Japan, Nagatomi (1972) found the rate of parasitism by the larval–pupal parasite *Enicospilus sakaguchii* (Matsumura et Uchida) to be as high as 67% in September, 1968, and 40% in September, 1969. They regarded this ichneumonid the only highly effective parasite of *Sesamia inferens*.

Apanteles flavipes (Cameron), although common in southeast Asia, does not control rice stem borers. Its parasitism of *S. inferens* in Sarawak never exceeded 2% (Rothschild, 1970), but, interestingly, when introduced from India to control *Diatraea saccharalis* (Fabricius) in sugarcane in Barbados, it built up an extraordinarily high population in a short period of time. Assessment showed that crop damage consequently was reduced considerably. The infestation, which fluctuated around 15% until 1966, decreased to less than 6% in 1970 (Alam *et al.*, 1971). Mohyuddin (1971) considers that, as a potential biological control agent of graminaceous borers, *A. flavipes* has the advantage of adaptation to drier situations, while *A. sesamiae* Cameron has the advantage of a wider host range. Rothschild (1970) also observed that in his Sarawak study, *Tetrastichus israeli* (Rohwer) attacked 8% of the pupae of *Chilo suppressalis* and 18% of those of *S. inferens*.

In Australia, Li (1971) reported high parasitism of aestivating rice stem borers by

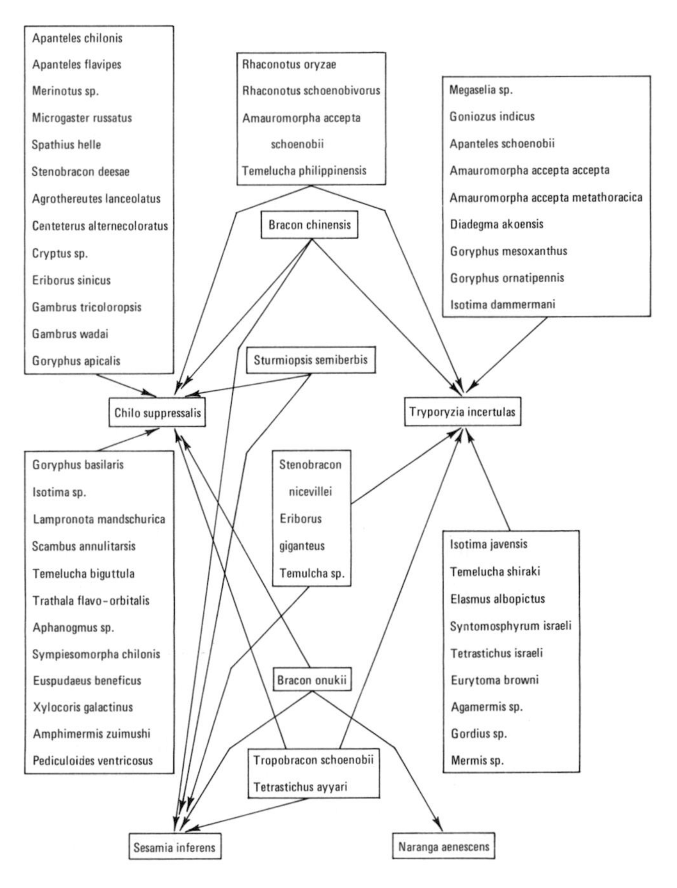

Fig. 2. Relationship between natural enemies and rice stem borers in the larval stage (After Yasumatsu, 1967).

Bracon sp. *Itoplectis narangae* (Ashmead) is also a common pupal parasite of both rice stem borers and certain rice defoliators.

In subtropical and tropical paddies, species of the wasp genus *Ropalidia* are important in carrying away rice-defoliating Lepidoptera.

3. *Predators of Rice Pests*

The importance of spiders as predators of leafhoppers and plant hoppers in paddies was first studied by Kobayashi (1961) in Japan. Among about 70 species of spiders, 50–80% of the spider populations was represented by *Oedothorax insecticeps* Boesenberg et Strand. About 81% of the leafhopper populations in paddies was controlled by this species from mid-July to early August. *Lycosa pseudoannulata* (Boesenberg et Strand), another important species, reduced the same leafhopper populations by about 9% during the same period. However, where the population densities of the leafhoppers and plant hoppers were high, the spiders could not suppress them. Spiders may also prevent resurgence of leafhopper populations after their suppression by use of selective insecticides, of which the latter improves the balance between spider and leafhopper populations (Sasaba and Kiritani, 1972). At present, 15 families, 57 genera, and 60 species in Taiwan, and 13 families, 28 genera, and 46 species are found in Thailand.

B. Future Use of Natural Enemies for Rice Pest Control

So long as traditional, primitive rice cultivation lasts and only the older "classical" insecticides are used, natural control of rice pests would likely continue. However, with the recent trend of growing high yielding varieties over extensive areas and substantial use of chemical fertilizers and machines, some modern agricultural pesticides are required. It then becomes more essential to learn how to conserve the important natural enemies (above). The following measures may be considered helpful: (1) using more selective insecticides, applied at the best time, and only when needed; (2) introducing resistant varieties; (3) providing hibernacula, shelters, or nesting sites to natural enemies in and around the paddy fields; (4) providing nectar- and pollen-producing plants in the rice environment (for such natural enemies as braconids and ichneumonids); (5) preserving some reservoir ratoon or volunteer rice plants, or weed hosts (Graminaceae and Cyperaceae) which harbor the rice pests and their natural enemies; (6) inducing double or triple cropping of rice to permit continuous use of some populations of rice pests so the natural enemies can persist adequately and may in turn regulate the populations of the rice pests below economic levels.

Interareal or regional introduction and colonization or exchange of natural enemies should be more aggressively pursued. The example cited above of the success in Barbados illustrates this potential. As there are indications of occurrence of strains in *Tetrastichus schoenobii* (Ferriere) and *Telenomus rowandi* Gahan (egg parasites), comparative studies are needed of the characteristics and utility of the different populations, as applicable to various geographic regions and conditions of culture. This approach should also be tried relative to other natural enemies and other rice pests.

Finally, mass rearing and strategic release (augmentation) of natural enemies may be found useful as a tactic of integrated control where chemicals are also used (Li, 1970, 1971; Lim, 1970; Rao, 1970, 1972; Yasumatsu, 1967, 1972; Yasumatsu and Torii, 1968). Encouraging for their utility in such programs, the following species of parasites and predators of rice pests have been mass reared either on alternate hosts or on primary hosts reared on semiartificial diets: *Trichospilus diatraea* (Chen, 1972) *Tetrastichus inferens* (Chen, 1972), *Apanteles chilonis* (Yasumatsu, 1967), *Apanteles flavipes* (Yasumatsu, 1967; Alam *et al.*, 1971; Mohyuddin, 1971), *Bracon chinensis* (Yasumatsu, 1967), *Tropobracon schoenobii* (Yasumatsu, 1967), *Itoplectis narangae* (Shin, 1970), *Lyctocoris beneficus*, *Xylocoris galactinus* (Chu, 1969), and *Trichogramma*.

III. CEREAL GRAINS OTHER THAN RICE

Cereals have ecological and economic peculiarities which affect the possibilities of utilizing biological control of their pests. Their short cropping period and the rotation practices commonly used in association with their culture create an unstable environment preventing substantial conservation and buildup of natural enemy populations. Moreover, in many Old-World countries the pest fauna of cereal crops is largely indigenous. This limits the prospects of finding and introducing effective exotic natural enemies. In addition, the rather low crop value per acre commonly makes it uneconomical to use expensive, repetitive releases of artificially produced parasites or predators. Perhaps this explains why there have been few attempts and few successes in the biological control of grain pests.

One of the claimed successes is the result of acclimatization in Canada of *Collyria calcitrator* (Grav.) introduced against the European wheat stem sawfly, *Cephus pygmeus* L. (Turnbull and Chant, 1961). This ichneumonid was introduced from Europe in 1937—1940. Parasitism by *C. calcitrator* increased to 47% in 1958 (Smith, 1959) when it became the most abundant parasite attacking this sawfly. Sawfly numbers decreased during this time to very low levels and have apparently remained so subsequently. However, according to Beirne (1972) there are no convincing reasons to believe that this project was, in fact, a success. Introduction of this parasite into the United States, beginning in 1935, was also followed by increase in host mortality. Parasitization during 1962—1963 ranged from 35—69% even in the low host populations then prevailing. The marked decline of the sawfly populations occurred slightly earlier than in Canada. Yet, the fact that the related *Cephus tabidus* (F.), which is not subject to attack by *Collyria calcitrator*, also declined in abundance during the same period (Streams and Coles, 1965) casts some doubt on the claim of success for this project.

Several species of parasites of Hessian fly, a serious pest of wheat, in the United States were imported in the 1890's. Clausen (1956) reported that only *Pleurotropis metallica* (Nees) became widely established and that it has exerted no substantial control. Perhaps an extended effort in the introduction of natural enemies of the Hessian fly would have been undertaken had not the problem been largely alleviated by the success of another form of biological control, i.e., the development and use of wheat varieties resistant to this insect (Chapter 25).

As noted previously, in recent years several natural enemies of the cereal leaf beetle, *Oulema melanopus* (L.), a recent arrival in this country, have been introduced in the eastern United States with some promise of success.

The value of the introduced rust *Puccinia chondrillina* in control of skeleton weed in wheat and other small grains in Australia is discussed in Chapter 19.

In the U.S.S.R. different species and strains of *Trichogramma* have long been used for control of various lepidopterous pests of grain, including *Euxoa segetum* Schiff.

Methods of mass rearing *Trichogramma* in the U.S.S.R. were first elaborated by Meyer (1941), substantially improved by Telenga and Schepetilnikova (1949), and still more recently, made largely automatic (see Lebedev, 1970). Important in the practical use of *Trichogramma* is the choice of species and strains best adapted to particular pests and ecological conditions.

Large-scale field releases of *Trichogramma* have been made in recent years against *E. segetum* and *P. nubulalis*. It is currently (1974) relied upon entirely for control of European corn borer in Moldavia (Maria Tuteovitch, personal communication). *Trichogramma* is now annually used for pest control over some 5 million hectares of grain and other crops in the Ukraine SSR, and extensively also in Moldavia and other provinces. Many authors in different countries have contributed to the elucidation of the complicated taxonomy of *Trichogramma* in recent decades (e.g., Telenga, 1956; Nagaraja and Nagarkatti, 1973). According to Telenga the most useful species for biological control of *E. segetum* is *T. evanescens* s.str., which is a common parasite of noctuid eggs in fields and grasslands.

Much effort has been spent on biological control of pest pentatomids (*Aelia* spp. and *Eurygaster* spp.) on cereal crops in Europe, North Africa, and the Near and Middle East. Significant loss in quality of the produce is caused by the first generation larval and adult pentatomids. The present high marketing standard for corn sets a very low economic threshold, creating difficulties in control of corn pests.

Eurygaster integriceps Put. is especially serious and has been the main object of biological control. This species, the sunn pest, has two important groups of natural enemies: (1) egg parasites (Scelionidae and Encyrtidae) and (2) tachinid flies. These parasites sometimes very strongly affect the population dynamics of *E. integriceps* (Viktorov, 1967).

Scelionids are promising for mass release purposes because of their high reproductive potential and rather simple methods of mass rearing.

The taxonomy of pentatomid egg parasites, especially the genus *Trissolcus* (= *Asolcus*), has been studied in recent years by several authors, and according to Viktorov (1967), both the species composition and their relative numbers among a complex are different in various parts of the range of *E. integriceps*. These differences are well correlated with differences in adaptations of the particular species to abiotic factors, especially humidity (Viktorov, 1960).

All the known egg parasites of these pentatomids are polyvoltine and use different host species during their various generations. Therefore, their abundance is low in regions where a single monoculture of cereal crop predominates and there are few alternative host species or suitable hibernating places. This is well demonstrated in the range of *E. integriceps* in some regions of the U.S.S.R. (e.g., Kamenkova, 1958; Viktorov, 1960, 1964).

Among the tachinids, besides polyphagous species, there are some host-specific ones. *Clytiomyia helluo* F. is closely associated only with species of *Eurygaster* (Dupuis, 1963). It has two generations per year, well synchronized with those of its hosts. Yet, the numerical response of this parasite to changes in the population density of *E. integriceps* is hindered by the very short periods of contact between the ovipositing parasite females of the summer generation and the adult hosts of the new generation prior to harvesting (Viktorov, 1967).

The first attempts to utilize natural enemies for control of *E. integriceps* were made in Russia at the end of the 19th and beginning of the 20th centuries (Saakov, 1903; Vassiliev, 1913). An extensive program of mass culture and release of scelionids was later conducted in the U.S.S.R., i.e., during World War II, against a severe outbreak. The main species used in this program was *Trissolcus grandis* Thoms. widely distributed in the U.S.S.R. Methods of culture have been elaborated by several authors (Kulakov, 1940; Schepetilnikova, 1942; and others). Wheat seedlings were the recommended food. However, egg production under the conditions described by the above authors was very low, only four to five eggs per female (Archangelskey and Polyakov, 1951). The same methods (Alexandrov, 1947—1949) were used in Iran where large-scale releases [mainly of *Trissolcus semistriatus* Nees and *T.* (= *M.*) *vassilievi* Mayr] were carried out for several years (Vaezi, 1950; Zomorrodi, 1959). However, the methods of egg production were then improved by Remaudière (1960), who suggested using wheat grains in water as food; the fecundity is double that when wheat seedlings are used.

To overcome the difficulty in collecting the hibernating bugs, Skaf *et al.* (1961) suggested collecting adults in the fields immediately before their emigration. Another possibility is the artificial rearing of a suitable polyvoltine pentatomid to serve as host of the egg parasites, e.g., species of *Aelia* or *Dolycoris* (Safavi, 1968). The most suitable culture host in the U.S.S.R. has been *Graphosoma italicum* Muell. (Gusev and Shmetzer, 1971; Popov, 1971).

Remaudière (1961) considered the biological control of sunn pest to cost only one-fifth as much as chemical control. However, there are still few data on the influence of releases of egg parasites upon the actual qualitative deterioration of grain caused by the pest, even though this is most essential to an evaluation of this method of control of *Eurygaster*. Release of egg parasites in the spring does not reduce the numbers of hibernated adult bugs whose feeding is especially damaging to seedlings of summer wheat. The release tactic is therefore incompletely effective.

There is a possibility of effective integrated control of *Eurygaster* (Viktorov, 1964, 1967) employing natural enemies in combination with other measures. Application of insecticides (organochlorines and organophosphates) to control the overwintered adults coincides with the time when adults of the very scarce overwintered populations of the scelionid and tachinid parasites are active, and their populations are also at their lowest ebb. The treatments greatly reduce the efficiency of these parasites (e.g., Viktorov, 1960, 1967; Kamenkova, 1971). Also these spring sprays are frequently not very effective against the pest itself (adults at this time). Consequently, with a minimum effect on the pest and a major harmful effect on its natural enemies, the nymphs of the new generation of *Eurygaster* in early summer are often more numerous in treated than in untreated fields (Viktorov, 1964, 1967; Zaeva, 1969). On the other hand, chemical control of these nymphs is not as detrimental to either group of parasites since many of them at this time are either within host eggs (the scelionids) or in the soil (pupae of the tachinids). Maximal restriction of spring chemical treatments can be compensated for by later treatments of the nymphs and this is very important for the increase in efficiency of the natural enemies.

This integrated approach is becoming popular in the U.S.S.R. Chemical controls in the spring are now much restricted, e.g., to densities above 2 hibernated adults/m^2 in winter wheat (Areshnikov *et al.*, 1971). Some recent studies involve a search for selective insecticides such as juvenile hormone analogs (Burov *et al.*, 1971).

Eurygaster integriceps is an example of a pest which has lost its natural regulating mechanism(s) under conditions of modern, large-scale single crop agriculture. Rather different is the human influence on the population dynamics of *Hadena sordida* Bkh., another important wheat pest in the U.S.S.R., especially in the steppe zone of North Kasakhstan and West Siberia. Severe outbreaks of this species immediately follow periods of extensive cultivation of virgin soils (Shek, 1962; Grigor'eva, 1965). *Hadena sordida* has a large complex of natural enemies, foremost of which is a host-specific parasite *Lampronota agnatha* Grav. (Shapiro, 1965). Their efficiency is greatly reduced during the first years of wheat cultivation but then increases gradually. As a rule, chemical control, due to the adverse effects on the natural enemies, prolongs the duration of outbreaks (Grigor'eva, 1965). The last outbreak of *Hadena* collapsed under the influence of a complex of natural mortality factors, including a virus disease and parasites. At this point adequate natural control was reestablished and *Hadena* has since been maintained at a low endemic population density (Shek, 1962; Grigor'eva, 1965).

IV. RANGE AND PASTURE GRASSLANDS

Grasslands used for grazing are grouped on the basis of growth habit and growth form of the plants. They fall roughly into open grasslands, dense grasslands, and various herb, grass, shrub associations (Brown, 1954).

In their native "climax" form, open grasslands, found on all continents, usually experience a distinct dry period during part of the year and are sufficiently adapted to present a high degree of stability year after year; hence, they should be conductive to biological control. The bunch-growth habit and occurrence of (sometimes bare) interspaces are characteristic.

Dense grasslands have an even cover of contiguous plants which leave virtually no interspaces, and may consist of annual or perennial grasses or both.

Dense annual grasslands are usually found in areas having a Mediterranean climate, and the associated plant species are also annuals. The composition of annual grasslands is very unstable year to year and no attempts to control the pests (except weeds) of annual grasslands by biological control have been made. It is noteworthy that the biological control of Klamath weed, *Hypericum perforatum* L., in California primarily restored a dense annual grass condition to the ranges (Chapters 11 and 19).

Dense perennial grasslands approach the stability of the open grasslands, and occur extensively in humid temperate regions where the climax forests are removed, or they may be artificially maintained by irrigation in areas of lesser rainfall.

Shrubs consisting of various woody plants may occur with open or dense grasslands in varying degrees.

Natural grasslands are the final or stable community in a successional series (sere). However, most grasslands are not in their most stable state due to actions by man and domestic animals and may be said to be disclimax except for a few relict areas. The full diversity and stability of climax vegetation promotes stability at the consumer level, particularly in a stable and favorable climate, but in agroecosystems, including many grasslands, this diversity and stability has often been reduced. Grasslands would fall between modified rain forest and subsistence agriculture with respect to isolation, diversity, and permanence in the scheme used by Southwood and Way (1970) to indicate their suitability to integrated control. The stability of the climate, however, would be less in many areas because the grasslands there are a result of climate unfavorable to more mesophytic vegetation.

A. Insects, Management Practices, and the Ecology of Grasslands

Little attention has been given the role of insects in determining the composition and structure of vegetation (Weaver and Clements, 1938). Range ecologists such as Stoddart and Smith (1943) also failed to appreciate the full impact of insects on range condition. Brues (1946, p. 90), however, recognized that insects are a prime factor in regulating the abundance of plants. Huffaker (1957, 1964, p. 643) and others have given clear examples of the importance of insects in relation to weed abundance and the composition of vegetation.

One might ask, "What kind and number of insects would one expect to find in grasslands which could influence grassland ecology and evolution?" Hendrickson (1930) found some 1175 species and varieties of insects in grassland communities in Iowa, and in 1932 found 144 additional species, belonging principally to the orders Orthoptera, Hemiptera, Homoptera, Coleoptera, and Diptera. In grassland pastures and grassy meadows in New York, invertebrates averaged 777 individuals/m^2 (Walcott, 1937), of which ants made up 26%, leafhoppers 15%, other insects 34%, and spiders 9%. Kelsey (1957) listed 45 species of insects that feed on the roots or aerial portions of native tussock grasses in New Zealand. Only a few species caused severe damage, and this was commonly localized and periodic. Sometimes they dealt the final death blow to tussocks already heavily injured by grazing animals and their trampling. Cumber (1958) found an average of 230 species and 20,000 specimens of insects per one-fifth acre of New Zealand grassland. One sampling indicated approximately 1,500,000 collembolans per acre. Populations of 6,500,000 grasshoppers per acre were reported by Ball (1915). Ricou (1967) concluded from a study in Normandy (France) that the study area was supporting an average minimum population of some 7,500,000 arthropods in the soil and 280,000 on the vegetation. Of this population, 50% was saprophytic, 36% phytophagous, and 12% parasitic or predatory.

The relationship between grazing practices and other management practices and grasshopper populations in Kansas bluestem pastures was investigated by Arnett (1960). Grasshopper populations (and damage?) were found to be largest in pastures where there were such practices as early spring burning and heavy grazing, and smallest in unburned pastures, with light grazing or deferred rotation grazing. Of the ten most abundant grasshopper species, nine were most abundant in either the spring burned or the heavily grazed pastures.

Spiders may help to regulate the numbers of insects on grasslands. In Nebraska, Muma and Katherine (1949) found 111 species of spiders in high open grassland sites, making up 7% of the arthropod population, but no relationship to control of any phytophagous species was shown.

The complexity of prey regulation by predators in a Normandy pasture was suggested by Ricou (1967). At times predators reacted in a density-dependent way primarily to the phytophagous insects, but at other times mainly to the saprophagous insects. Parasites were apparently regulating the numbers of the frit fly, *Oscinella frit* (L.), until September when spiders became quite abundant. The fly then reached outbreak proportions, suggesting the possibility that the spiders may have selectively limited the effectiveness of the parasites much as ants do (e.g., Bartlett, 1961).

These examples present a brief look into the complex interrelations of grasslands and grassland insects. There are many secondary consumers, i.e., parasites and predators of the primary consumers in natural grasslands which may play significant roles in the balance normally prevailing. However, the specific reasons for the usual lack of insect outbreaks and for the outbreaks that do occur are little understood.

Chada and Wood (1960) reported that the introduced rhodesgrass scale, *Antonina graminis* (Mask.), was not a problem in the Rio Grande area of Texas except on rhodesgrass. Schuster (1967), however, found that 38 species of grasses in this area were significantly damaged. Of importance was the fact that two of the climax dominants were not damaged. When the range lands are not overgrazed to the extent that these dominants are removed the overall damage would be much less. Schuster *et al.* (1971) showed that in an area where one of the two species of dominants was not present, a yield increase of 44% was obtained by the effects of the introduced parasite, *Neodusmetia sangwani* (Rao), which suppressed the scale.

B. Biological Control Introductions

Introduction of natural enemies for control of grassland insects has been attempted for Orthoptera, Lepidoptera, and Homptera. Of these, completely successful results have only been attained for the rhodesgrass scale, although for several of the latest introductions against other pests the ultimate results are as yet unknown. *Blaesoxipha lineata* (Fallen), introduced from France for control of grasshoppers in the United States, has a good chance to become established (Rees, 1970); only time will afford an answer as to its ultimate value.

In addition to introduction of natural enemies against rhodesgrass scale in Texas, several species have been introduced into other areas for control of grassland pests, as follows: (1) against rhodesgrass scale: *Neodusmetia sangwani* into Florida (with complete success), Arizona, California, Bermuda, Mexico, Brazil (Machado da Costa *et al.*, 1970), and Israel (many with unknown or undetermined outcome as yet); *Anagyrus antoninae* into Florida (partial success, Questel and Genung, 1957), Mexico (partial success, Dean, 1960), and Israel (unknown outcome as yet); (2) against *Thymelicus lineola* (Ochs): *Stenichneumon scutellator* (Grav.) into Canada (unknown outcome); (3) against *Oncopera: Hexamera signata* (Walk.) and *H. acis* (Walk.) into Australia (unknown outcome, Wilson, 1960); *Ichneumon suspiciosus* (Wesm.) into Australia (unknown outcome, Wilson, 1960); (4) against locusts and grasshoppers: two species of *Sacrophaga* into the United States (unknown outcome, Clausen, 1956), *Acridomyia sacharovi* (Stackelberg), *Blaesoxipha lineata* (Fallen), and *B. filipjevi* (Rohd.) into the United States (unknown outcome, Rees, 1970; Coulson, personal communication); *Acridiophaga caridei* (Brethes), *Protodexia australis* (Bl.), and *Tephromviella neuquenensis* (Bl.) into Canada (failure, Smith, 1939); *Systoechus vulgaris* (Loew), *Sarcophaga hunteri* (Hough.), *S. opifera* (Coq.), *S. atlanis* (Ald.), and *S. acudesta* (Ald.) into Argentina (unknown outcome, Smith, 1939); *Kellymyia kellyi* (Ald.) within Canada (failure, Wilson, 1960); and (5) against *Hemileuca oliviae* (Ckh.): *Ooencyrtus kuwanai* (Howard), *Compsilura concinnata* (Meig.), *Calosoma sycophanta* (L.), *C. lugubre* (Le Conte), and *C. calidum* (Fab.) into Arizona (United States) (failure, Wildermuth and Frankenfeld, 1933; Clausen, 1956).

Rhodesgrass scale is a cosmopolitan insect, feeding on over 100 hosts (e.g., Chada and Wood, 1960). The introduced parasite *Anagyrus antoninae* (Timb.) was studied by Dean and Schuster (1958) in Texas grasslands. They found this parasite to be adversely affected by the hot dry summers in Texas, except in some ecologically moderated areas around lakes and canals of the lower Rio Grande Valley, where it has been partially successful. In addition to this introduction from Hawaii, also introduced from 1954 to 1959 were *Xanthoencyrtus phragmitis* Ferr., *Boucekiella antoninae* (Ferr.), *Timberlakia europaea* (Mercet), *Anagyrus diversicornis* Mercet, and *Neodusmetia sangwani* (Rao), the latter alone becoming established.

Introduction of *N. sangwani* was made in 1959 (Dean *et al.*, 1961), and two supporting programs were pursued. The first was a study of the resistance of the native Texas range grasses and promising plant introductions (Schuster, 1967). The second was development of a rhodesgrass scale-tolerant variety of rhodesgrass which raised the damage threshold and increased the utility of biological control (Schuster and Dean, 1973). The effectiveness of *N. sangwani* was established by 1968 (Schuster *et al.*, 1971). Scale reductions of 68% during a year were recorded. The two normal scale population peaks (June and November) were reduced by 50%. Parasitized scales produced 33.7% fewer crawlers on rhodesgrass and 90.1% fewer on paragrass.

Neodusmetia sangwani eliminated losses due to the scale in a *Paspalum*-fringed signal grass stand. This field control potential was substantiated by experiments in greenhouses with rhodesgrass, fringed signal grass, Texas grass, and cane sour grass. Yield comparisons between treatments was the best measure of control and laboratory data gave results similar to, and as good as, field data. Other studies showed that rhodesgrass was not killed by the scales even under severe grazing conditions. The crucial period for control occurred during the June peak at which time the parasite was just recovering from suppressed populations due to general host unsuitability during May. The resistant rhodesgrass variety "Bell" was found to tolerate twice the population of scales without appreciable damage and where used, it provided relief until the parasites had effectively reduced the scale populations (Schuster and Dean, 1973).

The area of Texas to benefit most from the establishment of effective biological control embraced 45,000 square miles, in an area south of a line from San Antonio to Victoria. The problem was complicated because the female parasite was functionally wingless and had a very short life-span of about 24 hours. Several experiments were conducted to develop a practical means of parasite distribution to supplement its poor dispersal powers.

Experiments showed that release of adults or distribution of parasitized scales on grass cuttings gave about equal results in establishing colonies. August was the best month for releases. From 100 to 200 female *N. sangwani* were required per release site to ensure about 64% establishment. A colony was found to spread naturally about one-half mile per year, females mostly being carried by wind.

For distribution of *N. sangwani*, infected rhodesgrass was cut off and distributed, at first by truck at frequent points along predetermined lines, and later by airplane (Schuster *et al.*, 1971).

Even with this scheme it was found that 4 years were required for the parasites to inhabit 80% of the range. Natural colony spread was restricted mostly by barriers of vegetation that did not support host scales. Airplane distribution proved both practical and inexpensive. Machado da Costa *et al.* (1970) have described methods of distribution of *N. sangwani* used in Brazil, where the parasite has also been introduced.

C. Future Prospects

In spite of the poor record thus far, much of the effort having been on Acridoidea, grasslands provide a relatively stable environment which should be conductive to biological control. Moreover, the low value of grasslands and the dangers of upsetting various biological balances involving both wildlife and pest systems by use of chemicals greatly increases the attractiveness of using biological control. Little insight into which grassland insects might be most promising for biological control is afforded by the few attempts, but there is no a priori reason why grassland pests should not prove amenable to this approach.

The "probability" of biological control is greater with indirect pests, i.e., ones that do not attack the marketed product (Turnbull and Chant, 1961; DeBach *et al.*, 1971). Since animal products, meat, wool, or hides constitute the marketable product of grasslands, all grass or forb-feeding insects are "indirect" pests. Thus, we may conclude that the competition of the pests with man's animals must be measured as the effect on net primary aerial production eaten by these animals and expressed in their biomass. Huffaker (1969) has made clear the competition and interaction that exist between the large herbivores and insects. The indirect role of selective phytophagous insects in altering plant composition by reversing competitive advantages of one species over another may be important far beyond the amount of forage consumed. Biological control would be of value, therefore, in two respects, (1) direct forage yield loss, and (2) when ecological succession detrimental to yield is involved. In most cases, however, we would expect that the reduction of a phytophagous species would be reflected in greater forage productivity and, hence, animal yields.

The enhancement of the habitats of native predators and parasites has received little attention. The value of flowering and nectar-bearing forbs in the conservation of parasites and predators should be investigated for grasslands. The biology of natural predators and parasites of many range insects has been viewed only casually by observation. Range scientists must learn to integrate the management of insects and other range pests, as well as management of the large herbivores, into range management programs.

V. FORAGE LEGUMES

Although there are many forage plants, i.e., ones whose stems and leaves are consumed by farm livestock, including grasses and various forbs, in this section we discuss principally alfalfa, and to some extent the clovers and peas. The perennial grasses and legumes which make up the flora of ranges, pastures, and meadows, as well as cultivated alfalfa and clover fields, commonly abound with a diversity of insects. These field "insectaries" are important sources of natural enemies of pests attacking other, especially annual crops. These extensive field insectaries need as much protection as possible from various disruptions.

Here we touch on but a few cases of naturally occurring biological control in forage, since only a few indigenous phytophagous insects and their enemies have been well studied (Hagen *et al.*, 1971). However, there are some introduced forage pests that have been controlled by indigenous predators.

The alfalfa or lucerne (*Medicago sativa* L.) community can support a large complex of arthropods. In New York state, for example, Pimentel and Wheeler (1973) collected 591 arthropod species from alfalfa plots. The most serious insect pests are mainly aphids and weevils, but outbreaks of some lepidopterous and homopterous species do occur. In alfalfa grown for seed, lygus bugs and seed chalcids often attain damaging densities. Classical biological control has been undertaken in many parts of the world against alfalfa pests, and the supervised control and integrated control approaches were developed initially for managing insect pests in alfalfa, a crop that tolerates substantial insect feeding. Since the "clovers," i.e., all *Medicago* spp., *Trifolium* spp., and *Melilotus* spp., share a number of pests with alfalfa, many natural enemies that are active against alfalfa pests are also effective in protecting clovers.

A. Aphids

There are primarily three important aphid pests of alfalfa: the pea aphid, spotted alfalfa aphid, and cowpea aphid. These aphids attain outbreak proportions if their natural enemies are insufficiently present or differentially hindered by climatic conditions, agricultural practices, etc., or where they enter reproductive diapause while the aphids are still active. The impact of natural enemies of aphids is reviewed by Hagen and van den Bosch (1968). Starý (1968) discussed the principles for use of parasites for biological control of aphids. Included were the geographical distribution and faunistic complexes of aphids and their aphidiid parasites. The taxonomy, biology, and ecology of Aphidiidae are extensively covered by Starý (1970).

1. The Pea Aphid

Acyrthosiphon pisum (Harris) is an Old-World species, occurring across the Palearctic and Oriental regions. It has invaded parts of Africa, the entire Nearctic region, and, recently, Argentina and Peru, where it has become a serious pest of alfalfa. In its endemic areas, aphidiid parasites and a variety of predators appear to be the most important natural enemies. In the Nearctic region, native predators often prevent outbreaks in the spring and fall, and with the introduction of two aphidiid parasites (below), a greater degree of control has been obtained. Throughout the world, 20 aphidiid species are recorded as parasites of the pea aphid (Mackauer and Starý, 1967).

In Europe serious outbreaks of pea aphid occur some years. Starý (1968) correlated outbreaks in central Europe with years having hot weather in late April. Starý (1971) also concluded that migration and parasitization (predators and fungi not considered) were the two main factors affecting the population dynamics of the pea aphid in Czechoslovakia. The relation between pea aphid migration and parasitization by *Aphidius ervi ervi* Haliday is dynamic and mainly affected by the weather. Under favorable conditions for the parasite, parasitization is the main controlling factor of the fundatrices whereas the virginoparae are regulated through emigration as well as parasitization. In unfavorable conditions for the parasite, the parasitization of fundatrices is negligible, and if an outbreak follows in the F_1 and F_2 of the virginoparae, it is regulated through emigration.

Aphidius smithi was introduced into Poland in 1960 and Czechoslovakia in 1967 from California (originally from India); the impact is yet unknown (Starý, 1970). Further importations of *A. smithi* into other parts of Europe are under consideration, but Campbell and Mackauer (1973) believe that *A. smithi* could be significant in pea aphid control only in Steppe- or Mediterranean-type climates but not in cooler, wet localities. In the U.S.S.R., Voronina (1971) notes that abundance of pea aphids varies widely year to year and seasonally, the causes being poorly understood. Various U.S.S.R. authors consider meteorological factors and predators to be the key factors retarding the aphid's buildup, but Voronina and several other workers have noted the role of fungi (Entomophthoraceae) in causing epizootics. Significant, perhaps controlling, fungus epizootics frequently remain unevaluated because of the primitive state of investigation on these fungi. Declines in abundance of pea aphids due to fungi are frequently ascribed to weather conditions, which, in reality, may have acted only to make conditions conducive to an epizootic. After studying pea aphid dynamics in U.S.S.R. alfalfa for 8 years, Voronina (1971) was able to correlate the frequency of pea aphid epizootics caused by *Entomophthora thaxtoriana* Petch with increasing rainfall or humidity.

In England, alfalfa and trefoil, *Lotus corniculatus*, are subject to damage by pea aphid, and Dunn and Wright (1955) observed parasites to be important in limiting some infestations, but attributed much of the natural control to weather.

In India, pea aphid is of little economic concern, and is attacked by *Aphidius* sp., *A. smithi*, *Ephedrus plagiator* (Nees), and *Praon myzoaphagum* Mackauer (Rao, 1969).

In Pakistan, the pea aphid attacks 17 legume species, seems to prefer pea, and only occasionally reaches damaging densities. *Aphidius smithi* is the most important parasite in Pakistan but is confined to the hill regions, rarely extending into the plains. Polyphagous predators, coccinellids, and syrphids were considered important in controlling legume aphids (Ghani, 1971).

In North America, the first damaging populations of pea aphid were observed in the late 1800's. By 1900 it had spread from the eastern seaboard to Wisconsin, and by 1926 to the Pacific Coast and into Canada and Mexico. (Campbell, 1926). Alfalfa and canning peas are the crops most harmed in North America.

Many indigenous natural enemies now exploit the pea aphid in North America. In coastal California, Campbell (1926) found 20 predators, 4 parasites, and 2 fungi attacking it on pea. He concluded that these enemies had little importance in checking infestations. However, in Wisconsin, Fluke (1929) considered natural enemies, 22 predaceous (syrphids, coccinellids, chrysopids), and 3 parasitic insects (aphidiids) and a fungus largely responsible for keeping the aphid under control. In Utah, Knowlton *et al.* (1937) found that, although coccinellids and syrphids restricted a 1936 aphid outbreak, the infestation still markedly decreased yield and quality of peas. Noteworthy, in these various studies the pea aphid was not mentioned as a problem in alfalfa.

In central California, predators (mainly *Hippodamia convergens* Guerin (Fig. 3) and *H. quinquesignata* Kirby) usually prevent the pea aphid's attaining damaging levels during the spring and fall. Parasitization by aphidiids and aphelinids during 1956—1959, prior to the spread of introduced parasites, was negligible (Smith and Hagen, 1966; Neuenschwander *et al.*, 1975). The indigenous parasites attacking pea aphid in California are *Aphidius ervi pulcher* Baker (= *A. pisivorus*), *Praon pequodorum* Viereck, *Monoctonus paulensis* Ashmead, *Ephedrus californicus* Baker, and *Aphelinus howardi* Della Torre (Hagen and Schlinger, 1960; Mackauer and Campbell, 1972). Occasionally, *M. paulensis* and *P. pequodorum* have become fairly common in coastal

Fig. 3. *Hippodamia convergens*, a common North American natural enemy of aphids in alfalfa and grain. Photography by F. E. Skinner.

areas but rarely important as control factors (Hagen and Schlinger, 1960; Calvert and van den Bosch, 1972). In Oregon and Washington, these native parasites are also considered to be of minor importance (Cooke, 1963).

The introduction of the exotic parasite *Aphidius smithi* from India in 1958 by G. Angalet has greatly altered the picture in California. By 1960, this USDA-University of California introduction had spread over much of the alfalfa-growing area of California and had become an important controlling agent of the pea aphid in the coastal valleys (Hagen and Schlinger, 1960). In the Central Valley, pea aphid outbreaks have occurred even in recent years during midsummer but only in solid-cut alfalfa fields. If the alfalfa is strip-cut, giving habitat continuity, *A. smithi* survives sufficiently and controls the pea aphid (van den Bosch *et al.*, 1967). *Aphidius smithi* exhibits a clear numerical density-dependent response to the pea aphid increase in alfalfa, but this relationship is interrupted by harvesting practices or fungus epizootics.

Aphidius smithi (Fig. 4) has spread naturally from California into central Mexico, a distance of 1500 miles (Clancy, 1967). It also became quickly established in Hawaii after its intentional introduction (Beardsley, 1961).

From 1959 through 1964, *A. smithi*, as well as the introduced European *A. ervi ervi* and the indigenous *A. ervi pulcher* and *Praon pequodorum* were reared and released by federal and state entomologists in Idaho, Oregon, and Washington. *Praon pequodorum* and *A. ervi pulcher* became more effective, and in only 5 years, *A. smithi* spread into nearly all alfalfa and peagrowing areas of these states (Halfhill *et al.*, 1972), and apparently into Canada, where by 1971 it was the most common parasite of this aphid in the interior of the country (Mackauer and Campbell, 1972).

In eastern North America, *A. smithi* was released in 1958 and in 1964–1965 in some North Atlantic states and Nova Scotia. Mackauer and Bisdee (1965) collected it in Ontario, indicating immigration from the eastern United States. Angalet and Coles (1966) recovered it in New Jersey, Pennsylvania, and Delaware, where parasitization averaged 18%. Pass and Parr (1971) recorded peak parasitizations of 46 and 69% in 1968 and 1969 respectively, for pea aphids in clover and alfalfa in Kentucky.

Recently, the pea aphid invaded Argentina, Chile, and Peru, and *A. smithi* was imported into these countries from California. By 1974, *A. smithi* had become established in Argentina (D. Calvert, personal communication) and Peru (O. Beingolea, personal communication).

Campbell and Mackauer (1973) concluded from laboratory and field evaluations that in hot, dry areas *A. smithi* exerts the most effective control of pea aphids whereas under mild, wet conditions *Aphidius ervi ervi* is predominant and has the controlling effect.

The European *A. ervi ervi* was introduced (1959–1965) into Arizona, California, Oregon, Washington, Idaho, and British Columbia. It became established in Oregon, Washington (Halfhill *et al.*, 1972), and British Columbia. In British Columbia it has been a negligible factor (Mackauer and Campbell, 1972). On the other hand, introduced into Delaware and New Jersey in 1959, it has provided excellent control (*USDA Coop. Insect. Quart. Rep.* **2**, 1973).

Fig. 4. Mummified pea aphids containing the parasite pupae of *Aphidius smithi*. Photography by F. E. Skinner.

2. *The Spotted Alfalfa Aphid (SAA)*

Therioaphis maculata (Buckton) is indigenous to the Old World where it attacks alfalfa and some clovers (Manglitz and Russell, 1974). Harpaz (1955) observed coccinellids, syrphids, and *Chrysopa carnea* Stephens preying on this aphid in Israel where it is an occasional pest of alfalfa. A parasite, *Praon* sp. [now known to be *P. exsoletum* (Nees)], was also noted. In Pakistan, this aphid attacks nine different legumes, but occurs at extremely low populations. Polyphagous predators were considered by Ghani (1971) to be important in maintaining it at low levels; however, three parasites were also found to attack it: *Aphelinus* sp., *Praon exsoletum*, and *Trioxys complanatus* Quilis. *Trioxys complanatus* constituted about 75% of the total parasitism.

The SAA invaded the New World through the southwestern United States in the early 1950's. It swept through California causing devastating damage (Smith, 1959), and spread through most of the United States in the 1960's (Angalet, 1970). The same native predators that adapted to the pea aphid (mainly *Hippodamia* spp.) also adapted to the SAA in California, and were mainly responsible for its control during the spring and fall in the late 1950's (e.g., Neuenschwander *et al.*, 1975; Smith and Hagen, 1966). *Hippodamia convergens* also became the most important predator in Arizona (Nielson and Currie, 1960), Kansas (Simpson and Burkhardt, 1960), and Utah (Goodarzy and Davis, 1958).

Three exotic parasites were introduced from the Old World into California in 1955–1956: *Aphelinus asychis* Walker, *Praon exsoletum*, and *Trioxys complanatus*, at first in southern and later in northern California. By 1959, all three species were established in most of the alfalfa districts, and were contributing significantly to biological control of the aphid (van den Bosch *et al.*, 1959). *Trioxys complanatus* was generally dominant over *P. exsoletum* and *A. asychis*. Phenologically, *Trioxys* is active during spring, fall, and winter while *P. exsoletum* largely confines its activity to the spring and fall. *Aphelinus* is most active during the cooler, more humid times of the year (van den Bosch *et al.*, 1964). In total they greatly supplement one another, relative to the whole range of the SAA in California. Bioclimatic studies (Messenger, 1972) suggest that climate is the key feature in their distributional patterns. In areas of overlap, competition between the parasites can occur, the outcome of which can be influenced by climate.

These three parasites were also released in Arizona between 1955 and 1957. Recoveries of each species have been made and *T. complanatus* became well established in southern Arizona where it is a substantial control factor (Barnes, 1960).

In Kansas, either the native *A. semiflavus* or *A. asychis* from France and India was colonized in 1956. SAA on alfalfa in greenhouses became heavily parasitized by this culture. It is not clear if the *Aphelinus* involved was the native *A. semiflavus* or the introduced *A. asychis* (Simpson *et al.*, 1959).

In eastern United States, these three parasite species were cultured in the United States Department of Agriculture (USDA) Moorestown laboratory in 1956 and released against the yellow clover aphid, *Therioaphis trifolii* (Monell), which mainly attacks red clovers. The original stock of *P. exsoletum* came from France, *A. asychis* from Israel, and *T. complanatus* from the yellow clover aphid in New Jersey. It is thought that the latter parasite had already been accidentally introduced into the United States (Mackauer and Stary, 1967). Angalet (1970) reported all three species as parasitizing

SAA in New Jersey, Delaware, and Maryland. Maximum parasitism in the fall was 51%. Predators were observed but not listed, and no fungus disease was noted on SAA although 52–98% of the pea aphids were recorded infected with *Entomophthora aphidis* Hoffman. Alfalfa did not appear to be damaged by aphids although SAA-susceptible varieties were being grown.

The SAA in southern California was found to be susceptible to infection by five species of *Entomophthora*, two of which also occur in the Old World. Epizootics in high populations were widespread in the fall of 1955 (Hall and Dunn, 1957).

During the late 1950's an integrated control program was developed for the SAA, based on monitoring of aphid populations and using a selective insecticide when necessary. A low dosage of demeton killed enough aphids to prevent damage and left enough aphids to retain sufficient natural enemies to give a favorable controlling ratio of predators to aphids during the summer (Stern *et al.*, 1959). This system was effective with greatly reduced use of insecticides until, in the 1960's, SAA-resistant varieties of alfalfa had largely replaced the susceptible ones. Since then, SAA only rarely has been a pest in alfalfa hay fields in the western United States. On the other hand, in seed alfalfa where heavy insecticide applications are often made against *Lygus*, SAA frequently erupts severely, especially in susceptible varieties.

3. Sweet Clover Aphid

Therioaphis riehmi (Börner) is common in Europe and North America on *Melilotus* spp. (Manglitz and Hill, 1964). It is apparently controlled by native predators and introduced parasites in the United States. Methoxychlor check method treatments (Chapter 11) were used in Nebraska to inhibit or destroy these natural enemies to assess their impact on the aphid in untreated plots. The aphids became much more abundant in the treated plots throughout most of the season. Coccinellids appeared to be most important. During October and November, the fungus, *Entomophthora sphaerosperma*, appeared in the methoxychlor-treated plots but not in the untreated ones. The higher aphid populations in the treated areas probably enhanced the fungus epizootic (Manglitz and Hill, 1964). Two parasites, *Praon exsoletum* and *Trioxys complanatus*, introduced to control the SAA (above), also attack *T. riehmi* in Nebraska and California.

4. The Cowpea Aphid

Aphis craccivora Koch is a cosmopolitan species attacking many legumes and also species in other families (Bodenheimer and Swirski, 1957), and is believed to have originated in the Mediterranean Basin (Gutierrez *et al.*, 1971). In Israel and Egypt it mainly damages peas, beans, and broad beans in the spring. In Pakistan, it attacks annual clovers and alfalfa (Ghani, 1971). It migrates widely and transmits clover stunt virus to crop and pasture legumes in southeastern Australia (Gutierrez *et al.*, 1971). In

North America, it is apparently not an important pest perhaps because of naturally occurring biological control.

Among the natural enemies of *A. craccivora*, aphidiid parasites appear to be most important, some sixteen species in eight genera being listed by Mackauer and Starý (1967). In Pakistan, five aphidiids attack this aphid, with *Lysiphlebus fabarum* Marshall being the most common (Ghani, 1971). *Lysiphlebus testaceipes* (Cresson), a polyphagous species, was introduced into Hawaii in 1923 (Fullaway, 1924). It is an excellent parasite of *A. craccivora* in California (Schlinger and Hall, 1960), now occurs widely in North and South America, is reported to occur in Nairobi (Ramaseshiah *et al.*, 1969), and has recently been sent to Bangalore, India, but its establishment is not yet confirmed (Ramaseshiah *et al.*, 1969). The host plant supporting the cowpea aphid can influence the efficiency of *L. testaceipes* (Sekhar, 1960).

Since *L. testaceipes* is an effective parasite of *A. craccivora*, at least in western North America, it is worth consideration for importation into other parts of the world. Since the natural enemies associated with *A. craccivora* in Australia are rather few and are poorly adaptable, the use of *L. testaceipes* is suggested (Gutierrez *et al.*, 1974).

B. Weevils

Several *Hypera* spp. alien to the New World have become serious pests of alfalfa and clovers. Importations of parasites began in the early 1900's against *Hypera postica* (Gyllenhal) in Utah and in California in the early 1930's. Complete biological control was achieved by an ichneumonid, *Bathyplectes curculionis* (Thomson), in some areas of California. Invasion of another strain of *H. postica* in the northeastern United States in the early 1950's initiated new parasite introductions. A complex of five introduced parasites thus far has resulted in various degrees of biological control of the new invader in several northeastern states, and parasites are continuing to be released and are spreading throughout the infested areas in the eastern United States. Invasion of the Egyptian alfalfa weevil into Arizona in the late 1930's and California in the early 1940's stimulated new parasite introductions. By 1974, three parasites had been established in California. Several other *Hypera* spp. and *Sitona* spp. have received some attention by workers and are under biological control in some areas (see below).

1. The Alfalfa Weevil

Hypera postica has a wide distribution in the Old World and occurs wherever alfalfa and certain clovers are grown throughout Europe, western and central Asia, the Middle East, and northern India. It reached North America in Utah in 1904 and by 1933 had spread to many other western states (Essig and Michelbacher, 1933). From a separate invasion found in Maryland in 1951, it had spread by 1967 into most of the eastern half of the country and rapidly became the most destructive pest of alfalfa (Blickenstaff *et al.*, 1972). By 1972 (USDA, 1972 distribution map) all central 48 states were infested.

The eastern *H. postica* apparently is a different strain than the western *H. postica*, as indicated by their differential degree of encapsulation of the eggs of the parasite *Bathyplectes curculionis* (Thomson), and a degree of mating incompatibility between them (Armbrust *et al.*, 1970).

A few native predators in North America and some native parasites were observed to attack *H. postica*, but apparently had little control effect (Essig and Michelbacher, 1933).

Bathyplectes curculionis, an ichneumonid larval parasite of *H. postica*, was first established in the New World in Utah between 1911 and 1914. The material came mostly from Italy and Switzerland. By 1914, parasitism of 22% was found near the release sites, and by 1920 the parasite had spread into Idaho and Wyoming. By 1924, parasitization of 30 to 100% occurred, but Chamberlin (1926) considered the parasite of questionable value in Utah for reasons of inadequate synchrony. Hamlin *et al.* (1949) concluded that early cutting inhibited the parasite.

Bathyplectes curculionis from Utah was introduced and established in California in 1933—1934, first in the area of the San Francisco Bay, and then in the San Joaquin Valley in 1935. By 1938 it had greatly reduced the weevil population in the Bay Area and Livermore Valley, and substantially so in the San Joaquin Valley (Michelbacher, 1940). Michelbacher (1940) found that the parasite's active period is limited by the hot climate of the San Joaquin Valley where it is less effective, whereas it remains active throughout the spring and summer months in the cooler coastal areas, where it has been completely successful.

Bathyplectes curculionis from Utah was introduced in the eastern United States between 1953—1955 but did not become established, or only poorly so. In 1959—1960, *B. curculionis* from *H. brunneipennis* in southern California and from France resulted in recoveries in 1960 in New Jersey, Delaware, and Virginia (Puttler *et al.*, 1961). By 1962, this parasite had been released in ten eastern states and was generally established near the release sites (Brunson and Coles, 1968). By 1965 it was widespread in New Jersey (Blickenstaff *et al.*, 1972), and in Pennsylvania by 1970 (Smilowitz *et al.*, 1972) and had moved naturally into Quebec, Canada (Mailloux and Pilon, 1970b), and Illinois (Armbrust *et al.*, 1967). The effectiveness of *B. curculionis* against the eastern *H. postica* is reduced because of encapsulation of its eggs by the host unless superparasitism occurs (Smilowitz *et al.*, 1972).

Bathyplectes anurus (Thomson), a univoltine larval parasite native to Eurasia, was introduced against *H. postica* in eastern United States in the early 1960's. It was established in New Jersey and Pennsylvania in 1963 (Brunson and Coles, 1968), and two other states by 1970. Sailer (1972) believes this species will be most effective in the southern United States where it has recently been released. It also has been released and recovered in California on the Egyptian alfalfa weevil (van den Bosch, personal communication).

Bathyplectes stenostigma (Thomson) is also a univoltine larval parasite of *H. postica* in northern Europe, and has been released in the United States from 1961 to 1970

(Dysart and Coles, 1971). It has been recovered recently in New Hampshire and Colorado (Eklund, 1972).

Microctonus aethiops (Nees) is an endoparasite of adult *Hypera* that kills the female weevil upon emergence, and whose larvae castrate diapausing males (Drea, 1968). It was first introduced from France in 1948 against the sweet clover weevil (Clausen, 1956), and in 1958 the USDA released it against *H. postica* in eastern United States. It has been established since 1961 (Brunson and Coles, 1968), and by 1971 in eleven northeastern states (Day *et al.*, 1971; Stehr and Casagrande, 1971). Sailer (1972) considers it the key parasite responsible for reducing *H. postica* populations below economic levels in an area some 200 miles radius from Moorestown, New Jersey. It has also been introduced into California against *H. brunneipennis*.

Microctonus colesi Drea oviposits in weevil larvae and emerges from adults. The origin of this species in the United States is uncertain, but Day *et al.* (1971) believes it may have come in with *H. postica* when it inadvertently arrived in the eastern United States. It has been found in eleven eastern states (Brunson and Coles, 1968; Day *et al.*, 1971).

Tetrastichus incertus (Ratzeburg), a multivoltine endoparasite of *Hypera* larvae, was obtained from Italy and released in Utah (Chamberlin, 1925). It was sent from France to California, Oregon, and Utah in 1935–1939, but apparently was not established there (Streams and Fuester, 1967). *Tetrastichus incertus* from France was propagated and over 12,000 released in six northeastern states in 1960–1961 (USDA). By 1968 it had become established in ten eastern states (Schroder *et al.*, 1969), and parasitization averaged 71% in Pennsylvania during 1964–1965 (Brunson and Coles, 1968). It had spread into Quebec by 1969 and parasitization attained 100% in certain instances (Mailloux and Pilon, 1970a). Horn (1971) found that a ratio of one *T. incertus* adult parasite to eight alfalfa weevil larvae is necessary to achieve 50% parasitism, and he did not find such a ratio in New York when hosts were common because of asynchrony between host and parasite caused by differential incidence of diapause. In competition with *B. curculionis*, parasitism by *T. incertus* decreased but parasitism by *B. curculionis* increased since this parasite prefers to oviposit in earlier larval stages of the host.

Dibrachoides dynastes (Förster), an ectoparasitic pteromalid which attacks prepupae and pupae of several *Hypera* spp., was introduced into Utah from Italy in 1911 (Webster, 1912). It was later found parasitizing *Hypera nigrirostris* (F.) in Washington (Rockwood, 1920), and *H. rumicis* (L.) in Oregon (Chamberlin, 1933). It was released in California in 1933–1936 but did not become established (Clausen, 1956). An importation from Iran did become established against *H. brunneipennis* in southern California (Gonzalez *et al.*, 1969). It was released against *H. postica* in New Jersey, Pennsylvania, and Virginia in the early 1960's, but apparently not established by 1968 (Brunson and Coles, 1968).

Peridesmia discus (Walker) is a pteromalid egg predator of *Hypera* in Europe and was introduced into Utah in 1920's, California in the 1930's (Clausen, 1956), and in eastern United States mostly in the 1960's (Brunson and Coles, 1968), but it did not become established. Brunson and Coles (1968) believe that *Hypera* eggs serve as the winter host and that some other necessary alternate host is not available in the United States where it has been released, perhaps explaining its not becoming established in the New World.

Trichomalus inops (Walker), another pteromalid, was released in the eastern United States from 1959 to 1967, but it too has not been recovered (Brunson and Coles, 1968).

Mymar pratensis (Forster), a European mymarid egg parasite of *Hypera* spp., was introduced into Utah from Italy in 1911–1918 (recorded as an *Anaphoidea* sp.). It apparently did not become established, and new importations were made in 1925–1928. It was then established and a greater degree of parasitization of *H. punctata* eggs than of *H. postica* was indicated (Clausen, 1956). From Utah it was sent to Indiana for use against *H. nigrirostris*, but it is unknown if it was established (Brunson and Coles, 1968).

Campogaster exigua (Meigen) is a European species that was released in Delaware and New Jersey in 1957, but no recoveries have been made up to 1968 (Brunson and Coles, 1968).

A few native parasites now attack *H. postica* in the United States. The mymarid *Patasson luna* Girault is an egg parasite of several weevil genera in eastern United States. The tachinid *Hyaolomyodes triangulifera* (Loew) has been reared from adult weevils in eastern United States (Brunson and Coles, 1968), and in low numbers in Georgia (Miller *et al.*, 1972). Another native tachinid, *Leucostoma simplex* (Fallon), and a nematode, *Hexamermis arvalis* Poinar and Gyrisco, have been reared from *H. postica* in New York (Richardson *et al.*, 1971).

The native primary and hyperparasites that attack endemic North American species of *Hypera* and other studies of hyperparasites of *Hypera* are reviewed by Puttler *et al.* (1973). Apparently there was an inadvertent introduction of the European hyperparasite *Mesochorus nigripes* Ratz. which has become established on *Bathyplectes stenostigma* in Colorado, and efforts are underway in the attempt to eradicate it. Furthermore, three pteromalids, parasitic on *B. curculionis*, have been discovered in Wyoming, and appear to be recent inadvertent introductions from Europe (Pike and Burkhardt, 1974).

The biological control programs against *H. postica* demonstrate the importance, not only of releasing a complex of suitable parasite species against a widespread introduced pest, but also of releasing a complex of biotypes of single species from appropriate climatic zones and of biotypes closely attuned physiologically with the populations of the pest at the respective sources of origin. Introduction of only a single parasitic species, *B. curculionis*, which is feebly encapsulated in the egg stage in the host (van den Bosch, 1964, 1971), gave complete control of *H. postica* in central coastal California but only partial control in the hot Central Valley (Michelbacher, 1940). This parasite often achieved high parasitization in Utah but the degree of parasitization can be adversely influenced by alfalfa cutting practices (Hamlin *et al.*, 1949). Cutting alfalfa timed to accumulated 507 day-degrees (C°) (base 8.0°C) in Michigan gave a 79%

reduction in production of weevils and a 57% reduction of *B. curculionis*. Earlier and later times of cutting caused different reductions in the respective populations and different levels of crop damage (Casagrande and Stehr, 1973).

In the eastern United States, *B. curculionis* eggs are consistently encapsulated by the weevil except for those that have been superparasitized (Puttler, 1967; Smilowitz *et al.*, 1972). Here, the parasite has become coextensive with the weevil and has reduced weevil populations, but by itself is unable to control the weevil satisfactorily. The establishment of *Tetrastichus incertus* increased but did not substantially improve the biological control of the weevil (Blickenstaff *et al.*, 1972; Miller, 1970; Miller *et al.*, 1972; Smilowitz *et al.*, 1972). However, in some areas of New York, *T. incertus* is emerging earlier than in the past, indicating possible natural adjustment of its life cycle (Richardson *et al.*, 1971). A complex of at least four parasites have, at present, become established (*Microctonus aethiops*, *Bathyplectes anurus*, and the above two species), and now have a distinct and increased impact upon *H. postica*, especially in the northeastern states (Richardson *et al.*, 1971; Sailer, 1972). *Microctonus aethiops* appears to be a key parasite species responsible for reducing the weevil to subeconomic status (Sailer, 1972). Sailer (1972) predicted that the problem will be greatly reduced throughout eastern United States in 5 years. It was calculated that since 1957 the entire biological control program against the eastern invasion of *H. postica* cost $600,000, and savings to farmers during 1970 alone in the area where the weevil had been reduced to subeconomic levels amounted to more than $3,000,000. This estimate accounted only for the savings in insecticide costs and omitted prospective gains from the increased acreage of alfalfa planted because farmers no longer had to contend with the weevil. It also omitted all external cost savings from reduced insecticide contamination of the environment (Sailer, 1972).

2. *The Egyptian Alfalfa Weevil*

Hypera brunneipennis (Boheman) is indigenous to the Middle East, but was inadvertently introduced into Arizona in the 1930's and had spread widely into California by 1971 (van den Bosch and Marble, 1971). It is much more serious now than *H. postica* in California and is not suppressed by the natural enemies that suppress *H. postica*. Although introductions against it were soon begun, only in 1973 did a promising parasite become established.

Bathyplectes curculionis from *H. postica* in Utah was first released and apparently established against *H. brunneipennis* in Arizona in 1941 (McDuffie, 1945). It probably came into California along with the weevil in baled hay from Arizona, for it was found to be established in southern California in 1950 (Dietrick and van den Bosch, 1953). Larval parasitization of *H. postica* by *B. curculionis* during 1964—1967 ranged from 12 to 40% in southern California and 13 to 72% in Arizona (Clancy, 1969).

The efficacy of this Utah strain of *B. curculionis* attacking *H. brunneipennis* in southern California is much reduced and is limited in two ways. First, a degree of encapsulation of the parasite's eggs, and second, the unfavorable hot, dry summers limit the effectiveness of the parasite (van den Bosch *et al.*, 1971). Limitation in effectiveness against *H. postica* of this *B. curculionis* strain by the hot summers in California's Central Valley was mentioned earlier. The Middle East has a climate similar to California's Central Valley, so that area has been explored since 1961 for biotypes of *B. curculionis* and other parasites of *Hypera* weevils for shipment to California.

Van den Bosch (1964) found no encapsulation of *B. curculionis* eggs in *H. brunneipennis* in Egypt, but when tested in the laboratory against southern California *H. brunneipennis*, from 15 to 50% of the Egyptian *B. curculionis* eggs were encapsulated; the strain was not released (van den Bosch, 1964). Utah strain *B. curculionis* attacking *H. postica* in northern California are even more severely affected when parasitizing *H. brunneipennis*, since 94% of its eggs are encapsulated and killed (Salt and van den Bosch, 1967).

Both *B. curculionis* and *B. anurus* were collected in hot lowland plateau and mountain areas of Iran and introduced into California in 1971 (van den Bosch *et al.*, 1971), and importations are continuing. A small release of European *B. anurus* had been made in southern California in 1967 by the USDA but not established (Clancy, 1969). Finally, establishment of the promising *B. anurus* was achieved, resulting in high parasitization of *H. brunneipennis* in a release area in the San Joaquin Valley in 1973. The parasite is now being widely distributed in California (van den Bosch, personal communication). The northern European *B. stenostigma* was also released in California in 1970, but no recoveries have been made (van den Bosch *et al.*, 1971).

Over the past decade both the University of California and the USDA have introduced other parasitic Hymenoptera into Arizona and California against the Egyptian alfalfa weevil. The mymarid, *Patasson* sp., and two pteromalids, *Habrocytus*, sp. and *Dibrachoides dynastes* (Foerster), were introduced from Iran in 1960–1962 (Fisher *et al.*, 1961; Clancy, 1969). *Dibrachoides dynastes* became widely established in southern California but is not particularly effective (Gonzalez *et al.*, 1969). Large numbers of European *Tetrastichus incertus* were released in Arizona and California but establishment was not apparent by 1968 (Clancy, 1969). European and Iranian *T. incertus* have been extensively colonized in central and northern California. It was recovered at Albany in 1970, with up to 70% parasitization, but since then has not been recovered (van den Bosch, personal communication). *Microctonus aethiops*, which is effective in the eastern United States on *H. postica* (above), was released in Arizona and California against *H. brunneipennis* by the USDA in 1965 but apparently not established (Clancy, 1969). It had also been released widely in northern California in the 1960's, but not detected to be established by 1971 (van den Bosch *et al.*, 1971). However, in 1973 and 1974, both the European and Iranian strains appeared to be established in montane northeastern California (van den Bosch, personal communication).

Intensive research is still being carried out against this most serious pest of alfalfa, pasture clovers, and range *Medicago hispida* in California. Success would reduce the need for spring insecticide treatments. These treatments not only cause disruption of natural biological control of aphids, Lepidoptera, and mites in alfalfa, which induces outbreaks of these pests in alfalfa, but they also deplete the reservoir of natural enemies that help suppress pests of a wide variety of other crops.

3. The Clover Leaf Weevil

Hypera punctata (F.) is indigenous to Palearctica where it occasionally reaches injurious densities. It was first noted to cause damage to alfalfa in New York in 1881 (Titus, 1911), and to alfalfa and clovers in California in the early 1920's, but in the late 1920's no extensive serious damage to either of these crops was noted (Essig, 1931). Smith and Michelbacher (1944) believed a fungus disease and the limited acreage of clover accounted for the weevil's not being an important pest in California. A number of authors cited by Titus (1911) and Puttler and Coles (1962) credited the fungus, *Entomophthora sphaerosperma* (Fres.) Thax., as the main controlling agent in both eastern and northwestern United States.

Bathyplectes tristis (Graven.), a Palearctic parasite of *H. punctata*, was introduced into Utah and Washington, D.C. in 1912 (Puttler and Coles, 1962). It was recovered in Virginia in 1935 (Dicke, 1937), and is considered by Puttler and Coles (1962) to be more important than the fungus in certain areas. The latter authors found parasitization of *H. punctata* by *B. tristis* in northeastern states during the early 1960's to range from 37 to 92%. *Bathyplectes tristis* is commonly collected in California alfalfa fields but its impact on *H. punctata* in alfalfa or clovers has not been evaluated in the western states. The egg parasite, *Mymar pratensis*, introduced against *H. postica* in North America, also attacks *H. punctata* (Clausen, 1956).

4. The Lesser Clover Leaf Weevil

Hypera nigrirostris (F.), a European species attacking red and white clovers, was first recorded in North America in 1876. By 1910 it was found only in the northeastern United States and southeastern Canada (Titus, 1911). In the midwest states, it became the most important pest of red clover. It was also later found in the Pacific northwest states along the coast, spreading eastward and southward (Rockwood, 1920). Today it occurs throughout North America where clovers occur. The same fungus disease found associated with *H. punctata* also attacks. *H. nigirostris* in the United States (Titus, 1911). Detwiler (1923) believed *Bathyplectes exiguus* (Grav.) and a native *Bracon* sp., probably *B. tenuiceps* (Muesebeck), to be important in control of *H. nigrirostris* in New York. Rockwood (1920) found *Dibrachoides dynastes* parasitizing the weevil in the northwest states. These two latter parasites had been introduced into the United States in the early 1900's against *H. postica*.

5. Sweetclover Weevil

Sitona cylindricollis Fahr. is a Palearctic species, accidentally introduced into North America, and is a pest of sweet clovers. The first attempt to introduce parasites against it in the United States was in 1948. The braconid, *Microctonus aethiops*, and the tachinid, *Campogaster exigua* (Meig.), were sent by the USDA to North Dakota. Early recoveries were made but permanent establishment was not determined (Clausen, 1956). The greatest effort against this pest was made in Canada, beginning in 1952, when three braconid species were introduced from Europe: *M. aethiops*, *Perilitus rutilus* (Nees), and *Pygostolus falcatus* (Nees). Later the tachinid, *C. exigua*, was released in Manitoba. Only the euphorine *P. falcatus* from Sweden had become established on *S. cylindricollis* and *S. scissifrons* by 1969 near the 1960 release sites in Ontario (Loan, 1971).

C. Lepidoptera

Very few natural enemies have been introduced against lepidopterous pests of legume crops. Sporadic outbreaks of these pests occur but generally they are under natural biological control. At least this is the case in western United States where indigenous moth and butterfly species that moved into alfalfa were accompanied by their natural enemies (Hagen *et al.*, 1971). The use of nonselective pesticides can create outbreaks of noctuids and *Colias* and the frequent cutting or cropping of hay fields can disengage these pests from their natural enemies, causing outbreaks. Therefore strip harvesting of alfalfa hay fields is advocated (Bashir and Venkatraman, 1968; van den Bosch and Stern, 1969).

In northern Chile (county of Arica) there is a small valley where pesticides have never been used and alfalfa has been grown for many years. Entomologists are comparing the alfalfa fauna with nearby valleys where insecticides are frequently employed to determine what biotic factors are operating to maintain a favorable status for all the potential pests in the valley not receiving insecticides (Cortes *et al.*, 1972).

D. Collembola

Sminthurus viridis Lubbock is a Palearctic species which rarely becomes abundant enough in France and Italy to cause any damage to alfalfa (Bonnemaison, 1962), but it became a serious pest of this crop in Australia and South Africa where it was accidentally introduced. In Australia, a predatory European bdellid mite, *Bdellodes lapidaria* (Kramer), apparently introduced accidentally from Europe was found to prey on *S. viridis* and was dispersed widely in that country. It accounts for considerable biological control of this collembolan. This came to light mainly when it was found that use of DDT in alfalfa caused an increase in the lucerne flea populations. The predatory mites but not the pest were killed by DDT (Wilson, 1960). Another predatory bdellid mite,

Neomolgus capillatus (Kramer), collected in Morocco and southern France, was released in 1969 and recovered in 1970 and 1971 in southwestern Australia (Wallace, 1971). Both of these bdellid predators have been introduced into South Africa. *Bdellodes lapidaria* alone proved insufficient in controlling *S. viridis* there, and in 1969 *N. capillatus* was also introduced and is apparently established (Greathead, 1971).

REFERENCES

Alam, M. M., Bennett, F. D., and Carl, K. P. (1971). Biological control of *Diatraea saccharalis* (F.) in Barbados by *Apanteles flavipes* Cam. and *Lixophaga diatraeae* T. T. *Entomophaga* **16**, 151–158.

Alexandrov, N. (1947–1949). *Eurygaster integriceps* Put. à Varamine et ses parasites. *Entomol. Phytopathol. Appl.* **5**, 29–41; **6–7**, 28–47; **8**, 16–52.

Andres, L. A., and Goeden, R. D. (1971). The biological control of weeds by introduced natural enemies. *In* "Biological Control" (C. B. Huffaker, ed.), pp. 143–164. Plenum, New York.

Angalet, G. W. (1970). Population, parasites, and damage of the spotted alfalfa aphid in New Jersey, Delaware and the eastern shore of Maryland. *J. Econ. Entomol.* **63**, 313–315.

Angalet, G. W., and Coles, L. W. (1966). The establishment of *Aphidius smithi* in the eastern United States. *J. Econ. Entomol.* **59**, 769–770.

Archangelskey, N. N., and Polyakov, I. M. (eds.). (1951). "Sunn Pest Control." Selhozguiz, Moscow.

Archer, T. L., Cate, R. H., Eikenbary, R. D., and Starks, K. J. (1974). Parasitoids collected from green bugs and corn leaf aphids in Oklahoma in 1972. *Ann. Entomol. Soc. Amer.* **67**, 11–14.

Areshnikov, B. A., Rogochaya, L. G., and Feshin, D. M. (1971). Chemical protection of wheat against sunn pest. *Zashch. Rast.* **6**, 13–14.

Armbrust, E. J., Banerjee, A. C., Petty, H. B., and White, C. E. (1967). Distribution of the alfalfa weevil in Illinois and spread of *Bathyplectes curculionis*. *J. Econ. Entomol.* **60**, 604–605.

Armbrust, E. J., White, C. E., and Roberts, S. J. (1970). Mating preference of eastern and western United States strains of alfalfa weevil. *J. Econ. Entomol.* **63**, 674–675.

Arnett, W. H. (1960). Responses of Acridid populations to rangeland practices and range sites. Ph.D. Dissertation, Department of Entomology, Kansas State University, Manhattan, Kansas.

Ball, E. D. (1915). How to control the grasshopper. *Utah Agr. Coll. Exp. Sta. Bull. No.* **B8**, 79–116.

Barnes, O. L. (1960). Establishment of imported parsites of the spotted alfalfa aphid in Arizona. *J. Econ. Entomol.* **53**, 1094–1096.

Bartlett, B. R. (1961). The influence of ants upon parasites and predators and scale insects. *Ann. Entomol. Soc. Amer.* **54**, 543–551.

Bashir, M. O., and Venkatraman, T. V. (1968). Insect parasite complex of Berseem armyworm *Spodoptera exigua* (Hubn.) (Lepidoptera: Noctuidae). *Entomophaga* **13**, 151–158.

Beardsley, J. W. (1961). A review of the Hawaiian Braconidae (Hymenoptera). *Proc. Hawaii. Entomol. Soc.* **17**, 333–366.

Beirne, B. P. (1972). The biological control attempt against the European wheat stem sawfly, *Cephus pygmaeus* (Hymenoptera: Cephidae), in Ontario. *Can. Entomol.* **104**, 987–990.

Bess, H. A. (1967). Feasibility and problem of chemical control and biological control of rice stem borers (Research on the natural enemies of rice stem borers). *Mushi* (*Suppl.*) **39**, 45–50.

Bess, H. A. (1972). Lepidopterous stem borers in different rice growing areas. *Mushi* **46**, 65–80.

Blickenstaff, C. C., Huggans, J. L., and Schroder, R. W. (1972). Biology and ecology of the alfalfa weevil, *Hypera pastica*, in Maryland and New Jersey, 1961–1967. *Ann. Entomol. Soc. Amer.* **65**, 336–349.

Bodenheimer, F. S., and Swirski, E. (1957). "The Aphidoidea of the Middle East," 378 pp. Weizmann Sci. Press, Jerusalem.

Bonnemaison, L. (1962). "Les Ennemis Animaux et des Forets," 599 pp. Editions Sep, Paris.

Brown, D. (1954). Methods of surveying and measuring vegetation. *Commonw. Bur. Pastures Field Crops Bull.* **42**, 1–223.

Brues, C. T. (1946). "Insect Dietary," 146 pp. Harvard Univ. Press, Cambridge, Massachusetts.

Brunson, M. H., and Coles, L. W. (1968). The introduction, release, and recovery of parasites of the alfalfa weevil in eastern United States. *U.S. Dep. Agr. Prod. Res. Rep. No.* **101**.

Burnov, V. N., Gamper, N. M., and Sazonov, A. P. (1971). Comparative evaluation of activity of some synthetic juvenile hormone analogues for sunn pest. *Byull. Vses. Nauch. Issled. Inst. Zashch. Rast.* **19**, 62–66.

Calvert, D. J., and van den Bosch, R. (1972). Host range and specificity of *Monoctonus paulensis* (Hymenoptera: Braconidae), a parasite of certain Dactynotine aphids. *Ann. Entomol. Soc. Amer.* **65**, 422–432.

Campbell, A., and Mackauer, M. (1973). Some climatic effects on the spread and abundance of two parasites of the pea aphid in British Columbia (Hymenoptera: Aphidiidae — Homoptera: Aphididae). *Z. Angew Entomol.* **74**, 47–55.

Campbell, R. E. (1926). The pea aphis in California. *J. Agr. Res.* **32**, 861–881.

Casagrande, R. A., and Stehr, F. W. (1973). Evaluating the effects of harvesting alfalfa on alfalfa weevil (Coleoptera: Curculionidae) and parasite populations in Michigan. *Can. Entomol.* **105**, 1119–1128.

Chada, H. L., and Wood, E. A. Jr. (1960). Biology and control of the rhodesgrass scale. *U.S. Dep. Agr. Tech. Bull.* **1221**.

Chamberlin, T. R. (1925). A new parasite of the alfalfa weevil. *J. Econ. Entomol.* **18**, 597–602.

Chamberlin, T. R. (1926). The introduction and establishment of the alfalfa weevil parasite, *Bathyplectes curculionis* (Thoms.), in the United States. *J. Econ. Entomol.* **19**, 302–316.

Chamberlin, T. R. (1933). Some observations on the life history and parasites of *Hypera rumicis* (L.) (Coleoptera: Curculionidae). *Proc. Entomol. Soc. Wash.* **35**, 101–109.

Chen, C. B. (1972). Artificial propogation of *Trichogramma australicum*, *T. japonicum*, *Trichospilus diatraeae* and *Tetrastichus inferens* for the control of rice and sugar cane borers in Taiwan. *Mushi (Suppl.)* **45**, 47–49.

Chu, Y. (1969). On the bionomics of *Lyctocoris beneficus* (Hiura) and *Xylocoris galactinus* (Fieber) (Heteroptera: Anthocoridae). *J. Fac. Agr. Kyushu Univ.* **15**, 1–136.

Chu, Y., and Okuma, C. (1970). Preliminary survey on the spider-fauna of the paddy fields in Taiwan. *Mushi* **44**, 65–88.

Clancy, D. W. (1967). Discovery of the pea aphid parasite *Aphidius smithi* in central Mexico. *J. Econ. Entomol.* **60**, 1743.

Clancy, D. W. (1969). Biological control of the Egyptian alfalfa weevil in California and Arizona. *J. Econ. Entomol.* **62**, 209–213.

Clausen, C. P. (1956). Biological control of insect pests in the continental United States. *U.S. Dep. Agr. Tech. Bull.* **1139**.

Cooke, W. C. (1963). Ecology of the pea aphid in the Blue Mountain area of eastern Washington and Oregon. *U.S. Dep. Agr. Tech. Bull.* **1287**.

Cortes, R., Aguilera, P. A., Vargas C. H., Hichins, N., Campos, L. E., and Pacheco W., J. (1972). Las "cuncunillas" (Noctuidae) de la alfalfa en Lluta y Camarones, Arica-Chile. Un problema bio-ecologico de control (Resumen). *Rev. Perv. Entomol.* **15**, 253–266.

Cumber, R. A. (1958). The insect complex of sown pastures in North Island. *N. Z. J. Agr. Res.* **1**, 719–749.

Davidson, W. M. (1924). Observations and experiments on the dispersion of the convergent lady beetle in California. *Trans. Amer. Entomol. Soc.* **50**, 163–175.

Davis, C. J. (1967). Progress in the biological control of the southern green stink bug, *Nezara viridula* variety *smaragdula* (Fabricius) in Hawaii (Heteroptera: Pentatomidae). *Mushi Suppl.* **39**, 9–16.

Day, W. H., Coles, L. W., Stewart, J. A., and Fuester, R. W. (1971). Distribution of *Microctonus aethiops* and *M. colesi*, parasites of the alfalfa weevil, in the eastern United States. *J. Econ. Entomol.* **64**, 190–193.

Dean, H. A. (1960). Introduction and establishment of *Anagyrus antoninae* on rhodesgrass scale in Mexico. *J. Econ. Entomol.* **53**, 694.

Dean, H. A., and Schuster, M. F. (1958). Biological control of rhodesgrass scale in Texas. *J. Econ. Entomol.* **51**, 363–366.

Dean, H. A., Schuster, M. F., and Bailey, J. C. (1961). The introduction and establishment of *Dusmetia sangwani* on *Antonina graminis* in South Texas. *J. Econ. Entomol.* **54**, 952–954.

DeBach, P., Rosen, D., and Kennett, C. E. (1971). Biological control of coccids by introduced natural enemies. *In* "Biological Control" (C. B. Huffaker, ed.), pp. 165–194, Plenum, New York.

Detwiler, J. D. (1923). Three little-known clover insects: The clover-head weevil, (*Phytonomus meles* Fab.), the lesser clover-leaf weevil (*P. nigrirostris* Fab.) and the clover-seed weevil (*Tychius picirostris* Fab.). *Cornell Univ. Agr. Exp. Sta. Bull.* **420**, 28 pp.

Dicke, F. F. (1937). First record of *Bathyplectes tristis* (Grav.), a parasite of the clover leaf weevil in the United States. *J. Econ. Entomol.* **30**, 375–376.

Dietrick, E. J., and van den Bosch, R. (1953). Further notes on *Hypera brunneipennis* and its parasite, *Bathyplectes curculionis*. *J. Econ. Entomol.* **46**, 1114.

Drea, J. J. (1968). Castration of male alfalfa weevils by *Microctonus* spp. *J. Econ. Entomol.* **61**, 1291–1295.

Dunn, J. A., and Wright, D. W. (1955). Population studies of the pea aphid in East Anglia. *Bull. Entomol. Res.* **46**, 369–387.

Dupuis, C. (1963). Essai monographique sur les Phasiinae (Diptères Tachinaires parasite d'Hétéroptères). *Mem. Mus. Hist. Nat. Ser. Zool.* **A26**, 1–461.

Dysart, R. J., and Coles, L. W. (1971). *Bathyplectes stenostigma*, a parasite of the alfalfa weevil in Europe. *Ann. Entomol. Soc. Amer.* **64**, 1361–1367.

Eklund, L. R. (1973). Preliminary report of *Bathypletectes stenostigma* and *Mesochorus nigripes* in western Colorado. *Proc. Forage Insect Conf., Ottawa, Canada, July 10–12, 1972*, **16**, 24–27.

Essig, E. O. (1931). "A History of Entomology," 1029 pp. Macmillan, New York.

Essig, E. O., and Michelbacher, A. E. (1933). The alfalfa weevil. *Univ. Calif. Coll. Agr. Bull.* **567**, pp. 99.

FAO (1970). Report to the Government of Burma on the study of agricultural pests and their control, based on the work of B. M. Lazarevic. *U.N. Dev. Progr. Rep. TA* **2837**.

Fernando, H. E. (1972). Ecological studies on the rice gall midge, *Pachydiplosis oryzae* (Wood-Mason) (Diptera: Cecidomyidae), in Ceylon. *Jap. Pesticide Inform.* **10**, 121–122.

Fisher, T. W., Schlinger, E. I., and van den Bosch, R. (1961). Biological notes on five recently imported parasites of the Egyptian alfalfa weevil, *Hypera brunneipennis*. *J. Econ. Entomol.* **54**, 196–197.

Fluke, C. L. (1929). The known predacious and parasitic enemies of the pea aphid in North America. *Univ. Wis. Agr. Exp. Sta. Res. Bull.* **93**.

Fullaway, D. T. (1924). *Lysiphlebus* sp. from California. *Proc. Entomol. Soc. Hawaii* **5**, 345.

Ghani, M. A. (1971). Natural enemies of forage and grain legume aphids in Pakistan. *CIBC Pakistan Sta. Rep. Jan. 1967-Dec. 1971, Rawalpindi*, pp. 34.

Glenn, P. A. (1909). The influence of climate upon the green bug and its parasite. *Bull. Univ. Kans.* **9**, 165–200.

Gonzalez, D., van den Bosch, R., and Dawson, L. H. (1969). Establishment of *Dibrachoides druso* on the Egyptian alfalfa weevil in southern California. *J. Econ. Entomol.* **62**, 1320–1322.

Goodarzy, K., and Davis, D. W. (1958). Natural enemies of the spotted alfalfa aphid in Utah. *J. Econ. Entomol.* **51**, 612–616.

Greathead, D. J. (1963). A review of the insect enemies of Acridoidea (Orthoptera). *Trans. Roy. Entomol. Soc.* **114**, 437–517.

Greathead, D. J. (1971). A review of biological control in the Ethiopia Region. *Commonw. Inst. Biol. Control Tech. Commun.* **5**, pp. 162.

Grigor'eva, T. G. (1965). On the cause for the origin and fading out of *Hadena sordida* Bkh. out-breaks in virgin areas of Kasakhstan and Siberia. *Horae. Soc. Entomol. Rossicae (Unionis Sovet.) Moscow* **50**, 146–169.

Gusev, G. V., and Shmetzer, N. V. (1971). A possibility of utilizing different pentatomid species in the artificial breeding of *Trissolcus grandis* and *T. semistriatus*. *Byull. Vses. Nauch. Issled. Inst. Zashch. Rast.* **20**, 3–5.

Gutierrez, A. P., Morgan, D. J., and Havenstein, D. E. (1971). The ecology of *Aphis craccivora* Koch and subterranean clover stunt virus. I. The phenology of aphid populations and the epidemiology of virus in pastures in south-east Australia. *J. Appl. Ecol.* **8**, 699–721.

Gutierrez, A. P., Havenstein, D. E., Nix, H. A., and Moore, P. A. (1974). The ecology of *Aphis craccivora* Koch and subterranean clover stunt virus in south-east Australia. II. A model of cowpea aphid populations in temperate pastures. *J. Appl. Ecol.* **11**, 1–20.

Hagen, K. S. (1962). Biology and ecology of predaceous Coccinellidae. *Annu. Rev. Entomol.* **7**, 289–326.

Hagen, K. S., and Schlinger, E. I. (1960). Imported Indian parasite of pea aphid established in California. *Calif. Agr.* **14**, 5–6.

Hagen, K. S., and van den Bosch, R. (1968). Impact of pathogens, parasites, and predators on aphids. *Annu. Rev. Entomol.* **13**, 325–384.

Hagen, K. S., Sawall, E. F., and Tassan, R. L. (1971). The use of food sprays to increase effectiveness of entomophagous insects. *Proc. Tall Timbers Conf. Ecol. Anim. Contr. Manage.* **2**, 59–80.

Halfhill, J. E., and Featherston, P. E. (1973). Inundative releases of *Aphidius smithi* against *Acyrthosiphon pisum*. *Environ. Entomol.* **2**, 469–472.

Halfhill, J. E., Featherston, P. E., and Dickie, A. G. (1972). History of the *Praon* and *Aphidius* parasites of the pea aphid in the Pacific Northwest. *Environ. Entomol.* **1**, 402–405.

Hall, I. M., and Dunn, P. H. (1957). *Entomophthorous* fungi parasitic on the spotted alfalfa aphid. *Hilgardia* **27**, 159–181.

Hamlin, J. C., Lieberman, F. V., Bunn, R. W., McDuffie, W. C., Newton, R. C., and Jones, L. J. (1949). Field studies of the alfalfa weevil and its environment. *U.S. Dep. Agr. Tech. Bull.* **975**.

Hardy, D. E. (1972). Studies on Pipunculidae (Diptera) of the Oriental Region. I. *Oriental Insects Suppl.* **2**, 1–76.

Harpaz, I. (1955). Bionomics of *Therioaphis maculata* (Buckton) in Israel. *J. Econ. Entomol.* **48**, 668–671.

Haynes, D. L., Gage, S. H., and Fulton, W. (1974). Management of the cereal leaf beetle pest ecosystem. *Quaest. Entomol.* **10**, 165–176.

Helfer, J. R. (1953). "How to Know the Grasshoppers, Cockroaches and Their Allies," 353 pp. Wm. C. Brown, Dubuque, Iowa.

Hendrickson, G. O. (1930). Studies on the insect fauna of Iowa prairies. *Iowa State Coll. J. Sci.* **4**, 49–179.

Hight, S. C., Eikenbary, R. D., Miller, R. J., and Starks, K. J. (1972). The green bug and *Lysiphlebus testaceipes*. *Entomol.* **1**, 205–209.

Horn, D. J. (1971). The relationship between a parasite, *Tetrastichus incertus* (Hymenoptera: Eulophidae), and its host, the alfalfa weevil, *Hypera postica* (Coleoptera: Curculionidae), in New York. *Can. Entomol.* **103**, 83–94.

Huffaker, C. B. (1957). Fundamentals of biological control of weeds. *Hilgardia* **27**, 101–157.

Huffaker, C. B. (1964). Fundamentals of biological weed control. *In* "Biological Control of Insect Pests and Weeds" (P. DeBach, ed.), pp. 631–649. Reinhold, New York.

Huffaker, C. B. (1969). Insects as competitive herbivores. *In* "A Practical Guide to the Study of the Productivity of Large Herbivores" (F. B. Golley and H. K. Buechner, eds.), pp. 267–271. Blackwell, Oxford.

Hunter, S. J. (1909). The green bug and its enemies. *Bull. Univ. Kansas* **9**, 1–163.

Jackson, H. B., Rogers, C. E., and Eikenbary, R. D. (1971). Colonization and release of *Aphelinus asychis*, an imported parasite of the green bug. *J. Econ. Entomol.* **64**, 1435–1438.

Kajita, H., and Drake, E. F. (1969). Biology of *Apanteles chilonis* and *A. flavipes* (Hymenoptera: Braconidae), parasites of *Chilo suppressalis*. *Mushi* **42**, 163–179.

Kamenkova, K. V. (1958). Causes of high efficiency of sunn pest egg parasites in the pre-mountainous zone of Krasnodar Reon. *Tr. Vses. Nauch. Issled. Inst. Zashch. Rast.* **9**, 285–311.

Kamenkova, K. V. (1971). An evaluation of resistance of sunn pest egg parasites to insecticides. *Byull. Vses. Nauch. Issled. Inst. Zashch. Rast.* **21**, 7–10.

Kelsey, J. M. (1957). Insects attacking tussock. *N. Z. J. Sci. Technol.* **38**, 638–643.

Kieckhefer, R. W., and Olson, G. A. (1974). Dispersal of marked adult coccinellids from crops in South Dakota. *J. Econ. Entomol.* **67**, 52–54.

Knowlton, G. F., Smith, C. F., and Harmston, F. C. (1937). Pea aphid investigations. *Utah Acad. Sci. Arts Lett.* **15**, 11–80.

Kobayashi, T. (1961). The effect of insecticidal applications to the rice stem borer on the leaf-hopper populations. Special Rep. on Prediction of Pests. *Min. Agr. For. Jap.* **3**, pp. 126.

Kulakov, M. F. (1940). Sunn pest (*Eurygaster integriceps* Put.). *Sb. Nauch. Issled. Rab. Azovo Chernomor. Selśkokhoz. Inst.* **2**, 103–135.

Lebedev, G. I. (1970). Utilization des méthodes biologique de lutte biologique contre les insects nuisibles et les mauveses herbes en Union Sovietique. *Ann. Zool. Ecol. Amin. Hors Ser.*, pp. 17–23.

Li, C. S. (1970). Some aspects of the conservation of natural enemies of rice stem borers and the feasibility of harmonizing chemical and biological control of these pests in Australia. *Mushi* **44**, 15–23.

Li, C. S. (1971). Integrated control of the white rice borer, *Tryporyza innotata* (Walker) (Lepidoptera: Pyralidae) in northern Australia. *Mushi Suppl.* **45**, 51–59.

Lim, G. S. (1970). Some aspects of the conservation of natural enemies of rice stem borers and the feasibility of harmonizing chemical and biological control of these pests in Malaysia. *Mushi* **43**, 127–135.

Loan, C. C. (1971). *Sitona cylindricollis* Fahr., sweetclover weevil, and *Hypera postica* Gyll., alfalfa weevil. *Commonw. Inst. Biol. Control Tech. Commun.* **4**, 43–46.

McDuffie, W. C. (1945). The legume weevil *Hypera brunneipennis* (Boh.) during the season 1941–1942. *Spec. Publ. Calif. Dep. Agr. No.* **209**, 27–28.

McLeod, J. H. (1962). Biological control of pests of crops, fruit trees, ornamentals and weeds in Canada up to 1959. *Commonw. Inst. Biol. Control Tech. Commun.* **2**, 1–33.

Machado da Costa, J., Williams, R. N., and Schuster, M. F. (1970). Cochonilha dos capins, (*Antonina graminis*) no Brasil. II. Introducao de *Neodusmetia sangwani* inimigo natural da cochnilha. *Pesqui. Agropec. Bras.* **5**, 339–343.

Mackauer, M., and Bisdee, H. E. (1965). *Aphidius smithi* Sharma and Subba Rao, a parasite of the pea aphid, new in southern Ontario. *Proc. Entomol. Soc. Ont.* **95**, 121–124.

Mackauer, M., and Campbell, A. (1972). The establishment of three exotic aphid parasites (Hymenoptera: Aphidiidae) in British Columbia. *J. Entomol. Soc. Brit. Columbia* **69**, 54–58.

Mackauer, M., and Starý, P. (1967). "Hym. Ichneumonoidea: World Aphidiidae. Index of Entomophagous Insects." LeFrancois, Paris, pp. 195.

Mailloux, G., and Pilon, J.-G. (1970a). *Tetrastichus incertus* (Ratzeburg) (Hymenoptera: Eulophidae), au Quebec. *Ann. Soc. Entomol. Quebec* **15**, 123–127.

Mailloux, G., and Pilon, J.-G. (1970b). *Patassan luna* (Girault) (Hymenoptera: Mymaridae) and *Bathyplectes curculionis* (Thomson) (Hymenoptera: Ichneumonidae), two parasites of *Hypera postica* (Gyllenhal) (Coleoptera: Curculionidae) in Quebec. *Can. J. Zool.* **48**, 607–608.

Manglitz, G. R., and Hill, R. E. (1964). Seasonal population fluctuations and natural control of the sweetclover aphid. *Nebr. Univ. Agr. Res. Sta. Bull.* **217**.

Manglitz, G. R., and Russell, L. M. (1974). Cross matings between *Therioaphis maculata* (Buckton) and *T. trifolii* (Monell) (Hemiptera: Homoptera: Aphididae) and their implications in regard to the taxonomic status of the insects. *Proc. Entomol. Soc. Wash.* **76**, 290–296.

Messenger, P. S. (1972). Climatic limitations to biological controls. *Proc. Tall Timbers Conf. Ecol. Anim. Contr. Habitat Manage.* **3**, 97–114.

Meyer, N. F. (1941). "Trichogramma." Selhozhuiz, Leningrad.

Michelbacher, A. E. (1940). Effect of *Bathyplectes curculionis* on the alfalfa weevil population in lowland middle California. *Hilgardia* **13**, 81–99.

Miller, M. C. (1970). Biological control of the alfalfa weevil in Massachusetts. *J. Econ. Entomol.* **63**, 440–443.

Miller, M. C., Smith, C., and White, R. (1972). Activity of the alfalfa weevil and its parasites in north Georgia. *Environ. Entomol.* **1**, 471–473.

Mohyuddin, A. I. (1971). Comparative biology and ecology of *Apanteles flavipes* (Cam.) and *A. sesamiae* Cam. as parasites of graminaceous borers. *Bull. Entomol. Res.* **61**, 33–39.

Mohyuddin, A. I., and Greathead, D. J. (1970). An annotated list of the parasites of graminaceous stem borers in East Africa, with a discussion of their potential in biological control. *Entomophaga* **15**, 241–274.

Moutia, L. A., and Mamet, R. (1946). A review of twenty-five years of economic entomology in the island of Mauritius. *Bull. Entomol. Res.* **36**, 439–472.

Muma, M. H., and Katherine, E. (1949). Studies on populations of prairie spiders. *Ecology* **30**, 485–503.

Nagaraja, H., and Nagarkatti, S. (1973). A key to some new world species of *Trichogramma* (Hymenoptera: Trichogrammatidae), with descriptions of four new species. *Proc. Entomol. Soc. Wash.* **73**, 288–297.

Nagatomi, A. (1972). Parasites of *Sesemia inferens* in sugar cane fields in Kagoshima Pref., Japan (Lepidoptera: Noctuidae). *Mushi* **46**, 81–105.

Neuenschwander, P., Hagen, K. S., and Smith, R. F. (1975). Predation on aphids in California's alfalfa fields. *Hilgardia*, **43**, 53–78.

Nielson, M. W., and Currie, E. (1960). Biology of the convergent lady beetle when fed a spotted alfalfa aphid diet. *J. Econ. Entomol.* **53**, 257–259.

Nishida, T., and Wongsiri, T. (1972). Rice stem borer population and biological control in Thailand. *Mushi Suppl.* **45**, 25–37.

Okuma, C. (1968). Preliminary survey on the spider-fauna of the paddy fields in Thailand. *Mushi* **42**, 89–118.

Pass, B. C., and Parr, J. C. (1971). Seasonal occurrence of the pea aphid and braconid parasite, *Aphidius smithi*, in Kentucky. *J. Econ. Entomol.* **64**, 1150–1153.

Pemberton, C. E. (1948). History of the entomology department experiment station, H.S.P.A., 1904–1945. *Hawaii. Plant. Rec.* **52**, 53–90.

Pike, K. S., and Burkhardt, C. C. (1974). Hyperparasites of *Bathyplectes curculionis* in Wyoming. *Environ. Entomol.* **3**, 953–956.

Pimentel, D., and Wheeler, A. G., Jr. (1973). Species diversity of arthropods in the alfalfa community. *Environ. Entomol.* **2**, 659–668.

Popov, G. A. (1971). Breeding of pentatomids (Hemiptera: Pentatomidae) for sunn pest egg parasites. *Byull. Vses. Nauch. Issled. Inst. Zashch. Rast.* **19**, 3–10.

Puttler, B. (1967). Interrelationship of *Hypera postica* and *Bathyplectes curculionis* in the eastern United States with particular reference to encapsulation of the parasite eggs by the weevil larvae. *Ann. Entomol. Soc. Amer.* **60**, 1031–1038.

Puttler, B., and Coles, L. W. (1962). Biology of *Biolysia tristis* (Hymenoptera: Ichneumonidae) and its role as a parasite of the clover leaf weevil (*Hypera punctata*). *J. Econ. Entomol.* **55**, 831–833.

Puttler, B., Jones, D. W., and Coles, L. W. (1961). Introduction, colonization and establishment of *Bathyplectes curculionis*, a parasite of the alfalfa weevil in eastern United States. *J. Econ. Entomol.* **54**, 878–880.

Puttler, B., Thewke, S. E., and Warner, R. E. (1973). Bionomics of three Nearctic species, one new, of *Hypera* (Coleoptera: Curculionidae), and their parasites. *Ann. Entomol. Soc. Amer.* **66**, 1299–1306.

Questel, D. D., and Genung, W. G. (1957). Establishment of the parasite *Anagyrus antoninae* in Florida for control of rhodesgrass scale. *Fla. Entomol.* **40**, 123–125

Ramaseshiah, G., Bhat, K. V., and Dharmadhikari, P. R. (1969). Influence of host aphid, host plant and temperature on the laboratory breeding of *Lysiphlebus testaceipes*. *Indian J. Entomol.* **30**, 281–285.

Rao, V. P. (1969). Survey for natural enemies of aphids in India. *Final Tech. Rep. U.S. PL-480 Proj. Indian Sta. Commonw. Inst. Biol. Control. Bangalore, India.*

Rao, V. P. (1970). Aims, objectives, and future of biological control of rice stem borers. *Mushi* **44**, 11–14.

Rao, V. P. (1972). Rice stem borers and their natural enemies in India, Pakistan, Ceylon and Malaysia. *Mushi (Suppl.)* **45**, 7–23.

Rao, V. P., Ghani, M. A., Sankaran, T., and Mathur, K. C. (1971). A review of the biological control of insects and other pests in south-east Asia and the Pacific region. *Commonw. Inst. Biol. Control Tech. Commun.* **6**.

Rees, N. E. (1970). Suitability of selected North American grasshopper species as hosts for Eurasian parasites, *Acridomyia sachovovi* and *Blaesoxepha lineata* (Diptera). *Ann. Entomol. Soc. Amer.* **63**, 901–903.

Remaudière, G. (1960). Project d'installation et de fonctionnement d'une unite de production d'oeufs d'*Eurygaster integriceps. Sunn Pest Circ.* **5**. FAO Sunn Pest Inform. Rome.

Remaudière, G. (1961). "Sunn Pest in the Middle East," Draft report at termination of mission. FAO Sunn Pest Inform. Rome.

Richardson, R. L., Nelson, D. E., York, A. C., and Gyrisco, G. G. (1971). Biological control of the alfalfa weevil *Hypera postica* (Coleoptera: Curculionidae) in New York. *Can. Entomol.* **103**, 1653–1658.

Ricou, G. (1967). Etude biocoenotique d'un milieu "natural": la prairie permanete pasturee. *Ann. Epiphyt.* **18**.

Riley, C. V., Packard, A. S., and Thomas, C. (1878). "First Annual Report of the United States Entomological Commission for the year 1877, Relating to the Rocky Mountain Locust," 249 pp. U.S. Gov. Print. Off., Wash. D.C.

Rockwood, L. P. (1920). *Hypera nigrirostris* Fab. in the Pasific Northwest. *Can. Entomol.* **52**, 38–39.

Rothschild, G. H. L. (1970). Parasites of rice stem borers in Sarawak (Malaysian Borneo). *Entomophaga* **15**, 21–51.

Saakov, A. I. (1903). Artificial breeding of egg parasites of grain bug. *Tr. Bur. Entomol.* **4** (2), 1–12.

Safavi, M. (1968). Étude biologique et écologique des Hyménoptères parasites des oeufs des punaises des céréales. *Entomophaga* **13**, 381–495.

Sailer, R. I. (1972). A look at USDA's biological control of insect pests: 1888 to present. *Agr. Sci. Rev.* **10**, 15–27.

Salt, G., and van den Bosch, R. (1967). The defense reactions of three species of *Hypera* (Coleoptera: Cruculionidae) to an ichneumon wasp. *J. Invertebr. Pathol.* **9**, 164–177.

Sasaba, T., and Kiritani, K. (1972). Evaluation of mortality factors with special reference to parasitism of the green rice leafhopper, *Nephatettix cincticeps* Uhler (Hemiptera: Deltocephalidae). *Appl. Entomol. Zool.* **7**, 83–93.

Sasaji, H. (1968). Coccinellidae collected in the paddy fields of the Orient, with descriptions of new species (Coleoptera). *Mushi* **42**, 119–132.

Schepetilnikova, V. A. (1942). Distribution, biology and application of various sunn pest egg parasites in different life conditions. *Dokl. Vses. Akad. Selŝkokhoz. Nauk* **5** (6), 20–28.

Schlinger, E. I., and Hall, J. C. (1960). Biological notes on Pacific coast aphid parasites, and lists of California parasites (Aphidiinae) and their aphid hosts (Hymenoptera: Braconidae). *Ann. Entomol. Soc. Amer.* **53**, 404–415.

Schroder, R. F. W., Huggans, J. L., Horn, D., and York, G. T. (1969). Distribution and establishment of *Tetrastichus incertus* in the eastern United States. *Ann. Entomol. Soc. Amer.* **62**, 812–815.

Schuster, M. F. (1967). Response of forage grasses to rhodesgrass scale. *J. Range Manage.* **20**, 307–309.

Schuster, M. F., and Dean, H. A. (1973). Rhodesgrass scale resistance studies in rhodesgrass. *J. Econ. Entomol.* **66**, 467–469.

Schuster, M. F., Boling, J. C., and Morony, J. J., Jr. (1971). Biological control of rhodesgrass scale by airplane releases of an introduced parasite of limited dispersing ability. *In* "Biological Control" (C. B. Huffaker, ed.), pp. 227–250. Plenum, New York.

Sekhar, P. S. (1960). Host relationships of *Aphidius testaceipes* (Cresson) and *Praon aguti* (Smith), primary parasites of aphids. *Can. J. Zool.* **38**, 593–603.

Shapiro, V. A. (1965). On the formation of the parasite fauna of *Hadena sordida* Bkh. on the wheat fields in the virgin areas of Kasakhstan. *Horae Soc. Entomol. Unionis Sovet.* **50**, 193–217.

Shek, G. H. (1962). Regularities in outbreaks of *Hadena sordida*. *Tr. Kaz. Nauch. Issled. Inst. Zashch. Rast.* **7**, 334–341.

Shin, Y. H. (1970). On the bionomics of *Itoplectis narangae* (Ashmead) (Hymenoptera: Ichneumonidae). *J. Fac. Agr. Kyushu Univ.* **16**, 1–75.

Silveira-Guido, A., and Conde-Jahn, E. (1945). El pulgon verde de los cerales del Uruguay. *Rev. Fac. Agron. Univ. Repub. Montevideo* **41**, 35–86.

Simpson, R. G., and Burkhart, C. C. (1960). Biology and evaluation of certain predators of *Therioaphis maculata* (Buckton). *J. Econ. Entomol.* **53**, 89–94.

Simpson, R. G., Burkhardt, C. C., Maxwell, F. G., and Ortman, E. E. (1959). A chalcid parasitizing spotted alfalfa aphids and green bugs in Kansas. *J. Econ. Entomol.* **52**, 537–538.

Skaf, R., Ibara, W., Remaudière, G., Safavi, M., and Zommorrodi, A. (1961). First test in the collection of *Eurygaster* adults of the young generation in the cultivated fields. *Sunn Pest Circ.* **6**, FAO Sunn Pest Inform. Rome.

Smilowitz, Z., Yendol, W. G., and Hower, A. A., Jr. (1972). Population trends of the alfalfa weevil larvae and the distribution of its parasites in Pennsylvania. *Environ. Entomol.* **1** , 42–48.

Smith, C. W. (1939). An exchange of grasshopper parasites between Argentina and Canada with notes on parasitism of native grasshoppers. *70th Rep. Entomol. Soc. Ont.*, pp. 57–62.

Smith, R. F. (1959). The spread of the spotted alfalfa aphid, *Therioaphis maculata* (Buckton), in California. *Hilgardia* **28**, 647–685.

Smith, R. F., and Hagen, K. S. (1966). Natural regulation of alfalfa aphids in California. *In* "Ecology of Aphidophagous Insects" (I. Hodek, ed.), pp. 297–315. Proc. Symp. Liblice near Prague, Academia, Prague.

Smith, R. F., and Michelbacher, A. E. (1944). Clover leaf weevil in California. *Pan Pac. Entomol.* **20**, 120.

Smith, R. W. (1959). Status in Ontario of *Collyria calcitrator* (Grav.) (Hymenoptera: Ichneumonidae) and of *Pediobius beneficus* (Gahan) (Hymenoptera: Eulophidae) as parasites of the European wheat stem sawfly, *Cephus pygmaeus* (L.) (Hymenoptera: Cephidae). *Can. Entomol.* **91**, 697–700.

Southwood, T. R. E., and Way, M. J. (1970). Ecological background to pest management. *In* "Concepts of Pest Management" (R. L. Rabb and F. E. Guthrie, eds.), pp. 6–28. North Carolina State Univ., Raleigh, North Carolina.

Starý, P. (1970). "Biology of Aphid Parasites (Hymenoptera: Aphidiidae) with Respect to Integrated Control," Ser. Entomol. 6. Dr. W. Junk Publ., The Hague.

Starý, P. (1971). Migration and parasitization as factors of population regulation of the pea aphid, *Acyrthosiphon pisum* (Harris) in alfalfa fields in central Europe. *Acta Entomol. Bohemoslov.* **68**, 353–364.

Stehr, F. W., and Casagrande, R. A. (1971). Establishment of *Microctonus aethiops*, a parasite of adult alfalfa weevils, in Michigan. *J. Econ. Entomol.* **64**, 340–341.

Stern, V. M., Smith, R. F., van den Bosch, R., and Hagen, K. S. (1959). The integration of chemcial and biological control of the spotted alfalfa aphid. *Hilgardia* **29**, 81–101.

Starý, P. (1968). Population dynamics of alfalfa pest aphids (Homoptera: Aphidoidea) in Czechoslovakia. *Boll. Lab. Entomol. Agr. Portici* **26**, 271–292.

Stoddart, L. A., and Smith, A. D. (1943). "Range Management," 547 pp. McGraw-Hill, New York.

Streams, F. A., and Coles, L. W. (1965). Wheat stem sawfly parasites, *Collyria calcitrator* and *Pediobius nigritarsis*, in eastern United States. *J. Econ. Entomol.* **58**, 303–306.

Streams, F. A., and Fuester, R. W. (1967). Biology and distribution of *Tetrastischus incertus*, a parasite of the alfalfa weevil. *J. Econ. Entomol.* **60**, 1574–1579.

Telenga, N. A. (1956). Studies of *Trichogramma evanescens* Westw. and *T. pallida* Meyer (Hymenoptera: Trichogrammatidae) and their use for pest control in the U.S.S.R. *Rev. Entomol. URSS* **35**, 599–610.

Telenga, N. A., and Schepetilnikova, V. A. (1949). "A Manual for Breeding and Application of *Trichogramma* in Agricultural Pest Control." Izdatelstvo Akademie Nauk Ukranian SSR, Kiev.

Titus, E. G. (1911). Genera *Hypera* and *Phytonomus* in America, north of Mexico. *Ann. Entomol. Soc. Amer.* **4**, 383–473.

Turnbull, A. L., and Chant, D. A. (1961). The practice and theory of biological control of insects in Canada. *Can. J. Zool.* **39**, 697–753.

Uvarov, B. P. (1928). "Locusts and Grasshoppers," 352 pp. Imp. Bur. Entomol., London.

Vaezi, M. (1950). Rapport du laboratoire d'elevage des parasites d'*Eurygaster integriceps* Put. *Entomol. Phytopathol. Appl.* **2**, 12–18, 27–41.

van den Bosch, R. (1964). Observations on *Hypera brunneipennis* (Coleoptera: Curculionidae) and certain of its natural enemies in the Near East. *J. Econ. Entomol.* **57**, 194–197.

van den Bosch, R. (1971). Biological control of insects. *Annu. Rev. Ecol. Syst.* **2**, 45–66.

van den Bosch, R., and Marble, V. L. (1971). Egyptian alfalfa weevil—the threat to California alfalfa. *Calif. Agr.* **25**, 3–4.

van den Bosch, R., and Stern, V. M. (1969). The effect of harvesting practices on insect populations in alfalfa. *Proc. Tall Timbers Conf. Ecol. Anim. Contr. Habitat Manage.* **1**, 47–54.

van den Bosch, R., Schlinger, E. I., Dietrick, E. J., and Hall, J. C. (1959). The role of imported parasites in the biological control of the spotted alfalfa aphid in southern California. *J. Econ. Entomol.* **52**, 142–154.

van den Bosch, R., Schlinger, E. I., Dietrick, E. J., Hall, J. C., and Puttler, B. (1964). Studies on succession, distribution, and phenology of imported parasites of *Therioaphis trifolii* (Monell) in southern California. *Ecology* **45**, 601–621.

van den Bosch, R., Lagace, C. F., and Stern, V. M. (1967). The interrelationships of the aphid *Acyrthosiphon pisum* and its parasite *Aphidius smithi* in a stable environment. *Ecology* **48**, 993–1000.

van den Bosch, R., Finney, G. L., and Legace, C. F. (1971). Egyptian alfalfa weevil—biological control possibilities. *Calif. Agr.* **5**, 6–7.

Vassiliev, I. V. (1913). Sunn pest (*Eurygaster integriceps* (Osch.) Put.) and new methods of its control by parasites from the insect world. *Trudy. Bureau. Entomol.* **4**, 1–31.

Viktorov, G. A.(1960). Factors of sunn pest (*Eurygaster integriceps* Put.) population dynamics in Kuban in 1956–1958, *In* "The Nofious Pentatomid *Eurygaster intergriceps* Put. Collection of Works of the Laboratory of Invertebrate Morphology" (D. M. Fedotor, ed.), pp. 226–236. Moscow, Akad Nauk. (In Russian).

Viktorov, G. A. (1964). Factors of population dynamics of *Eurygaster integriceps* Put. in Saratov District in 1961–1962. *Zool. Zh.* **43**, 1317–1334.

Viktorov, G. A. (1967). "Problems of Insect Population Dynamics Exemplified by Sunn Pests." Nauka, Moscow.

Voronina, E. G. (1971). Entromophthorosis epizootics of the pea aphid *Acyrthosiphon pisum* (Homoptera: Aphidoidea). *Entomol. Rev.* **50**, 444–453.

Walcott, G. N. (1937). An animal census of two pastures and a meadow in northern New York. *Ecol. Monogr.* **7**, 1–90.

Walker, A. L., Bottrell, D. G., and Cate, J. R., Jr. (1973). Hymenopterous parasites of Biotype C greenbug in the High Plains of Texas. *Ann. Entomol. Soc. Amer.* **66**, 173–176.

Wallace, M. M. H. (1971). Lucerne flea. *CSIRO Div. Entomol. Ann. Rep. 1970–1971*, p. 53.

Weaver, J. E., and Clements, F. C. (1938). "Plant Ecology," 520 pp. McGraw-Hill, New York.

Webster, F. M. (1912). A preliminary report on the alfalfa weevil. *U.S. Dep. Agr. Bur. Entomol. Bull.* **112**, 47.

Wildermuth, V. L., and Frankenfeld, J. C. (1953). The New Mexico range caterpillar and its natural control. *J. Econ. Entomol.* **26**, 794–798.

Wilson, F. (1960). A review of the biological control of insects and weeds in Australia and Australian New Guinea. *Commonw. Inst. Biol. Control Tech. Commun.* 1.

Yano, K. (1970). Preliminary key to the Sciomyzid flies occurring in rice fields. *IBP Handb.* **14**, 35–36.

Yasumatsu, K. (1967). Distribution and bionomics of natural enemies of rice stem borers (research on the natural enemies of rice stem borer). *Mushi Suppl.* **39**, 33–44.

Yasumatsu, K. (1972). Activity, scope and problems in rice stem borer research. *Mushi Suppl.* **45**, 3–6.

Yasumatsu, K., and Torii, T. (1968). Impact of parasites, predators and diseases on rice pests. *Annu. Rev. Entomol.* **13**, 295–324.

Zaeva, I. P. (1969). Comparative role of spring chemical treatments and predators and parasites complex in population dynamics of sunn pest. *Zool. Zh.* **48**, 1652–1660.

Zomorrodi, A. (1959). La lutte biologique contre la punaise du blé *Eurygaster integriceps* Put. par *Microphanurus semistriatus* Nees, en Iran. *Rev. Pathol. Veg. Entomol. Agr. Fr.* **38** (3), 167–174.

17

BIOLOGICAL CONTROL OF INSECT PESTS OF ROW CROPS

R. van den Bosch, Oscar Beingolea G., Mostafa Hafez, and L. A. Falcon

I. INTRODUCTION

There has been but limited successful employment of classic biological control of insect and mite pests of row crops. For example, of the 110 species under partial to complete control by introduced natural enemies reported by DeBach (DeBach, 1964), only 13 involved pests of row crops (including sugarcane). The poor record probably relates largely to the instability of the row crop environment, which presumably does not permit establishment of the effective host—natural enemy relationships which often characterize more stable environments. The row crop only persists for a short period of time, often less than 1 year, during which time the natural enemy must discover and move into the crop, find, attack, and begin to build up in numbers on the host, then be subjected to abrupt habitat destruction at the end of the crop season. Because of this record a defeatist attitude appears to have developed concerning biological control in row crops. Thus, it may well be that in a number of cases where introduction programs might have been undertaken they were not, simply because the chances for success were deemed minimal. It is impossible to even surmise just how great a deterrent this negative attitude has been, but it would be most unfortunate if it were to dominate our future thinking and activities in biological control of pests of row crops. The several substantial successes scored against such pests are proof enough that important natural enemies can be established and be effective even under the basically unfavorable circumstances of the temporary crop environment. Furthermore, our developing knowledge of mass production and artificial manipulation of natural enemies (e.g., by periodic colonization, nutritional augmentation, use of selective insecticides or selective use of conventional ones, and habitat improvement) should permit increasingly greater benefits from the use of introduced natural enemies in row crop pest control. This also seems true for the use of naturally occurring biological control agents. Indigenous natural enemies play an immensely important role in the restraint of pest and potential pest arthropods in row crops, and under modern pest control philosophy this naturally occurring biological control becomes an important element in the developing pest management system.

The importance of these naturally occurring biological control agents can be seen rather clearly when they are removed from an agricultural habitat. A very good example of the removal of such species and the chaotic consequences which result from the loss of the biological control which they effect is seen in the insecticide problems associated in recent years with cotton (Adkisson, 1971; Smith and Reynolds, 1972; and see below). One of the most striking examples of such a pest control breakdown in cotton occurred in northeastern Mexico, where the tobacco budworm, *Heliothis virescens* Fabricius, destroyed extensive cotton industries at Matamoras-Reynosa, and later at Tampico-Mante. The budworm, a secondary pest in cotton, frequently erupts to great abundance where early season insecticide treatments for boll weevil and plant bug control strip the fields of predators and parasites. This, then, permits the bollworm to ravage the crop since it is highly resistant to all available insecticides and thus uncontrollable (Adkisson, 1971). Such debacles present a clear signal that similar breakdowns may occur, and probably are already developing, in other row crops.

It behooves us, then, to direct our thoughts and efforts to the development of pest management systems which take all possible advantage of naturally occurring biological control. Indeed, the future of biological control in row crops may well lie more in the preservation, augmentation, and manipulation of naturally occurring parasites, predators, and pathogens than in the introduction of exotic species. This is especially true for pest species that are in reality pesticide-induced ones. For those that have never been under satisfactory biological control, importations of exotic natural enemies is indicated. Whatever the case, each of these aspects of biological control has an important future in row crop pest management.

II. CLASSIC BIOLOGICAL CONTROL

Introductions of natural enemies resulting in successful control have been made against a number of row crop pests: *Aphidius salicis* Haliday against the aphid *Cavariella aegopodii* (Scop.) on carrot in Australia and Tasmania (M. Day, personal communication; L. Stubbs, personal communication); *Cyrtorhinus fulvus* Knight against *Tarophagus proserpina* (Kirk) on taro in Hawaii and other Pacific islands (Fullaway, 1940; Matsumoto and Nishida, 1966; Pemberton, 1954); *Encarsia formosa* Gah. against the greenhouse whitefly, *Trialeurodes vaporariorum* (Westw.), on a variety of vegetable crops in Australia and Tasmania (Wilson, 1960); *Ascogaster quadridentata* Wesm. and *Glypta hesitator* Grav. against the pea moth, *Laspeyresia nigricana* (Steph.), in vegetable crops in British Columbia (Baird, 1956; McLeod, 1961); *Apanteles glomeratus* (L.), *Apanteles rubecula* (Marsh.), and *Pteromalus puparum* (L.) against the imported cabbageworm, *Pieris rapae* (L.), on cruciferous crops in New Zealand and Australia (Miller *et al.*, 1936; Clausen, 1958; Thompson, 1958; Todd, 1959); *Angitia cerophaga* (Grav.), *Thyraella collaris* Grav., and *Apanteles plutellae* Kurj. against the diamondback moth, *Plutella maculipennis* (Curt.), on cruciferous crops in New Zealand and Australia (Wilson, 1960); *Trissolcus basalis* (Wall.) against the southern green stink bug (green tomato bug), *Nezara viridula smaragdula* (L.), on various vegetable crops in Australia, New Zealand, and Hawaii (Wilson, 1960; Davis, 1967; R. A. Cumber, personal communication); and various species against a variety of sugarcane pests in Hawaii, the West Indies, Florida, Saipan, and Mauritius (Clausen, 1956; DeBach, 1964; Greathead, 1971; Pemberton, 1948; Simmonds, 1958; Yasumatsu *et al.*, 1953; and Chapter 3, this treatise).

As has so often been the case in the employment of classic biological control, the documentation of many of these successes has been scanty, for when the problem has been suddenly alleviated the support required for good evaluation has commonly been withheld. However, several of the successful cases in row crops have been reasonably well documented, and these are discussed below.

A. Sugarcane Pests

The considerable success in the classic biological control of sugarcane pests is almost certainly related to the stability of the cane field environment. In tropical areas, cane is allowed to grow for 18 to 24 months before being harvested. Furthermore, cane has a relatively high tolerance for insect damage before yields are significantly affected. Thus, the lengthy period of stability of the sugarcane environment enables a more effective population development and sustained activity of natural enemies to occur, and pest suppression by these enemies need not be as great as in crops with lower economic injury levels.

Biological control in sugarcane has reached its greatest efficiency in Hawaii, where

TABLE 1
Hawaiian Sugarcane Pests and Their Natural Enemies

Scientific name	Common name	Natural enemies	Control
Aphis sacchari Zhnt.	Sugarcane aphid	Complex of parasites and predators	Substantial
Perkinsiella saccharicida Kirk.	Sugarcane leafhopper	*Tytthus mumdulus* Bredd. predator	Complete
Tyttius mundulus Bredd.	Pink sugarcane mealybug	*Anagyrus saccharicola* Timb. parasite	Substantial
Pseudaletia unipuncta (Haw.)	Armyworm	A complex of parasites	Partial
Spodoptera exempta (Wlk.)	Nut grass armyworm	A complex of parasites	Partial
Anomela orientalis Waterh.	Oriental beetle	*Compsomeris marginella modesta* Sm.; *Tiphia segregata* Cwfd. parasites	Substantial
Rhabdoscelus obscurus (Bdv.)	Sugarcane weevil	*Ceromasia sphenophori* Vill. parasite	Substantial
Gryllotalpa africana P. DeB.	African mole cricket	*Larra luzonensis* Roh. parasite	Partial
Oxya chinensis (Thunb.)	Chinese grasshopper	*Scelio pembertoni* Timb. parasite	Substantial

for decades the crop has been kept largely free of serious pest problems through the action of imported natural enemies (Pemberton, 1948). Successful natural enemy introductions into Hawaii have been made against no less than nine pest insects, with substantial to complete control being effected against six of these (Table 1; DeBach, 1964). The overall impact of the introduced natural enemies has been such that for most of this century biological control has been the mainstay of pest control in Hawaiian sugarcane (Pemberton, 1948).

In the Caribbean area, particular success has been attained against the sugarcane borer, *Diatraea saccharalis* (F). In this area the borer problem has been significantly reduced by introduction of the parasites *Lixophaga diatraea* (Tns.) and *Agathis stigmaterus* (Cress.), and *Apanteles flavipes* Cam., into several of the islands and into southern Florida (Alam *et al.*, 1971; Simmonds, 1958; Clausen, 1956). This has not been true, however, in Louisiana, where the more severe climate, and shorter crop cycle appear to have prevented effective parasite activity.

In Mauritius, the scarabaeid beetles, *Clemora smithi* (Arrow) and *Oryctes tarandus* (Oliv.), have been significantly affected by the introduction of the parasite *Campsomeris phalerata* Sauss. and *Tiphia parallela* Sm. versus the former and *Scolia oryctophaga* Coq. versus the latter, while damage caused by the spotted borer, *Chilo saccariphagus* Bojer, was substantially reduced following introduction of the parasite *Apanteles flavipes* (Cam.) (Greathead, 1971). Finally the beetle *Anomala sulcatula* (Burm.) was completely controlled on the island of Saipan by introduction of the scoliid parasite, *Campsomeris annulata* Fabr (Yasumatsu *et al.*, 1953; see also Chapter 3).

B. Cabbageworms

Two lepidopterous pests of Cruciferae, the imported cabbageworm (cabbage white butterfly), *Pieris rapae* (L.), and the diamondback moth (cabbage moth, *Plutella maculipennis* (Curt), have been the objects of biological control efforts in several countries.

Biological control of *P. rapae* is of special historical significance because it involved the first successful transfer of a natural enemy from one continent to another, i.e., the introduction in 1883 of *Apanteles glomeratus* (L.) from Europe into the United States. The parasite did not have a measurable effect on *P. rapae* but, despite this, its introduction was of great significance because it demonstrated the feasibility of intercontinental transfer of natural enemies. It is also noteworthy that in this landmark endeavor, entomologists committed their first major misjudgment in natural enemy introductions. This was revealed by recent investigations which show *A. glomeratus* to have been the wrong species to import since it is primarily a parasite of *Pieris brassicae* and rather poorly adapted to *P. rapae* (Wilkinson, 1966; Puttler *et al.*, 1970). The more effective *P. rapae* parasite, *Apanteles rubecula*, was somehow overlooked in the early biological control effort. Recently, *A. rubercula* was discovered as an accidental invader in British Columbia and material from this source was imported into the United States where it has played an important role in experimental integrated control studies in Missouri. In those tests where the cabbage was deliberately "salted" with *P. rapae* eggs and larvae and *Trichogramma evanescens* Westw. and *A. rubecula* in the early season the *P. rapae* population was satisfactorily controlled, but where there was no early season "salting," injurious infestations developed in late summer and fall (Parker, 1971; Parker *et al.*, 1971).

The imported cabbageworm has also been the object of biological control efforts in New Zealand, Australia, and Tasmania. In New Zealand, *A. glomeratus* is reported to have effected substantial control of the pest (Miller *et al.*, 1936; Thompson, 1958; Todd, 1959). In Australia, three introduced parasites, *A. glomeratus*, *A. rubecula*, and *Pteromalus puparum* (L.) have provided a "useful" degree of control of *P. rapae* in the Australian Capital Territory, where observations were probably more intense, while varying degrees of control appear to have been effected in South Australia, New South Wales, and Victoria and in Tasmania by *A. glomeratus* and *P. puparum* singly and/or in combination (Wilson, 1960).

C. Diamondback Moth

The diamondback moth, *Plutella maculipennis*, like the imported cabbageworm, is a Palearctic species which has been widely distributed by the activities of man. In New Zealand, the introduced parasite *Angitia cerophaga* Grav. is reported to have effected substantial control of this pest in Australia dating back to the early 1900's. Unidentified parasites were introduced into Western Australia from 1902 to 1907 with claims of considerable success. However, these claims must be doubted since the species remained an important pest.

Later (1936–1951), another group of parasites was introduced into Australia from New Zealand and Europe. Of these, *A. cerophaga*, *Thyraella collaris* Grav., and *Apanteles plutellae* Kurj. became established. These parasites are reported to have caused a marked reduction in the abundance of *P. maculipennis*. The greatest benefit seems to have been in control of the diamondback moth on fodder crucifers, but Wilson (1960) noted that reduction in the use of insecticides in cabbage and cauliflower in South Australia subsequent to parasite introduction indicates that there has been some benefit in these crops too.

D. Greenhouse Whitefly

Beginning in 1934, the parasite *Encarsia formosa* Gahan was introduced from New Zealand into Tasmania and parts of Australia against *Trialeurodes vapororiorum*. Its colonization over a period of several years on infested tomato and other vegetable crops in both greenhouses and the open field resulted in such effective reduction of the whitefly populations that chemical control of the pest was thereafter rarely needed, the program being considered a success.

E. Southern Green Stink Bug (Green Tomato Bug)

In Australia, the egg parasite *Trissolcus basalis* (Woll.) obtained from Egypt and the Caribbean area produced outstanding control of *Nezara viridula smaragdula* in Western Australia, South Australia, New South Wales, Victoria, and other areas (Wilson, 1960; D. F. Waterhouse and E. McC. Callan, personal communication with C. B. Huffaker). At first the parasite was, at best, only partially effective in interior areas of New South Wales and Queensland and in the Australian Capital Territory, but in recent times effective control has occurred in these places. The Australians speculate that either hybridization between certain of the introduced *T. basalis* ecotypes, or preadaptation of one or more of these ecotypes, resulted in a strain or strains that could perform effectively in the colder interior areas.

In Hawaii, where *N. viridula smaragdula* is a relatively recent invader, biological control effected by *T. basalis* and the tachinid *Trichopoda pennipes* var. *pillipes* Fabr. has been a general success (Davis, 1967). It is noteworthy that in Louisiana where a strain of *T. basalis* is indigenous, *N. viridula* is sometimes heavily attacked by this species in pastures but not in soybean.

F. Carrot Aphid

The successful program against the formerly epidemic carrot aphid, *Cavariella aegopodii*, in Australia is one of the most noteworthy in the field of biological control, for it is the only case in which a serious crop disease has been controlled by an introduced natural enemy of the transmitting insect. In this case, the parasite *Aphidius salicis* was introduced from California into Australia in 1960 and quickly became established after very limited colonization. Following establishment, *A. salicis* spread rapidly, even crossing from the Australian mainland to Tasmania unaided by deliberate human effort. The parasite drastically reduced the aphid to a generally low level of abundance in all areas. As a result, carrot motley dwarf, a virus disease which is transmitted by *C. aegopodii*, has virtually disappeared from carrot plantings in Australia and Tasmania (M. Day, personal communication).

G. Taro Leafhopper

Tarophagus proserpina, a native of the southwest Pacific and Indo-Malayan areas, has been accidentally introduced into other areas of the Pacific Basin where in some cases it has caused serious damage. In 1930 it was found in damaging abundance on the island of Oahu in the Hawaiian group, and subsequently spread to all other taro growing areas in Hawaii. In 1938 the mirid, *Tytthus fulvus* (Knight), an egg predator, was introduced into Hawaii from the Philippines. It quickly became established and spread rapidly over the leafhopper-infested areas. The spread and increase in abundance of *T. fulvus* was followed by a corresponding and general decline in abundance of the leafhopper, which has remained below damaging levels ever since. *T. fulvus* was credited with complete biological control by Fullaway (1940), a conclusion strongly supported by more precise studies of Matsumoto and Nishida (1966). In 1947, *T. fulvus* was successfully introduced against the taro leafhopper in Guam, where it also effected complete biological control, and in Ponape, where it is reported to have had a substantial effect (Pemberton, 1954).

III. NATURALLY OCCURRING BIOLOGICAL CONTROL AND INTEGRATED CONTROL

The recurrent insecticide-associated target pest resurgences and secondary pest outbreaks in such crops as cotton, strawberry, tobacco, and various vegetables provide abundant testimonial to the importance of naturally occurring biological control in these plantings. Indeed, much of the future of insect control in row crops should involve ways and means of preserving and augmenting naturally occurring biological control. Integrated control programs based largely on utilization of natural enemies are already in operation in cotton in several countries, while pilot programs, or avenues indicating potentials, have been developed in such crops as strawberry, tobacco, and cole crops (Huffaker and Kennett, 1956; Rabb, 1971; Parker, 1971; Parker *et al.*, 1971). The potential for further exploitation of naturally occurring biological control in row crops seems substantial when we consider the scope of crops and the extensiveness of some of them, e.g., sugarbeet, beans, and other pulse crops, peanut and tomato, and such specialty crops as cucurbits, peppers, crucifers, pea, green beans, and leafy vegetables. Some of these are grown as row crops in glasshouses where employment of biological control has developed extensively in Europe and currently is being explored in North America (Hussey and Bravenboer, 1971; McClanahan 1971a,b).

At present very little progress has been made in integrated control in these crops (except for the special glasshouse situation). Chemicals are still largely relied upon, especially in the short-cycle crops. Growers remain generally satisfied with this approach and unless there is a pesticide-related crisis in a given crop they are difficult to budge from their established pattern. However, such crises are developing, either directly in the crops due to development of resistance, resurgence of target pests or induction of new pests, or because of public concern over, and action against, the use of pesticides.

The widespread development of insecticide-associated problems in cotton exemplifies the direct problem, while the increasing legal restrictions placed on pesticides point up the crisis created by public reaction. In the first instance, target pest resurgence following treatment, induced secondary pest outbreaks, and the development of pest resistance to the pesticides has led to the use of heavier dosages of given pesticides, combinations of materials, and increased frequency of treatments. This in turn has led to increased control costs and not infrequently to reduced control efficacy. Classic breakdowns of this sort have been experienced in Peru, Central America, Mexico, and the United States (Texas) and have forced the development of integrated control programs in several of these countries (Smith and Reynolds, 1972; Adkisson, 1971). The banishment of the residual organochlorine insecticides is forcing the development of additional programs in the southeast United States and California.

Patterns of control in both the established programs and those that will develop in the future can perhaps be best depicted by a summary of what has already been accomplished in Peru, Egypt, and California.

A. Integrated Control of Cotton Pests in Peru

The pest management system developed in cotton in Peru's Cañete Valley has been one of the most widely discussed integrated control programs of the synthetic organic insecticide era. What is not generally known is that equally successful programs were developed and have been maintained in several other of Peru's agricultural valleys. Among these are the valleys of Ica, Pisco, Chincha, Rimac, Chancay, Carobayllo, Huaura, Supe, and Pativilca, in addition to Cañete.

Several of these valleys developed crisis pest problems in cotton following the introduction of synthetic organic insecticides in the late 1940's and early 1950's. As these problems increased in gravity one after another of the valleys abandoned the modern insecticides, adopting legal restrictions on the use of such materials and relying to an increased extent on cultural practices. Among the latter were the adoption of fixed planting dates, clean fallow periods, and an active campaign to increase the natural enemy faunas. The techniques employed to increase these natural enemy populations included the interplanting of corn in cotton fields, planting of corn and wheat acreages in the more heavily infested areas, and transportation of corn tassels bearing various stages of the entomophagous species from heavily populated areas to impoverished ones. The corn interplantings and corn and wheat acreage intermixtures permit the early increase of a complex of entomophagous species, several of which readily move into the developing cotton nearby.

The key to integrated control in Peruvian cotton has been the maintenance of the natural enemies which suppress populations of bollworms, principally *Heliothis virescens* (Fab.), which can explode to devastating levels when released from their biotic suppressants. However, other pest problems are also aggravated by the interference of the synthetic organic insecticides with naturally occurring biological control. For example, in the Cañete Valley the complex of damaging species in cotton virtually doubled in numbers (from 7 to 13 species) following the adoption of the synthetic organic insecticides for use in cotton. Following the establishment of integrated control the numbers of pest species reverted to their former level.

Integrated control programs have been in effect in Peru's cotton growing valleys for nearly a quarter century now, and stand as strong testimonial to the utility of this pest control philosophy in row crops.

B. Integrated Control of Cotton Pests in the San Joaquin Valley of California

There is really only one perennially serious pest of cotton in the San Joaquin Valley, the mirid, *Lygus hesperus* Knight, known colloquially as lygus bug or simply lygus. Lygus is a much feared pest, and justly so, for when it occurs in sufficient abundance at a critical time in plant development it can cause severe damage. Consequently, cotton is heavily treated with insecticides for lygus control. Unfortunately, these treatments are often applied at the wrong times or against infestations which are not injurious. This is not only economically wasteful and ecologically disruptive, but it leads to an aggravated secondary pest outbreak problem. The latter is directly related to the destruction of natural enemies which normally suppress such potentially serious pests as bollworm, *Heliothis zea* Boddie, cabbage looper, *Trichoplusia ni* (Hubner), and beet armyworm, *Spodoptera exigua* (Hubner) (Falcon *et al.*, 1968, 1971; Ehler *et al.*, 1973; Eveleens *et al.*, 1973; van den Bosch *et al.*, 1971). For this reason, the integrated control program in the San Joaquin Valley is primarily geared to the preservation of the natural enemy complex in the fields, and the key to this lies in the limitation of insecticide applications against lygus to those situations, both in time and place, where this pest poses a real threat to the crop. Knowledge of the phenology of the noctuid species has also been of critical importance in this program. The following are salient aspects: (1) the virtual restriction of lygus control to the time of flower budding (squaring) of the Acala cotton varieties, grown in the valley, (2) use of carefully determined economic thresholds as the basis for chemical control decision-making for the two most dangerous pests, *Lygus* and bollworm, and (3) discouragement of applications of broadly toxic insecticides in late July—early August, usually a period of heavy noctuid oviposition, because these insecticides destroy the predators which normally consume most of the eggs and small larvae of bollworm, cabbage looper, and beet armyworm. Where the predators are destroyed by such pesticides, severe outbreaks of one or another or all three of these pest noctuids often occur (Ehler *et al.*, 1973; Eveleens *et al.*, 1973; van den Bosch *et al.*, 1971).

The integrated control program in San Joaquin Valley cotton has not yet been fully implemented. However, where it has been adopted growers have strikingly reduced insecticide usage (hence costs) without reductions in yield or quality, and their fields have been virtually free of lepidopterous pest problems (L. Chrisco, personal communication).

C. Integrated Control of Cotton Pests in Egypt

Egyptian cotton is attacked by more than a dozen pest arthropod species, with the pink bollworm, *Pectinophora gossypiella* (Saunders), and the cotton leafworm, *Spodoptera littoralis* (Boisd.), being the key ones. At the outset of the synthetic organic insecticide era, Egyptian entomologists, like their colleagues elsewhere, undertook an intensive chemical control program in cotton and just as elsewhere they soon encountered the three pesticide-associated problems of target pest resurgence, secondary pest outbreaks, and pest resistance to insecticides.

It became apparent that insecticide disruption of natural enemy populations in the cotton fields was a key contributor to the resurgence and secondary pest problems. Assessments of predator populations at certain localities in 1939—1941 and in 1959—1961 indicated that there had been a drastic decrease in predator numbers, especially in lower Egypt where pesticides had been used most heavily. This effect on natural enemy populations led to studies and practices aimed at understanding and maximizing the role of naturally occurring biolgoical control in the overall pest control program. In other words, an integrated control program based largely on utilization of naturally occurring biological control was perfected and widely implemented.

This program is mainly oriented to the utilization of natural enemies as suppressants of the pest complex as long as possible into the growing season. This is accomplished by minimizing insecticide treatments until the time of predator population breakdown which usually occurs in early July. To help accomplish this, populations of such early-season pests as thrips, aphids, and spider mites are closely monitored and insecticides used only where there is a definite need. Formerly, mere presence of these pests in the fields was deemed sufficient reason for application of insecticides. This, of course, led to exessive insecticide usage. Now the establishment of a valid economic threshold for thrips has permitted a reduction in treated cotton from about one million acres in the early 1960's to approximately 100,000 acres in 1972. Furthermore, with *S. littoralis*, early season insecticidal treatment has been largely replaced as a control measure by hand picking of egg masses at least through late May and June and, in some cases, part of July. This measure is quite feasible under Egypt's social and economic conditions, and is rather efficient as is indicated by the fact that even where infestations are heavy, an estimated 80% of the egg masses can be removed and destroyed. This either eliminates the need for chemical control or delays the applications until after the breakdown of the predator populations.

With pink bollworm, establishment of an economic threshold has permitted a considerable delay in treatment over what was the case when controls were invoked upon detection of the pest in the fields. It has also been found that the pink bollworm moths which infest the new cotton crop originate overwhelmingly from diapausing larvae which survive the winter in dried bolls attached to cotton stalks stored on village rooftops to be used as fuel. Moths from this source deposit the eggs of the first generation in cotton fields occurring in 150- to 250-meter wide belts adjacent to the villages. Fields outside these belts remain virtually free of infestation until they are invaded by moths, which disperse from the infested areas (Hosny and Metwally, 1967). Accordingly, a pink bollworm control system was devised to take advantage of the restricted area of early infestation. Under this system, which involved 700,000 acres in 1968, three insecticidal sprays are applied only to the village bordering belts, starting in early June and covering the period of emergence of the bulk of the first generation moths. This program has resulted in the achievement of efficient pink bollworm control with the treatment of only about 25% of the cotton acreage. Furthermore, it has permitted survival and maximum impact of predators and parasites in the bulk of the acreage which lies outside the treated belts (Hafez, 1968; Abdel-Kawi, 1971).

As a result of the practices just described, considerable benefit appears to have been realized in the Egyptian cotton ecosystem. Among other things, there has been a trend toward the restoration of natural balance in the cotton fields, pest control has been more efficient, there has been a reduction in the use of pesticides, pest population densities have been lower, and there has been an associated reduction in insect injury for the crop with a resultant increase in yield (Hafez, 1972).

IV. CONCLUSION

Success in classic biological control of row crop pests can quite probably be increased through more intensive effort. The results will almost certainly continue to be less rewarding than in the more stable situations afforded by horticultural crops, pastures, forests, and the like. Technological advances in mass propagation of natural enemies and the augmentation of their populations in the field will permit effective, economically feasible manipulation of row crop pest natural enemies. This should be particularly true in high value crops such as strawberry and many kinds of vegetables. Knowledge of agroecosystems and the biotic elements in them, coupled with the development of valid economic thresholds, selective pesticides, and better pest and natural enemy manipulation, will almost certainly permit increasingly effective utilization of naturally occurring biological control in both high value crops and the more extensive field crops such as cotton, field tomato, sugar beet, and soybean.

REFERENCES

Abdel-Kawi, F. (1971). Studies on different predators of certain economic pests. M.Sc. Thesis, Plant Protection Department, Faculty of Agriculture, Assiut University, Egypt.

Adkisson, P. L. (1971). Objective uses of insecticides in cotton. *In* "Agricultural Chemicals—Harmony or Discord for Food People Environment" (J. E. Swift, ed.), pp. 43–51. Univ. Calif. Div. Agr. Sic., California.

Alam, M. M. Bennett, F. D., and Carl, K. P. (1971). Biological control of *Diatraea saccharalis* (F) in Barbados by *Apanteles flavipes* Cam. and *Lixophaga diatrae* T. T. *Entomophaga* **16**, 151–158.

Baird, A. B. (1956). Biological control of insect and plant pests in Canada. *Proc. 10th Int. Congr. Entomol.* **4**, 483–485.

Clausen, C. P. (1956). Biological control of insect pests in the continental United States. *U.S. Dep. Agr. Tech. Bull.* **1139**, 151 pp.

Clausen, C. P. (1958). Biological control of insects. *Annu. Rev. Entomol.* **3**, 291–310.

Davis, C. J. (1967). Progress in the biological control of the southern green stink bug, *Nezara viridula* variety *smaragdula* (Fabricius) in Hawaii (Heteroptera: Pentatomidae). *Mushi* **39**, 9–16.

DeBach, P. (ed.). (1964). "Biological Control of Insect Pests and Weeds." Reinhold, New York.

Ehler, L. E., Eveleens, K. G., and van den Bosch, R. (1973). An evaluation of some natural enemies of cabbage looper on cotton in California. *Environ. Entomol.* **2**, 1009–1015.

Eveleens, K. G., van den Bosch, R., and L. E. Ehler. (1973). Secondary outbreak induction of beet armyworm by experimental insecticide applications in cotton in California. *Environ. Entomol.* **2**, 497–503.

Falcon, L. A., van den Bosch, R., Ferris, C. A., Etzel, L. K., Stinner, R. E., and Leigh, T. F. (1968). A comparison of season-long cotton pest control programs in California during 1966. *J. Econ. Entomol.* **61**, 633–642.

Falcon, L. A., van den Bosch, R., Gallagher, J., and Davidson, A. (1971). Investigation of the pest status of *Lygus hesperus* in cotton in central California. *J. Econ. Entomol.* **64**, 56–61.

Fullaway, D. T. (1940). An account of the reduction of the immigrant taro leafhopper (*Megamelus prosperina*) population to insignificant numbers by the introduction and establishment of the egg-sucking bug *Cyrtorhinus fulvus*. *Proc. 6th Pac. Sci. Congr.* **4**, 345–346.

Greathead, D. J. (1971). A review of biological control in the Ethiopian region. *Commonw. Inst. Biol. Control Tech. Commun. No.* **5**.

Hafez, M. (1968). Effect of controlling the pink bollworm by spraying belts of cotton on the abundance of predators in cotton fields in U.A.R. *13th Int. Congr. Entomol.* **9**, 97. (Moscow, Abstr.)

Hefez, M. (1972). Statement of Arab Republic of Egypt. Methods of integrated insect control in cotton. *31st Plen. Meet. Int. Cotton Advisory Committee, Managua, Nicaragua*, pp. 30–58.

Hosney, M. M., and Metwally, A. G. (1967). New approaches to the ecology and control of three major cotton pests in U.A.R.—A new approach to the problem of pink bollworm control in U.A.R. *U.A.R. Min. Agr. Tech. Bull.* **1**, 37–54.

Huffaker, C. B., and Kennett, C. E. (1956). Experimental studies on predation: (I). Predation and cyclamen mite populations on strawberries in California. *Hilgardia* **26**, 191–222.

Hussey, N. W., and Bravenboer, L. (1971). Control of pests in glasshouse culture by the introduction of natural enemies. *In* "Biological Control" (C. B. Huffaker, ed.), pp. 195–216. Plenum, New York.

McGugan, B. M., and Coppell, H. C. (1962). Biological control of forests insects—1910–1958. *Commonw. Inst. Biol. Control Tech. Commun.* **2**, 35–216, 215 pp.

McClanahan, R. J. (1971a). *Tetranychus urticae* (Koch), two-spotted spider mite (Acarina: Tetranychidae). *Commonw. Inst. Biol. Control Tech. Commun.* **4**, 49–50.

McClanahan, R. J. (1971b). *Trialeurodes vaporariorum* (Westwood), greenhouse whitefly (Homoptera: Aleyrodidae). *Commonw. Inst. Biol. Control Tech. Commun.* **4**, 57–59.

McLeod, J. H. (1961). Biological control of pests of crops, fruit trees, ornamentals and weeds in Canada up to 1959. *Commonw. Inst. Biol. Control Tech. Commun.* **2**, 1–33.

Matsumoto, B. M., and Nishida, T. (1966). Predator-prey investigations on the taro leafhopper and its egg predator. *Hawaii. Agr. Exp. Sta. Tech. Bull.* **64**, 32 pp.

Miller, D., Clark, A. F., and Dumbleton, L. J. (1936). Biological control of noxious insects and weeds in New Zealand. *N.Z. J. Sci. Technol.* **18**, 579–593.

Parker, F. D. (1971). Management of pest populations by manipulating densities of both hosts and parasites through periodic releases. *In* "Biological Control" (C. B. Huffaker, ed.), pp. 365–376. Plenum, New York.

Parker, F. D., Lawson, F. R., and Pennell, R. E. (1971). Suppression of *Pieris rapae*: Mass releases of both the pest and its parasite. *J. Econ. Entomol.* **64**, 721–735.

Pemberton, C. E. (1948). History of the entomology department experiment station, H.S.P.A., 1904–1945. *Hawaii. Plant. Rec.* **52**, 53–90.

Pemberton, C. (1953). Biological control of insects in Hawaii. *Proc. 7th Pac. Sci. Congr.* **4**, 220–223.

Pemberton, C. (1954). Invertebrate Consultants Commission for the Pacific Report for 1949–54. Pacific Science Board, Nat. Acad. Sci. Nat. Res. Council.

Puttler, B., Parker, F. D., Pennell, R. E., and Thewke, S. E. (1970). Introduction of *Apanteles rubecula* into the United States as a parasite of imported cabbageworm. *J. Econ. Entomol.* **63**, 304–305.

Rabb, R. L. (1971). Naturally occurring biological control in the eastern United States with particular reference to tobacco. *In* "Biological Control" (C. B. Huffaker, ed.), Chapter 12, pp. 294–311. Plenum, New York.

Simmonds, F. J. (1958). Recent work on biological control in the British West Indies. *Proc. 10th Int. Congr. Entomol.* **4**, 475–478.

Smith, R. F., and Reynolds, H. T. (1972). Effects of manipulation of cotton agro-ecosystems on insect pest populations. *In* "The Careless Technology" (M. J. Farvar and J. P. Milton, eds.), pp. 373–406. Natural History Press, Garden City, New York.

Thompson, W. R. (1958). Biological control in some Commonwealth countries. *Proc. 10th Int. Congr. Entomol.* **4**, 479–482.

Todd, D. H. (1959). Incidence and parasitism of insect pests of cruciferous crops in the North Island. Evaluation of data, 1955–58 seasons. *N.Z. J. Agr. Res.* **2**, 859–869.

van den Bosch, R., Leigh, T. F., Falcon, L. A., Stern, V. M., Gonzales, D., and Hagen, K. S. (1971). The developing program of integrated control of cotton pests in California. *In* "Biological Control" (C. B. Huffaker, ed.), pp. 377–394. Plenum, New York.

Wilkinson, A. T. S. (1966). *Apanteles rubecula* Marsh. and other parasites of *Pieris rapae* in British Columbia. *J. Econ. Entomol.* **59**, 1012–1018.

Wilson, F. (1960). A review of biological control of insects and weeds in Australia, and Australian New Guinea. *Commonw. Inst. Biol. Control Tech. Commun. No.* **1**, 102 pp.

Yasumatsu, K., Nomura, K., Utida, S., and Yamasaki, T. (1953). "Applied Entomology," 266 pp. Asakura Publ., Tokyo. (In Jap.)

18

BIOLOGICAL CONTROL OF MEDICAL AND VETERINARY PESTS

E. C. Bay, C. O. Berg, H. C. Chapman, and E. F. Legner

I. INTRODUCTION

Insects that directly attack man and his animals differ significantly in two ways from those affecting his crops, with respect to biological control strategy. There are exceptions in both cases. On the positive side, the most common and vexatious pests of man, including mosquitoes, flies, ants, and wasps, are only injurious as adults, whereas most crop damage by insects is inflicted by larvae. Thus, within any particular pest generation, time is on our side for parasites or predators to deplete the immature population and reduce the number of adults free to attack man and his animals. On the negative side, man's tolerance for arthropods attacking his person is often so low that the best natural enemies cannot reduce populations of these pests to his satisfaction. For instance, no matter what the mosquito breeding potential of a given area is, with or without natural controls, one mosquito bite per person per night would often be regarded as unacceptable. This is particularly true if the mosquito happens to be a disease vector. To look at it another way, an insect like a scale or an aphid that might be reduced from thousands to tens per square meter of leaf surface by a parasite would be regarded under outstanding biological control, but vector mosquito adults resting among foliage at the same density, or less, would not be considered under control at all.

Progress in the biological control of medically significant arthropods, except for the use of fish in mosquito control, has been notably slow for a number of reasons. Perhaps primary among these is that man, in altering his environment for his convenience, has also provided a more favorable environment for many of his direct insect pests than for their natural enemies. Mosquitoes will often breed in habitats that are too heavily polluted to sustain their natural enemies. Poor irrigation practices create new mosquito breeding sources faster than natural enemies can respond to them. Fly larvae and pupae that would be exposed to natural enemies in small scattered animal droppings find more protection in steeped manure.

Although snails are not arthropods, these molluscs, particularly those that are vectors of schistosomiasis, share this chapter with the housefly and the mosquito as among those organisms most harmful and annoying to man that have been investigated as subjects for biological control. These snails, too, have benefited from man's manipulation of his environment. As inhabitants of quiet water, their breeding grounds have been immeasurably increased in recent years by the construction of new dams and irrigation canals in tropical and semitropical countries. Schistosomiasis, or bilharzia, now vies with malaria as the world's most debilitating disease of humans. Moreover, unlike *Anopheles* mosquitoes that are vectors for malaria, there is no known effective or economic control for the snail host of schistosomiasis. For this reason, among others, the search for biological control agents of snails is given special attention. Ironically, whereas flies are the targets for control in the case of the housefly and the mosquito, it is another fly, of the family Sciomyzidae, that functions as a prominent natural enemy under investigation as a control agent for snails.

II. MOSQUITOES

Perhaps few other animals have so many natural enemies as the mosquito. Jenkins (1964) lists 212 pathogens and parasites that affect mosquito larvae, and more than 500 predators that attack the larvae and adults. With such an array of enemies, and with mosquitoes the problem that they are, the question logically arises whether natural enemies of mosquitoes can be of any practical value. The answer is that some, at least, particularly certain larvivorous fishes, are of demonstrated value where it is practical to employ them. Others, including many aquatic beetle larvae, seem completely ineffectual from the practical standpoint as far as has been determined and reported. Still others, including notonectid bugs (Bay, 1967; Kühlhorn, 1965) and *Toxorhynchites* mosquitoes (Trpis, 1970), whose adults do not feed on blood and whose larvae destroy the larvae of species that do, are strikingly effective in some field situations but are insufficiently reliable, and so far too expensive for effective mass rearing and release.

An especially important consideration in the biological control of mosquitoes is the diversity of habitats in which they breed. Mosquito breeding sites run the gamut of rain puddles, salt and freshwater marshes, tree holes, rock pools, wheel ruts, hoof prints, cisterns, tin cans and tires, snowmelt pools, leaf axils, rat-gnawed coconuts, crab holes, snail shells, flood plains, sewage oxidation lagoons, privys, and even the hollow interiors of some carnivorous plants that have evolved to lure and destroy other insects. Clearly, no single parasite or predator could be expected to find application against a group of insects with such an array of breeding sources and protection. To complicate matters, some mosquito larvae inhabit a variety of breeding sites at a common location, so that an effective predator in one of these might not have a noticeable effect on the adult mosquito population as a whole.

Other problems include the relatively short life cycle of most mosquitoes. This may be as brief as $3\frac{1}{2}$ to 4 days for *Psorophora confinnis* (Lynch Arribálzaga) in the Coachella Valley of California, and for certain species of floodwater *Aedes*. Most established predators have little chance to discover and respond to such "instantaneous" mosquito breeding. Man has even less time to mobilize for mass release with those parasites or predators that he may be successful in rearing. Furthermore, mosquito predators of which we are aware are not dependent on mosquito larvae, and are seldom if ever reciprocally regulating in the sense we commonly think of as biological control. Some of these predators, particularly fish, however, are often destructive to mosquito larvae and can totally suppress larval populations to which they have access in permanent standing water. Unfortunately, as suggested earlier, "permanent" water constitutes only a fraction of the available mosquito breeding sites in many localities. Even in permanently marshy areas much exposed water is actually temporary or intermittent, dependent upon seasonal drought and flooding. In those areas where water is relatively permanent, many mosquito larvae are protected amid grass hummocks and in breeding pockets among debris-congested shallows.

Given the circumstances described, it is understandable that insecticides have been so heavily relied upon for mosquito control in recent years. Their advantages of economy, ready availability, storage, and applicability over discontinuous and obstructed larval habitats present formidable competition to known measures of biological control. Microbial pathogens, nematodes, and hydra are among the few organisms that conceivably could be produced, stored, and applied in the manner of insecticides. For this reason, except for refinements in the use of fish, the biological control research that has involved mosquitoes has been mostly with pathogens, including certain nematodes. Mosquito control prospects of hydra are discussed by Qureshi and Bay (1969).

A. Fish

The practice of biological control of mosquitoes is still essentially restricted to use of fish. There are literally hundreds of species of fish that will feed on mosquito larvae, as any tropical fish fancier should be aware, but very few of these are practical in mosquito control. The most noted exception is the mosquito fish, *Gambusia affinis* Baird and Girard, that is native to the southeastern and gulf coast United States. It is a top minnow (family Poeciliidae) which bears living young. *Gambusia* has been explored and used successfully for the control of pool-breeding mosquitoes since the early 1900's. During this period, it has been spread to virtually all parts of the world in which it can survive, and has established an excellent reputation for mosquito control. Hildebrand (1921) reported success with *Gambusia* in providing rather permanent anopheline control at reduced cost in several states of the southern United States. Also, during the 1920's *Gambusia* hatcheries were established in Richmond, Virginia, and Sacramento, California, from where fish were distributed to communities requesting them within those states. In several states, and especially in California, mosquito abatement districts and health departments today maintain stocks of *Gambusia* and use them on a regular basis.

Gambusia has been credited with particular success in reducing malaria in Istria (Hackett, 1931). More recently, Tabibzadeh *et al.* (1970) reported that *Gambusia* has served as a supplemental control of *Anopheles* in Iran where it has helped reduce exophilic species that are not exposed to residual DDT house sprays. It is pointed out that because of the combined malarial control measures employed in all extensive areas, the effects of individual approaches cannot be identified. However, it was noted that after *Gambusia* introduction *Anopheles* larvae disappeared from numerous sites where they previously had occurred for many years.

In Hawaii, where *Gambusia* was first introduced in 1905, this fish continues to be an important mosquito control measure. It is often not sufficient that fish be introduced into breeding sites and left ignored. Because of seasonal and physical alterations to their environment, fish populations must sometimes be augmented or replaced. This is the case in Hawaii's 700-acre Kawainui swamp where the surface area fluctuates drastically between the wet and dry seasons. Nakagawa and Ikeda (1969) have found that fish seeded at a rate of 1000 to 2000/acre reproduce sufficiently within 6 to 8 weeks to establish full mosquito control, eliminating the need for further chemical applications that are sometimes required while fish are becoming established. Tabibzadeh *et al.* (1970) found 2 female and 1 male *Gambusia*/m^2 to be adequate for small unobstructed mosquito breeding sites, and 15 females/m^2 required for extensive sites. These rates extrapolate to roughly 12,000 and 60,000 fish/acre, respectively. Hoy and Reed (1971), working in carefully managed California rice fields, obtained excellent mid- and late-season control of *Culex tarsalis* Coquillett from *Gambusia* stocked at only 100 and 200 gravid females/acre in April and May. Bay (unpublished data) found that in small freshly filled shallow ponds at Riverside, California, the equivalent of 12,000 female *Gambusia*/acre gave immediate and complete control of *Culex peus* Speiser and destroyed 94% of *C. peus* egg rafts before they could hatch. In similar ponds where *G. affinis* was stocked at a rate of 1200 females/acre, mosquito larvae were eliminated 20 days after filling and stocking with fish. The differences in effective stocking rates for

G. affinis reported by various workers can easily be reconciled by considering the reproductive capability of this fish. This is enhanced by its ovoviviparous habit; its newborn young have a high survival. This is especially so in newly flooded habitats where piscivorous fish and birds are few or absent. Gravid *Gambusia* may give birth at any time after stocking and bear up to 100 or more young which will mature and themselves become gravid within 4 to 6 weeks. The parent fish, meanwhile, produces additional broods at 3- to 6-week intervals until she dies at the end of the summer. Estimating brood survival at 50 young per fish, the 100 gravid females/acre released by Hoy and Reed (1971) could easily have produced over 5000 adult fish/acre in less than 2 months. Even this figure is apt to be conservative, since Hoy and O'Grady (1971) found evidence of a leveling off before midseason for populations begun at either 100 or 200 fish/acre. One of us (Bay) has found population limits equal to approximately 125,000 *Gambusia*/acre (250 lbs) within 60 days for fish stocked at a 1200/acre rate in experimental ponds in Riverside. Terminal population levels will vary dependent upon the carrying capacity of individual ponds.

There are several factors that contribute to the extraordinary capability of *G. affinis* as a mosquito larvivore. The advantages of ovoviviparity, early maturity, seasonal brood repetition, and high density distribution have been discussed. Other attributes include its 3—6 cm size coupled with its shoaling behavior which enables it to penetrate and negotiate pool and marsh mosquito larval habitats, a salinity tolerance up to 5% (Ahuja, 1964), organic waste tolerance, a relatively wide temperature range, and a highly omnivorous as well as larvivorous habit. Cold tolerance is the most limiting factor in the practical range of *Gambusia* utilization. Paradoxically, its very efficiency limits its full acceptance as a mosquito larvivore in amenable habitats. Because of its combined characteristics of voracity, omivorousness, rapid reproduction, high survival, and pervasiveness it sometimes displaces or eliminates other small fish species either directly or by its superior competition for food. For this reason its use as a larvivore in some areas is discouraged. The fact that *G. affinis* so completely eliminates other forms of zooplankton in addition to mosquito larvae from shallow ornamental pools sometimes causes another annoyance. Without the zooplankters to feed on phytoplankton these pools often become an unaesthetic dense murky green (Hurlbert *et al.*, 1972). When considering the objections to *Gambusia*, however, one must think of the alternatives where mosquito control is demanded or desired. Any other predator that might be as capable would likely evoke the same criticisms. The only other options presently available are chemicals and habitat modification or destruction.

Other fishes that have been studied and found to be occasionally successful for mosquito control include the common guppy, *Poecilia reticulata* Peters (Bay and Self, 1972), and certain of the annual fishes that occur in parts of Africa and South America. These latter fishes are cyprinodontid killifishes that depend upon alternating seasons of flooding and drought in order to complete their life cycles and reproduce. Therefore, hatches that produce insufficient fish populations for mosquito control cannot improve within a particular flood season.

Among the many fishes that feed upon mosquito larvae, many species have a far higher per capita larva consumption than do *Gambusia*, but occur at disproportionately low characteristic density. In order that a nonselective predator can have a measurable effect on a given prey density it must have both a good per capita rate of consumption for that prey, and also an effectively high characteristic density (see further Chapter 20).

Larvivorous fishes, where these can be used, will likely remain the most effective and practical biological controls of mosquitoes for some time to come. Mosquito pathogens and nematodes, however, if these can be propagated and made to respond on demand, should have much broader utilization.

B. Pathogens

Among the 212 mosquito pathogens listed by Jenkins (1964) the most numerous were protozoa (83), followed by fungi (57), bacteria (22), nematodes (20), and viruses (2). Examination of the bibliography updated through 1967 by Laird in Burges (1971) showed that the most common pathogens mentioned in the titles were fungi (29), protozoa (23), nematodes (12), bacteria (6), and viruses (6). Most of these reports concern laboratory studies, host–parasite relationships, distribution records, and new host records, with relatively little on actual use of these pathogens in field studies.

This chapter treats the present status of certain pathogens in these five common groups with special emphasis on those that seem to offer the most hope for imminent field use (see further Chapter 7).

1. Viruses

Only a single virus was listed by Jenkins (1964). This was a cytoplasmic polyhedrosis virus (CPV) from California that exhibited large tetragonal inclusion bodies in the wing and leg buds of larval *Culex tarsalis* (Kellen *et al.*, 1963, 1966). An identical pathogen was also reported from larvae of *Culex salinarius* Coquillett in Louisiana (Clark and Chapman, 1969) and from larvae of *Anopheles crucians* Wiedemann (Chapman *et al.*, 1970).

Two occluded viruses, a nuclear polyhedrosis virus (NPV) and a CPV, that infect the gut and gastric cecal cells of *Aedes sollicitans* (Walker) and *Culex salinarius* in Louisiana were reported by Clark *et al.* (1969). Similar NPV and CPV infections have been seen in at least 18 other species of mosquito larvae.

An epizootic involving a CPV and a NPV, predominately the latter, produced combined infections of more than 71% in a natural larval population of *A. sollicitans* in Louisiana (Clark and Fukuda, 1971). A similar epizootic infecting over 65% of larval *A. sollicitans* in another site was recently seen (Chapman, unpublished). However, when both sites were reflooded after a short period of drying, infection levels were extremely low.

CPV strains most commonly occurring in the field in *A. sollicitans* and *C. salinarius* were not cross-transmissible in the laboratory. Levels of infection, both in the field and laboratory, are usually below 5% and such infections are seldom lethal.

On the other hand, observed infections of NPV are normally lethal. Infection levels, however, from the transmission of NPV in the laboratory, have never approached those measured in the two epizootics. Since we have never seen 100% infection in the field or laboratory, it is possible that a segment of the population is resistant; higher levels of infection might occur in mosquito species not previously exposed to such a virus.

The mosquito iridescent virus (MIV) was first reported by Clark *et al.* (1965) and Weiser (1965). Reported hosts of MIV are larvae of *Aedes annulipes* (Meigen) (Czech.), *A. cantans* (Meigen) (Czech.), *A. detritus* (Haliday) (Tunisia), *A. dorsalis* (Meigen) (Nevada), *A. fulvus pallens* Ross (Louisiana), *A. sticticus* (Meigen) (Louisiana), *A. stimulans* (Walker) (Connecticut, New Jersey), *A. taeniorhynchus* (Wiedemann) (Florida, Louisiana), *A. vexans* (Meigen) (Louisiana), *Psorophora ferox* (Humboldt) (Louisiana), *P. horrida* (Dyar & Knab) (Louisiana), and *P. varipes* (Coquillett) (Louisiana). MIV was also recently noted in *P. confinnis* (Lynch Arribálzaga) in Louisiana. Infection levels in the field have seldom exceeded 1%.

Since the yield of virus particles from *in vivo* cultures of mosquitoes is ridiculously small, future field releases will undoubtedly depend on the successful culturing of viruses *in vitro* or *in vivo* in larger insects (see further Chapter 7).

2. *Bacteria*

Few observations have been recorded of bacterial epizootics in field populations of mosquitoes. Also, laboratory studies have been few and except for one, commercial strains of bacilli have shown little promise against mosquitoes. The exception, a bicrystalliferous strain of *Bacillus thuringiensis* isolated from *Culex tarsalis*, has shown promise against some aedine species in California (Reeves and Garcia, 1970). It is readily mass produced *in vitro* and has been limitedly successful in small field experiments against *A. nigromaculis* (Ludlow) (see further Chapter 7).

3. *Fungi*

One of the most promising pathogens of mosquitoes would seem to be *Coelomomyces* since it has been reported to infect more than 63 species of mosquitoes involving 11 genera, the most common being *Anopheles*, followed by *Aedes*, *Culex*, *Culiseta*, *Psorophora*, *Aedomyia*, *Opifex*, *Armigeres*, *Uranotaenia*, *Toxorhynchites*, and *Tripteroides*. Although at least 25 species and 4 subspecies of *Coelomomyces* have been reported from mosquitoes, the true number is unknown and for most, information is sadly lacking on host specificity.

The effect of various species of *Coelomomyces* against specific mosquitoes in the field has been documented. Muspratt (1963) reported up to 100% infection of some populations of the important *Anopheles gambiae* Giles in Zambia; in a 3-year study in a lake in North Carolina the average fungal infections in *Anopheles quadrimaculatus* Say was 24% (Umphlett, 1970); Chapman and Glenn (1972) reported average weekly infections of 33 and 48% by two species of *Coelomomyces* over $4\frac{1}{4}$ and $2\frac{1}{4}$ years, respectively, in *Anopheles crucians* in two ponds in Louisiana. The highest population parasitism noted was 95% in *Culiseta inornata* Williston and *Psorophora howardii* Coquillett by *Coelomomyces psorophorae* Cough, 97% in *Aedes triseriatus* (Say) by *C. macleayae* Laird, and 85% in *Culex peccator* Dyar & Knab by *C. pentangulatus* Couch (Chapman, unpublished data). The successful introduction and establishment of a *Coelomomyces* was made by Laird (1967) on a small Pacific Isle against *Aedes polynesiensis* Marks, a vector of filariasis, and represents one of the few attempts to establish pathogens of mosquitoes into other areas or countries.

While both Madelin (1966) and Couch (1967) have had some success in obtaining laboratory infections in certain mosquitoes with *Coelomomyces*, the expertise is still lacking for production of large amounts of inoculum either by *in vitro* or *in vivo* means.

Other fungi such as *Beauveria bassiana* (Balsamo), *Metarrhizium anisopliae* (Metsch.), and *Entomophthora* spp. have been tried against mosquitoes in the laboratory with varying results. This is summarized by Roberts (1970) (see further Chapter 7).

4. Protozoa

Many groups of protozoans attack mosquito larvae, including flagellates (*Blastocrithidia* and *Crithidia*), eugregarines (*Lankesteria*), schizogregarines (*Caulleryella*), and ciliates (*Vorticella* and *Tetrahymena*), but so far only one group, the Microsporida, has shown sufficient pathogenicity to warrant study.

Microsporidan species are often highly pathogenic and occur in nature in a variety of mosquitoes. However, levels of infection observed in the field have mostly been very low (usually no more than 1%). Less than 35 microsporidan species have been described, mostly *Thelohania* species, followed by *Pleistophora*, *Nosema*, *Stempellia*, and *Toxoglugea*; the true number of species is uncertain since some have been synonymized and others remain undescribed. Worldwide, only about 43 mosquito species representing four genera (*Anopheles*, *Aedes*, *Culex*, and *Culiseta*) were listed as hosts of Microsporida by Jenkins (1964). Now, in the United States alone, an additional 34 mosquitoes are known to be hosts. New generic hosts are *Psorophora*, *Orthopodomyia*, *Mansonia*, *Toxorhynchites*, and *Uranotaenia* (Chapman *et al.*, 1967, 1969).

Despite more recorded mosquito hosts and microsporidan species, the problems with the use of most microsporidan pathogens (some were known in the 1920's) have not been resolved. Most microsporidan pathogens of mosquitoes, unlike those that occur in nonaquatic insects, cannot be transmitted by feeding infected material to healthy specimens in the laboratory, nor can per os Microsporida transmission be demonstrated in the field. Apparently only about three microsporidan species (*Nosema algerae* Vavra & Undeen, *Pleistophora culicis* (Wieser), and *Stempellia Miller*: Hazard and Fukuda) can be consistently transmitted per os in the laboratory. Mosquito infection by most Microsporida is either transovarian (*Thelohania*, *Pleistophora*, and *Stempellia*) or transovum (*Nosema*).

The only attempt to release a microsporidan against mosquitoes was with *Pleistophora culicis* on the Pacific Island of Nauru and the assessment of the attempt is incomplete (Laird, 1971). Worthwhile experiments in the near future will have to be accomplished with *P. culicis, Nosema stegomyiae,* or *N. algerae* because inadequate amounts of inoculum for other microsporidan species cannot yet be produced (see further Chapter 7).

5. Nematodes

In this group, a number of rhabditoid and mermithid nematodes appear to offer promise as control agents of mosquitoes.

A member of the first group, *Neoaplectana carpocapsae* Wieser (DD-136), was tested with some success against mosquitoes in the laboratory (Welch and Bronskill, 1962) and in the field (Briand and Welch, 1963). Since DD-136 is easily mass reared and released, additional research is imperative to better understand its potential.

Many reports mention the sometimes high level of infection of mosquitoes in nature by mermithids (Jenkins, 1964; Welch, 1960; Petersen *et al.*, 1968) but most investigators also mention the absence of the parasite in most similar adjacent sites. Nickle (1972) revised some mermithid genera and renamed some nema parasites of mosquitoes.

Apparently Muspratt (1965) cultured the first mermithid parasite of mosquitoes, a *Romanomermis* sp. from a Zambian tree hole; this nematode was later released (on a small scale) against mosquitoes in 1967 on Nauru Island in the Pacific with the assessment incomplete at present (Laird, 1971).

Three mermithid nematodes have been cultured in the Lake Charles, Louisiana (United States) laboratory: (1) *Perutilimermis culicis* (Stiles) (= *Agamomermis culicis*) specific to *Aedes sollicitans* (Petersen *et al.*, 1967); (2) *Diximermis peterseni* (= *Gastromermis* sp.) specific to anopheline larvae (Petersen and Chapman, 1970); and (3) *Reesimermis nielseni* Tsai & Grundmann (= *Romanomermis* sp.) known to infect larvae of 22 mosquito species in the field and another 33 species in the laboratory (Petersen *et al.*, 1968). *Reesimermis nielseni* has been much investigated prior to its release in the field. The most important result was its successful mass culture *in vivo* using the southern house mosquito (*Culex pipiens quinquefasciatus* Say) by Petersen and Willis (1972). A number of field sites was sprayed with the infective stage of the nematode in 1971 and treatments produced an average parasitism of 65% in anopheline larvae (inundative effect). Only time will demonstrate the extent of the inoculative effect of these applications; up to 47% of some anopheline populations were infected when the nematode recycled in the habitats (Petersen, unpublished data).

Careful research must precede the use of this nematode against mosquitoes in the field since some mosquito species exhibit physical and/or physiological resistance to it;

in addition, certain environments of mosquitoes (salt marshes and tree holes) negate the use of this parasite because of the susceptibility of the infective stage to salts (see further Chapter 7).

6. Appraisal

We have some idea of the extent of infections in mosquitoes caused by various pathogens in specific habitats. Known sites are few, and expertise is needed to produce sufficient inoculum to treat new sites. Naturally this has to follow the establishment of such parameters as the susceptibility of the target mosquitoes and the tolerance of the pathogens to the water, which would include the effect of various salts as well as micropredators.

Also, we now have considerable knowledge concerning those parasites and pathogens that are host-specific for mosquitoes and those that are not. Mermithid nemas and *Coelomomyces* that have been studied from mosquitoes are definitely host-specific. The same appears to be true for Microsporida and while few nontarget tests have been made, the viruses found in mosquitoes appear to fall in this same category. Too little is known of the bacilli, notably BA-068, but like other variants of *Bacillus thuringiensis* it does affect certain Lepidoptera as well as mosquitoes. Pathogens such as *Beauveria*, *Metarrhizium*, and *Entomophthora* are not specific to mosquitoes. This includes Dutky's nema DD-136. It is doubtful, however, that these organisms will ever show sufficient promise to be recommended in the field.

Eventually, if pathogens are to be released on a large scale against any mosquito, relatively economical methods of mass *in vitro* culturing and dispersal must be available. We are only fooling ourselves if we believe that this can be brought to fruition by the monies and research personnel presently available for biological control studies of mosquitoes. The cream has been skimmed—now the seemingly insurmountable problems that have plagued researchers of mosquito pathogens for too long must be attacked and overcome. Only then can we hope to attain the successes in biological control with pathogens that are presently available against some arthropod pests of field crops and forests.

III. MEDICALLY IMPORTANT SNAILS*

The biologists primarily responsible for the control of medically important snails have made very little study of biological control methods. Many are content to depend on chemical control methods. Others have advocated attempts at biological control, but without any adequate study of the natural enemy suggested as a control agent. Many of these have merely been observed to feed on snails in aquariums under con-

*Strategies for the biological control of snail-borne diseases that bypass the snail hosts and aim directly at the trematode parasites are reviewed by Berg (1973).

ditions that could not be duplicated in nature. For example, Lo (1967) attributed the heavy densities of Ostracoda that frequently jeopardize laboratory snail colonies to continuously rich food supply, optimal temperature, and lack of predators in the aquariums. He wisely suggested that ostracod predation may offer little hope as a method of snail control, because such densities probably can be produced only in laboratory snail tanks.

Examples of ill-considered candidate agents of biological control are found especially among the fish, frogs, turtles, waterfowl, and mammals. Although some species of fishes are exceptions, the diets of most vertebrates that feed on snails are so varied that they prove unreliable as mortality factors in the population dynamics of the pest, and also endanger harmless and valuable non-target species.

Certain leeches and competing snails of the genus *Physa* eliminate medically important snails in laboratory aquariums but prove completely ineffective in nature (Wright, 1968). Some of the water bugs (Belostomatidae) discussed by Voelker (1968), dragonfly naiads (Odonata), and the crayfish (Deschiens and Lamy, 1955) and crabs (Deschiens *et al.*, 1955) probably would act in the same way. Like the vertebrate predators (Chapter 20), these invertebrates eat too many different things in nature to exert much effect on snail populations. These animals fail as agents of biological control, but they teach the important lesson that succeess in the laboratory often is remote indeed from success in nature.

Successful biological control depends on the selection of a natural enemy having high searching ability, a high degree of host specificity or preference, good reproductive capacity relative to the host, and good adaptation to a wide range of environmental conditions (De Bach, 1974). It also depends on studies of the ecology and population biology of both the natural enemy and the pest and on realistic quantitative assessments of the impact of the natural enemy on pest populations. As emphasized by Thomas (1973), "The lack of success in the malacological field can be mainly attributed to the dearth of workers and to the fact that few are population biologists. Further work on biological control of molluscs is clearly indicated. Potential predators such as sciomyzid flies should be screened by using the experimental approach . . . ." In addition to the sciomyzid flies, certain competing and predatory snails have shown considerable promise as agents of biological control.

A. Competing and Predatory Snails

A large, aquatic snail that competes with and preys on other snails, *Marisa cornuarietis* (L.), is being studied in Puerto Rico for the control of *Biomphalaria glabrata* (Say), the major snail host of schistosomiasis in the Western Hemisphere. *Marisa* feeds voraciously, devouring *B. glabrata* and its eggs and also consuming submerged aquatic weeds on which *Biomphalaria* depends for food, cover, and oviposition sites. Radke *et al.* (1961) reported demonstrated control of the schistosome host snail in small ponds.

Ruiz-Tibén *et al.* (1969) concluded that *Marisa* is effective in irrigation ponds (night storage reservoirs) as well as farm ponds. Nine years after introductions into irrigation ponds began in 1956, *B. glabrata* "had been displaced in 89 out of 97 ponds (92%). . . ."

Jobin (1970) analyzed environmental parameters and determined survival, growth, and reproduction rates of snails in three farm ponds containing *B. glabrata*. *Marisa* was introduced into one pond, but the other two were maintained as controls. Although this study was not conclusive in all respects, *Marisa* apparently was successful. All *B. glabrata* were eliminated in the experimental pond, but remained in the control ponds.

The cost-efficiency data of Ruiz-Tibén *et al.* (1969) indicate that it costs several times as much per year to obtain comparable control with the molluscicides most used in Puerto Rico (sodium pentachlorophenate and Bayluscide) as to use *Marisa*. (See further Jobin *et al.*, 1970.)

Following the promising reports in Puerto Rico, similar results of trials of *Marisa* against other medically important snails in Egypt were reported (Demian and Lutfy, 1966; Demian and Kamel, 1973).

In recent laboratory tests, ampullariid snails belonging to a species of *Pomacea* have shown predatory capabilities comparable to those of *Marisa* against *B. glabrata*. Like *Marisa*, these Brazilian snails also are voracious herbivores (see Chapter 19). By destroying aquatic "weeds," *Marisa* and *Pomacea* modify environments to the disadvantage of the trematode host snails while also preying on them and competing with them for the reduced food supply (Ferguson and Ruiz-Tibén, 1971).

Weed-consuming snails may be welcome in natural bodies of water and irrigation canals, but *Pomacea* and *Marisa* can be destructive in rice paddies and aquatic gardens of watercress or water chestnuts. *Marisa* is known to damage rice seedlings (Ortíz-Torres, 1962). *Pomacea lineata* Spix inflicts serious damage on rice in Surinam, where it is the object of considerable search for effective control measures (van Dinther, 1956).

Marisa and *Pomacea* perhaps could be dangerous also as alternative hosts of helminth parasites of man and domestic animals. Before these snails are introduced anywhere, they should be tested for susceptibility to all local, snail-transmitted parasites of medical or veterinary importance.

B. Sciomyzid Flies as Snail Predators

Insects that feed principally or exclusively on snails (e.g., Voelker, 1968; Maillard, 1971) should be studied thoroughly before their possibilities for biological control are dismissed. The larvae of all marsh flies (Sciomyzidae) reared in the laboratory (about 200 species) feed on molluscs, and most have never been known to accept any other food. Differences in habitat selection, prey preferences, and modes of attack indicate that some species of Sciomyzidae may be useful for the control of snail hosts of trematode parasites and others may be effective against land snails or slugs destructive to agricultural crops. Larvae classified as "aquatic predators" (Berg, 1961) attack aquatic, pulmonate snails of various families, and a single larva may kill from 8 (Geckler, 1971) to 24 (Eckblad, 1973) snails during its development. When foraging in the water, they kill snails that come up to the surface, and some species also descend to search for prey.

Adult flies rest, mate, and lay their eggs on vegetation, especially in marshes and swamps and at borders of ponds and slow streams. They feed primarily on decaying animal matter, are not attracted to man or to domestic animals, and are seldom seen outside of larval breeding sites (Neff and Berg, 1966).

The first series of laboratory trials of sciomyzid larvae against medically important snails showed great differences in snail vulnerability. Hosts of *Schistosoma mansoni* (Sambon) (*Biomphalaria* spp.) were very vulnerable; hosts of *S. haematobium* (Bilharz) (*Bulinus* spp.) were somewhat less so; and hosts of *S. japonicum* Katsurada (*Oncomelania* spp.) seemed completely invulnerable (Berg, 1964).

A central American sciomyzid was introduced experimentally into Hawaii in 1959 to control the snail host of the giant liver fluke of cattle. Larvae of *Sepedon macropus* Walker attacked the target snail in the laboratory and in nature, and populations of this sciomyzid soon became established on all four major islands of Hawaii. However, there was no immediate evidence that transmission rates had been reduced (Berg, 1964). Breeding stock of another species of *Sepedon*, *S. sauteri* Hendel, was obtained from Japan and released in Hawaii primarily in 1967. The Hawaiian biologists most closely associated with this experimental introduction " . . . feel that on the basis of slaughterhouse records between 1967 and 1972 the purposely introduced sciomyzids . . . are making important contributions in reducing transmission rates" (C. J. Davis, *in literature*). Graphs of liver fluke incidence for that period show downward trends on all four islands.

C. Prospects for the Future

The costs and dangers of biological control programs should be understood, stated with realism, and weighed against the costs and dangers inherent in alternative courses of action. Quick, positive chemical methods unfortunately are prohibitively expensive for use over large areas. Where used, molluscicides have killed fish and other animals exposed to them. This probably includes key organisms in essential freshwater food chains and natural enemies of snails that are already limiting snail populations. Many insect pest problems have worsened because natural enemies were thus eliminated by broad-spectrum insecticides. Programs of snail control should not repeat those mistakes.

The use of a snail predator that might also kill some harmless snails clearly poses less of a threat to the ecosystem. Predators that kill nothing except snails are more selective and specific than any chemical method of snail control. Perhaps it was these considerations that led Hairston *et al.* (1975) to recommend that, "The Organization [W.H.O.] encourage search in the tropics . . . for sciomyzid flies or other obligatory snail predators . . . , The Organization encourage field trials of appropriate sciomyzids and of *Limnogeton fieberi* . . . , [and] The Organization encourage interaction between workers in non-chemical means of snail control and those experts in biological control in the field of agriculture. The experience of those experts is contradictory to the untested opinions of some schistosome biologists. . . ."

As suggested by the last recommendation given above, programs of biological control of medically important snails lag far behind such programs for insect pests of farm crops. Unfortunately, no one combines a detailed knowledge of all aspects of this complex problem with an understanding of the organisms and methodology available for its solution. The best qualified experts on biological control methods lack the required knowledge of the biology and ecology of both trematodes and intermediate host snails. The most knowledgeable specialists on the trematodes and snails do not really understand the theory and practice of biological control. When the urgent need for better snail control methods is more widely recognized, the interdisciplinary communication and cooperation needed for successful biological control of snails perhaps can be attained.

IV. SYNANTHROPIC DIPTERA—VECTOR AND NOXIOUS FLY SPECIES

The flies treated here belong to a group generally referred to as synanthropic, or those associated with man in various degrees over varying periods of time. The efficacy of a natural enemy associated with one of these fly species is influenced to a large extent by host behavior and habitat, especially the character of larval and pupal breeding sites. Endophilous flies are those most dependent upon man and domestic animal habitations (Povolny, 1971).

The housefly, *Musca domestica* L., as well as certain species of *Drosophila* and *Psychoda*, and to a lesser extent *Stomoxys calcitrans* (L.), *Fannia canicularis* (L.), and *F. femoralis* (Stein) are all species that depend heavily upon man's waste management and disposal practices to produce and maintain the population densities that we often consider pestiferous in these Diptera. Populations of these species would be largely inconspicuous were man's accumulated waste habitats greatly reduced, or properly managed.

Musca domestica and associated flies have been targets for biological control around the world for the better part of this century. In Hawaii alone, between 1909 and 1967, sixteen direct or indirect "natural enemy" species have been released of which twelve are known to be established (Legner, in press). Unfortunately, as with many biological control attempts, preintroduction surveys of the natural enemies involved have been few and existing species have often been reintroduced (Legner and Olton, 1968).

In Fiji, a scarabaeid dung beetle, *Copris prociduuis* Say (which destroys the fly's habitat and is native to Mexico), was introduced and established via Hawaii in 1928 (Simmonds, 1929). The dipterous predator, *Mesembrina meridiana* (L.), was imported from England in 1931, but disappeared after initial "establishment." Also unsuccessful was an attempt to establish *Dirhinus* sp., a chalcidid parasitoid of *M. domestica* puparia, and native to Africa. However, the coprophagous dung beetle, *Hister chinensis* Quensel, was imported from Java in 1938 and established (Lever, 1938).

Simmonds (1958) attributed what he observed and interpreted as successful biological control of muscids in Fiji and Samoa to the ant *Pheidole megacephala* (F.), an accidental introduction around 1910, and to *H. chinensis*.

On Guam, *Spalangia endius* Walker was imported from Hawaii in 1928, and later implicated as responsible for significant fly reductions, with parasitism reported at 75–80% (Vandenburg, 1931, 1933).

Biological control efforts were temporarily halted during the 1940's with the advent of DDT and other effective insecticides. However, coincident with resistance problems during the last decade there has emerged an era of renewed and concentrated research in this field. Work begun in California in the early 1960's has extended to diverse parts of the world. Largely through the dissemination efforts of the Commonwealth Institute of Biological Control, and other cooperating agencies, an array of parasitoid and predator species was introduced into Mauritius, the West Indies, Pakistan, and India. Independent organizations in Chile, El Salvador, New Zealand, the Pacific Islands, Europe, and throughout the United States have acquired key natural enemies from original stock secured by University of California researchers.

Emphasis in the 1960–1970 period was placed on a group of parasitoid and predator species that demonstrated a high host destructibility. The encyrtid *Tachinaephagus zealandicus* Ashmead has adapted to attacking dipterous larvae, except for Tachinidae with which its name is misassociated, in animal excrement, plant refuse, and in carrion (Legner and Olton, 1968; Olton, 1971). Five separate species of the pteromalid genus *Muscidifurax* are recognized for their parasitization capabilities near the surface of breeding sites, areas not searched thoroughly by other parasitoids (Legner, 1969; Kogan and Legner, 1970).

The nearly ubiquitous *Spalangia* species occur as many strains which differ both behaviorly, and in climatic preference (Legner, 1969, 1972). *Spalangia longipetiolata* Bŏucek from East Africa, for instance, appeared to be most active in poultry manure and at cooler temperature (70°F) in Africa (Legner and Greathead, 1969).

Chicken cockerels have been used successfully as direct predators of flies in poultry and rabbit manure in California (Rodriguez and Riehl, 1962).

Other approaches to fly control that have met with only limited success include seasonal inundations with parasites and predators (Legner and Brydon, 1966) and the use of predatory Acarina. Predatory Acarina associated with certain fly breeding sites have received more attention from the standpoint of their conservation than their introduction (Axtell, 1970; Rodriguez and Wade, 1960). Insect pathogens used for fly control have also met with only slight success (Greenberg, 1969).

Exophilic flies that thrive well in the wild but whose populations often increase in association with man include species in the genera *Phaenicia*, *Calliphora*, *Muscina*, and *Stomoxys*, and also *Musca sorbens* (Wiedemann) (Povolny, 1971).

At very high latitudes some of these exophilic forms become endophilic to the extent that they depend upon man for their survival. Where exophilic species share the same habitat with endophilic flies they are often attacked by the same natural enemies (Legner, 1971).

Although many studies have been made of the natural enemies of calliphorid species and biological control attempts have been made in various parts of the world, including the United States (e.g., Holdaway, 1930; Froggatt and McCarthy, 1914; Ullyet, 1950), Australia (e.g., Morgan, 1929; Wilson, 1963), New Zealand (Wilson, 1960), and South Africa (D. P. Annecke, personal communication), only those in California and Hawaii are considered reasonably successful (C. P. Davis, personal communication; Legner, 1967).

The braconid parasite *Alysia ridibunda* Say, native to southern Arizona, New Mexico, and eastern North America, was mass released as 50,000 adults in Uvalde, Texas, in 1934 and 1935, resulting within a short time in up to 100% parasitization of *Phaenicia sericata* (Meigen) and *Sarcophaga* species. By 1938, however, *A. ridibunda* was considered rare (Lindquist, 1940).

Hemisynanthropic flies, those which exist independent of man's environment (Povolny, 1971), occasionally become nuisances during man's interference with nature. These species include *Lucilia caesar* (L.), *L. illustris* (Meigen), *Calliphora vomitoria* (L.), *Ophyra leucostoma* (Wiedemann), *Hydrotaea irritans* (Fallen), *Hylemya* spp., and *Muscina pabulorum* (Fallen) (Povolny, 1971). Some species of *Glossina* and *Hippelates*, as well as *Musca sorbens*, also fall within this group. In recent years the gregarious larval parasite *Tachinaephagus zealandicus* has been stressed for biological control of hemisynanthropic flies, as its ability to range in a wide array of habitats is unparalleled by any known fly parasitoid.

There have been no biological control efforts specifically against the Sarcophagidae, but some indirect effects have been reported when the target species were calliphorids or endophilous muscoid flies. Especially significant has been the high recovery of *Tachinaephagus zealandicus* from *Sarcophaga* species in southern California by one of the authors (E. F. Legner).

Eye gnats of the genus *Hippelates* are of widespread concern. Only in California, however, have biological control efforts been attempted against this genus (i.e., *Hippelates collusor* Townsend and *H. pusio* Loew). Following the discovery of a native cynipid parasitoid (Mulla, 1962) a concerted effort was made in search of other indigenous parasitoids, in which four additional species were found that attack the pupae (Bay *et al.*, 1964). Foreign explorations, mostly in the West Indies, produced an additional eight species and strains, some of which became established, but were not demonstrated to affect eye gnat population densities (Legner, 1970; Legner *et al.*, 1966). Bay and Legner (1963), in discussing the prospects for the biological control of *Hippelates* eye gnats in California, proposed various factors that undoubtedly relate to the failure of subsequently introduced parasitoids to be more successful. Cultivation, in particular, favored increased survival of *Hippelates* by burying larvae and pupae out of reach of the parasitoids and by removing vegetation that afforded protection, and presumably pollen and nectar, for the adult wasps. Gnats emerging from buried puparia were able to work their way through several centimeters of soil by means of their ptilinum. Most parasitoid activity was usually restricted to within a few millimeters of the surface.

The only known successful biological control experiment against tabanids was an inundative release with the egg parasitoid *Phanurus emersoni* Girault in Texas. This produced an estimated 50% population reduction, but only during the season of its release (Parman, 1928).

Only recently have biological control efforts been directed at the *Musca sorbens* complex of flies in Australia and the Pacific Islands. Waterhouse (personal communication) writes:

> Australia lacks an ecological counterpart of the large domestic herbivores in its native fauna, and the native dung-beetles are not adapted to exploit large masses of dung such as cowpads. As a result, dung pads often remain on the ground surface for long periods. There they provide a breeding medium for two insect pests, the bushfly (*Musca vetustissima*) and the buffalo fly (*Haematobia exigua*). Perhaps more importantly the pads cover large areas of pasture and, because they dry out *in situ*, prevent the recycling of plant nutrients.
>
> To overcome these disadvantages CSIRO is importing from Africa species of dung beetles which specialize in burying large wet dung pads. The following species have already been introduced:
> *Onthophagus gazella* F., *O. sagittarius* F., *O. binodis* Thunberg, *O. bubalus* Harold, *Euoniticellus africanus* Harold, *E. intermedius* Reiche, *Onitis alexis* Klug, *O. westermanni* Lansberg, *O. caffer* Boheman, *Sisyphus spinipes* Thunberg, *S. mirabilis* Arrow, *S. rubripes* Boheman, *Chironitis scabrosus* F., *Canthon humectus* Say, *Copris incertus* Say, *Heliocopris andersoni* Bates, *Liatongus militaris* Cast.
>
> The earliest releases were made in the tropical north, and dung dispersal has been markedly accelerated since the establishment of the exotic beetle species. Dung pads formerly remained on the ground surface for 4 to 12 months or even more, but during the wetter times of year are now buried completely within about 30 hours. This rate of activity has caused a drop in buffalo fly numbers on livestock, but as yet the beetle activity does not span the full breeding season of the fly, and a wider range of beetle species is to be introduced.
>
> Currently species of beetles suited to subtropical and temperate climates are being imported from Africa. The first of these are already established in some areas, where the rate of dung dispersal has increased, and the effect of these beetles on bushfly populations is being evaluated.
>
> Laboratory studies with grass plants and arid zone herbage seedlings have shown that dung burial by beetles produces faster and better germination and plant growth than thorough mechanical mixing of dung and soil. The tunnelling of the beetles aerates the soil and improves water penetration, as well as recycling nitrogen, much of which is lost into the air if the pad is not buried while still fresh.

Symbovine Diptera, associated with pastured cattle (Povolny, 1971), include among others the horn fly *Haematobia irritans* (L.), *H. exigua* (de Meijere), and the face fly *Musca autumnalis* De Geer. Parasitoids (mostly Braconidae and Diapriidae) commonly associated with field droppings of cattle differ from those visiting accumulated manure heaps. Some coprophagous and insectivorous Coleoptera that are natural enemies of symbovine Diptera in scattered droppings also inhabit accumulated manure while others, e.g., some Staphylinidae, do not (Legner and Olton, 1970). Errors of interpreting parasitoid activity continue as investigators, eager to concentrate pupating hosts for ease of collection, deliberately gather field excrement to consolidate into localized heaps. A misrepresented natural enemy biota is thus often attracted [e.g., *Haematobia irritans* (L.) and *Musca autumnalis* De Geer].

Among the many attempts at biological control of symbovine Diptera in Australia since 1932, the most successful to date have been introductions of coprophagous dung beetles to reduce animal excrement in pastures (Bornemissza, 1960; Snowball, 1941).

In Hawaii, 48 species of natural enemies of *Haematobia irritans* (L.) have been introduced within this century, with 10 known to be established, their effect on horn fly populations resulting in a substantial reduction (Fullaway, 1923; Pemberton, 1948; C. J. Davis, personal communication).

In North America, biological control attempts against the imported face fly of cattle, *Musca autumnalis*, by introduced *Aleochara tristis* Gravenhorst, *Hister nomas* Erichson, *H. caffer* Erichson, and *Heterotylenchus autumnalis* in California, and elsewhere with the nematode *H. autumnalis* and *A. tristis*, have so far been unsuccessful, or have not had time to be properly evaluated.

In conclusion, existing natural and biological control of endophilous flies may be enhanced through cultural modifications of the breeding habitat. This task becomes more difficult with the other groups discussed, since their populations often lie outside the range of management practices. For these species, further search and introduction of predatory and scavenging species that forage through the breeding habitat would seem most promising.

REFERENCES

Ahuja, S. K. (1964). Salinity tolerance of *Gambusia affinis*. *Indian J. Exp. Biol.* **2**, 9–11.

Axtell, R. C. (1970). Integrated fly-control programs for caged-poultry houses. *J. Econ. Entomol.* **63**, 400–405.

Bay, E. C. (1967). Potential for naturalistic control of mosquitoes. *Proc. Pap. Annu. Conf. Calif. Mosq. Contr. Ass.* **35**, 34–37.

Bay, E. C., and Legner, E. F. (1963). The prospect for the biological control of *Hippelates collusor* (Townsend) in southern California. *Proc. Pap. Annu. Conf. Calif. Mosq. Contr. Ass.* **31**, 76–79.

Bay, E. C., and Self, L. S. (1972). Observations of the guppy, *Poecilia reticulata* Peters, in *Culex pipiens fatigans* breeding sites in Bangkok, Rangoon, and Taipei. *Bull. W.H.O.*, **46**, 407–416.

Bay, E. C., Legner, E. F., and Medved, R. (1964). *Hippelates collusor* (Diptera: Chloropidae) as a host for four species of parasitic Hymenoptera in southern California. *Ann. Entomol. Soc. Amer.* **57**, 582–584.

Berg, C. O. (1961). Biology of snail-killing Sciomyzidae (Diptera) of North America and Europe. *Verh. XI Int. Kongr. Entomol. Wien, 1960* **1**, 197–202.

Berg, C. O. (1964). Snail control in trematode diseases: the possible value of sciomyzid larvae, snail-killing Diptera. *In* "Advances in Parasitology" (B. Dawes, ed.), Vol. 2, pp. 259–309. Academic Press, New York.

Berg, C. O. (1973). Biological control of snail-borne diseases: a review. *Exp. Parasitol.* **33**, 318–330.

Bornemissza, G. F. (1960). Could dung eating insects improve our pastures? *J. Aust. Inst. Agr. Sci.* **26**, 54–56.

Briand, L. J., and Welch, H. E. (1963). Use of entomophilic nematodes for insect pest control. *Phytoprotection* **44**, 37–41.

Chapman, H. C., and Glenn, F. E., Jr. (1972). Incidence of the fungus *Coelomomyces punctatus* and *C. dodgei* in larval populations of the mosquito *Anopheles crucians* in two Louisiana ponds. *J. Invertebr. Pathol.* **19**, 256–261.

Chapman, H. C., Woodard, D. B., and Petersen, J. J. (1967). Pathogens and parasites in Louisiana Culicidae and Chaoboridae. *Proc. N. J. Mosq. Exterm. Ass.* **54**, 54–60.

Chapman, H. C., Clark, T. B., Petersen, J. J., and Woodard, D. B. (1969). A two-year survey of pathogens and parasites of Culicidae, Chaoboridae, and Ceratopogonidae in Louisiana. *Proc. N. J. Mosq. Exterm. Ass.* **56**, 203–212.

Chapman, H. C., Clark, T. B., and Petersen, J. J. (1970). Protozoans, nematodes, and viruses of anophelines. *Misc. Publ. Entomol. Soc. Amer.* **7**, 134–139.

Clark, T. B., and Chapman, H. C. (1969). A polyhedrosis in *Culex salinarius* of Louisiana. *J. Invertebr. Pathol.* **13**, 312.

Clark, T. B., and Fukuda, T. (1971). Field and laboratory observations of two viral diseases in *Aedes sollicitans* (Walker) in southwestern Louisiana. *Mosquito News* **31**, 193–199.

Clark, T. B., Chapman, H. C., and Fukuda, T. (1969). Nuclear-polyhedrosis and cytoplasmic-polyhedrosis infections in Louisiana mosquitoes. *J. Invertebr. Pathol.* **14**, 284–286.

Clark, T. B., Kellen, W. R., and Lum, P. T. M. (1965). A mosquito iridescent virus (MIV) from *Aedes taeniorhynchus* (Wiedemann). *J. Invertebr. Pathol.* **7**, 519–521.

Couch, J. N. (1967). Sporangial germination of *Coelomomyces punctatus* and the conditions favoring the infection of *Anopheles quadrimaculatus* under laboratory conditions. *U.S.–Jap. Semi. Microb. Contr. Insect Pests, Fukuoka*, pp. 93–105.

DeBach, P. (1974). "Biological Control by Natural Enemies," xi + 323 pp. Cambridge University Press, London and New York.

Demian, E. S., and Kamel, E. G. (1973). Biological control of *Bulinus truncatus* under semifield conditions using the snail *Marisa cornuarietis*. *Ninth Inter. Congr. on Tropical Medicine and Malaria, Athens*, (G. J. Papaevangelou, ed.), Vol. 2, 77–78.

Demian, E. S., and Lutfy, R. G. (1966). Factors affecting the predation of *Marisa cornuarietis* on *Bulinus* (*B.*) *truncatus*, *Biomphalaria alexandrina* and *Lymnaea caillaudi*. *Oikos* **17**, 212–230.

Deschiens, R., and Lamy, L. (1955). Préhension et ingestion des mollusques vecteurs des bilharzioses par les écrevisses du genre *Cambarus*. *Bull. Soc. Pathol. Exot.* **48**, 201–203.

Deschiens, R., Dechancé, M., and Vermeil, C. (1955). Action prédatrice des crabes d'eau douce du genre *Potaman* sur les mollusques vecteurs des bilhazioses. *Bull. Soc. Pathol. Exot.* **48**, 203–207.

Eckblad, J. W. (1973). Experimental predation studies of malacophagous larvae of *Sepedon fuscipennis* (Diptera: Sciomyzidae) and aquatic snails. *Exp. Parasitol.* **33**, 331–342.

Ferguson, F. F., and Ruiz-Tibén, E. (1971). Review of biological control methods for schistosome-bearing snails. *Ethiop. Med. J.* **9**, 95–104.

Froggatt, W. W., and McCarthy, T. (1914). The parasite of the sheep-maggot fly (*Nasonia brevi-*

cornis). Notes and observations in the field and laboratory. *N.S.W. Agr. Gaz.* **25** (9), 759–764.

Fullaway, D. T. (1923). Report of the entomologist. *Hawaii. Board Comm. Agr. For. Bien. Rep.*, pp. 53–68.

Geckler, R. P. (1971). Laboratory studies of predation of snails by larvae of the marsh fly, *Sepedon tenuicornis* (Diptera: Sciomyzidae). *Can. Entomol.* **103**, 638–649.

Greenberg, B. (1969). *Salmonella* suppression by known populations of bacteria in flies. *J. Bacteriol.* **99**, 629–635.

Hackett, L. W. (1931). Recent developments in the control of malaria in Italy. *J. Southern Med. Ass.* **24**, 426–430.

Hairston, N. G., Wurzinger, K. H., and Burch, J. B. (1975). "Non-chemical Methods of Snail Control." *WHO/VBC/75.573, WHO/SCHISTO/75.40*, 1–30.

Hildebrand, S. F. (1921). Suggestions for a broader application of *Gambusia* for the purpose of mosquito control in the south. *Pub. Health Rep. Wash.* **36**, 1460–1461.

Holdaway, F. G. (1930). Field populations and natural control of *Lucilia sericata*. *Nature* (*London*) **126**, 648–649.

Hoy, J. B., and O'Grady, J. J. (1971). Populations of mosquito fish in rice fields. *Proc. Pap. Annu. Conf. Calif. Mosq. Contr. Ass.* **39**, 107.

Hoy, J. R., and Reed, D. E. (1971). The efficacy of mosquitofish for the control of *Culex tarsalis* in California rice fields. *Mosquito News* **31**, 567–572.

Huffaker, C. B., Messenger, P. S., and DeBach, P. (1971). The natural enemy component in natural control and the theory of biological control. *In* "Biological Control" (C. B. Huffaker, ed), pp. 16–67. Plenum, New York.

Hurlbert, S. H., Zedler, J., and Fairbanks, D. (1972). Ecosystem alteration by mosquitofish (*Gambusia affinis*) predation. *Science* **175**, 639–641.

Jenkins, D. W. (1964). Pathogens, parasites and predators of medically important arthropods, annotated list and bibliography. *Bull W.H.O.* pp. 1–50. (Suppl. to Vol. **30**.)

Jobin, W. R. (1970). Population dynamics of aquatic snails in three farm ponds of Puerto Rico. *Amer. J. Trop. Med. Hyg.* **19**, 1038–1048.

Jobin, W. R., Ferguson, F. F., and Palmer, J. R. (1970). Control of schistosomiasis in Guayoma and Arroyo, Puerto Rico. *Bull. W.H.O.* **42**, 151–156.

Kellen, W. R., Clark, T. B., and Lindegren, J. E. (1963). A possible polyhedrosis in *Culex tarsalis* Coquillett (Diptera: Culicidae). *J. Invertebr. Pathol.* **5**, 98–103.

Kellen, W. R., Clark, T. B., Lindegren, J. E., and Sanders, R. B. (1966). A cytoplasmic-polyhedrosis virus of *Culex tarsalis* (Diptera: Culicidae). *J. Invertebr. Pathol.* **8**, 390–394.

Kogan, M., and Legner, E. F. (1970). A biosystematic revision of the genus *Muscidifurax* (Hymenoptera: Pteromalidae) with descriptions of four new species. *Can. Entomol.* **102**, 1268–1290.

Kühlhorn, F. (1965). "An Investigation of the Natural Enemies of *Anopheles* Larvae (Diptera: Culicidae) in Different Areas at Varying Altitudes in West Germany," pp. 1–18 (*WHO*/EBL/37.65) (Mimeo.).

Laird, M. (1967). A coral island experiment. *World Health. Organ. Chron.* **21**, 18–26.

Laird, M. (1971). Microbial control of insects of medical importance. *In* "Microbial Control of Insects and Mites" (H. D. Burges, ed.), pp. 387–406. Academic Press, New York.

Legner, E. F. (1967). Two exotic strains of *Spalangia drosophilae* merit consideration in biological control of *Hippelates collusor* (Diptera: Chloropidae). *Ann. Entomol. Soc. Amer.* **60**, 458–462.

Legner, E. F. (1969). Reproductive isolation and size variation in the *Muscidifurax raptor* Girault and Sanders complex. *Ann. Entomol. Soc. Amer.* **62**, 382–385.

Legner, E. F. (1969) Adult emergence interval and reproduction in parasitic Hymenoptera influenced by host size and density. *Ann. Entomol. Soc. Amer.* **62** (1), 220–26.

Legner, E. F. (1970). Advances in the ecology of *Hippelates* eye gnats in California indicate means for effective integrated control. *Proc. 38th Annu. Conf. Calif. Mosq. Contr. Ass.* pp. 89–90.

Legner, E. F. (1971). Some effects of the ambient arthropod complex on the density and potential parasitization of muscoid Diptera in poultry wastes. *J. Econ. Entomol.* **64**, 111–115.

Legner, E. F. (1972). Observations on hybridization and heterosis in parasitoids of synanthropic flies. *Ann. Entomol. Soc. Amer.* **65** (1), 254–63.

Legner, E. F. (1976). Diptera. Medical and veterinary pests. *In* "Introduced Parasites and Predators of Arthropod Pests and Weeds: A World Review" (C. P. Clausen, ed.). USDA Handb.

Legner, E. F., and Brydon, H. W. (1966). Suppression of dung-inhabiting fly populations by pupal parasites. *Ann. Entomol. Soc. Amer.* **59**, 638–651.

Legner, E. F., and Greathead, D. J. (1969). Parasitism of pupae in East African populations of *Musca domestica* and *Stomoxys calcitrans*. *Ann. Entomol. Soc. Amer.* **62**, 128–133.

Legner, E. F., and Olton, G. S. (1968). Activity of parasites from Diptera: *Musca domestica, Stomoxys calcitrans*, and species of *Fannia, Muscina*, and *Ophyra* II. At sites in the Eastern Hemisphere and Pacific area. *Ann. Entomol. Soc. Amer.* **61**, 1306–1314.

Legner, E. F., and Olton, G. S. (1970). Worldwide survey and comparison of adult predator and scavenger insect populations associated with domestic animal manure where livestock is artificially congregated. *Hilgardia* **40**, 225–266.

Legner, E. F., Bay, E. C., and Farr, T. H. (1966). Parasitic and predacious agents affecting the *Hippelates pusio* complex in Jamaica and Trinidad. *Can. Entomol.* **98**, 28–33.

Lever, R. J. A. W. (1938). Entomological notes 3. A Javanese beetle to combat houseflies. *J. Dep. Agr. Fiji* **9**, 15–18.

Lindquist, A. W. (1940). The introduction of an indigenous blowfly parasite, *Alysia ridibunda* Say, into Uvalde County, Texas. *Ann. Entomol. Soc. Amer.* **33**, 103–112.

Lo, C.-T. (1967). The inhibiting action of ostracods on snail cultures. *Trans. Amer. Microsc. Soc.* **86**, 402–405.

Madelin, M. F. (1966). Fungal parasites of insects. *Annu. Rev. Entomol.* **11**, 423–448.

Maillard, Y.-P. (1971). La malacophagie dans le genre *Hydrophilus* Geoffroy (Ins. Coléoptères Hydrophilidae); son intérêt dans le contrôle naturel des hotês intermédiares d'helminthiases. *C. R. Acad. Sci.* (Paris) **272**, 2235–2238.

Morgan, W. L. (1929). *Alysia manducator* Pz., an introduced parasite of the sheep blowfly maggot. *Agr. Gaz. N.S.W.* **16**(11), 818–829.

Mulla, M. S. (1962). Recovery of a cynipoid parasite from *Hippelates* pupae. *Mosquito News* **22**, 301–302.

Muspratt, J. (1963). Destruction of the larvae of *Anopheles gambiae* Giles by a *Coelomomyces* fungus. *Bull. W.H.O.* **29**, 81–86.

Muspratt, J. (1965). Technique for infecting larvae of the *Culex pipiens* complex with a mermithid nematode and for culturing the latter in the laboratory. *Bull. W.H.O.* **33**, 140–144.

Nakagawa, P. Y., and Ikeda, J. (1969). "Biological Control of Mosquitoes with Larvivorous Fishes in Hawaii," pp. 1–25 (WHO/VBC/69.173) (Mimeo.).

Neff, S. E. (1964). Snail-killing sciomyzid flies: application in biological control. *Verh. Int. Verein. Theor. Angew. Limnol.* **15**, 933–939.

Neff, S. E., and Berg, C. O. (1966). Biology and immature stages of malacophagous Diptera of the genus *Sepedon* (Sciomyzidae). *Bull. Agr. Exp. Sta., Va. Poly. Inst.* **566**, 1–113.

Nickle, W. R. (1972). A contribution to our knowledge of the Mermithidae (Nematoda). *J. Nematol.* **4**, 113–146.

Olton, G. S. (1971). Bioecological studies of *Tachinaephagus zealandicus* Ashmead (Hymenoptera: Encyrtidae), parasitoid of synanthropic Diptera. Ph.D. Thesis, University of California, Riverside, California.

Ortíz-Torres, E. (1962). Damage caused by the snail, *Marisa cornuarietis*, to young rice seedlings in Puerto Rico. *J. Agr. Univ. Puerto Rico* **46**, 241.

Parman, D. C. (1928). Experimental dissemination of the tabanid egg parasite, *Phanurus emersoni* Girault, and biological notes on the species. *U.S. Dep. Agric. Circ.* **18**, 1–16.

Pemberton, C. E. (1948). History of the Entomology Dept. Expt. Sta. H. S. P. A. 1904–1945. *Hawaii. Plant. Rec.* **52**, 53–90.

Petersen, J. J., and Chapman, H. C. (1970). Parasitism of *Anopheles* mosquitoes by a *Gastromermis* sp. (Nematodae: Mermithidae) in southwestern Louisiana. *Mosquito News* **30**, 420–424.

Petersen, J. J., and Willis, O. R. (1972). Procedures for the mass rearing of a mermithid parasite of mosquitoes. *Mosquito News* **32**, 226–230.

Petersen, J. J., Chapman, H. C., and Woodard, D. B. (1967). Preliminary observations on the incidence and biology of a mermithid nematode of *Aedes sollicitans* (Walker) in Louisiana. *Mosquito News* **27**, 493–498.

Petersen, J. J., Chapman, H. C., and Woodard, D. B. (1968). The bionomics of a mermithid nematode of larval mosquitoes in southwestern Louisiana. *Mosquito News* **28**, 346–352.

Povolny, D. (1971). Synanthropy. Definition, Evolution and Classification. *In* "Flies and Disease, Ecology, Classification and Biotic Associations" (B. Greenberg, ed.), Vol. I, pp. 17–54. Princeton Univ. Press, Princeton, New Jersey.

Qureshi, A. H., and Bay, E. C. (1969). Some observations on *Hydra americana* Hyman as a predator of *Culex peus* Speiser mosquito larvae. *Mosquito News* **29**, 465–471.

Radke, M. G., Ritchie, L. S., and Ferguson, F. F. (1961). Demonstrated control of *Australorbis glabratus* by *Marisa cornuarietis* under field conditions in Puerto Rico. *Amer. J. Trop. Med. Hyg.* **10**, 370–373.

Reeves, E. L., and Garcia, C., Jr. (1970). Pathogenicity of bicrystalliferous *Bacillus* isolate for *Aedes aegypti* and other aedine mosquito larvae. *Proc. 4th Int. Colloq. Insect Pathol.*, pp. 219–228.

Roberts, D. W. (1970). *Coelomomyces, Entomopthora, Beauveria*, and *Metarrhizium* as parasites of mosquitoes. *Misc. Publ. Entomol. Soc. Amer.* **7**, 140–154.

Rodriguez, J. G., and Wade, C. F. (1960). Preliminary studies on biological control of housefly eggs using macrochelid mites. *Proc. 11th Int. Congr. Entomol., Vienna.*

Rodriguez, J. L., and Riehl, L. A. (1962). Control of flies in manure of chickens and rabbits by cockerels in south California. *J. Econ. Entomol.* **55**, 473–477.

Ruiz-Tibén, E., Palmer, J. R., and Ferguson, F. F. (1969). Biological control of *Biomphalaria glabrata* by *Marisa cornuarietis* in irrigation ponds in Puerto Rico. *Bull. W.H.O.* **41**, 329–333.

Simmonds, H. W. (1929). Introduction of natural enemies against the house fly in Fiji. *Agr. J. Dep. Agr. Fiji* **2**, 46.

Simmonds, H. W. (1958). The housefly problem in Fiji and Samoa. *S. Pac. Comm. Quart. Bull.* **8** (2), 29–30, 47.

Snowball, G. J. (1941). A consideration of the insect population associated with cow dung at Crawley. *J. Roy. Soc. West Aust.* **28**, 219–245.

Tabibzadeh, I., Behbehani, G., and Nakhai, R. (1970). Use of *Gambusia* Fish in the Malaria Eradication Programme of Iran, pp. 1–13 (WHO/MAL/70.716, WHO/VBC/70.198.) (Mimeo.).

Thomas, J. D. (1973). Schistosomiasis and the control of molluscan hosts of human schistosomes with particular reference to possible self-regulatory mechanisms. *In* "Advances in Parasitology" (B. Dawes, ed.), Vol. 11, pp. 307–394. Academic Press, New York.

Trpis, M. (1970). "Adult Population Estimate of *Toxorhynchites brevipalpis* Breeding in Man-Made Containers in Dar-Es-Salaam, Tanzania," pp. 1–7 (WHO/VBC/7.231) (Mimeo.).

Ullyett, G. C. (1950). Pupation habits of sheep blowflies in relation to parasitism by *Mormoniella vitripennis* Wlk. (Hymn., Pteromalidae). *Bull. Entomol. Res.* **40**, 533–537.

Umphlett, C. J. (1970). Infection levels of *Coelomomyces punctatus*, an aquatic fungus parasite, in a natural population of the common malarial mosquito, *Anopheles quadrimaculatus*. *J. Invertebr. Pathol.* **15**, 299–305.

Vandenburg, S. R. (1931). Report of the entomologist. *Guam Agr. Exp. Sta. Rep.*, pp. 23–25.

Vandenburg, S. R. (1933). Report of the entomologist. *Guam Agr. Exp. Sta. Rep.*, pp. 20–22.

van Dinther, J. B. M. (1956). Control of *Pomacea* (*Ampullaria*) snails in rice fields. *Landbouwproefsta. Suriname Bull.* **68**, 1–20.

Voelker, J. (1968). Untersuchungen zu Ernährung, Fortpflanzungsbiologie und Entwicklung von *Limnogeton fieberi* Mayr (Belostomatidae, Hemiptera) als Beitrag zur Kenntnis von natürlichen Feinden tropischer Süsswasserschnecken. *Entomol. Mitt.* **3** (60), 1–31.

Weiser, J. (1965). A new virus infection of mosquito larvae. *Bull. W.H.O.* **33**, 586–588.

Welch, H. E. (1960). *Hydromermis churchillensis* n. sp. (Nematoda: Mermithidae) a parasite of *Aedes communis* (DeG.) from Churchill, Manitoba, with observations on its incidence and bionomics. *Can. J. Zool.* **38**, 465–474.

Welch, H. E., and Bronskill, J. F. (1962). Parasitism of mosquito larvae by the nematode DD 136 (Nematoda: Neoaplectanidae). *Can. J. Zool.* **40**, 1263–1268.

Wilson, F. (1960). A review of the biological control of insects and weeds in Australia and Australian New Guinea. *Commonw. Inst. Biol. Control Tech. Commun.* **1**.

Wilson, F. (1963). The results of biological control investigations in Australia and New Guinea. *Proc. 9th Pac. Sci. Congr.* **9**, 112–123.

Wright, C. A. (1968). Some views on biological control of trematode diseases. *Trans. Roy. Soc. Trop. Med. Hyg.* **62**, 320–329.

Yasuraoka, K. (1970). Some recent research on the biology and control of *Oncomelania* snails in Japan. *In* "Recent Advances in Researches on Filariasis and Schistosomiasis in Japan" (M. Sasa, ed.), pp. 291–303. Univ. of Tokyo Press, Tokyo; Univ. Park Press, Baltimore, Maryland.

19

BIOLOGICAL CONTROL OF WEEDS

L. A. Andres, C. J. Davis, P. Harris, and
A. J. Wapshere

I. INTRODUCTION

All plants, including weeds, have natural enemies. In some instances these enemies can be manipulated to influence the abundance of their host plants. The cactus-feeding moth, *Cactoblastis cactorum* (Berg), was transported from its native home in Argentina to Australia where it greatly reduced the abundance of the prickly pear cactus, *Opuntia* spp. (Dodd, 1940). The leaf-feeding chrysomelid beetle, *Chrysolina quadrigemina* (Suffrian), which was imported from Europe to the United States, via Australia, successfully controlled the poisonous range weed, *Hypericum perforatum* L. (Holloway, 1964; Huffaker and Kennett, 1959). All in all the transfer and manipulation of the natural enemies of weeds, primarily weed-feeding insects, have resulted in the control of a wide range of weed pests in many parts of the world (Anon., 1968) and at present over 70 species of weedy plants are under study or consideration for control by biological means.

Insects have received particular attention for the biological control of weeds, partly because of their size, their high rate of reproduction, and their high degree of host specificity. However, studies are broadening to include the use of plant pathogens (Wilson, 1969; Inman, 1971), nematodes (Ivannikov, 1969a), parasitic plants (Rudakov, 1961), competing plants (Yeo and Fisher, 1970), and other organisms. The growing interest in the biological control of weeds is evident from the number of reviews that have appeared (Huffaker, 1957, 1959; Huffaker and Andres, 1970; Wilson, 1960, 1964; Holloway, 1964; Anon., 1968; Zwölfer, 1968; Harris, 1971; Andres and Goeden, 1971; Goeden, in press; Blackburn *et al.* 1971). In addition, several papers have appeared reviewing topics of specific importance to the biological control of weeds, e.g., host specificity determination of weed-feeding insects (Harris and Zwölfer, 1968; Zwölfer and Harris, 1971; Wapshere, 1974) and the strategy of control (Harris, 1973; Zwölfer, 1974).

II. WEEDS AND NATURAL CONTROL

Plants are able to establish in almost every conceivable habitat. Wherever they compete with plants of importance to mankind or are in some way discomforting to man, either physically or economically (e.g., the production of allergenic pollens, poisonous range plants), they are declared weedy. The losses from weeds run upward from $5 billion annually for the United States alone (USDA, ARS, 1965) and are second only to losses caused by soil erosion (Saunders *et al.*, in King, 1966). These losses far exceed those caused by insect pests.

No two weed problems are alike. The presence and abundance of a weed in a particular area are a product of that area's history and the weed's ability to reproduce under existing climatic, edaphic, and biotic limitations. Variations in soil, water, disturbance of the habitat, and cropping practices, all influence the abundance of weeds and the species involved. In some instances alterations of the environment can be planned to regulate the abundance of particular plant species. The biological control practitioner strives to reduce the abundance of a weed species by introducing or augmenting the weed's natural enemies.

The influences of natural enemies on plant abundance has been well demonstrated. Outbreaks of populations of the moth *Aroga websteri* Clarke severely denude big sagebrush plants, *Artemisia tridentata* Nutt., over wide areas of its native range in the northwestern United States, and in some years destroy thousands of acres, permitting the return of native grasses (Ritcher and Dickason, 1964). The combined action of the root-infesting scale insect, *Orthezia annae* Cockerell, and the crown moth, *Eumysia idahoensis* Mackie, leads to the death of the native shad scale, *Atriplex confertifolia* (Torr. and Frem) S. Wats., in central Idaho, with subsequent invasion by other plants (Mackie,

1957). Huffaker (1959) lists other instances in which naturally occurring insects have played an important role in affecting the abundance of particular plant species.

Where effective indigenous natural enemies are lacking, introduced species may hold the plants in check. Segments of the cactus *Opuntia vulgaris* Miller infested with *Dactylopius ceylonicus* (Green) were distributed to control naturalized stands of the plant in southern India as early as 1863–1868. This scale was specifically introduced from India to Ceylon in 1865 for the control of this same pest [Goeden, in press (citing Tryon, 1910)]. Hawaiian ranchers, observing the devastation that the accidentally introduced scale, *Orthezia insignis* Browne, was causing to the weedy *Lantana camara* L., spread the scale to uninfested stands of the plant (Perkins and Swezey, 1924). An entomologist subsequently went to Mexico, where *Lantana* is native, in search of other *Lantana*-feeding insects (loc. cit.). Efforts have since been made to introduce phytophagous insects to control weeds in over 50 projects throughout the world (Goeden, loc. cit.). Although success has not automatically followed the introduction of a weed's natural enemies, reduction and effective control of the target weeds have been demonstrated in sufficient cases to establish biological control as a valuable method of weed suppression.

III. THE DEVELOPMENT OF A PROGRAM OF BIOLOGICAL WEED CONTROL

The threshold of economic damage varies with the weed species, the crop with which it competes, the habitat, and other factors. To meet these varying requirements an array of control methods, tailored to the problem, are needed. Biological control, like any other method, is better suited to some weed problems than others and depends upon the availability of host-specific biological control agents, and the ease and safety with which they can be handled. Of the several approaches that can be followed in developing a biological control program, the introduction of natural enemies has received most emphasis.

The introduction technique hinges on the availability of organisms that will suppress the weed but will not harm nonpest plants. The work generally progresses according to protocol, the steps of which were described by Harris (1971): (1) determine the suitability of the weed for biological control, (2) survey the plant's natural enemies, (3) study and evaluate the ecology of the several natural enemies, (4) conduct host specificity studies of the organisms to ascertain the safety of their introduction in the weed area, (5) introduction and establishment, and (6) evaluation studies.

A. Suitability of a Weed for Biological Control—The Plant, The Problem

All too often biological control is considered a last resort method of weed control. In most instances, the plants studied have been introduced perennial species on little disturbed, low value land. Recent developments indicate that biological control of weeds can have wider application.

Two considerations are of importance in determining whether a weed is suited for biological control—the plant and the problem. Whether the plant is a native or introduced species and whether it has close relatives of economic importance bear on the possibility of finding suitable natural enemies. Whether these enemies can provide relief depends on the problem: the number of weed species involved, the type and stability of the habitat, and the degree and urgency of control.

The closer the weed is related to plants of economic or "ecological" value, the more difficult it generally becomes to find host-specific natural enemies which will not attack valued species. Current work on the biological control of various composite thistles in North America is hampered by the presence there of cultivated artichoke (*Cynara scolymus* L.) and safflower (*Carthamus tinctorius* L.) which belong to the same family and plant tribe (Compositae; Tribe Cynareae). The use of biological control against weedy grasses has never seriously materialized because of closely related grain crops.

Despite the several instances of successful control of native weeds with introduced insects [e.g., *Leptospermum scoparium* Forster in New Zealand by the scale insect *Eriococcus orariensis* Hoy (Hoy, 1949); *Opuntia* spp. on Santa Cruz Island, California by *Dactylopius* sp. (Goeden, in press)], the likelihood of finding natural enemies capable of controlling introduced weeds with a minimum of manipulation and attention is greater than in the case of indigenous weed species. The use of plant pathogens and the possibility of finding or creating more virulent strains of existing pathogens may well improve the biological control outlook for native weeds (Sands and Rovira, 1971). The control of northern jointvetch [*Aeschynomene virginica* (L.) B.S.P.] with sprays of *Colletotrichum gloeosporioides* (Penz.) Sacc: f. sp. *aeschynomene*, an indigenous pathogen, supports this viewpoint (Daniel *et al.*, 1973). Natural enemies of indigenous weeds may also be obtained from other areas of the weed's range and from related plant species in other areas.

Another point that continues to pose a serious obstacle to the use of biological control organisms is the question of conflicting interests. Even important weeds have virtues, in certain seasons and locales. For example, johnsongrass, *Sorghum halepense* (L.) Pers., is a weed throughout most of the United States but has some forage value in at least one southern state. The introduced phreatophyte, saltcedar, *Tamarix pentandra* Pall., forms dense stands over washes and stream beds in southern areas of Arizona, New Mexico, and parts of Texas where it impedes water flow, causing flooding during the rainy season and high transpirational losses for the remainder of the year. However, saltcedar is valued as a nesting site for the white-winged dove, *Zenaida asiatica* L., a favorite game bird in the area, and is also an important source of nectar.

Since it is almost impossible to limit the distribution of weed-feeding agents, once they have been introduced to an area, the value of a plant and its ecological significance must be carefully weighed against its potential for causing loss. In resolving conflicts it should be kept in mind that biological control, unlike chemical or mechanical controls, effects a gradual reduction in weed numbers, and eradication over large areas is rarely, if ever, achieved. Thus, if the virtues of the weed can be retained at a lower level of abundance, and as part of a more diversified plant community, the conflict of interest is resolved.

A weed problem can be characterized by the number of plant species involved, the type and stability of the habitat, and the level and timing of control necessary to minimize loss. Host-specific weed-feeding organisms are useful for the control of only closely related weed species. When the problem involves a complex of two or more plant species, the nonhost weeds will develop unabated, and may even increase in abundance, requiring the use of other control measures. One remedy might be the introduction of additional natural enemies to attack the other plants of the problem complex. In some areas where *Agasicles hygrophila* Selman and Vogt has reduced the abundance of alligatorweed, *Alternanthera philoxeroides* (Mart.) Griseb (Maddox *et al.*, 1971), waterhyacinth, *Eichhornia crassipes* (Mart.) Solms, has increased. Several insects are being considered for the control of this latter plant (Coulson, 1971; Perkins, 1973).

Most biological control efforts have been against terrestrial weeds. This may have been due to the feeling that there are fewer host-specific insects on aquatic plants than on terrestrial species. This may prove true regarding submerged hydrophytes [e.g., *Myriophyllum spicatum* L. (Lekić and Mihajlović, 1970)], but relatively rich complexes of insect species are closely associated with alligatorweed, an emersed plant (Maddox *et al.*, 1971) and the floating waterhyacinth (Bennett, 1970b).

The value of using biological control in areas of high disturbance (e.g., cultivated cropland) remains questionable, especially if the inoculative approach is used (i.e., the natural enemies are released and left to increase to effective levels on their own). If the life cycle of the natural enemy is long, the chance for interruption of development is high and achievement of control is low. However, the control of the annual *Emex spinosa* Campd. in Hawaii (Anon., 1968) and *Tribulus terrestris* L. in Hawaii and areas of California and Arizona (Andres and Goeden, 1971) suggest that biological control may have some application against weedy annuals of disturbed areas. Also, the suppression of weeds in low disturbance situations (e.g., fence rows, roadsides) may provide some relief in adjacent disturbed sites. In crop areas, quick and effective weed control is required if loss is to be avoided. Although weed-feeding insects have provided dramatic control in some instances, at best a period of at least 1 year has been required to achieve control and, in most cases, it has been 3—10 years before the weeds were suppressed below the level of economic importance.

B. The Search for Natural Enemies

"Any organisms which curtail plant growth or reproduction may be used as biological weed control agents. Such could potentially include animals, other than the insects, and as well, parasitic higher plants, fungi, bacteria and viruses" (Huffaker, 1964). Thus, the search for natural enemies should encompass all organisms associated with the target plant.

Organisms other than insects have been used in the control of weeds, although their use has been negligible. In the case of plant pathogens and nematodes, one major obstacle has been the difficulty of determining with certainty the species involved. Also, their damage is less visible than that of insects and they frequently attack stages of the plant which are less readily investigated or only occur for a short time in any one year (e.g., seedlings).

However, the use of plant pathogens has increased, especially where the insect species on a particular weed have not had sufficient potential or host specificity to be used as biological control agents. This was true for rush skeletonweed, *Chondrilla juncea* L., in the western Mediterranean where fungi, particularly the rust *Puccinia chondrillina* Bubak and Syd., are considerably more effective in reducing the plant than the associated arthropod species (Wapshere, 1970; Hasan and Wapshere, 1973). Hasan (1972) found *P. chondrillina* to be specific to *Chondrilla juncea* and has introduced it into Australia (Hasan, 1974a) where it has spread rapidly and is already causing considerable damage to long established populations of *C. juncea* (Cullen *et al.*, 1973; Cullen, 1974). Hasan (1974b) is also conducting similar ecological and host specificity studies with two powdery mildews, one a form of *Erysiphe cichoracearum* D. C. and the other a form of *Laveillula taurica* Arnaud, both of which attack *Chondrilla juncea*. Inman (1971) studied the rust *Uromyces rumicis* (Schiem) Wint., which attacks *Rumex crispus* L. (curly dock) in Europe. In Russia, *Alternaria cuscutacidae* Rudak was used to control *Cuscuta* spp. in alfalfa (*Medicago sativa* L.) (Rudakov, 1961). Attempts have also been made to use viruses to control algal blooms in sewage ponds (Jackson, 1967). A review of the role of plant pathogens in weed control was made by Wilson (1969). Zettler and Freeman (1972) discuss the role of pathogens in the biological control of aquatic weeds.

Despite their close similarity to insects in general structure and the type of damage they cause, plant mites (acarians) have been used rarely for weed control. *Tetranychus opuntiae* Banks (= *T. desertorum* Banks) was of initial importance in controlling *Opuntia inermis* (DC.) DC. and *O. stricta* (Haw.) Haw. in Australia, until displaced by the introduced moth, *Cactoblastis cactorum* (Dodd, 1940). The only other instance where a mite species has been used in the biological control of weeds is that of an eriophyid gall mite, *Aceria chondrillae* Can., which was recently introduced into Australia against *Chondrilla juncea* (Caresche and Wapshere, 1974) where it has become well established (Cullen, 1974).

Where control of a single weed species is desired, host specificity is important to protect surrounding plants from damage. Where control of a complex of plant species or total vegetation control is the goal, little host specificity on the part of the control organisms is needed as long as their distribution can be controlled and useful plants remain unharmed. This has frequently been the case with aquatic weeds, permitting the use of a large range of organisms against these plants. Blackburn *et al.* (1971) summarizes the use of a number of relatively polyphagous organisms for control of aquatic weeds [e.g., weed-eating fish, particularly *Tilapia* spp. and carp, *Cyprinus* sp., ducks, manatees, snails, and crayfish (see Chapter 18)]. More recently, detailed studies of the use of the polyphagous white amur grass carp, *Ctenopharyngodon idella* (Val.), have been made in Austria (Liepolt and Weber, 1971), Holland (Van Zon, 1974), and the United States (Sneed, 1971).

Only recently in the control of alligatorweed with insects has there been an attempt to control a specific aquatic weed while conserving some of the vegetation for bank protection and food for wildlife (Maddox *et al.*, 1971).

Despite recent studies with organisms other than insects as biological control agents, their use is still a little-explored field. In the future, plant mites, nematodes, and fungi are likely to be used since their host specificity can be readily determined. However, host specificity determination problems with the plant pathogenic bacteria, viruses, and similar microorganisms still pose a problem. With increasing use of these and other organisms, biological control of weeds will no longer be an "entomological preserve."

C. Ecological Observations

Observations on the ecology of weed-feeding organisms are not only important in strengthening the case for host specificity, but also valuable in assuring the selection of organisms that have a high control potential.

The concern for specificity has often preempted evaluation studies on the potential effectiveness of prospective candidate agents. For this reason, many host-specific insects may be introduced, but only a few prove of value. For example, only 5 of the 51 cactus insects imported to Australia against prickly pear were of value for its control (Wilson, 1960). Since each insect released represents a substantial cost in time and money, it is of advantage to select those organisms of greatest control value.

Introduced organisms have to tolerate or adapt to the environmental conditions of the problem area if they are to reproduce and attain controlling levels. This is generally considered during preliminary studies when an effort is made to select natural enemies from areas ecologically analagous to the problem site (Wapshere, 1970). Harris and Peschken (1971) describe the climatic requirements of several species of *Chrysolina* beetles, noting where each is, or would be, most effective within the range of the weedy *Hypericum* in Canada. *Chrysolina hyperici* (Förster) is thriving in those areas characterized by vigorous stands of Douglas fir, while on the drier sites, or those where ponderosa pine is dominant, *C. hyperici* has been completely or partially replaced by *C. quadrigemina*. Since neither beetle has been effective in the driest sites occupied by the weed, the moth, *Anaitis plagiata* (L.), adapted to dry conditions, was released in 1967. In Hawaii, on the other hand, neither beetle is causing much damage to the plant and control is being attributed to year-around high populations of the gall midge, *Zeuxidiplosis giardi* (Kieffer), which prefers more moist situations.

Biological control organisms, once introduced and established to a new area, should become better adapted to their new environment and may gradually extend their range and improve their control of the weed. *Chrysolina quadrigemina* has apparently adapted to the climatic conditions of British Columbia and is now providing improved control of *Hypericum* (Harris and Peschken, 1971).

Of considerable importance is the synchronization of the damage caused by natural enemies to the plant's developmental cycle and to the particular conditions or limiting factors under which the plant is developing. Defoliation of the *Hypericum* rosettes in the fall and winter by the larvae of *Chrysolina quadrigemina* contributed to the death of the plant during the dry summer season in California. The defoliated plants did not have time to redevelop an adequate root system before the summer drought (Huffaker, 1953). *Teleonemia scrupulosa* Stål causes defoliation to *Lantana* during the summer in Hawaii but the plant then recovers during the remainder of the year. When this period of defoliation was extended by introducing several lepidopterans that fed during the cooler months, many of the plants in the lower rainfall areas were killed (Andres and Goeden, 1971). Other examples in which the weed was unable to compensate for the damage inflicted by a well-synchronized attack include the defoliation of *Senecio jacobaea* by *Tyria jacobaeae* in Nova Scotia, about 2 months before frost. This does not allow the plants time to accumulate sufficient root reserves before winter. They are cold tender and killed (Harris, 1974).

Observations of the plant's growth cycle may indicate the type and time of attack to which the plant will be vulnerable. The level of carbohydrate reserves in the storage organs of perennial weeds is a guide to the application of sprays, better control generally being obtained when herbicide application is timed to low carbohydrate reserve levels. Conversely, ranchers often limit livestock grazing when food plant carbohydrate reserves are low so as not to damage the plants. Harris (1973) concluded that biological control agents selected to attack the weed at a time of low carbohydrate reserves would be preferred, but noted that this period will vary from plant to plant and cautioned that a poorly timed attack may even stimulate the weed. Defoliation of alligatorweed by *Agasicles hygrophila* at a time when the carbohydrate reserves of the plants were at their lowest level (March–June) may have been instrumental in the control of the plant at the original Jacksonville, Florida release site (Andres and Davis, 1974).

However diligent the ecological observations on the plant and its natural enemies, it is still difficult to predict which agent has the greatest control potential. Harris (1973) suggested a possible solution in the form of a scoring system involving 12 criteria for comparing the effectiveness of possible agents before host specificity tests are initiated (e.g., type of damage inflicted, fecundity of insect, number of generations, mortality factors).

D. Determination of Host Specificity

The use of exotic or indigenous biological control agents is practical only if they can be relied upon not to damage desirable plants. Thus, the evidence used to determine the host specificity of the agent and the reliability that can be placed on it are of overriding importance. Much of this evidence is normally provided by laboratory tests. However, it should not be forgotten that a parasite's ability to take nourishment from a plant is only one aspect necessary for its survival; it must also be able to find the plant and complete its life cycle in the environment involved. Thus, a complete investigation of host specificity, which is not always necessary, is tailored to the agent. Following is an adaptation and summary of several steps considered in the host specificity studies of weed-feeding insects from papers by Harris and Zwölfer (1968), Zwölfer and Harris (1971), and Wapshere (1974). These steps need modification for work with pathogens or nonarthropod organisms.

1. Biology and Host-Determined Adaptations

The first requirement is to determine the biology of the insect, paying special attention to morphological, physiological, behavioral, and other adaptations that restrict its host plant range (e.g., the length of the ovipositor in many insects that lay in flower heads determines the size of the heads, and consequently the plant species, susceptible to attack; intricate interaction between the behavior of gall-forming insects and the specific physiological responses of the plant determine host suitability). These studies are normally reinforced with a broad field and literature survey to determine the actual host plant range. The previous safe use of the insect for biological control is the best proof, but additional tests may be needed if the plant associations are different in the new environment.

2. Plants Attacked by Related Insects

A high degree of reliability can be assigned to the host specificity of an insect if it belongs to a taxonomic group (subgenus, genus, etc.) which is restricted to a single plant taxon. This suggests that the insect has speciated on the plant group concerned. Hence, over a long time period and usually a wide geographical area, they have not exploited any other plants. The host plants of sibling species are of particular interest in this regard.

3. Host Specificity Tests

There are two basic strategies to testing: (a) determination of the plants that cannot be utilized (crop testing method) and (b) determination of the range of plants that can be utilized (biologically relevant method). The methods are not mutually exclusive and most investigators use both. The former method is emphasized when the insect is from a region where little is known about the taxonomy and host ranges of the insects or about the pests of particular crops; but the latter method is to be preferred where there is adequate background information. The tests are usually done in the laboratory but field cages or even uncaged plants may be used.

The emphasis in the crop testing method is to show that important crop plants would not be damaged by the insect. It is particularly important to test economic plants that may grow with the target weed, and those species to which the insect has had little or no previous exposure or if exposed, about which there is little knowledge of its pests (Wapshere, 1974). It is unlikely that these plants will have the stimuli to attract the insect (unless they are related to the host), but they may lack inhibitors preventing attack by the insect. This renders them susceptible to casual transfer which is likely to occur immediately following deterioration or disappearance of the weed. Under these conditions temporary crop damage can occur.

Unfortunately in the crop testing method, the results may be misleading as insects in small cages or feeding under forced conditions commonly eat and sustain themselves on plants that they do not attack in nature. The consequence is either rejection of a safe and potentially useful control agent or intuitive interpretation of the results.

The reasoning behind the biologically relevant method is that if the host range of the agent is known, and if possible the basis of its host specificity, then a priori, all other plants are immune. The test plants are all selected on the basis of their risk to attack. The difficulties with the method are that quarantine authorities tend to find the results unconvincing in comparison to the crop testing method and the plants must be selected individually for each agent, although they fall into general categories (e.g., plant species related to the weed host, plants recorded as hosts, plants attacked by insects related to the agent or by other insects primarily associated with the same weed host, and plants unrelated to the host that have morphological or biochemical similarities to it).

4. Causal Analysis of Host Plant Specificity

The host specificity of stenophagous insects is primarily determined by secondary plant substances plus visual and tactile features, and their determination provides a rational method of assessing an insect's host specificity and for revealing taxonomically unrelated hosts. Unfortunately, the investigation of host-recognition tokens is involved and difficult, partly because most insects require several factors to be present simultaneously for acceptance of a plant.

E. Liberation and Establishment

Despite careful planning in the selection of weed control candidates their establishment in the problem areas may fail. Parasites or predators indigenous to the release areas may attack the weed control agent [e.g., *Harpobitticus* on *Tyria* in Australia (Bornemissza, 1966)]. Nocturnal insects released during daytime may be especially vulnerable to predation. Disease organisms unknowingly introduced with weed-feeding insects may cause failure of establishment (Bucher and Harris, 1961). Catastrophic destruction of the release areas by herbicides, grazing animals, and flooding prevents establishment. Differences in the myriad of environmental factors (e.g., competing plants, climatic and edaphic conditions) between the release area and the source of the weed control agent may retard or negate control action.

F. Evaluation Studies

Evaluation studies are not essential to the success of a project, but do have a bearing on the implementation and success of future projects.

Ideally, the populations of natural enemies should be monitored and correlated with damage to the host plant. The impact of the control organisms may range from gross, rapid destruction of the weed to a subtle lessening of competitiveness with other plants in the community. In the latter instances, studies on the productivity of infested and uninfested plants will be needed. "Before and after" photos are quite striking and provide documentation of success or failure, but often little else. Studies by Huffaker (1967) and Huffaker and Kennett (1959) on the control of *Hypericum perforatum* form one of the better documentations of the results of a "biological-control-of-weeds" project.

IV. THE UTILIZATION OF BIOLOGICAL CONTROL

The introduction of exotic organisms has proved valuable where effective natural enemies were absent. Where potentially effective natural enemies of a weed are already present, the possibility of conserving or augmenting their action should not be overlooked.

Annecke *et al.* (1969) noted that, due to predation by coccinellids, the imported cochineal insect *Dactylopius* sp. was ineffective in the control of the pricklypear cactus (*Opuntia megacantha* Salm. Dyck) in certain areas of the eastern Cape Province, South Africa. By applying a low dosage of DDT (about 2 oz. actual/acre) directly to the pear plants, the predators were reduced sufficiently to conserve the scale, allowing it to destroy the plants. Wilson (1960) described how the scale *Dactylopius* sp. was periodically distributed to stands of *Opuntia aurantiaca* Lindl. in areas of Australia to augment its action in controlling this weed. Efforts are underway to evaluate the mass production of an indigenous moth, *Bactra verutana* Zeller, for early spring release against *Cyperus rotundus* L. in the United States (Frick and Garcia, 1975).

Approximately 75 species of weedy plants have been or are being considered for biological control. The actual number of projects will be higher than this, since once a biological control agent has proved successful in one area, it is often transferred to other areas to control the same or closely related plants. Thus the biological control of *Opuntia* spp. has been attempted in at least ten different areas of the world; *Lantana*, eleven areas; *Hypericum*, six areas; and *Senecio jacobaea* L., four areas (Goeden, in press). The success of a number of these projects was rated in the NAS Handbook on weed control (Anon., 1968) and also by Goeden (in press). It is now apparent that insects have provided excellent control to some otherwise uncontrollable weed problems.

A. Perennial Weeds

1. Opuntia spp. (Pricklypear Cacti)

Various species of *Opuntia* occur as weeds in range and cropland areas of the world, crowding out the more desirable forage plant species. The successful reduction of *Opuntia inermis* and *O. stricta* from 60 million acres to a fraction of this amount by the Argentine phycitid moth, *Cactoblastis cactorum*, demonstrated conclusively the value of the biological method of weed control [see Dodd (1940) and Wilson (1960)]. Goeden (in press) details the Australian work in addition to similar attempts elsewhere.

2. Hypericum perforatum (St. Johnswort, Klamath weed)

This perennial European native is spread throughout much of the temperate areas of the world but has attracted attention as a major weed pest primarily in Australia and northwestern North America where attempts were made to control it with insects. Following surveys of the insects associated with *H. perforatum* in England and southern France by Australian entomologists (1928—1940), eight species were introduced into Australia. Some control was achieved with two beetles, *Chrysolina hyperici* and *C. quadrigemina* (Wilson, 1960; Huffaker, 1967).

In 1944—1945 American entomologists imported these beetles to the United States with remarkable success (e.g., in California over 2 million acres of infested land was reduced to less than 1% of this amount). For detailed accounts see Huffaker and Kennett (1959); Holloway (1964); Huffaker (1967) (see also Chapter 11).

In Canada, the *Chrysolina* beetles have brought the weed under control in most of the interior of British Columbia, reducing the plant by as much as 98% of its former density [Harris and Peschken (1971)]. Control has also been attempted in New Zealand (some success), South Africa (some success, Stephan Neser, personal communication, 1974), Chile (Goeden, in press) and recently Hawaii (some success).

3. Lantana camara (Lantana)

Lantana, a native to subtropical and tropical Central and South America, infested over 443,000 acres of land in Hawaii in 1962. In 1902 a tingid, *Teleonemia scrupulosa*, was introduced from Mexico and reduced the spread of this plant. In the 1950's improved control was obtained in the drier areas with the introduction of three leaf-feeding lepidopterans, *Catabena esula* (Druce), *Syngamia haemorrhoidalis* Guenée, and *Hypena strigata* F. (Krauss, 1962). A stem-boring cerambycid, *Plagiohammus spinipennis* (Thom.), and two leaf-feeding beetles, *Octotoma scabripennis* Guer. and *Uroplata girardi* Pic, have been introduced to aid control in the wetter areas (Andres and Goeden, 1971). The two leaf beetles are gradually displacing the lepidopterans in some areas. The effectiveness of the latter has also been reduced by parasitism of the eggs, larvae, and pupae. Goeden (in press) has reviewed the biological control of lantana in other areas.

4. Senecio jacobaea (Tansy ragwort)

This poisonous European weed is a problem in pastures of the northwestern United States, parts of Canada, New Zealand, Tasmania, Australia, South Africa, and South America (Frick and Holloway, 1964). The moth, *Tyria jacobaeae* (L.), whose larvae are foliage and flower feeders, has been introduced to many of the ragwort infestations with varying effect. Remarkable control has been achieved in areas of California (Hawkes, 1968) and also in Canada (Harris *et al.*, 1971). Two other insects, a seed fly, *Hylemya seneciella* (Meade), and a crown-infesting chrysomelid, *Longitarsus jacobaeae* (Waterhouse), have been recently introduced to the United States to enhance control. The latter has increased to high levels at one site with marked effect on the plant (R. B. Hawkes, USDA, ARS, Albany, Calif., personal communication).

Other successful perennial weed projects include *Cordia*, *Eupatorium*, and *Clidemia*.

B. Annual Weeds

Projects are underway to develop biological control agents for several annual weeds. *Emex spinosa* and *E. australis* Steinh., pasture weeds in Hawaii, have been substantially controlled by a small weevil, *Apion antiquum* Gyll., from South Africa (Anon., 1968). Two weevils, *Microlarinus lareynii* (Jacq. du Val) and *M. lypriformis* (Woll.), have been introduced to control the annual weed *Tribulus terrestris* in the United States and other areas. Although these weevils are of value, clear-cut documentation of control of this sporadic annual is difficult. The control of *T. cistoides* L., a perennial, and *T. terrestris* in Hawaii has been reported (Andres and Goeden, 1971).

C. Aquatic Weeds

Alternanthera philoxeroides (alligatorweed)

In 1964, a South American chrysomelid beetle, *Agasicles hygrophila*, was introduced to the United States for the control of floating and emergent mats of this plant. A second insect, the stem mining phycitid, *Vogtia malloi* Pastrana, was released in 1971 (Brown and Spencer, 1973).

Defoliation by the *Agasicles* larvae and adults has had its greatest impact in reducing alligatorweed in northern Florida, southern Alabama, Louisiana, and Texas (Maddox *et al.*, 1971).

Programs to control other aquatic weeds are now underway. *Neochetina eichhorniae* Warner has been released against *Eichhornia crassipes* in the United States. The grasshopper, *Paulinia acuminata* De Geer, the weevil, *Cyrtobagous singularis* Hulst., and the moth, *Samea multiplicallis* Guenée, are under consideration for the control of *Salvinia auriculata* Aubl. in Africa (Bennett, 1966). (See also *Marisa*, Chapter 18.)

V. SUMMARY OF BIOLOGICAL CONTROL OF WEEDS PROJECTS

The projects listed below were compiled from the literature and progress reports of research laboratories. The status of each project is indicated by the numbers: *1* (survey for control organisms, feeding tests, etc.), *2* (introduction and/or establishment of exotic organisms), and *3* (evaluation of results).

Some references of special interest are: (1) Goeden (in press) reviews most projects reported in the literature until 1968, noting weed species, insects used for control, and results obtained in various parts of the world; (2) The Commonwealth Institute of Biological Control Technical Communication No. 4 (see Harris, 1971, and others for summary of weed work in Canada from 1959 to 1968, with references to work in other countries) and Technical Communication No. 5 (see Greathead, 1971, for summary of biological control projects in the Ethiopian Region and other areas of Africa south of the Sahara); (3) Andres and Davis (1974) list the more important insects released to control weeds in the Continental United States and Hawaii, with brief notes on the status of the work; (4) Simmonds (1970a) and earlier reports annually summarizes the projects of the Commonwealth Institute of Biological Control; (5) The Commonwealth Scientific and Industrial Research Organization (CSIRO 1970—1971 and earlier reports) annually summarizes biological control projects in Australia; (6) Proceedings of the First, Second, and Third International Symposia on Biological Control of Weeds, which appears as Commonwealth Institute of Biological Control Miscellaneous Publications (see Simmonds, 1970b); (7) Zwolfer (1968) summarizes projects underway in Europe.

As an aid to the reader, the organization performing the work has often been cited [e.g., (USDA) United States Department of Agriculture, Albany, California; (CDA) Canada Department of Agriculture, Regina, Saskatchewan; (CIBC) Commonwealth Institute of Biological Control, Curepe, Trinidad; (CSIRO) Commonwealth Scientific and Industrial Research Organization, Canberra, Australia; (Hawaii) Hawaii Department of Agriculture, Honolulu, Hawaii; (Univ. Calif.) University of California, Berkeley and Riverside, California].

Amaranthaceae: *Alternanthera philoxeroides* (Mart.) Griseb, United States, *3* (Maddox *et al.*, 1971); India, *2* (CIBC)

Anacardiaceae: *Rhus diversiloba* T. and G., United States, *1* (Univ. Calif.); *Schinus terebinthifolius* Raddi, Hawaii, *3* (Goeden, in press)

Boraginaceae: *Cordia macrostachya* (Jacquin) Roemer & Schultes, Mauritius, *3* (Bennett, 1970a; Goeden, in press)

Cactaceae: *Opuntia* spp., India, Ceylon, Australia, South Africa, Mauritius, New Caledonia, Celebes, Java, Hawaii, United States, Leeward Island (West Indies), East Africa, *3* (Goeden, in press; Bennett, 1970a)

Chenopodiaceae: *Halogeton glomeratus* (M. Bieb.) C. A. Mey, United States, *1, 2* (Simmonds, 1970a; USDA); *Salsola iberica* Sennen and Pau (= *S. pestifer* A. Nelson = *S. kali* var. *tenuifolia* G. F. W. Mey.), United States, *1, 2* (Simmonds, 1970a; USDA, Univ. Calif.)

Compositae: *Acanthospermum hispidum* DeCandolle, Nigeria, inactive (Huffaker, 1959); *Ambrosia artemisiifolia* L., *A. trifida* L., Canada, *1* (CDA); USSR, *1* (Kovalev and Runeva, 1970); *Baccharis halimifolia* L., Australia, *1, 2, 3* (CIBC, Queensland Dept. of Lands); *Bidens pilosa* L., Papua, New Guinea, *1* (Simmonds 1970a); *Carduus nutans* L., *C. acanthoides* L., Canada *3* (CDA, CIBC); United States, *1,2, 3* (USDA); *Carduus pycnocephalus* L., *C. tenuiflorus* Curt., United States, *1* (Univ. Calif., USDA); *Centaurea diffusa* Lam., *C. maculosa* Lam., Canada, *2, 3* (CDA, CIBC); *Centaurea repens* L., United States, *1* (Univ. Calif.); USSR, *1* (Kovalev, 1968); *Centaurea solstitialis* L., United States, *1, 2* (CIBC, Univ. Calif.); *Centaurea* spp., Canada, *1* (Zwölfer, 1968); *Chondrilla juncea* L., Australia, *1, 2, 3* (Goeden in press; CDA, USDA); *Silybum marianum* (L.) Gaertn., United States, *1, 2* (USDA); England, *1* (Zwölfer, 1968); *Cirsium vulgare* (Savi) Ten., Canada, *1* (Zwölfer, 1968); *Elephantopus mollis* H.B.K., Hawaii, *3*; Fiji, *3* (Goeden, in press); *Eupatorium adenophorum* Spr., Hawaii, *1, 2, 3*, Australia, New Zealand, *3* (Goeden, in press; Hoy, 1964); *Eupatorium odoratum* L., Nigeria, Malaysia, *1, 2* (CIBC); *Pluchea odorata* (L.) Cassini, Hawaii, *3* (Goeden, in press); *Senecio jacobaea* L., Australia, New Zealand, Canada, United States, *1, 2, 3* (Goeden, in press; CDA, USDA); *Silybum marianum* (L.) Gaertn., United States, *1, 2, 3* (USDA, Univ. Calif.); *Sonchus* spp. (CIBC), *1* (Schroeder, 1974; CIBC); *Xanthium strumarium* L., Australia, Fiji, *2, 3* (Goeden, in press; CIBC, CSIRO, Queensland Dept. of Lands)

Convolvulaceae: *Convolvulus arvensis* L., North America, *1* (USDA, Univ. Calif.); *Cuscuta* spp., Barbados, *1, 2, 3* (CIBC); United States, *1* (CIBC); USSR, *1* (Ivannikov, 1969b)

Cruciferae: *Cardaria draba* (L.) Desv. and other cruciferous weeds, United States, *1* (USDA).

Cyperaceae: *Cyperus rotundus* L., Hawaii, *3* (Goeden, in press); North America, *1* (CIBC, USDA)

Euphorbiaceae: *Euphorbia esula* L., *E. cyparissias* L., Canada, *2, 3* (CDA, CIBC); *Euphorbia geniculata* Ort., Papua, New Guinea, *1* (CIBC)

Graminae: *Sorghum halepense* (L.) Pers., United States, *1* (USDA)

Haloragidaceae: *Myriophyllum spicatum* L., United States, *1* (USDA, CIBC)

Hydrocharitaceae: *Hydrilla verticillata* (L.P.) L. C. United States, *1* (CIBC) (Baloch and Sana-Ullah, 1974; CIBC)

Hypericaceae: *Hypericum perforatum* L., Australia, United States, Canada, Chile, Hawaii, New Zealand, South Africa, *3* (Goeden, in press; CDA, Univ. Calif., Hawaii)

Labiatae: *Salvia aethiopis* L., United States, *1, 2, 3* (USDA)

Leguminosae: *Cytisus scoparius* (L.) Link, United States, *3* (USDA); *Mimosa pudica* L., Papua, New Guinea, *1* (CIBC); *Prosopis juliflora* (Swartz) D. C., United States *1* (USDA); *Ulex europaeus* L., United States, Hawaii, Tasmania, *3* (Goeden, in press); New Zealand, *1, 3* (CIBC)

Loranthaceae: *Arceuthobium* sp., United States, inactive (Holloway, 1964)

Melastomataceae: *Clidemia hirta* D. Don, Fiji, Hawaii, *3* (Goeden, in press; CIBC, Hawaii); *Melastoma malabathricum* L., Hawaii, *3* (Goeden, in press; Hawaii)

Myricaceae: *Myrica faya* Aiton, Hawaii, *3* (Goeden, in press)

Myrtaceae: *Leptospermum scoparium* Forster, New Zealand, *3* (Hoy, 1964; Holloway, 1964)

Onagraceae: *Ludwigia* spp., United States, *1* (Simmonds, 1970a)

Orobanchaceae: *Orobanche cumana* Waller, Yugoslavia, *3* (Lekic, 1970); *Orobanche ramosa* L., United States, *1* (Lekic, 1970); *Orobanche* spp., USSR, *3* (Kovalev, 1968)

Polygonaceae: *Emex australis* Steinh., *E. spinosa* Campd., Hawaii, *3* (Goeden, in press); *Rumex crispus* L., United States, *1* (USDA); *Rumex obtusifolius* L., Nat. Grassland Inst., Japan, *1*, *2* (Miyazaki and Naito, 1974)

Pontederiaceae: *Eichhornia crassipes* (Mart.) Solms, United States, India, Africa, *1*, *2* (USDA, CIBC)

Portulacaceae: *Portulaca oleracea* L., Papua, New Guinea, *1* (Simmonds, 1970a)

Potamogetonaceae: *Potamogeton nodosus* Poir., United States, *1* (Simmonds, 1970a)

Rhamnaceae: *Rhamnus cathartica* L., Canada, *1* (CDA, CIBC)

Rosaceae: *Acaenia sanguisorbae* Vahl., New Zealand, inactive (Goeden, in press); *Rosa rubiginosa* L., New Zealand, *1* (Zwölfer, 1968); *Rubus penetrans* Bailey, Hawaii, New Zealand, *3* (Goeden, in press)

Salvinaceae: *Salvinia auriculata* Aubl., Kenya, Botswana, Zambia, Kariba Lake, *3* (CIBC)

Scrophulariaceae: *Linaria dalmatica* (L.) Mill., *L. vulgaris* Mill., Canada, *3* (CDA); United States, *1*, *2* (USDA); *Striga lutea* Lour., United States, *1* (CIBC)

Solanaceae: *Solanum elaeagnifolium* Cav., United States, *1* (Goeden, 1970)

Tamaricaceae: *Tamarix pentandra* Pall., United States, *1* (USDA)

Verbenaceae: *Lantana camara* L., Australia, Caroline Islands, East Africa, Fiji, Hawaii, Hong Kong, India, Indonesia, New Caledonia, Mauritius, South Africa, *2*, *3* (Goeden, in press; CIBC, CSIRO); *Stachytarpheta jamaicensis* (L.) Vahl., Fiji, Pacific Islands, inactive (Huffaker, 1959)

Zygophyllaceae: *Tribulus cistoides* L., Hawaii, West Indies, *3* (Goeden, in press; USDA, CIBC); *Tribulus terrestris* L., United States, Hawaii, *3* (Goeden, in press; USDA, Univ. Calif., Hawaii)

Miscellaneous: aquatic weeds, *Ceratophyllum*, *Najas*, *Potamogeton*, *Elodea*, *Hydrilla*, etc. (Blackburn *et al.*, 1971).

REFERENCES

Andres, L. A., and Davis, C. J. (1974). The biological control of weeds with insects in the United States. *Commonw. Inst. Biol. Control Misc. Publ.* **6**, 11–25.

Andres, L. A., and Goeden, R. D. (1971). The biological control of weeds by introduced natural enemies. *In* "Biological Control" (C. B. Huffaker, ed.), pp. 143–164. Plenum, New York.

Annecke, D. P., Karney, M., and Burger, W. A. (1969). Improved biological control of the pricklypear, *Opuntia megacantha* Salm-Dyck, in South Africa through the use of an insecticide. *Phytophylactica* **1**, 9–13.

Anon. (1968). Principles of plant and animal pest control. Vol. 2, Weed Control. *Nat. Acad. Sci. Wash. D.C. Publ.* **1597**, 471 pp.

Baloch, G. M., and Sana-Ullah. (1974). Insects and other organisms associated with *Hydrilla verticillata* (L.f.) L.C. (Hydrocharitaceae) in Pakistan. *Commonw. Inst. Biol. Control, Trinidad, Misc. Publ.* **8**, 61–66.

Bennett, F. D. (1966). Investigations on the insects attacking aquatic ferns, *Salvinia spp.* in Trinidad and northern South America. *Proc. S. Weed Conf.* **19**, 497–504.

Bennett, F. D. (1970a). Recent investigations on the biological control of some tropical and subtropical weeds. *Proc. 10th Brit. Weed Contr. Conf.*, pp. 660–668.

Bennett, F. D. (1970b). Insects attacking waterhyacinth in the West Indies, British Honduras and the USA. *Hyacinth Contr. J.* **8**(2), 10–13.

Bennett, F. D. (1974). Biological control. *In* "Aquatic Vegetation and Its Use and Control" (D. S. Mitchell, ed.), pp. 99–106, UNESCO, Paris.

Blackburn, R. D., Sutton, D. L., and Taylor, T. (1971). Biological control of aquatic weeds. *J. Irrig. Drain. Div. A.S.C.E.* **97**(IR3), 421–432.

Bornemissza, G. F. (1966). An attempt to control ragwort in Australia with the Cinnabar moth, *Callimorpha jacobaeae* (L.) (Arctiidae: Lepidoptera). *Aust. J. Zool.* **14**, 201–243.

Brown, J. L., and Spencer, N. R. (1973). *Vogtia malloi*, a newly introduced phycitid to control alligatorweed. *Environ. Entomol.* **2**, 521–523.

Bucher, G. E., and Harris, P. (1961). Food plant spectrum and elimination of disease of cinnabar moth larvae, *Hypocrita jacobaeae* (L.) (Lep.: Arctiidae). *Can. Entomol.* **93**, 931–936.

Caresche, L., and Wapshere, A. (1974). Biology and host specificity of the *Chondrilla* gall mite *Aceria chondrillae*. *Bull. Entomol. Res.* **64**, 183–192.

Coulson, J. R. (1971). Prognosis for control of water hyacinth by arthropods. *Hyacinth Control J.* **9**, 31–34.

Commonwealth Scientific and Industrial Research Organization. (1971). *Div. Entomol. Annu. Rep. 1970–1971.* Canberra, A.C.T., Australia.

Cullen, J. M. (1974). Seasonal and regional variation in the success of organisms imported to combat skeleton weed, *Chondrilla juncea* L., in Australia. *Commonw. Inst. Biol. Control, Trinidad, Misc. Publ.* **8**, 111–117.

Cullen, J. M., Kable, P. F., and Catt, M. (1973). Epidemic spread of a rust imported for biological control. *Nature (London)* **244**, 462–464.

Daniel, J. T., Templeton, G. E., Smith, R. J., Jr., and Fox, W. T. (1973). Biological control of northern jointvetch in rice with an endemic fungal disease. *Weed Sci.* **21**, 303–307.

Dodd, A. P. (1940). "The Biological Campaign against Prickly Pear," 177 pp. Commonw. Prickly Pear, Brisbane, Australia.

Frick, K. E., and C. Garcia, Jr. (1975). *Bactra verutana* as a biological agent for purple nutsedge. *Ann. Entom. Soc. Amer.* **16**, 7–14.

Frick, K. E., and Holloway, J. K. (1964). Establishment of the cinnabar moth, *Tyria jacobaeae*, on tansy ragwort in the western United States. *J. Econ. Entomol.* **57**, 152–154.

Goeden, R. D. (1970). Current research on biological weed control in Southern California. *Commonw. Inst. Biol. Control, Trinidad, Misc. Publ.* **1**, 25–28.

Goeden, R. D. Biological control of weeds. *In* "Introduced Parasites, Predators, and Pathogens of Harmful Insects and Weeds—A Review," (C. P. Clausen, ed.), in press. USDA Handb.

Greathead, D. J. (1971). A review of biological control in the Ethiopian Region. *Commonw. Inst. Biol. Control Tech. Commun.* **5**, 162 pp.

Harris, P. (1971). Current approaches to biological control of weeds. *Commonw. Inst. Biol. Control Tech. Commun.* **4**, 67–76.

Harris, P. (1973). The selection of effective agents for biological control of weeds. *Can. Entomol.* **105**, 1495–1503.

Harris, P. (1974). The impact of the cinnabar moth on ragwort in eastern and western Canada and its implication for biological control strategy. *Commonw. Inst. Biol. Control, Trinidad, Misc. Publ.* **8**, 119–123.

Harris, P., and Peschken, D. P. (1971). *Hypericum perforatum* L., St. John's Wort (Hypericaceae). *Commonw. Inst. Biol. Control, Trinidad, Tech. Commun.* **4**, 89–94.

Harris, P., Wilkinson, A. T. S., Neary, M. E., and Thompson, L. S. (1971). *Senecio jacobaea* L., Tansy Ragwort (Compositae). *Commonw. Inst. Biol. Control, Trinidad, Tech. Commun.* **4**, 97–104.

Harris, P., and Zwölfer, H. (1968). Screening of phytophagous insects for biological control of weeds. *Can. Entomol.* **100**, 295–303.

Hasan, S. (1972). Specificity and host specialisation of *Puccinia chondrillina*. *Ann. Appl. Biol.* **72**, 257–263.

Hasan, S. (1974a). First introduction of a rust fungus in Australia for the biological control of skeleton weed. *Phytopathology* **64**, 253–254.

Hasan, S. (1974b). The powdery mildews as potential biological control agents of skeleton weed, *Chondrilla juncea* L. *Misc. Publ. Commonw. Inst. Biol. Control* **6**, 114–119.

Hasan, S., and Wapshere, A. J. (1973). The biology of *Puccinia chondrillina*, a potential biological control agent of skeleton weed. *Ann. Appl. Biol.* **74**, 325–332.

Hawkes, R. B. (1968). The cinnabar moth, *Tyria jacobaeae*, for control of tansy ragwort. *J. Econ. Entomol.* **61**, 499–501.

Huffaker, C. B. (1964). Fundamentals of biological weed control. *In* "Biological Control of Insect Pests and Weeds" (P. DeBach, ed.), pp. 631–649. Reinhold, New York.

Hoy, J. M. (1949). Control of manuka by blight. *N. Z. J. Agr.* **79**, 321–324.

Hoy, J. M. (1964). Present and future prospect for biological control of weeds. *N. Z. Sci. Rev.* **22**(2), 17–19.

Huffaker, C. B. (1953). Quantitative studies on the biological control of St. John's wort (Klamath weed) in California. *Proc. 7th Pac. Sci. Congr.* **4**, 303–313.

Huffaker, C. B. (1957). Fundamentals of biological control of weeds. *Hilgardia* **27**, 101–157.

Huffaker, C. B. (1959). Biological control of weeds with insects. *Annu. Rev. Entomol.* **4**, 251–276.

Huffaker, C. B. (1964). Fundamentals of biological weed control. *In* "Biological Control of Insect Pests and Weeds" (P. DeBach, ed.), pp. 631–649. Reinhold, New York.

Huffaker, C. B. (1967). A comparison of the status of biological control of St. Johnswort in California and Australia. *Mushi Suppl.* **39**, 51–73.

Huffaker, C. B., and Andres, L. A. (1970). Biological weed control using insects. *Tech. Pap. FAO Int. Conf. Weed Contr., June 22-July 1, 1970, Davis, California*, pp. 436–449. Weed Sci. Soc. Amer. Champaign, Illinois.

Huffaker, C. B., and Kennett, C. E. (1959). A ten-year study of vegetational changes associated with biological control of Klamath weed. *J. Range Manage.* **12**, 69–82.

Inman, R. E. (1971). A preliminary evaluation of Rumex rust as a biological control agent for Curly Dock. *Phytopathology* **61**, 102–107.

Ivannikov, A. I. (1969a). A nematode controlling *Acroptilon picris*. *Zaschch. Rast.* (*Moscow*), pp. 54–55.

Ivannikov, A. I. (1969b). The fly *Melanagromyza cuscutae* Hg.—a pest of dodder in Kazakhstan. *Vestn. Sel'skokhoz. Nauki* (*Alma Ata*) **12**, 85–88. (In Russian.)

Jackson, D. F. (1967). Interaction between Algal populations and viruses in model pools—a possible control of Algal blooms. Presented at *ASCE Ann. Nat. Meet. Water Resources Eng., Statler Hilton, New York, 1967.*

King, L. J. (1966). "Weeds of the World, Biology and Control," 526 pp. Interscience, New York.

Kovalev, O. V. (1968). Perspectives of weed biological control in the U.S.S.R. *Proc. 13th Int. Congr. Entomol. Moscow* **II**, 158–159 (Abstr.).

Kovalev, O. V., and Runeva, T. D. (1970). *Tarachidia candefacta* Hübn. (Lep: Noctuidae), a promising phytophage in the biological control of weeds of the genus *Ambrosia* L. *Entomol. Rev.* **49**, 8–16 (Transl. of *Entomol. Obozr.*, pp. 23–36).

Krauss, N. L. H. (1962). Biological control investigations on lantana. *Proc. Hawaii. Entomol. Soc.* **18**, 134–136.

Lekić, M. (1970). The role of the dipteron, *Phytomyza orobanchia* Kalt. (Agromyzidae), in reducing parasitic phanerogam populations of the *Orobanche* genus in Vojvodina. *Contemp. Agr.* **18**(7–8), 59–68.

Lekić, M., and Mihajlović, Lj. (1970). Entomofauna of *Myriophyllum spicatum* L. (Halorrhagidaceae), and aquatic weed on Yugoslav territory. *J. Sci. Agr. Res.* **23**(82), 59–74.

Liepolt, R., and Weber, E. (1971). Bischerige erfahrungen mit den weissen amur (*Ctenopharyngodon idella*) in Osterreich. *Oesterr. Fisherei* **24**, 159–162.

Mackie, R. (1957). The biology of *Eumysia idahoensis* Mackie. Unpublished M.S. Thesis, Univ. of Idaho, Moscow, Idaho.

Maddox, D. M., Andres, L. A., Hennessey, R. D., Blackburn, R. B., and Spencer, N. R. (1971). Insects to control alligatorweed. *Bioscience* **21**, 985–991.

Miyazaki, M., and Naito, A. (1976). Studies on the host specificity of *Gastrophysa atrocyanea* Mot. (Col: *Chrysomelidae*), a potential biological control agent against *Rumex obtusifolius* L. (Polygonaceae) in Japan. *Commonw. Inst. Biol, Trinidad, Misc. Publ.* **8**, 97–107.

Perkins, B. D. (1973). Potential for waterhyacinth management based on studies in Argentina. *Proc. Tall Timbers Conf. Ecol. Anim. Contr. Habitat Manage.* **4**, 53–64.

Perkins, R. C. L., and Swezey, O. H. (1924). The introduction into Hawaii of insects that attack Lantana. *Hawaii. Sugar Plant. Ass. Entomol. Ser. Bull.* **16**, 1–53.

Ritcher, P. O., and Dickason, E. A. (1964). The *Aroga* epidemic in Oregon. *Proc. Range Insect Meet. Albany, Calif.*, 4 pp. (mimeo.).

Rudakov, O. L. (1961). Pervye resultaty biologicheskoy borby s povilikoy. *Zashch. Rast.* (*Moscow*) **6**, 23–24.

Sands, D. C. and Rovira, A. D. (1971). Modifying the virulence and host range of weed pathogens. *Proc. 3rd Int. Conf. Plant Pathogenic Bacteria, Wageningen, The Netherlands*, p. 50. (Abstr.).

Schroeder, D. (1974). The phytophagous insects attacking *Sonchus* spp. (Compositae) in Europe. *Commonw. Inst. Biol. Control, Trinidad, Misc. Publ.* **8**, 89–96.

Simmonds, F. J. (1970a). Commonwealth Agricultural Bureaux, Commonw. Inst. of Biol. Control. Annual report of work carried out during 1970.

Simmonds, F. J. (ed.). (1970b). *Commonw. Inst. Biol. Control, Trinidad, Misc. Publ.* **1**, 110 pp.

Sneed, K. E. (1971). The white amur: A controversial biological control. *Amer. Fish Farmer* **2**(6), 6–9.

U.S. Department of Agriculture. (1965). A survey of extent and cost of weed control and specific weed problems. *Agr. Res. Serv. ARS 34–23–1, August*, 78 pp.

Van Zon, J. C. J. (1974). Studies on the biological control of aquatic weeds in the Netherlands. *Commonw. Inst. Biol. Control, Trinidad, Misc. Publ.* **8**, 31–38.

Wapshere, A. J. (1970). The assessment of biological control potential of the organisms attacking *Chondrilla juncea* L. *Commonw. Inst. Biol. Contr. Trinidad, Misc. Publ.* **1**, 81–89.

Wapshere, A. J. (1974). A comparison of strategies for screening biological control organisms for weeds. *Misc. Publ. Commonw. Inst. Biol. Contr.* **6**, 151–158.

Wilson, C. L. (1969). Use of plant pathogens in weed control. *Annu. Rev. Phytopathol.* **7**, 411–434.

Wilson, F. (1960). A review of the biological control of insects and weeds in Australia and Australian New Guinea. *Commonw. Inst. Biol. Contr. Trinidad Tech. Commun.* **1**, 102 pp.

Wilson, F. (1964). The biological control of weeds. *Annu. Rev. Entomol.* **9**, 225–244.

Yeo, R. R., and Fisher, T. W. (1970). Progress and potential for biological weed control with fish, pathogens, competitive plants and snails. *Tech. Pap. FAO Int. Conf. Weed Control, June 21-July 1, 1970, Davis, Calif.*, pp. 450–463. Weed Sci. Soc. Amer. Champaign, Illinois.

Zettler, F. W., and Freeman, T. E. (1972). Plant pathogens as biocontrols of aquatic weeds. *Annu. Rev. Phytopathol.* **10**, 455–470.

Zwölfer, H. (1968). Some aspects of biological weed control in Europe and North America. *Proc. 9th Brit. Weed Control Conf.*, pp. 1147–1156.

Zwölfer, H. (1974). Competitive coexistence of phytophagous insects in the flower heads of *Carduus nutans* L. *Commonw. Inst. Biol. Control, Misc. Publ.* **6**, 74–80.

Zwölfer, H., and Harris, P. (1971). Host specificity determination of insects for biological control of weeds. *Annu. Rev. Entomol.* **16**, 157–178.

20

BIOLOGICAL CONTROL AMONG VERTEBRATES

D. E. Davis, Kenneth Myers, and J. B. Hoy

I. INTRODUCTION

Application of biological control either of vertebrates or by vertebrates has, with a few conspicuous exceptions, generally failed. Although the attempts have not been numerous, unpromising results have discouraged more intensive development. This chapter, by contrasting the vertebrates and the invertebrates, will examine the basic reasons for this history of failures, thereby perhaps pointing the way to success. Examples of attempts at biological control will be presented, and the conspicuous success (myxomatosis) will be described in detail. Unfortunately, the long history of the use of various bacteria in Russia for the control of rats and other rodents cannot be described because records are not available. The scope of the chapter will include vertebrates on vertebrates, various microorganisms on vertebrates, and also vertebrates preying upon insects. It will exclude discussion of accidentally introduced predators or diseases (Warner, 1968) and also discussion of habitat changes which have affected vertebrate populations. A specialized and rare extension of biological control is the use of vertebrates to control vegetation which is analogous to a predator–prey system. The discussion will also exclude the specialized biological control consisting of man's trapping and poisoning of pest species, which is beyond the scope of this book.

II. PRINCIPLES

The principles of biological control which have been described in considerable detail in earlier chapters are inseparable from predation principles. In general, the relationship depends on the details of the ratios of birth, death, and movement rates on the part of predator and prey species. Thus, a predator with a high birth rate has a chance to increase rapidly and through such high numerical responsiveness to bring increasing prey populations under control. However, aspects such as learning and escape behavior will affect the rate of responsiveness and the result. Similarly, death rates will be found to lie within particular ratios, otherwise the predator would tend to overexploit or exterminate the prey population and thereby itself. Also, the existence of alternate prey (buffer species) will influence the result. Last, the nature of movements either as escape or as colonization of new areas will also influence the result. To summarize these complicated relationships, it can be said that no rigid answer is available concerning the appropriate rates, but that a fluctuating system is constantly in action; and under certain circumstances the predator and the prey populations will come to a balance involving some variation but which may be maintained so that the pest is no longer a problem.

Over the centuries the predator and prey adapt to each other's existence and habits so that a constant evolution of the system occurs, although not always at the same level of abundance. When either one or both of the species become extinct, the system, of course, breaks down. Some aspects of this evolution differ considerably among vertebrates in contrast to the situation among many insects. Generally speaking, vertebrates are relatively omnivorous in contrast to insects. For example, many birds and mammals can catch and eat a wide range of prey. A list of the food of foxes would contain dozens of species and differ greatly in different seasons. Also, a single prey species is captured by a large number of vertebrates. For example, field voles are captured by snakes, birds, and mammals among others. While some arthropods are somewhat omnivorous, nevertheless, many have very rigid feeding habits, and these are the ones considered to be the most efficient biological control agents.

Another aspect is the development of various homeostatic mechanisms so that the predator and prey adjust to each other. For example, the existence of advanced forms of learning is a common behavioral device among vertebrates that produces a potential adaptation not available to insects. Furthermore, a number of physiological mechanisms, such as warm bloodedness and highly organized nervous systems, help to allow a close adjustment of certain vertebrate predators and their prey. An example of the behavioral intricacies is seen in the existence of social hunting by wolves and hyenas. Under these circumstances members of the pack cooperate to capture the prey. Another behavioral mechanism is the territorial behavior of both predator and prey species. The division of land into territories prevents the prey from becoming excessively abundant and thus prevents the predator from exhausting the prey in a particular area. Thus, territoriality presents the potential for a more delicate balance between certain vertebrate predator and vertebrate prey species.

A somewhat different feature is the extensive development of immunity mechanisms in vertebrates so that pathogens have less chance of overcoming their hosts (the prey), as contrasted to insect parasitoids and their hosts. These immunity mechanisms again provide opportunities for close adjustments of the predator and prey. A final adaptation is the existence of self-limitation of populations below the resources of the habitat. It seems likely that many vertebrates, with a high reproductive rate, have escaped from the control of predators, and have in contrast developed some self-limiting device which prevented destruction of the habitat. Those species or groups which failed to limit their numbers destroyed their habitat or became poor competitors in that habitat, and became extinct. The existence of these intrinsic mechanisms (for example, pituitary-adrenal negative feedback) prevents the population of prey from becoming overly abundant and thus susceptible to severe predation. None of the above adaptations is rigid or universal; all of them exist to a greater or lesser extent in different species or within the same species in different localities. Thus, while these statements are true as generalities, conspicuous exceptions could be cited for each one of them.

In terms of birth rates (b) and death rates (d) most species maintain stability for many centuries, or $b = d$. Now a temporary or local change in predation would require a compensation by some other mortality factor (unless the change in predation itself is a compensation for a change in such other factor) so that d would not change (for more than a trivial period of time). Vertebrate prey in contrast to arthropods have more flexibility in the aspects mentioned above and hence more opportunities to compensate for a change in predation. Therefore, d changes very little and the population remains more nearly stationary.

III. ATTEMPTS AT BIOLOGICAL CONTROL

The possibility of biological control of vertebrates has been recognized for a long time (Shelford, 1942), and recently revived by Howard (1962, 1967). Naturally, the spectacular success of myxomatosis of the European rabbit in Australia, in particular, has stimulated renewed interest in the possibility of using some microorganisms for control of vertebrate pests. However, the evidence available indicates that successes have been rare. The following examples summarize the available information on the number of attempts.

A. Myxomatosis

The introduction of myxomatosis into the Australian rabbit population in 1950 remains the classic example of the biological control of a mammal. The early experiments leading to its introduction, the ecological consequences, and the ensuing evolutionary adjustments between virus and host have been well documented (Fenner and Ratcliffe, 1965).

Today, some 20 years later, rabbit populations throughout the continent continue to persist at significantly lower densities than those which existed prior (Myers, 1954) to the release of the virus (Table 1). Measurements taken in 1970 in the same manner show that the present numbers of rabbits in those areas are less than 1% of their 1950 levels. This degree of control has not been achieved in all areas, especially on the cooler tablelands or in wet coastal habitats, where rabbit populations still reach high densities on occasion. However, the overall reduction in numbers in those areas has also been significant.

1. Recent Field Observations

Recent work helps to explain these differences. In those areas where the highest degree of control has been achieved (Table 1) myxomatosis occurs as an annual,

TABLE 1
Rabbit Population Indexes in Relation to Introduction of Myxoma Virus

Area	Before (1950)	After (1970)
Rutherglen, Vic. 52.6 hectares (13 acres)		
Number of warrens	15	0
Number of rabbits counted	1000	3
Active burrows	850	4
Coreen, N.S.W. 68.8 hectares (17 acres)		
Number of warrens	26	0
Number of rabbits counted	400	2
Active burrows	600	2
Balldale, N.S.W. 68.8 hectares (17 acres)		
Number of warrens	150	5
Number of rabbits counted	300	20
Active burrows	1250	17
Urana, N.S.W. 121.3 hectares (30 acres)		
Number of rabbits counted	5000	11

mosquito-borne, epizootic during the spring and summer months. The regularity of the outbreaks, and the initially high mortalities, have altered the biological system to the extent that despite a fall in virus fatality rate to 30% or less, predators are now exerting tremendous effect on population numbers. In a recent intensive study it has been shown that at Urana, New South Wales (Table 1), predators (mainly the feral cat and introduced European fox) account for 85% of the young of each year, leaving 15% to be exposed to myxomatosis in the ensuing epizootic (Parer, unpublished observations). This situation illustrates the principle that a vertebrate predator may hold a pest at low levels even though it cannot control the pest if it escapes to high levels.

Outbreaks during the cooler months of the year have been observed since 1952 on the tablelands and slopes of eastern Australia and in other areas in Victoria and Tasmania. The ecology of those epizootics suggest an ectoparasite as the most likely vector. An initial survey of the regional distribution of ectoparasites of the rabbit in New South Wales has shown that the mite *Listrophorus gibbus* Pagenstecher is the only ectoparasite with the necessary geographical distribution and seasonal behavior patterns. The mite is confined to the cooler eastern areas and increases in number during the winter months (Williams, 1972). The same species occurs commonly on rabbits throughout Britain and may also be implicated in some of the transmission of myxomatosis which has been attributed to the rabbit flea, *Spilopsyllus cuniculi* (Dale).

The so-called "winter outbreaks" are less predictable, and usually transmission extends for long periods. Dunsmore *et al.* (1971) describe such an epizootic in subalpine New South Wales in which an attenuated field strain was transmitted slowly through the population for 5 months resulting in a case-mortality rate of 90%. Williams *et al.* (1971) describe an even more dramatic winter outbreak in which the case-mortality rate reached 100%. The results conform with Marshall's (1959) prediction of higher fatality rates in winter, based on his findings that rabbits maintained under hot conditions exhibited lower mortality rates and less marked symptoms and serological changes than those kept under cold conditions. The fact that rabbit populations maintain themselves at generally higher levels results from the more sporadic nature of the outbreaks and the less intense predation pressure (Myers, 1971).

2. New Vectors

The European rabbit flea, *S. cuniculi*, was first introduced into Australia in 1966 (Sobey *et al.*, 1969) in an attempt to improve the rate of transmission of myxomatosis in areas where outbreaks of the disease occurred sporadically and where numbers remained high. It was generally assumed that the insect was sedentary, that it would readily transmit the virus throughout the year, and it would preferentially carry the more virulent virus strains, since lower survival rates in the host would shift fleas about more quickly.

Work in Australia suggests that the flea is not sedentary; it frequently moves from host to host (Williams, 1971). It is strictly seasonal, with peak numbers occurring in spring (Williams, 1973), and its dispersal is a direct result of rabbit movements (Williams and Parer, 1971). The flea is far more mobile than had been supposed (Mead-Briggs, 1964). Furthermore, British workers (Vaughan and Vaughan, 1968) report that the flea transmits attenuated strains of virus in preference to virulent ones. The hopes that the insect may thus improve transmission rates significantly may not be fully borne out.

3. Field Strains

Interesting and probably significant changes have occurred in the virulence of the virus strains collected in the field during the past 20 years in Australia. Following the release of the virulent Brazilian virus in 1950, a mixture of strains of high to very low virulence (based on mean survival times in laboratory rabbits) became endemic by 1959 (Marshall and Fenner, 1960). A similar pattern of change occurred in Britain (Fenner and Chapple, 1965), although the level of virulence tended to drop somewhat lower in Australia.

During the past decade in Australia, further changes in virulence appear to have occurred. Strains of very high virulence, and of severely attenuated strains have disappeared. The initial variability now appears to be giving way to a more stable distribution in which strains of intermediate virulence predominate. Similar observations, although of a more limited nature, have been reported in other parts of Australia, especially in Tasmania (Johnson, unpublished observations) and Western Australia (Oliver, unpublished observations).

A further interesting observation has been that a regional difference appears to be developing in the virulence of the strains collected. Thus in Victoria the most virulent strains are being collected in the northwestern corner of the state where annual epizootics are common and ambient temperatures high. This change agrees with the prediction that increasing genetic resistance of the rabbit might favor viruses of higher virulence (Fenner, 1959).

4. Genetic Resistance

Measurement of the rate of development of true genetic resistance in the rabbit continues to be difficult. Since Fenner and Marshall (1957) showed a negative correlation between mortality rate, mean survival time, and symptomatology, survival time has become the main factor for consideration. Measures of selection using survival time, of course, are not necessarily related to the capacity of the virus to survive in nature.

Survival time has been shown to vary with dose (Fenner and Marshall, 1957), age of the rabbit infected (Sobey *et al.*, 1970), the ambient temperature (Marshall, 1959; Sobey *et al.*, 1968), and route of infection (Mykytowcyz, 1956). The measure of survival time is also influenced by how the virus is collected.

Despite problems in recognizing the qualities of the test animals, some degree of genetic resistance has occurred. Following Marshall and Fenner's (1960) and Marshall and Douglas' (1961) demonstrations of the development of innate resistance to myxomatosis in wild rabbits, Sobey (1969) has recently shown that in a white, domestic breed, heritability of resistance, using survival time as an index, was of the order of 35 to 40%. The work revealed an elevated and inexplicable dam component in heritability. Douglas (unpublished observations) reports that research data from Victoria clearly indicates that the level of genetic resistance is still fairly low and has not progressed much in the last decade. Resistance is highest in the warm northwest, where the virulent virus, "Glenfield" strain, kills 94–98% of susceptible rabbits, and an attenuated strain KM13 kills 45–82%. In the cooler coastal areas of Gippsland, recovery from Glenfield virus is rare, and KM13 kills 91–94%. In Britain, Vaughan and Vaughan (1968) have reported that, despite the activity of attenuated virus strains, genetic resistance has not developed to any extent.

5. *New Virus Strains*

The production of new strains of virus has been the concern of the Division of Animal Genetics, Commonwealth Scientific and Industrial Research Organization (CSIRO). Workers in that Division early rejected the use of mutagens in order to avoid alteration of the virus/host specificity. Their first approach was to create as big a pool of genetic variation as possible by recombining the most varied virus strains available. New strains from recombinants of North American and South American strains of virus were selected on the basis of plaque variants. Another approach has been to put different selection pressures on virus strains and note alterations in their characteristics. Despite their efforts, no strains giving promise of better performance in the field have yet been produced. Inoculation of wild rabbits is now confined mainly to the Glenfield strain.

6. *Evolution*

The key to understanding the kinds of change occurring in the host/virus relationship in Australia lies in a knowledge of the evolutionary background of the disease in the Americas. Myxomatosis is a disease of rabbits and hares, the Leporidae. Immunologically, the many hundreds of strains of myxoma virus obtained in America, Australia, and Europe do not show marked variation, although modern workers can distinguish between strains from California and Brazil, and at least two variants have been recognized from Colombia (Regnery, personal communication). The antigenic stability of the virus is characteristic of the poxvirus group to which it belongs.

Two types of myxoma virus occur, the Californian, maintained in *Sylvilagus bachmani*, and the South American, maintained in *S. brasiliensis*. The types occur in their respective regions to the virtual exclusion of other variants, within the limits of the biological and immunological properties detectable by routine tests. The lack of variability of virus strains within each region has been accompanied by a remarkable stability of strain characteristics over a 10 year period and hundreds of miles apart (Marshall *et al.*, 1963). Although it may not be strictly true for the Colombia-Panama region (Regnery, personal communication), the evidence suggests one of static balance, as a result of long association between parasite and host, in strong contrast to the wide variety of strains recovered in Australia and Europe.

Marshall and Regnery (1963), in an important set of observations, indicate the subtle host/parasite relationships in operation. They show that the attenuated Australian KM13 strain could not become enzootic in *S. bachmani* in California, since between 100 and 1000 times as many particles of the virus are necessary to infect *S. bachmani* as *O. cuniculus*. Doses of such magnitude cannot be transmitted by insects. In addition, KM13 forms very transient tumors in *S. bachmani*. The brush rabbit is an insensitive, recipient host for this strain. With South American strains, *S. bachmani* does not produce a high enough titre of virus particles in the skin to ensure efficient transmission by mosquito vectors. The Californian virus differs vitally from both the other strains. More recently, Regnery (in prep.) has shown by serological analysis of rabbit skins from museum collections that the Californian virus was present in California in 1909 and was thus not introduced into California with sick *Oryctolagus* in 1928, as has been suggested in the literature.

7. *Myxomatosis in* Oryctolagus

Myxomatosis has been successful as an agent of biological control in Australia because of the following set of circumstances:

a. The rabbit is an introduced species. No other species in Australia is remotely related to it, except for the introduced European hare (*Lepus europaeus*). Myxomatosis does not affect hares adversely (Vaughan and Vaughan, 1968), and the species does not need protection.

b. The virus is specific to the Leporidae, and possesses a remarkable antigenic stability.

c. Australia contains a rich and varied fauna of biting insects which have adapted readily to the introduced rabbit as a source of food. The life history of mosquito and rabbit in both California and in large areas of southern Australia are somewhat similar. Many of the required ecological factors, both physical and biological, are thus present.

d. The ecological system is sufficiently elastic to permit a swing from a high density (food-limiting) situation to a sensitive biologically controlled system in which predation has become very important.

e. The European rabbit has not diverged far from its American counterpart. If, as seems possible, *Oryctolagus* spread westward and *Sylvilagus* spread eastward from an Asiatic birthplace (now inhabited by *Caprolagus*) and *Sylvilagus* met up with myxomatosis, *Oryctolagus* may never have been exposed to the myxoma virus; its genetic defenses were thus wide open.

The rabbit in Australia represents a perfect situation for biological control of a pest mammal by microorganisms. Myxomatosis is only a beginning. Many other species of rabbit in other parts of the world, inhabit a wide variety of habitats and carry a broad array of endoparasites. The rabbit cannot be safely left to its own devices in the altered systems it now inhabits (Myers, 1971).

B. Mongoose and Weasels

The mongoose, which is a carnivore belonging to the family Viverridae, and the weasel, also a carnivore but belonging to the family Mustelidae, have been frequent candidates as biological control agents. The history of the mongoose (Hinton and Dunn, 1967) has been extensive but rarely have studies been adequate to measure its impact on field populations. For at least a century mongooses have been liberated in areas where rats were pests in the hope that the mongooses would reduce the rats. Some 36 species of true mongooses exist, but the most commonly introduced is *Herpestes auropunctatus*. The history has been the same whether the mongoose was introduced to get rid of Norway rats or of roof rats. The mongoose (de Vos and Manville, 1956) soon found other prey such as lizards and birds more attractive and fed on rats only occasionally (Pimentel, 1955). Thus, the mongoose has become a terrible pest, especially on islands where some endemic species were exterminated (Seaman and Randell, 1962; Tomich, 1969). Recently, Uchida (1969a, b) has tested the effect of the Japanese weasel (*Mustela sibirica*) on rats on some Pacific islands. While he feels that his results merit further work, the evidence does not indicate any obvious success. To summarize, the data on biological control of rats by mongooses and weasels are meager, but fail to suggest any substantial effect. However, introduction of weasels on an island in 1931 at a time of cyclic decline of the voles (Van Wyngaarden and Bruigns, 1961) was followed by extermination in a pine plantation within 4 years.

C. Cats

The presumed value of cats in rodent control is legendary, but, unfortunately, not supported by data. Several studies indicate some of the circumstances involved in the relationship. Elton (1953) found that cats will not rid a farm of rats, but if the rats have been removed from the farm, then cats will often prevent the reinvasion simply by catching the occasional individual that moves in. Furthermore, the cats did not stay on the farm without the provision of additional food such as milk. In another study (Davis, 1957) it was found that the provision of cat food kept four cats on a farm and these four cats killed a sufficient number of young rats to reduce the population substantially. However, in May the cats turned to squabs and allowed the young rats to survive and the rat population to increase. These studies illustrate some of the complexities and the relationships of such a predator to its prey and the importance of buffers. The widely publicized "cat drop" in Burma (Pomerantz, 1971) seems to have little support in fact (Harrison, 1965).

In contrast, cats, liberated on islands for rat control or simply escapes, have frequently exterminated birds. Thompson (1963) quotes Stonehouse as saying that cats liberated on Ascension Island in 1815 resulted in extirpation of 10 of 11 species of sea birds.

D. Shrews

A valiant attempt to control the larch sawfly by the introduction of shrews has provided interesting results (Buckner, 1966; Warren, 1970). The shrew, *Sorex cinereus*, was introduced into Newfoundland in 1960 and spread rapidly for a decade. However, it has not demonstrated any effect on the abundance of the larch sawfly (Annual report for Insect Disease Survey 1967, Department of Forestry Canada).

E. Bats

For many decades the encouragement of bats as a possible control for insects, especially mosquitos, has been suggested. In about 1910, in San Antonio, Dr. C. A. R. Campbell built bat towers that were supposed to attract bats. Evidence has not been presented that bats control any pest insects.

F. Foxes and Wolves

An interesting recent example is the use of foxes (and raccoons) to control gulls. Kadlec (1967) reports the introduction of these predators on islands where the gulls were nesting. This procedure for the past 2—4 years caused major declines in the number of gulls and even abandonment of some colonies. The predator has to be introduced each year since it cannot survive the winter without gulls for food. As another example, wolves apparently reduced the deer on an island to a level that no longer destroyed the habitat (Merriam, 1964).

G. *Salmonella*

For half a century various types of *Salmonella* (often called Danyz virus or Ratin) have been combined with red squill poison (ratinin) for the control of rats, especially in the Russian Empire. About 1900—1910 many claims were made, but the data are not adequate for confirmation. Apparently the material is no longer used, perhaps because the *Salmonella* is dangerous to humans. An attempt to reduce rat populations by the use of *Salmonella* (Davis and Jenson, 1952) was a failure. Possibly other strains of *Salmonella* would have had an effect.

H. Birds

The previous sections of this chapter have been discussions of control of vertebrate pests with both vertebrate and microbial agents. Most of the cases involved mammalian pests and mammalian agents. Documented cases of control of invertebrate pests or weeds with vertebrate agents, on the other hand, have been largely brought about by either birds or fish. It is perhaps noteworthy that of the vertebrates, birds have relatively low reproductive potentials, whereas fish have relatively high potentials. This difference suggests that encouragement such as nest site improvement may dramatically increase the population of a particular bird species and result in controlling action by those birds — continuous maintenance of an adequate predator population. In contrast, properly timed inoculative release of agents with a high reproductive potential allows use of species that cannot maintain themselves through crises such as a cold season or adverse cultural practices. (However, when inoculative releases are made each season, proper timing, quality stock, and reliable supplies of stock are very important.)

Birds have long been given credit for controlling insects. However, that credit has been largely based on the observations of feeding habits of birds rather than an evaluation of the impact of bird predation on insect populations. For example, the purple martin has been widely publicized and commercially promoted as a mosquito control agent. Yet critical review of the scientific literature reveals no evidence of reduction of populations and very little evidence that purple martins feed on mosquitoes (Kale, 1968). Forest entomologists and ornithologists have developed a large part of the quantitative evidence that birds are controlling and/or regulating factors of insect populations. Betts (1955) concluded that female winter moths, *Operophtera brumata* (L.), were heavily preyed upon by titmice. Mountain chickadee populations were found to respond both functionally and numerically to outbreaks of lodgepole needle miners, *Coleotechnites milleri* (Busek), where nest boxes were provided (Dahlsten and Herman, 1965). In Canada, R. F. Morris and others have elucidated the importance of avian predators in control of spruce budworm, *Choristoneura fumiferana* (Clemens). Morris (1972) confirmed the impact of avian predators on fall webworm. Bird predators can have an indirect effect on a pest beyond that of preying on individuals. For example Moore (1972) discovered that removal of outer bark by woodpeckers allowed drying of the inner bark, thus increasing the impact of disease and cold weather on bark beetles.

Woodpeckers have been found to have significant impact on overwintering populations of codling moth (Chapter 14) and of three important pests of maize. For example, Black *et al.* (1970) found that the yellow-shafted flicker had great impact on southwestern corn borer, *Diatraea grandiosella* (Dyar).

The above examples, like those that follow, are selected to represent successes with respect to such factors as geographical area, pest species, etc. Space neither permits analysis of failures nor comprehensive review of successes. These examples illustrate the possibility that natural populations of birds can reduce a pest insect. Some efforts in Europe encourage biological control by provision of nest boxes (Creutz, 1949) but while the birds increased, evidence was not presented that insects decreased. (see also Chapter 12).

I. Fish

Fish have been used for control of insects, mollusks (see Chapter 18), and aquatic weeds. Many species of fish have short generation times, rapid growth, and large broods, all of which contribute to a rapid numerical response to a food resource. Transport and culture methods are well developed for many species of food and game fish. Furthermore, fish have relatively limited powers of dispersal. Therefore, they tend to remain in the area with the pest problem. These factors combine to make many fish species attractive theoretical candidates for employment as control agents.

Among the disadvantages of fish are that those species that are the most promising candidates are not generally available in numbers adequate for conclusive testing of their abilities, and field tests of efficacy risk establishment of exotic species that can be detrimental to endemic species. The aforementioned limited powers of dispersal have resulted in many fish species that are endemic to one or a few lakes or drainage systems, and therefore particularly vulnerable to competition. These endemic species are threatened by introduction of species such as *Cyprinus carpio* L., the common carp, which is locally prized as a food and widely despised as a trash fish.

A diverse scientific literature is available to those interested in fish. Jenkins (1964) listed many species of fish known to feed on medically important arthropods, and in a more specialized vein, Gerberich and Laird (1966) have published an annotated bibliography of papers "relating to the control of mosquitos (sic) by the use of fish." Blackburn *et al.* (1971) reviewed the state-of-the-art of biological control of aquatic weeds in canals, devoting more space to the potential of fish than to any other taxon.

Mosquito control has been the goal of much of the utilization of fish as biological control agents. *Gambusia affinis* (Baird and Girard) bears the common name mosquito fish and has been widely used especially in the malarious areas of the world. Krumholz (1948) reviewed the use of *G. affinis* and provided detailed data on the biology and efficacy of that species for mosquito control in the midwestern United States. The moquito fish has been found to be effective in both permanent and temporary bodies of water. Notable examples of the impact of mosquito fish on larval mosquito populations in permanent waters are found in the reports by Hess and Tarzwell (1942), Krumholz (1948), and Nakagawa and Ikeda (1969), whose studies were done in a large reservoir, several ponds, and a swamp, respectively.

Temporary bodies of water offer even greater challenge than permanent waters when management of mosquito populations is attempted. Tidal marshes and seasonally flooded waterfowl marshes produce mosquitoes regularly and are candidates for scheduled acts of management. Rees and co-workers (1969) have investigated the use of *G. affinis* in conjunction with a second fish, *Lucania parva* (Baird and Girard), in a Utah waterfowl management area. Mosquito control was achieved with *G. affinis* alone and also with *G. affinis* in conjunction with *L. parva*. (Another example of multispecies stocking of fish for mosquito control is provided in Chapter 18.)

The temporary nature of rice fields invites exploitation by opportunistic aquatic insects such as the mosquitoes, a group noted for short generation times and high reproductive potential. Many researchers have investigated the use of *G. affinis* for moquito control in rice fields (Craven and Steelman, 1968; Hoy *et al.*, 1971, 1972; Hoy and Reed, 1970, 1971; Washino, 1969). Varied results have been reported, but on balance it is clear that *G. affinis* plays a great part in controlling several species of rice field mosquitoes when the fish are stocked in adequate numbers. Fields in California stocked with 0.5 pounds or more of fish per acre rarely produce unacceptable numbers of mosquitoes (Hoy and Reed, 1970, 1971; Hoy *et al.*, 1971, 1972).

Mosquito fish are used extensively by a few agencies responsible for mosquito control and they are used nominally by many. Perhaps the greatest obstacle to full utilization of this control agent is the limited supply of stock at the proper time. Sewage oxidation ponds as sources of mosquito fish vary greatly, both in time and by site (Fisher *et al.*, 1972). Yet, such ponds are the major source of stock in California.

Both common carp (*Cyprinus carpio* L.) and goldfish (*Carassinus auratus* L.) have been shown to be effective for control of pestiferous chironomid gnats in California (Bay and Anderson, 1965; Mezger, 1967). In both studies cited, stocking rates were quite high (150—1400 lb/acre); however, it should be pointed out that control was very rapidly achieved, and that both cases were essentially inundative releases.

Fish have been studied extensively in the search for effective means of controlling aquatic weeds. The white amur, *Ctenopharyngodon idella* Val., has received much attention because it is a hardy species with a voracious appetite for a wide range of aquatic plants (Michewicz *et al.*, 1972). The hearty appetite is no doubt at least in part due to poor utilization of what is eaten. According to Hickling (1966), "Much of the material passes through the gut unchanged." Other fish that show potential for control of aquatic weeds include four species of *Tilapia*, *Metynnis roosevelti* Eig., *Mylossoma agrenteum* E. Ahl., and *Cyprinus carpio* L., the common carp (Blackburn *et al.*, 1971).

J. Miscellaneous Examples

A curious case of biological control resulted from the introduction of the coho salmon into the Great Lakes. The history in brief is that when the lamprey (Eschmeyer, 1955) reduced the lake trout the alewife subsequently increased. Then coho salmon were introduced primarily as a source of sport fishing recreation. As a consequence of this introduction, the alewife has declined in numbers, and is no longer a pollution problem on the beaches. Apparently this history is one of the best cases of biological control, although the salmon was not introduced for this purpose.

Finally, some very special cases of attempts at biological control exist. The Mandan Indian tribe in the Dakotas was exterminated by smallpox which they obtained from blankets purposefully given to them, although DeVoto (1947) denies the tradition of purpose. Also rumors exist that the distemper virus was scattered throughout a county in Georgia in an attempt to reduce the numbers of rabid foxes. Hog cholera has been surreptitiously used to eradicate wild pigs (Herman, 1964). Also, toads have been introduced in some places in an attempt to control insects.

IV. SUCCESSES AND FAILURES

The examples show that in one situation (myxomatosis) success has been spectacular, but in other cases the results have not been sufficiently promising to encourage substantial additional research. Several reasons exist and are listed below:

1. In many cases the pest presents a hazard to human health and hence must be kept at very low levels or exterminated (rats). A biological control agent might not maintain itself on so few individuals unless a good buffer existed.
2. In other cases the agent, if effective against a mammalian pest, may be a hazard to humans (e.g., *Salmonella*).
3. Many pests are carnivores (foxes, skunks, etc.) which exist at low densities and thus provide little chance for an agent to maintain itself by transmission.
4. Carnivores, introduced to reduce herbivores (rodents), feed on so many species that the impact on the pest may be trivial (mongooses on rats).
5. While on islands a species (cats) may eradicate a pest, the system is so small that the predator cannot maintain itself after eradication and dies out.

V. UNTESTED IDEAS

In spite of the lack of success some ideas merit further trial. Of course, there *always* is the possibility that a different species (e.g., mongoose) or a different place (e.g., tropics) or a different season (e.g., winter) will produce biological control, but we have neither time nor funds to try everything. Hence, some likely prospects might be tried first.

One scheme is to reduce the pest by some campaign (e.g., poison) and then expect a predator to maintain control. This sequence occurred in rabbits (see above) where the virus reduced the population and cats maintained control. It also occurred naturally (Pearson, 1964) where the dry season reduced breeding of voles and the local feral cats supplied artificially held down the population.

Another scheme is to introduce several predators simultaneously. Apparently this idea has not seriously been tried.

A third procedure is to introduce the predator at intervals as, for example, foxes to control gulls on islands (Kadlec, 1971).

A fourth procedure is to supply a buffer (as for cats) that itself is no hazard.

Cases of biological control (i.e., the planned introduction of a predator or pathogen) merge imperceptibly into unplanned situations and then into natural predator-prey systems. In its narrow sense, biological control has meant a planned introduction of an agent that may reduce the problem (e.g., myxomatosis). However, in many situations, introduction by accident (cats in New Zealand) or for other purposes (Coho salmon) sets up a predator-prey system that results in reduction of a species that happens to be a pest (rabbits and alewives). While the results are beneficial no planning was involved in respect to the pest. In other situations a natural predator-prey system may occur that results in a balance so that prey scarcity limits the number of predators and the predators prevent the prey from destroying the habitat.

An excellent example is the story (Mech, 1966) of the relation of wolves and moose on Isle Royal. The moose were increasing to a level that was destroying the vegetation when the wolves came onto the island over the ice. A good balance was achieved in a few years so that the vegetation is good and wolves usually have enough food. In another case, Schnell (1968) showed that natural predators could hold a population of cotton rats (which are rarely a pest) at levels of 6 per hectare. However, the generality of this example is not known. Woodpeckers (MacLellan, 1970) destroy large numbers of codling moths. However, at high densities the woodpeckers cannot keep up and the moths cause severe damage. (see also Chapter 14).

Predation presents a spectrum of intensity ranging from failure to prevent a species from destroying its habitat to extermination of the prey. Perhaps a consideration of the reasons that vertebrate predators rarely control their prey will serve as a guide to future efforts. (1) The ratio of reproductive rates is such that the prey increases more rapidly than does the predator. (2) The predator may shift to a buffer species for food. (3) The predator may be eliminating only a surplus of prey. (4) The prey may adapt its behavior to avoid the predator. (5) Habitat changes may protect the prey from the predator. (6) The predator may change the age composition rather than the total number. Information about these possibilities may indicate whether or not a plan for biological control will succeed.

REFERENCES

Bay, E. C., and Anderson, L. D. (1965). Chironomid control by carp and goldfish. *Mosquito News* **25**, 310–316.

Betts, M. M. (1955). The food of titmice in the oak woodland. *J. Anim. Ecol.* **24**, 282–323.

Black, E. R., Davis, F. M., Henderson, C. A., and Douglas W. A. (1970). The role of birds in reducing overwintering populations of the southwestern corn borer, *Diatracea grandiosella* (Lepidoptera: Crambidae), in Mississippi. *Ann. Entomol. Soc. Amer.* **63**, 701–706.

Blackburn, R. D., Sutton, D. L. and Taylor, T. (1971). Biological control of aquatic weeds. *J. Irrig. Drain. Div., ASCE Vol.* **97**, (No. IR3), 421–432.

Buckner, C. H. (1966). The role of vertebrate predators in the biological control of forest insects. *Annu. Rev. Entomol.* **11**, 449–470.

Craven, B. R., and Steelman, C. D. (1968). Studies on a biological and a chemical method of controlling the dark rice field mosquito in Louisiana. *J. Econ. Entomol.* **61**, 1333–1336.

Creutz, G. (1949). Die Entwicklung zweier Populationen des Trauerschnappers *Muscicapa h. hypoleuca* (Poll) nach Herkunft and Alter. *Beitr. z. Vogelkunde*, pp. 27–53.

Dahlsten, D. L., and Herman, S. G. (1965). Birds as predators of destructive forest insects. *Calif. Agr.* **19**, 8–10.

Davis, D. E. (1957). The use of food as a buffer in a predator-prey system. *J. Mammal.* **38**, 466–472.

Davis, D. E., and Jensen, W. L. (1952). Mortality in an induced epidemic. *Trans. N. Amer. Wildl. Conf.* **17**, 151–160.

DeVoto, B. (1947). "Across the Wide Missouri," 483 pp. Houghton Mifflin, Boston, Massachusetts.

de Vos, A., and Manville, R. H. (1956). Introduced mammals and their effect on native biota. *Zoologica* **41**, 163–194.

Dunsmore, J. D., Williams, R. T., and Price, W. J. (1971). A winter epizootic of myxomatosis in sub-alpine southeast Australia. *Aust. J. Zool.* **19**(3), 275–286.

Elton, C. S. (1953). The use of cats in farm rat control. *Brit. J. Anim. Behav.* **1**(4), 151–155.

Eschmeyer, P. H. (1955). The near extinction of lake trout in Lake Michigan. *Trans. Amer. Fish. Soc.* **85**, 102–119.

Fenner, F. (1959). Myxomatosis in Australian wild rabbits—evolutionary changes in an infectious disease. *Harv. Lect.*, pp. 25–55.

Fenner, F., and Chapple, P. J. (1965). Evolutionary changes in myxoma virus in Britain. *J. Hyg.* **63**, 175–185.

Fenner, F., and Marshall, I. D. (1957). A comparison of the virulence for European rabbits (*Oryctolagus cuniculus*) of strains of myxoma virus recovered in the field in Australia, Europe and America. *J. Hyg.* **55**, 149–191.

Fenner, F., and Ratcliffe, F. N. (1965). "Myxomatosis," 379 pp. Cambridge Univ. Press, Cambridge.

Fisher, J. L., Washino, R. K., and Fowler, J. (1972). Populations of *Gambusia affinis* in a cline of oxidation ponds. *Proc. & Pap. Annu. Conf. Calif. Mosq. Contr. Ass.* **40**, 120–121.

Gerberich, J. B., and Laird M. (1966). An Annotated Bibliography of Papers Relating to the Control of Mosquitos by the Use of Fish. (Revised and enlarged to 1965.) WHO/EBL/66.71. (Mimeo.)

Harrison, T. (1965). Operation catdrop. *Animals* **5**, 512–513.

Herman, C. (1964). Disease as a Factor in Bird Control. *Second Bird Contr. Sem. Bowling Green State Univ.*, pp. 112–120. (Mimeo.)

Hess, A. D., and Tarzwell, C. M. (1942). The feeding habits of *Gambusia affinis affinis* with special reference to the malaria mosquito, *Anopheles quadrimaculatus*. *Amer. J. Hyg.* **35**, 142–151.

Hickling, C. F. (1966). On the feeding process in the White Amur (*Ctenopharyngodon idella* Val.). *J. Zool. (London)* **148**, 408–419.

Hinton, H. E., and Dunn, A. M. S. (1967). "Mongooses," 144 pp. Univ. Calif. Press, Berkeley, California.

Howard, W. E. (1962). Means of improving the status of vertebrate pest control. *Trans. N. Amer. Wildl. Nat. Res. Conf.* **27**, 139–150.

Howard, W. E. (1967). Biological control of vertebrate pests. *Proc. 3rd Vertebr. Pest Conf.*, pp. 137–157.

Hoy, J. B., and Reed, D. E. (1970). Biological control of *Culex tarsalis* in a California rice field. *Mosquito News* **30**, 222–230.

Hoy, J. B., and Reed, D. E. (1971). The efficacy of mosquitofish for control of *Culex tarsalis* in California rice fields. *Mosquito News* **31**, 567–572.

Hoy, J. B., O'Berg, A. G., and Kauffman, E. E. (1971). The mosquitofish as a biological control agent against *Culex tarsalis* and *Anopheles freeborni* in Sacramento Valley rice fields. *Mosquito News* **31**, 146–152.

Hoy, J. B., Kauffman, E. E., and O'Berg, A. G. (1972). A large-scale field test of *Gambusia affinis* and Chlorpyrifos for mosquito control. *Mosquito News* **32**, 161–171.

Jenkins, D. W. (1964). Pathogens, parasites and predators of medically important arthropods. *Bull. WHO Suppl.* **30**, 150 pp.

Kadlec, J. A. (1971). Effect of introducing foxes and raccoons on Heering Gull colonies. *J. Wildl. Manage.* **35**, 625–636.

Kale, H. W. (1968). The relationship of Purple Martins to mosquito control. *Auk* **85**, 654–661.

Krumholz, L. A. (1948). Reproduction in the western mosquito fish, *Gambusia affinis affinis* (Baird and Girard), and its use in mosquito control. *Ecol. Monogr.* **18**, 1–43.

MacLellan, C. R. (1970). Woodpecker ecology in the apple orchard environment. *Proc. Tall Timbers Conf. Ecol. Anim. Contr. Habitat Manage.* **2**, 273–284.

Marshall, I. D. (1959). The influence of ambient temperature in the course of myxomatosis in rabbits. *J. Hyg.* **57**, 484–497.

Marshall, I. D., and Douglas, G. W. (1961). Studies in the epidemiology of infectious myxomatosis of rabbits. VIII. Further observations on changes in the innate resistance of Australian wild rabbits exposed to myxomatosis. *J. Hyg.* **59**, 117–122.

Marshall, I. D., and Fenner F., (1960). Studies in the epidemiology of infectious myxomatosis of rabbits. VII. The virulence of strains of myxoma virus recovered from Australian wild rabbits between 1951 and 1959. *J. Hyg.* **58**, 485–488.

Marshall, I. D., and Regnery, D. C. (1963). Studies in the epidemiology of myxomatosis in California. III. The response of brush rabbits (*Sylvilagus bachmani*) to infection with exotic and enzootic strains of myxoma virus, and the relative infectivity of the tumors for mosquitoes. *Amer. J. Hyg.* **77**, 213–219.

Marshall, I. D., Regnery, D. C., and Grodhaus, G. (1963). Studies in the epidemiology of myxomatosis in California. I. Observations on two outbreaks of myxomatosis in coastal California and the recovery of myxoma virus from a brush rabbit (*Sylvilagus bachmani*). *Amer. J. Hyg.* **77**, 195–204.

Mead-Briggs, A. R. (1964). Some experiments concerning the interchange of rabbit fleas, *Spilopsyllus cuniculi* (Dale), between living rabbit hosts. *J Anim. Ecol.* **33**, 13–26.

Mech, L. D., 1970. "The Wolf," 384 pp. Amer. Mus. Nat. Hist., New York.

Merriam, H. R. (1964). The wolves of Coronation Island. *Proc. Alaska Sci. Conf.* **15**, 27–32.

Mezger, E. G. (1967). Insecticidal and naturalistic control of chironomid larvae in Lake Delwigk, Vallejo, California. *Proc. & Pap. Annu. Conf. Calif. Mosq. Contr. Ass.* **35**, 125–128.

Michewicz, J. E., Sutton, D. L. and Blackburn, R. L., (1972). The white amur for aquatic weed control. *Weed Sci.* **20**, 106–110.

Moore, G. E. (1972). Southern pine beetle mortality in North Carolina caused by predators and parasites. *Environ. Entomol.* **1**, 58–65.

Morris, R. F. (1972). Predation by wasps, birds, and mammals on *Hyphantria cunea*. *Can. Entomol.* **104**, 1581–1591.

Myers, K. (1954). Studies in the epidemiology of infectious myxomatosis of rabbits, II. Field experiments August-November 1950, and the first epizootic of myxomatosis in the riverine plain of southeastern Australia. *J. Hyg.* **52**, 47–59.

Myers, K. (1971). The rabbit in Australia. *In* "Dynamics of Populations" (P. J. den Boer and G. R. Gradwell, eds.), pp. 478–506. Wagenigen, The Netherlands.

Mckytowycz, R. (1956). The effect of season and mode of transmission on the severity of myxomatosis due to an attenuated strain of the virus. *Aus. J. Exp. Biol. Med. Sci.* **34**, 121–7.

Nakagawa, P. Y., and Ikeda, J. (1969). "Biological Control of Mosquitoes with Larvivorous Fish in Hawaii." WHO/VBC/69.173. (mimeo.)

Pearson, O. P. (1964). Carnivore-mouse predation: an example of its intensity and bioenergetics. *J. Mammal.* **45**, 177–188.

Pimentel, D. (1955). Biology of the Indian mongoose in Puerto Rico. *J. Mammal.* **36**, 62–68.

Pomerantz, C. (1971). The Day they Parachuted Cats on Borneo, 122 pp. Young-Scott, New York.

Rees, D. M., Bown, D. N., and Winget, R. N. (1969). Mosquito larva control with *Gambusia* and *Lucania* fish in relation to water depth and vegetation. *Proc. & Pap. Annu. Conf. Calif. Mosq. Control. Ass.* **37**, 110–114.

Seaman, G. A., and Randall, J. E. (1962). The mongoose as a predator in the Virgin Islands. *J. Mammal.* **43**, 544–546.

Schnell, J. H. (1968). The limiting effects of natural predation on experimental cotton rat populations. *J. Wildl. Manage.* **32**, 698–711.

Shelford, V. E. (1942). Biological control of rodents and predators. *Sci. Mon.* **55**(4), 331–341.

Sobey, W. R. (1969). Selection for resistance in myxomatosis in domestic rabbits (*Oryctolagus cuniculus*). *J. Hyg.* **67**, 743–754.

Sobey, W. R. and Menzies, W. (1969). Myxomatosis: The introduction of the European rabbit flea *Spilopsyllus cuniculi* (Dale) into Australia. *Aust. J. Sci.* **31**, 404–405.

Sobey, W. R., Conolly, D., Haycock, P., and Edmonds, J. W. (1970). Myxomatosis. The effect of age upon survival of wild and domestic rabbits (*Oryctolagus cuniculus*) with a degree of genetic resistance and unselected domestic rabbits infected with myxoma virus. *J. Hyg.* **68**, 137–149.

Sobey, W. R., Menzies, W., Connolly, D., and Adams, K. M. (1968). Myxomatosis: The effect of raised ambient temperature on survival time. *Aust. J. Sci.* **30**, 322–324.

Thompson, H. V. (1963). The limitations of control measures. *Ann. Appl. Biol.* **51**, 326–329.

Tomich, Q. (1969). Mammals in Hawaii. *Spec. Publ. Bishop Mus.* **17**, 1–238.

Uchida, T. (1969a). Rat-control procedures on the Pacific Islands, with special reference to the efficiency of biological control agents. I. Appraisal of the monitor lizard, *Varamus indicus* (Daudin), as a rat-control agent on Ifaluk, Western Caroline Islands. *J. Fac. Agr. Kyushu Univ.* **15**, 311–330.

Uchida, T. (1969b). Rat-control procedures on the Pacific Islands, with special reference to the efficiency of biological control agents. II: Efficiency of the Japanese weasel, *Mustela sibirica itatsi* Temminck and Schlegel, as a rat-control agent in the Ryukus. *J. Fac. Agr. Kyushu Univ.* **15**, 355–385.

Van Wyngaarden, A. and Bruigns, M. F. M. (1961). De Hermlynen, *Mustela erminea* L. van Terschelling. *Lutra* **3**, 35–42.

Vaughan, H. E. N., and Vaughan, J. A. (1968). Some aspects of the epizootiology of myxomatosis. *Symp. Zool. Soc. London* **24**, 289–309.

Warner, R. E. (1968). The role of introduced diseases in the extinction of the endemic Hawaiian Avifauna. *Condor* **70**(2), 101–120.

Warren, G. L. (1970). Introduction of the Masked Shrew to improve control of forest insects in Newfoundland. *Proc. Tall Timbers Conf. Ecol. Anim. Contr. Habitat Manage* **2**, 185–202.

Washino, R. K. (1969). Progress in biological control of mosquitoes-Invertebrate and vertebrate predators. *Proc. Pap. Annu. Conf. Calif. Mosq. Contr. Ass.* **37**, 16–18.

Williams, R. T. (1971). Observations on the behaviour of the European rabbit flea, *Spilopsyllus cuniculi* (Dale), on a natural population of wild rabbits, *Oryctolagus cuniculus* (L.), in Australia. *Aust. J. Zool.* **19**, 41–51.

Williams, R. T. (1972). The distribution and abundance of the ectoparasites of the wild rabbit, *Oryctolagus cuniculus* (L.), in New South Wales, Australia. *Parasitology* **64**, 321–330.

Williams, R. T. (1973). Establishment and seasonal variations in abundance of the European rabbit flea, *Spilopsyllus cuniculi* (Dale), on wild rabbits in Australia. *J. Entomol.* **A 48**, 117–127.

Williams, R. T., Fullagar, P. J., Davey, C. C., and Kogan, C. (1971). Factors affecting the survival time of rabbits in a winter epizootic of myxomatosis at Canberra, Australia. *J. Appl. Ecol.* **9**, 399–410.

Williams, R. T., and Parer, I. (1971). Observations on the dispersal of the European rabbit flea, *Spilopsyllus curiculi* (Dale), through a natural population of wild rabbits, *Oryctolagus cunicolus* (L.) *Aust. J. Zool.* **19**, 129–140.

21

BIOLOGICAL CONTROL OF PLANT PATHOGENS

William C. Snyder, G. W. Wallis, and Shirley N. Smith

I. INTRODUCTION

The control of plant pathogens by chemicals can be spectacular. Unfortunately, costs are often uneconomic for the relatively short term of effective control. Furthermore, accumulation of chemical residues can be dangerous to humans and other animals. Steaming and flooding of soil to eliminate pathogens also have shortcomings. All these treatments affect the entire biophase of the soil rather than just the plant pathogens and reinvasion by both desirable and undesirable microorganisms occurs quickly. Some effects of such alterations of ecosystems on plant responses which are phytopathological in nature are discussed by Kreutzer (1965). "Disease trading," in which a dominant pathogen is controlled by soil treatment, and a minor pathogen not controlled becomes the new dominant pathogen, is one such effect. Another, called the "boomerang" effect, involves the disappearance of the dominant pathogen after treatment, which is soon followed by its reappearance in even greater quantities.

Control of plant diseases through genetic resistance of the host is often very successful. However, pathogens can and do mutate. Many of the older varieties listed as resistant or tolerant to a certain disease are now observed to succumb to the disease in the field. There is a different race of *Fusarium oxysporum* Schlecht. f. sp. *vasinfectum* Atk. for each of the cultivated cotton species, upland, Egyptian, and Indian (Armstrong and Armstrong, 1960) and we now have numerous races described of *F. oxysporum* f. sp. *pisi* (F. R. Jones) Snyd. and Hans. (Haglund and Kraft, 1970). The perfect fungi exhibit even greater variations in nature, hence, the wheat breeders are constantly busy keeping ahead of the races of rusts. In the area of Mexico where the perfect stage of *Phytophthora infestans* (Mont.) d. By. exists in nature, potato varieties developed elsewhere for tolerance to late blight come down very quickly with the blight because of presence of many races of the fungus.

As more and more land is intensively cultivated, the buildup and the genetic variation of plant pathogens becomes larger and larger. In many instances, the buildup of soil-borne pathogens is not reversible. Once established, certain of the wilt fungi remain in the soil for at least 20 years and a susceptible crop planted in soil once contaminated will invariably succumb again. Such establishment of pathogens in soil constitutes a kind of pollution of one of our most valuable resources.

Because of its potential in solving some of these problems, biological control methods have been investigated in the field of plant pathology. A detailed review has recently been published by Baker and Cook (1974).

II. BIOLOGICAL CONTROL OF PLANT PATHOGENS IN NATURE

Biological control of plant pathogens has been observed under field conditions as well as in laboratory and greenhouse experiments. Many types of biotic agencies are known.

A. Viruses

Plants may become infected with viruses which are only weakly pathogenic to them but which at the same time protect them from more virulent, closely related viruses. Stout (1950) discussed the finding of strains of peach mosaic virus which caused only very slight symptoms in infested trees, but when these trees were subsequently inoculated with severe strains of the virus, the tree continued to show only slight symptoms. Another example of virus "cross-protection" has been demonstrated by Gäumann (1950) within the X-group of potato viruses. Potato plants can be preinfected with the H-strain virus of the X-group without showing any disease symptoms, while still transmitting H virus to their vegetative progeny. If they are subsequently inoculated with the vigorous N-strain, which produces interveinal necrosis in severe cases, they do not become infected by it; they are protected against all X-viruses.

It was hoped that by prior inoculation with these cross-protecting avirulent virus strains, an effective method of control against severe strains might be obtained. Unfortunately, there are too many disadvantages to this form of protection. In the case of the peach mosaics the protection extends only to local forms. A mild form from one area does not protect against a severe form from a distant area. Furthermore, inoculation of peach trees with a virus, even though mild, causes a disease with concurrent losses, even though in the normal course of events such a tree would never have become infected with a severe peach mosaic strain.

The protecting mild or avirulent viruses do not impart immunity to other viruses except those closely related, and in fact their presence may make the host plant even more susceptible to unrelated viruses. Also, it is possible that an innocuous virus strain may mutate to a virulent one. These factors have thus far prevented practical applications of the cross-inoculation principle in controlling virus diseases.

Phages that attack plant pathogenic bacteria occur in nature and therefore probably afford protection to plants in some instances. Soil close to diseased hosts has been found to be a reservoir for these phages (Stolp, 1956; Cross, 1959; Fulton, 1950).

In addition to phages, Stolp and Petzold (1962) have reported tiny "microbacteria" that are obligate parasites and lyse bacteria in soil, including some that are plant pathogenic.

Recently, Boyd *et al.* (1971) reported that crown galls produced on tomato plants were smaller if an *Agrobacterium tumefaciens* (E. F. Sm. and Town.) Conn. bacteriophage was present in the plant. However, only when the phages were absorbed 12 hours prior to bacterial inoculation was the gall growth retarded.

Although there have been numerous reports concerning viruses associated with fungi, until recently plant pathogenic fungi were seldom implicated. Zoospores of the water mold, *Olpidium brassicae* (Woron.) Dang., infected with tobacco necrosis virus (TNV), were found to encyst on and enter roots of lettuce plants and carry TNV into the plants with them. Lettuce plants were subsequently infected with the virus by this means of transmission (Teakle, 1962). Yarwood (1971) and Nienhaus (1971) have found that certain members of the Erysiphaceae carry tobacco mosaic virus and that their conidia can transmit the virus to host plants. The effect of this virus on the powdery mildew is unknown.

French workers (Lemaire, *et al.*, 1970) have implicated the presence of a virus in *Ophiobolus graminis* Sacc. with the decrease in pathogenicity of that parasite and suggest that this fact might be useful for biological control.

So far, practical use of viruses for biological control has not been achieved. More research is necessary to determine effective techniques that can be employed for their use, but up to the present, attempts to control plant diseases by application of phages or other viruses have been disappointing.

B. Fungi

In addition to their direct action as pathogens of higher plants, certain fungi have long been known to be associated with other fungi and other plant pathogens or parasites in a variety of ways. Potentialities for use of these organisms as biological control agents are discussed below.

1. Fungi Parasitic on Plant Pathogenic Fungi

The terms hyperparasite and mycoparasite are used interchangeably to refer to the parasitism of one fungus by another. Unlike antibiotic-producing microorganisms, mycoparasites do not initiate their parasitic activity at a distance but require an intimate association of the host and parasite. Some mycoparasites obtain nutrients from living host cells, causing little apparent harm; this is "balance" or "biotrophic" parasitism (Barnett, 1963). When the action of the parasite kills the host, the phenomenon is called "necrotrophic" parasitism. It is this latter mode of action which is of most interest to those concerned with the control of plant pathogens. The nature of mycoparasitism is outlined by Barnett (1963) and the topic comprehensively reviewed by DeVay (1956) and Madelin (1968).

Mycoparasitism of plant pathogens has been known for many decades. Hino and Kato (1929) reported a *Ciccinobolus* sp. parasitizing *Oidium* species.

Parasitism by *Trichoderma,* as described by Weindling (1932), typifies the "hyphal interference" phenomenon now known to be common in a range of coprophilous fungi (Ikediugwu and Webster, 1970) and as the means by which *Peniophora gigantea* (Fr.) Mass. parasitizes *Fomes annosus* (Fr.) Cooke (Ikediugwu *et al.*, 1970). This phenomenon implies an encircling or coiling around and constriction of the parasitized hyphae, followed by penetration, lysis of the hyphal membrane, and disintegration of the protoplasm.

Numerous plant rusts also serve as hosts for a variety of mycoparasites. *Eudarluca caricis* Erik. in both the perithecial and pycnidial [= *Darluca filum* (Fr.) Cast.] stages attacks many rusts (Eriksson, 1966). Fruiting occurs for the most part in the uredosori but also may be found in the aecidia and teleutosori.

Tuberculina maxima Rostr. is a known mycoparasite of first the aecia and later the pycnia of *Cronartium* spp., particularly *Cronartium ribicola* Dietr., the white pine blister rust. A number of workers have attributed some degree of control to the mycoparasite but the ability of the American strain to inactivate cankers has not been encouraging. More virulent strains of *T. maxima* are being sought from those occurring in Europe and Asia.

Numerous mycoparasites are harbored in the soil. *Dactylella spermatophaga* Drechsl. and *Trinacrium subtile* Drechsl. produce hyphae capable of penetrating the walls of oospores of root-rotting fungi, forming internal haustoria, and significantly reducing their numbers (Drechsler, 1938). As well as through the production of inhibitory amino acids, Siegle (1961) found that *Didymella exitialis* (Mor.) Müller effectively destroyed *Ophiobolus graminis* by directly penetrating and killing the hyphae. Boosalis (1956) and Butler (1957) noted that the rate and intensity of infection of host hyphae by *Rhizoctonia solani* Kuehn was profoundly affected by temperature and nutrition. Parasitism of *R. solani* by *Penicillium vermiculatum* Dang. was severe when the concentration of dextrose was high; the susceptibility of the host was apparently increased in the high dextrose concentration (Boosalis, 1954, 1956). Parasitism of *Armillaria mellea* (Fr.) Kumm. by *Trichoderma* spp. was substantially less when the pH of the medium was adjusted below 5.1 and parasitism was inhibited at pH 7 (Aytoun, 1953).

From the foregoing rather extensive studies it is apparent that, although the use of mycoparasites in the biological control of plant parasites would appear promising, successes have been extremely limited. What appeared in the first instances to be a simple association of microorganisms, has turned out, in many cases, to be a very complex symbiotic association extremely difficult to manipulate.

For reasons best stated by Wood and Tveit (1955) following their work on antagonists, the use of mycoparasites to control pathogens of aerial plant parts must remain highly speculative. To be successful the mycoparasite would require an extremely high reproductive capacity, persistence under unfavorable environmental and nutritional conditions, and extreme aggressiveness.

Although the soil environment is more amenable to manipulation than is that aboveground, the level of control of root pathogens achieved with mycoparasites has been most disheartening. Successes achieved in the laboratory have usually met with failure when applied to the field or have proved impractical on a commercial basis. Until our knowledge of conditions governing the actions and interactions of the extremely complex soil microorganism community is múch more advanced, the possibility of controlling root parasites with mycoparasites will undoubtedly remain only an interesting possibility. Intensive study of these relationships in undisturbed soils could be enlightening.

2. Fungi Predacious on Nematodes

Predacious fungi form a well-defined ecological group in the soil. For the most part, they fall into the Zoopagales and the Hyphomycetes; the predacious nematophagous fungi occur primarily in the latter group.

Nematophagous fungi were known as early as 1877 but until recent years have received little attention. A number of trapping structures have been studied, probably the most common being the adhesive network as found in *Arthrobotrys oligospora* Fres. (Drechsler, 1937). The sticky, short lateral branches typical of *Dactylella lobata* Duddington (Duddington, 1951) and knobs of *Monacrosporium ellipsosporum* Grove (Drechsler, 1937), however, also appear quite effective in trapping the prey. The most complex of the trapping devices studied to date are the nonconstricting rings as formed in *Dactylaria candida* (Nees) Sacc. and constricting rings found in *Dactylaria bembicoldes* Drechs. (Drechsler, 1937; Duddington, 1968). The action of these rings is mechanical rather than sticky. Nonconstricting rings depend on the forward movement of the nematode after it enters the ring to wedge it in a position from where it cannot escape. The constricting ring, on the other hand, captures the nematodes after they enter the ring by an extremely rapid swelling of the ring cells, grasping the nematode so firmly that there is no possibility of escape. Sticky spores and ingestion of spores by the host are two other common ways in which nematophagous fungi gain a foothold in or on the host. Dollfus (1946) reported Phycomycetes to attack helminth eggs in soil and Ellis and Hesseltine (1962) noted Rhopalomyces parasitizing nematode eggs.

Unlike the Zoopagales, the nematode-trapping Hyphomycetes grow vigorously and sporulate abundantly in the absence of nematodes. Their numbers in the soil appear to be correlated with the size of the nematode population (Linford, 1937; Duddington, 1956), and ensure the balance of nature.

C. Insects Predacious on Fungus Parasites and Nematodes

Many insects are predacious on plant pathogenic fungi but the possibility of biological control by this means has not been encouraging. The inoculum of the smuts, rusts, and mildews may be reduced by feeding Coleoptera and Diptera. The larvae and adults of *Phalacrus caricis* feed on the spores of *Cintractia subinclusa* (d'Aguilar, 1944) and when the insect is present in large numbers it probably helps to significantly reduce the inoculum. In 1922, Grassé observed *Deuterosminthus bicinctus* var. *repanda* devouring conidiophores and conidia of *Plasmopara viticola* Berl. and de Toni, the grape downy mildew.

The predation by insects on plant parasitic nematodes has been reported by a number of workers (Hutchinson and Streu, 1960; Murphy and Doncaster, 1957; Brown, 1954). Gilmore (1970) has shown that where populations of nematodes are high, a Collembolan may consume one every $4\frac{1}{2}$ seconds and the effect in reducing the nematode population, when considered in conjunction with other parasitizing soil agents, may be significant.

D. Nematodes Parasitic on Plant Parasitic Nematodes

In some California citrus orchard soils *Thornia* sp. were observed feeding on *Tylenchus semipenetrans* Cobb larvae and *Aphelenchus avenae* Bastian (Boosalis and Mankau, 1965). Although unable to prove it statistically in the laboratory, the authors were of the opinion that under certain conditions predation by *Thornia* sp., when combined with the feeding of other parasites, may account for a significant degree of natural control of citrus nematodes.

Opinions regarding the use of predacious nematodes in the control of plant parasitic nematodes are based, for the most part, on observations. This most interesting field has the potential for yielding results of great practical value but still remains largely unexplored.

E. Microbial Antagonism

Bacteria, actinomycetes, and fungi that produce substances antibiotic to plant pathogens *in vitro* have been isolated from soil. One of the earliest reports was that of Sanford (1926) who suggested that control of potato scab obtained by plowing under a green rye crop was due to the antibiotic qualities of certain predominant soil bacteria enhanced by the green rye residue. He arrived at this concept because cultures obtained from the soil were inhibitory to *Streptomyces scabies* (Thaxt.) Waks. and Henrici, some by substantially lowering the pH and others without altering the pH. Since that report, there have been many others demonstrating that cultures or culture filtrates inhibited plant pathogens. Many of the early papers dealt with antibiosis of Actinomycete cultures or culture filtrates on petri plates. Such antibiosis was demonstrated against *Streptomyces scabies* (Millard and Taylor, 1927), *Colletotrichum lindemuthianum* (Sacc. and Magn.) Briosis and Cav. (Alexopoulos *et al.*, 1938), *Fusarium oxysporum* f. sp. *cubense* E. F. Sm. (Meredith, 1944), *Pythium arrhenomanes* Drechs. (Cooper and Chilton, 1950), and other pathogens. Sanford and Broadfoot (1931) demonstrated that pathogenicity of *Ophiobolus graminis* may be modified and at times controlled by co-culture association with several soil-inhibiting bacteria, fungi, and actinomycetes in greenhouse experiments. They recognized that the extent of effectiveness of this control under field conditions would depend on environmental factors, including availability of nutrients, and that the extreme effects observed in the greenhouse may have been due to the use of sterilized soil.

It was observed by Warcup (1955) that although some fungi existed in soil as hyphal fragments or sclerotia, the majority of the colonies obtained on an agar dilution plate were derived from spores. Garrett (1956) suggested that although sporing fungi are favored on dilution plates the mycelial forms which are not isolated are active in the soil. The observation that spores incorporated into soil from laboratory cultures did not germinate readily led to the discovery of "widespread fungistasis in soil." Dobbs and Hinson (1953) described an inhibitory factor in soils which prevented spores of "sugar fungi" (those unable to utilize cellulose or lignin) from germinating. However, sterile mycelia, some with clamp connections, grew on or near the buried cellophane. The identity of the fungistatic factors in soil is still a matter of controversy. Dobbs and Hinson believed that the inhibition factors differed from antibiotics produced by microorganisms, the presence of which is stimulated by carbohydrates. In their experiments, when glucose as dilute as 0.1% was added to the soil, germination of the spores occurred. The inhibition effect was not removed by repeated leaching or percolation and the water used did not acquire the effect.

The production of antibiotics which were antagonistic to *Pythium* species was detected in unsterilized soil amended with organic matter by Gregory *et al.*, (1952a,b). They also temporarily controlled damping-off in natural soil infested with *Pythium* plus antagonistic organisms, *Penicillium patulum* Bain, *Trichoderma lignorum* Harz, and a Streptomycete, but the control rapidly declined after 10 days. This finding supports Garrett's statement (1965), pertaining to the proposal of soil inoculations with antagonistic organisms as a means of biological control. He contends that microflora of a soil are selected from among the species currently available by environmental conditions. By artificially augmenting the population of a species already present, the balance is upset only temporarily.

Most of the antibiotics synthesized by *Streptomyces* and *Bacillus* species were short-lived in the soil, but phytotoxic Actidione and Fradicin remained for at least 14 days (Gregory *et al.*, 1952b).

Siminoff and Gottlieb (1951) concluded that basic antibiotics such as streptomycin do not play an active biological role in soil because they are strongly absorbed by the soil colloidal complexes.

According to Brian's (1957) discussion of the ecological significance of antibiotic production in normal soils the shortage of carbon sources limits the production of antibiotics to soils which are heavily enriched by organic matter, and then only small amounts may be present. Not only are the antibiotic producers inhibited, but biodegradation quickly occurs. He later allows (1960) that there is evidence for antibiotic production in localized environments, such as on fragments of plant material, seed coats, or in the rhizosphere. A fungus able to produce such substances in food substrates will be able to reduce the growth rate of susceptible organisms. This becomes an aspect of competition for foodstuffs. Resistance to antibiotics is a factor determining success in competition. He also suggests that complex antifungal materials, such as certain polypeptides and polyenic antibiotics, might be involved in soil inhibitions.

Griffin (1964) questioned the importance of antibiotic production. He observed that soil-borne organisms need not be antagonistic in culture or produce specific antibiotic substances to cause fungistasis of *Gliocladium fimbiatum* Gilman & Abbott conidia and hyphae. He postulated that ordinary culture metabolites and accumulating staling products are fungistatic.

Lingappa and Lockwood (1961) failed to demonstrate any antifungal antibiotics, volatile or nonvolatile, in soil extractions. In fact soil extracts were stimulatory rather than inhibitory. They concluded that the soil was not a reservoir for such substances, but that they could exist in microenvironments. They also thought that the spores may act as substrate for bacteria and actinomycetes because a small amount of nutrient leaks out. They point out that spores and hyphae often lyse in soil.

F. Resting Structures of Fungi in Soils

In most of the earlier work on antagonism and fungistasis, conidial germination or hyphal growth was measured and often saprophytic fungi were used. When pathogens were used the inoculum came from spores or mycelium from laboratory cultures. In many plant pathogenic fungi, conidia may play a role in dispersal of the fungus and the ease of their formation in culture tubes has contributed to their use in laboratory studies, but the principal surviving and infecting structure in nature is a less vulnerable body. Such structures as chlamydospores, resting sporangia, oospores, and sclerotia survive in nature. They not only germinate less readily than conidia, or sporangiospores, but they are more difficult to kill by chemical treatments.

In the genus *Fusarium*, only species which are capable of forming chlamydospores survive in soil (Jackson, 1957; Nash *et al.*, 1961). Other *Fusaria* produce ascospores or sporodochia under moist conditions, on above-ground substrates, and are air- or water-borne. Conidia of soil-borne *Fusaria* when placed in moist soil soon convert to the double-walled chlamydospore. These resting structures germinate less readily than conidia do, needing both a nitrogen and a carbon source (Cook and Schroth, 1965). In soil, carbon is more apt to be in short supply than is nitrogen, but carbon and nitrogen compounds are exuded from most roots and released from decomposing plant residues. The bean foot and root rot organism, *Fusarium solani* (Mart.) Sacc. f. sp. *phaseoli* (Burke.) Snyd. and Hans., attacks hypocotyls in preference to roots, even though hypocotyls secrete less amino acids. Therefore, in nitrogen-poor soils, germination of chlamydospores near hypocotyls is slow (Toussoun and Snyder, 1961), except if inorganic nitrogen is supplied to the soil (Weinke, 1962; Maurer and Baker, 1965).

Cook and Snyder (1965) found that although chlamydospores contiguous to bean seeds in soil germinated more rapidly than those contiguous to the below-ground hypocotyl, their hyphae also lysed rapidly. Lysis of germ tubes also occurred rapidly in soil in the presence of amino acid and sugars, which are exuded in soil surrounding germinating seeds. Although both carbon and nitrogen are needed by the germinating chlamydospore, the more nitrogen that is present, the sooner lysis occurs. This suggests that the phase between germination of the chlamydospore and infection is most critical, the organism having changed from a resistant, inactive state to a vulnerable, active state where it is subject to lysis. If the pathogen's germ tube is lysed after the reserve energy of the resting structure is exhausted but before it can either form a parasitic relationship with a host or make a replacement resting structure, a kind of biological control is achieved.

Burke (1965) found that *F. solani* f. sp. *phaseoli* formed only very few tiny chlamydospores in certain soils in the Columbia River basin. Consequently, low pathogen populations and little bean root rot occurred in these fields in spite of the long history of bean crops. Ford *et al.* (1970) suggested that distinct chlamydospore-inducing substances are elaborated by soil bacteria. In addition to the chlamydospore-inducing substance the number of chlamydospores formed depends on an energy source such as glucose.

Fusarium oxysporum Schlecht. forms chlamydospores more easily than the bean root rot fungus, and is capable itself of synthesizing chlamydospore-inducing substance (Ford *et al.*, 1970). Fusarium wilts, however, do not establish in all soils, even though chlamydospores form in many soils in which wilts are slow to develop. In these soils, the chlamydospores of the "formae speciales" which cause wilts germinate and grow less easily in response to nutrients than in the soils which are prone to fusarium wilts (Smith and Snyder, 1972). Thus, there are soil types that are difficult for establishment of *Fusarium solani* f. sp. *phaseoli*, a cortical rot organism, and for *Fusarium oxysporum* formae speciales which cause wilting. The mechanisms involved appear to be different, but both may involve other soil flora and represent a natural biological control.

Thielaviopsis basicola (Berk. and Br.) Ferraris forms chlamydospores in plant material and soil (Patrick *et al.*, 1965) that require an ageing period for germination and that the soil be moist for several days before they germinate (Linderman and Toussoun,

1968). A few chlamydospores (10–15%) may germinate when kept for weeks in saturated natural soil without nutrients, but the germ tubes are thin and light staining as compared with the robust hyphae produced when 10% carrot juice is present.

Another type of resting structure formed by fungi is the sclerotium, which is a compact mass of fungal hyphae varying greatly in size. It may or may not produce a rind. They are produced on or within the host or saprophytically on plant refuse. Some of the larger ones are carpogenic (Garrett, 1956), i.e., they produce a fruiting body and ascospores which attack fruit and flowers. Usually light is required for their fruiting. The microsclerotia of *Macrophomina* are myceliogenic, but *Verticillium* microsclerotia are capable of producing conidiophoric mycelium. The smaller sclerotia often are produced by fungi attacking the below-ground parts of plants (Garrett, 1956). Some sclerotia germinate repeatedly and some only once, some require specific stimulants for germination, and some germinate with nonspecific nutrients. Soil is the reservoir of all sclerotia (Coley-Smith and Cooke, 1971).

Inorganic nitrogen compounds in the soil reduce infection by *Sclerotium rolfsii* Sacc. According to Henis and Chet (1968) ammonia affects sclerotial viability, but other nitrogenous compounds and amendments act by increasing antagonistic activity. Boyle (1961) found that soil organic matter appeared to be necessary for a parasitic relationship to occur. Linderman and Gilbert (1969) found that the volatile components of an alfalfa hay, which was used as a soil amendment, stimulated the sclerotia to germinate. Such compounds as acetaldehyde, ethanol, and methanol derived from alfalfa residues also stimulated germination of *Verticillium* microsclerotia (Owens *et al.*, 1969). A significant reduction in *Verticillium* wilt can be achieved by removal of much of the crop residue before it is added to the soil organic matter, and by the use of inorganic nitrogenous fertilizers.

Sclerotium cepivorum Berk. which attacks members of the genus *Allium*, is an example of a sclerotium which germinates only once and requires very specific stimulants for germination (Coley-Smith and King, 1969)—certain volatile alkyl sulfides. King and Coley-Smith (1969) showed that intact *Allium* plants exude into soil small quantities of alkyl cysteine sulfoxides, which are thermostable, water soluble, and diffusible, but are nonstimulatory to the sclerotia. These compounds are broken down by soil bacteria to yield the stimulatory mixture of volatile alkyl sulfides.

Microsclerotia of *Verticillium albo-atrum* Reinke and Berth. germinate in the presence of roots and the basic portion of root exudates of host plants such as tomatoes and, to a lesser extent, those of such plants as wheat (Schreiber and Green, 1963) and resistant varieties of peppermint (Lacey and Horner, 1966), whose vascular systems are not invaded. Moreover, their population can increase in the presence of these wilt-resistant crops because they are able to form new microsclerotia in the roots. They also continue to increase in the crop residue of these infected nonsusceptible plants (Martinson and Horner, 1962; Lacey and Horner, 1966).

Green and Papavizas (1968) found that *Verticillium albo-atrum* microsclerotia in soil responded to available carbon sources such as glucose by germination and the production of conidia. Such conidia can readily infect host roots. However, the rapid increase in population due to the conidia may soon fall again resulting in a net reduction of the pathogen propagules if there is not sufficient energy for production of secondary resting propagules. Nitrogen compounds or organic amendments little affect germination and survival of microsclerotia. Wilhelm (1951) found that nitrogen and barley straw amendments did reduce *Verticillium* wilt in tomato plants in greenhouse tests. Green and Papavizas (1968) concluded that this effect may have been more on host response to the pathogen than on the pathogen population itself.

G. Fungal Pathogens Which Grow Extensively through the Soil

Certain Basidiomycete pathogens have more mobility in soil than the fungi so far discussed. Blair (1943) demonstrated the ability of *Rhizoctonia solani* Kuehn hyphae to make continued growth through soil. It is capable of infecting only susceptible root tips and grows indiscriminately through soil and over the mature roots with which it comes into contact (Garrett, 1956). Mycelium of such other plant parasites as those which develop on woody hosts typically do not grow through soil as single hyphae, but instead form into mycelial sheets or strands or into rhizomorphs (Garrett, 1956, 1970). Strands develop most commonly acropetally by an ensheathing growth of branches of the parent hyphae. Rhizomorphs are morphologically akin to the roots of higher plants in that they grow from an apical meristem.

It has been demonstrated that a critical level of inoculum potential must be present for infection of roots to occur. Whereas this is met in most instances through single hyphal development on the small, simple roots of herbaceous plants, the large complex cork-covered woody roots require the greater inoculum potential associated with the sophisticated fungal aggregate strands and rhizomorphs for infection to be successful.

The aggregation of hyphae into rhizomorphs has also enabled these fungi to grow freely through the soil from a food base. Rhizomorphs of *Armillaria mellea* (Fr.) Kumm. 30 feet in length have been measured by Findlay (1951); Napper (1938) recorded rhizomorph growth of *Fomes lignosus* (Klotzsch) Bres. epiphytic on rubber tree roots 15 feet ahead of the last point of bark penetration. The radius of spread of rhizomorphs, however, is probably considerably longer than the distance at which they can successfully infect a host, for they must remain attached to their food base until they have successfully invaded another nutritionally suitable substrate or they will perish.

H. Fungal Pathogens with Limited Longevity in Soil

Besides the soil-borne fungi which are either (1) capable of forming long-lived resting structures in soil and germinate in response to host exudates, or (2) active saprophytes which grow extensively through soil from a food base, there are some pathogens which are limited in these capabilities. Their mycelium is harbored in plant residues and as the residues decay the amount of surviving inoculum falls below the level which affects the next susceptible crop. These pathogens, discussed by Glynne (1965), are the most easily controlled by rotations with nonsusceptible crops.

The eyespot fungus, *Cercosporella herpotrichoides* Fron, survives as mycelium on infected straws. During cool moist weather the straws brought to the soil surface in plowing produce conidia profusely. This inoculum is spread by splashes of rain. Therefore, the extent of eyespot depends on the amount of surface inoculum present and the weather. A 3-year interval under grass or alfalfa sufficed to make eyespot unimportant although it was not altogether eliminated.

Take-all, caused by *Ophiobolus graminis* Sacc., is usually controlled by just one year free from susceptible cereal crops. The fungus survives on infected root pieces in soil, which act as a food base prior to attacks on seminal roots of the next crop (Garrett, 1956).

Where consecutive wheat crops are grown, the disease is most severe on the second and third successive crops. The fourth, fifth, etc., crops after a break showed less take-all and larger yields than second or third crops, but yielded less than the first. The factors influencing this partial control are still obscure. Observations suggest that there is also a similar though smaller decline in eyespot and there is evidence that the cereal root eelworm, *Heterodera avenae* Woll., may increase and then decrease in successive oat crops (Glynne, 1965).

The importance of a fungal pathogen in utilizing a substrate such as straw residue in soil has been emphasized in *Cephalosporium gramineum* Nisikado and Ikata (Bruehl and Lai, 1968; Bruehl, 1970). This fungus must obtain a new lease on life every 2 or 3 years through parasitism or it is doomed to reduction to a harmless level. It persists in straw either dormant, when it is too dry to grow, or by production of small amounts of a wide-spectrum antibiotic when conditions favor its growth. Isolates that produce antibiotics possess straw more strongly, sporulate more profusely, and persist longer in straw buried in soil than do similar isolates incapable of antibiotic production.

III. CONCLUSIONS—THE COMPLEXITY OF RELATIONSHIPS AND FUTURE POSSIBILITIES

Changing crop sequences and the use of cover crops have long been recognized as beneficial in terms of yield of crops which follow. Some or perhaps most of the benefit may be attributed to the improved health of roots. The buildup of root pathogens is interrupted by such practices and the quality and quantity of elements of microbial populations are shifted drastically toward more natural balance. One of the major research goals in root disease research has been to better understand what the changes are and how to use such cultural practices as crop rotation and cover cropping knowledgeably. Some crop rotations may increase disease, i.e., a barley rotation may increase potato scab, while soybean may decrease it (Weinhold *et al.*, 1964). If *Rhizoctonia* is severe on a crucifer crop, little or no benefit is to be expected from planting another susceptible crucifer in rotation. A short rotation may accomplish the objective of lowering the population level of a pathogen, while in other cases it would not. *Ascochyta* blight pathogens of peas may be effectively reduced by 2—4 years of rotation with another crop such as wheat. Even *Fusarium* root rot of cucurbits may be controlled economically by a similar rotation of a noncucurbit crop. However, Panama disease of banana caused by *Fusarium oxysporum* f. sp. *cubense* may persist in the soil by invading grass roots and survive for decades (Waite and Dunlap, 1953) in the absence of its banana host. It is because the results from the use of organic residues have been unpredictable or at times even contradictory that plant pathologists have sought to unravel the principles involved through research.

Considerable stimulation for research in this area has come from the formation of active research organizations such as the Pacific Coast Research Conference on Soil Fungi (Soil Fungus Conference) created in 1953, the Western Regional Project No. 38 (Cook and Watson, 1969) on the nature of the influence of crop residues on soil-borne, fungus-induced root diseases initiated in 1955, and the International Symposium on the Ecology of Soil-Borne Plant Pathogens—Prelude to Biological Control (Baker and Snyder, eds., 1965). All these conferences remain active, and a second international symposium was held in 1968 (Toussoun *et al.*, 1970). Although a great deal of research has been published here and abroad in the past 15 years, and pathologists are able to advise growers much more knowledgeably on how to avoid disastrous root disease losses, much more research is required to evolve the principles involved in the biological control of root diseases. Other regional organizations have also been established to this general aim, such as S-26, NC-70, NE-45, and an Eastern Soil Fungus Conference.

Diverse mechanisms may be involved in the biological control of root diseases. Although direct predation of the pathogen may take place, as in the biological control of one insect by another, indirect phenomena are probably more common to plant pathogen control. For example, the competition for nutrients from the soil solution may deprive the pathogen of the energy it needs to germinate and infect living host tissues. In this latter case, exemplified by the control of *Fusarium* root rot of bean (above) by a barley crop, rotation is probably a more common and effective control mechanism than, for example, the direct invasion of *Rhizoctonia* hyphae by *Trichoderma* or other organisms. Antibiotics produced by one organism such as subtilin by *Bacillus subtilis* (Cohn) Praz may be operative against organisms such as the pathogen *Streptomyces scabies* (Thaxt.) Waks. and Henrici (Weinhold and Bowman, 1968), or may biologically protect the generative organism in the soil habitat as with *Cephalosporium* in invaded wheat straw (Bruehl, 1970). Biological control of soil-borne pathogens thus may be direct such as by predation, or indirect through a chain of biological events, but by definition must involve the biological activity of one or more organisms acting to control another.

In order to evaluate the phenomena of biological control obtained by a bean-barley rotation, in the economic control of root rot of bean, detailed studies were necessary on the interaction in field soils of the bean and barley plants, the *Fusarium* pathogen, the soil environment, and the associated microflora (Snyder *et al.*, 1959b).

It has been supposed earlier that fungus pathogens were growing through the soil in search of their hosts. It was some 20 years ago that Dobbs and Hinson (1953) showed that to the contrary, most fungi were in a spore stage in the soil, quiescent, and that it was in the presence of nutrients that spores germinated and resumed active growth. Different pathogens show different nutrient requirements for their germination. The fungistatic inhibition of germination of resting spores (chlamydospores) of *F. solani* f. sp. *phaseoli* is overcome by nonspecific carbohydrate and nitrogenous substances such as sugars and amino acids. Resting structures of other pathogens, such as *Scelerotium cepivorum* Berk., are rather specific in their nutrient requirements, in this case requiring volatile extracts or exudates derived from its onion host in order for the sclerotia to germinate in soil.

Nutrients serve more than just to germinate resting structures of fungus pathogens lying dormant in soil. It has been shown clearly that the pathogen needs proper nutrition in order to invade living host tissue. Pathogenesis by *F. solani* f. sp. *phaseoli* is favored by ammonium nitrogen (Toussoun *et al.*, 1960). Mycelium of *Rhizoctonia solani* may grow extensively over the surface of the below-ground portion of the hypocotyl yet not invade the tissue unless supplied adequately with nitrogen either from the soil solution or from a base substrate (Weinhold, *et al.*, 1969).

Environmental factors influence all elements of the host–pathogen relationship. Below optimum temperatures for germination of host seeds increases the exudation of nutrients essential to germination and vegetative growth of the pathogen (Hayman, 1970), while high CO_2 levels in the soil may inhibit chlamydospore formation in *Fusarium* but favor their germination (Louvet, 1970; Bourret, *et al.*, 1968). High levels of soil moisture favor some pathogens such as the water molds, but inhibit pathogenesis by *Fusarium roseum* (L. K. ex Fr.) amend. Snyd. and Hans. f. sp. *cereales* (Cke.) Snyd. and Hans. (Cook and Papendick, 1970), and *Streptomyces scabies* (Sanford, 1923). Volatile substances in the soil atmosphere serve either to inhibit growth of soil fungi, stimulate them, or both (Coley-Smith and King, 1969; Owens, *et al.*, 1969). Fusarium wilts are usually favored by light textured soils but usually not by heavy soils (Stover, 1962).

In essence, the attempt is made in biological control to keep the propagule population of the pathogen as low in the soil as possible, and/or to prevent them from germinating and functioning. Thus the biology of the pathogen in total relationship to the soil microbiology with which it is associated in the field must be understood—hence, the recent research emphasis on soil microbiology in the field.

Some generalizations may be made at this point. Most pathogenic fungi are at rest in the soil in a resistant structure, the germination of which is inhibited by the soil environment. Nutrients of a nonspecific or specific nature are necessary to overcome the inhibitory influences and enable germination. Growth in proximity to the host and penetration of its tissues requires proper nutrition of the germling. Successful penetration and pathogenic invasion of the host plant is influenced by factors of the soil environment such as temperature, moisture, levels of CO_2 and O_2, other volatiles, and the activity and makeup of the soil flora. In addition to these generalizations must be added the complex genetic nature of the pathogen and microbial flora. *Fusarium solani* f. sp. *phaseoli*, for example, may exist in the field in two dozen or more genetically distinct clones (Snyder *et al.* 1959a), each with its own capacity to adjust to the changes taking place in its local environment and survive. The genetic flexibility of these components of the soil flora makes most difficult manipulation of the favorable elements of the flora against the unfavorable elements (the pathogens). This is the nature of the problem of the biological control of plant diseases.

REFERENCES

Alexopoulos, C. J., Arnett, R., and McIntosh, A. V. (1938). Studies in antibiosis between bacteria and fungi. *Ohio J. Sci.* **38**, 221–235.

Armstrong, G. M., and Armstrong, J. K. (1960). American, Egyptian and Indian cotton-wilt Fusaria: their pathogenicity and relationship to other wilt Fusaria. *U.S. Dept. Agr. Tech. Bull.* **1219**, 19.

Aytoun, R. S. C. (1953). The genus *Trichoderma:* its relationship with *Armillaria mellea* and *Polyporus schweinitzii* together with preliminary observations on its woodland ecology. *Proc. Bot. Soc. Edinburgh* **36**, 99–114.

Baker, K. F., and Cook, R. J. (1974). "Biological Control of Plant Pathogens," 433 pp. Freeman, San Francisco, California.

Baker, K. F., and Snyder, W. C. (eds.). (1965). "Ecology of Soil-Borne Plant Pathogens," 571 pp. Univ. California Press, Berkeley, California.

Barnett, H. L. (1963). The nature of mycoparasitism by fungi. *Annu. Rev. Microbiol.* **17**, 1–14.

Blair, I. D. (1943). Behaviour of the fungus *Rhizoctonia solani* Kühn in the soil. *Ann. Appl. Biol.* **30**, 118–127.

Boosalis, M. G. (1954). *Penicillium* sp. parasitic on *Rhizoctonia solani*. *Phytopathology*, **44**, 482 (Abstr.)

Boosalis, M. G. (1956). Effect of soil temperature and green-manure amendment of unsterilized soil on parasitism of *Rhizoctonia solani* by *Penicillium vermiculatum* and *Trichoderma* sp. *Phytopathology* **46**, 473–478.

Boosalis, M. G., and Mankau, R. (1965). Parasitism and predation of soil microorganisms. *In* "Ecology of Soil-Borne Plant Pathogens" (K. F. Baker and W. C. Snyder, eds.), pp. 374–391. Univ. California Press, Berkeley, California.

Bourret, J. A., Gold, A. H., and Snyder, W. C. (1968). Effect of carbon dioxide on germination of chlamydospores of *Fusarium solani* f. sp. *phaseoli*. *Phytopathology* **58**, 710–711.

Boyd, R. J., Hildebrandt, A. C., and Allen, O. N. (1971). Retardation of crown gall enlargement after bacteriophage treatment. *Plant Dis. Rep.* **55**, 145–148.

Boyle, L. W. (1961). The ecology of *Sclerotium rolfsii* with emphasis on the role of saprophytic media. *Phytopathology* **51**, 117–119.

Brian, P. W. (1957). The ecological significance of antibiotic production. *Symp. Soc. Gen. Microbiol.* **7**, 168–188.

Brian, P. W. (1960). Antagonistic and competitive mechanisms limiting survival and activity of fungi in soil. *In* "The Ecology of Soil Fungi" (D. Parkinson and J. S. Waid, eds.), pp. 115–129. Liverpool Univ. Press, Liverpool.

Brown, W. L., Jr. (1954). Collembola feeding upon nematodes. *Ecology* **35**, 421.

Bruehl, G. W. (1970). Nature of the influence of crop residues on fungus-induced root diseases. Factors affecting the persistence of fungi in soil. *Wash. Agr. Exp. Sta. Bull.* **716**, 32 pp.

Bruehl, G. W., and Lai, P. (1968). The probable significance of saprophytic colonization of wheat straw in the field by *Cephalosporium gramineum*. *Phytopathology* **58**, 464–466.

Burke, D. W. (1965). Fusarium rot of beans and behavior of the pathogen in different soils. *Phytopathology* **55**, 1122–1126.

Butler, E. E. (1957). *Rhizoctonia solani* as a parasite of fungi. *Mycologia* **49**, 354–373.

Coley-Smith, J. R., and Cooke, R. C. (1971). Survival and germination of fungal sclerotia. *Annu. Rev. Phytopathol.* **9**, 65–92.

Coley-Smith, J. R., and King, J. E. (1969). The production by species of *Allium* of alkyl sulphides and their effect on germination of sclerotia of *Sclerotium cepivorum* Berk. *Ann. Appl. Biol.* **64**, 289–301.

Cook, R. J., and Papendick, R. I. (1970). Effect of soil water on microbial growth, antagonism, and nutrient availability in relation to soil-borne fungal diseases of plant. *In* "Root Diseases and Soil-Borne Pathogens" (T. A. Toussoun, R. V. Bega, and P. E. Nelson, eds.), pp. 81–88. Univ. California Press, Berkeley, California.

Cook, R. J., and Schroth, M. N. (1965). Carbon and nitrogen compounds and germination of chlamydospores of *Fusarium solani* f. *phaseoli*. *Phytopathology* **55**, 254–256.

Cook, R. J., and Snyder, W. C. (1965). Influence of host exudates on growth and survival of germlings of *Fusarium solani* f. *phaseoli* in soil. *Phytopathology* **55**, 1021–1025.

Cook, R. J., and Watson, R. D. (1969). Nature of the influence of crop residues on fungus-induced root diseases. *Wash. Agr. Exp. Sta. Bull.* **716**, 32 pp.

Cooper, W. E., and Chilton, S. J. P. (1950). Studies on antibiotic soil organisms. I. Actinomycetes antibiotic to *Pythium arrhenomanes* in sugar cane soils of Louisiana. *Phytopathology* **40**, 544–552.

Crosse, J. E. (1959). Plant pathogenic bacteria and their phages. *Commonw. Phytopathol. News* **5**, 17–32.

d'Aguilar, J. (1944). Contribution a l'etude des phalacridae. Note sur *Phalacrus caricis* Sturm. *Ann. Epiphyt.* **10**, 85–91.

De Vay, J. E. (1956). Mutual relationships in fungi. *Annu. Rev. Microbiol.* **10**, 115–140.

Dobbs, C. G., and Hinson, W. H. (1953). A widespread fungistasis in the soil. *Nature (London)* **172**, 197.

Dollfus, R. P. (1946). Parasites des helminthes. *In* "Encyclopedie Biologigue 27" (Paul Lechevalier, ed.), 481 pp. Paris.

Drechsler, C. (1937). Some hyphomycetes that prey on free-living terricolous nematodes. *Mycologia* **29**, 447–552.

Drechsler, C. (1938). Two hyphomycetes parasitic on oospores of root-rotting oomycetes. *Phytopathology* **28**, 81–103.

Duddington, C. L. (1951). *Dactylella lobata*, predacious on nematodes. *Trans. Br. Mycol. Soc.* **34**, 489–491.

Duddington, C. L. (1956). The predacious fungi: zoopagales and moniliales. *Biol. Rev.* **31**, 152–193.

Duddington, C. L. (1968). Fungal parasites of invertebrates. 2. Predacious fungi. *In* "The Fungi, An Advanced Treatise. III. The Fungal Population" (G. C. Ainsworth and Alfred S. Sussman, ed)., pp. 239–251. Academic Press, New York.

Ellis, J. J., and Hesseltine, C. W. (1962). *Phopalomyces* and *Spinellus* in pure culture and the parasitism of *Rhopalomyces* on nematode eggs. *Nature (London)* **193**, 699–700.

Eriksson, O. (1966). On *Eudarluca caricis* (Fr.) O. Eriks., comb. nov., a cosmopolitan uredinicolous Pyrenomycete. *Bot. Notis.* **119**, 33–69.

Findlay, W. P. K. (1951). The development of *Armillaria mellea* rhizomorphs in a water tunnel. *Trans. Br. Mycol. Soc.* **34**, 146.

Ford, E. J., Gold, A. H., and Snyder, W. C. (1970). Interaction of carbon nutrition and soil substances in chlamydospore formation by *Fusarium*. *Phytopathology* **60**, 1732–1737.

Fulton, R. (1950). Bacteriophages attacking *Pseudomonas tabaci* and *P. angulatum*. *Phytopathology* **40**, 936–949.

Garrett, S. D. (1956). "Biology of Root-Infecting Fungi," 293 pp. Cambridge Univ, Press, Cambridge.

Garrett, S. D. (1965). Towards biological control of soil-borne plant pathogens. *In* "Ecology of Soil-Borne Plant Pathogens" (K. F. Baker and W. C. Snyder, eds.), pp. 4–17. Univ. California Press, Berkeley, California.

Garrett, S. D. (1970). "Pathogenic Root-Infecting Fungi," 294 pp. Cambridge Univ. Press, Cambridge.

Gäumann, E. (1950). "Principles of Plant Infection," 543 pp. Crosby Lockwood and Sons, London. (Translated by W. B. Brierly.)

Gilmore, S. K. (1970). Collembola predation on nematodes. *Search-Agriculture* **1**, 1–12.

Glynne, M. D. (1965). Crop sequence in relation to soil-borne pathogens. *In* "Ecology of Soil-Borne Plant Pathogens" (K. F. Baker and W. C. Snyder, eds.), pp. 423–435. Univ. California Press, Berkeley, California.

Grasse, P. P. (1922). Notes sur la biologie d'un Collembole *Hypogastrura armata* (Nicolet). *Ann. Soc. Entomol. Fr.* **91**, 190–192.

Green, R. J., and Papavizas, G. C. (1968). The effect of carbon source, carbon to nitrogen ratios, and organic amendments on survival of propagules of *Verticillium albo-atrum* in soil. *Phytopathology* **58**, 567–570.

Gregory, K. F., Allen, O. N., Riker, A. J., and Peterson, W. H. (1952a). Antibiotics and antagonistic microorganisms as control agents against damping-off of alfalfa. *Phytopathology* **42**, 613–622.

Gregory, K. F., Allen, O. N., Riker, A. J., and Peterson, W. H. (1952b). Antibiotics as agents for the control of certain damping-off fungi. *Amer. J. Bot.* **39**, 405–415.

Griffin, G. J. (1964). Long-term influence of soil amendments on germination of conidia. *Can. J. Microbiol.* **10**, 605–612.

Haglund, W. A., and Kraft, J. M. (1970). *Fusarium oxysporum* f. *pisi*, Race 5. *Phytopathology* **60**, 1861–1862.

Hayman, D. S. (1970). The influence of cottonseed exudates on seedling infection by *Rhizoctonia solani*. *In* "Root Diseases and Soil-Borne Pathogens" (T. A. Toussoun, R. V. Bega, and P. E. Nelson, eds.), pp. 99–102. Univ. California Press, Berkeley, California.

Henis, Y., and Chet, I. (1968). The effect of nitrogenous amendments on the germinability of sclerotia of *Sclerotium rolfsii* and on their accompanying microflora. *Phytopathology* **58**, 209–211.

Hino, I., and Kato, H. (1929). Cicinnoboli parasitic on mildew fungi. *Bull. Miyazaki Coll. Agr. For.* **1**, 91–100.

Hutchinson, M. T., and Streu, H. T. (1960). Tardigrades attacking nematodes. *Nematologica* **5**, 149–150.

Ikediugwu, F. E. O., and Webster, J. (1970). Hyphal interference in a range of coprophilous fungi. *Trans. Br. Mycol. Soc.* **54**, 205–210.

Ikediugwu, F. E. O., Dennis, C., and Webster, J. (1970). Hyphal interference by *Peniophora gigantea* against *Heterobasidion annosum*. *Trans. Br. Mycol. Soc.* **54**, 307–309.

Jackson, R. M. (1957). Fungistasis as a factor in the rhizosphere phenomenon. *Nature* (*London*) **180**, 96–97.

King, J. E., and Coley-Smith, J. R. (1969). Production of volatile alkyl sulphides by microbial degradation of synthetic alliin and alliin-like compounds, in relation to germination of sclerotia of *Sclerotium cepivorum* Berk. *Ann. Appl. Biol.* **64**, 303–314.

Kreutzer, W. (1965). The reinfestation of treated soil. *In* "Ecology of Soil-Borne Plant Pathogens" (K. F. Baker and W. C. Snyder, eds.), pp. 495–508. Univ. California Press, Berkeley, California.

Lacy, M. L., and Horner, C. E. (1966). Behavior of *Verticillium dahliae* in the rhizosphere and on roots of plants susceptible, resistant, and immune to wilt. *Phytopathology* **56**, 427–430.

Lemaire, J. -M., Lapierre, H., Jouan, B., and Bertrand, G. (1970). Discovery of virus particles in certain strains of *O. graminis*, causal agent of take-all of cereals. Anticipated agronomic consequences. *Proc. Acad. Agr. Fr.* **56**, 1134–1137.

Linderman, R. G., and Gilbert, R. G. (1969). Stimulation of *Sclerotium rolfsii* in soil by volatile components of alfalfa hay. *Phytopathology* **59**, 1366–1372.

Linderman, R. G., and Toussoun, T. A. (1968). Predisposition to *Thielaviopsis* root rot of cotton by certain phytotoxins obtained from decomposing barley residues. *Phytopathology* **58**, 1571–1574.

Linford, M. B. (1937). Stimulated activity of natural enemies of nematodes. *Science* **85**, 123–124.

Lingappa, B. T., and Lockwood, J. L. (1961). The nature of the widespread soil fungistasis. *J. Gen. Microbiol.* **26**, 473–485.

Louvet, J. (1970). Effect of aeration and of concentration of carbon dioxide on the activity of plant pathogenic fungi in the soil. *In* "Root Diseases and Soil-Borne Pathogens" (T. A. Toussoun, R. V. Bega, and P. E. Nelson, eds.), pp. 89–91. Univ. California Press, Berkeley, California.

Madelin, M. F. (1968). Fungi parasitic on other fungi and lichens. *In* "The Fungi, An Advanced Treatise. III. The Fungal Population" (G. C. Ainsworth and A. S. Sussman, eds.), pp. 253–269. Academic Press, New York.

Martinson, C. A., and Horner, C. E. (1962). Importance of non-hosts in maintaining the inoculum potential of *Verticillium*. *Phytopathology* **52**, 742 (Abstr.).

Maurer, C. L., and Baker, R. R. (1965). Ecology of plant pathogens in soil. II. Influence of glucose, cellulose, and inorganic nitrogen amendments on development of bean root rot. Phytopathology **55**, 69–72.

Meredith, C. H. (1944). The antagonism of soil organisms to *Fusarium oxysporum cubense*. *Phytopathology* **34**, 426–429.

Millard, W. A., and Taylor, G. B. (1927). Antagonism of microorganisms as the controlling factor in the inhibition of scab by green manuring. *Ann. Appl. Biol.* **14**, 202–215.

Murphy, P. W., and Doncaster, C. C. (1957). A culture method for soil meiofauna and its application to the study of nematode predators. *Nematologica* **2**, 202–214.

Napper, R. P. N. (1938). Root disease and underground pests in new plantings. *Planter* **19**, 453–455.

Nash, S. M., Christou, T., and Snyder, W. C. (1961). Existence of *Fusarium solani* f. *phaseoli* as chlamydospores in soil. *Phytopathology* **51**, 308–312.

Neinhaus, F. (1971). Tobacco mosaic virus strains extracted from conidia of powdery mildews. Virology **46**, 504–505.

Owens, L. D., Gilbert, R. G., Griebel, G. E., and Menzies, J. D. (1969). Identification of plant volatiles that stimulate microbial respiration and growth in soil. *Phytopathology* **59**, 1468–1472.

Patrick, Z. A., Toussoun, T. A., and Thorpe, H. J. (1965). Germination of chlamydospores of *Thielaviopsis basicola*. *Phytopathology* **55**, 466–467.

Sanford, G. B. (1923). The relation of soil moisture to the development of common scab of potato. *Phytopathology* **13**, 231–236.

Sanford, G. B. (1926). Some factors affecting the pathogenicity of *Actinomyces scabies*. *Phytopathology* **16**, 525–547.

Sanford, G. B., and Broadfoot, W. C. (1931). Studies of the effects of other soil-inhabiting microorganisms on the virulence of *Ophiobolus graminis* Sacc. *Sci. Agr.* **11**, 512–528.

Schreiber, L. R., and Green, R. J., Jr. (1963). Effect of root exudgtes on germination of conidia and microsclerotia of *Verticillium albo-atrum* inhibited by the soil fungistatic principle. *Phytopathology* **53**, 260–264.

Siegle, H. (1961). Über mischinfektionen mit *Ophiobolus graminis* und *Didymella exitialis*. *Phytopathol. Z* **42**, 305–348.

Siminoff, P., and Gottlieb, D. (1951). The production and role of antibiotics in the soil. I. The fate of streptomycin. *Phytopathology* **41**, 420–430.

Smith, S. N., and Snyder, W. C. (1972). Germination of *Fusarium oxysporum* chlamydospores in soils favorable and unfavorable to wilt establishment. *Phytopathology* **62**, 273–277.

Snyder, W. C., Nash, S. M., and Trujillo, E. E. (1959a). Multiple clonal types of *Fusarium solani* f. *phaseoli* in field soil. *Phytopathology* **49**, 310–312.

Snyder, W. C., Schroth, M. N., and Christou, T. (1959b). Effect of plant residues on root rot of bean. *Phytopathology* **49**, 744–756.

Stolp, H. (1956). Bakteriophagenforschung und phytopathologie. *Phytopathol. Z.* **26**, 171–218.

Stolp, H., and Petzold, H. (1962). Untersuchungen uber einen obligat parasitischen mikroorganisms mit lytischer aktivitat fur *Pseudomonas*-bakterien. *Phytopathol. Z.* **45**, 364–390.

Stout, G. L. (1950). New methods of plant disease control. *Calif. Dep. Agr. Bull.* **39**, 129–136.

Stover, R. H. (1962). Fusarial wilt (panama disease) of bananas and other *Musa* species. *Commonw. Mycol. Inst. Phytopathol. Pap. No.* **4**, 117 pp.

Teakle, D. S. (1962). Transmission of tobacco necrosis virus by a fungus, *Olpidium brassicae*. *Virology* **18**, 224–231.

Toussoun, T. A., and Snyder, W. C. (1961). Germination of chlamydospores of *Fusarium solani* f. *phaseoli* in unsterilized soils. *Phytopathology* **51**, 620–623.

Toussoun, T. A., Nash, S. M., and Snyder, W. C. (1960). The effect of nitrogen sources and glucose on the pathogenesis of *Fusarium solani* f. *phaseoli*. *Phytopathology* **50**, 137–140.

Toussoun, T. A., Bega, R. V., and Nelson, P. E., (eds.). (1970). "Root Diseases and Soil-Borne Pathogens," 252 pp. Univ. California Press, Berkeley, California.

Waite, B. H., and Dunlap, V. C. (1953). Preliminary host range studies with *Fusarium oxysporum* f. *cubense*. *Plant Dis. Rep.* **37**, 79–80.

Warcup, J. (1955). Studies on the occurrence and activity of fungi in a wheat-field soil. *Trans. Br. Mycol. Soc.* **40**, 237–259.

Weindling, R. (1932). *Trichoderma lignorum* as a parasite of other soil fungi. *Phytopathology* **22**, 837–845.

Weinhold, A. R., and Bowman, T. (1968). Selective inhibition of the potato scab pathogen by antagonistic bacteria and substrate influence on antibiotic production. *Plant Soil* **28**, 12–24.

Weinhold, A. R., Oswald, J. W., Bowman, T., Bishop, J., and Wright, D. (1964). Influence of green manures and crop rotation on common scab of potato. *Amer. Potato J.* **41**, 265–273.

Weinhold, A. R., Bowman, T., and Dodman, R. L. (1969). Virulence of *Rhizoctonia solani* as affected by nutrition of the pathogen. *Phytopathology* **59**, 1601–1605.

Weinke, K. E. (1962). The influence of nitrogen on the root disease of bean caused by *Fusarium solani* f. *phaseoli*. *Phytopathology* **52**, 757 (Abstr.).

Wilhelm, S. (1951). Effect of various soil amendments on the inoculum potential of the verticillium wilt fungus. *Phytopathology* **41**, 684–690.

Wood, R. K. S., and Tveit, M. (1955). Control of plant diseases by use of antagonistic organisms. *Bot. Rev.* **21**, 441–492.

Yarwood, C. E. (1971). Erysiphaceae transmit virus to *Chenopodium*. *Plant Dis. Rep.* **55**, 342–344.

SECTION V

COMPONENTS OF INTEGRATED CONTROL AND ITS IMPLEMENTATION

22

THE IMPORTANCE OF NATURAL ENEMIES IN INTEGRATED CONTROL

P. S. Messenger, E. Biliotti, and R. van den Bosch

I. THE IMPORTANCE OF NATURAL ENEMIES

The ability of any living organism to increase is limited by the action of other living organisms. The level of population of any of them, in addition, is the result of a series of complex interactions.

In the case of insects, we find that their principal antagonists belong to the same zoological class. The entomophagous habit has been developed in 224 families belonging to 15 orders (Clausen, 1940). Every species of phytophagous insect is exploited by a series of parasites and predators. All the developmental stages are subject to attack; apart from general predators, there are species of parasites which specialize on the egg, the larva, the nymph, or the adult. Their ways of finding, selecting, and taking possession of their host are widely diversified, as are their types of reproduction; entomophagous insects constitute by themselves a fascinating world. Many other animals also feed upon insects. They include nematodes, spiders, fishes, frogs, lizards, birds, and mammals. To complete the picture we must add the wide variety of diseases caused by bacteria, fungi, protozoa, and viruses to which insects are subject.

For an entomologist, the only surprising thing is that some species are able to build up very large populations regardless of the pressure exerted by natural enemies.

It may be estimated that the total number of insect species is about 3,000,000 of which only 10,000 are considered economically important (Sailer, 1969) because they cause some type of injury to crops, man, or domestic animals. This number would certainly be very much higher if the natural enemies were missing. However, man does not generally recognize the importance of their action. In fact, the part played by insects is only seen in two situations, i.e., when they are suppressed (generally following man-made disturbance of the environment) and when they are purposely used as a specific means of controlling pests (biological control).

There are many known cases of species becoming pests after the suppression of their natural enemies. One example is that of certain *Lecanium* species on English walnut in California, developed by Michelbacher (1962). These scales presented no problem prior to the advent of DDT and were rarely encountered. They suddenly became major pests of walnuts when their natural enemies were destroyed by the chemical. The fact that these small hymenopterous parasites were able to maintain their host at a low level would probably have escaped detection if the insecticide had not been used.

The most striking example of appearance of new pests after the use of broad-spectrum insecticides is that seen with cotton in many areas of the world, mainly in Central and South America. In most cotton areas one may estimate that more than ten insect species have changed from an innocuous status to that of major pests because of such pesticide interference.

On the other hand, when the use of chemicals is restrained, parasites and predators become more efficient. In the Lauragais area in France where rape is subject to the attack of a complex of pest species, it has been shown (P. Jourdheuil, unpublished data) that a thorough reduction in the use of insecticides resulted in a growing efficiency of natural enemies. Populations of the key pest, *Ceuthorrhynchus assimilis* Payk. were reduced to such an extent by the action of larval parasites that they become insufficient for experimental purposes after 3 years.

The real importance of a natural enemy sometimes escapes attention in the beginning and is only recognized later on. The introduced San José Scale, *Quadraspidiotus perniciosus* Comst., was, for more than 40 years, one of the most important pests in the United States. Its population then decreased under the influence of a parasite *Prospaltella perniciosi* Tow. which very probably introduced itself, perhaps later than the pest (Sailer, 1972). Strains of this wasp coming from the United States, and others from China (through U.S.S.R.), have been introduced in western Europe and a very good control of the scale has been obtained in various areas of Germany, Switzerland and France (Benassy *et al.*, 1968). The spread of the pest in the United States was faster than that of the parasite, but the latter won the battle.

The most striking examples of the efficacy of natural enemies are to be found in the successes of biological control.

The analysis by DeBach (1971) of work done between 1888 and 1969 shows that, during this period, attempts have been made to control 223 species of pests. Among them, 42 have been completely controlled, that is to say that there is no need to undertake any type of treatment against them except in very special cases. Forty-eight other species are under substantial control. The economic savings are somewhat less pronounced because the pest or crop is less important or the crop area is restricted. Occasionally insecticidal treatments are needed. In 30 other cases the success has been only partial in that the chemical control measures are usually necessary though less frequently than before. Taking into consideration the small financial inputs that this represents, a considerable success with a return of 30 to 1 (compared with a ratio of 5 to 1 for chemical control) gives a good idea of the potentiality of the method using parasites and predators.

Another fact is also striking. These successes have been obtained against pest representatives of all the major insect orders, in a great variety of ecological situations, and under all types of climates throughout the world. We must also consider that, if the species of useful insects involved belong to the most important orders of entomophaga Diptera, Hymenoptera, Coleoptera, . . .), they represent a very small part of the total of parasites and predators available (about 200,000 species at least).

We often have a very limited knowledge of the potentialities of a given species. For example, *Chilocorus bipustulatus* (Linn.) is a well known coccinellid which is widely distributed; it is considered a good general predator of coccids, but is unable to control any species by itself. During the last few years a strain of *C. bipustulatus* collected in Iran was introduced in Mauritania and gave a very good biological control of the date palm scale in this country (Iperti *et al.*, 1970).

We also certainly have many things to learn concerning the role of natural enemies of weeds, although the successful control of prickly pear, lantana, hypericum, etc., is proof that other successes are possible.

One should also remember that importation of exotic species is not the only way of utilizing natural enemies; manipulation of indigenous predators, for example, as shown by Ridgway (1969) offers considerable promise.

The value of egg parasites of the genus *Trichogramma* as controlling agents has been frequently discussed, but entomologists of the U.S.S.R. have shown that mass propagation of selected strains well adapted to the different climatic areas provided a good control of various pests. This system has been applied to some 5 millions hectares in 1972 and a special type of factory producing daily 50 million of *Trichogramma* has been developed (Lebedev, 1970 and unpublished data).

The efficacy of natural enemies is often reduced in periods of low population levels of the host and they are unable to prevent a consecutive outbreak. This is frequently found in the "gradation" type of fluctuation of populations of forest pests.

Manipulations of pest numbers could provide means to avoid a pronounced reduction of natural enemies. In Yugoslavia, Maksimovic *et al.* (1970) have shown that, by distributing egg masses of *Porthetria dispar* (L.) in a forest at the period of lowest density of the pest, it was possible to maintain populations of parasites and predators avoiding the pest outbreak which was observed in the control area. Very promising results were obtained in the U.S.A. by periodic liberation of both host and parasites (Parker *et al.*, 1971) in the case of *Pieris*.

Aside from the world of insects, the past 20 years have also given ample evidence that diseases, either introduced in new areas (the use of viruses against sawflies in Canada) or industrially produced (such as *Bacillus thuringiensis* in U.S.A., France, and U.S.S.R.) are able to control pest populations in the field.

It is quite evident that natural enemies are not only one of the most important factors of natural population regulation, but also a powerful tool to control pests in ecosystems modified by human activity.

However, often the pest species, even in the presence of natural enemies, rises to densities sufficient to cause crop damage. In such cases two questions may be asked: Why are these natural enemies not able to keep the pest populations adequately suppressed? How can we apply additional control measures against the pest while still taking advantage of such benefits as are produced by the natural enemies? The first question is considered in Section II of this chapter, while the second question is dealt with in Section III.

II. FACTORS AFFECTING NATURAL ENEMY EFFICIENCY

In most agricultural environments the principal pests are attacked to a greater or lesser extent by natural enemies. The efficiency with which such natural enemies suppress pest populations is influenced, on the one hand, by their own intrinsic properties and limitations and, on the other hand, by environmental factors and conditions occurring in the agroecosystem under consideration. Where the natural enemies are able to act with high effectiveness, pest populations are prevented from reaching injurious levels, and little need for additional pest control measures exist. Where the natural enemies are less effective, supplemental pest control measures must be considered, i.e., an integrated control approach is indicated.

It is, therefore, of interest to consider as a matter of early priority the natural enemy complex in an agroecosystem, to determine the reasons why they are prevented from exerting effective control of the pests, and then to consider ways to augment, complement, or otherwise improve their action. Among those factors known to affect natural enemy effectiveness are climate, developmental or populational synchrony (phenology) between enemy and host, food for natural enemy adults, host or prey properties, presence of alternative hosts, competition or interference from other insect species, hyperparasites, agronomic or cultural practices utilized in growing the crop, use of pesticides, and other miscellaneous factors, capabilities, or requirements.

A. Climatic Factors

Climate is fundamentally influential on all elements in the agroecosystem, affecting crop, pests, and natural enemies together. In any given situation climate will usually be favorable for both crop and pest. The same may not be said for natural enemies, and indeed one or more climatic factors can often be the main limitations on the effectiveness of any given natural enemy.

For many natural enemies extremes of temperature and/or humidity restrict their distribution or suppress their abundance. Hence, regarding the citrus blackfly, *Aleurocanthus woglumi* Ashby, in tropical and subtropical Central America, the otherwise completely effective parasite *Eretmocerus serius* Silvestri, was ineffective against this pest after the latter invaded northwestern Mexico (Clausen, 1958). It was found to be unable to pass the winter due to excessive cold. A similar restriction resulting from winter cold prevents the very effective egg parasite, *Patasson nitens* Girault, from controlling its host, the eucalyptus snout weevil, *Gonipterus scutellatus* Gyllenhal, at the higher elevations (above 1250 m) in South Africa (Tooke, 1955). The relationship here is more an effect upon the host than on the parasite, with winter cold inhibiting the production of eggs such that the parasite is left for a period without suitable hosts for maintenance of a flourishing population.

In the case of the spotted alfalfa aphid, *Therioaphis trifolii* (Monell), in California, climate, particularly temperature, limits the distribution and effectiveness of the several parasites established for biological control (Messenger, 1971). The parasite *Praon exsoletum* (Nees) is restricted in numbers in the hot summer periods in the interior valleys because it has no capability to aestivate and high temperatures are severely limiting to adult survival or oviposition. On the other hand, it is capable of a hibernal diapause through severe winter cold and so it is the dominant parasite in the colder, more northerly alfalfa growing regions of the state. The parasite *Trioxys complanctus* Quilis, on the contrary, is able to withstand summer heat by virtue of an aestival diapause and therefore persists in high numbers in the southerly or central interior valleys. It is not, however, able to provide effective control of its host during the time it is in diapause. A third parasite *Aphelinus asychis* Walker, because it lacks the ability to diapause during either summer or winter period, is restricted in numbers to coastal and intermediate valley situations. Hence climate influences both the differential distribution of these three natural enemies and their seasonal effectiveness in controlling their host. Because of similar failures of other natural enemies (native predators), integrated control procedures must be applied against the spotted alfalfa aphid and other pests of alfalfa during the mid- and late-summer (Stern *et al.*, 1959).

Climate is a major factor influencing the distribution and effectiveness of the various parasites used to control the California red scale, *Aonidiella aurantii* (Maskell), in southern California citrus groves. During the 1950's *Aphytis lingnanensis* Compere was shown to be effective against the red scale in coastal and intermediate citrus regions but, because of excessive heat and aridity, ineffective in the interior regions (DeBach *et al.*, 1955). On the contrary, the red scale parasite *Comperiella bifasciata* Howard is restricted to interior and intermediate zones (DeBach, 1965). As will be discussed below, interrelations among red scale parasites themselves also have influenced their distribution and effectiveness.

Some strains of a natural enemy survive and flourish in one set of climatic conditions while another strain is better adapted to a different set of conditions. This was found to be the case with the parasite *Trioxys pallidus* (Haliday), imported into California to control the walnut aphid, *Chromaphis juglandicola* (Kaltenbach) (van den Bosch *et al.*, 1962; Messenger, 1970; Messenger and van den Bosch, 1971). A strain derived from southern France proved to be establishable and effective against the aphid in southern California where the coastal climate is milder. The same strain proved ineffective in central and northern California walnut growing regions because of excessive heat and aridity during the summer season. A strain of the parasite was then procured from Iran which proved to be well adapted to these limiting conditions, was easily established, and provided excellent biological control of the pest.

Sometimes the natural enemy proves to be very effective during the crop growing season, but unable to maintain suitable population numbers through the winter conditions. Such is the case with *Cryptolaemus montrouzieri* Mulsant, a predator of the citrus mealybug, *Planococcus citri* (Risso). Periodic colonization of the predator from insectary mass culture each growing season provides an acceptable solution to this limitation (Clausen, 1956, 1958).

The above examples involve exotic species which have been introduced for purposes of biological control. Often, however, otherwise effective native natural enemies are similarly restricted in their native homes. The native parasite *Aphidius testaceipes* (Cresson) is quite able to control the wheat pest *Schizaphis graminum* (Rondani) in mid- and late season in the central plains area of the U.S.A., but not in the early part of the season where it would be much more valuable (Clausen, 1956). Survival is not involved here, but rather the early spring temperatures are too cool for adequate oviposition. There are many more examples of this sort. The tachinid parasite *Chaetexorista javana* Brauer and Bergstrom, very effective in controlling the oriental moth, *Cnidocampa flavescens* (Walker), in most years, is killed by winters with exceptional cold such that control is lost in the next year or two after which it returns (Clausen, 1958). The cuban fly, *Lixophaga diatraeae* (Townsend), introduced into southern Florida to control the sugarcane borer, *Diatraea saccharalis* (Fabricus), is reduced in effectiveness because winter cold results in wide fluctuations in parasitization from season to season (Clausen, 1956). The purple scale parasite (*Aphytis lepidosaphes* Compere) is limited in effectiveness in southern California by winter cold and summer heat, though it does provide a sufficient mortality of its host, *Lepidosaphes beckii* (Newman), to provide a basis for integrated control (DeBach and Landi, 1961). The olive scale parasite *Aphytis maculicornis* (Masi) is quite effective in California on the spring brood of scales (nearly 90% parasitism) but is poor against the summer brood (20% parasitization) due to the hot, dry conditions. The addition of a second parasite, *Coccophagoides utilis* Doutt, more effective against the summer brood, gave the necessary additional parasitism required for fully effective biological control (Doutt, 1954; Clausen, 1958; Huffaker and Kennett, 1966). The ichneumonid *Bathyplectes curculionis* (Thomsen) is very effective against the alfalfa weevil, *Hypera postica* (Gyllenhal), in the mild coastal San Francisco Bay area, less effective in the more severe intermediate valleys, and ineffective in the very hot dry Central Valley (Michelbacher, 1943; Messenger, 1970). Differential overwintering sites, caused by differences in response to winter climate, reduce the effectiveness of the egg parasites, *Trissolcus* spp., which attack the sunn pest, *Eurygaster integriceps* Paton, in Iran (Safavi, 1968). Reassociation of the parasite and host was required at the beginning of each new season.

B. Asynchrony of Life Cycles

For many natural enemies only certain stages of the host are suitable for attack by searching adults or development by immature stages. This means that these two stages of the life cycles of enemy and host must occur at the same time if the enemy is to survive and bring about effective suppression of pest. For numerous natural enemies, asynchrony in life cycles between them and their hosts (or prey) constitutes a major limitation. Sometimes this asynchrony results from differential responses to climate. In other cases, it derives from intrinsic limitations. In a few interesting cases the asynchrony results from the interactive association between enemy and host.

The tachinid parasite *Hyperecteina aldrichi* Mesnil is not able to provide dependable control of its host, the Japanese beetle, *Popillia japonica* Newman, in the eastern U.S.A. because the adult flies emerge from the soil after winter diapause several weeks before the host does (Clausen, 1956; Messenger, 1970). What hosts are available (early emergers) are heavily parasitized, but most of the host brood escapes. A similar phenomenon has been considered to prevent the establishment of the pupal parasite *Phygadeuon wiesmanni* Sachtleben against the cherry fruit fly, *Rhagoletis indifferens* Curran, in California (Messenger, 1970). This parasite, which attacks *R. cerasi* (L.) in Central Europe, emerges in May and June of each year just several weeks after the emergence of the adult fly. However, in California, the adult *R. indifferens* emerge much later, in July and August, so that the pupal stages of the new host generation are not present until August through September. The parasite is unable to survive long enough to find pupae.

A well-known case of asynchrony in life cycles concerns the parasite *Metaphycus helvolus* (Compere) and its host, the black scale, *Saissetia oleae* (Oliv.), on citrus and olive in central and northern California (Clausen, 1956; Messenger and van den Bosch, 1971). The parasite which attacks the early nymphal stages of its host, is very effective in controlling the pest in coastal southern California. The difference in effectiveness is due to the continuous presence of suitable stages of the host in the milder southern region, a result of the lack of synchrony or the overlapping of generations in the host population. In the more interior or northerly regions the host generations are much more closely synchronized with each other such that for long periods each year suitable host stages for the parasite to attack and develop are absent.

C. Adult Food

Many adult natural enemies require food sources for survival and reproduction (Doutt, 1964b). When such food resources are lacking in the agroecosystem, the natural enemy population suffers as a consequence. For example, the parasite *Tiphia vernalis* Rohwer is effective against the Japanese beetle, *P. japonica*, mainly in environments where adult food (honeydew) is present (Clausen, 1956). In areas where such honeydew is absent, parasitism rates are lower. Adults of the egg parasites, *Trissolcus* spp., which attack the wheat pest *E. integriceps* in Iran require aphid honeydew for survival (Safavi, 1968). This factor then becomes a limiting one for these otherwise efficient parasites in the early spring in wheat fields before aphid populations increase in abundance.

Leius (1960) has shown that the parasite *Orgilus obscurator* Nees requires umbelliferous flower nectar for proper adult maturation and survival, preferring nectar of wild parsnip, *Pastinaca sativa*, above other sources. The same author showed that parasitism of tent caterpillar pupae, *Malacosoma americanum*, (Fabricius), could be increased eighteenfold in orchards where an abundant covercrop of wild flowers was maintained, as compared to orchards where clean culture was practiced (Leius, 1967). Tent caterpillar egg parasitism was increased fourfold. Parasitism of codling moth larvae was increased fivefold.

This limitation of adult food in the abundance of natural enemies has been exploited by Hagen *et al.* (1970), who applied artificial food sources to fields of cotton and alfalfa in order to retain and maintain populations of such important predators as *Chrysopa carnea* Stephens and *Hippodamia convergens* Guerin.

D. Host Suitability

For parasites particularly, the identity, nature, condition, and microdistribution of hosts are often very important. Some parasites are stenophagous, and even closely related hosts may prove unequally suitable. Of great historical interest is the fact that *Apanteles glomeratus* (L.), the first natural enemy successfully imported from Europe to North America in 1883 (Doutt, 1964a), is but poorly adapted to its intended host, *Pieris rapae* (L.). In fact, in Europe this parasite is mainly associated with the related host, *P. brassicae* (L.) and when it attacks *P. rapae* it often is encapsulated internally (Puttler *et al.*, 1970).

The genetic constitution of the host population can have important effects upon natural enemies. Chabora (1970) has shown that the pupal parasite *Nasonia vitripennis* (Walker) exhibits a higher fecundity and greater powers of numerical increase when attacking and developing on strains of hosts, *Musca domestica* L., from Florida than from New York. This is presumed to be a result of differences in the nutritive value of these strains.

Host size when attacked can influence natural enemy effectiveness. The Japanese beetle parasite, *Tiphia popilliavora* Rohwer, emerges in summer in the eastern U.S.A. and attacks host larvae when they are still quite small. This results in a change in sex ratio of progeny parasites, there being less females which reduces the control effectiveness of the parasite (Clausen, 1956).

Host density for one reason or another temporarily can become so low as to interfere with natural enemy effectiveness. At the beginning of the growing season the cyclamen mite, *Steneotarsonemus pallidus* (Banks), on strawberry is so scarce that the otherwise effective predatory mite *Metaseiulus occidentalis* (Nesbitt) is unable to find enough hosts to survive with the result that those cyclamen mites that are present soon increase to damaging levels (Huffaker and Kennett, 1956). When both pest and predator are added to the strawberry plants together, both increase somewhat in numbers but the pest is prevented reaching damaging levels.

This situation, found to occur also with *P. rapae* on cabbage, has been similarly exploited by Parker *et al.* (1971). Early releases of the parasites *Trichogramma evanescens* Westwood and *Apanteles rubecula* (Marshall) proved ineffective against this cabbage pest because host densities (eggs and larvae, respectively) are at this time too low. By releasing both host eggs and parasite adults provided sufficient resources so that the natural enemies survived and increased, with corresponding effective control of the pest.

In some cases, even though the natural enemy is physiologically well adapted to its host, its powers of numerical increase prevent it from overtaking and suppressing a pest population. Such is the case with *Trichogramma* sp. when released for control of the bollworm, *Heliothis* spp., on cotton (Lingren, 1969).

The microdistribution of the host can interfere with natural enemy effectiveness. Leaf and bark infestations of the woolly apple aphid, *Eriosoma lanigerum* (Hausman), are well controlled by the parasite *Aphelinus mali* (Haldeman), whereas subterranean root and bark infestations are not.

Host availability to natural enemy attack can be limiting. Codling moth larvae, once they penetrate into the fruit, are no longer accessible to larval parasites until they emerge for pupation. The ichneumonid pupal parasite *Itoplectis conquisitor* (Say) more effectively attacks infestations of the European pine shoot moth, *Rhyacionia buoliana* (Schiffermüller), on Scots pine than on red pine because the leaf buds of the former, within which the host pupates, are smaller and less protected by sheathes of needles (Arthur, 1962).

In some cases the host may exhibit, or develop, a capacity for resisting parasitic attack which then renders the enemy less effective. Such has been found to occur in the larch sawfly, *Pristiphora erichsonii* (Hartig), with respect to its parasite *Mesoleius tenthredinis* Morley in Canada (Muldrew, 1953; Kelleher, 1969; Pschorn-Walcher and Zinnert, 1971). A similar phenomenon has been found to restrain the effectiveness with which the alfalfa weevil parasite *Bathyplectes curculionis* is able to attack the related host, *Hypera brunneipennis* (Bohemann) (Dietrick and van den Bosch, 1953). In this particular case, the adaptive improvement of the natural enemy to this new host has also been observed (van den Bosch and Dietrick, 1959; Salt and van den Bosch, 1967).

The nature of the host plants upon which the host or prey develop can influence natural enemy effectiveness. The famous lady beetle *Rodolia cardinalis* (Mulsant) is ineffective against the cottony-cushion scale, *Icerya purchasi* Maskell, when growing on scotch broom (*Cytisus scoparius*) or maple (*Acer* sp.) (Clausen, 1956). The tachinid endoparasite *Lypha dubia* (Fallen) is more effective against larval infestations of the European pine shoot moth on Scots pine (*Pinus sylvestris*) than infestations on *P. contorta* (Schröder, 1969). The red scale strain of the parasite *Comperiella bifasciata* attacks its host, *Aonidiella aurantii* with facility on citrus, much less so on sago palm (Flanders, 1942), but very poorly on avocado (Flanders, 1966).

E. Alternate Hosts

In several cases what appeared to be an effective natural enemy failed to become established or to provide acceptable control of the pest because the agroecosystem lacked a necessary alternate host. This often occurs when the target host becomes unavailable for a period of time during the year. In some cases the alternate host is known, in others it is assumed because of the behavior of the natural enemy.

The parasite *Agathis diversus* (Muesebeck) is less effective against its principal host, the oriental fruit moth, *Grapholitha molesta* (Busck), in the eastern U.S.A. than it is in Japan where it is native. Circumstantial evidence suggests a poorer synchrony with an alternate host in the U.S.A. (Clausen, 1956). Against the same pest the native *Macrocentrus ancylivorus* Rohwer is quite effective in the eastern U.S.A. because of the presence of its native host *Ancylis comptana fragariae* Walsh & Riley. In other regions of the U.S.A. where this alternate host is absent, *Macrocentrus ancylivorus* is either unable to persist or can do so only in ineffective numbers (Clausen, 1956). Another parasite, *Horogenes molestae* (Uchida), imported from Japan to control the oriental fruit moth was unable to persist in effective numbers because of the lack of a suitable alternate host on which it could overwinter (Clausen, 1956).

The egg parasite *Mymar pratensis* (Förster) only produced low levels of parasitism of the alfalfa weevil because it cannot overwinter on this host since the latter passes the summer, autumn, and early winter as the adult. However, when its preferred host, *Hypera punctata* (Fabricius), is also present in the same fields the parasite thrives (Clausen, 1956). The clover leaf weevil, *H. punctata*, overwinters in the egg stage. In interior central California grape vineyards the egg parasite *Anagrus epos* Girault requires an alternate host for overwintering before it can increase to effective numbers against the grape leafhopper, *Erythroneura elegantula* (Doutt and Smith, 1971). The grape leafhopper overwinters as the adult. A suitable alternate host, *Dikrella* sp. grows on wild blackberry, *Rubus* sp. on which the parasite can maintain itself through the winter. To increase the effectiveness of *A. epos* in suppressing *E. elegantula* the deliberate planting of blackberry hedges in the vicinity of grape vineyards has provided a solution, experimentally.

In a similar way densities of the egg parasite *Trissolcus basalis*, which attacks the southern green stink bug *Nezara viridula*, are increased at the beginning of the season (spring) by the presence of alternate host eggs which support increased overwintering populations (Callen, 1969).

Populations of the Willamette mite, *Eotetranychus willamettei* Ewing, are not well controlled in weed-free grape vineyards by the predatory mite *Metaseiulus occidentalis* because prey at low densities became too evenly dispersed. However, when weedy grasses (Johnson grass) are allowed to grow in the vineyards, the alternate prey mites are maintained which support low numbers of predators and allow them to restrain Pacific mite to noneconomic numbers (Flaherty, 1969).

F. Natural Enemy Competition

When more than one natural enemy species attacks the same host (or prey) population, there is the possibility for the occurrence of interspecific competition between enemies. Such interaction can lead to the areal displacement of one natural enemy by another (DeBach, 1966), or the reduction in abundance of an inferior enemy by a superior one. In some cases the overall effectiveness of biological control of the common host is unaffected or even improved by the competition (DeBach, 1965, 1966). In other cases, control is diminished. In any event, competition is a factor which often affects natural enemies in agroecosystems.

An interesting case of a disturbing competitive interaction concerned the coconut leaf-mining beetle, *Promecotheca reichei* Baly, in Fiji (Taylor, 1937). In earlier years this native pest was under effective natural control by a complex of native enemies until the accidental invasion of the predatory mite *Pediculoides ventricosus* Newport. This new enemy, by virtue of its short developmental time and heavy predation rate caused wide fluctuations to occur in the density of the host. These wide fluctuations interfered with the effective action of the endemic enemies such that additional control measures were required, which eventually led to the successful introduction of a parasite from Java.

The series of introductions of parasites to control the California red scale, *Aonidiella aurantii*, in southern California produced a series of interesting competitive interactions (DeBach, 1965, 1969). When *Aphytis lingnanensis* was introduced in 1947 it soon displaced the earlier established *A. chrysomphali* (Mercet). Biological control was improved. When, in 1956, *A. melinus* DeBach was established, *A. lingnanensis* became restricted in its effectiveness to coastal areas, while *A. melinus* provided still better control in intermediate and interior regions. Attempts to establish still another species, *A. fisheri* DeBach, proved ineffective, presumably because of competition from *A. melinus*. Competition by *A. melinus* also restricted the red scale strain of *Prospaltella perniciosi* Tower to coastal areas.

Lypha dubia in Europe is a major parasite of the winter moth, *Operophtera brumata* (L.), but on occasion it also attacks European pine shoot moth larvae (Schröder, 1969). However, whenever it attacks a host already parasitized by another endoparasitic species it succumbs. *Lypha dubia* in such cases of multiple parasitism is intrinsically inferior.

Sometimes the inferiority of one parasite to another is based on female oviposition behavior. Such is the case of the two egg parasites *Asolcus mitsukurii* Ashmead and *Telenomus nakagawai* Watanabe which attack the southern green stink bug, *Nezara viridula*. The former parasite is the better competitor and displaces the latter by virtue of the more aggressive behavior of the ovipositing *Asolcus* female which fights off the *Telenomus* female (Hokyo and Kiritani, 1966).

The attack of larvae of *Pieris rapae* by *Apanteles glomeratus* and *A. rubecula* has already been discussed. Whenever both species attack the same host larva, the solitary *A. rubecula* survives and the gregarious *A. glomeratus* dies (Parker *et al.*, 1971).

G. Ants

Often certain species of ants can be found in close association with populations of honeydew-producing Homoptera (aphids, mealy bugs, soft scales), feeding on or gathering honeydew. In some cases these foraging ants protect the honeydew producers from certain of their natural enemies. Where such occurs, the homopteran populations may increase in numbers to damaging levels. Such has been found to be the case with the citrus mealy bug, *Planococcus citri* (Risso), black scale, *Saissetia oleae*, and yellow scale, *Aonidiella citrina* (Coquillett), on citrus in southern California, the ant in question most often being the Argentine ant, *Iridomyrmex humilis* Mayr (Flanders, 1951, 1958). A logical approach to prevent such ant-induced outbreaks is to eliminate the ant colonies from the orchard or apply barriers to ant foragers.

H. Hyperparasites

Many parasites of crop pests are themselves attacked by parasites. These hyperparasites then constitute mortality factors which may reduce the effectiveness of the primary parasites in suppressing the pest. Clausen (1956) notes that hyperparasitic destruction of *Meteorus versicolor* (Wesmael), parasitic on gypsy moth, satin moth, and brown-tail moth, reduces its abundance and effectiveness. The effectiveness of *Diaeretiella rapae* (M'Intosh) in parasitizing populations of the cabbage aphid, *Brevicoryne brassicae* L., is said to be inhibited by hyperparasitic attack which becomes particularly heavy late in the year (Sedlag, 1964; Paetzold and Vater, 1967). In southern California the black scale parasite, *Metaphycus lounsburyi* (Howard), is reduced in efficacy by the hyperparasite *Quaylea whittieri* (Girault) (Clausen, 1956). The endoparasite, *Olesi-*

campe benefactor Hinz, which attacks the larch sawfly, *Pristiphora erichsonii*, in central Europe undergoes "drastic reduction" (as much as 70%) by the hyperparasite *Mesochorus dimidiatus* Hlgr. (Pschorn-Walcher and Linnert, 1971). Parker *et al.* (1971) report that pupal hyperparasitism of *Apanteles rubecula*, particularly during winter, reduces its control of *P. rapae* in Missouri.

Natural enemies of insect predators, ecological though not taxonomic hyperparasites, are often observed. Clancy (1946) refers to the numerous parasites which attack species of Chrysopidae, and Iperti and van Waerebeke (1968) describe a nematode parasite of the aphidophagous lady beetles, *Harmonia* 14-*punctata* L. and *H. conglobata* L.

I. Cultural Practices

The manner or timing of certain cultural or agronomic practices used in growing the crop plant often interfere with natural enemies. The parasitic and predatory fauna associated with cotton are seriously reduced when the plants are destroyed and fields plowed in winter to control the pink bollworm after the cotton is harvested (Clausen, 1956). No alternate plants or alternate host insects are then present to maintain the natural enemies over the winter. In Louisiana, where sugarcane is grown as an annual crop and is harvested each autumn, both habitats and hosts of the several parasites (*Lixophaga*, *Paratheresia*, and *Metagonistylum*) of the sugarcane borer, *Diatraea saccharalis*, are eliminated which sharply reduces their abundance each year (Clausen, 1956). Early spring harvesting of the first cutting of alfalfa in central California inflicts heavy mortality on the alfalfa weevil parasite, *Bathyplectes curculionis* (Clausen, 1956). When alfalfa is harvested in large, solid blocks, the natural enemies of the pea aphid are decimated (van den Bosch and Stern, 1969). Harvesting the alfalfa in alternate, narrow strips protects and maintains parasite populations (short-term refuge effect) such that effective biological control is maintained throughout the season. Elimination of hedgerows, fencerows, turn alleys, and thickets in Arkansas, a consequences of land preparation for soybean production, eliminates refuges, food resources, and alternate hosts for natural enemies of pests of cotton and corn (Lincoln, 1969). Use of corn borders adjacent to cotton fields partially restores this beneficial fauna. The abundance of one of the principal predators of cotton pests, *Geocoris pallens* Stal, is influenced by the frequency with which experimental cotton fields are irrigated (Leigh *et al.*, 1969).

J. Pesticides

The harmful effect of pesticides on natural enemies is now well known. Reviews of the effects of pesticides on natural enemies of citrus pests are provided by DeBach (1969) and of phytophagous mites by McMurtry *et al.* (1970). The use of DDT for control of the codling moth in northwestern United States in the mid-1940's led to the local extermination of the very effective *Aphelinus mali*, with subsequent occurrences of outbreaks of woolly apple aphid *Eriosoma lanigerum* (Clausen, 1956). Wille (1951) has reported on the drastic effects of pesticides on the natural enemy fauna in cotton fields in Peru (see also Ridgway, 1969 for similar effects in Texas). The elimination of parasites of various *Lecanium* species on walnut, and the subsequent scale outbreaks which resulted, are described by Bartlett and Ortega (1952) and Michelbacher and Hitchcock (1958). Recently, Safavi (1968) has described how winter pesticide treatments for apple pests in Iran destroy *Trissolcus* spp. adults, egg parasites of the cereal pest, *Eurygaster integriceps*. The adult wasps utilize the trunks of apple trees as overwintering sites.

K. Miscellaneous Influences

A variety of other environmental factors also limit natural enemy effectiveness. Dust deposits interfere with parasite action in relation to citrus scales (DeBach, 1965, 1969). The provision of artificial shelters near tobacco fields provides nesting sites for *Polistes* spp., wasp predators of bollworms, budworms, and other lepidopterous larvae, such that increased predation results (Rabb and Lawson, 1957; Lawson *et al.*, 1961). The fostering of colonies of the ant, *Formica rufa* L., has proved useful in suppressing certain forest pests in Europe (Gösswald, 1951).

III. NATURAL ENEMIES AND INTEGRATED CONTROL

Emtomophagy has been a basic component of virtually all operational integrated control programs. Indeed, the widespread difficulties resulting from insecticide interference with biological control largely triggered the move toward the integrated control philosophy. Accordingly, the primary approach in a number of integrated control programs has been the development of selective pesticides or the selective use of these chemicals (Smith and van den Bosch, 1967; Huffaker, 1971; van den Bosch, 1971; Anonymous, 1974a).

The use of the microbial insecticide *Bacillus thuringiensis* Berliner against certain lepidopterous species in urban (i.e., city) pest control programs in California exemplifies the adaptation of a selective insecticide to integrated control needs. *B. thuringiensis*, which replaced organophosphate materials and carbaryl, has several advantages as an integrated control tool: (i) it is effective against the target pest but spares that pest's natural enemies, thereby augmenting its own efficacy; (ii) it spares the parasites and predators of other phytophagous species occurring in the treated plant community, permitting them to carry on their suppressive activities in full; (iii) it is completely innocuous to warm blooded animals. This microbial control agent is now of major importance in urban pest management in California where it is a key factor in integrated control programs developed for urban street trees in several cities, the San Francisco Bay Area Regional Park System, and State of California highway rights-of-way vegetation (Olkowsky, *et al.*, 1974).

A second biological control manipulation utilized in integrated control is the classic introduction of exotic natural enemies. Natural enemy introduction has been a key factor in a highly successful integrated control program developed for street tree pests in Berkeley, California, where parasites introduced from Europe have sufficiently depressed populations of the linden aphid, *Eucallipterus tiliae* (Linn.) and an elm aphid, *Tinocallis plantani* Kalt. to eliminate disruptive spraying with organophosphate insecticides (Olkowsky *et al.*, 1974). Introduced natural enemies have also played important roles in integrated control programs developed for alfalfa in California, citrus in Israel and California (Harpaz and Rosen, 1971; Huffaker, 1971) and pome fruits in Europe (Anonymous, 1974a).

Periodic colonization of natural enemies is another tactic which has played an important role in integrated control. Outstanding examples of this tactic include use of predatory mites against spider mites on glasshouse pests in Europe (Anonymous, 1974b), and the long-standing program of natural enemy colonization against citrus pests on several thousand acres of citrus in Southern California (Fisher and Finney, 1964).

Augmentation of natural enemies by cultural manipulations (e.g., harvesting manipulation, use of pollen and nectar plants), nutritional augmentation (e.g., spraying of artificial nutrients), and pheromone manipulation have been shown to be effective at a research level but have played only a very limited role in practice.

Indigenous natural enemies have been used in various ways in integrated control programs. For example, in California cotton timely and/or restricted spraying for control of lygus bug, *Lygus hesperus* Knight, permits maximum survival of generalist predators which maintain virtually complete control of lepidopterous pests such as the fruit infesting bollworm, *Heliothis zea* (Boddie), and the defoliators, *Trichoplusia ni* (Hübner), and *Spodoptera exigua* (Hübner) (van den Bosch *et al.*, 1971; Ehler *et al.*, 1973; Eveleens *et al.*, 1973). Other integrated control programs which have made effective use of naturally occurring biological control include those in alfalfa in California (Stern and van den Bosch, 1959), tobacco in North Carolina (Ellis *et al.*, 1972), oil palm and rubber in Malaysia (Wood, 1971), cotton in Latin America (Boza Barducci, 1969), apple in the U.S.A., Canada, and Europe (Pickett, 1960; Asquith and Colburn, 1971; Hoyt and Caltagirone, 1971; van den Bosch, 1971), and cling peach in California (Hoyt and Caltagirone, 1971).

There have been some particularly ingenious manipulations of natural enemies in the fruit tree programs. For example, in cling peach in California, reduced spraying against the virtually harmless peach silver mite, *Aculis cornutus* (Banks), has been found to favor maximum survival of the predatory mite *Metaseiulus occidentalis* (Nesbitt), a major enemy of the highly injurious two spotted spider mite, *Tetranychus urticae* Koch, a European red mite *Panonychus ulmi* (Koch). In former years, populations of the peach silver mite were routinely sprayed on the assumption that they were injurious. However, investigation revealed that the species was not only essentially harmless, but indeed often beneficial in that its populations often support *M. occidentalis* during periods of scarcity of the injurious *T. urticae* and *P. ulmi*. Where spraying for peach silver mite has been eliminated, sufficient numbers of *M. occidentalis* are maintained to provide highly satisfactory regulation of *T. urticae* and *P. ulmi* (Hoyt and Caltagirone, 1971).

Another example of the imaginative adaptation of a native natural enemy to a deciduous fruit integrated control program involves the utilization of insecticide tolerant *M. occidentalis* against the McDaniel spider mite, *Tetranychus mcdanieli* McGregor in Washington State. Here, the discovery that *M. occidentalis* had developed a tolerance (resistance?) to organophosphate insecticides was put to effective use when it was found that reduced dosages of azinphosmethyl, though adequate to control codling moth, permitted sufficient survival of *M. occidentalis* for it to effect satisfactory control of the McDaniel spider mite (Hoyt and Caltagirone, 1971).

In Pennsylvania there is another case of the use of a native predator against spider mites in apple. In this program, where the pest mites involved are *T. urticae* and *P. ulmi*, the predatory coccinellid, *Stethorus punctum* (LeConte), is the key to the system. The concern with *S. punctum* is that its populations not be severely reduced by sprays directed against the nonacarine pests of apple, i.e., insecticide selectivity is required. In this case selectivity has been accomplished by the use of reduced insecticide dosages in what are called "half sprays." Half spraying is accomplished by guiding the sprayer between alternate tree rows so that the off-side trees (those in the skipped rows) receive only that insecticide which drifts through the trees. This protects the *S. punctum* on the off-sides of the trees from the full dosage of the insecticides and permits the lady beetles to disperse into the heavily sprayed sides as the insecticide residue dissipates (Asquith and Colburn, 1971).

The previous comments have largely related to natural enemy utilization against individual pest species or to the manipulation of individual natural enemy species in integrated control programs. There is a tendency to consider natural enemies or their manipulation in this isolated way, when in actuality most integrated control programs involve the orchestration of natural enemy complexes and the multiplicity of factors which preserve or enhance them. This is illustrated by the spotted alfalfa aphid program which not only involved protection and augmentation of imported and indigenous entomophages and pathogens directly affecting the aphid, but also had to take into consideration the natural enemies of the entire pest arthropod fauna of alfalfa and to some extent of associated crops, particularly as these were affected by drift of insecticides directed against *T. trifolii*.

The within-crop complexity of natural enemy utilization in integrated control is further illustrated by the spider mite program in apple in Washington. Here, three phytophagous mites, the McDaniel spider mite, the European red mite, and the apple rust mite, as well as their natural enemies, are being successfully manipulated within a pest control framework that involves acaricides, insecticides, and fungicides as well as fruit thinning chemicals, all of which can have negative and/or positive effects on the spider mite-natural enemies relationship.

The cases just cited are but a sampling of the situations in which natural enemies have been effectively used in integrated control programs. However, they provide an overview of the role of natural enemies and the great relevance of biological control to integrated pest management.

REFERENCES

Anonymous (1974a). "Lutte Integree en Vergers, 5ᵉ Symposium," 369 pp. *OILB/SROP, Bozen-Bolzano* **IX**, 3–7.

Anonymous (1974b). "Integrated Control in Glasshouses," *OILB-SROP/WPRS, Bulletin*, 73 pp.

Arthur, A. P. (1962). Influence of host tree on abundance of *Itoplectis conquisitor* (Say) (Hymenoptera, Ichneumonidae) a polyphagous parasite of the European pine shoot moth, *Rhyacionia buoliana* (Schiff.) (Lepidoptera, Olethreutidae). *Can. Entomol.* **94**, 337–347.

Asquith, D., and Colburn, P. (1971). Integrated pest management in Pennsylvania apple orchards. *Bull. Entomol. Soc. Amer.* **17**, 89–91.

Bartlett, B. R., and Ortega, J. C. (1952). Relation between natural enemies and DDT-induced increases in frosted scale and other pests of walnut. *J. Econ. Entomol.* **45**, 783–785.

Benassy, C., Mathys, G., Neuffer, G., Milaire, H., Bianchi, H., and Guignard, E. (1968). L'utilisation pratique de *Prospaltella perniciosi* Tow. parasite du pou de San Jose *Quadraspidiotus perniciosus* Comst. *Entomophaga Mem. Hors. Ser.* **4**, 28 pp.

Boza Barducci, T. (1969). Ecological consequences of pesticides used for the control of cotton insects in Cañete Valley, Peru. *In* "The Careless Technology—Ecology and International Development" (M. T. Farvar and J. T. Milton, eds.), pp. 423–438. Natur. Hist. Press, Garden City, New York.

Callan, E. M. (1969). Ecology and insect colonization for biological control. *Proc. Ecol. Soc. Aust.* **4**, 17–31.

Chabora, P. C. (1970). Studies of parasite-host interaction. II. Reproductive and developmental response of the parasite *Nasonia vitripennis* (Hymenoptera: Pteromalidae) to strains of the house fly host *Musca domestica*. *Ann. Entomol. Soc. Amer.* **63**, 1632–1636.

Clancy, D. W. (1946). The insect parasites of the Chrysopidae (Neuroptera). *Univ. Calif. Publ. Entomol.* **7**, 403–496.

Clausen, C. P. (1940). "Entomophagous Insects," 688 pp. McGraw Hill, New York.

Clausen, C. P. (1956). Biological control of insect pests in the continental United States. *U.S. Dep. Agr. Tech. Bull.* **1139**, 151 pp.

Clausen, C. P. (1958). Biological control of insect pests. *Annu. Rev. Entomol.* **3**, 291–310.

DeBach, P. (1965). Some biological and ecological phenomena associated with colonizing entomophagous insects. *In* "The Genetics of Colonizing Species" (H. G. Baker and G. L. Stebbins, eds.), pp. 287–306. Academic Press, New York.

DeBach, P. (1966). The competitive displacement and coexistence principles. *Annu. Rev. Entomol.* **11**, 183–212.

DeBach, P. (1969). Biological control of diaspine scale insects on citrus in California. *Proc. 1st Intern. Citrus Symp.* **2**, 801–815.

DeBach, P. (1971). The use of imported natural enemies in insect pest management ecology. *Proc. Tall Timbers Conf. Ecol. Anim. Contr. Habitat Manage.* **3**, 211–233.

DeBach, P., and Landi, J. (1961). The introduced purple scale parasite *Aphytis lepidosaphes* Compere, and a method of integrating chemical with biological control. *Hilgardia* **31**, 459–497.

DeBach, P., Fisher, T. W., and Landi, J. (1955). Some effects of meteorological factors on all stages of *Aphytis lingnanensis*, a parasite of the California red scale. *Ecology* **36**, 743–753.

Dietrick, E. J., and van den Bosch, R. (1953). Further notes on *Hypera brunneipennis* and its parasite *Bathyplectes curculionis*. *J. Econ. Entomol.* **46**, 1114.

Doutt, R. L. (1954). An evaluation of some natural enemies of the olive scale. *J. Econ. Entomol.* **47**, 39–43.

Doutt, R. L. (1964a). The historical development of biological control. *In* "Biological Control of Insect Pests and Weeds" (P. DeBach, ed.), Chapter 2, 844 pp. Chapman and Hall, London.

Doutt, R. L. (1964b). Biological characteristics of entomophagous adults. *In* "Biological Control of Insect Pests and Weeds" (P. DeBach, ed.), Chapter 6, 844 pp. Chapman and Hall, London.

Doutt, R. L., and Smith, R. F. (1971). The pesticide syndrome—diagnosis and suggested prophylaxis. *In* "Biological Control" (C. B. Huffaker, ed.), Chapt. 1, 511 pp. Plenum, New York.

Ehler, L. E., Eveleens, K. G., and van den Bosch, R. (1973). An evaluation of some natural enemies of cabbage looper on cotton in California. *Environ. Entomol.* **2**, 1009–1015.

Ellis, H. E., Ganyard, M. C., Jr., Singletary, H. M., and Robertson, R. L. (1972). North Carolina pest management. Second Annual Rpt. (1972). *Agr. Ext. Serv., North Carolina St. Univ., Raleigh, North Carolina.* 67 pp.

Eveleens, K. G., van den Bosch, R., and Ehler, L. E. (1973). Secondary outbreak induction of beet armyworm by experimental insecticide applications in cotton in California. *Environ. Entomol.* **2**, 497–503.

Fisher, T. W., and Finney, G. L. (1964). Insectary facilities and equipment. *In* "Biological Control of Insect Pests and Weeds" (P. DeBach, ed.), Chapt. 13, pp. 381–401. Chapman and Hall, London.

Flaherty, D. L. (1969). Ecosystem trophic complexity and densities of the willamette mite, *Eotetranychus willamettei* (Acarina, Tetranychidae) *Ecology* **50**, 911–916.

Flanders, S. E. (1942). Abortive development in parasitic Hymenoptera induced by the food-plant of the insect host. *J. Econ. Entomol.* **35**, 834–835.

Flanders, S. E. (1951). The role of the ant in the biological control of homopterous insects. *Can. Entomol.* **83**, 93–98.

Flanders, S. E. (1958). The role of the ant in the biological control of scale insects in California. *Proc. 10th Int. Congr. Entomol.* **4**, 579–584.

Flanders, S. E. (1966). The circumstances of species displacement among parasitic Hymenoptera. *Can. Entomol.* **98**, 1009–1024.

Gösswald, K. (1951). "Die rote Waldameise im dienste der Waldhygiene," 160 pp. Metta Kinau Verlag, Wolfu. Täuber, Lüneburg, Germany.

Hagen, K. S., Sawall, E. F., Jr., and Tassan, R. L. (1970). The use of food sprays to increase effectiveness of entomophagous insects. *Proc. Tall Timber Conf. Ecol. Anim. Contr. Habitat Manage.* **2**, 59–82.

Harpaz, I., and Rosen, D. (1971). Development of integrated control programs for crop pests in Israel. *In* "Biological Control" (C. B. Huffaker, ed.), Chapt. 20, pp. 458–468. Plenum, New York.

Hokyo, N., and Kiritani, K. (1966). Oviposition behavior of two egg parasites, *Asolcus mitsukurii* Ashm. and *Telenomus nakagawai* Watanabe (Hymenoptera, Proctotrupoidea, Scelionidae), *Entomophaga* **11**, 191–201.

Hoyt, S. C., and Caltagirone, L. E. (1971). The developing programs of integrated control of pests of apples in Washington and peaches in California. In "Biological Control" (C. B. Huffaker, ed.). Chapt. 18, pp. 395–421. Plenum, New York.

Huffaker, C . B. (ed.) (1971). "Biological Control," 511 pp. Plenum, New York.

Huffaker, C. B., and Kennett, C. E. (1956). Experimental studies on predation: predation and cyclamen-mite populations on strawberries in California. *Hilgardia* **26**, 191–222.

Huffaker, C. B., and Kennett, C. E. (1966). Studies of two parasites of olive scale *Parlatoria oleae* (Colvée) 4. Biological control of *Paralatoria oleae* (Colvée) through compensatory action of two introduced parasites. *Hilgardia* **37**, 283–335.

Iperti, G. and van Waerebeke, D. (1968). Description, biologie et importance d'une nouvelle espèce d'Allantonematidae (Nématoda), parasite des Coccinelles aphidiphages: *Parasitylenchus coccinellinae*, n. sp. *Entomophaga* **13**, 107–119.

Iperti, G., Laudeho, Y., Brun, J., and Choppin de Janvry, E. (1970). The natural enemies of *Parlatoria blanchardi* Targ. in the palm groves of the Adrar region of Mauritania. III. Introduction, acclimatization, and effectiveness of a new predacious coccinellid, *Chilocorus bipustulatus* L., variety *iranensis* (var. nov.). *Ann. Zool. Ecol. Anim.* **2**, 617—638.

Kelleher, J. S. (1969). Introduction practices—past and present. *Bull. Entomol. Soc. Amer.* **15**, 235—236.

Lawson, F. R., Rabb, R. L., Guthrie, F. E., and Bowery, T. G. (1961). Studies on an integrated control system for horn worms on tobacco. *J. Econ. Entomol.* **54**, 93—97.

Lebedev, G. I. (1970). Colloquium franco-sovietique sur l'utilisation des entomophages, Antibe, 13—18 Mai 1968. *Ann. Zool. Ecol. Anim. Hors. Ser.* 17—23.

Leigh, T. F., Grimes, D. W., Yamada, H., Stockton, J. R., and Bassett, D. (1969). Arthropod abundance in cotton in relation to some cultural management variables. *Proc. Tall Timbers Conf. Ecol. Anim. Contr. Habitat Manage.* **1**, 71—83.

Leius, K. (1960). Attractiveness of different foods and flowers to the adults of some hymenopterous parasites. *Can. Entomol.* **92**, 369—376.

Leius, K. (1967). Influence of wild flowers on parasitism of tent caterpillars and codling moth. *Can. Entomol.* **99**, 444—446.

Lincoln, C. (1969). The effect of agricultural practices on insect habitats in a typical delta community. *Proc. Tall Timbers Conf. Ecol. Anim. Contr. Habitat Manage.* **1**, 13—18.

Lingren, P. D. (1969). Approaches to the management of *Heliothis* spp. in cotton with *Trichogramma* spp. *Proc. Tall Timbers Conf. Ecol. Animal Contr. Habitat Manage.* **1**, 207—217.

Maksimovic, M., Bjegovic, P., and Vasiljevic, L. J. (1970). Maintaining the density of the gypsy moth enemies as a method of biological control. *Zastita Bilja* **21**, 3—15.

McMurtry, J. A., Huffaker, C. B., and van de Vrie, M. (1970). Ecology of tetranychid mites and their natural enemies: a review. I. Tetranychid enemies: their biological characters and the impact of spray practices. *Hilgardia* **40**, 331—390.

Messenger, P. S. (1970). Bioclimatic inputs to biological control and pest management programs. *In* "Concepts of Pest Management" (R. L. Rabb and F. E. Guthrie, eds.), pp. 84—102. No. Carolina State Univ., Raleigh, North Carolina.

Messenger, P. S. (1971). Climatic limitations to biological control. *Proc. Tall Timbers Conf. Ecol. Anim. Contr. Habitat Manage.*, **3**, 97—114.

Messenger, P. S., and van den Bosch, R. (1971). The adaptability of introduced biological control agents. *In* "Biological Control" (C. B. Huffaker, ed.) Chapt. 3. Plenum, New York.

Michelbacher, A. E. (1943). The present status of the alfalfa weevil in California. *Univ. Calif. Agr. Exp. Sta. Bull.* **677**, 24 pp.

Michelbacher, A. E. (1962). Influence of natural factors on insect and spider mite populations. *Proc. 11th Intern. Congr. Entomol., Vienna (1960)* **2**, 694.

Michelbacher, A. E. and Hitchcock, S. (1958). Induced increase of soft scales on walnut. *J. Econ. Entomol.* **51**, 427—431.

Muldrew, J. A. (1953). The natural immunity of the larch sawfly (*Pristiphora erichsonii* Hartig) to the introduced parasite *Mesoleius tenthredinis* (Morley) in Manitoba and Saskatchewan. *Can. J. Zool.* **31**, 312—332.

Olkowsky, W., Pinnock, D., Toney, W., Mosher, G., Neasbitt, W., van den Bosch, R., and Olkowsky, H. (1974). A model integrated control program for street trees. *Calif. Agr.* **28(1)**, 3—4.

Paetzold, D., and Vater, G. (1967). Populationsdynamische Untersuchungen an den parasiten und hyperparasiten von *Brevicoryne brassicae* (L.) (Homoptera, Aphididae). *Acta Entomol. Bohemoslov.* **64**, 83—90.

Parker, F. D. Lawson, F. R., and Pennell, R. E. (1971). Suppression of *Pieris rapae* using a new control system: mass release of both the pest and its parasites. *J. Econ. Entomol.* **64**, 721—735.

Pickett, A. D. (1960). The ecological effects of chemical control practices on arthropod populations in apple orchards in Nova Scotia. *In* "The Ecological Effects of Biological and Chemical Control of Undesirable Plants and Animals. 8th Technical Meeting, Intern. Union Conserv. Nat Resources, Warsaw, 15—24 July 1960" (D. J. Kuenen, ed.), pp. 19—24. E. J. Brill, Leiden.

Pschorn-Walcher, H., and Zinnert, K. D. (1971). Investigations on the ecology and natural control of the larch sawfly (*Pristiphora erichsonii* Hartig) (Hymenoptera: Tenthredinidae) in Central Europe. Part II. Natural enemies: their biology and ecology and their role as mortality factors in *Pristiphora erichsonii*. *Commonw. Inst. Biol. Contr. Tech. Bull.* **14**, 1—50.

Puttler, B., Parker, F. D., Pennell, R. E., and Thewke, S. E. (1970). Introduction of *Apanteles rubecula* into the United States as a parasite of the imported cabbageworm. *J. Econ. Entomol.* **63**, 304—305.

Rabb, R. L., and Lawson, F. R. (1957). Some factors influencing the predation of *Polistes* wasps on the tobacco hornworm. *J. Econ. Entomol.* **50**, 778—784.

Ridgway, R. L. (1969). Control of the bollworm and tobacco budworm through conservation and augmentation of predaceous insects. *Proc. Tall Timbers Conf. Ecol. Anim. Contr. Habitat Manage.* **1**, 127—144.

Safavi, M. (1968). Etude biologique et écologie des Hyménopteres parasites des oeufs des punaises des céréales. *Entomophaga* **13**, 381—495.

Sailer, R. I. (1969). A taxonomist's view of environmental research and habitat manipulation. *Proc. Tall Timbers Conf. Ecol. Anim. Contr. Habitat Manage.* **1**, 37—45.

Sailer, R. I. (1972). Concepts, principles and potentials of biological control parasites and predators. *Proc. North Central Branch, Entomol. Soc. Amer.* **27**, 35—39.

Salt, G., and van den Bosch, R. (1967). The defense reactions of three species of *Hypera* (Coleoptera, Curculionidae) to an ichneumon wasp. *J. Invertebr. Pathol.* **9**, 164—177.

Schröder, D. (1969). *Lypha dubia* (Fall.) (Diptera: Tachinidae) as a parasite of the European pine shoot moth, *Rhyacionia buoliana* (Schiff.) (Lepidoptera: Eucosmidae) in Europe. *Commonw. Inst. Biol. Contr. Tech. Bull.* **12**, 43—60.

Sedlag, U. (1964). Zur Biologie und Bedeutung von *Diaeretiella rapae* (M'Intosh) als Parasit der Kohlblattlaus (*Brevicoryne brassicae*) (L.). *Z. Nachr. Deut. Pflanz.* **18**, 81—86.

Smith, R. F., and van den Bosch, R. (1967). Integrated control. *In* "Pest Control—Biological, Physical and Selected Chemical Methods" (W. W. Kilgore and R. L. Doutt, eds.), Chapt. 9, pp. 295—340. Academic Press, New York.

Stern, V. M., and van den Bosch, R. (1959). The integration of chemical and biological control of the spotted alfalfa aphid. Part 2. Field experiments on the effects of insecticides. *Hilgardia* **29**, 103—130.

Stern, V. M., Smith, R. F., van den Bosch, R., and Hagen, K. S. (1959). The integration of chemical and biological control of the spotted alfalfa aphid. *Hilgardia* **29**, 81—154.

Taylor, T. H. C. (1937). The biological control of an insect in Fiji. An account of the coconut leaf-mining beetle and its parasite complex. *Imp. Inst. Entomol. London*, 239 pp.

Tooke, F. G. C. (1955). The eucalyptus snout-beetle *Gonipterus scutellatus* Gyll. A study of its ecology and control by biological means. *Union S. Afr. Dep. Agr. Entomol. Mem.* **3**, 1—282.

van den Bosch, R. (1971). Biological control of insects. *Ann. Rev. Ecol. Syst.* **2**, 45—66.

van den Bosch, R., and Dietrick, E. J. (1959). The interrelationships of *Hypera brunneipennis* (Coleoptera, Curculionidae) and *Bathyplectes curculionis* (Hymenoptera, Ichneumonidae) in Southern California. *Ann. Entomol. Soc. Amer.* **52**, 609—616.

van den Bosch, R., and Stern, V. M. (1969). The effect of harvesting practices on insect populations in alfalfa. *Proc. Tall Timbers Conf. Ecol. Anim. Contr. Habitat Manage.* **1**, 47—54.

van den Bosch, R., Schlinger, E. I., and Hagen, K. S. (1962). Initial field observations in

California on *Trioxys pallidus* (Haliday), a recently introduced parasite of the walnut aphid. *J. Econ. Entomol.* **55**, 857–862.

van den Bosch, R., Leigh, T. F., Falcon, L. A., Stern, V. M., Gonzalez, D., and Hagen, K. S. (1971). The developing program of integrated control of cotton pests in California. *In* "Biological Control" (C. B. Huffaker, ed.). Chapt. 17, pp. 377–394. Plenum, New York.

Wille, J. E. (1951). Biological control of certain cotton insects and the application of new organic insecticides in Peru. *J. Econ. Entomol.* **44**, 13–18.

Wood, B. J. (1971). Development of integrated control programs for pests of tropical perennial crops in Malaysia. *In* "Biological Control" (C. B. Huffaker, ed.). Chapt. 19, pp. 422–457. Plenum, New York.

23

SELECTIVE PESTICIDES AND SELECTIVE USE OF PESTICIDES

L. D. Newsom, Ray F. Smith, and W. H. Whitcomb

I. INTRODUCTION

Conventional chemical pesticides remain one of the most powerful and dependable tools available for the management of pest populations. They are more effective, dependable, economical, and adaptable for use in a wide variety of situations than any other proved tool for controlling pest populations at subeconomic levels. Indeed, use of conventional chemical pesticides is the only known method for control of many of the world's most important pests of agriculture and public health. No other agent lends itself to such comparative ease of manipulation and none can be brought to bear so quickly on outbreak populations.

For more than 15 years, entomologists have been shifting research from chemical pesticides to various alternatives. The shift has accelerated rapidly during the last decade. Consequently, a tremendous amount of data has been accumulated on alternative techniques, many of which show great promise, particularly in limited or specific situations or as part of an integrated control system. However, when promise is measured against performance the alternate approaches have failed to provide a solution to meet the demands of present day agriculture in most cases. The performance of these alternatives has been disappointing. As far into the future as can be reasonably projected, chemical pesticides will continue to be valuable assets in pest management and will be an integral component of most animal and plant protection schemes.

During the current period of heavy reliance upon chemical pesticides their undesirable side effects have demonstrated convincingly that their continued use in the patterns of the past is no longer permissible (e.g., Chapters 14—17). However, despite the many serious problems arising from their use, it is becoming increasingly apparent that these useful chemicals must be retained in our arsenal of weapons for pest management. The question is not whether their use should be continued, rather it is a question of how they may be used with the minimum of undesirable side effects and complications. The most appropriate answer to this question is in the use of selective insecticides and in the selective use of insecticides. The principle of selectivity must be used as effectively in pest management as it has been used in the medical and veterinary professions.

Selectivity, in the broad sense, is accounted for by differences in physiological sensitivity (physiological selectivity) after contact between organisms and toxic chemical and by differences in exposure or behavior of the fauna (ecological selectivity) (Winteringham, 1969). Winteringham (1969) described selective physiological toxicity as injury of one kind of living matter without harming some other kind with which the first is in intimate contact. Albert (1965) defined selectively toxic chemicals as those "substances which injure certain cells without harming others, even when the two kinds of cells are close neighbors." He listed three main principles by which selective agents may exert their favorable effects.

1. Through accumulation, e.g., the comparatively large surface area per unit of weight of an insect pest resting on a mammal may result in the intake of a proportionally greater amount of the applied chemical by the pest than by the mammal.
2. Through comparative cytology—cells are full of component parts, organelles—and each kind of these organelles displays strong differences.
3. Through comparative biochemistry—the most striking difference between species is found in the choice and biosynthesis of enzymes and similar substances used in growth and development.

Unlike physiological selectivity which stresses activity of the chemical, ecological selectivity emphasizes application. Ecological selectivity is secured by applying compounds having broad spectra of activity but in a manner to ensure contact of a toxic dose with the target species while avoiding completely, or greatly minimizing, contact of a toxic dose with nontarget species (Unterstenhofer, 1970).

The objective of this chapter is to examine the role of physiological and ecological selectivity in the use of insecticides in pest management schemes where biological control agents, especially, are useful and require this protection.

II. PHYSIOLOGICAL SELECTIVITY

Considering the remarkable degree of success of chemotherapy in human and veterinary medicine based on the principle of physiological selectivity, surprisingly little effort has been devoted to this area of research in entomology. Obviously, a potential for developing highly selective chemicals for control of insect pests does exist (Bailenger *et al.*, 1970; Metcalf, 1964, 1971; O'Brien, 1961, 1967; Winteringham, 1969). If the medical and veterinary professions can find and develop drugs that injure certain cells without seriously harming others in the same organism, surely it is possible to find chemicals that are capable of injuring some species or taxa without seriously harming other species or higher taxa which it is desirable to protect.

The necessity for achieving physiological selectivity in order to protect the host organisms of our target insect pests has been recognized as long as chemicals have been used for insect control. This degree of selectivity was comparatively easy to achieve. The recent wide use of the synthetic organochlorine insecticides has, however, emphasized the need for much more narrow differential toxicity within the arthropods. The broad spectrum of activity of these chemicals, especially their effects on insect predators and parasites (e.g., Chapters 3, 10, 11, 14–17, and 22) and on other nontarget species has further proved the importance of predators and parasites, including native species, for regulation of pest populations. The previous use of the comparatively selective inorganic and botanical materials had not resulted in such widely occurring, dramatic examples of disruption: resurgence of treated populations and elevation of occasional and secondary species to the status of key pests. These phenomena became commonplace following wide-scale use of the broad-spectrum synthetics and entomologists began to plead for development of narrowly selective chemicals.

For two decades the pesticide chemical industry has been increasingly challenged to produce selective pesticides. In managing the complex of pest populations on the respective crops, having available chemicals that are toxic to no other species except boll weevil, *Anthonomus grandis* Boh. (in southern United States) and *Lygus* spp. (in western United States) for cotton, and codling moth, *Laspeyresia pomonella* (L.), for apples, for example, would be of incalculable value. There are enough examples of narrowly selective insecticides to justify the belief that chemicals that are toxic to single families, genera, and even species can be developed.

Unterstenhofer (1970) noted that 1,3,6,8-tetranitrocarbazole was so effective for control of grape-berry moths, *Clysia ambiguella* (Hubner) and *Polychrosis botrana* Schiffermuller, that further use of arsenicals in vineyards in Germany was banned in 1942 and it became the leading insecticide in wine growing for about a decade. Other potential uses for this narrowly selective chemical could not be found due to lack of, or inadequate, insecticidal and acaricidal activity! This seemingly ideal insecticide had a short life in spite of possessing such desirable properties as narrow selectivity among the arthropods, harmlessness to mammals, and lack of phytotoxicity. It was superseded by broad-spectrum insecticides because other major pests were not affected by it. The pear leaf roller, *Byctiscus betulae* (L.), a species of little previous importance, became a serious pest. This development of a replacement pest caused Klett (1945) to make the important point that selective pesticides are not inherently different from broad-spectrum materials in causing a change in status of secondary pests.

Jung and Scheinpflug (1970) described the specificity of 2-isopropoxyphenyl *N*-methylcarbamate for control of *Nephotettix cincticeps* (Uhler) in rice. In field experiments in Japan this chemical achieved better than 95% control of the leafhopper population with no effect on the important predators *Oedothorax insecticeps* Bosenberg & Strand and *Nabis ferus* (L.) contrasted to 56% control of the pest species and 100% kill of the two predator species by the previously used standard, lindane.

Fraser *et al.* (1967) examined some 120 substituted aryl *N*-methylcarbamates and their *N*-aryl derivatives in their search for a compound effective against the sheep blowfly before discovering 3,5-di-*t*-butylphenyl *N*-methylcarbamate, which proved to be outstanding in its toxicity to blowfly larvae, persistence in fleece, and low mammalian toxicity. In spite of its powerful anticholinesterase activity *in vitro*, it proved to have little toxicity except for sheep blowfly larvae.

Binns (1969) and Gould (1971) demonstrated that the selective aphicide, pirimicarb, could be used to control the melon aphid, *Aphis gossypii* Glover, on cucumber without adversely affecting populations of the predatory mite *Phytoseiulus persimilis* Athias-Henriot. However, Hussey and Bravenboer (1971) reported that the significantly wide differences between the LD_{50}'s of this predator (used for control of spider mites) and of *Aphis gossypii* in laboratory tests did not hold up commercially. Discriminating dosages of the chemicals, established in laboratory tests, failed to control the aphid effectively in commercial glasshouses. Simmonds (1970) also reported that pirimicarb applied to strawberries after fruiting controlled aphids without harming predatory insects or mites.

Although the discovery of a sufficient number of monotoxic compounds to have a significant impact on pest management, generally, is highly unlikely for the forseeable future, compounds with limited spectrum activity are discovered more frequently than any other class of toxic chemical. It should be remembered that we do not need the ultimate in specificity that would permit us to prescribe a specific chemical for each pest species. However, we do need materials that are specific for groups of pests, such as aphids, locusts, lepidopterous larvae, and weevils. The degree of selectivity provided by such compounds has already proved useful in many cases. If exploited fully, they could make many more significant contributions to pest control while lessening the undesirable side effects of currently used materials. The wide-scale use of a variety of selective acaricides illustrates the potential of such oligotoxic compounds (Smith and van den Bosch, 1967).

The physiological selectivity principle has been used with success in a number of integrated control programs, one of the first and most successful being conducted in Nova Scotia orchards (MacPhee and MacClellan, 1971). Much of the information needed had been developed by entomologists seeking to use selective pesticides to help preserve the effectiveness of natural enemies for regulating populations of mites, aphids, cankerworms, scales, and bugs. They had found nicotine sulfate and lead arsenate to be moderately selective, some of the synthetic acaricides to be highly selective, and the organochlorine, organophosphate, and carbamate insecticides to be so broad in their activity that they were of very limited use in the integrated control program. The key pest was the codling moth. It was not until Patterson and MacClellan (1954) had demonstrated the effectiveness of ryania for codling moth control and its relative innocuousness to natural enemies that a commercial integrated control program could be developed. This program was accepted by a high percentage of the growers in the late 1950's and has been practiced in the area since that time.

Hoyt (1969) found that a wide variety of pesticides, including most of the organophosphorus compounds, endosulfan, Morestan, Omite, the dithiocarbamate fungicides, dodine, and captan, can be used in apple orchards in Washington with little adverse effect upon the important predacious mite, *Metaseiulus occidentalis* (Nesbitt). However, he was uncertain whether the selectivity of these compounds results from natural tolerance of *M. occidentalis* or from the development of resistant strains in the population.

The above examples show that rather narrowly selective insecticides exist and that they have been used effectively in pest management systems. However, there appear to be serious handicaps to their substantial and speedy development. The discovery of physiologically selective pesticides is a difficult undertaking. Metcalf (1971), one of the foremost proponents of designing selective insecticides based on an understanding of the relationships between chemical structure and biochemical action, stated, "It is evident that after 40 years of intensive study we have only an elementary understanding of the forces and reactions involved in the inhibition of AChE by both carbamates and organophosphates. Improved understanding of these problems will certainly lead to the design of interesting new insecticides." Ludwig and Potterfield (1971) also concluded that it is "difficult to predict biological activity on the basis of chemical structure . . . consequently, the testing and evaluation of drugs remains pragmatic."

Unterstenhofer (1970) pointed out that the incorporation in a single molecule of all the properties desired in a pesticide is a chance occurrence that will be all the rarer the greater the number and complexity of the requirements. He concluded that strictly selective "monotoxic" compounds cannot yet be developed systematically. He also noted that testing for effects on beneficial insects is at least as expensive as testing for effects on pest species.

Now, when the need for narrow selectivity has become increasingly acute, industry appears to have become reluctant to accelerate, indeed even to continue, their search for these desirable chemicals. Under present commercial development the selective material is not favored (Von Rumker *et al.*, 1970). Rather, the companies are forced to develop only those compounds which can be marketed on a very large scale (Persing, 1965). This emphasizes the source of one of our difficulties; it stems from the manner in which these compounds are developed commercially. First, only limited ecological considerations enter into the search for new compounds. Candidate materials are laboratory screened on the basis of high percentage kill to a small select group of pest species. A few companies may use more than 20 insects in their screen but the average is about 5 or 6 (Persing, 1965; Von Rumker *et al.*, 1970). Promising chemicals are then field tested on a small scale against a wide variety of important crop arthropods. If a given candidate is effective against major pests on important crops, it may be able to compete against the established pesticides. Once such a compound is developed for major markets, it is relatively easy to expand those markets by including minor ones. The resultant pesticide registered for use for a variety of crops has a broad toxicity spectrum and is precisely the type of compound which is so disruptive ecologically.

Von Rumker *et al.* (1970) concluded that the following were the most important obstacles to the development of more selective pesticides: (1) high cost of research and development; (2) lack of knowledge of basic plant and animal biochemistry; (3) competition from existing, nonselective, relatively inexpensive pesticides; (4) lack of grower interest resulting from the high cost of selective products which the grower is unwilling to pay; (5) cumbersome government registration procedures; (6) fear of consumer complaints and litigation; and (7) lack of interest, support, and experience on the part of agricultural workers. They also point out that the research and development costs of bringing a single compound to market has risen from $1,196,000 in 1956 to $4,060,436 in 1969, while the chance of success for an experimental product has dropped from 1 in 1800 to 1 in 5040.

The ideal selective material is not one that eliminates all individuals of the pest species while leaving all of the natural enemies. The maintenance of low populations of pest species is, of course, essential to the continuity of predators or parasites. In Washington apple orchards, the apple rust mite, *Aculus schlechtendali* Nalepa, can be tolerated at moderate populations without economic loss, and populations of this mite are a good food source for mite predators which not only prevent eruptions of apple rust mite but also regulate McDaniel mites at low densities and help to control European red mites (Hoyt and Caltagirone, 1971). Flaherty and Huffaker (1970) also described a similar situation for grapes in California.

After a shift to pesticide chemicals which are more selective in their action, it may be some time before balance is restored. The effects of the previous treatments may last several years. Biological control agents may even need to be reestablished or a key alternate host or prey species may need to be reestablished. This may be a slow process (Stern *et al.*, 1959; Flaherty and Huffaker, 1970). It took 4 years to reestablish normal predator—prey relationships for mites in apple orchards in Washington after mite control sprays were modified (Hoyt and Caltagirone, 1971).

Perhaps the most important deterrent of all to proper emphasis on selective pesticides is the difficulty experienced by industry, growers, and even many entomologists in substituting a philosophy of pest control based on regulation of pest populations below economic injury thresholds for that of maximum kill based on repetitive applications of broad-spectrum, highly effective, relatively cheap insecticides, without regard to pest population assessment and side effects. It has not proved easy to quickly change a philosophy that has been effective and has prevailed for a quarter-century. A generation of entomologists has developed under the strong influence of a philosophy that has advocated the complete elimination of every individual of a pest population wherever possible.

Clearly, the research required for development of selective chemicals will not be achieved in the near future without substantial financial support from the government. Industry is unconvinced that concentrating their research toward the discovery and development of narrowly selective (monotoxic or oligotoxic) compounds is the proper step to take. Moore (1970) has fairly stated the attitude of a majority of the industry as follows: "For example, it does not pay industry to look for and market specific pesticides, or to study new methods of integrated pest control, when these reduce the amounts of pesticides sold, and therefore reduce profits. So, only governments can sponsor adequate research on specific pesticides and integrated control effectively." Few, if indeed any, universities or governmental agencies have adequate personnel or the facilities required to mount research programs of the required magnitude that would offer a reasonable chance of significant success during the forseeable future. Therefore, it appears that the most appropriate procedure would be government subsidy support to industry to help finance the required research and development costs.

III. ECOLOGICAL SELECTIVITY

Fortunately, the development of effective and economical systems of pest management for many major species is not dependent upon physiological selectivity which could only be provided by the availability of a large number of narrowly selective pesticides.

Pesticides having a broad spectrum of activity may be used in an ecologically selective manner. The selective use of insecticides (ecological selectivity) will probably continue to be far more important in pest management than the use of selective insecticides (physiological selectivity).

Selective action of nonselective chemicals can be obtained by manipulating dosage, formulations, timing of applications, method of application, and localization of area to be treated (Ripper, 1944, 1956; Ripper *et al.*, 1948, 1951; Bartlett, 1964; Stern *et al.*, 1959; Lean, 1965).

A. Dosage

From the time of Paracelsus it has been known that dose alone determines whether or not a chemical is poisonous. The remarkable success of chemotherapy in human and veterinary medicine is based on differences in species-specific response to dosage (Albert, 1965). One of the most elegant examples to illustrate this principle is a recent one from veterinary medicine provided by Thompson *et al.* (1972) in their description of a method for control of the vampire bat, *Desmodus rotundus* E. Geoffroy Saint Hilaire, on cattle in Mexico. They reported that a single intraruminal dose of the anti-coagulant diphenadione (2-diphenylacetyl-1,3-indandione) at 1 mg/kg to cattle gave effective control of the bats that fed on the treated animals. Field tests showed 93% reduction in fresh bites 2 weeks after the cattle were treated. Laboratory studies had shown an LD_{50} for diphenadione of 0.91 mg/kg for vampire bats but that a single oral dose of up to 5 mg/kg produced no observable signs of intoxication in cattle beyond a moderate increase in the clotting time of prothrombin. The method depends on: (1) the daily requirements of vampire bats for blood; (2) the tendency of anticoagulants temporarily to bind to blood protein; and (3) differential sensitivity of cattle and vampire bats to diphenadione.

Although not so spectacular as this control of vampire bats, use of organophosphorus systemic insecticides for control of cattle grubs, *Hypoderma* spp., in oral or dermal applications and control of the house fly, *Musca domestica* L., by oral application of Rabon to contaminate the feces of treated animals at levels toxic to larvae of the pest also demonstrate that dosage can be used to achieve ecological selectivity (Khan, 1969; Skaptason and Pitts, 1962; Miller and Gordon, 1972).

In agricultural entomology, Ripper (1956) demonstrated the value of reduced dosages of the systemic insecticide schradan for control of the cabbage aphid, *Brevicoryne brassicae* (L.). In the integrated control of the spotted alfalfa aphid, *Therioaphis maculata* (Buckton), in California a key element is the use of demeton at a discriminating dosage that gives adequate kill of the aphid while only minimally affecting its predators and parasites (Stern and Van den Bosch 1959).

The importance of dose in the toxicity of a compound was emphasized by Schulz (1888) who published data indicating that poisonous substances have a stimulatory effect when given in small doses. His work and that of the German psychiatrist Arndt became the basis for the Arndt-Schulz law which holds that weak stimuli excite physiological activity, moderately strong ones favor it, strong ones retard it, and very strong ones arrest it. Many apparently unrelated stimuli, including chemical pesticides, produce this series of responses, i.e., stimulation upon exposure to subinhibitory doses, and depression, then arrest, with increased doses. That hormoligosis (the phenomenon of the stimulatory effect of a small amount of an agent on living organisms) may be more important in insect control by chemicals than has been generally recognized is indicated by work of Huffaker and Spitzer (1950), Hueck *et al.* (1952), Kuenen (1958), and Luckey (1968). Hueck *et al.* found that exposure of *Metatetranychus ulmi* Koch to DDT at subinhibitory concentrations had a stimulating effect on egg production. Kuenen found that the granary weevil, *Sitophilus granarius* (L.), exposed to very low doses of DDT mixed with the culture medium (wheat) produced about 20% more offspring than unexposed weevils. More recently, Luckey (1968) found that growth was increased in the house cricket, *Acheta domesticus* (L.), when reared at suboptimum conditions and exposed to low concentrations of most of the fourteen pesticides tested.

The significance of hormoligosis for pest management is unknown. It deserves much more study. If it occurs widely, as is suggested by the few studies made, another selectivity avenue could be opened to pest management by manipulation of pesticide dosages such that, in some cases at least, the pest could be suppressed and the natural enemies stimulated.

The philosophy of "overkill" so widely prevalent for a quarter-century has resulted in use of far more pesticides than required for optimum insect control. Not only has this philosophy resulted in excessive pollution of the environment, rapid development of insecticide-resistant populations, rapid increase in status of secondary and occasional pests, and unnecessarily severe effects on nontarget species, but it has also been economically unsound. Establishment of minimum dosages required to hold pest populations just below economic injury levels for the minimum period of time needed for optimum crop production, and acceptance of these levels by growers is an objective of highest priority.

B. Confining Insecticides to Restricted Areas

There are many opportunities for restricting the total amount of pesticides used for control of pest species. Thorough knowledge of the biology, ecology, and behavior of a pest often makes possible its control by application of insecticides to very restricted areas. The effectiveness of such a technique was demonstrated by Isely (1926) almost 50 years ago by his "spot-dusting" method, which consisted of dusting with calcium arsenate only those areas of cotton fields that were infested by overwintered adults of the target species, the boll weevil. It took advantage of an element of behavior of the pest that results in areas of fields adjacent to favorable overwintering quarters being invaded first by adult weevils emerging from hibernation. Such areas are characterized by differences in soil or fertility levels that stimulate cotton to more vigorous early growth and development. Isely found that overwintered weevils were invariably attracted to, and concentrated in, such areas and that often only a very small percentage of the total area in a field was infested.

By taking advantage of this behavioral trait, a high percentage of the total boll weevil population that had survived the winter in a given locality could be destroyed by treating only a small percentage of the crop acreage. Such a method had the advantage of preventing, or reducing substantially, the need for additional insecticide treatments. Consequently, both costs and adverse effects on nontarget organisms were held to a minimum.

Isely (1934) advocated a refinement of the method based on early planting of an early maturing variety in areas found by experience to be those most attractive to overwintered boll weevils in order to attract and confine the pests, as they emerged from hibernation, to smaller and more specific areas.

More recently, L. D. Newsom (unpublished data) has found that the bean leaf beetle, *Cerotoma trifurcata* (Forster), has a behavioral trait very similar to that of the boll weevil. Relatively small areas of a field planted near favorable hibernation quarters 2 weeks or more before the main crop attract and hold a high percentage of the overwintered beetles. Populations concentrated by this method can be controlled by insecticide applications timed so as to prevent egg deposition, or to control teneral adults of the first generation before they have time to disperse to other areas of the field. In addition to reducing the amount of acreage required to be treated and the amount of insecticide released into the environment, thus reducing costs and environmental pollution and conserving populations of predators and parasites, the method also controls bean pod mottle mosaic of soybean when it is a problem. This important virus disease of soybean overwinters in its perennial wild hosts and is transmitted to soybean almost exclusively by the bean leaf beetle. Concentrating the overwintered beetles in small trap plots of early planted soybeans and destroying them by application of insecticide prior to dispersion to adjoining areas resulted in excellent control of the virus. Selective action on a pest-parasite complex can also be obtained by restricted area usage, i.e., by treating only those areas where the pest-parasite ratio is unfavorable. Thus, on an area-wide basis, the pest-parasite ratio was shifted in favor of the hymenopterous parasite, *Apanteles medicaginis* Muesebeck, as contrasted to its host caterpillar, in the supervised control program for alfalfa butterfly control in California, even though many parasites were destroyed in the treated areas (Smith, 1970).

Other behavioral traits have been found to occur in various species which allow high percentages of a population to be concentrated in restricted places. Nishida and Bess (1950) and Nishida (1954) showed that the melon fly, *Dacus cucurbitae* Coquillett, in Hawaii spent a relatively small part of each day in cultivated crops, and the major part in bordering or adjacent vegetation. The females move into the fields to oviposit and then return to these outside areas. They control this pest more effectivly by treating these adjacent areas than by treating the fields themselves.

Control of the dimpling bug, *Rhodolygus milleri* Ghauri, a serious pest of apple in Tasmania, was unsuccessful when the orchards were treated (Terauds, 1970). An investigation of the biology of the pest revealed that its primary host was *Cupressus macrocarpa* Hartweg, a tree planted around the orchards for windbreaks. Properly timed applications of 0.1% DDT to the windbreak trees alone gave good control of the bug in the orchards and reduced crop damage to negligible levels.

An excellent example of selective action of pesticides is the discriminative use of insecticides for control of tsetse flies, *Glossina* spp. (Chadwick *et al.*, 1964; Hocking *et al.* 1966; Davies, 1971; Park *et al.*, 1972). By taking advantage of the tsetse's highly selective choice of resting sites, degrees of control usually described as "eradication" have been achieved by application of insecticides to extremely localized sites. In the bushland of northern Tanganyika "eradication" from a 35 square mile area was achieved by treating only the undersides of tree branches with a diameter of 1 to 4 inches, between 4 and 9 feet from the ground, and inclined less than 35 degrees from the horizontal, with 3% dieldrin or endosulfan. In northeastern Nigeria, tsetse-infested country has been reclaimed by treating narrow strips of vegetation along the edges that border rivers and streams. In the most frequently occurring situations where narrow fringing forests border well-defined river banks the width of the sprayed strips of vegetation required for control is 10 yards for *Glossina morsitans* Westwood and 5 yards for *G. tachinoides* Westwood. The cost per square mile of these methods range from one-sixth to one-half that for treatment by aircraft (Davies, 1971; Chadwick *et al.*, 1964; Hocking *et al.*, 1966; Park *et al.*, 1972).

Control of migratory locusts has evolved toward emphasis on preventing these pests from moving away from their breeding grounds to form migratory swarms. An international cooperative organization exists to combat these locusts wherever they are observed, regardless of distance from agricultural areas, in order to stop potential outbreaks at their source. Control is based on use of chemicals that has evolved from the poison baits with sodium arsenite, sodium silicofluoride, or benzene hexachloride as the toxicants to low volume spraying by aircraft with synthetic organic insecticides such as dinitrocresol, benzene hexachloride, or aldrin. It culminated during the 1950's when "vegetation baiting" with dieldrin sprays was adopted. Strips of vegetation as much as 1 km apart are treated using dieldrin applied at 1 oz/acre in a nonvolatile oil. Such sprayed strips remain toxic to all hopper bands that attempt to move across them for as long as 1 month (Lean, 1965).

Anderson (1965) and Anderson and Poorbaugh (1964) reported that flies and mosquitoes were also controlled effectively by confining insecticide treatment to their resting sites rather than to the entire surfaces of poultry houses. Lice and mites of poultry have for many years been controlled effectively by painting the roosts with an appropriate insecticide.

The importance of narrowly restricting the placement of insecticides to get maximum effectiveness as well as differences in species-specific effects of a given toxicant is well illustrated by Way's (1959) studies on the mode of action of insecticidal seed dressings. He found that upon hatching larvae of the wheat bulb fly, *Leptohylemyia coarctata* (Fallen), move upward through the soil to near the surface where they search for host plants, entering the shoots at a depth of $\frac{1}{4}$ to 1 inch. About one-half of the larvae were killed by contact with dieldrin-treated seeds sown $\frac{1}{4}$ inch deep, contrasted with few or none killed by similarly treated seeds sown at a depth of 3 inches.

Commercial damage to tobacco by the tobacco hornworm, *Manduca sexta* (L.), could be prevented with one-fifth the recommended rate of insecticide if treatments were directed to the upper five or six leaves at the time the oldest larvae were in the 3rd and 4th instars (Guthrie *et al.*, 1956). Restricted application cost less and resulted in less insecticide residue at harvest.

A behavioral pattern of the May beetle, *Melolantha vulgaris* F., has been used successfully in restricting insecticide applications for its control to very localized areas (Schneider, 1958). Adults of this species fly in "swarm paths" from larval breeding places to the bordering woods where they congregate in very definite places. The beetles were found to orient by optical stimuli to the section of the horizon with the greatest average silhouette height, the attraction decreasing with distance and vanishing at about 10,000 ft. By using this and other information, systematized, selective control measures were devised that permit the treatment of minimal areas.

Thus, through the understanding of insect behavior, broad-spectrum insecticides may be used in ways to attain high degrees of selectivity. Ecological selectivity so obtained deserves much more attention than it has received to date. It reduces the amount of pesticides introduced into the environment and restricts the amount of acreage treated. This lowers costs, conserves populations of beneficial insects, minimizes the selection pressure toward pesticide resistance, and reduces the hazard to other nontarget species. These advantages far outweigh any inconveniences caused in obtaining the biological knowledge needed for restrictive application, or by planting earlier than at optimal time, using less desirable varieties, or less then optimal cultural practices.

C. Formulation and Methods of Application

Methods of formulating and applying broad-spectrum insecticides can often be varied so as to obtain a high degree of selectivity. An outstanding example is provided by a method developed for control of peanut pests. The primary pest of peanuts in Texas is the lesser cornstalk borer, *Elasmopalpus lignosellus* (Zeller), which feeds below the soil surface on developing nuts. Until recently, peanut growers applied foliar sprays, mainly by airplane, for control of this pest. The fallacy of this type of application is that peanuts have a large number of secondary pests, including many foliage-feeding caterpillars and spider mites (J. W. Smith, unpublished data). The foliar sprays applied for control of the borer killed the beneficial arthropods and unleashed the secondary pests. As a result, many producers were forced to treat the crop eight to ten times per season. The placement of a granular insecticide in the soil zone where the borers were feeding was highly effective for controlling this pest without disrupting the arthropod fauna of the foliage. This has allowed many growers to reduce the number of treatments to one or two per season.

Lepidopterous pests of maize, especially the fall armyworm, *Spodoptera frugiperda* (J. E. Smith), are effectively controlled in areas of South America, Central America, and Mexico by applying insecticides formulated as granules to the central whorl of the plants. Insecticides applied as granules exert minimal adverse effects on populations of natural enemies on other parts of the plants.

Dust formulations of carbaryl effectively control the southern chinch bug, *Blissus insularis* Barber, on pasture and forage grasses in southern Louisiana and Mississippi (E. H. Floyd, unpublished data). Granular and sprayable formulations of carbaryl are not effective because they do not penetrate the thick mat of vegetation under which *B. insularis* is found on these crops.

D. Baits

Poison baits have provided effective and long-used means of obtaining ecological selectivity of broad-spectrum insecticides. Arsenical baits have been effective for control of various species, e.g., grasshoppers, locusts, Mormon crickets, armyworms, and cutworms; they provided the only effective, economical control for many of these pests prior to the synthetic organic insecticide era. These highly effective and relatively cheap modern insecticides have now generally replaced the arsenicals. In fact, poison baits were essentially abandoned for several years. Recently, this tactic appears to be making a comeback as serious problems have developed in the use of these modern insecticides and as the value of selectivity is becoming widely recognized.

A striking example of achieving ecological selectivity by use of poison bait is that of mirex to control imported fire ants, *Solenopsis* spp., in southern United States. This broad-spectrum, highly persistent organochlorine dissolved in soybean oil and impregnated on finely ground corn cob "grits" at 0.3% of the finished bait broadcast at 1.25 lb/acre gives excellent control of both *Solenopsis saevissima* (F. Smith) and *S. invicta* Buren. This dosage is an almost unbelievably small amount (1.7 gm of mirex/acre/application). Three applications 4—6 months apart provide highly effective control (Anonymous, 1972).

Selectivity is due to the specialized behavior of the ants and to several features of the pesticide itself. Also it is unattractive to the great majority of nontarget species. Mirex is relatively slow acting; this allows the workers to carry the bait into the nests to the queen and immature forms. The foragers remove bait particles from the soil surface so rapidly that it is exposed to nontarget species for a relatively short period of time. Although highly selective when used at such a low dosage and in such a highly selective formulation, this use of mirex is not without undesirable side effects. Several species of oil-loving ants are killed. Other nontarget arthropods are also affected (Harris and Burns, 1972). Thus applied mirex enters the ecosystem food chains and because of its extreme stability it is highly concentrated in a number of species of insects, birds, fish, and mammals. However, this usage is remarkably more selective than the previously used organochlorine insecticides heptachlor and dieldrin applied at rates of active ingredient far greater than required for mirex.

Several fruit fly species have been effectively controlled by using poisoned baits. A striking example is the eradication of the Oriental fruit fly, *Dacus dorsalis* Hendel, from the island of Rota by male annihilation using the highly effective male attractant, methyl eugenol. The bait, consisting of small cane fiberboard wafers saturated with methyl eugenol and naled (95:5), was distributed over the island by airplanes (Steiner *et al.*, 1965, 1970). Generally, use of such baits means much reduced amounts of insecticide per unit area and substantial reductions in the total area requiring treatment.

E. Seed Treatments

Treating of seeds provides another means of achieving selectivity. The synthetic organic insecticides, for the first time, made this usage available for control of many pests, including especially seed maggots, wireworms, rootworms, and thrips. Often, seed dressings at very low amounts of insecticide per acre have given comparable control to that obtained by broadcast or band applications of the same chemical at much higher rates of application. In southern United States a complex of seedling pests of corn composed of the seed-corn maggot, *Hylemya platura* (Meigen), sand wireworm, *Horistonotus uhlerii* Horn, and southern corn rootworm, *Diabrotica undecimpunctata howardi* Barber, is effectively controlled using seed dressings. Dieldrin, applied at 32 gm of active ingredient per bushel of seed corn and seeded at 10 to 15 lb/acre, 5—8 gm/acre of toxicant, gives control equal to 100 times this amount of insecticide applied broadcast. Use of seed treatment rather than broadcast application for control of such a complex has the following advantages: reduction in costs resulting from reduced amounts of pesticide, no additional costs for application equipment or application, reduced levels of environmental contamination, minimum contact with nontarget species, and reduced selective pressures on the pest species to develop resistance to the chemical because only a small percentage of the population is exposed in contrast to broadcast application. It has some disadvantage in not giving effective control of the cutworm pests of seedling corn, *Agrotis ypsilon* (Hufnagel) and *Felita subterranea* (F.); however, they are relatively minor, highly localized pests of sporadic occurrence.

Unfortunately, seed treatments do not control seedling corn pests prevalent in the large corn producing areas of midwestern United States. Comparatively large amounts, 1 to 2 lb/acre/year of persistent organochlorine insecticides applied broadcast or in bands, have been required. Consequently, there has developed the well-known syndrome of resistance to the chemicals in several species, especially *Diabrotica* spp., adverse effects on nontarget organisms, and unacceptably high levels of environmental pollution.

F. Proper Timing of Insecticide Application

Timing of applications is often the most effective and economical way of achieving differential insecticide selectivity on the pest—natural enemy complex. Detailed knowledge of the biology and ecology of the species involved and of economic injury relationships is required. Too often insufficient biological data are available. The enormous initial successes with the synthetic organic insecticides, contrasted to the insecticides previously used, severely reduced research on these basic principles of applied ecology. Work in these areas was largely ignored by a generation of entomologists. Only recently, after these initial successes have been eclipsed by failures resulting from the well-known syndrome of unwanted side effects (above), has the profession become increasingly aware that pest management must be based on better information on agroecosystem biology, ecology, and economic injury levels.

Excellent results have often been obtained by simply delaying initial insecticide applications until pest populations have reached economic injury thresholds and discontinuing applications when the cost of further protection would equal or exceed the increased gain. Long and Concienne (1964) demonstrated that treatment was unnecessary for control of the sugarcane borer, *Diatraea saccharalis* (F.), during the 1st generation in June and early July and the 4th and 5th generations during September and October. Restricting treatment to 2nd and 3rd generations gave adequate protection to the crop. Control of earlier or later generations had no effect on yield or quality. Their findings allowed a reduction by more than one-half in the number of insecticide applications required. Similarly, Keriem *et al.* (1971) reported that two applications of carbaryl were as effective for control of stem borers of maize when applied at 20 and 45 days after planting, as three at 20, 30, and 45 days, or four at 20, 30, 45, and 60 days after planting. Yields were not different between schedules of applications under levels of infestation that reduced yields in the control plots by about 60%.

Cotton is attacked by a large complex of serious pests and profitable production in many parts of the world has required heavy use of pesticides. In some countries, e.g., United States, the amount of chemicals used for control of insects and related pests of cotton has been about one-third to one-half of the total used on all crops. Effective methods must be developed to obtain greater selectivity with reference to pesticide action on pests and natural enemies. It was thought that this could be realized by the application of insecticides timed to coincide with the beginning of fruiting in order to destroy overwintered boll weevils before they could reproduce and then omitting further applications in order to allow the above-mentioned predator and parasite populations to increase to effective controlling levels (Ewing and Parencia, 1949). Suppression of boll weevil populations was not satisfactorily achieved with the two or three "early season" applications as initially envisioned. The method quickly developed into a "full-season" program of applications, representing a 180-degree change from the previously practiced methods based on delay of insecticidal treatment until weevil infestations had reached levels of 10 to 25% of the flower buds (squares) infested, or plant bug populations had reached densities of 25 to 40 per 100 plant terminals, or both. The latter method frequently permitted delay of insecticide applications until early July or later.

In the "full-season" method, applications are started as soon as the seedling plants emerge to a stand and are continued at weekly intervals until the crop is mature and safe from insect attack. Such a program decimated the predator—parasite complex. Previously, these natural control agents had exerted such consistent and heavy pressure on the tobacco budworm, *Heliothis virescens* (F.), that it was virtually unknown as a pest of cotton in the United States. Under "full-season" treatment of cotton, its release from the pressure of predators and parasites during the 2nd generation has converted cotton to a major host. This has resulted in changing the status of the tobacco budworm from a species rarely reported from cotton (Brazzel *et al.*, 1954) to one that has destroyed a 500,000-acre cotton industry in northeast Mexico and is currently posing a serious threat to cotton production in much of south, central, and eastern Texas and the mid-south cotton-producing states. It has also made the species a greater threat to other crops in cotton growing areas. (See also Chapter 17.)

Recently, cotton insect control has undergone another drastic change in many of the areas where the boll weevil is a serious pest. Problems that have developed from misuse of insecticides, i.e., problems inherent in methods that rely on fixed schedules of application without regard to pest population assessment or economic injury thresholds, have become so acute that there must be substantial reductions made in the amount of insecticide applied. In these areas there has been a return to the philosophy of delaying the start of treatments as long as possible (based on pest population assessment). This is the same method that was prevalent prior to the synthetic organic insecticide era, but a unique component has been added. This component consists of applying an organophosphorus insecticide to control the boll weevil in a series of applications scheduled to destroy the potential diapausing generation(s) which alone is capable of overwintering successfully. The method has been termed "diapause control" or "reproduction-diapause control."* It is based on continuing applications after the crop has matured until it is harvested and until the plants are destroyed mechanically or killed by frost so as to prevent the boll weevil from entering hibernation quarters in a physiological state that will make it capable of overwintering successfully. Usually, one application added to the defoliant application plus two or three additional ones made at approximately 10-day intervals is all that is required in most areas. This method suppresses the weevil population to the point that applications the following year may often be delayed for a month or more beyond the normal starting time. Thus, the amount of insecticide required to control cotton pests is reduced by one-third to one-half. This allows maximum predator and parasite pressure on *Heliothis* spp., *Trialeurodes abutilonea* (Haldeman), *Aphis gossypii*, *Lygus* spp., and *Pseudatomoscelis seriatus* (Reuter). It also reduces the selective pressure toward development of resistance by these species by exposing fewer generations to the heavy dosages of insecticides required in the full-season program. Thus, it may halt the rapidly accelerating rate of development of resistance that has become widespread in *Heliothis virescens* and *Trialeurodes abutilonea*; it may even permit some loss of already attained resistance.

*Neither "diapause control" nor "reproduction-diapause control" describes accurately what happens when insecticides are applied during late season and postharvest for control of weevils that are in the physiological state of diapause. Photoperiod is the major factor involved in the induction of diapause in the boll weevil and induction takes place during the immature stages (Earle and Newsom, 1964). Obviously diapause cannot be controlled by destroying with insecticides adults in which diapause has already been induced. Weevils that are in this physiological state are incapable of reproducing until they have undergone diapause development that takes place during their overwintering period. Those that are not in diapause will continue to reproduce within limits imposed by environmental conditions. Thus, reproduction can be prevented by applying insecticides timed in such a manner that the insects will be killed before they can reproduce but this does not mean "reproduction control." A more accurate and properly descriptive term would be "late season—postharvest control."

This method appears to be the only currently available tactic that offers hope of delaying the rapid spread in the southern United States of what appropriately has been termed the *Heliothis* disaster syndrome. Nevertheless, Newsom (1972) called attention to the weaknesses and potential hazards involved in use of this method: Its possible adverse effects on populations of nontarget species has not been assessed; the insecticides are applied on fixed schedules without regard to pest population assessment; it exerts such heavy selective pressure upon the boll weevil that the possibility of its developing resistance to the organophosphorus (O-P) insecticides is greatly increased; and the possibility that late season—postharvest applications of the O-P insecticides may release 3rd and 4th generation larvae of *Heliothis virescens* from the regulatory effects of parasites and predators, is especially disturbing. Such an eventuality could result in an enormous increase in overwintering populations of this increasingly serious pest. (See also Chapter 17.)

Unfortunately, the information required for using proper timing and other principles, as described above, is much too infrequently available. However, in some cases the information required for timing application of insecticides in such a manner as to obtain a high degree of selectivity is both simple and available. The honeybee is highly susceptible to some of the insecticides widely used for control of cotton and soybean pests in southern United States, e.g., to methyl parathion and carbaryl. By taking advantage of the short residual activity of methyl parathion and the fact that foraging bees discontinue their activity at about 5 to 6 P.M., this highly toxic chemical can be used with minimum adverse effects. Restricting applications to late afternoon hours after the foraging bees have returned to the hives prevents unacceptably high levels of mortality in honeybee populations in treated areas.

Selective timing of the extremely broad-spectrum O-P insecticide parathion in the form of a bed drench for control of the French fly, *Tyrophagus longior* Gervais, is an important component of a promising integrated control program being developed for glasshouse pests in England (Gould, 1971; Parr and Scopes, 1971; Binns, 1969; Hussey and Bravenboer, 1971).

G. Systemic Insecticides

Development of systemic insecticides (Ripper, 1956) made possible a previously unknown technique of using insecticides in a selective way. Schradan was one of the first of these new materials to be developed and used widely. Discovery of its unique and useful properties stimulated a great deal of research and many organophosphorus and carbamate insecticides were quickly made available for evaluation. Many of these chemicals proved to have useful properties for obtaining selective action. They are absorbed rapidly by foliage thus becoming essentially unavailable directly to all except those species feeding on the treated plants. They may also be applied to the soil or to seeds in seed dressings, enter the plants through the roots, and be translocated to aerial portions in amounts that are selectively toxic to some phytophagous species.

Such qualities have resulted in overoptimism in evaluating the role of systemics in pest management. The relative safety of systemic insecticides to predators and parasites when applied in a manner not producing contact effects has usually been overestimated (Ripper, 1956; Bartlett, 1964; O'Brien, 1961; Metcalf, 1964; Ridgway, 1969; Cate *et al.*, 1972).

Ahmed *et al.* (1954) demonstrated that some species of predators are killed by feeding upon prey poisoned with systemic insecticides. The significance of such secondary poisoning has not been determined. Some facultatively phytophagous predator species may be severely affected by certain systemic insecticides. However, these effects are probably far less important than the indirect hazards from elimination of hosts that are the targets of the applications. Very early, Boyce (1936) recognized the danger in reducing population densities of pest species that serve as hosts of predators and parasites to such low levels that the control agents are either decimated by lack of prey (or hosts) or are forced to emigrate. The significance of such an indirect hazard to predator and parasite complexes of cotton pests has been stressed (Newsom, 1972; Newsom and Brazzel, 1968).

Techniques for evaluating the impact of destruction of such pests as thrips, spider mites, aphids, and plant bugs by use of systemic insecticides to soil, seeds, or plants usually have involved small plot field tests as the final step before recommendations for commercial use. This experimental procedure is quite inadequate for evaluating responses of parasite and predator populations to a toxicant. Such plots are usually surrounded by untreated areas, or areas treated with some other chemical. In either case results obtained are likely to be badly biased. Where the untreated areas surrounding treated plots are large they may serve as a refuge for predators and parasites and allow rapid recolonization of the treated areas even though the latter may have been totally defaunated initially by the chemical. The results obtained may be grossly misleading. A good example is provided by research with aldicarb applied to cotton in furrow at planting, or as a side dressing later, for control of thrips, plant bugs, and overwintered boll weevils. Early field experiments with this chemical were conducted in which the ratio of untreated to treated area was very large. Thus, rapid recolonization of treated areas by natural enemies was possible and the extreme hazard of this material to such nontarget species was masked. Consequently, the manufacturer was encouraged to proceed with the very expensive process of establishing tolerances and obtaining registration for use of aldicarb on cotton at heavy dosage rates. When large contiguous areas of cotton have been treated with aldicarb at dosages of 1 to 2 lb, or more, per acre, populations of many species of predators and parasites have been severely depressed. Ridgway (1969) reported that side-dress applications of aldicarb for control of boll weevil showed almost 80% reduction in populations of predators, with a corresponding sixfold increase in populations of *Heliothis* larvae, compared with populations in control plots.

Earlier, Cowan *et al.* (1966) and Coppedge *et al.* (1966) had shown the severe effects of aldicarb on thrips, aphids, and cotton fleahoppers, the principal hosts of members of the predator complex that regulate populations of *Heliothis* spp. in cotton. Ridgway *et al.* (1967) found a close correlation between reduction in populations of predator species in cotton previously treated with aldicarb and subsequent buildup in populations of *Heliothis*. Cate *et al.* (1972) reported that soil applications of aldicarb as a side dress at 2 lb/acre resulted in sufficient translocation to the extrafloral nectar to kill 100% of adults of the parasite *Campoletis perdistinctus* Viereck if fed the nectar collected 3 days after treatment, and 13% if fed the nectar collected 22 days after treatment. These authors arrived at the curious conclusion that ". . . soil applications of aldicarb in the field did not appear to harm adult *C. perdistinctus*, though nectar collected from treated plants killed some parasites." The demonstrated severe effects of aldicarb upon the natural enemy complex of cotton ecosystems in the southern United States should preclude its recommendation for use on the crop there.

Although systemic insecticides may be less specific than is often claimed, they are nevertheless very useful. They are especially effective in situations where aphids and spider mites are members of a pest complex. In such cases they often may be applied at dosages so low that aphid and spider mite populations can be regulated at subeconomic levels with little harm to predators and parasites.

IV. INSECT GROWTH REGULATORS AND PHEROMONES

These two relatively new approaches are here considered together although there is no implication that they are related.

A. Insect Growth Regulators

An intensive effort to explore the potential for pest control of insect growth regulators (IGRs) and their analogs has been underway for almost a decade. These chemicals have generated an interest and excitement in pesticide research that is in sharp contrast to the current attitude toward research on conventional pesticides. Only the juvenile hormone analogs have thus far been demonstrated to have any immediate promise for use in pest management systems (Staal, 1972, 1975).

Bagley and Bauernfeind (1972) reported the results of extensive laboratory and small-scale field trials with three juvenile hormone analogs: methyl 10,11-epoxy-7-ethyl-3,11-dimethyltridecadienoate (mixture of eight isomers of JH 1, the "Roeller compound"); ethyl *trans*-7,11-dichloro-3,7,11-trimethyl-2-dodecenoate, "Romanuk compound"; and 6,7-epoxy-3,7-dimethyl-1,3,4-(methylenedioxy) phenoxy-2-nonene (mixture of four isomers), "Bowers compound." A large number of pest species representing several orders was included in the tests. Varying degrees of activity of the following types were exhibited: sterility, reduced reproduction, ovicidal effects, reduced emergence, deformity due to disturbed metamorphosis, mortality, and abnormalities in larval pigmentation.

A large volume of similar research has been subsequently published. Many additional species have been found to be susceptible to the effects of these insect growth regulators. However, progress has been very slow in developing any of them to the point that they may be expected to become widely useful in pest management systems in the near future. In general, it has been found that they possess such favorable characteristics as high biological activity against insects, including strains that are resistant to conventional insecticides, low toxicity to vertebrates, short persistence in the field, and intermediate to high selectivity. Predators and parasites are generally not so adversely affected directly, because of the greater selectivity, as with conventional pesticides. Their greatest weaknesses appear to be that they have no effect on young larvae or nymphs, are effective only during critically short periods in the life cycle of most species, often require long periods of time between application and expression of effects, and are degraded so rapidly that repeated applications at short intervals would be required for control of multivoltine species whose generations or broods are not closely synchronized.

Some early workers had been enthusiastic in their convictions that insect pests would not be able to become resistant to hormone materials because they play an indispensable role in the normal development of insects. This enthusiasm may be ill-founded, however, because Staal (1975) cited several references that reported some insecticide-resistant strains of insects showed cross-resistance to IGRs. He suggested the probability that the same classes of enzymes responsible for the metabolism of conventional insecticides may be involved in the breakdown and regulation of hormone titers in insects. Schneiderman (1972) expressed the view that the wide differences in sensitivity of different insect families to specific juvenile hormone analogs ensured the occurrence of insects resistant to these chemicals. Although there are some who oppose this view, it is supported by the relatively rapid development of resistance to methoprene in the mosquito *Culex pipiens pipiens* L. and the housefly, *Musca domestica* L. Brown and Brown (1974) reported that they had induced 13-fold resistance in a strain of *C. p. pipiens* by selection with methoprene. According to E. C. Burns (personal communication) measurable levels of resistance have been induced in a strain of housefly by selection with methoprene in "feed through" treatments for control of dung breeding flies.

In spite of the intensive research effort devoted to them for almost a decade, only one juvenile hormone analog, i.e., methoprene, has been registered for use. Its immediate use is limited to control of floodwater mosquitoes. The promise it shows for certain other Diptera, particularly for use in "feed-through" treatments of cattle and poultry for control of flies breeding in manure, is clouded by the development of resistance to it in the housefly.

The status of IGRs from the standpoint of possible carcinogenic, teratogenic, and mutagenic activity has not yet been clarified.

Much additional research remains to be done before the final role of insect growth regulators in insect pest control can be determined. Results obtained thus far indicate that widespread utility in the near future for specific uses in integrated pest management programs cannot be anticipated.

B. Pheromones

The potential value for pest control of synthetic pheromones and related chemicals has been the subject of extensive research for more than a decade. Although results were disappointing initially, research during the last few years indicates that some of these chemicals may soon become powerful tactical weapons in the strategy of integrated pest management. They show much promise for direct control and more accurate monitoring of pest populations (Birch *et al.*, 1974).

Two approaches have been taken for employing the synthetic pheromones in direct control of pests:

1. *Disruption* of pheromone communication by permeating the environment with the pheromone of the target species to the extent required for disrupting mating. Mating disruption has been demonstrated for major lepidopterous pests including the following: *Trichoplusia ni* (Hübner) (Shorey *et al.*, 1972); *Pectinophora gossypiella* (Saunders) (Shorey *et al.*, 1974); *Argyrotaenia velutinana* (Walker) and *Paralobesia viteana* (Clemens) (Taschenberg *et al.*, 1974); *Porthetria dispar* (L.) (Beroza *et al.*, 1974); and *Grapholitha molesta* (Busck) (Gentry *et al.*, 1974). Population control *has not been* clearly demonstrated to date. There is promise, however, that such potentials for control may soon be realized.

2. *Trapping*. This method has proved to be particularly promising for control of *A. velutinana* (Roelofs *et al.*, 1970); and *Dendroctonus brevicomis* LeConte (Birch *et al.*, 1974). Trapping has also been proved to be highly effective for monitoring populations of *Anthonomus grandis* Boh. *A. velutinana*, *D. brevicomis*, and *P. gossypiella*. Toscano *et al.* (1974) demonstrated that timing applications of conventional insecticides by monitoring populations of *P. gossypiella* with traps baited with Hexalure made possible reduction in numbers of applications by about one-third and reduction in costs of about $8.00/acre compared to conventional methods. They listed the disadvantages of the method as being daily inspection of the traps, necessity for making insecticide applications within 24 hours of the time for treatment, and adverse effects of wind and rain on trap catch. However, they considered these disadvantages to be far outweighed by the advantages of reductions in costs, risks of inducing outbreaks of other pests, and probability of developing resistant populations.

Most studies indicate the need for suppression of populations of the target species to very low levels by use of conventional pesticides or other appropriate methods, before beginning applications of the synthetic pheromones in order for the method to work most effectively (Beroza and Knipling, 1972; Beroza *et al.*, 1974; Mitchell and Hardee, 1974). This may offset to a considerable extent any potential gains from decreased use of conventional insecticides when the method is employed. A unique feature claimed for the confusion method is that it becomes progressively more efficient as the population is reduced (Beroza *et al.*, 1974). The synthetic pheromones are not as narrowly specific as they were originally considered to be (Kaae *et al.*, 1972), but they have a very narrow spectrum of activity. All available evidence suggests that they are environmentally safe.

Much additional research remains to be done before their promise can be realized. Answers must be obtained to such basic questions as number of applications required for control, most effective formulations, time of application, and dosage rates, and mechanisms of mating disruption, i.e., confusion, habituation, arrestment, and repellency. Taschenberg *et al.* (1974) have called attention to the possibility that habituation of males to the chemicals, as well as confusion, may be a mechanism responsible for disruption of mating. If so, lower rates of application than required for confusion may be desirable. Finally, the important question of their possible carcinogenic and teratogenic activity must be resolved.

The excellent results obtained with these chemicals in large-scale field trials for control of such important pests as *Anthonomus grandis*, *Argyrotaenia velutinana*, and *P. gossypiella* is reason for optimism that synthetic pheromones will become important components of integrated pest management systems of the future. It should be emphasized, however, that these chemicals will be most useful in such systems and not as the panacea that they were envisioned to become by some workers.

V. CONCLUSIONS

Research on alternatives to conventional chemical insecticides has yielded an overwhelming body of data which shows that nothing is yet available that can take the place of these indispensable chemicals. Equally conclusive data have been obtained during the last two decades which show that society cannot and will not tolerate the profligate manner in which these potent chemicals have been used for a quarter of a century. Moreover, their use in this way has often proved self-defeating for the grower himself, as the above accounts describe. The only way out of the impasse is the development of pest management systems that make intelligent use of chemical pesticides. Although much too slow, substantial and rapidly accelerating progress has been made toward achieving this objective. Selective pesticides and the selective use of pesticides are now, and must continue to be for the forseeable future, one of the foundations upon which pest management programs are constructed. No other method is available that can so quickly be brought to bear for controlling populations of a wide spectrum of pests and pest complexes with such predictably reliable results. Indeed, the intelligent use of conventional chemical pesticides provides the only currently known effective means for controlling a large number of the world's most important pests.

Narrowly selective (monotoxic) chemicals appear to offer an almost ideal means of pest control. However, only a very few such chemicals have been discovered and future prospects for additional discoveries have become very dim. The chemical industry, for the most part, has historically had little interest in finding and developing such useful compounds. The financial return upon investments in research and development of truly physiologically selective insecticides is very small when compared with that for the broad-spectrum compounds so widely used for insect control. Except for a relatively few key pests of major crops. e.g., boll weevil, rice stem borer, and codling moth, industry would be hard pressed to recover research and development costs of monotoxic compounds for pest control.

Problems involved with development of resistance to insecticides in many pest species, with resultant rapid obsolescence of the chemical, unwanted side effects, high costs of securing tolerances and registrations for use, and a society that has become increasingly critical of chemical pesticide use make it highly unlikely that industry will be willing to make a substantial effort to discover new selective compounds. In fact, the prospects are so unattractive it is unlikely that industry will attempt to develop compounds previously synthesized and known to possess selective properties but that are now sitting on the shelves of chemical laboratories.

Fortunately, it is not necessary to rely upon the physiological selectivity of chemicals to obtain the effects required. Ecological selectivity obtained by discriminating use of even the most broad-spectrum insecticides can be employed in many cases for the development of effective, economical, and ecologically sound pest management systems. Development of such systems is presently limited to some extent by a lack of knowledge of the biology and behavior of pest—natural enemy—crop complexes. A more serious limiting factor is a shortage of properly trained, imaginative, and capable entomologists dedicated to the development of pest management systems based on the principles of integrated control. Nevertheless, there are many encouraging examples of progress. Recognition of the need for improved pest management has finally resulted in substantial financial support to a large number of institutions for both research and extension in this field. The most encouraging development of all, however, is the accelerating change taking place in the entomological profession that is orienting applied entomologists toward the philosophy of integrated control.

REFERENCES

Ahmed, M. K., Newsom, L. D., Emerson, R. B., and Roussel, J. S. (1954). The effect of systox on some common predators of the cotton aphid. *J. Econ. Entomol.* **47**, 445–459.

Albert, A. (1965). "Selective Toxicity," 394 pp. Methuen, London.

Anderson, J. R. (1965). A preliminary study of integrated fly control on northern California poultry ranches. *Proc. Calif. Mosq. Contr. Ass.* **33**, 42–44.

Anderson, J. R., and Poorbaugh, J. H. (1964). Observations on the ethology and ecology of various Diptera associated with northern California poultry ranches. *J. Med. Entomol.* **1**, 131–147.

Anonymous. (1972). "Report of the Mirex Advisory Committee to William D. Ruckelshaus, Administrator," 70 pp. Environmental Protection Agency. Washington, D.C. (Revised March 1, 1972.)

Bagley, R. W., and Bauernfeind, J. C. (1972). Field experiences with juvenile hormone mimics. *In* "Insect Juvenile Hormones Chemistry and Action" (J. J. Menn and M. Beroza, eds.), pp. 113–151. Academic Press, New York.

Bailenger, J., Rouge, M., and Tribouley, J. (1970). Etude comparie de l'activite insecticide de contact de plusieurs derives du diphenyltrichlorethane. *Bull. WHO* **43**, 827–840.

Bartlett, B. R. (1964). Integration of chemical and biological control. *In* "Biological Control of Insect Pests and Weeds" (P. DeBach, ed.), pp. 489–511. Reinhold, New York.

Beroza, M., and Knipling, E. F. (1972). Gypsy moth control with a sex attractant pheromone. *Science* **117**, 19–27.

Beroza, M., Hood, C. S., Trefrey, D., Leonard, D. E., Knipling, E. F., Klassen, W., and Stevens, L. J. (1974). Large field trial with microencapsulated sex pheromone to prevent mating of the gypsy moth. *J. Econ. Entomol.* **67**, 659–664.

Binns, E. S. (1969). Integrated control of *Tetranychus urticae* and *Aphis gossypii* on cucumbers. *Rep. Glasshouse Crops Res. Inst.*, pp. 87–88.

Birch, M., Trammel, K., Shorey, H. H., Gaston, L. K., Hardee, D. D., Cameron, E. A., Sanders, C. J., Bedard, W. D., Wood, D. L., Burkholder, W. E., and Muller-Schwarze, D. (1974). Programs utilizing pheromones in survey or control. *In* "Pheromones" (M. C. Birch, ed.), pp. 411–461. American Elsevier, New York.

Boyce, A. M. (1936). The citrus red mite *Paratetranychus citri* McG. in California and its control. *J. Econ. Entomol.* **29**, 125–130.

Brazzel, J. R., Newsom, L. D., Roussel, J. S., Lincoln, C., Williams, F. J. and Barnes, C. (1954). Bollworm and tobacco budworm as cotton pests in Louisiana and Arkansas. *La. Agr. Exp. Sta. Tech. Bull.* **482**, 47.

Brown, T. M., and Brown, A. W. A. (1974). Experimental induction of resistance to a juvenile hormone mimic. *J. Econ. Entomol.* **67**, 799–801.

Cate, J. R., Jr., Ridgway, R. L., and Lingren, P. D. (1972). Effects of systemic insecticides applied to cotton on adults of an ichneumonid parasite, *Campoletis perdistinctus*. *J. Econ. Entomol.* **65**, 484–488.

Chadwick, P. R., Beesley, J. S. S., White, P. J., and Matechi, H. T. (1964). An experiment on eradication of *Glossina swynnertoni* Aust. by insecticidal treatment of its resting sites. *Bull. Entomol. Res.* **55**, 411–419.

Coppedge, J. R., Lindquist, D. A., Ridgway, R. L., Cowan, C. B., and Bariola, A. L. (1969). Sidedress applications of Union Carbide UC-21149 for control of overwintered boll weevils. *J. Econ. Entomol.* **62**, 558–565.

Cowan, C. B., Jr., Ridgway, R. L., Davis, J. W., Walker, J. K., Watkins, W. C., Jr., and Dudley, R. F. (1966). Systemic insecticides for control of cotton insects. *J. Econ. Entomol.* **59**, 958–961.

Davies, H. (1971). Further eradication of the tsetse in the Chad and Gongala river systems in northeastern Nigeria. *J. Appl. Ecol.* **8**, 563–578.

Earle, N. W., and Newsom, L. D. (1964). Initiation of diapause in the boll weevil. *J. Insect Physiol.* **10**, 131–139.

Ewing, K. P., and Parencia, C. R., Jr. (1949). Experiments in early season application of insecticides for cotton insect control in Wharton County, Texas, during 1948. *U.S. Dep. Agr. Bur. Entomol. Plant Quar.* **E-772**.

Flaherty, D. L., and Huffaker, C. B. (1970). Biological control of Pacific mites and Willamette mites is San Joaquin Valley vineyards. I. Role of *Metaseiulus occidentalis*. *Hilgardia* **40**, 267–308.

Fraser, J., Greenwood, D., Harrison, I. R., and Wells, W. H. (1967). The search for a veterinary insectide. II. Carbamates. *J. Sci. Food Agr.* **18**, 372–376.

Gentry, C. R., Beroza, M., Blythe, J. L., and Bierl, B. A. (1974). Efficacy trials with the pheromone of the oriental fruit moth and data on the lesser appleworm. *J. Econ. Entomol.* **67**, 605–606.

Gould, H. J. (1971). Large-scale trials of an integrated control programme for cucumber pests on commercial nurseries. *Plant Pathol.* **20**, 194–196.

Guthrie, F. E., Rabb, R. L., Bowery, T. G., Lawson, F. R., and Baron, R. L. (1956). Control of hornworms and budworms on tobacco with reduced insecticide dosage. *Tobacco Sci.* **3**, 65–68.

Harris, W. G., and Burns, E. C. (1972). Predation on the lone star tick by the imported fire ant. *Environ. Entomol.* **1**, 362–365.

Hocking, K. S., Lee, C. W., Beesley, J. S. S., and Matechi, H. T. (1966). Aircraft applications of insecticides in East Africa. XVI. Airspray experiment with endosulfan against *Glossina morsitans* Westw., *G. swynnertoni* Aust. and *G. pallidipes* Aust. *Bull. Entomol. Res.* **56**, 737–744.

Hoyt, S. C. (1969). Integrated chemical control of insects and biological control of mites on apple in Washington. *J. Econ. Entomol.* **62**, 74–86.

Hoyt, S. C., and Caltagirone, L. E. (1971). The developing programs of integrated control of pests of apples in Washington and peaches in California. *In* "Biological Control" (C. B. Huffaker, ed.), pp. 395–421. Plenum, New York.

Hueck, H. J., Kuenen, D. J., den Boer, P. J., and Jaeger-Draafsel, E. (1952). The increase of egg-production of the fruit-tree spider mite (*Metatetranychus ulmi* Koch) under influence of DDT. *Physiol. Comp. Oecol.* **2**, 371–377.

Huffaker, C. B. and Spitzer, C. H., Jr. (1950). Some factors affecting red mite populations on pears in California. *J. Econ. Entomol.* **43**, 819–831.

Hussey, N. W., and Bravenboer, L. (1971). Control of pests in glasshouse culture by the introduction of natural enemies. *In* "Biological Control" (C. B. Huffaker, ed.). pp. 195–216. Plenum, New York.

Isely, D. (1926). Early summer dispersion of boll weevils. *Arkansas Agri. Exp. Sta. Bull.* **204**, 17.

Isely, D. (1934). Relationship between early varieties of cotton and boll weevil injury. *J. Econ. Entomol.* **27**, 762–766.

Jung, H. F., and Scheinpflug, H. (1970). Rice growing in Japan with special emphasis on problems of crop protection. *Pflanzenschutz-Nachr.* **23**, 235–263.

Kaae, R. S., McLaughlin, J. R., Shorey, H. H., and Gaston, L. K. (1972). Sex pheromones of Lepidoptera. XXXII. Disruption of intraspecific pheromone communication in various species of Lepidoptera by permeation of the air with Looplure or Hexalure. *Environ. Entomol.* **1**, 651–653.

Keriem, Saad, A. S. A., Zeid, M., and El-Sebae, A. H. (1971). Uber Versuche zur Bekampfung von *Sesamia cretica* Led., *Chile agamemnon* Bles. und *Ostrinia nubilalis* Hbn. (Lepidoptera) *Z. Angew. Entomol* **69**, 91–98.

Khan, M. A. (1969). Systemic pesticides for use on animals. *Annu. Rev. Entomol.* **14**, 369–386.

Klett, W. (1965). Integrieter und praktischer Pflanzenschutz. *Mitt. Biol. Bundesanst. Land. Forstwirt. Berlin Dahlem* **115**, 8–13.

Kuenen, D. J. (1958). Influence of sublethal doses of DDT upon the multiplication rate of *Sitophilus granarius* (Coleoptera: Curculionidae). *Entomol. Exp. Appl.* **1**, 147–152.

Lean, O. B. (1965). FAO's contribution to the evolution of international control of the desert locust 1951–1963. *FAO Desert Locust Newslett. Spec. Issue, Rome, 1965*, 142.

Long, W. H., and Concienne, E. J. (1964). Critical period for controlling the sugarcane borer in Louisiana. *J. Econ. Entomol.* **57**, 350–353.

Luckey, T. D. (1968). Insecticide hormoligosis. *J. Econ. Entomol.* **61**, 7–12.

Ludwig, D. J., and Potterfield, J. R. (1971). The pharmacology of propanediol carbamates. *Advan. Pharm. Chemother.* **9**, 173–240.

MacPhee, A. W., and MacClellan, C. R. (1971). Ecology of apple orchard fauna and development of integrated pest control in Nova Scotia. *Proc. Tall Timbers Conf. Ecol. Anim. Contr. Habitat Manage.* **3**, 197–208.

Metcalf, R. L. (1964). Selective toxicity of insecticides. *World Rev. Pest Contr.* **3**, 28–43.

Metcalf, R. L. (1971). Structure activity relationships for insecticidal carbamates. *Bull. WHO* **44**, 43–78.

Miller, R. W., and Gordon, C. H. (1972). Technical Rabon for larval house fly control in cow manure. *J. Econ. Entomol.* **65**, 1064–1066.

Mitchell, E. B., and Hardee, D. D. (1974). In-field traps: a new concept in survey and suppression of low populations of boll weevils. *J. Econ. Entomol.* **67**, 506–508.

Moore, N. W. (1970). Implications of the pesticide age. *Ceres* **3**, 26–29.

Newsom, L. D. (1972). Theory of population management for *Heliothis* spp. in cotton. *S. Coop. Ser. Bull.* **169**, 92.

Newsom, L. D., and Brazzel, J. R. (1968). Pests and their control. *In* "Advances in Production and Utilization of Quality Cotton: Principles and Practices" (Elliot, F. C., Hoover, M., and Porter, W. K. Jr., eds.), pp. 367–405. Iowa State Univ. Press. Ames, Iowa.

Nishida, T. (1954). Further studies on the treatment of border vegetation for melon fly control. *J. Econ. Entomol.* **47**, 226–229.

Nishida, T., and Bess, H. A. (1950). Applied ecology in melon fly control. *J. Econ. Entomol.* **43**, 877–883.

O'Brien, R. D. (1961). Selective toxicity of insecticides. *Advan. Pest Contr. Res.* **4**, 75–116.

O'Brien, R. D. (1967). "Insecticides: Action and Metabolism," 378 pp. Academic Press, New York.

Park, P. O., Gledhill, J. A., Alsop, N., and Lee, C. W. (1972). A large-scale scheme for the eradication of *Glossina morsitans morsitans* Westw. in the western Province of Zambia by aerial ultra-low volume application of endosulfan. *Bull. Entomol. Res.* **61**, 373–384.

Parr, W. J., and Scopes, N. E. E. (1971). Recent advances in the integrated control of glasshouse pests. *NAAS Quart. Rev.* **3**, 101–108.

Patterson, N. A., and MacLellan, C. R. (1954). Control of the codling moth and other orchard pests with ryania. *85th Rep. Entomol. Soc. Ont.*, pp. 25–32.

Persing, C. O (1965). Problems in the development of tailor-made insecticides. Specific insecticides. *Bull. Entomol. Soc. Amer.* **11**, 72–74.

Ridgway, R. L. (1969). Control of the bollworm and tobacco budworm through conservation and augmentation of predaceous insects. *Proc. Tall Timbers Conf. Ecol. Anim. Contr. Habitat Manage.* **1**, 127–144.

Ridgway, R. L., Lingren, P. D., Cowan, C. B., and Davis, J. W. (1967). Populations of arthropod predators and *Heliothis* spp. after application of systemic insecticides to cotton. *J. Econ. Entomol.* **60**, 1012–1016.

Ripper, W. E. (1944). Biological control as a supplement to chemical control of insects. *Nature (London)* **153**, 448–551.

Ripper, W. E. (1956). Effect of pesticides on balance of arthropod populations. *Annu. Rev. Entomol.* **1**, 403–438.

Ripper, W. E., Greenslade, R. M., Heath, J., and Barker, K. (1948). New formulation of DDT with selective properties. *Nature (London)* **161**, 484.

Ripper, W. E., Greenslade, R. M., and Hartley, G. S. (1951). Selective insecticides and biological control. *J. Econ. Entomol.* **44**, 448–459.

Roelofs, W. L., Glass, E. H., Tette, J., and Comeaux, A. (1970). Sex pheromone trapping for red-banded leaf roller control: theoretical and actual. *J. Econ. Entomol.* **63**, 1162–1167.

Schneider, F. (1958). Sinnesphysiologie Untersuchungen in Dienste der landwirtschaftlichen Entomologie. *Mitt. Schweiz. Entomol. Ges.* **31**, 146–153.

Schneiderman, H. A. (1972). Insect hormones and insect control. *In* "Insect Juvenile Hormones Chemistry and Action" (J. J. Menn and M. Beroza, eds.), pp. 3–27. Academic Press, New York.

Schulz, H. (1888). Uber Hefegifte. *Arch. Ges. Physiol.* **42**, 517–541.

Shorey, H. H., Kaae, R. S., Gaston, L. K., and McLaughlin, J. R. (1972). Sex pheromones of Lepidoptera. XXX. Disruption of sex pheromone communication in *Trichoplusia ni* as a possible means of mating control. *Environ. Entomol.* **1**, 641–645.

Shorey, H. H., Kaae, R. S., and Gaston, L. K. (1974). Sex pheromones of Lepidoptera. Development of a method for pheromonal control of *Pectinophora gossypiella* in cotton. *J. Econ. Entomol.* **67**, 347–350.

Simmonds, S. P. (1970). The possible control of *Steneotarsonemus pallidus* on strawberries by *Phytoseiulus persimilis. Plant Pathol.* **19**, 106–107.

Skaptason, J. S., and Pitts, C. W. (1962). Fly control in feces from cattle fed Co-Ral. *J. Econ. Entomol.* **55**, 404–405.

Smith, R. F. (1970). Pesticides: Their use and limitations in pest management. *In* "Concepts of Pest Management" (R. L. Rabb and F. E. Guthrie, eds.), pp. 103–113. North Carolina State Univ. Raleigh, North Carolina.

Smith, R. F., and van den Bosch, R. (1967). Integrated control. *In* "Pest Control, Biological, Physical, and Selected Methods" (W. W. Kilgore and R. L. Doutt, eds.), pp. 295–340. Academic Press, New York.

Staal, G. B. (1972). Biological activity and bio-assay of juvenile hormone analogs. *In* "Insect Juvenile Hormones Chemistry and Action" (J. J. Menn and M. Beroza, eds.), pp. 69–94. Academic Press, New York.

Staal, G. B. (1975). Insect growth regulators with juvenile hormone activity. *Annu. Rev. Entomol.* **20**, 417–460.

Steiner, L. F., Mitchell, W. C., Harris, E. J., Kozuma, T. T., and Fujimoto, M. S. (1965). Oriental fruit fly eradication by male annihilation. *J. Econ. Entomol.* **58**, 961–964.

Steiner, L. F., Hart, W. G., Harris, E. J., Cunningham, R. T., Ohinata, K. and Tamakhi, D. C. (1970). Eradication of the oriental fruit fly from the Mariana Islands by the methods of male annihilation and sterile insect release. *J. Econ. Entomol.* **63**, 131–135.

Stern, V. M., Smith, R. F., van den Bosch, R., and Hagen, K. S. (1959). The integrated control concept. *Hilgardia* **29**, 81–101.

Taschenberg, E. F., Carde, R. T., and Roelofs, W. L. (1974). Sex pheromone mass trapping and mating disruption for control of red-banded leaf roller and grape berry moths in vineyards. *Environ. Entomol.* **3**, 239–242.

Terauds, A. (1970). Evolution of methods for the control of damage to apples by dimpling bug in Tasmania. *Aust. J. Exp. Agr. Anim. Husb.* **10**, 647–650.

Thompson, R. D., Mitchell, G. C., and Burns, R. J. (1972). Vampire bat control by systemic treatment of livestock with an anticoagulant. *Science* **177**, 806–807.

Toscano, N. C., Mueller, A. J., Sevacherian, V., Sharma, R. K., Nilus, T., and Reynolds, H. T. (1974). Insecticide application based on Hexalure trap catches versus automatic schedule treatments for pink bollworm moth control. *J. Econ. Entomol.* **67**, 522–524.

Unterstenhofer, G. (1970). Integrated pest control from the aspect of industrial research on crop protection chemicals. *Pflanzenschutz-Nachr.* **23**, 264–272. (Engl. Ed.)

Von Rumker, R., Guest, H. R., and Upholt, W. M. (1970). The search for safer, more selective, and less persistent pesticides. *Bioscience* **20**, 1004–1007.

Way, M. J. (1959). Experiments on the mode of action of insecticidal seed-dressings especially against *Leptohylemia coarctata* Fall., Muscidae, the wheat bulb fly. *Ann. Appl. Biol.* **47**, 783–801.

Winteringham, F. P. W. (1969). Mechanisms of selective insecticidal action. *Annu. Rev. Entomol.* **14**, 409–442.

24

CULTURAL CONTROLS

Vernon M. Stern, Perry L. Adkisson, Oscar Beingolea G., and
G. A. Viktorov

I. INTRODUCTION

Agroecosystems vary widely in stability, complexity, and the area they occupy. The kinds of crops, agronomic practices, changes in land use, and weather are important elements affecting the degree of stability of an agroecosystem. All of these, except perhaps weather, are subject to manipulations to influence pest or natural enemy populations (Smith and van den Bosch, 1967). Oftentimes, the agroecosystem may lack only a minor key factor or feature to adversely affect a pest or favorably modify the environment to increase the effectiveness of its natural enemies. These needs can often be met by proper use of cultural controls.

The two basic principles in the cultural control of arthropod pests are (1) manipulation of the environment to make it less favorable to the pest and (2) manipulation to make it more favorable to their natural enemies. Both may be used together to prohibit, reduce, or delay pest population increase.

Classical use of cultural practices has involved such measures as stalk destruction, plowing and tillage, crop rotation, timely harvesting, selected planting dates, use of trap crops, barriers, flooding, and the planting of special plants to increase natural enemies. Recently, such techniques as selective use of preharvest chemicals, modification of crop variety, manipulation of pest populations using pheromones, and selective employment of insecticides also have helped suppress the pests and preserve the natural enemies. Modification of crop varieties will not be covered in this chapter but is treated separately in Chapter 25. Certain aspects of using selectivity in pesticide technology are cultural in nature but these are treated in Chapter 23 and are not dealt with here.

Cultural methods require a thorough knowledge of production of the crop and the biology and ecology of the pest and its natural enemies in order to integrate the techniques for pest control into proved agronomic procedures for crop production. This knowledge permits assessment of the agronomic procedures that favor a particular pest and enables agronomic changes to be made to reduce pest population numbers and damage.

However, pest control is only one factor in crop production. Changes in agronomic procedures should not lead to other problems. If this occurs the most valuable practice should be followed. For example, the last sustained and widespread grasshopper outbreak in the United States occurred in the 1930's, a period of severe drought combined with high winds and temperatures. Wind erosion was also a serious problem and the United States Department of Agriculture (USDA) and state agencies initiated strip cropping to reduce erosion on the western Great Plains and elsewhere. However, in certain years the strip cropping favored buildup and damage from grasshoppers, mainly *Melanoplus mexicanus* Saus. (Wilber *et al.*, 1942). In this case, wind erosion was by far the more serious long-term problem and strip-cropping prevailed as an agronomic procedure because, in most cases, grasshoppers could be controlled with baits and their outbreaks were sporadic.

Many cultural measures for pest control are closely associated with ordinary farm, forest, or water management practices. They can be simple and inexpensive because they can often be carried out with only slight modification of routine management operations. In addition, they do not contribute to pest resurgence, resistance to pesticides, undesirable residues, and contamination of the environment. Moreover, some insects are easier to control by cultural methods at a stage when they are doing no damage or at the end of a season than in a stage at which they cause serious damage. This is especially true where the pest has no important natural enemies and is very difficult to control with available chemical methods, such as the pink bollworm, *Pectinophora gossypiella* (Saunders) (Noble, 1969), or where the pest has developed a resistance to pesticides, such as the cotton leaf perforator, *Bucculatrix thurberiella* Busck (Reynolds, 1971).

There are various reasons why growers may not use or have abandoned cultural controls to reduce pest populations. These include: domestic use of the infested crop residues combined with primitive agriculture, as in the case of lepidopterous stem borers attacking cereal crops in Africa, south of the Sahara (Harris, 1962; Swaine, 1957); educational problems in the case of the sorghum midge, *Contarinia sorghicola* (Coq.), in Nigeria (Harris, 1961) and in Ghana (Bowden, 1965); similar reasons for weevils and soil pests attacking a variety of crops in Rhodesia (Rose and Hodgson, 1965); advancements in farm technology in the case of corn rootworms in midwest United States (i.e., mechanized corn pickers, readily available fertilizers, insecticides, and herbicides) (Chiang, 1973; Patel and Apple, 1967; Peterson, 1955, 1957; USDA, 1971; and many others); lack of enforcement of laws requiring cotton stalk destruction for the pink bollworm in parts of Turkey, and for spiny bollworm, *Earias insulana* Boisduval, in Iraq (Stern, 1968).

Other reasons for growers' disinterest in cultural control methods are that these preventive measures are usually applied long in advance of the actual pest outbreak. Growers may not accept them for this reason, or if they are accepted, they may use them too late or improperly. Moreover, there is often a reluctance to use them because they may not provide the complete economic control that is often attainable by use of an insecticide even though a population reduction can delay buildup to damaging levels and reduce the number of insecticide applications necessary for crop protection.

Cultural practices can be used by individual growers for controlling pests such as the meadow spittlebug, *Philaenus spumarius* (L.) (King and Weaver, 1953), or corn rootworms (Chiang, 1973); or, they may be used to achieve partial suppression of pests in adjunct with other control measures, especially chemicals (Gentry *et al.*, 1967; Rabb, 1971). In other cases, cultural practices require community or area-wide adaptation to achieve population suppression over a relatively large geographical area. This is especially important for insects that migrate or disperse a considerable distance, such as the pink bollworm (Noble, 1969).

When direct suppression of the pest is involved, a vulnerable stage in its life cycle, or, one of its special behavioral patterns in relation to the habitat is often singled out for manipulation. For example, corn rootworms have one generation per year and lay their eggs in corn fields in late summer and fall; the larvae hatch the next spring. The life cycle can be broken by crop rotation (Forbes, 1883, 1891–1892; Patel and Apple, 1967; Chiang, 1973). Pink bollworm larvae enter diapause in October and November. Early defoliation or desiccation of cotton in late August and early September causes high larval mortality because most of their food is destroyed. Shredding the stalks, plowing them under, and winter weather kills most of the remaining larvae (Fenton and Owen, 1953; Adkisson *et al.*, 1960; Noble, 1969). The Hessian fly, *Mayetiola destructor* (Say), emerges from the summer generation in the fall and must oviposit within 4 to 5 days because they soon die. Oviposition can be essentially eliminated by delaying the planting of the winter wheat until after the fall emergence (Painter, 1951). Long-term ecological research may be necessary before such cultural methods can be implemented.

The first consideration in any insect pest management program should be given to the "key" pest species, including their biological and behavioral characteristics, natural enemies, their main and alternative food supplies, and the direct and indirect influence of other environmental factors. Key pests are serious, perennially persistent species that dominate control practices, and which, in the absence of deliberate control by man, reach or exceed their economic thresholds one or more times a year (Smith and Reynolds, 1966; Stern, 1973).

While the total number of potential pest species in a crop may be high, the number of key pests involved is low, usually only one or two in an area. For example, in the San Joaquin Valley of California the key pests of cotton are lygus bugs, *Lygus hesperus* Knight and *L. elisus* van Duzee. On grapes in the same area the key pest is the grape leafhopper, *Erythroneura elegantula* Osborn.

In both cases, untimely, too frequent, and often unnecessary chemical treatments have created secondary pest problems. Cultural controls have been developed for lygus bugs (Stern *et al.*, 1967; Stern, 1969) and are being studied for the grape leafhopper (Doutt and Nakata, 1973).

II. CULTURAL PRACTICES TO REDUCE OVERWINTERING PEST POPULATIONS

Many pest species remain in or on the stalks, stems, and other parts of their host plant during the winter and destruction of these parts by shredding, burning, and plowing, etc. can greatly reduce the overwintering populations. A stalk destruction program followed by plowing was developed in Texas for control of the pink bollworm, *Pectinophora gossypiella* (Adkisson and Gaines, 1960; Noble, 1969). This pest lays its eggs on the fruiting forms of cotton and, immediately after hatching, the larvae burrow into the flower buds and bolls. This behavior makes chemical control expensive and unsatisfactory. In addition, no effective natural enemies have been found.

Early research (Ohlendorf, 1926) showed that the pink bollworm diapauses in the last larval instar, mainly in the seeds of bolls that remain in the field after harvest. Later studies showed that diapause is controlled by photoperiod (Lukefahr *et al.*, 1964; Adkisson *et al.*, 1963) and that induction occurs in early September when day length becomes less than 13 hours (Adkisson, 1964). Diapause incidence then increases rapidly and attains a maximum in mid-October and early November and the seasonal onset of diapause can be predicted with precision at any location (Adkisson, 1966).

This discovery of the place and timing of diapause in the pink bollworm was then used to achieve heavy suppression of overwintering populations by modification of certain cotton cultural practices (Adkisson, 1962). Until this time cotton growers usually allowed the plants to grow in a way that resulted in a lengthy period of boll opening and harvesting; plants often continued to grow and bolls to open after harvest was completed; plants were left undisturbed in the fields through the winter; and, next year's cotton planting times were varied at will of the growers. Such practices provided excellent overwintering conditions for the pest. Anti-pink bollworm cultural controls eliminated or modified all these conventional practices. The mature cotton plants were managed by use of defoliants and desiccants so that bolls opened at nearly the same time and were promptly harvested. Soon after harvest the plants were then shredded mechanically and plowed deeply into the soil so as to prevent growth of new fruiting forms that provided food and diapause sites for the overwintering pest generation. New cotton crops were not allowed to be planted in the spring until after a designated time, which time was set well after adult moths from the overwintering generation emerged and died (Adkisson and Gaines, 1960).

Early defoliation provides the initial step for pink bollworm suppression, followed by the remaining cultural procedures because most of the overwintering generation comes from eggs laid after mid-September. If a preharvest defoliant or desiccant is applied in late August or early September (before days are short enough to induce diapause) the mature bolls open and immature fruiting forms either shed or dry up and little suitable food is left for larval development. However, if the application of the desiccant or defoliant is delayed until early October it is relatively ineffective in reducing potential overwintering larval numbers because most of the larvae have already reached the diapause stage by this time.

When cotton is mechanically stripped, virtually all pink bollworm larvae are carried to the gin since the stripper leaves almost no bolls in the field. Almost 100% of the larvae are killed by the ginning process (Robertson *et al.*, 1959). When cotton is harvested by a spindle picker, some immature bolls are left in the field and these may constitute a source of infestation the next season. However, a rotary stalk cutter may kill from 50 to 85% of the larvae that remain on standing stalks after harvest (Wilkes *et al.*, 1959). When the stalks are plowed under immediately after shredding, the combined mortality from stalk shredding, plowing, and winter weather may exceed 90% (Adkisson *et al.*, 1960; Noble, 1969; Fenton and Owen, 1953).

These cultural procedures have been so successful in Texas that insecticides are seldom needed for the control of this pest. The effectiveness of this program has largely depended on legislation that prohibits planting before an established date in the spring and requires plowing up of the crop by another date in the fall. Growers who do not comply with these dates are subject to a fine. There is also a "social" aspect to the success of this program because few if any growers care to be accused of contaminating their neighbors' fields.

A similar type of cultural program utilizing stalk destruction and black-light trapping of the tobacco hornworm, *Manduca sexta* (Johannson), and tomato hornworm, *M. quinquemaculata* (Haworth), attacking tobacco has been developed in North Carolina (Gentry *et al.*, 1967; Rabb *et al.*, 1964; Rabb, 1971). Insecticidal treatments on tobacco for these two pests in a 113 square mile study area were reduced more than 90% following wide-scale implementation of the program, and there was a reduction by 60% in applications for all tobacco pests.

III. USE OF A HOST-FREE SEASON

Where the winters are not cold enough to kill cotton roots, the land may be left unplowed after the stalks are shredded, and the old plant stubs will grow new sprouts for producing the next crop. This is referred to as "ratooning" or "stub cotton." This practice has been discouraged and prohibited by law in many areas of the world because it can provide a continuous source of food for pest species and it results in higher winter survival of pests, such as the pink bollworm, when the infested crop residue is not plowed under. Rigid enforcement can be of tremendous benefit even where the pest can survive on native host plants (Pearson, 1958; Ripper and George, 1965), which in some areas may not be common enough to pose a problem (Pearson, 1958). This occurs with the cotton leaf perforator in southern California (H. T. Reynolds, personal communication).

The elimination of ratooning cotton or its restriction to 1 year in dry areas has reduced successfully a number of pests in Peru. The scale, *Pinnaspis minor* Mask., established in Peru in 1905 (Townsend, 1912), had, by 1920, spread into the entire coastal area where cotton was grown. Several parasites, *Aspidiotiphagus citrinus* Crawf., *Prospaltella aurantii* (How.), *P. berlesei* (How.), *Aphelinus fuscipennis* How., and *Signiphora* sp., and predators, *Microweisia* sp. and *Scymnus* sp., attacked this scale. However, the practice of ratooning cotton favors the pest over its natural enemies, particularly in the hotter and drier areas (Wille, 1943). The scale also has a number of other hosts (*Gossypium raimondii*, *Ricinus communis*, *Manihot utilissima*, *Tessaria integrifolia*, *Cassia* sp., *Sida panniculata*, *Malachra* spp., *Malvastrum* spp., etc.) which favor its persistence. Satisfactory control of *Pinnaspis minor* was achieved by prohibiting ratoon cotton in the more northern valleys and by reducing the number of ratoon seasons, and/or reducing the height of the ratoon stumps and eliminating wild hosts within and around the fields in the middle coastal area (Townsend, 1912, 1913, 1924; Wille, 1943).

The weevil, *Eutinobrothrus gossypii* Pierce, girdles cotton at the base of the stem and root. This wilts or kills the plants if girdling is complete. Laws prohibiting ratooning decreased populations below the economic level when combined with complete cleaning of the fields of hosts. This includes uprooting the plants within the row, then plowing as many as three times crosswise, and hand removal of any remnants of roots. Attempts to simplify the operation have usually resulted in increased weevil damage (Wille, 1943).

Experience has also shown that when ratooning is suppressed both the square weevil, *Anthonomus vestitus* Boh., and the lesser bollworm, *Mescinia peruella* Schaus., are reduced to negligible numbers. In the Cañete Valley, the incidence of damage from the square weevil dropped to 1% or less, and the lesser bollworm, which ordinarily destroys more than 10% of the late formed bolls, essentially disappeared. The reason for this is that cotton is the only known host of this latter pest.

In Peruvian valleys where there is insufficient irrigation water to germinate and sustain new seed plantings, ratooning cannot be suppressed as a regular practice.

Several devices have been used to keep *E. gossypii*, *A. vestitus*, and *M. peruella* at low levels. In the case of *E. gossypii*, ratooning is limited to 1 year and the plants selected for ratooning are those which show little or no damage. The overwintering populations of *A. vestitus* which can be found as adults and larvae, and *M. peruella* as larvae on terminal cotton buds, can be drastically reduced by goat or sheep browsing on the sprouts. A measure to enhance natural enemy effectiveness is to hand pick the early-attacked squares and place them in rearing chambers. As natural enemies emerge they are collected and returned to the fields whereas unparasitized larvae are destroyed.

Populations of the red stainer, *Dysdercus peruviansus* Guerin, in Peruvian cotton come from two sources: (1) remnant populations favored by ratooning, abandoned fields, or lack of clean fallowed fields, and (2) mass immigrant populations from wild hosts. In valleys with adequate water, the problem has been solved through suppression of ratooning and the destruction of the more important host plants (*Sida panniculata*, *Malachra* spp., etc.) in and near cotton fields. Where water is scarce and ratooning is permitted for 1 year, alternate host destruction is combined with control measures using trap plants (*Urocarpidium* spp.) which are much more attractive to the red stainer than cotton.

Urocarpidium is planted in strips which can be treated to eliminate the red stainer without disturbing the natural enemies of other pests in the cotton. In addition, cotton seed is soaked in a pesticide solution and distributed in small heaps spaced 25 m along the cotton row in every fifth row. This apparently attracts the red stainer and they are killed by the poisoned seed.

IV. USE OF CROP ROTATION

The cultural practice of rotating a crop that is attacked by an insect with another crop that is not a host is most effective against pest species having a restricted host range, limited power of dispersal, or certain special behavior.

The larvae of the northern corn rootworm, *Diabrotica longicornis* (Say), and western corn rootworm, *D. vigifera* LeConte, feed mainly and preferably, but not exclusively, on corn roots and can be serious pests of corn in the midwest United States (Chiang, 1973; Branson and Ortman, 1970, 1971). These two rootworms have one generation a year and their eggs are laid in corn fields in late summer and early fall (Lawson, 1964; Patel and Apple, 1967). The adults soon die after oviposition and the diapausing eggs usually hatch in the following late spring, although Chiang (1965) found that some eggs of the northern corn rootworm will go through two winters before hatching.

Using knowledge of the single generation per year and the preference for ovipositing in corn fields, Forbes (1883, 1891–1892) recommended crop rotation to curtail the life cycle of the northern corn rootworm. A 4-year crop rotation of corn, corn oats, and clover was generally adopted for the level lands of northern Illinois and Iowa. In other areas (ca. 1900) sorghum, alfalfa, soybean (Weiss, 1949), and other crops were used in rotation with corn. Although few growers maintained a systematic consistent rotation pattern, corn was seldom planted consecutively for more than 2 or 3 years (K. J. Frey, personal communication; J. B. Peterson, personal communication). The rotation program maintained the northern corn rootworm as a minor pest until the late 1940's (Patel and Apple, 1967). The same situation applied to both rootworm species in Nebraska (Burr, 1944).

This corn rootworm situation changed after World War II. Before the development of the organochlorine insecticides, soil insect control was virtually impossible except in limited cases where cultural methods (e.g., for wireworms, Landis and Onsager, 1966; Lane, 1941) and a biological method (for Japanese beetle, *Popillia japonica* Newman, Steinhaus, 1949) were effective. When DDT became available in the late 1940's, followed by the cyclodiene insecticides a few years later, general soil insect control became practical (Harris, 1972). Aldrin and heptachlor were particularly effective. These chemicals were quickly and extensively adopted for control of soil insects. They were often misused, resulting in more rapid development of resistance to them in many species, and needless contamination of large acreages of agricultural land (Harris, 1972). In addition, these insecticides, absorbed by soil particles, have been transported by erosion into streams and other water systems, causing significant pollution (Newsom, 1967).

With effective soil insecticides, herbicides, cheap commercial fertilizers, new high-yielding hybrids, and the development of efficient farm machinery, farmers in the corn belt began to grow corn continuously in the same field and usually on their best non-erodible land. The three classes of chemicals mentioned required a small outlay of money and profits were greater from continuous planting of corn than from rotating with other crops (Peterson, 1955). Thus, changes in farm technology, economic, and social factors contributed to general abandonment of a once successful cultural control program for corn rootworms (USDA, 1973). However, continuous corn cropping has now led to increased rootworm populations in all corn growing areas of the midwest. The use of insecticides has also increased markedly. In 1966, 33% of the United States corn acreage was treated, an increase of 50% from 1964 (USDA, 1970). Moreover, while these pests were of minor importance under cultural control up to the late 1940's and early 1950's, they are now major and disturbing corn pests because in many areas they now have resistance to the formerly effective materials used against them.

V. USE OF HARVESTING PROCEDURES

Harvesting procedures involving crop maturity, time of harvest or cutting practices, selective harvesting, and strip-harvesting can be of considerable assistance in suppressing a variety of insect pests, increasing their natural enemies, thus affecting yields. A delay in harvesting cotton can reduce both its quality and quantity due to adverse weather factors and a subsequent delay in the shredding and plowing under of the cotton stalks favors overwintering populations of the pink bollworm (see Section II).

A. Time of Harvest

Harvesting can have a marked effect on the insects in certain crops. In contrast to cereals and other seed crops which are usually dry at harvest time, forage crops such as alfalfa are harvested green and may be cut several times a year. Harvesting suddenly changes the insects' physical environment. It generally becomes much hotter and drier. Insects such as lygus bug adults and various parasites are caused to leave, seek shelter, or die (Stern *et al.*, 1967; Stern, 1969; van den Bosch *et al.*, 1967). With regrowth of the alfalfa, many insects such as the alfalfa caterpillar, *Colias eurytheme* Boisduval, return to the field. This species prefers to oviposit on short regrowth alfalfa stems (Smith and Allen, 1954; Smith *et al.*, 1949). Following heavy butterfly flights, it is these fields where larval populations often reach damaging numbers.

Another example of harvesting that affects future pest numbers concerns the meadow spittlebug, *Philaenus spumarius* (L.). Damaging infestations of nymphs vary from field to field. King and Weaver (1953) were able to explain this in Ohio by (1) the time of cutting alfalfa fields, (2) the time of first egg development and maximum oviposition, and (3) the fall behavior of the adult spittlebugs. When adults emerge from the spittle masses, their dispersal is brought about gradually by a "hardening off" of the maturing plants where fields are left uncut, or abruptly by the first cutting in early June. Some adults return to these fields as alfalfa regrowth occurs but they will leave when the plants mature or the field is cut a second time. The adult dispersals proceed through the season, but decrease when the first fully developed eggs appear in the females during the latter part of August (Weaver, 1951). A preoviposition period of about 3 to 4 weeks occurs between the time of full development of the first eggs and maximum egg deposition in September and October (Weaver, 1952; Weaver and King, 1954).

When comparing adult spittlebug numbers with alfalfa conditions during the preoviposition period, the number is greatest in fields with the most succulent foliage. When the last cutting is removed in early August, plant regrowth is rapid. Large numbers of adults accumulate in these fields during the preoviposition period and large nymphal infestations occur the next spring. On the other hand, when the last cutting occurs in July, the plants attain advanced maturity by late August and these fields are unattractive to the adults. Likewise, when the last alfalfa cutting occurs in early September the majority of adults leave the field before ovipositing. Thus, a harvesting schedule that produces succulent foliage in September generally promotes large spittlebug infestations the following spring.

B. Selective Harvesting

Each year bark beetles, mainly the western pine beetle, *Dendroctonus brevicomis* LeConte, destroy millions of board feet of lumber in western United States (Miller and Keen, 1960). In contrast to many pests which feed on their host plant but do not kill it, the western pine beetle, when successfully established in a tree, changes it from a living to a dying and then a dead organism; decay follows (Stark, 1970).

Early forest entomologists and silviculturists observed that the western pine beetle has a tendency to select mature slow-growing or decadent trees on poor sites and seemingly avoid younger, vigorously growing trees. Denning (1928) used this idea and described seven categories for ponderosa pine based on their silvical characteristics. Keen (1936, 1943) expanded the categories to 16, based on four age groups and four degrees of crown vigor. Records over a 7-year period involving approximately 39,000 beetle-killed trees showed that they were largely mature and overmature trees with poor to very poor crown vigor (Keen and Salman, 1942).

The value of the tree susceptibility classifications was that if the type of tree most likely to be attacked and killed could be recognized, it would be possible to make a light-cut of beetle-susceptible trees to improve sanitation, reduce subsequent pine beetle infestations, and salvage valuable trees. However, Salman and Bongberg (1942) commented that the application of Keen's (1936, 1943) system would have entailed removal of a considerable percentage of the standing virgin timber along the eastern slopes of the Sierra and Cascade mountains.

Salman and Bongberg (1942) designed a more practical risk-rating system for selecting individual trees, based on four categories: low risk, moderate risk, high risk, and very high risk. Trees in the high and very high risk categories were those most likely to be attacked and die in the near future. This system proved to be very successful in terms of timber management objectives and is still used by foresters in sanitation and salvage logging on the drier eastern side of California, Oregon, and Washington (Miller and Keen, 1960). For unknown reasons, in mixed tree stands on the western slopes of the Sierra and Cascade mountain ranges which receive more rainfall, the risk-rating system for selective timber harvesting has not been too reliable.

C. Strip-Harvesting

As noted above, lygus bugs, primarily *Lygus hesperus*, are a key pest of cotton in the San Joaquin Valley of California. During May, most cotton fields are still in the seedling stage and unattractive to *Lygus*. By mid-June the situation changes and often large numbers of adults can be found in the fields (Stern *et al.*, 1967).

The adults come from two sources, alfalfa and safflower (Mueller, 1971). Thousands of acres of alfalfa are grown and adjoin cotton fields through the middle part of the Valley (N-S) and across the southern part of the Valley (E-W). When an entire alfalfa field is suddenly mowed, *Lygus* is drastically affected. The humidity drops and the temperature quickly rises. This sudden change destroys the lygus bug's habitat, food, shelter, and oviposition sites. During hot weather, nearly all the adults leave solid cut fields within 24 hours after cutting (Stern *et al.*, 1964). The problem then is to stabilize the alfalfa hay environment to keep the lygus bug adults in the alfalfa habitat where they do little or no damage.

This can be done by strip-cutting the alfalfa fields. Under the strip-cutting technique, the alfalfa is harvested in alternate strips so that two different aged hay growths occur in a field simultaneously. When one set of strips is cut, the alternate strips are about half-grown. The field becomes a rather stable habitat because the lygus bugs move from the cut strips to the half-grown strips instead of flying to adjoining crops, such as cotton, as they do when the entire field is cut at one time.

Since natural enemies (predators) of the *Lygus* also move from strip to strip, there is no increase in the lygus bug population in the alfalfa. Moreover, when the *Lygus* adults move into the uncut strips they deposit eggs in the half-grown hay. However, these strips mature and are cut in about 2 weeks. The time required for *Lygus* development (egg to adult) is much longer and most of the nymphs and the unhatched eggs are destroyed by high temperatures and the drying of the mowed alfalfa.

Strip-harvesting also reduces other pest problems in alfalfa itself, particularly as a result of more effective biological control. Van den Bosch *et al.* (1967) found a reduction in pea aphid, *Acyrthosiphon pisum* (Harris), populations because strip-harvesting favors its parasite *Aphidius smithi* Sharma and Rao. In strip-cut fields both the aphid and the parasite persist during mid-summer, a time when populations of both species are disrupted in solid-cut alfalfa. In the latter fields, *A. smithi*, being host-density dependent, is particularly hard hit, not only by the adverse physical conditions but also by the protracted scarcity of its host. It is virtually eradicated from the fields and does not show vigorous activity again until autumn. By contrast, in strip-cut alfalfa the parasite remains abundant, in continuous interaction with its' host, and quickly responds to the aphid upsurge in late summer.

Although intensive analyses have not been made of other parasite—host relationships, there are some indications that parasites of lepidopterous pests are also favored by strip-cutting. This appears to be true for parasites of the alfalfa caterpillar, *Colias eurytheme*, the beet armyworm, *Spodoptera exigua* (Hübner), and the western yellow-striped armyworm, *Spodoptera praefica* (Grote). There is also evidence that spotted alfalfa aphid, *Therioaphis trifolii* (Monell), populations are reduced because of more effective biological control in strip-cut fields (Schlinger and Dietrick, 1960).

Bashir and Venkatraman (1968) found that the efficiency of *Zelomorpha sudanensis* Gahan (an important parasite of the beet armyworm, *Spodoptera exigua*, attacking alfalfa near Khartoum, Sudan) is high only during the winter. The frequent early cutting of alfalfa and the summer "dead season" cause unfavorable changes in the microclimate and deprive the adult parasites of nectar-bearing plants. They suggested that parasite efficiency could be increased by strip-cutting alfalfa throughout the growing season. The alternating uncut strips would serve as a favorable habitat for the natural enemies and increase their efficiency.

VI. USE OF HABITAT DIVERSIFICATION

The replacement of natural communities with monocultures of agricultural crops has caused general faunal impoverishment, while certain species of phytophagous arthropods have become extremely abundant (Smith, 1956; Cole, 1964). Many of these are pests and many have a high degree of vagility (Southwood, 1966), often colonizing the disrupted agroecosystems ahead of their natural enemies (van den Bosch and Telford, 1964).

Burnett (1960), Odum (1971), and many others comment that as biotic complexity increases, particularly with reference to the number and kinds of trophic interactions, the stability of the agroecosystem will increase. An opposing viewpoint of this ecological "dogma" is taken by van Emden and Williams (1973). They argue that species diversity does not necessarily cause greater stability. However, there are examples of diversification of the crop environment which do make it more favorable to natural enemies and less favorable to a pest.

An example is the sunn pest, *Eurygaster integriceps* Puton, and its scelionid egg parasites, mainly *Trissolcus* spp. (= *Asolcus*). Kamenkova (1958), Viktorov (1962, 1967), and others report that sunn pest egg parasites are very efficient in areas with small wheat fields surrounded by diverse vegetation. Under these conditions (e.g., Armenia) the polyvoltine egg parasites have a number of other pentatomid hosts and favorable hibernating places. On the other hand, in areas with extensive wheat monocultures (North Caucasus, lower Volga region), the density of scelionids is usually low. However, their numbers and efficiency increase near forests and in winter wheat fields following corn and sunflower which are inhabited by other pentatomid hosts (Shapiro, 1959). Thus, it is possible to increase the efficiency of sunn pest parasites by changes in the rotation practice and the spatial distribution of crops.

Doutt and Nakata (1973) studied vineyard ecosystems in California to determine the reason for different population levels of the grape leafhopper, *Erythroneura elegantula*. *Anagrus epos* Girault was found to be an effective egg parasite of this leafhopper in the north coastal region, but the parasite is not effective in most of the San Joaquin Valley and the grape leafhopper is a chronic pest there.

During the winter the grape leafhoppers are all adults in reproductive diapause while *A. epos* has no diapause. In the north coastal region and similar areas, *A. epos* can breed through the winter on eggs of another leafhopper, *Dikrella cruentata* (Gillette), that occurs on native and introduced blackberries (*Rubus* spp.). In these wooded areas there is sufficient water for trees to grow along the stream beds to provide shade for these host plants.

On the other hand, the southern San Joaquin Valley was virtually reclaimed from a desert. It lacks the trees to provide shade for the wild grape (original host of the grape leafhopper) and blackberries except along the banks of major, continuously flowing streams, and populations of *D. cruentata* exist in very isolated spots. R. L. Doutt (personal communication) has suggested planting trees around water-holding irrigation ponds to provide shade for the blackberries to increase populations of *D. cruentata* as a host for *A. epos* during the winter months. The growers will not do this because they fear these trees will provide nesting sites for birds which feed on the grapes. Nevertheless, research is continuing to find possible ways to diversify the vineyard areas to provide a suitable habitat for *D. cruentata*.

In East Germany (near Leipzig) a tachinid, *Campogaster exigua* (Meig.), and a braconid, *Pygostolus falcatus* Nees, are usually unimportant in control of monovoltine *Sitona* species. Parasitization of adult beetles rarely attains 80% (Müller, 1963). Among several *Sitona* species, *S. humeralis* Steph. is the most abundant on alfalfa (79.6%) while *S. flavescens* Msh. is the most abundant on red clover (51.9%). Both *C. exigua* and *P. falcatus* have three or four generations per year and are therefore adapted to natural meadow habitats populated by several *Sitona* species having different reproductive periods. Thus, the efficiency of these parasites can be increased by planting both alfalfa and red clover in the same field.

Considerable work in the U.S.S.R. has been devoted to the use of nectar-bearing plants as a source of adult food for entomophagous insects to increase their effectiveness. This subject has been reviewed by several authors (Shepetilnikova, 1963; Shepetilnikova *et al.*, 1968; and others). Stark (1940) conducted some of the original work on species of Scoliidae. However, there is still very little quantitative data to explain clearly how the cultivation of nectar-bearing plants affects pest numbers on various crops.

Field experiments of Chumakova (1960) in the North Caucasus have shown that the growing of *Phacelia* spp. in orchards greatly increases the parasitization of *Prospaltella perniciosi* Tower by its parasite *Aphytis proclia* (Walker). Studies of tachinid and ichneumonid parasites attacking *Barathra brassicae* (L.) and *Plutella xylostella* (Curtis) were conducted near Moscow and the data show that parasite efficiency was substantially higher in cabbage fields when they were grown near flowering umbelliferous plants. Similar results were reported by Allen and Smith (1958) who found that parasitization of the alfalfa caterpillar, *C. eurytheme*, by *Apanteles medicaginis* Muesebeck was far greater in California's San Joaquin Valley where weeds were in bloom along irrigation canals in contrast to areas where the weeds were destroyed.

Many experiments have been conducted using trap crops to attract the pest species or to provide a more favorable habitat to increase natural enemies. An example is the interplanting of alfalfa strips in cotton (20-ft strips every 250–500 ft of cotton row). Stern (1969) and Sevacherian (1970) showed that *Lygus hesperus* prefers alfalfa over cotton as long as the alfalfa remains in a lush crowing condition.

In Peru, the planting of corn either in neighboring fields or rows of corn interplanted in cotton fields (one row to every fifth or seventh row of cotton) favors reproduction of many natural enemies of cotton pests; they build up earlier in corn than in cotton (Christidis and Harrison, 1955; Combe and Lamas, 1954; Combe and Gamero, 1954; Simón, 1954; Beingolea, 1957). Beingolea commented that the benefits are real although not easy to establish.

As a substitute for, or supplement to interplanting special plants within a main crop to increase natural enemies, the transfer of plants densely populated with natural enemies can often be used to advantage. The introduced parasites, *Praon palitans* Mues. and *Trioxys utilis* Mues., of the spotted alfalfa aphid, *Therioaphis trifolii*, were rapidly spread throughout California by cutting and spreading infested hay (van den Bosch *et al.*, 1959). This technique, along with (1) the use of resistant alfalfa varieties, (2) the use of a selective insecticide (Demeton) to permit maximum survival of the parasites and lady beetles, and (3) the aid of a fungus disease, reduced the annual cost of damage and chemical control from about $13 million in 1955 to about 1 million by 1958. Thereafter this insect became a very minor alfalfa pest (DeBach, 1964).

Similar parasite movements have been used in U.S.S.R. Kolobova (1959) reported that *Bruchophagus roddi* Gussakov., an important pest of seed alfalfa in the Ukraine, has several species of parasites, especially *Habrocytus medicaginis* Gahan, *Liodontomerus perplexus* Gahan, and *Tetrastichus bruchophagi* Gahan. When the alfalfa seed crop is harvested in early August, up to 50% of the pest population is in diapause. However, the majority of the parasites (99% of *H. medicaginis* and 82% of *T. bruchophagi*) complete their development and many then die without parasitizing new hosts because the early harvesting precludes oviposition by any *B. roddi* that have not entered diapause.

The situation is quite different when the alfalfa seed crop is harvested at the end of August. At this time, parasitization of *B. roddi* is about 75% and most of the parasites are then in diapause. Thus, the chaff from alfalfa seed harvested in late August contains a great many hibernating parasites. By saving the chaff these hibernating parasites can be easily moved to other fields where their benefits can be gained the following year.

Beingolea (1957, 1962) reported similar movements of plants containing natural enemies in the Cañete Valley of Peru. Corn plants carrying great numbers of cotton green leaf roller egg masses per plant, and heavily parasitized by *Trichogramma* sp. and *Prospaltella* sp., were moved by truck loads. Corn tassels carrying anthocorids and mirids were also moved into crops to increase the local populations of these parasites.

Studies on the population dynamics of the weevils *Apion apricans* Hbst. and *A. aestivum* Germ. on wild and cultivated clovers in U.S.S.R. (Telenga, 1954) revealed a low pest infestation and high parasite [mainly *Spintherus dubius* (Nees) (= *S. linearis* Kurdjumov)] activity in the wild habitats as contrasted to the cultivated clover fields. *Apion* spp. hibernate as adults under the litter in forests and forest "windbreak" belts. *Spintherus dubius*, which has two generations per year, hibernates as full-grown larvae in the clover heads. Thus, the parasite population is substantially reduced by seed harvesting. However, a large number of hibernating parasites remain in the chaff. Therefore, it is possible to increase substantially the parasite abundance by leaving the chaff in piles in the clover fields to permit parasite emergence the following spring (Diadetchko, 1970).

VII. USE OF PLANTING TIME

Control of some insect pests can be achieved by manipulating the planting date so that the most susceptible stage of the crop coincides with the time of the year when the pest is least abundant.

The Hessian fly, *Mayetiola destructor* (Say), has been a severe pest of wheat, its most important host. It normally has two full generations per year but under certain weather conditions there may be more (Painter, 1951). When there are two generations, the eggs of one generation are laid in the fall and the winter is passed in the maggot stage in a brown puparium, the "flaxseed" stage. In the spring the maggot pupates and most of the adults emerge in April. These adults give rise to the 2nd generation and much of the summer is also passed in the "flaxseed" stage. Adults from the summer generation usually begin to emerge in late August.

Wheat varieties with variable degrees of resistance to the Hessian fly are now available (Painter, 1951; Chapter 25, herein) but for many years delayed planting, coupled with measures to destroy volunteer wheat on which the fall generation might deposit their eggs, and the plowing under of infested wheat stubble after harvest was the only effective control.

The adult Hessian flies from either generation emerge over a 30-day period but the individuals live only 3–4 days. Thus, the planting date for winter wheat in a given climatic zone can be delayed so that most of the adults have emerged and died before the wheat has grown to a stage that can be attacked (Cartwright and Jones, 1930).

There is some variation in time of adult emergence in different years because development is largely dependent on the weather. However, tests have shown that it is nearly always possible to plant late enough for fly control yet early enough to secure sufficient plant growth for the wheat to withstand winter cold. Eight years of experiments in Illinois on planting before and after the "safe" planting date gave an average gain of 5.8 bushels per acre when the wheat was planted after the "safe" date of planting (Metcalf and Flint, 1962).

VIII. USE OF PLOWING AND TILLAGE

Soil tillage is used in nearly all commercial crop production and is probably the most widely used method of suppressing pest species, excluding chemical control. A number of examples of effects of plowing on insect pests have already been discussed. However, plowing can be used to affect differentially the pest species and their natural enemies. Telenga (1950) and Telenga and Zhigaev (1959) found that deep plowing may destroy up to 95% of the egg parasite *Xenocrepis* (= *Caenocrepis*) *bothynoderes* Gromakov which hibernates in surface soil layers in the eggs of its weevil host, *Bothynoderes punctiventris* Germer. The replacement of deep plowing by surface disc tillage greatly increased parasite survival and was twice as effective in reducing the pest population as was deep plowing (Telenga and Zhigaev, 1959; Diadetchko, 1970).

Bobinskaya (1959) and Aleynikova and Utrobina (1960) showed that such surface cultivation greatly reduces the numbers of elaterid larvae in field crops. Studies in the Kourgan District (Bobinskaya, 1959) showed that under surface tillage soil-inhabiting insects build up mainly in the upper soil layer (0–10 cm) and are easily accessible to carabids. Under deep plowing they are turned into the soil (up to 30 cm) where the activity of such predators is greatly reduced.

IX. MANAGEMENT OF DRIFT OF CHEMICALS AND ROAD DUST

The problems associated with drift of agricultural chemicals and of road dust are not clearly within the scope of cultural controls. Yet both deserve mention because they are serious problems and are not covered elsewhere in this treatise.

The problems arising from drift of agricultural chemicals have been revived by Akesson and Yates (1964). In addition to direct kill of predators and parasites in the treated area there may be other side effects. There is good evidence that drift of certain pesticides can cause pest outbreaks on property adjacent to the treated area. Harpaz and Rosen (1971) reported that in certain areas of Israel citrus groves lie adjacent to cotton fields. The integrated control program on citrus is very restricted as to the chemicals used, but aerial sprays of organophosphorus and chlorinated-hydrocarbon pesticides may be repeatedly applied to the adjacent cotton. Drift of these materials into the citrus groves often upsets the biological equilibrium, resulting in serious outbreaks of California red scale, *Aonidiella aurantii* (Maskell). Legislation was enacted forbidding aerial spraying of cotton with nonselective pesticides within a distance of some 200 yards from a citrus grove.

Adkisson (1971) reported that methyl parathion applied at 1 to 2 lb/acre for control of bollworm, *Heliothis zea*, and tobacco budworm, *H. virescens*, in cotton in Texas has caused upsets in adjacent citrus, mainly of brown soft scale, *Coccus hesperidum* L. (H. A. Dean, personal communication).

Another hazard due to improper timing and/or drift of insecticides is the effect on honeybees and other pollinating insects. The effects on honeybees have been reviewed by Anderson and Atkins (1968). The effect on honeybee kill at various distances from a treated area has been discussed by Anderson *et al.* (1968).

Road dust is an inevitable companion of cultivation and of unpaved access roads for farm machinery. DeBach (1958) reported the inhibition of effective parasitism and predation of the California red scale, *A. aurantii*, by dust and he has artificially increased California red scale populations by purposeful road dust applications. As a means of reducing dust in citrus groves, DeBach suggested overhead sprinkling, cover crops, and dust-reducing road surfaces.

Das (1960) also commented that tea bushes (*Camellia sinensis*) along dusty roads are often severely damaged by the red spider, *Oligonychus coffeae* (Nietner). He noted a low incidence of predators on dusty leaves. Similar situations have been noted on cotton and other crops in many countries. A common practice among growers in California is to sprinkle daily all dirt roads commonly used by automobiles, trucks, and other farm equipment.

REFERENCES

Adkisson, P. L. (1962). Timing of defoliants and desiccants to reduce populations of the pink bollworm in diapause. *J. Econ. Entomol.* **55**, 949–951.

Adkisson, P. L. (1964). Action of the photoperiod in controlling insect diapause. *Amer. Natur. XCVIII* **902**, 357–374.

Adkisson, P. L. (1966). Internal clocks and insect diapause. *Science* **154**, 234–241.

Adkisson, P. L. (1971). Objective uses of insecticides in agriculture. *In* "Agricultural Chemicals—Harmony or Discord for Food, People and Environment" (J. E. Swift, ed.), pp. 43–51. Univ. Calif., Div. Agri. Sci., Berkeley, California.

Adkisson, P. L. and Gaines, J. C. (1960). Pink bollworm control as related to the total cotton insect control program of Central Texas. *Texas Agr. Exp. Sta. Misc. Publ.* **444**, 7 pp.

Adkisson, P. L., Wilkes, L. H., and Cochran, B. J. (1960). Stalk shredding and plowing as methods for controlling the pink bollworm, *Pectinophora gossypiella*. *J. Econ. Entomol.* **53**, 436–439.

Adkisson, P. L., Bell, R. A., and Wellso, S. G. (1963). Environmental factors controlling the induction of diapause in the pink bollworm, *Pectinophora gossypiella* (Sanders). *J. Insect Physiol.* **9**, 299–310.

Akesson, N. B., and Yates, W. E. (1964). Problems relating to application of agricultural chemicals and resulting drift problems. *Annu. Rev. Entomol.* **9**, 285–318.

Aleynikova, M. M., and Utrobina, N. M. (1960). Influence of soil cultivation according to T. S. Malzev's method on soil fauna. *Tr. Kazan. Filiala Akad. Nauk S.S.S.R. Ser. Biol. Sci.* **5**, 19–29.

Allen, W. W., and Smith, R. F. (1958). Some factors influencing the efficiency of *Apanteles medicaginis* Muesebeck (Hymenoptera: Braconidae) as a parasite of the alfalfa caterpillar, *Colias philodice eurytheme* Boisduval. *Hilgardia* **28**, 1–42.

Anderson, L. D., and Atkins, E. L., Jr. (1968). Pesticide usage in relation to beekeeping. *Annu. Rev. Entomol.* **13**, 213–238.

Anderson, L. D., Atkins, E. L., Jr., Todd, F. E., and Levin, M. D. (1968). Research on the effect of pesticides on honey bees. *Amer. Bee J.* **108**, 277–279.

Bashir, M. O., and Venkatraman, T. V. (1968). Insect parasite complex of beerseem armyworm, *Spodoptera exigua* (Hübn.) (Lepidoptera: Noctuidae). *Entomophaga* **13**, 151–158.

Beingolea, O. G. (1957). El Sembrío del Maíz y la Fauna Benéfica del Algodonero. *Estac. Exp. Agr. La Molina. Inf. No. 104, Julio, 1957.* Lima-Perú.

Beingolea, O. G. (1962). Empleo de Insecticidas Orgánicos en el Perú y Posibilidades de Reducirlo por medio del control Integrado. *Rev. Peru. Entomol. Agr.* **5** (No 1).

Bobinskaya, S. G. (1959). The influence of soil cultivation and sowing according to T. S. Malzev's method of the development and survival of harmful and beneficial insects. *Zool. Zh.* **38**, 1601–1611.

Bowden, J. (1965). The sorghum midge, *Contarinia sorghicola* (Coq.) and other causes of grain-sorghum loss in Ghana. *Bull. Entomol. Res.* **56**, 169–189.

Branson, T. F., and Ortman, E. E. (1970). The host range of larvae of the western corn rootworm: Further studies. *J. Econ. Entomol.* **63**, 800–803.

Branson, T. F., and Ortman, E. E. (1971). Host range of larvae of the northern corn rootworm: Further studies. *J. Kans. Entomol. Soc.* **44**, 50–52.

Burnett, T. (1960). Control of insect pests. *Fed. Proc. Fed. Amer. Soc. Exp. Biol.* **19**, 557–561.

Burr, W. W. (1944). *58th Annu. Rep. Agr. Exp. Sta. Univ. Nebr.*, 125 pp.

Cartwright, W. B., and Jones, E. T. (1930). The Hessian fly—and how losses from it can be avoided. *U.S. Dep. Agr. Farmer's Bull.* **1627**, 14 pp.

Chiang, H. C. (1965). Survival of northern corn rootworm eggs through one and two winters. *J. Econ. Entomol.* **58**, 470–472.

Chiang, H. C. (1973). Bionomics of corn rootworms. *Annu. Rev. Entomol.* **18**, 47–72.

Christidis, B., and Harrison, G. J. (1955). "Cotton Growing Problems," 633 pp. McGraw-Hill, New York.

Chumakova, B. M. (1960). Adult feeding as a factor of increase in efficiency of insect pests parasites. *Tr. Vses. Inst. Zashch. Rast.* **15**, 57–70.

Cole, L. D. (1964). *Amer. J. Publ. Health* **54**(1), 24–31.

Combe, I., and Gamero, O. (1954). Resultados del Experimento. b. Investigaciones sobre la protección del sembrío del Maíz Intercalado con el algodonero, en el Valle del Rímac. *Estac. Exp. Agr. La Molina, 1953, Inf. Mens Mayo, 1954.*

Combe, I., and Lamas, J. M. (1954). Resultados del Experimento. a. Investigaciones sobre la protección del sembrio de Maíz Intercalado en el algodonero en el Valle de Cara bayllo. *Estac. Exp. Agr. La Molina, 1953, Inf. Mens Mayo, 1954.*

Das, G. M. (1960). Occurrence of the red spider, *Oligonychus coffeae* (Nietner) on tea in northeast India in relation to pruning and defoliation. *Bull. Entomol. Res.* **51**, 415–426.

DeBach, P. (1958). Application of ecological information to control of citrus pests in California. *Proc. 10th Int. Congr. Entomol. Montreal, 1956.* **3**, 187–194.

DeBach, P. (1964). The scope of biological control. *In* "Biological Control of Insect Pests and Weeds" (P. DeBach, ed.), pp. 3–20. Reinhold, New York.

Denning, D. (1928). A tree classification for the selection forests of the Sierra Nevada. *J. Agr. Res.* **36**, 755–771.

Diadetchko, M. D. (1970). Role des procédés agrotechniques et agro nomiques dans la multiplication naturelle des entomophages. *Ann. Zool. Ecol. Anim. Hors Ser.*, pp. 39–44.

Doutt, R. L., and Nakata, J. (1973). The *Rubus* leafhopper and its egg parasitoid: An endemic biotic system useful in grape-pest management. *Environ. Entomol.* **2**, 381–386.

Fenton, F. A., and Owen, W. L. (1953). The pink bollworm of cotton in Texas. *Texas Agr. Exp. Sta. Misc. Publ. No.* **100**, 39.

Forbes, S. A. (1883). Noxious and beneficial insects of Illinois. *Ill. State Entomol. Rep.* **12**, 10–31.

Forbes, S. A. (1891–1892). Noxious and beneficial insects of Illinois. *Ill. State Entomol. Rep.* **18**, 135–145.

Gentry, C. R., Lawson, F. R., Knott, C. M., Stanley, J. M., and Lam, J. J., Jr. (1967). Control of hornworms by trapping with blacklight and stalk cutting in North Carolina. *J. Econ. Entomol.* **60**, 1437–1442.

Harpaz, I., and Rosen, D. (1971). Development of integrated control programs for crop pests in Israel. *In* "Biological Control" (C. B. Huffaker, ed.), pp. 458–468. Plenum, New York.

Harris, C. R. (1972). Factors influencing the effectiveness of soil insecticides. *Annu. Rev. Entomol.* **17**, 177–198.

Harris, K. M. (1961). The sorghum midge, *Contarinia sorghicola* (Coq.) in Nigeria. *Bull. Entomol. Res.* **52**, 129–146.

Harris, K. M. (1962). Lepidopterous stem borers of cereals in Nigeria. *Bull. Entomol. Res.* **53**, 139–171.

Kamenkova, K. V. (1958). The causes of high efficiency of sunn pest egg parasites in premountainous zone of Krasnodar Region. *Tr. Vses. Inst. Zashch. Rast.* **9**, 285–311.

Keen, F. P. (1936). Relative susceptibility of ponderosa pines to bark-beetle attack. *J. Forest.* **34**, 919–927.

Keen, F. P. (1943). Ponderosa pine tree classes redefined. *J. Forest.* **41**, 249–253.

Keen, F. P., and Salman, K. A. (1942). Progress in pine beetle control through tree selection. *J. Forest.* **40**, 854–858.

King, D. R., and Weaver, C. R. (1953). The effect of meadow management on the abundance of meadow spittlebugs. *J. Econ. Entomol.* **46**, 884–887.

Kolobova, A. N. (1959). An ecological background for the method of reproduction stimulation of parasites of *Bruchophagus roddi* in natural conditions. *Vop. Ekol. (Kiev)* **3**, 25–34.

Landis, B. J., and Onsager, J. A. (1966). Wireworms on irrigated lands in the west: How to control them. *U.S. Dep. Agr. Farmer's Bull.* **2220**, 14 pp.

Lane, M. C. (1941). Wireworms and their control in irrigated lands. *U.S. Dept. Agr. Farmer's Bull.* **1966**, 22 pp.

Lawson, D. E. (1964). Rootworm egg distribution in corn fields. *Proc. N. Cent. Br. Entomol. Soc. Amer.* **19**, 92.

Lukefahr, M. J., Noble, L. W., and Martin, D. F. (1964). Factors inducing diapause in the pink bollworm. *U.S. Dep. Agr. Tech. Bull.* **1304**, 17 pp.

Metcalf, C. L., and Flint, W. P. (1962). "Destructive and Useful Insects: Their Habits and Control," 4th ed., 1087 pp. McGraw-Hill, New York. (Revised by R. L. Metcalf).

Miller, J. M., and Keen, F. P. (1960). Biology and control of the western pine beetle. *U.S. Forest. Serv. Misc. Publ.* **800**, 381 pp.

Mueller, A. J. (1971). The bionomics of lygus bugs in the safflower-cotton-alfalfa seed cropping system in California. Ph.D. Thesis, 141 pp. Univ. California, Riverside, California.

Müller, H. (1963). Zur Populationsdynamik von *Sitona* Germar (Curculionidae) auf Luzerne und Rotklee unter besonderer Berücksichtigung entomophager Parasiten. *Zool. Jahrb. Abt. Syst. Okol. Geog. Tiere* **90**, 659–696.

Newsom, L. D. (1967). Consequences of insecticide use on nontarget organisms. *Annu. Rev. Entomol.* **12**, 257–286.

Noble, L. W. (1969). Fifty years of research on the pink bollworm in the United States. *U.S. Dep. Agr. Handb. No.* **357**, 62 pp.

Odum, E. P. (1971). "Fundamentals of Ecology," 574 pp. Saunders, Philadelphia, Pennsylvania.

Ohlendorf, W. (1926). Studies of the pink bollworm in Mexico. *U.S. Dep. Agr. Bull.* **1374**, 64 pp.

Painter, R. H. (1951). "Insect Resistance in Crop Plants," 520 pp. Univ. Press of Kansas, Lawrence, Kansas.

Patel, K. K., and Apple, J. W. (1967). Ecological studies on the eggs of the northern corn rootworm. *J. Econ. Entomol.* **60**, 496–500.

Pearson, E. O. (1958). "The Insect Pests of Cotton in Tropical Africa." 355 pp. Emp. Cotton Grow. Corp. Commonw. Inst. Entomol., Eastern Press Ltd., London and Reading.

Peterson, J. B. (1955). Continuous cultivation—Its conservation significance. *J. Soil Water Conserv.* **10**, 281–285, 302.

Peterson, J. B. (1957). Examples of influence of soil science on agronomic practice. *Agron. J.* **49**, 651–659.

Rabb, R. L. (1971). Naturally-occurring biological control in the eastern United States, with particular reference to tobacco insects. *In* "Biological Control" (C. B. Huffaker, ed.), pp. 294–311. Plenum, New York.

Rabb, R. L., Neunzig, H. H., and Marshall, H. V., Jr. (1964). Effect of certain cultural practices on the abundance of tobacco hornworms, tobacco budworms, and corn earworms on tobacco after harvest. *J. Econ. Entomol.* **57**, 791–792.

Reynolds, H. T. (1971). Resistance of insects and mites to insecticides and acaricides and the future of pesticide chemicals. *In* "Agricultural Chemicals—Harmony or Discord for Food, People and the Environment" (J. E. Swift, ed.), pp. 108–112. Univ. Calif. Div. Agr. Sci., Berkeley, California.

Ripper, W. E., and George, L. (1965). "Cotton Pests of the Sudan—Their Habits and Control," 345 pp. Blackwell, Oxford.

Robertson, O. T., Stedronsky, V. L., and Currie, D. H. (1959). Kill of pink bollworms in the cotton gin and the oil mill. *U.S. Dep. Agr. Prod. Res. Rep.* **26**, 22 pp.

Rose, D. J. W., and Hodgson, C. J. (1965). *Systates exaptus* and related species as soil pests in Rhodesia. *Bull. Entomol. Res.* **56**, 303–318.

Salman, K. A., and Bongberg, J. W. (1942). Logging high-risk trees to control insects in the pine stands of northeastern California. *J. Forest.* **40**, 533–539.

Schlinger, E. I., and Dietrick, E. J. (1960). Biological control of insect pests aided by strip-farming alfalfa in experimental program. *Calif. Agr.* **14**, 8–9, 15.

Sevacherian, V. (1970). Spatial distribution pattern, sequential sampling and host preference of

Lygus hesperus Knight and *L. elisus* Van Duzee in California cotton fields. Ph.D. Thesis, 176 pp. Univ. California, Riverside, California.

Shapiro, V. A. (1959). The influence of agricultural and silvicultural methods on the efficiency of sunn pest egg parasites. *In* "Biologicheskdy Method Bor'by s Vreditelyamie Rasteniy," Akad. Nauk Ukrain. SSR. pp. 182–191. Kiev.

Shepetilnikova, V. A. (1963). Problems of biological control methods in plant protection. *Tr. Vses. Inst. Zashch. Rast.* **17**, 162–185.

Shepetilnikova, V. A., Fedorinchik, N. S., Kolmakova, V. D., and Kapustina, O. V. (1968). A biological control program as a basis of fruit orchard protection against pests in the area with one codling moth generation. *Tr. Vses. Inst. Zashch. Rast.* **31**, 21–62.

Simón, J. E. (1954). Algunas experiencias en el control de las plagas del algodonero: Maíz Intercalado e Insecticidas Sistémicos. *Estac. Exp. Agr. La Molina, Bolet. No. 90, Junio, 1954.*

Smith, R. C. (1956). Upsetting the balance of nature, with special reference to Kansas and the Great Plains. *Science* **75**, 649–654.

Smith, R. F., and Allen, W. W. (1954). Insect control and the balance of nature. *Sci. Amer.* **190**, 38–42.

Smith, R. F., and Reynolds, H. T. (1966). Principles, definitions, and scope of integrated pest control. *Proc. FAO Symp. Integrated Pest Contr.* **1**, 11–17.

Smith, R. F., and van den Bosch, R. (1967). Integrated Control. *In* "Pest Control, Biological, Physical, and Selected Methods" (W. W. Kilgore, R. L. Doutt, eds.), pp. 295–340. Academic Press, New York.

Smith, R. F., Bryan, D. E., and Allen, W. W. (1949). The relation of flights of *Colias* to larval population density. *Ecology* **30**, 288–297.

Southwood, T. R. E. (1966). "Ecological Methods," 391 pp. Chapman & Hall, London.

Stark, V. N. (1940). Use of scoliids for the control of scarabaeid larvae. *Vestn. Zashch. Rast.* **1**, 120–122.

Stark, R. W. (1970). Section 1. General. *In* "Studies on the Population Dynamics of the Western Pine Beetle, *Dendroctonus brevicomis* LeConte" (R. W. Stark and D. L. Dahlsten, eds.), pp. 1–5. Univ. California Press, Berkeley, California.

Steinhaus, E. A. (1949). "Principles of Insect Pathology," 757 pp. McGraw-Hill, New York.

Stern V. M. (1968). "Mission Rep. Turkey, Cyprus, Israel, Iraq—Development of an integrated pest control program on cotton," 39 pp. FAO of the United Nations, Rome, Italy.

Stern. V. M. (1969). Interplanting alfalfa in cotton to control lygus bugs and other insect pests. *Proc. Tall Timbers Conf. Ecol. Anim. Contr. Habitat Manage.* **1**, 55–69.

Stern, V. M. (1973). Economic Thresholds. *Annu. Rev. Entomol.* **18**, 259–280.

Stern, V. M., van den Bosch, R., and Leigh, T. F. (1964). Strip cutting alfalfa for lygus bug control. *Calif. Agr.* **18**, 4–6.

Stern, V. M., van den Bosch, R., Leigh, T. F., McCutcheon, O.D., Sallee, W. R., Houston, C. E., and Garber, M. J. (1967). *Lygus* control by strip cutting alfalfa. *Univ. Calif. Agri. Ext. Serv. AXT-241,* 13 pp.

Swaine, G. (1957). Maize and sorghum stalk borer, *Busseola fusca* (Fuller) in peasant agriculture in Tanganyika Territory. *Bull. Entomol. Res.* **48**, 711–722.

Telenga, N. A. (1950). *Caenocrepis bothynoderis* Grom. and its importance in reproduction of beet weevil. *Nauch. Tr. Inst. Entomol. Fitopatol. Akad. Nauk Ukr. SSR.* **2**, 142–170.

Telenga, N. A. (1954). Parasites of clover *Apion*, their importance and methods of use. *In* "Biologicheskiy Method Bor'by s Vreditelyamie Rasteniy, Kiev," Akad. Nauk. Ukrain. SSR. pp. 55–72.

Telenga, N. A., and Zhigaev, G. N. (1959). The influence of different soil cultivation on reproduction of *Caenocrepis bothynoderis*, an egg parasite of beet weevil. *Nauch. Tr. Ukr. Nauch. Issled. Inst. Zashch. Rast.* **8**, 68–75.

Townsend, C. H. T. (1912). The work in Perú against the white scale of cotton. *J. Econ. Entomol.* **5**, 256–263.

Townsend, C. H. T. (1913). A brief report on the piojo blanco of cotton. *J. Econ. Entomol.* **6**, 318–327.

Townsend, C. H. T. (1924). El gorgojo de la Chupadera del algodonero, *Gasterocercodes gossypii* Pierce. *Vida Agr.* **1**, 257–263.

U.S.Dep. Agr. (1970). Economic consequences of restricting the use of organochlorine insecticides on cotton, corn, peanuts and tobacco. *Econ. Res. Serv. Agr. Econ. Rep.* **178**, 52 pp.

U.S. Dep. Agr. (1971). Midwestern corn farms: Economic status and the potential for large and family-sized units. *Econ. Res. Serv. Agr. Econ. Rep.* **216**, 53 pp.

U.S. Dep. Agr. (1973). "Monoculture in Agriculture: Extent, Causes, and Problems—Report of Task Force on Spatial Heterogeneity in Agricultural Landscapes and Enterprises" (H. J. Hodgson, Chairman), 64 pp. Coop. State Res. Serv., Wash., D.C.

van den Bosch, R., and Telford, A. D. (1964). Environmental modification and biological control. *In* "Biological Control of Insect Pests and Weeds" (P. DeBach, ed.), pp. 459–488. Reinhold, New York.

van den Bosch, R., Schlinger, E. I., Dietrick, E. J., Hagen, K. S., and Holloway, J. K. (1959). The colonization and establishment of imported parasites of the spotted alfalfa aphid in California. *J. Econ. Entomol.* **52**, 136–141.

van den Bosch, R., Lagace, C. F., and Stern, V. M. (1967). The interrelationship of the aphid, *Acyrthosiphon pisum*, and its parasite, *Aphidius smithi*, in a stable environment. *Ecology* **48**, 993–1000.

van Emden, H. F., and Williams, G. F. (1973). Insect stability and diversity in agro-ecosystems. *Annu. Rev. Entomol.* **19**, 455–475.

Viktorov, G. A. (1962). Causes of low *Eurygaster integricips* numbers in some regions of Transcaucasia. *Zool. Zh.* **41**, 63–76.

Viktorov, G. A. (1967). "Problems of Insect Population Dynamics Exemplified by Sunn Pest *Eurygaster integricips*." Nauka, Moscow.

Weaver, C. R. (1951). The seasonal behavior of the meadow spittlebug and its relation to a control method. *J. Econ. Entomol.* **44**, 350–353.

Weaver, C. R. (1952). Fall applications of insecticides to control spittlebug nymphs. *J. Econ. Entomol.* **45**, 238–241.

Weaver, C. R., App, B. A., and King, D. R. (1954). Field trials of fall insecticide treatments in Ohio to control the meadow spittlebug. *Ohio Agr. Exp. Sta. Res. Bull.* **741**, 99 pp.

Weiss, M. G. (1949). Soybeans. *Advan. Agron.* **1**, 77–157.

Wilbur, D. A., Fritz, R. F., and Painter, R. H. (1942). Grasshopper problems associated with strip cropping in western Kansas. *J. Amer. Soc. Agron.* **34**, 16–29.

Wilkes, L. H., Adkisson, P. L., and Cochran, B. J. (1959). Stalk shredder tests for pink bollworm control. *Texas Exp. Sta. Progr. Rep.* **2095**, 2 pp.

Wille, J. E. (1952) "Entomología Agrícola del Perú," 2nd ed. Edit. por Est. Exp. Agr. La Molina, Dir. Gral. Agr. Min. Agr. Lima-Perú, 543 pp.

25

USE OF PLANT RESISTANCE

Stanley D. Beck and Fowden G. Maxwell

I. INTRODUCTION

An ideal method of preventing insect depredation of agricultural and forest plantings would be to cultivate plant varieties that are highly resistant or immune to insect attack. Pest populations could thereby be suppressed well below the level of economic damage with neither disturbance nor pollution of the ecosystem, and at no additional cost to the producer. Although such a goal may not always be attainable, progress has been made in that direction in many instances. Resistance to pathogens and insects are characteristics being taken into account in the selection and development of an increasing number of modern crop plants. Where the use of resistant plants may be combined with other forms of biological and integrated control, efficient pest population management is fully feasible.

Plant resistance is defined as the collective heritable characteristics by which a plant species or genetic line reduces the probability of its successful utilization as the host of an exploiting species or biotype. Nearly all plants display some degree of resistance toward phytophagous insects, and the selection of resistant genetic lines for agricultural use more frequently involves augmentation of existing defenses than the creation of additional or novel resistance mechanisms.

Some general characteristic that would render a crop variety resistant to all plant-feeding insects is clearly impossible, because of the simple fact that a host plant is more than an inert nutritional substrate for the complex of insects that live on it. In each case, the host plant is the central component of the environment of the insect, but each insect species may be highly adapted to specific physical, chemical, developmental, and phenological characteristics of its host. The insect's host specificity, the identity of the plant characteristics that influence host utilization, and the nature of the plant's natural defenses against damage, must all be taken into account in any attempt to develop genetic lines of a crop plant that will display economically significant resistance (see Section IV).

Plant-feeding insects vary widely in the range of plant species utilized. There are general feeders (polyphagous), such as the desert and migratory locusts, *Schistocerca gregaria* (Forsk) and *Locusta migratoria migratoriodes* R & F, at one extreme, and highly specific forms (monophagous) such as the silkworm, *Bombyx mori* L., and certain sawflies, at the other. Most pest insects display relatively narrow ranges of host plants, and are termed "oligophagous." Even polyphagous species do not attack every plant species, and within the broad restrictions of their host range, not all plants are equal in terms of preference and utilization. The evolutionary history of insect—plant interactions has been one of complementary adaptations of both plants and insects, in which plant species have evolved defense mechanisms and insects have evolved counteracting adaptations enabling them to maintain or improve host utilization (Fraenkel, 1959, 1969). There is little evidence that the evolution of insect phytophagy was from a primitive polyphagy toward a highly specialized monophagy; a coevolution involving a continuum of interacting plant and insect adaptations is a more likely interpretation (Ehrlich and Raven, 1964; Breedlove and Ehrlich, 1968; Dethier, 1970). Although a female insect may deposit up to several hundred eggs, usually, only a very few of her progeny will reach adulthood and reproduction. This severe attrition is due, in part, to the hazards of specialized phytophagous ways of life and to the effectiveness of host plant resistance.

II. Types of Resistance

A phytophagous insect must be able to identify and orient successfully to the utilizable hosts among the multitude of plant species available within its temporal and spatial environment. In these behavioral patterns, the insect is dependent on its ability to respond to physical and biochemical stimuli emanating from the plant itself. Plant resistance may thus be based on characteristics exerting adverse effects on the insect's behavior, thereby lessening the probability of successful orientation, and subsequent host plant utilization. Such behavior-based plant resistance was termed "nonpreference" by Painter (1951).

The feeding stages of an insect (all larval or nymphal stages and frequently young adults) are dependent on the host plant for a nutritionally adequate and nontoxic substrate. In addition, the host plant tissues must be physically available, and in quantities allowing successful completion of development. Plant resistance based on the plant's exerting adverse effects on the physiological maturation of the insect was classified as "antibiosis" by Painter (1951).

Although convenient designations, "nonpreference" and "antibiosis" are arbitrary categories that cannot always be clearly separated, and the two types of resistance very frequently show a high degree of interdependence (Beck, 1965; Kennedy, 1965). Plant resistance to the cereal leaf beetle, *Oulema melanopus* (L.), for example, is associated with the density of pubescence on the leaves of the host plant, in that the beetles deposit far fewer eggs on the more densely haired leaves; this is an example of non-preference (Schillinger and Gallun, 1968). However, Schillinger and Gallun also observed that eggs laid on densely pubescent leaves tended to become desiccated, and fewer than 10% hatched. Of newly hatched larvae confined to such leaves, only 20% survived 72 hours; these effects are clearly examples of antibiosis.

A. Resistance to Oviposition

Most phytophagous insects deposit their eggs on or near the host plant to be utilized by their progeny. Oviposition is the culmination of a catena of behavioral events, the first component of which (for the purpose of this discussion) is the orientation of the gravid female to the prospective host plant, in which it is selected from among the array of plants available. Following orientation to the plant as a whole, the insect orients to different plant parts in the selection of a specific oviposition site. Plant characteristics tending to prevent oviposition may do so either by failing to provide the appropriate releasing stimuli, or by providing stimuli that inhibit behavioral release.

Orientation to a prospective host plant may involve visual as well as chemosensory stimuli. In studies of the ovipositional behavior of the tobacco hornworm, *Manduca sexta* (Johan.), two phases in the orientation of the moth to the plant were observed—the approach and the landing (Yamamoto *et al.*, 1969; Sparks and Cheatham, 1970). The approach appeared to be based on visual clues, but the landing was elicited in response to olfactory stimuli. After landing, contact chemosensory factors released actual egg deposition. The importance of plant-borne attractants in insect orientation has been demonstrated in many studies, and the role of olfactory stimuli in some instances of plant resistance has also been confirmed (Thorsteinson, 1960). An upwind orientation of desert locusts in response to grass odors was reported by Kennedy and Moorhouse (1969); alfalfa weevils, *Hypera postica* (Gyllenhal), were attracted to alfalfa and alfalfa extracts (Byrne and Steinauer, 1966; Pienkowski and Golik, 1969); many other similar examples might be cited. The odors of nonhosts may fail to evoke any orientation response, or they may elicit a negative reaction in which the insect moves away from the odor source (de Wilde *et al.*, 1969; Maxwell *et al.*, 1969). Host plants may be encountered during the course of random, undirected locomotor activity. In such cases the insect tends to resume locomotion after encountering a nonhost plant, but to remain or display a behavioral change upon contacting an acceptable host.

Upon arriving at the prospective host plant, the insect is responsive to stimuli that will release the subsequent components of the ovipositional behavioral pattern. Oviposition is seldom indiscriminate over the surface of the plant or in the surrounding soil, but most frequently on selected plant parts. The specific sites selected may vary according to leaf maturity and the physiological state of the plant (Goeden and Norris, 1965; Miller and Hibbs, 1963; Gara *et al.*, 1971). Specific ovipositional stimulants may be involved; such a stimulant for the carrot rust fly, *Psila rosae* F., was found in carrot leaves and identified as *trans*-1,2-dimethoxy-4-propenylbenzene (Berüter and Städler, 1971). Matsumoto and Thorsteinson (1968) found that oviposition by the onion maggot was stimulated by a number of organic sulfur compounds, most effective of which was *n*-proply disulfide and *n*-propyl mercaptan, both of which are normal constituents of onion.

Tactile and proprioceptive stimuli have been shown to be important components of the ovipositional sequence. As mentioned earlier, the cereal leaf beetle prefers glabrous rather than densely pubescent leaves; the soybean pod borer, *Grapholitha glicinivorella* Matsumura, on the other hand, tends to lay more eggs on pubescent than on glabrous varieties of soybean (Nishijima, 1960). Tactile as well as chemotactic stimuli were shown to play a role in the ovipositional behavior of the diamondback moth, *Plutella maculipennis* (Curtis) (Gupta and Thorsteinson, 1960). Proprioceptive stimuli associated with the specific oviposition site, such as with insects that deposit their eggs only on the stalk, or underside of the leaf, etc., are relatively important in many insect species.

B. Resistance to Feeding

Insect larvae issuing from eggs deposited on a host plant are usually confined to that plant for the whole of their immature feeding stages. Adult forms that feed and the more mobile of immature insects may feed on a few or a great many plants. In either case, feeding involves a sequence of stereotyped behavioral components that parallels that described for oviposition. The sequential steps are: (1) host plant recognition and orientation (whole plant or plant part); (2) initiation of feeding (biting or piercing); (3) maintenance of feeding; and (4) cessation of feeding, followed by dispersal, at least in the case of some very mobile species (Dethier, 1954; Thorsteinson, 1960). Each behavioral pattern is manifested only in response to the appropriate combination of external releaser stimuli. The host plant is normally the source of releasing stimuli, and resistance may result from failure of the plant to provide the stimuli required for one or more components of the sequence, or because the plant provides adverse stimuli that tend to prevent the release of the behavior.

The range of food plants of an oligophagous insect has, in a number of instances, been shown to be characterized by the presence of identical or related chemicals that stimulate the insect's feeding. The classical demonstration of the relationship of feeding specificity and taxa-correlated chemical factors was the study of larval feeding of *Papilio ajax* by Dethier (1941). He showed that the Umbelliferae, on which this butterfly feeds almost exclusively, contained a group of related essential oils that attract the larvae. Plants not containing these substances were not attractive, nor would the larvae attempt to feed on them.

Feeding of larvae of the diamondback moth, *Plutella maculipennis* (Curtis), is confined almost entirely to the plant family Cruciferae; a number of mustard oil glycosides that commonly occur in cruciferous species were shown to be feeding stimulants for the larvae (Thorsteinson, 1953; Nayar and Thorsteinson, 1963). The vegetable weevil, *Listroderes costirostris obliquus* (Klug.), feeds on both Umbelliferae and Cruciferae. Study of the larval feeding behavior disclosed that many of the essential oils of Umbelliferae and the mustard oil glycosides of Cruciferae act as attractants and feeding incitants (Sugiyama and Matsumoto, 1959).

The relationship of feeding factors to plant resistance has been studied in a few cases, although much more work is needed in this area. The Colorado potato beetle, *Leptinotarsa decemlineata* Say, utilizes solanaceous host plants of the genus *Solanum* almost exclusively. *Solanum* species differ greatly in their suitability as hosts, however, with the common potato, *S. tuberosum*, being the most efficiently utilized. Foliage of *S. luteum* contains a feeding deterrent that tended to prevent larval feeding, and *S. lycopersicum* foliage was consumed only if no other food was available (Bongers, 1970).

Adults of the beetle *Scolytus multistriatus* (Marsham) feed on bark of the American elm *Ulmus americana*, and its feeding behavior is controlled by attractants, incitants, and stimulants present in the bark of elm twigs (Baker and Norris, 1968). Extracts of the bark of some nonhosts (poplar, oak, hickory) contained feeding deterrents. Hickory bark was found to contain 5-hydroxy-1,4-naphthoquinone, which strongly deterred the feeding of *Scolytus multistriatus* but not that of the hickory bark beetle, *S. quadrispinosus* Say; when this substance was removed from the hickory bark extracts, both beetle species fed readily (Gilbert and Norris, 1968). Feeding deterrents may play a part in host utilization among even the most polyphagous species. The desert locust, *Schistocerca gregaria*, for example, does not feed on seeds of the neem tree; this was shown to be due to the presence of the terpenoid azadirachtin, which acts as a powerful feeding deterrent (Butterworth and Morgan, 1971).

The effects of plant chemicals on insect feeding behavior, chemosensory processes, and on host plant relationships have been widely studied. The details cannot be pursued here and the reader is referred to the excellent reviews of Schoonhoven and Dethier (1966), Fraenkel (1969), and Dethier (1970) and for further information.

C. Resistance to Development and Survival

A plant may resist the successful development and survival of insects feeding on it by: (1) physical barriers, (2) unfavorable nutritional properties, (3) presence of toxins and inhibitors, or (4) presence of hormonal antagonists. In most cases such resistance mechanisms are directed against the immature stages of the insect, but they may also influence the fecundity and longevity of adult stages.

1. Physical Barriers

The physical form and tissue structure of plants undoubtedly influence their utilizability as host plants. Extensive testing of rice varieties for relative resistance to the Asiatic rice stem borer, *Chilo suppressalis* (Walker), led Patanakamjorn and Pathak (1967) to conclude that resistance tended to be associated with hairy upper lamina, tight leaf-sheath wrapping, small stems with ridged surfaces, and thick hypodermal layers. Cucurbit varieties that were resistant to the squash vine borer, *Melittia cucurbitae* (Harris), were found to have hard, woody stems with closely packed, tough vascular bundles (Howe, 1950). Hooklike trichomes were observed to be a very effective defense mechanism of *Passiflora adenopoda* against larvae of several species of heliconiine butterflies, the larvae becoming entrapped on the hooks and thereby perishing (Gilbert, 1971).

Physical defenses against insects may, in some cases, be evoked by the invasion of plant tissue by the insect. This is the case with some leaf miners whose larvae may be crushed by proliferating wound tissue, or which become isolated by a ring of necrotic and drying host cells surrounding the larvae causing death (Oatman, 1959). Some pine varieties have been shown to be resistant to needle miners by virtue of a high rate of resin flow, tending to drive the larvae out of the needles (Bennett, 1954; Harris, 1960).

2. Nutrition and Resistance

The role of the nutrient content of the host plant in its resistance to insects is currently unclear. A number of modifying factors make the question quite difficult to study experimentally. If such considerations as sensory influences on feeding behavior, the presence or absence of toxic or growth-inhibiting substances, and the digestibility of the plant tissue, are arbitrarily excluded from the question of nutritional adequacy, the problem becomes one of whether or not a plant can be resistant to insects by virtue of deficiencies in the specific nutrients (amino acids, vitamins, etc.) required by the insect. Although there have been instances in which the adequacy of different host plants was shown to be related to the digestibility of the plant tissues, no clear demonstration has been made of a causal relationship between plant resistance and insect nutritional requirements. There is, however, some evidence that suggests that amino acid balance in plant tissues might influence developmental rates of phytophagous insects (McGinnis and Kasting, 1961; Maltais and Auclair, 1962). The influence of the proportions or ratios of required nutrients in the plant tissues merits much further research. House (1969), for example, found that insect growth rates and survival were markedly influenced by rather subtle differences in the proportions of required nutrients in chemically defined diets that were qualitatively complete.

3. Toxins and Physiological Inhibitors

Not all of the chemical constituents of plant tissues are conducive to insect growth and well-being. Plant defense against utilization by herbivores, including insects, includes an array of alkaloids, phenols, flavonoids, and other substances. In their adaptation to a range of host plants, insects have necessarily evolved some degree of efficiency in detoxifying these inhibiting resistance factors (Krieger *et al.*, 1971; Self *et al.*, 1964; Kircher *et al.*, 1967). Nevertheless, many instances of varietal differences in resistance of a plant species to an insect species have been shown to be closely associated with varietal differences in the tissue content of insect-inhibiting chemical factors.

The resistance of maize varieties to the growth and survival of European corn borer larvae, *Ostrinia nubilalis* (Hübner), was shown to be the result of feeding deterrency and growth inhibition exerted by benzoxazolinone compounds in certain tissues of young plants (Beck, 1957, 1965; Smissman *et al.*, 1957). The resistance of maize to the corn borer will be discussed in greater detail in a later section. Plant constituents that are toxic to insects have been detected in many agronomic plant species and varieties including cabbage, cotton, oats, alfalfa, barley, turnips, beets, kale, parsnips, and wheat; only a few of these have been chemically characterized, however (Beck and Smissman, 1960; Swailes, 1960; Lichtenstein and Casida, 1963; Lichtenstein *et al.*, 1962).

Insect adaptations and plant defenses are interestingly illustrated by the interactions between the green peach aphid, *Myzus persicae* (Sulzer), and tobacco, *Nicotiana tabacum*, which is one of its many hosts. The aphids feed in the phloem, and not in the nicotine-translocating xylem; they thereby avoid ingesting this toxic alkaloid (Guthrie *et al.*, 1962). The closely related plant, *Nicotiana gossei*, is highly resistant to the green peach aphid. The resistance is caused by a toxic exudate from leaf hairs, which evokes nicotinelike symptoms in contacted aphids (Thurston and Webster, 1962).

4. Insect Hormone Analogs

The recent findings of analogs of insect hormones in plant tissues (Slama and Williams, 1966; Bowers *et al.*, 1966; Heinrich and Hoffmeister, 1967; Carlisle and Ellis, 1968) has added an exciting new dimension to the field of insect–plant interactions. The significance of plant-borne hormonal activity to practical aspects of host plant resistance and integrated control systems has not been determined. Its potential importance seems likely, because of the growth-disrupting effects of hormonal imbalance.

The juvenile hormone (JH) is the methyl ester of epoxyfarnesoate: *trans,trans*-10-epoxy-7-ethyl-3,11-dimethyl-2,6-tridecadienoate (Röller *et al.*, 1967). A few very closely related molecules have also been shown to occur in insects and to have comparable hormonal activity. The presence of biologically active JH analogs in plant tissues was first reported by Slama and Williams (1966), who found that some commercial papers contained a factor that would prevent the metamorphosis of the plant bug *Pyrrhocoris apterus* L. Its source was identified as a number of paper-pulp conifer species, especially balsam fir, *Abies balsamea*, but its activity was highly specific and no other insect species tested was so influenced. The factor was isolated and characterized as a methyl todomatuate and given the common name "juvabione" (Bowers *et al.*, 1966). The closely related dehydro ester "dehydrojuvabione" has also been identified from conifer tissues (Cerny *et al.*, 1967). In a search for other plant-borne JH analogs, Bowers (1968, 1969) found that compounds containing a methylenedioxyphenyl grouping are not only common in plant material but also that many such compounds display appreciable JH activity. The possible role of such substances in plant resistance to insects has yet to be determined.

The molting hormone, ecdysone, is a complex steroid, of which a number of active isomers and closely related analogs have been reported in various arthropods and plants. Steroids of ecodysone activity occur in many species of ferns of the family Polypodiaceae, and in nearly twenty families of gymnosperms and angiosperms. The role of these substances in plant resistance to insects has not been determined (see recent reviews by Williams, 1970; Slama, 1969).

III. Selected Examples of Resistance

Painter (1951) provided an excellent history and numerous examples of general resistance to insects by field crops. Additional reviews by Painter (1958, 1968), Beck (1965), and Maxwell *et al.* (1972) have also provided additional examples of resistance. The reader is referred to these publications for more extensive information. The purpose in the limited space available here is to present information specifically on four major insects and crops: (1) Hessian fly [*Mayetiola destructor* (Say)] on wheat; (2) European corn borer [*Ostrinia nubilalis* (Hübner)] on corn; (3) spotted alfalfa aphid [*Therioaphis maculata* (Buckton)] on alfalfa; and (4) boll weevil (*Anthonomus grandis* Boh.) on cotton. These outstanding examples are sufficient to illustrate the success and potential of the field of host plant resistance.

A. Hessian Fly

The Hessian fly occurs in most parts of the holarctic region where winter wheat is grown and in many localities where spring wheat occurs. Painter (1951) described the biology and the history of resistance to this pest, including the programs in California and Kansas which led to the development and release of the resistant varieties, Poso 42 and Big Club 43 in California in 1942 and 1944, and Pawnee in Nebraska and Kansas in 1942–1943. The resistant varieties released in California eliminated the Hessian fly as a problem by 1945.

By 1945, Pawnee was planted on 5 million or more acres in Kansas and over 1 million acres in Nebraska. The Hessian fly, which had been a major pest, was reduced in a period of 2 to 3 years to nonsignificant levels. Ponca, a wheat resistant to Hessian fly in eastern Kansas was released in 1951 (Painter, 1958, 1968), and was followed by the development of the resistant variety, Dual, released in 1956 by workers in Indiana. Since 1956 several additional varieties of resistant wheat were developed and released for use in the states of Wisconsin, Kentucky, Indiana, Arkansas, Nebraska, Georgia, Oklahoma, Missouri, and Illinois.

It is currently estimated that over 10 million acres are planted with 23 different Hessian fly-resistant varieties of wheat in 34 states. Luginbill (1969) reported that the increased yield resulting from resistant varieties in the United States was valued at about 238 million dollars. Losses by Hessian fly in wheat have been reduced to less than 1% because of resistant varieties.

The genetics of Hessian fly resistance was studied by several investigators (Painter, 1951; Shands and Cartwright, 1953; Allan *et al.*, 1959); six genes were found to be responsible for resistance. The primary mechanisms of resistance to the fly controlled by these genes are antibiosis and tolerance. Refai *et al.* (1955), in studying possible biochemical causes of resistance to Hessian fly, found no differences in protein, ash cellulose, silica, and other trace mineral contents between resistant and susceptible varieties. Hemicellulose content and degree of resistance were correlated, and larvae were demonstrated *in vitro* to secrete hemicellulase as well as a material which inactivated or inhibited wheat plant phosphorylase. Miller *et al.* (1960) and Lanning (1966) conducted studies on the relation of plant pigments and silica to resistance and found that silica content was related to resistance in some varieties. There was no relation between plant pigments and resistance. These studies constitute essentially all the biochemical work known on the basic causes of resistance in wheat to the Hessian fly.

B. European Corn Borer

Ostrinia nubilalis is a lepidopterous species of Eurasian origin. The insect was introduced in North America on shipments of broom corn about 1912. In this new environment, its increase was greatly facilitated by the absence of its natural diseases, parasites, and predators, and by the presence of extensive plantings of its currently preferred host.

Soon after the discovery of the borer in North America, entomologists observed that corn planted early in the season was much more likely to be heavily damaged by first-generation borers than were later plantings. The larger, more mature plants were not only more attractive to ovipositing borer moths, but larval survival on them was more successful. The suitability of the corn plant as a host for the corn borer tends to increase as the plant grows, and reaches a maximum suitability after the tassel has fully emerged from the whorl. Plant resistance that is attenuated by plant growth is of limited utility, since it is not always economically feasible to plant corn late in the season to avoid borer infestation. Even such limited resistance indicates that resistance to the borer might be greatly improved by selective plant breeding. Varietal differences in susceptibility to borer infestation were observed as early as 1928, and it was expected that plant breeders would soon develop highly resistant varieties. Great difficulty was encountered, however, in the selection of genetic lines that displayed borer resistance that was more than the transitory resistance associated with early growth, and the selection of varieties showing economically significant resistance to second-generation borer attack has thus far been relatively unsuccessful. However, genetic lines which are resistant for a sufficiently long period to be effective against first-generation borers have been developed, and most widely used hybrids now carry these resistance genetic factors.

Resistance of corn to European corn borer is based on a low attraction of the plants for oviposition and on adverse effects on survival of the newly hatched larvae. Manifestations of plant resistance against larval survival are consistent with the hypothesis that the plant tissues contained toxic or growth-inhibiting chemicals. Detailed studies of the effects of addition of corn plant part extracts to artificial diets fed to newly hatched borer larvae showed that such was the case (Beck and Stauffer, 1957). Chemical studies showed that at least three substances contributed to the resistance. Two of these compounds proved to be closely related benzoxazolinones: 6-methoxybenzoxazolinone (Smissman *et al.*, 1957) and 2,4-dihydroxy-7-methoxy-1,4-benzoxazin-3-one (Kristianson, unpublished thesis, 1960). Of these two factors, the latter appeared to be the more effective (Klun *et al.*, 1967) and to be responsible for the resistance of internodes and ear husks at the time of second-generation borer attack, although this resistance was of a generally low order of effectiveness in the inbreds studied (Beck, 1957). The third resistance chemical factor (Beck, 1957) was not identified.

Tissue concentrations of the resistance factors were found to change as the corn plant matured, but varietal differences were found, not only in the amounts of the benzoxazolinones present but also in respect to the relative concentrations in different plant tissues at different stages of growth (Beck, 1957; Klun *et al.*, 1967). The leaves of a borer-susceptible corn variety were found to contain large amounts of the benzoxazolinones, but at a growth stage in which leaf-feeding by borer larvae does not occur. Similarly, very young tassel buds were found to contain high titers of resistance chemicals, but at the stage of growth when borer larvae invade the tassels these structures contained only very small amounts. Borer resistance is obviously dependent upon the presence of effective concentrations of resistance chemicals in the tissues on which the borer feeds. Because the biology of the borer is such that its feeding sites change as the plant grows, and because the resistance of the plant changes as it grows, one of the problems associated with the development of highly resistant corn plants is the relationship of the biologies of the two organisms.

C. Spotted Alfalfa Aphid

The spotted alfalfa aphid, *Therioaphis maculata* (Buckton), was first found in North America in New Mexico on alfalfa in the spring of 1954. It now occupies most all the alfalfa producing areas in the United States and Mexico (Smith, 1959; Sprague and Dahms, 1972). It is now considered one of the most widespread and serious pests of alfalfa, causing millions of dollars damage annually. In host range studies, Peters and

Painter (1957) found a total of 23 species in the genera *Medicago*, *Melilotus*, *Trifolium*, and *Trigomella* to be favorable for the reproduction and development of this aphid. A considerable difference in resistance to the aphids by some species and by varieties of single species was noted. Resistant plants were noted to occur frequently in plants of Turkestan origin. Howe and Smith (1957) reported that the variety Lahontan and its parental clones had a high degree of resistance in the seedling stage, showing both antibiosis and tolerance. Howe and Pesho (1959) reported on the performance of a large number of varieties in mature field stands; Lahontan, Moapa, Zia, Bam, and Sirsa No. 9 were rated resistant. The synthetic variety, Zia, carrying resistance to the aphid, was released in New Mexico in 1959. The following resistant synthetics were released in California, Arizona, Nevada, and Utah: Moapa, Sonora, Caliverde 65, and Mesa-Sirsa.

In Kansas, individual resistant plants were observed to occur in the common susceptible variety Buffalo (Harvey and Hackerott, 1956). These individually resistant clones served as the basis for the development of Cody alfalfa which was officially released in 1961. Currently, Cody is grown over large acreages in Kansas, Nebraska, Oklahoma, and Missouri. At present there are seventeen varieties carrying resistance to the spotted alfalfa aphid that have been released in the United States (Sprague and Dahms, 1972).

The genes controlling the inheritance of resistance in alfalfa to the spotted alfalfa aphid are unknown. It is of special significance that breeding for resistance has been accomplished through phenotypical recurrent selection without knowing the genetics or the basic causes of resistance. Antibiosis and preference are the main mechanisms of resistance present in alfalfa. Tolerance in some instances is probably also involved, but to a much lesser degree.

D. Boll Weevil

The boll weevil, *Anthonomus grandis* Boh., is a very destructive insect on cotton, *Gossypium hirsutum*, in the United States cotton belt. The weevil first entered the country from Mexico in 1892 and by 1922 had covered almost the entire southern cotton belt, stretching from Texas to North Carolina. Knipling (1966) estimated that the boll weevil causes a $300 million annual loss, plus insecticide costs estimated at $70 million.

Webber (1903) made the earliest observation on the extent of injury by the boll weevil to several varieties of Egyptian cotton. Cook (1904) was the first to investigate the possibility of plant resistance to the weevil. He pointed out that plant characteristics such as large leafy involucres, hairiness of stems, gelatinization in young bud tissues, and proliferation of tissue, crushing eggs and young larvae, might be useful. Proliferation was studied by Hinds (1907), who found that 13.5% of the squares and 6.3% of the bolls proliferated after oviposition punctures occurred. He concluded that the weevil larvae were killed by mechanical pressure from the proliferated cells and not from any toxic properties.

Cook (1911) reported that weevil-resistant characteristics such as earliness, quick fruiting, and determinate habits of growth are affected by arid and humid conditions. The importance of earliness to escape weevil damage was emphasized by several early workers. Without the development of early varieties, most of the cotton industry in the United States probably would have been eliminated by the boll weevil.

Wannamaker *et al.* (1957) reported on 2 years of work at North Carolina which showed that two types of hairy cotton, "Pilose" and "MU-8," were less damaged by weevils than standard commercial varieties of Upland cotton. Wannamaker (1951) further reported on the possibility of using hairy types of Upland cotton as a source of resistance to the boll weevil. Hunter *et al.* (1965) reported a major screening program involving a cross section of the old-time commercial varieties of cottons. The only significant resistance found was due to a mutant gene called *Frego*. This gene causes a rolled bract condition and apparently disrupts weevil behavior. Jenkins *et al.* (1969) examined five different genetic lines, each carrying the *Frego* gene, and concluded that the *Frego* condition in all backgrounds reduced seasonal oviposition significantly. Jenkins and Parrott (1971) demonstrated in large field plots that *Frego* was responsible for up to 90% reduction in number of eggs oviposited. The weevils were observed to be nervous because of the exposed environment and moved around more, often leaving the plant.

Buford *et al.* (1967, 1968) found that Sea Island seaberry (*Gossypium barbadense*) suppressed boll weevil oviposition by approximately 40% compared with commercial Uplands. Maxwell *et al.* (1969) described the breeding program to incorporate this factor into a basic breeding line. It is anticipated that future work will embrace the combining of the oviposition suppression factor with *Frego* for boll weevil resistance, possibly in a fast fruiting, early, determinate-type cotton that can be utilized in a total pest management system for cotton insects.

Probably more is currently known about biologically active materials in the cotton plant that affect the behavior of the boll weevil than for any other insect–host situation. A program was begun in 1962 at the USDA Boll Weevil Research Laboratory at State College, Mississippi, to determine some of the factors involved in the selection (host preference) of cotton and other alternate hosts of the boll weevil. The following briefly describes progress made by a number of researchers over the past 9 years. (See also Table 1.)

1. Feeding Stimulant

Keller *et al.* (1962) reported the detection and extraction of a feeding stimulant present in cotton squares. Maxwell *et al.* (1963b) and Jenkins *et al.* (1963) reported on the occurrence and distribution of the feeding stimulant in cotton and other plants, utilizing a bioassay method. Their results showed that the stimulant was specific to hosts of the boll weevil, was present in all parts of the cotton plant including the roots, was highest in the squares, and that the calyx (exterior covering of the square) contained the highest concentrations. They also demonstrated the powerful nature of the feeding stimulant by showing that treated cork, meat, beans, and nonhost plants were readily consumed by the weevil when treated with stimulant extract. Hedin *et al.* (1966) and Struck *et al.* (1968) reported the chemical identification research on the feeding stimulant.

2. Attractant

Keller *et al.* (1963) discovered that chloroform extracts of ice water obtained in the process of freeze-drying cotton squares were attractive to boll weevils. Hedin *et al.* (personal communication) steam distilled 3 tons of cotton squares to obtain approximately one pound of essential oil for chemical identification work. Minyard *et al.* (1965, 1966, 1967, 1968) identified many constituents of this oil. Most boll weevil activity was associated with esters and alcohols. Minyard *et al.* (1969) reported the following identified compounds were attractive to the boll weevil: (−)-pinene, (−)-limonene, (−)-caryophyllene, (+)-bisabolol, caryophyllene oxide, and two unidentified sesquiterpenoids. Thompson *et al.* (1970, 1971) studied plant volatiles collected from cotton in field cages; many of the same compounds found by Minyard *et al.* (1969) were present. The attractiveness of cotton to the boll weevil undoubtedly results from a complex mixture of many chemical compounds in variable concentrations.

3. Repellent

In attempts to purify the crude attractant extract, it was found that the residue left after passing nitrogen through the crude extract and retrapping the vapors in cold traps, was strongly repellent to the weevils (Maxwell *et al.*, 1963a). Squares treated with the repellent material were not touched by weevils for 48 hours or more. Seedling cotton sprayed with the material was also protected, while untreated seedlings were readily destroyed by the weevils. Preliminary chemical work indicated that it may be a mixture of terpenes.

4. Feeding Deterrent

Boll weevils are known to feed on *Hibiscus syriacus* (Rose of Sharon) near cotton fields, in times of food stress late in the season when the cotton has matured (Parrott *et al.*, 1966). In observing boll weevil feeding and oviposition on Rose of Sharon buds it was noticed that the insect strongly preferred the buds where the calyx had split prior to blooming; if the calyx was mechanically removed, weevils fed and oviposited on this host as well as on cotton. Weevils would not feed through the calyx when intact on the bud except on rare occasions. Calyxes were found to contain a nonvolatile, water-soluble material which prevented feeding (Maxwell *et al.*, 1965b) of the adult as well as inhibiting larval development. *Hibiscus syriacus* was bioassayed for boll weevil attractant. No activity was detected with the techniques employed. Significant levels of boll weevil repellent were found. It was concluded that the nonpreference shown in nature by the boll weevil to *H. syriacus* results from a low level of attractant, high level of repellent, and a highly active feeding deterrent in the calyx which effectively masks feeding and oviposition stimulants in the bud.

IV. PROBLEMS ASSOCIATED WITH BREEDING FOR RESISTANCE TO INSECTS

There are a number of problems that occur at different stages in the development of any host plant resistance program. Usually one of the primary limiting factors is adequate financing over a long period of time. Considerable work has been done over the years on a piecemeal type approach. Many times, initial effectiveness was lost because of the inability to follow through with a sustained program after the initial investigation.

The importance of a multidisciplinary team approach should be recognized; efforts by entomologists and plant breeders have often been thwarted by their fragmented or isolated endeavors without proper coordination and cooperation. A close-knit team is absolutely essential to good progress and success in a host plant resistance program. Not only does the entomologist and the plant breeder need to have extremely close working relations, but they must also be willing to bring in plant pathologists, biochemists, plant physiologists, and other disciplines at various stages of the project to help solve complex problems which will undoubtedly arise.

The entomological problems connected with breeding plants for resistance to insects differ from other insect control in several ways. First, a very intimate knowledge of the biology (life table) and feeding habits of the insect is absolutely essential. Insect variability is also a problem that must be closely monitored. Biotypical differences within insect species may cause different behavior in laboratory-reared colonies than in field-selected populations. Natural variability in growth, fecundity, and other parameters of insect development must be carefully determined and accounted for in tests for resistance. Second, a population of insects must be built up with field or laboratory cultures for use in resistance testing rather than having the investigation subject to the vagaries of natural populations. Third, methods must be designed to examine large numbers of plants for insect infestation or damage, frequently in a very short period of time. Seedling screening techniques for mass selection becomes a very valuable time-saving tool and should be developed wherever possible. Careful consideration must be given to correlation of resistance in the seedling stage to that in mature plants. The success in developing, in a short period of time, synthetic resistant varieties of alfalfa to the spotted alfalfa aphid and pea aphid can be attributed largely to seedling rating techniques coupled with a simple recurrent selection procedure.

Resistance is often found in off-types or exotic wild species, and involves considerable crossing and selection to transfer the resistant genome into a desirable agronomic background. This, unfortunately, requires the time-consuming process of discarding all undesirable characters. For this reason, it is extremely important that plant pathologists and agronomists also be involved early in the program to help decide the characters that are to be retained or disposed of. In some cases, concessions may have to be made in levels of insect or disease resistance or agronomic qualities in order to provide the best overall variety for a particular geographical area.

Another problem frequently encountered is that a line or variety developed for resistance to one pest may in fact be more susceptible to another insect that previously had not been of economic significance. Excellent recent examples included the glanded (gossypol glands) and glandless condition in cotton. Removal of the glands through breeding [desired to open up potential markets for cotton seed meal (40% protein) for human and livestock consumption] caused greater susceptibility to bollworms, *Heliothis zea* and *H. virescens* (Lukefahr *et al.*, 1966; Oliver *et al.*, 1970, 1971). In addition, a number of leaf-feeding insects not previously known to cause damage to cotton became serious pests (Maxwell *et al.*, 1965). The *Frego* character in cotton imparts high field resistance to the boll weevil but causes increased susceptibility to the tarnished plant bug, *Lygus lineolaris* (Lincoln *et al.*, 1971). The glabrous condition in cotton which contributes resistance to bollworms causes greater susceptibility to thrips, leafhoppers, aphids, and certain other pests. Table 1 summarizes the situation on cotton in which various characters for resistance have been isolated. A character highly desirable for one pest may contribute to greater susceptibility for another. Each has to be weighted individually based on economics and other feasibility factors and a first decision made to utilize or not to utilize the character in question. Plants found resistant to a pest should be tested early in a breeding program for susceptibility to other insects and diseases. Such procedures may necessitate changes in approaches, but would possibly prevent loss of many years' work and expenditures of large amounts of money. In some cases it may not be economically feasible to proceed with a program if there is a high probability of creating an equal or worse problem with another insect.

V. UTILIZATION OF RESISTANT VARIETIES

In some cases where resistance is at a high level, it may be sufficient in itself to effect control (Painter, 1968). In areas where resistant varieties are planted over large areas the effect on the population of the insect resisted has been observed to be specific, persistent, and cumulative. The result is often a drastic reduction to noneconomic levels of the insect species. Holmes and Peterson (1957), Painter (1958, 1968), Dahms (1969), Chaing (1968), Pathak *et al.* (1970), Luginbill (1969), and Luginbill and Knipling (1969) have provided statistical information on the impact of resistant varieties where known on certain insect populations where they have been employed as the principal method of control. This type of information is scarce and more care should be taken in the future to document the affect of resistant varieties on total insect populations and crop yields.

The greatest use of resistant varieties for the future will undoubtedly be as one component part of a pest management system. In this type of management system the value of even low levels of resistance can be high because the resistance works as just one of

TABLE 1

Genetic Sources of Insect Resistance Found in Cotton and Currently Being Utilized in Breeding Programs (Texas, Mississippi, USDA, and Cooperating States of Louisiana and Missouri)[a]

Morphological or chemical characters identified	Major cotton insects					
	Boll weevil	*Heliothis* complex	Plant bugs *Lygus lineolaris* *L.* and *hesperus*	Cotton fleahopper	Spider mites	Whiteflies
Frego	R (60–90% suppression)	N (insecticide coverage increased)	S	S	N	N
Nectariless	N	R (20–50% egg suppression)	R	R	N	N
Smoothleaf (glabrous)	N	R (60% egg suppression)	S	S	R–	N
High gossypol	N	R	N	R–	N	N
Pilose (pubescence)	R–	S	R?	R	N	S
Okra leaf	N (better insecticide coverage increased kill in squares)	N	N	N	N	R–
Red color	R (choice situation)	N	N	N	N	N
"X" Factor (*G. hirsutum* wild races)	N	R	N	N	N	N
Oviposition suppression factor (*G. hirsutum* wild races)	R (40% + suppression of egg)	N	N	N	N	N
Plant bug suppression factor (Stoneville, wild races and other sources)	N	N	R	R	N	N
G. barbadense (Pima S-2)	N	R–	N	N	R	N
Earliness of maturity	R (escape)	R (escape)	N	N	N	R?

[a]R, resistant; N, no effect; S, increased susceptibility.

many suppressant factors integrated to prevent the pest population from attaining an economic threshold level. The resistant crop may in fact enhance the operation of natural biological control organisms. Greenbug [(*Schizaphis graminum* (Rondani)] damage to barley and sorghum was greatly reduced by the combined effects of resistant varieties and the parasite *Lysiphelebus testaceipes* (Cresson), compared to either control agent by itself (Starks *et al.*, 1972). Resistant varieties used in combination with other suppressant measures (parasites, pathogens, and phermone trapping) may be sufficient to reduce the pest population to such a low level that the sterile male technique may become feasible. *Frego* cotton, resistant to the boll weevil, is being used as one of several biological measures for reducing boll weevil populations to the level of possible susceptibility to eradication by the sterile male method. Once a pest population has been severely suppressed, or eradicated, in an area, the continued use of resistant crop varieties may serve as a major deterrent to the insect's reestablishing its status.

REFERENCES

Allan, E. R., Heyne, E. G., Jones, E. T., and Johnston, C. O. (1959). Genetic analyses of ten sources of Hessian fly resistance, their interrelationships and association with leaf rust reaction in wheat. *Kans. Agr. Exp. Sta. Tech. Bull.* **104**, 51 pp.

Baker, J. E., and Norris, D. M. (1968). Behavioral responses of the smaller elm bark beetle, *Scolytus multistriatus*, to extracts of non-host tree tissues. *Entomol. Exp. Appl.* **11**, 464–469.

Beck, S. D. (1957). The European corn borer, *Pyrausta nubilalis* (Hubn.), and its principal host plant. VI. Host plant resistance to larval establishment. *J. Insect Physiol.* **1**, 158–177.

Beck, S. D. (1965). Resistance of plants to insects. *Annu Rev. Entomol.* **10**, 207–232.

Beck, S. D., and Smissman, E. E. (1960). The European corn borer, *Pyrausta nubilalis* (Hubn.), and its principal host plant VIII. Laboratory evaluation of host resistance to larval growth and survival. *Ann. Entomol. Soc. Amer.* **53**, 755–762.

Beck, S. D., and Stauffer, J. F. (1957). The European corn borer, *Pyrausta nubilalis* (Hubn.), and its principal host plant. III. Toxic factors influencing larval establishment. *Ann. Entomol. Soc. Amer.* **50**, 166–170.

Bennett, W. H. (1954). The effect of needle structure upon the susceptibility of hosts to the pine needle miner *Exoteleca pinefoliella* (Chamb.) (Lepidoptera: Gelechiidae). *Can. Entomol.* **86**, 49–54.

Berüter, J., and Städler, E. (1971). An oviposition stimulant for the carrot rust fly from carrot leaves. *Z. Naturforsch.* **26**, 339–340.

Bongers, W. (1970). Aspects of host-plant relationship of the Colorado beetle. *Meded. Landbouwhogesh. Wageningen, Ned.* **70–10**, 1–77.

Bowers, W. S. (1968). Juvenile hormone: Activity of natural and synthetic synergists. *Science* **161**, 895.

Bowers, W. S. (1969). Juvenile hormone: Activity of aromatic terpenoid ethers. *Science* **164**, 323.

Bowers, W. S., Fales, H. M., Thompson, M. J., and Uebel, E. C. (1966). Juvenile hormone: Identification of an active compound from balsam fir. *Science* **154**, 1020–1021.

Breedlove, D. E., and Ehrlich, P. R. (1968). Plant-herbivore coevolution: Lupines and Lycaenids. *Science* **162**, 671–672.

Buford, W. T., Jenkins, J. N., and Maxwell, F. G. (1967). A laboratory technique to evaluate boll weevil oviposition preference among cotton lines. *Crop Sci.* **7**, 579–581.

Budford, W. T., Jenkins, J. N., and Maxwell, F. G. (1968). A boll weevil oviposition suppression factor in cotton. *Crop Sci.* **8**, 647–649.

Butterworth, J. H., and Morgan, E. D. (1971). Investigation of the locust feeding inhibition of the seeds of the neem tree, *Azadirachta indica*. *J. Insect Physiol.* **17**, 969–977.

Byrne, H. D., and Steinauer, A. L. (1966). The attraction of the alfalfa weevil, *Hypera postica*, to alfalfa. *Ann. Entomol. Soc. Amer.* **59**, 303–309.

Carlisle, D. B., and Ellis, P. E. (1968). Bracken and locust ecdysones: Their effects on molting in the desert locust. *Science* **159**, 1472–1474.

Cerny, V., Dolejs, L. Labler, L., Sorm, F., and Slama, K. (1967). Dehydrojuvabione—a new compound with juvenile hormone activity from balsam fir. *Tetrahedron Lett.*, pp. 1053–1057.

Chiang, H. C. (1968). Host variety as an ecological environmental factor in the population dynamics of the European corn borer, *Ostrinia nubilalis*. *Ann. Entomol. Soc. Amer.* **61**, 1521–1523.

Cook, O. F. (1904). Evolution of weevil resistance in cotton. *Science* **20**, 666–670.

Cook, O. F. (1911). Relation of drought to weevil resistance in cotton. *U.S. Dep. Agr. Bur. Plant Ind. Bull.* **220**, 30 pp.

Dahms, R. G. (1969). Theoretical effects of antibiosis on insect population dynamics. *U.S. Dept. Agr. Entomol. Res. Div.*, 5 pp.

Dethier, V. G. (1941). Chemical factors determining the choice of food plants by papilio larvae. *Amer. Natur.* **75**, 61–73.

Dethier, V. G. (1954). Evolution of feeding preferences in phytophagous insects. *Evolution* **8**, 33–54.

Dethier, V. G. (1970). Chemical interactions between plants and insects. *In* "Chemical Ecology" (E. Sondheimer and J. B. Simeone, eds.), pp. 83–102. Academic Press, New York.

de Wilde, J., Ris-Lambers-Suverkropp, K. H., and Van Tol, A. (1969). Responses to air flow and airborne plant odour in the Colorado beetle. *Neth. J. Plant Pathol.* **75**, 53–57.

Ehrlich, P. R., and Raven, P. H. (1964). Butterflies and plants: A study in coevolution. *Evolution* **18**, 586–608.

Fraenkel, G. (1959). The *raison d'être* of secondary plant substances. *Science* **129**, 1466–1470.

Fraenkel, G. (1969). Evaluation of our thoughts on secondary plant substances. *Entomol. Exp. Appl.* **12**, 473–486.

Gara, R. I., Carlson, R. L., and Hrutfiord, B. F. (1971). Influence of some physical and host factors on the behavior of the Sitka spruce weevil, *Pissodes sitchensis*, in southwestern Washington. *Ann. Entomol. Soc. Amer.* **64**, 467–471.

Gilbert, B. L., and Norris, D. M. (1968). A chemical basis for bark beetle (*Scolytus*) distinction between host and non-host trees. *J. Insect Physiol.* **14**, 1063–1068.

Gilbert, L. E. (1971). Butterfly-plant coevolution: Has *Passiflora adenopoda* won the selection race with Heliconince butterflies? *Science* **172**, 585–586.

Goeden, R. D., and Norris, D. M. (1965). Some biological and ecological aspects of ovipositional attack in *Carya* spp. by *Scolytus quadrispinosus*. *Ann. Entomol. Soc. Amer.* **58**, 771–777.

Gupta, P. D., and Thorsteinson, A. J. (1960). Food plant relationships of the diamondback moth [*Plutella maculipennis* (Curt.)]. I. Gustation and olfaction in relation to botanical specificity of the larva. *Entomol. Exp. Appl.* **3**, 241–250.

Guthrie, F. E., Campbell, W. V., and Baron, R. L. (1962). Feeding sites of the green peach aphid with respect to its adaptation to tobacco. *Ann. Entomol. Soc. Amer.* **55**, 42–46.

Harris, P. (1960). Production of pine resin and its effect on survival of *Rhyacionia buoliana* (Schiff) (Lepidoptera: Olethreutidae). *Can. J. Zool.* **38**, 121–130.

Harvey, T. L., and Hackerott, H. L. (1956). Apparent resistance to the spotted alfalfa aphid selected from seedlings of susceptible alfalfa varieties. *J. Econ. Entomol.* **49**, 289–291.

Hedin, P. A., Thompson, A. C., and Minyard, P. J. (1966). Constituents of the cotton bud. III. Factors that stimulate feeding by the boll weevil. *J. Econ. Entomol.* **59**, 181–185.

Heinrich, G., and Hoffmesiter, H. (1967). Ecdyson als Begleitsubstanz des Ecdysterons in *Polypodium vulgare* L. *Experientia* **23**, 995.

Hinds, W. E. (1907). Proliferation as a factor in the natural control of the Mexican cotton boll weevil. *U.S. Dep. Agr. Bur. Entomol. Bull.* **59**, 45 pp.

Holmes, N. D., and Peterson, L. K. (1957). Effect of continuous rearing on Rescue wheat on survival of the wheat stem sawfly, *Cephus cinctus* Nort. *Can. Entomol.* **89**, 363–365.

House, H. L. (1969). Effects of different proportions of nutrients on insects. *Entomol. Exp. Appl.* **12**, 651–669.

Howe, W. L. (1950). Biology and host relationships of the squash vine borer. *J. Econ. Entomol.* **43**, 480–483.

Howe, W. L., and Pesho, G. R. (1959). Spotted alfalfa aphid resistance in mature growth of alfalfa varieties. *J. Econ. Entomol.* **53**, 234–238.

Howe, W. L., and Smith, O. F. (1957). Resistance to the spotted alfalfa aphid in Lahontan alfalfa. *J. Econ. Entomol.* **50**, 320–324.

Hunter, R. C., Leigh, T. F., Lincoln, C., Waddle, B. A., and Bariola, L. (1965). Evaluation of a selected cross section of cottons for resistance to boll weevil. *Arkansas Agr. Exp. Sta. Bull.* **700**, 70 pp.

Jenkins, J. N., and Parrot, W. L. (1971). Effectiveness of Frego bract as a boll weevil resistance character in cotton. *Crop. Sci.* **11**, 739–743.

Jenkins, J. N., Maxwell, F. G., Keller, J. C., and Parrott, W. L. (1963). Investigations of the water extracts of *Gossypium*, *Abelmoschus*, *Cucumis*, and *Phaseolus* for an arrestant and feeding stimulant for *Anthonomus grandis*. *Crop Sci.* **3**, 215–219.

Jenkins, J. N., Maxwell, F. G., Parrott, W. L., and Buford, W. T. (1969). Resistance to the boll weevil, *Anthonomus grandis*. *Crop Sci.* **9**, 369–372.

Keller, J. C., Maxwell, F. G., and Jenkins, J. N. (1962). Cotton extracts as arrestants and feeding stimulants for the boll weevil *J. Econ. Entomol.* **55**, 800–801.

Keller, J. C., Maxwell, F. G., Jenkins, J. N., and Davich, T. B. (1963). A boll weevil attractant from cotton. *J. Econ. Entomol.* **56**, 110–111.

Kennedy, J. S. (1965). Coordination of successive activities in an aphid. Reciprocal effects of settling on flight. *J. Exp. Biol.* **43**, 489–509.

Kennedy, J. S., and Moorhouse, J. E. (1969). Laboratory observations on locust responses to wind-borne grass odour. *Entomol. Exp. Appl.* **12**, 487–503.

Kircher, H. W., Heed, W. B., Russell, J. S., and Grove, J. (1967). Senita cactus alkaloids: Their significance to Sonoran Desert *Drosophila* ecology. *J. Insect Physiol.* **13**, 1869–1874.

Klun, J. A., Tipton, C. L., and Brindley, T. A. (1967). 2,4-Dihydroxy-7-methoxy-1,4-benzoxazin-3-one (DIMBOA), an active agent in the resistance of maize to the European corn borer. *J. Econ. Entomol.* **60**, 1529–1533.

Knipling, E. F. (1966). Some basic principles in insect population suppression. *Bull. Entomol. Soc. Amer.* **12**, 7–15.

Krieger, R. I., Feeny, P. P., and Wilkinson, C. F. (1971). Detoxication enzymes in the guts of caterpillars: An evolutionary answer to plant defense? *Science* **172**, 579–581.

Lanning, F. C. (1966). Relation of silicon in wheat to disease and pest resistance. *J. Agr. Food Chem.* **14**, 350–352.

Lichtenstein, E. P., and Casida, J. E. (1963). Myristicin, an insecticide and synergist occurring naturally in the edible parts of parsnips. *J. Agr. Food Chem.* **11**, 410–415.

Lichtenstein, E. P., Strong, F. M., and Morgan, D. G. (1962). Identification of 2-phenylethylisothiocyanate as an insecticide occurring naturally in the edible part of turnips. *J. Agr. Food Chem.* **10**, 30–33.

Lincoln, C., Dean, G., Waddle, B. A., Yearins, W. C., Phillips, F. R. and Roberts, L. (1971). Resistance of Frego-type cotton to boll weevil and bollworm. *J. Econ. Entomol.* **64**, 1326–1327.

Luginbill, P. (1969). Developing resistant plants—the ideal method of controlling insects. *U.S. Dep. Agr., Agr. Res. Serv. Prod. Res. Rep.* **111**, 14 pp.

Luginbill, P., and Knipling, E. F. (1969). Suppression of wheat stem sawfly with resistant wheat. U.S. Dep. Agr. Agr. Res. Serv. Prod. Res. Rep. **107**, 9 pp.

Lukefahr, M. J., Noble, L. W., and Houghtaling, J. E. (1966). Growth and infestation of bollworms and other insects on glanded and glandless strains of cotton. *J. Econ. Entomol.* **59**, 817–820.

McGinnis, A. J., and Kasting, R. (1961). Comparison of tissues from solid- and hollow-stemmed spring wheats during growth. I. Dry matter and nitrogen contents of pith and wall and their relation to sawfly resistance. *Can. J. Plant Sci.* **41**, 469–478.

Maltais, J. B., and Auclair, J. L. (1962). Free amino acid and amide composition of pea leaf juice, pea aphid haemolymph, and honeydew, following the rearing of aphids on single pea leaves treated with amino compounds. *J. Insect Physiol.* **8**, 391–400.

Matsumoto, Y., and Thorsteinson, A. J. (1968). Effect of organic sulfur compounds on oviposition in onion maggot, *Hylemya antiqua* (Meigen) (Diptera: Anthomyiidae). *Appl. Entomol. Zool.* **3**, 5–12.

Maxwell, F. G., Jenkins, J. N., and Keller, J. C. (1963a). A boll weevil repellent from the volatile substance of cotton. *J. Econ. Entomol.* **56**, 894–895.

Maxwell, F. G., Jenkins, J. N., Keller, J. C., and Parrott, W. L. (1963b). An arrestant and feeding stimulant for the boll weevil in water extracts of cotton plant parts. *J. Econ. Entomol.* **56**, 449–454.

Maxwell, F. G., Lafever, H. N., and Jenkins, J. N. (1965a). Blister beetles on glandless cotton. *J. Econ. Entomol.* **58**, 792–793.

Maxwell, F. G., Parrott, W. L., Jenkins, J. N., and Lafever, H. N. (1965b). A boll weevil feeding deterrent from the calyx of an alternate host, *Hibiscus syriacus*. *J. Econ. Entomol.* **58**, 985–988.

Maxwell, F. G. Hardee, D. D., Parrott, W. L., Jenkins, J. N., and Lukefahr, M. J. (1969). *Hampea* sp., host of the boll weevil. I. Laboratory preference studies. *Ann. Entomol. Soc. Amer.* **62**, 315–318.

Maxwell, F. G., Jenkins, J. N., and Parrott, W. L. (1972). Resistance of plants to insect. *Advan. Agron.* **24**, 187–265.

Miller, B. S., Robinson, R. J., Johnson, J. A., Jones, E. T., and Ponnaiya, B. W. X. (1960). Studies on the relation between silica in wheat plants and resistance to Hessian fly attack. *J. Econ. Entomol.* **56**, 995–999.

Miller, R. L., and Hibbs, E. T. (1963). Distribution of eggs of the potato leafhopper, *Empoasca fabae* on *Solanum* plants. *Ann. Entomol. Soc. Amer.* **56**, 737–740.

Minyard, J. P., Tumlinson, J. H., Hedin, P. A., and Thompson, A. C. (1965). Constituents of the cotton bud. Terpene hydrocarbons. *J. Agr. Food Chem.* **13**, 599–602.

Minyard, J. P., Tumlinson, J. H., Thompson, A. C., and Hedin, P. A. (1966). Constituents of the cotton bud. Sesquiterpene hydrocarbons. *J. Agr. Food Chem.* **14**, 332–336.

Minyard, J. P., Tumlinson, J. H., Thompson, A. C., and Hedin, P. A. (1967). Constituents of the cotton bud. The carbonyl compounds. *J. Agr. Food. Chem.* **15**, 517–524.

Minyard, J. P., Thompson, A. C., and Hedin, P. A. (1968). Constituents of the cotton bud. VIII. β-bisabolol, a new sesquiterpene alcohol. *J. Org. Chem.* **33**, 909–911.

Minyard, J. P., Hardee, D. D., Gueldner, R. C., Thompson, A. C., Wiygul, G., and Hedin, P. A. (1969). Constituents of the cotton bud. Compounds attractive to the boll weevil. *J. Agr. Food Chem.* **17**, 1093–1097.

Nayar, J. K., and Thorsteinson, A. J. (1963). Further investigations into the chemical basis of insect-host plant relationships in an oligophagous insect, *Plutella maculipennis* (Curtis). *Can. J. Zool.* **41**, 923–929.

Nishijima, Y. (1960). Host plant preference of the soybean pod borer, *Grapholitha glicinivorella* Matsumura. 1. Oviposition site. *Entomol. Exp. Appl.* **3**, 38–47.

Oatman, E. R. (1959). Host range studies of the melon leaf miner, *Liriomyza pictella* (Thomson). *Ann. Entomol. Soc. Amer.* **52**, 739–741.

Oliver, B. F., Maxwell, F. G., and Jenkins, J. N. (1970). A comparison of the damage done by the bollworm to glanded and glandless cotton. *J. Econ. Entomol.* **63**, 1328–1329.

Oliver, B. F., Maxwell, F. G., and Jenkins, J. N. (1971). Growth of the bollworm on glanded and glandless cotton. *J. Econ. Entomol.* **64**, 396–398.

Painter, R. H. (1951). "Insect Resistance in Crop Plants." Macmillan, New York.

Painter, R. H. (1958). Resistance of plants to insects. *Annu. Rev. Entomol.* **3**, 267–290.

Painter, R. H. (1968). Crops that resist insects provide a way to increase world food supply. *Kans. State Agr. Exp. Sta. Bull.* **520**.

Parrott, W. L., Maxwell, F. G., and Jenkins, J. N. (1966). Feeding and oviposition of the boll weevil, *Anthonomus grandis*, on Rose-of-Sharon, an alternate host. *Ann. Entomol. Soc. Amer.* **59**, 547–550.

Patanakamjorn, S., and Pathak, M. D. (1967). Varietal resistance of rice to the Asiatic rice borer, *Chilo suppressalis* (Lepidoptera: Crambidae), and its association with various plant characters. *Ann. Entomol. Soc. Amer.* **60**, 287–292.

Pathak, M. D., Andres, F., Galacgac, N., and Raros, R. (1970). Striped borer, *Chilo suppressalis*, resistance in rice varieties. *Int. Rice Res. Inst.*, 50 pp.

Peters, D. C., and Painter, R. H. (1957). A general classification of available small seeded legumes as hosts for three aphids of the "yellow clover aphid complex." *J. Econ. Entomol.* **50**, 231–235.

Pienkowski, R. L., and Golik, Z. (1969). Kinetic orientation behavior of the alfalfa weevil to its host plant. *Ann. Entomol. Soc. Amer.* **62**, 1241–1245.

Refai, F. Y., Jones, E. T., and Miller, B. S. (1955). Some biochemical factors involved in the resistance of the wheat plant to attack by the Hessian fly. *Cereal Chem.* **32**, 437–451.

Röller, H., Dahm, K. H., Sweely, C. C., and Trost, B. M. (1967). The structure of juvenile hormone. *Angew. Chem. Int.* Ed. **6**, 179–180.

Schillinger, J. A., and Gallun, R. L. (1968). Leaf pubescence of wheat as a deterrent to the cereal leaf beetle, *Oulema melanoplus*. *Ann. Entomol. Soc. Amer.* **61**, 900–903.

Schoonhoven, L. M., and Dethier, V. G. (1966). Sensory aspects of host plant discrimination by lepidopterous larvae. *Arch. Neerl. Zool.* **16**, 497–530.

Self, L. S., Guthrie, F. E., and Hodgson, E. (1964). Metabolism of nicotine by tobacco-feeding insects. *Nature* (*London*) **204**, 300.

Shands, R. G., and Cartwright, W. B. (1953). A fifth gene conditioning Hessian fly response in common wheat. *Agron. J.* **45**, 302–307.

Slama, K. (1969). Plants as a source of materials with insect hormone activity. *Entomol. Exp. Appl.* **12**, 721–728.

Slama, K., and Williams, C. M. (1966). "Paper factor" as an inhibitor of the embryonic development of the European bug *Pyrrhocoris apterus*. *Nature* (*London*) **210**, 329.

Smissman, E. E., Lapidus, J. B., and Beck, S. D. (1957). Isolation and synthesis of an insect resistance factor from corn plants. *J. Amer. Chem. Soc.* **79**, 4697–4698.

Smith, R. F. (1959). The spread of the spotted alfalfa aphid, *Therioaphis maculata* (Buckton), in California. *Hilgardia* **28**, 647–685.

Sparks, M. R., and Cheatham, J. S. (1970). Responses of a laboratory strain of the tobacco hornworm, *Manduca sexta*, to artificial oviposition sites. *Ann. Entomol. Soc. Amer.* **63**, 428–431.

Sprague, G. F., and Dahms, R. G. (1972). Development of crop resistance to insects. *Environ. Qual.* **1**, 28–34.

Starks, K. J., Muniappan, R., and Eikenbary, R. D. (1972). Interaction between plant resistance and parasitism against the greenbug on barley and sorghum. *Ann. Entomol. Soc. Amer.* **65**, 650–655.

Struck, R. F., Frye, J., Shealy, Y. F., Hedin, P. A., Thompson, A. C., and Minyard, J. P. (1968). Constituents of the cotton bud. IX. Further studies on a polar boll weevil feeding stimulant complex. *J. Econ. Entomol.* **61**, 270–274.

Sugiyama, S., and Matsumoto, Y. (1959). Studies on the host plant determination of the leaf-feeding insects. III. Attractiveness of umbelliferous plants to the vegetable weevil. *Nogaku Kenkyu* **47**, 141–148.

Swailes, G. E. (1960). Laboratory evaluation of resistance in rutabaga varieties to the cabbage maggot, *Hylemya brassicae* (Bouche). *Can. Entomol.* **92**, 958–960.

Thompson, A. C., Wright, B. J., Hardee, D. D., Gueldner, R. C., and Hedin, P. A. (1970). Constituents of the cotton bud. XVI. The attractancy response of the boll weevil to the essential oils of a group of host and nonhost plants. *J. Econ. Entomol.* **63**, 751–753.

Thompson, A. C., Baker, D. N., Gueldner, R. C., and Hedin, P. A. (1971). Identification and quantitative analysis of the volatile substances emitted by maturing cotton in the field. *Plant Physiol.* **48**, 50–52.

Thorsteinson, A. J. (1953). The chemotactic responses that determine host specificity in an oligophagous insect (*Plutella maculipennis* (Curt.) Lepidoptera). *Can. J. Zool.* **31**, 52–72.

Thorsteinson, A. J. (1960). Host selection in phytophagous insects. *Annu. Rev. Entomol.* **5**, 193–218.

Thurston, R., and Webster, J. A. (1962). Toxicity of *Nicotiana gossei* Domin. to *Myzus persicae* (Sulzer). *Entomol. Exp. Appl.* **5**, 233–238.

Wannamaker, W. K. (1951). The effect of plant hairiness of cotton strains on boll weevil attack. *J. Econ. Entomol.* **50**, 418–423.

Wannamaker, W. K., Wessling, W. H., and Stephens, S. G. (1957). Boll weevil resistant cotton? *N.C. Agr. Exp. Sta. Res. Farm.* **16** (1), 14.

Webber, H. J. (1903). Improvement of cotton by seed selection. *U.S. Dep. Agr. Yearb.*, pp. 384–386.

Williams, C. M. (1970). Hormonal interactions between plants and insects. *In* "Chemical Ecology" (E. Sondheimer and J. B. Simeone, eds.), pp. 103–132. Academic Press, New York.

Yamamoto, R. T., Jenkins, R. Y., and McClusky, R. K. (1969). Factors determining the selection of plants for oviposition by the tobacco hornworm. *Manduca sexta. Entomol. Exp. Appl.* **12**, 504–508.

26

USE OF AUTOCIDAL METHODS

D. F. Waterhouse, L. E. LaChance, and M. J. Whitten

I. INTRODUCTION

Earlier chapters in this treatise have dealt with various aspects of the relationship between a pest species and its natural enemies and how this can be exploited for the practical control of the pest. Here we review possible uses of an insect pest for its own control, that is to say, autocidal methods of pest management.

Many biological control successes have been obtained in certain orders (e.g., Hemiptera and Coleoptera) but the record is not nearly as satisfactory for some other orders, for example, for the Diptera where species of veterinary or medical importance (e.g., houseflies, blowflies, mosquitoes, and sand flies) are normally attacked by organisms incapable of maintaining host abundance at levels that can be tolerated. Because of the very great importance of such Diptera it is natural that they should be among the first explored for possibilities of autocidal control. Studies are already in progress in other groups, e.g., mites (Van Zon and Overmeer, 1972), crickets

(Hogan, 1971), Coleoptera (Klassen and Earle, 1970), and Lepidoptera (North and Holt, 1971).

Genetic and other aspects of autocidal control are still largely unexplored and our account will place emphasis on the principles of genetic control, with a brief review of factors determining the degree of success of the sterile insect release method.

In its simplest and best known form, autocidal control involves the mass release of sterilized individuals of a species and this is dealt with first. If we accept the possibility of releasing fertile individuals which have a genetic constitution that will lead to a reduction of the pest population, the opportunities become more varied, although the genetics becomes much more complex. Discussion of this approach forms the basis of the subsequent section.

II. STERILE INSECT RELEASE METHOD

Without the successful large-scale application of the sterile insect release method (SIRM) against the screwworm fly, *Cochliomyia hominivorax* (Coquerel), there would have been little impetus for the development of the other autocidal methods discussed in this chapter. In the largest autocidal control program ever conducted, sterile screwworm flies have been released continually since 1962 in the southwestern United States and northern Mexico at rates ranging from 50 to 200 million per week in an immensely successful program that has provided 10 years of effective control and suppression of a major insect pest for the area. This accomplishment constitutes ample proof that the sterile insect technique does work on an area-wide basis. The program suffered a major setback in the 1972 season. There were 90,000 cases in Texas, 1000 cases in New Mexico, 1000 cases in Oklahoma, and 2000 cases in Arizona (W. H. Newton, personal communication). These figures are the highest reported for any year since 1962 and can be compared with the 1971 case report of 444 for Texas, 0 for Oklahoma and New Mexico, and 4 for Arizona. It is too early to identify the causes for this serious setback in the program, but unusually favorable weather for screwworm breeding in northern Mexico and the southwestern United States during the winter of 1971–1972 and changes in the competitiveness of the released flies are possibilities.

The substantial success of the screwworm programs has, however, diverted attention from the fact that SIRM is likely to be applicable far more frequently in practice to the *control* of pests than to their *eradication*. There are, so far, no comparable examples of eradication, such as the screwworm program in the southwestern United States, whereas there are many programs demonstrating that SIRM can substantially reduce pest populations. Such programs also provide us with a measure of the considerable difficulty of achieving eradication, even with relatively small, somewhat isolated populations (International Atomic Energy Agency (IAEA) Symposia, 1968, 1971b; and IAEA Panels, 1969, 1970, 1971a).

There has been extensive research aimed at applying SIRM against additional insect pests (for reviews, see Knipling, 1966, 1970; LaChance *et al.*, 1967; Proverbs, 1969;

Bushland, 1971). It is currently being tested in the field against several species of mosquitoes and fruit flies, the boll weevil, *Anthonomus grandis* Boh., and several agriculturally important lepidopterous pests. The genetic basis of SIRM and the radiobiological and chemical events affecting insect reproductive cells and leading to sterility are discussed elsewhere (LaChance and Riemann, 1964; LaChance *et al.*, 1967, 1968). In this section we will discuss some of the requirements, restrictions, problems, and misconceptions associated with this method.

Theoretically, SIRM is applicable to the control of any sexually reproducing insect. Reduction of birth rate below the replacement level decreases population size and, when "all" the native females mate with sterile males (zero birth rate), the population rapidly becomes extinct. Short of this goal, it should be possible to reduce the population of many pests below economically damaging levels at a reasonable cost.

A. Requirements of the Sterile Insect Release Method

The principal requirements for SIRM were listed by Knipling (1955) and they remain substantially the same today:

1. A method for mass rearing the pest must be available or capable of development.

2. Adequate dispersion of the released sterile males must be attainable.

3. The area inundated with treated insects must be large enough or sufficiently isolated to render unlikely the immigration of fertilized females in significant numbers.

4. The rearing and sterilization procedures must result in adequately competitive males.

5. The females of the pest must mate only once or, if more frequent matings occur, sperm from sterilized males must be competitive with those from fertile males.

6. The cost of production of the sterile insects must be such that it is economically feasible to maintain in the field the required ratio of sterile males to normal males for extended periods; it may sometimes be possible to reduce the number of normal males by other means prior to using SIRM.

Factors that contribute to the success and economy of SIRM programs include:

1. Mechanization of mass rearing.

2. Maintenance of a highly competitive insect strain for release.

3. A laboratory system for monitoring quality of insects produced (fertility, fecundity, effectiveness of sterilization treatment, vigor, longevity, etc.).

4. Effective methods of packaging, distributing, and releasing sterile insects.

5. Data on ecology and population dynamics, including accurate delineation of the distribution of the species and evaluation of factors, especially density-dependent factors, that affect population size, e.g., natural enemies, competition, and migration.

6. Adequate methods for progressive assessment of the results of releases and the effectiveness of the released insects.

7. An organization capable of monitoring population trends in the field, especially for the detection of pockets of heavy infestation.

8. An organization capable of effecting rapid changes in all phases of a large-scale operation.

Nor surprisingly, most of the requirements which contribute to the success of SIRM programs are also necessary for the application of most of the autocidal methods discussed later in this chapter.

B. Mass Rearing

Progress in the mass rearing of insects has been truly phenomenal in the past decade. When the book, "Insect Colonization and Mass Production" (Smith, 1966) appeared, only three species of insects could be reared "by the millions." Currently, there are at least six laboratories capable of producing over 10 million "medflies," *Ceratitis capitata* (Wiedemann), per week, and the majority of these are in developing countries. In laboratories of the United States Department of Agriculture (USDA), the production of boll weevils is presently 1 million/week in a plant with a 15-million/week capability; pink bollworm, *Pectinophora gossypiella* (Saunders), production exceeds 1.3 million/day; codling moths, *Laspeyresia pomonella L.*, are produced at a rate of 3 million/year; and tobacco budworm, *Heliothis virescens* (F.), production is 70,000/day. The "screwworm plant" located at Mission, Texas, has a 200-million/week capability. Canada Department of Agriculture laboratories can produce 2 million codling moths per month and World Health Organization (WHO) laboratories in New Delhi are presently producing nearly 5 million *Culex fatigans* Wiedemann per week. Mass production could easily be increased in many of these facilities. Even the tsetse fly, *Glossina morsitans* (Westwood), is being produced today in numbers not considered possible only a few years ago (Nash *et al.*, 1971). It is most noteworthy that the cost of production, in most instances, has decreased as the number produced increased and as the techniques have improved. Rearing costs are less frequently the major obstacle they once were in launching autocidal programs (Knipling, 1970).

Once mass rearing is feasible, it is important that the fitness of the released insects approaches that of the native insects. In the absence of evidence to the contrary, we should endeavor to release insects that genetically resemble the wild ones. However, for some species this is difficult to accomplish as certain characteristics important for survival in the field may lose their selective advantage under laboratory conditions. It is difficult to predict whether the problem is alleviated or compounded by rearing insects under optimal or suboptimal conditions.

Several groups involved in insect mass rearing are currently modifying insectary conditions to simulate natural conditions (day length, temperature, etc.) and are also replacing or infiltrating the cultured strain regularly with newly collected field insects in an attempt to maintain high fitness in the release strain. These procedures are troublesome and not without risk. For example, wild insects sometimes introduce diseases as well as new genes. However, efforts expended to maintain fitness in the laboratory strain should pay handsome dividends.

C. Competitiveness

In addition to the possible deleterious effects of laboratory handling and rearing, reduced competitiveness can also result from the sterilizing treatment. Adult male house flies, *Musca domestica* L., can withstand doses of radiation that exceed the sterilizing dose by a factor of 10 before their competitiveness under laboratory conditions is affected (LaChance and Richard, unpublished). On the other hand, "medflies" show reduced competitiveness before the sterilizing dose has been attained (Hooper, 1971; Wakid, 1973). No way has been found of sterilizing the boll weevil by radiation that does not drastically reduce its competitiveness and longevity (Riemann and Flint, 1967), but chemicals have been found that produce the desired sterility without undue side effects (Klassen and Earle, 1970). Most, if not all lepidopterous species require exceedingly high radiation doses to induce complete sterility (LaChance *et al.*, 1967). In many Lepidoptera, the fully sterilizing dose so severely reduces competitiveness that it is doubtful whether Lepidoptera so treated can be used effectively in autocidal programs (North and Holt, 1971; IAEA, 1971a). However, the use of substerilizing doses which nevertheless result in high levels of sterility in the offspring may prove a more useful approach (see Section II,E).

Two points to be stressed are: (1) complete sterility is not necessarily required in SIRM programs, and (2) the sterilizing treatment should be administered to the appropriate life stage under conditions such that the detrimental effect on the organism is minimized.

D. Desired Level and Kind of Sterility

Autocidal control encompasses methods wherein the reproductive capacity of the released insects ranges from total sterility to complete fertility (see Section III). With SIRM, it was formerly thought that the released insects should be fully sterile. However, this is not necessary. As long as females are not released or, if released, they are fully sterile, the released males cannot immediately increase the size of the native population regardless of their fertility, although the qualities of the hybrids may influence the situation subsequently. Thus the level of male sterility can be adjusted to maximize their impact.

In many instances, it will be advantageous to adjust the level of sterility in the released males to increase competitiveness, even if some of their progeny are viable. Using the medfly as an example, LaChance (1970) estimated that a dose of 7 krad which induces approximately 95% sterility and allows a 50% level of competitiveness would be more effective than other available combinations in initiating a downward trend in the population. In this regard, two possible pitfalls should be noted. Measurements of competitiveness (Fried, 1971) are generally derived from cage tests performed in the laboratory and cannot be related directly to competitiveness in a field situation. It is improbable that released insects would be *more* competitive in a field situation although they might very well be *less* so.

Second, the permissible amount of residual fertility in the released males will depend upon the level of density-dependent mortality in the population and whether this mortality occurs before or after the genetic death that is induced by the sterility treatment.

If the environmental resistance to population growth consists mainly of density-independent factors, then partially sterile males could have a substantial effect on the size of the population of the next generation. We need to know a great deal more about environmental resistance factors, and particularly whether they are density-dependent. Climatic factors, such as temperature and soil moisture, would eliminate all or a part of the population regardless of its size; whereas factors such as host and food availability, and parasites and predators, tend to have proportionately less effect as density is lowered, especially at very low densities. Each insect species has its own characteristic potential rate of increase per generation but this may vary greatly according to population density and environmental conditions. The biotic potential of a pest is very important in considering the permissible level of fertility in males released for its control. A knowledge of the natural factors regulating the abundance of pests has been lacking in many autocidal programs and has contributed to the problems encountered.

The physiological and genetic basis of sterility in the male can be quite important in determining the female's future reproductive success and, therefore, the success of the program. Although sperm containing dominant lethal mutations usually provide the base of sterility, it is not necessary that sperm be transferred to the female. In species where the female only mates once, it is sufficient that the sterile male elicit this monocoitic behavior. In females that mate repeatedly, it becomes important that the males not only transmit lethal-bearing sperm that can compete with sperm contributed by untreated males, but should also elicit the same responses from the females as do the untreated males, such as termination of pheromone release.

E. Delayed Sterility

Many Lepidoptera are important pests and some have been studied intensively with the aim of control by SIRM. Moths are remarkably resistant to sterilization (LaChance *et al.*, 1967). For example, compared with a sterilizing dose of 6 to 8 krad for the screwworm fly, 30–40 krad are needed to sterilize male codling moths, cabbage loopers, [*Trichoplusia ni* (Hübner)], and tobacco budworms, and as much as 100 krad are needed for some other Lepidoptera. Such high levels of irradiation cause physiological and somatic damage, generally rendering the sterilized males insufficiently competitive with wild males.

Radiation resistance among Lepidoptera has been attributed to the chromsomes having diffuse cintromeres (LaChance *et al.*, 1967; North and Holt, 1968; LaChance *et al.*, 1970). The induction of dominant lethal mutations in the sperm and eggs is thought to require much higher doses of radiation because many of the genetic events that

TABLE 1

Reproductive Trends in Partially Sterilized Lepidoptera and Their Untreated Progeny[a]

Radiation dose to released male (krad)	Percentage egg hatch		
	Released male × normal female	F_1 male × normal female	F_1 female × normal male
Trichoplusia ni			
0	85.6	83.4	85.6
10	46.0	30.7	33.8
15	39.5	11.5	19.5
20	15.7	3.8	6.7
Heliothis virescens			
0	74.0	66.6	66.6
7.5	64.3	35.0	32.0
15.0	51.0	3.0	9.0
22.5	20.5	1.0	3.7

[a]Based on data from North and Holt (1968a,b) and Proshold and Bartell (1972).

would produce a dominant lethal in species with monocentric chromosomes are not lethal in Lepidoptera. For example, the irradiation of lepidopterous gametes does not produce dicentric elements (i.e., carrying two centromeres) and acentric fragments (without centromeres). In Diptera these products are the primary source of mitotic instability in developing embryos and of the ensuing dominant lethality. In Lepidoptera, chromosome rearrangements and fragments tend to be mitotically stable so that a high proportion of the F_1's derived from an irradiated parent survives to the adult stage. However, the surviving F_1 progeny are heterozygous for induced chromosomal rearrangements. At meiotic segregation in the adult F_1, a high proportion of the gametes receive genomes containing chromosomal deficiencies or duplications, and these are the cause of dominant lethality and sterility in the next generation (Table 1).

The role of diffuse centromeres as the explanation of radiation insensitivity in Lepidoptera has to be accepted with caution since Hemiptera are not especially resistant to radiation, yet they too have diffuse centromeres (LaChance *et al.*, 1970).

Males partially sterilized with a lower dose of radiation suffer less somatic damage and are more competitive. Thus it should be possible to regulate the dosage given to Lepidoptera to maximize the genetic death in the second generation after release. What appears initially as a disadvantage might, in fact, be used to some advantage, since the cumulative effect of releasing partially sterile insects whose progeny inherit sterility can ultimately exceed the effect of releasing fully sterile males (Knipling, 1970; Toba *et al.*, 1972). A comparison of the effects of releasing fully sterile males with the effects of releasing males whose major sterilizing influence operates one generation later will illustrate this point. If the release ratio is $r:1$ (released to wild), then, in the former instance, the genetic mortality is $r/(r + 1)$, whereas in the latter instance, it is $r(r + 2)/(r + 1)^2$ which is considerably higher for likely values of r. The outcome is more complex where both sexes are released because we have to consider the impact of the released semisterile females. Dosage and stage of irradiation can be adjusted so females are almost completely sterile and males only partially sterile. Models have been developed by Knipling (1970), but we need a much better understanding of the population dynamics of the pest species, as well as of the radiation biology of lepidopterans, before we can develop programs that give maximum genetic death.

F. Sterilization

During its life cycle each insect species has an optimum stage for application of the sterilizing treatment and this may differ from species to species. Irradiation at or after the late pupal stage generally produces more vigorous insects for release than irradiation at earlier stages. For species like *Culex fatigans* that have a brief pupal period, special measures might be required, such as slowing development by low temperature to allow restriction of irradiation treatment to late pupal stages, if adverse effects are to be minimized.

In many field programs with Diptera, the radiation treatment has been applied during the last 1–2 days of the pupal period rather than to the adults, although this choice is often for convenience of handling. There has been a general reluctance to irradiate adults, but in some cases this may be advantageous. Recent developments in chilling, handling, and distribution procedures for insects suggest that treatment of adults can be effective and economical (White *et al.*, 1970; White and Hutt, 1972; Hooper, 1970; Curtis and Langley, 1971).

In addition to determining the optimal stage for administering the sterilizing treatment, a number of other factors should be investigated with a view to increasing competitiveness of the treated insects. For example, the debilitating effects of ionizing radiation on the medfly were dramatically reduced when nitrogen was introduced into the irradiation containers before and during the radiation treatment (Hooper, 1971; Wakid, 1973). Chilling and uses of gases (Hooper, 1970; Curtis and Langley, 1971), irradiation in carbon dioxide plus air (LaChance, 1963), aeration (Baumhover, 1963), radioprotective agents in the diet (H. M. Flint, unpublished), and dose-fractionation (Ducoff, 1972) are but a few of the possibilities for improving the fitness of irradiated insects.

Although it is not possible to deal with chemosterilants in this chapter (for review see LaBrecque and Smith, 1968), in certain species the use of these materials either alone or in combination with radiation may provide the required degree of sterility with minimal interference with fitness.

When there is reason to suspect that the competitiveness of the released insects is reduced, there is a temptation to rectify the situation by increasing the number of released insects. This will suffice in some situations, although competitiveness can be reduced to the point where additional insects will not compensate, no matter how many are released.

G. Number of Sterile Insects Required

The number of insects required in a SIRM program depends on (1) the density of native insects per unit area and the size of the infested area; (2) the level to which the pest can be reduced by other control methods before initiation of releases or even during the releases if the method of reduction only affects the eggs, larvae, or pupae; (3) the capacity for increase of the species; (4) the nature of the factors regulating the pest population, and (5) whether control or eradication is the goal. The cost of rearing and releasing the insects will also have a major effect on the size of a program and its goal. For some species, there may be restrictions regarding the number of insects that can be added to the environment. This is particularly true for insects that are disease vectors or that have a high nuisance potential (houseflies, horn flies, stable flies).

In some species, the females may be more difficult to sterilize than the males or may constitute the pest component of the adult phase. The possibility of releasing male insects only has often been considered. Techniques for the mechanical separation of the sexes have become available for a number of important pest species. Size difference is exploited for *Aedes aegypti* (L.) and *Culex fatigans* and linkage of a pupal color mutant with the maleness chromosome provides the same opportunity for the Australian sheep blowfly (Whitten, 1969). Undoubtedly the most practical approach to separation of the sexes is the exploitation of male-killing or female-killing genetic systems. The achievement of McDonald (1971) in synthesizing a strain of housefly which produces both sexes at 25°C but only males when reared at 33°C demonstrates admirably what is possible once the need to separate sexes is realized.

In our discussions so far, we have assumed that SIRM acts solely through the presence of sterile males. Generally the release of sterile females is immaterial to the success of the program and consequently little thought is given to providing them with an effective role. Conditions under which their separate release would be beneficial are considered by a number of authors (Ailam and Galun, 1967; Lawson, 1967; Whitten and Norris, 1967; Whitten and Taylor, 1970). If we could exploit the female's ability to attract males and so permit the selective dispersal of minute quantities of sterilizing agent, such that each liberated booby-trap female was capable of sterilizing even one field male, then the release of one such treated female for each native female could produce an effect equivalent to swamping the natural population with sterile males. This possibility alone should stimulate serious consideration of ways of exploiting both sexes in SIRM programs.

H. Assessment of Results

Once the problems associated with mass rearing, sterilization without undue loss in competitiveness, and adequate release procedures have been resolved, and the necessary information on population ecology and dynamics has been acquired, serious consideration must be given to how the results will be monitored (IAEA Panels, 1969, 1970, 1971a).

The ultimate proof that the sterile males are mating with native females is the occurrence of infertile eggs in field collections. When the viability of the eggs from wild females progressively decreases, then a successful operation is under way. However, for some pests, collection of egg-hatch data may be extremely time-consuming or difficult.

Many field trials have involved marking the insects before release with dyes or radioactive tracers and later trapping to determine the ratio of released to native insects. From release–recapture data, it is possible to establish that the released insects appear to outnumber the native insects by a ratio of 30:1, 10:1, etc. Also information may be obtained on approximately how long they are surviving and how well they are dispersing. Such data do not measure the effectiveness of the released insects. It is easy to be misled by release–recapture data because the released and native insects may differ both in their distribution and in their response to attractants. The methods used in marking insects before release are not always completely effective, and it is difficult to maintain a consistent level of marking efficiency in a large-scale operation. For this reason, where trapping is used to monitor the program, marking efficiency should be continually monitored so that the limits to the interpretation of recapture data are known. Even very small fluctuations in marking efficiency may render the interpretation of recapture data rather difficult (LaChance, 1970).

Field data on percentage infestation of a crop and yield per unit area, disease incidence, and trap catches are useful adjuncts to the assessment of an autocidal program. However, these parameters are also subject to normal population fluctuations and seasonal factors that are independent of the effects of the sterile insects. Comparative data from untreated areas are desirable but there are frequently problems in selecting ecologically comparable areas.

In species with marked seasonal fluctuations in abundance the timing of the initial release should, if possible, coincide with the time of first appearance of native virgin females. This time is not always simple to determine. Insects may first appear in traps several weeks after they emerge or move into an area from elsewhere so it is not always possible to rely on trapping data. Obviously, data from host infestations are even less reliable for timing initial releases. In practice, release of sterile insects should probably commence as soon as they can survive in nature, and preferably even before the appearance of the first virgin females. In areas where reproduction is continuous, releases should commence during any seasonal low or should follow population suppression by other means. For some species it may be possible to continue selective population suppression against the larval stage during the release of sterile adults. For example, viruses might be used against lepidopterous larvae since they have no adverse effects on the liberated adults.

I. Autocidal Methods Similar to the Sterile Insect Release Method

In the next section, we discuss other autocidal methods. Some of these techniques mimic SIRM if only males are released. For example, methods such as the use of compound chromosomes, cytoplasmic incompatibility, and hybrid sterility have effects comparable with SIRM since the released males are fully sterile or produce sterile progeny when they mate with native females. On the other hand, if fertile males and females are released, other population manipulations (such as population-genotype replacement) become possible.

III. GENETIC TECHNIQUE

In Section II, except for the special case of Lepidoptera, we restricted our consideration to the release of semisterilized or fully sterilized insects in which the measure of control achieved is a direct product of the interaction of the released insects with the field population. In this section we examine the ways in which we might use a species for its own control by liberating partially or fully fertile members of the same species. Two major possibilities exist.

1. The released insects, in addition to mating among themselves, could mate with the field population, producing inviable or sterile hybrids. The release would result in population reduction for some generations subsequent to the actual liberation because the progeny produced from matings among released individuals also carry the condition that would cause the continued production of inviable or sterile hybrids.
2. The released strain could be used to manipulate the genetic composition of the natural population in a way that would ultimately lead to satisfactory control, either by altering the characteristics of the population, for example, by removing vector capacity, or by spreading some conditional lethal factor.

A. Control by Production of Sterile Offspring

1. Inherited Sterility

Hybrids arising from a cross between two strains differing for one homozygous reciprocal translocation (interchange of chromosomal material) have normal complements of genes (see Fig. 1) and consequently tend to be fully viable. However, in these hybrids, difficulties of chromosome pairing and segregation arise during meiosis, generating a proportion of genetically unbalanced gametes. This imbalance is passed on to the progeny and manifests itself in the form of dominant lethality. Thus a translocation heterozygote is viable but partially sterile. A single translocation in an individual can cause the death of 20 to 80% of its progeny, commonly around 50%. As more interchanges involving a larger number of chromosomes are incorporated into the heterozygote, sterility can increase without much adverse effect on viability in some cases. Translocation homozygotes, on the other hand, do not form unbalanced gametes (see Fig. 1) although position effects of the new gene arrangements or direct damage to genes in the neighborhood of the breakpoints often cause recessive lethality, sterility, or at least reduced viability (Patterson *et al.*, 1934).

There is evidence that some homozygous translocations are not disadvantageous: the variation in chromosome number and morphology throughout both animal and plant kingdoms is believed to be due, in large measure, to replacement of old gene sequences by new translocation forms (see Patterson and Stone, 1952; White *et al.*, 1967).

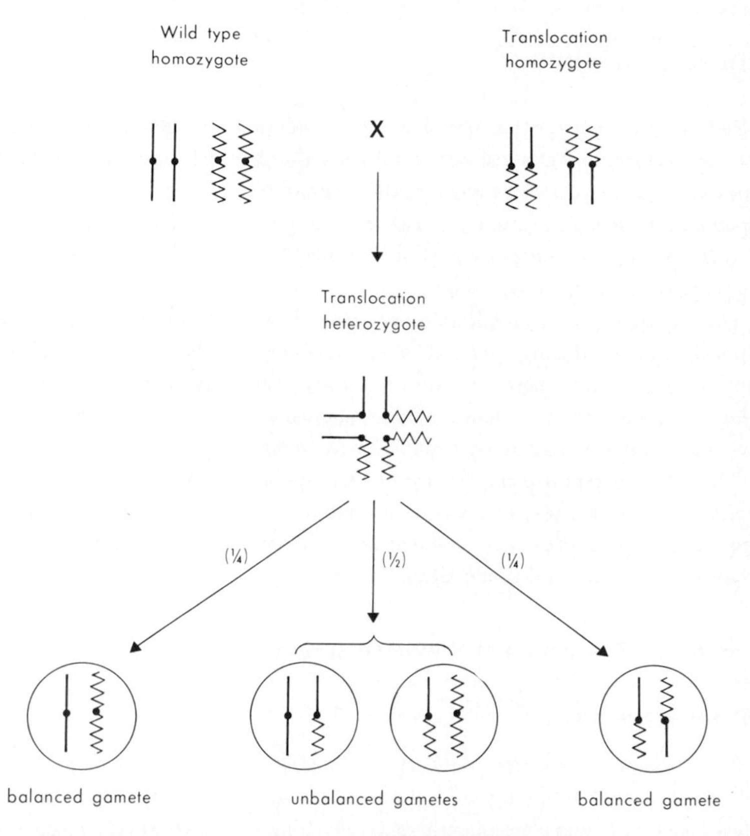

Fig. 1. Diagrammatic representation of a normal karyotype, translocation heterozygote, and translocation homozygote. About 50% of gametes from a translocation heterozygote are imbalanced. These function as normal gametes but ensure the death of ensuing zygotes.

It is not possible to deduce from this information what percentage of irradiation-induced translocations will prove suitable, but the work of Lorimer *et al.* (1972) on *Aedes aegypti* and Whitten (unpublished) on *Lucilia cuprina* suggest a figure of around 5%. We can conclude that viability, rather than fertility in translocation homozygotes, tends to be impaired, whereas in translocation heterozygotes the converse tends to be true.

2. Translocation Heterozygotes

The exploitation of translocation-induced sterility for insect control was suggested in a remarkable paper by Serebrovskii (1940) long before radiation-induced sterility was ever considered. This work went unnoticed until Curtis (1968a) made a similar suggestion during genetic studies on tsetse flies. Both workers postulated the release of strains homozygous for a single translocation. Partially sterile hybrids are produced for some generations following the release, and their continued presence may cause a population decline and eventual collapse. Some workers (Laven *et al.*, 1971; McDonald and Rai, 1971) have advocated the direct release of partially sterile translocation heterozygotes, as an alternative to sterilization by irradiation or chemosterilants. The release of partially sterile heterozygotes or their production following the release of fertile translocation homozygotes may appear equivalent procedures but their implications in population genetics are significantly different and must be examined carefully. Some difficulties arising from the direct release of translocation heterozygotes have been discussed by Whitten (1971a).

Since the release of translocation heterozygotes appears to be superior to the conventional sterile male approach only where higher levels of induced sterility impair competitiveness, it would seem as rewarding to pursue means of minimizing the adverse effects of radiation or chemical sterilization as to pursue this particular means of utilizing sterility associated with heterozygous translocations.

3. Translocation Homozygotes

The theory propounded by Serebrovskii (1940) for releasing translocation homozygotes is basically correct for single translocations but is erroneous for multiple translocations (Whitten, 1971b; Curtis and Hill, 1971). Curtis and Robinson (1971) have investigated the theory for single or double translocations and Whitten (1971b) the more extreme case where sufficient translocations are assumed to be present to render the hybrid fully sterile. The theory for the latter is simpler and therefore will be considered first.

Assume that a multiple translocation strain (TT) is released into a wild-type population (AA) at a 1:1 ratio and that the hybrid AT is fully sterile. With random mating we have in the F_1 generation $1AA{:}2AT{:}1TT$ and 50% of the individuals (all AT) are sterile. The only mating pair combinations to produce F_2 offspring and their frequencies are tabulated on the next page.

Mating pair	Frequency	Type of progeny	Genotype frequency of progeny
$AA \times AA$	1/16	All AA	$1AA$
$AA \times TT$	1/8	All AT	$2AT$
$TT \times TT$	1/16	All TT	$1TT$
Total	1/4		

All other matings involve one or both parents who are AT and, hence, are sterile (Serebrovskii erroneously estimated that all matings would be sterile if TT was a multiple translocation). Moreover, as in the F_1, 50% of the F_2 adults are AT and are therefore sterile. Thus it would appear that a single release of a multiple translocation homozygote at a 1:1 ratio would induce 75% genetic death in the second generation and for an indefinite time thereafter. Further, if we suppose that, instead of releasing one multiple translocation strain, we liberate $(n - 1)$ translocation strains, then the genetic death will be $(n^2 - 1)/n^2$ in every generation after the F_1 following the release. A release of four such translocation strains, for example, would generate 96% genetic death in the population indefinitely. Clearly, if this occurred we would have a powerful tool for insect control.

Unfortunately the effects of a single release of this nature cannot be permanent since the equilibrium between AA, AT, and TT is unstable, as demonstrated by the following continuity argument. Suppose one form (i.e., TT) is infrequent. TT will tend to mate with AT or AA because these types constitute the majority. With AT no offspring are produced, and with AA no F_2 arise. AA females, on the other hand, will tend to mate with AA or AT and, in the former instance, but not in the latter, will produce fertile offspring. Thus, in situations where the hybrid is less viable or fertile, either homozygote enjoys a mating advantage if it is frequent and a disadvantage if it is infrequent. At some intermediate frequency it has no advantage and an unstable equilibrium must obtain. In finite populations, genotype frequencies are never at the position of unstable equilibrium and so it is inevitable that the released translocation will displace the native genotype or, alternatively, be displaced by it.

When more than one multiple translocation strain is released the strains are still unable to coexist. Whichever strain is least frequent will be at a mating disadvantage and so, one by one, each strain is eliminated until a single strain remains and the genetic deaths from this source, cease. Table 2 gives a computer simulation of the percentage of matings, where at least one parent is sterile, in the seven generations following the "release" of 1, 4, and 8 multiple translocation strains. During the displacement, sterile hybrids are produced at a fairly constant frequency for some five generations. If we assume that the hybrid is fully sterile, these percentages would also indicate the level of genetic death. This level of genetic load may be sufficient to prevent the population from reaching nuisance levels and it may even cause a total population collapse. In insects with five to six generations per year, a single release each year might

TABLE 2
Estimated Genetic Death in the Eight Generations following the Release of 1,4, and 8 Homozygous Translocation Strains Each in Equal Frequency with the Native Population[a]

Generation	Number of translocation strains released		
	1	4	8
1	74.8	95.9	98.7
2	74.5	96.1	98.8
3	73.4	95.7	98.6
4	68.2	94.8	98.1
5	52.6	89.1	91.8
6	26.8	64.0	74.2
7	6.9	24.6	57.7
8	2.2	7.7	36.5

[a]After Whitten, 1971b.

be all that would be required. Thus, in cases where the liberation of sterilized males would prove too costly, the increased efficiency of the multiple translocation approach may create a much more favorable cost/benefit ratio. Multiple translocation strains have arisen often enough in the course of evolution but their synthesis in the laboratory may be a difficult problem . It might be noted that the release of strains, each homozygous for two different translocations will produce hybrids heterozygous for four translocations when intermating between homozygous strains occurs (these hybrids would be virtually sterile and so satisfy the conditions stipulated in Table 2).

An interesting fact to emerge from the work of Curtis and Hill (1971) is that the accumulated genetic death in a population is constant regardless of the complexity of the translocation. Single translocations induce progressively less genetic death each generation but they persist longer in the population. It becomes important, then, to have a thorough understanding of the population dynamics of each pest. For example, a genetic death of 96% for five generations might be preferable for a mosquito species, whereas a genetic death of 50% for 20 generations could be more effective for tsetse flies whose reproductive capacity is much lower. Fortunately, those species with a need for more complex rearrangements tend to be more amenable to laboratory manipulation. Attempts to synthesize suitable strains are under way for the tsetse fly (Curtis, 1969), the housefly (Wagoner *et al.*, 1969), for several mosquito species, namely, *Culex fatigans* (Laven *et al.*, 1971), *C. tritaeniorhynchus* Giles, (Sakai *et al.*, 1971), and *A. aegypti* (McDonald and Rai, 1971), and the Australian sheep blowfly (Whitten, 1971b). Translocation homozygotes already exist in most of these species.

B. Control through Genetic Manipulations

The autocidal techniques considered so far derive their regulating effects from the sterile individuals which are either released directly into the field or arise in time through hybridization between the field and released strains. There is also a quite different method of genetic control, namely manipulation of the genetic structure of the population in a manner that will lead to control. The discussion of this topic can be divided into two sections: first, the selection or production of genetically determined characteristics which can be utilized for the insects' destruction, and second, the development of an effective transporting system for introducing these genetic conditions into natural populations.

1. Suitable Characteristics for Introduction

Two kinds of desired genotype may be introduced: (1) those rendering a population vulnerable to subsequent control measures, i.e., conditional lethal, and (2) those nullifying the noxious characteristics of the pest thus making control unnecessary.

Genetically determined characteristics of the first kind already exist in a number of important pest species or could be synthesized by radiation or other means. For example, insecticide susceptibility is generally still present somewhere in the distribution of each species even though resistance might be widespread. Lethal response to either high or low temperature or the inability to diapause could be useful in controlling multivoltine species in temperate regions. Variation in ability to diapause occurs naturally among geographical races (Klassen *et al.*, 1970) and temperature-sensitive mutations are inducible with mutagens (Suzuki, 1970).

The second type of genotype suitable for introduction is exemplified by disease vectors, where a satisfactory solution would be the interruption of disease transmission by replacing vector forms of the species by nonvector forms (Foster *et al.*, 1972).

Conditional lethality and vector refractoriness exemplify the sorts of genotype that could be employed in pest management if we had at our disposal some effective means (a transporting mechanism) of introducing them into the natural population. It is reasonable to assume that, if the above kinds of desirable genotype were simply released into a population without the aid of a genetic transporting mechanism, they would be quickly rejected in most cases by natural selection. Strains specially synthesized in the laboratory for this purpose are likely to contain deleterious characteristics and, since flooding cannot by itself completely fix the genotype, it is probable that the original genetic structure of the population would be rapidly restored. A solution to this problem might be to release strains heterozygous for a number of recessive lethal factors (LaChance and Knipling, 1962) or carrying several conditional dominant lethal traits, such as the putative multiple-locus system controlling diapause in the boll weevil (Klassen *et al.*, 1970). Segregation of such genes would simulate a diffusion process such that most individuals in later generations would inherit at least one dose of the conditional lethal. Here a transporting mechanism is not required. The usefulness of such an approach is limited only by our ability to detect or synthesize complex genetic systems of the appropriate type.

2. *Transporting Mechanisms*

Except for the case where segregation suffices to spread the genetic conditions through the population, a genetic process is required which ensures the continued spread and fixation of the released material regardless of any relatively minor inaptitudes it may possess. In other words, a transporting mechanism is virtually essential.

Two basic kinds of genetic transporting mechanism have been suggested:

1. "Meiotic drive" (MD) where heterozygotes for a particular chromosome transmit that chromosome to more than the usual 50% of the offspring. According to the system, driven chromosomes would spread through the population regardless of their initial frequencies until they were fixed, along with genes inseparably linked to the driving element. An attractive feature of this scheme is the prospect of releasing relatively small numbers of insects. A thorough review of the nature of drive mechanisms, their modes of synthesis, their application to insect control either as transport mechanisms, or to distort sex ratios, and the likely response of natural selection to the introduction of such systems into field populations, is clearly necessary before a proper evaluation of this approach to insect control can be provided.

2. Genetic situations in which negative heterosis (i.e., reduced fitness of the heterozygote) leads to elimination of one or other homozygote. The movement of gene frequencies away from an intermediate unstable equilibrium point provides the transporting force. Translocations and pericentric inversions, interracial hybrid sterility, cytoplasmic incompatibility, and compound chromosomes (see Fig. 2) are all possibilities since hybrids in each case have a fitness lower than either "homozygote," due either to sterility or lethality. Unlike meiotic drive, the frequencies of these rearrangements will move in either direction away from an unstable equilibrium point. This behavior, in turn, necessitates an initial release of larger numbers than involved in merely seeding the population as in meiotic drive, but the process of fixation is much faster than in meiotic drive.

The rate of fixation by the more frequent homozygote increases as hybrid fitness decreases. Figure 3 shows how rapidly replacement occurs in cage experiments using *D. melanogaster* when two equally fit strains compete and when their hybrid is completely inviable. This example demonstrates the frequency-dependent nature of the

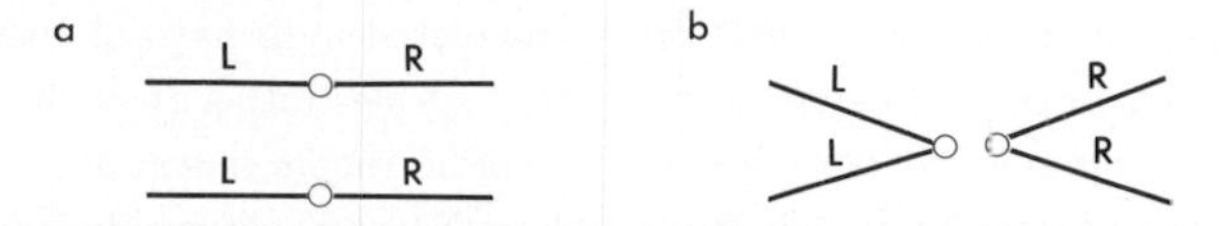

Fig. 2. Compound chromosomes are derived from a normal pair of chromosomes (a) by attachment of left and right arms to separate centromeres (b). (After Foster *et al.*, 1972.)

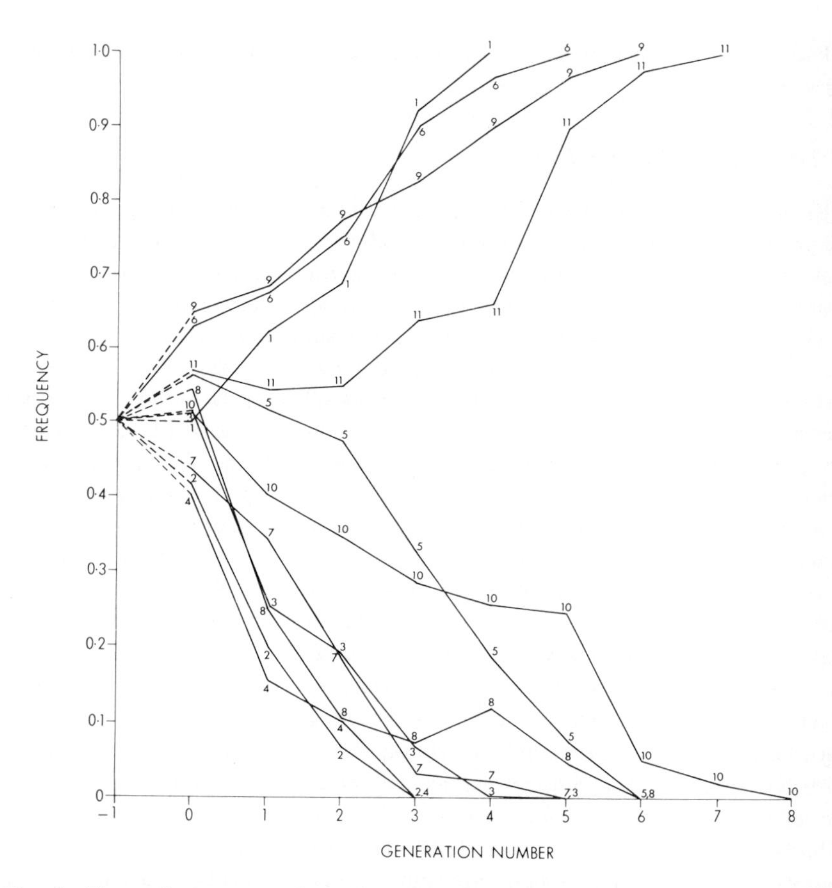

Fig. 3. Change in frequency of a strain of *D. melanogaster* in eleven population cages after the introduction of another equally fit strain. Hybrids between the two strains are inviable. (After Foster *et al.*, 1972.)

phenomenon. One strain's initial frequency exceeded 50% in roughly half of the cages and it continued to increase until it eliminated the other. In the remainder its initial frequency was less than 50% and it was eliminated. This illustrates a type of competitive displacement novel to biological control. Normally competitive displacement occurs when groups of organisms that do not interbreed compete in some ecological manner for a limiting resource (DeBach and Sundby, 1963) or are regulated by a natural enemy. In such cases it is the innate properties of the competing species rather than their initial frequencies that determine the outcome. With genetic displacement it is the relative frequencies of the competing strains that regulate the course of events. Innate properties of the competitors determine the location of the instability point, but beyond that they have no bearing on the direction or speed of the actual displacement process.

It might be possible to alter the equilibrium by giving the released strain a temporary genetic advantage. For example, insecticide resistance might be incorporated in its genome and the insecticide then applied (Whitten, 1971a) until such time as the released strain has a frequency advantage. Substantially fewer insects would need to be released to exceed the equilibrium point in these circumstances.

Since translocations generate negative heterosis they have been suggested as a transporting mechanism (Curtis, 1968b; Whitten, 1971b). If the translocation strain is not complex enough to ensure hybrid sterility, the displacement process might be too slow to be practical and difficulties could arise through leakage of genes from the displaced strain into the displacing strain. This latter problem would be serious if, for example, restoration of insecticide susceptibility were the objective of the displacement operation. There are various cytogenetic methods available to minimize the problems of leakage and speed of transport. For example, chromosome inversions which suppress crossing-over are easily isolated, as evidenced by the recent work of Baker *et al.* (1971) on *C. tritaeniorhynchus*, and may be introduced into the genome for this purpose (Curtis, 1968b).

If the hybrid is fully sterile or inviable the problem of leakage does not arise. Compound chromosomes satisfy this condition. They also permit the incorporation of large segments of the genome from a chromosomally normal strain into the compound strain in a single step (Foster *et al.*, 1972). This possibility, which is not available if multiple translocations provide the transporting force, could also allow the release of the same compound strain into many different environments, through the incorporation of local genomes into the compound strain in each instance before release.

An important question which must be considered both for control by inherited sterility and for genetic manipulation as described above, is the sensitivity of these techniques to population structure and to migration. If their application were restricted to species occurring in discrete populations the restriction would be severe indeed. Where the population is large or very widespread and logistics prevent suppression of the entire population, eradication of the pest over a portion of its range would be satisfactory only if migration were sufficiently slow for the period of respite to justify the expenditure. Otherwise it would seem preferable, in those cases where it applies, to displace the pest by a strain ecologically equivalent but devoid of the noxious features of the pest and competitively preventing reinvasion by the pest. There are known examples of natural occurrence of this kind of chromosomal displacement which could allow a corresponding displacement of characteristics (White, 1970). Furthermore, the displacement can be quite stable. White *et al.* (1969) have estimated that the hybrid zones in the wingless morabine grasshopper of the "viatica" group have occupied the same general areas for more than 10,000 years. The width of the hybrid zone is probably determined by the hybrid fitness and the mobility of the insect, the zone becoming more restricted as hybrid sterility increases or as mobility decreases. Applying this model to the displacement of certain mosquito pests we might envisage the stable displacement of vector forms over large areas by genetically isolated nonvector forms.

IV. CONCLUSIONS

With the exception of the sterile insect release method, this chapter has necessarily emphasized the developing theory concerning autocidal methods, since there is little practical experience to draw upon. It cannot be overemphasized, however, that the prospects for success of any autocidal method are greatly enhanced by the acquisition of comprehensive ecological information about the natural pest population. Fundamental to the effective planning of a program is knowledge of the absolute size of the pest population, its fluctuations, the components of the environment that are primarily responsible for the regulation of numbers, capacities for increase (especially at low levels of abundance), and, particularly, the occurrence of high local densities and their effect on the behavior of the pest, its natural enemies, and its food supply. Further necessary information includes the capacity of both released and wild insects to disperse, the pattern of distribution and seasonal abundance of both cultivated and wild hosts, the mating competitiveness of the released insects, and the longevity of released and wild insects under field conditions.

The collection of the necessary information may involve years of careful and demanding ecological research and few projects yet attempted have commenced with such comprehensive background information. Without such information, however, the inherent great promise of several autocidal approaches will continue to be severely handicapped.

REFERENCES

Ailam, G., and Galun, R. (1967). Optimal sex ratio for the control of insects by the sterility method. *Ann. Entomol. Soc. Amer.* **60**, 41–43.

Baker, R. H., Sakai, R. K., and Mian A. (1971). Linkage group-chromosome correlation in a mosquito. *J. Hered.* **62**, 31–36.

Baumhover, A. H. (1963). Influence of aeration during gamma irradiation of screw-worm pupae. *J. Econ. Entomol.* **56**, 628–631.

Bushland, R. C. (1971). Sterility principle for insect control: Historical development and recent innovations. *In* "Sterility Principle for Insect Control of Eradication," pp. 3–14. Int. Atomic Energy Agency, Vienna.

Curtis, C. F. (1968a). A possible method for the control of insect pests with special reference to tsetse flies (*Glossina* spp.) *Bull. Entomol. Res.* **57**, 509–523.

Curtis, C. F. (1968b). Possible use of translocations to fix desirable genes in insect pest populations. *Nature* (*London*) **218**, 368–369.

Curtis, C. F. (1969). The production of partially sterile mutants in *Glossina austeni*. *Genet. Res.* **13**, 289–301.

Curtis, C. F., and Hill, W. G. (1971). Theoretical studies on the use of translocations for the control of tsetse flies and other disease vectors. *Theoret. Popul. Biol.* **2**, 71–90.

Curtis, C. F., and Langley, P. A. (1971). The use of chilling and nitrogen in the radiation sterilization of *Glossina morsitans*. *Trans. Roy. Soc. Trop. Med. Hyg.* **65**, 230.

Curtis, C. F., and Robinson, A. S. (1971). Computer simulation of the use of double translocations for pest control. *Genetics* **69**, 97–113.

DeBach, P., and Sundby, R. A. (1963). Competitive displacement between ecological homologues. *Hilgardia* **34**, 105–166.

Ducoff, H. S. (1972). Causes of death in irradiated adult insects. *Biol. Rev.* **47**, 1–47.

Foster, G. G., Whitten, M. J., Prout, T., and Gill, R. (1972). Chromosome rearrangements for the control of insect pests. *Science* **176**, 875–880.

Fried, M. (1971). Determination of sterile-insect competitiveness. *J. Econ. Entomol.* **64**, 869–872.

Hogan, T. W. (1971). An evaluation of a genetic method for population suppression of *Telleogryllus commodus* (Walk.) (Orth. Gryllidae) in Victoria. *Bull. Entomol. Res.* **60**, 383–390.

Hooper, G. H. S. (1970). Use of carbon dioxide, nitrogen, and cold to immobilize adults of the Mediterranean fruit fly. *J. Econ. Entomol.* **63**, 1962–1963.

Hooper, G. H. S. (1971). Competitiveness of gamma-sterilized males of the Mediterranean fruit fly: Effect of irradiating pupal or adult stage and of irradiating pupae in nitrogen. *J. Econ. Entomol.* **64**, 1364–1368.

International Atomic Energy Agency (1968). Isotopes and radiation in entomology. *Proc. Symp. Vienna, Dec. 4–8, 1967*, 428 pp. Int. Atomic Energy Agency, Vienna.

International Atomic Energy Agency (1969). Sterile-male technique for eradication or control of harmful insects. *Proc. Panel, Vienna, May 27–31, 1968*, 142 pp. Int. Atomic Energy Agency, Vienna.

International Atomic Energy Agency (1970). Sterile-male technique for control of fruit flies. *Proc. Panel, Vienna, Sept. 1–5, 1970*, 175 pp. Int. Atomic Energy Agency, Vienna.

International Atomic Energy Agency (1971a). Application of induced sterility for control of lepidopterous populations. *Proc. Panel, Vienna, June 1–5, 1970*, 169 pp. Int. Atomic Energy Agency, Vienna.

International Atomic Energy Agency (1971b). Sterility principle for insect control or eradication. *Proc. Symp. Athens, Sept. 14–18, 1970*, 542 pp. Int. Atomic Energy Agency, Vienna.

Klassen, W., and Earle, N. W. (1970). Permanent sterility induced in boll weevils with busulfan without reducing production of pheromone. *J. Econ. Entomol.* **63**, 1195–1198.

Klassen, W., Knipling, E. F., and McGuire, J. U., Jr. (1970). The potential for insect-population suppression by dominant conditional lethal traits. *Ann. Entomol. Soc. Amer.* **63**, 238–255.

Knipling, E. F. (1955). Possibilities of insect control or eradication through the use of sexually sterile males. *J. Econ. Entomol.* **48**, 459–462.

Knipling, E. F. (1966). Some basic principles in insect population suppression. *Bull. Entomol. Soc. Amer.* **12**, 7–15.

Knipling, E. F. (1970). Suppression of pest Lepidoptera by releasing partially sterile males. A theoretical appraisal. *Bioscience* **20**, 465–470.

LaBrecque, G. C., and Smith, C. N. (eds.). (1968). "Principles of Insect Chemosterilization." 354 pp. North-Holland, Amsterdam.

LaChance, L. E. (1963). Enhancement of radiation-induced sterility in insects by pretreatment in CO_2 + air. *Int. J. Radiat. Biol.* **7**, 321–331.

LaChance, L. E. (1970). Problems and programs in the application of the sterility principle for the control of *Ceratitis capitata*. *Bol. Asoc. Nac. Ing. Agron.* **209**, 493–495.

LaChance, L. E., and Knipling, E. F. (1962). Control of insect populations through genetic manipulations. *Ann. Entomol. Soc. Amer.* **55**, 515–520.

LaChance, L. E., and Riemann, J. G. (1964). Cytogenetic investigations on radiation and chemically induced dominant lethal mutations in oocytes and sperm of the screw-worm fly. *Mutat. Res.* **1**, 318–333.

LaChance, L. E., Schmidt, C. H., and Bushland, R. C. (1967). Radiation-induced sterility. *In* "Pest Control: Biological, Physical, and Chemical Methods" (W. W. Kilgore and R. L. Doutt, eds.), pp. 147–196. Academic Press, New York.

LaChance, L. E., Degrugillier, M. E., and Leverich, A. P. (1970). Cytogenetics of inherited partial sterility in three generations of the large milkweed bug as related to holokinetic chromosomes. *Chromosoma* **29**, 20–41.

LaChance, L. E., North, D. T., and Klassen, W. (1968). Cytogenetic and cellular basis of chemically-induced sterility to insects. *In* "Principles of Insect chemosterilization" (G. C. LaBrecque and C. N. Smith, eds.), pp. 99–157. Appleton-Century-Crofts, New York.

Laven, H., Jost, E., Meyer, H., and Selinger, R. (1971). Semisterility for insect control. *In* "Sterility Principle for Insect Control or Eradication," pp. 415–423. Int. Atomic Energy Agency, Vienna.

Lawson, F. R. (1967). Theory of control of insect populations by sexually sterile males. *Ann. Entomol. Soc. Amer.* **60**, 713–722.

Lorimer, N., Hallinan, E., and Rai, K. S. (1972). Translocation homozygotes in the yellow fever mosquito, *Aedes aegypti*. WHO/VBC/72.355.

McDonald, I. C. (1971). A male-producing strain of the house fly. *Science* **172**, 489.

McDonald, P. T., and Rai, K. S. (1971). Population control potential of heterozygous translocations as determined by computer simulations, *Bull. WHO* **44**, 829–845.

Nash, T. A. M., Jordan, A. M., and Trewern, M. A. (1971). Mass rearing of tsetse flies (*Glossina* spp.): Recent advances. *In* "Sterility Principle for Insect Control or Eradication," pp. 99–110. Int. Atomic Energy Agency, Vienna.

North, D. T., and Holt, G. G. (1968a). Inherited sterility in progeny of irradiated male cabbage loopers. *J. Econ. Entomol.* **61**, 928–931.

North, D. T., and Holt, G. G. (1968b). The genetic and cytogenetic basis of radiation-induced sterility in the adult male cabbage looper, *Trichoplusia ni*. *In* "Isotopes and Radiation in Entomology," pp. 391–403. Int. Atomic Energy Agency, Vienna.

North, D. T., and Holt, G. G. (1971). Inherited sterility and its use in population suppression of Lepidoptera. *In* "Application of Induced Sterility for Control of Lepidopterous Populations," pp. 99–111, Int. Atomic Energy Agency, Vienna.

Patterson, J. P. Stone, W., Bedichek, and Suche, M. (1934). The production of translocations in *Drosophila*. *Amer. Natur.* **68**, 359–369.

Patterson, J. T., and Stone, W. S. (1952). "Evolution in the Genus *Drosophila*," 610 pp. MacMillan, New York.

Proshold, F. I., and Bartell, J. A. (1972). Inherited sterility and postembryonic survival of two generations of tobacco budworms, *Heliothis virescens* (Lepidoptera: Noctuidae) from partially sterile males. *Can. Entomol.* **104**, 221–230.

Proverbs, M. D. (1969). Induced sterilization and control of insects. *Annu. Rev. Entomol.* **14**, 81–102.

Riemann, J. G., and Flint, H. M. (1967). Irradiation effects on midguts and testes of the adult boll weevil, *Anthonomus grandis*, determined by histological and shielding studies. *Ann. Entomol. Soc. Amer.* **60**, 298–308.

Sakai R. K., Baker, R. H., and Mian, A. (1971). Linkage group-chromosome correlation in a mosquito. *J. Hered.* **62**, 90–100.

Serebrovskii, A. S. (1940). On the possibility of a new method for the control of insect pests. *Zool. Zh.* **19**, 618–630.

Smith, C. N. (ed.). (1966). "Insect Colonization and Mass Production," 618 pp. Academic Press, New York.

Smith, R. H. (1971). Induced conditional lethal mutations for the control of insect populations. *In* "Sterility Principles for Insect Control or Eradication," pp. 453–465. Int. Atomic Energy Agency, Vienna.

Suzuki, D. (1970). Temperature-sensitive mutations in *Drosophila melanogaster*. *Science* **170**, 695–706.

Toba, H. H., Kishaba, A. N., and North, D. T. (1972). Reduction of populations of caged cabbage loopers by the release of irradiated males. *J. Econ. Entomol.* **65**, 408–411.

Van Zon, A. Q., and Overmeer, W. P. J. (1972). Induction of chromosome mutations by X-irradiation in *Tetranychus urticae* with respect to a possible method of genetic control. *Entomol. Exp. Appl.* **15**, 195–202.

Wagoner, D. E., Nickel, C. A., and Johnson, O. A. (1969). Chromosome translocation heterozygotes in the house fly. *J. Hered.* **60**, 301–304.

Wakid, A. M. (1973). Effects of nitrogen during gamma irradiation of puparis and adults of the Mediterranean fruit fly on emergence, sterility, longevity and competitiveness. *Environ. Entomol.* **2**, 37–40.

White, L. D., and Hutt, R. B. (1972). Effects of treating adult codling moths with sterilizing and substerilizing doses of gamma irradiation in a low-temperature environment. *J. Econ. Entomol.* **65**, 140–143.

White, L. D., Hutt, R. B., and Onsager, J. A. (1970). Effects of CO_2, chilling, and staining on codling moths to be used for sterile releases. *J. Econ. Entomol.* **63**, 1775–1777.

White, M. J. D. (1970). Cytogenetics of speciation. *J. Aust. Entomol. Soc.* **9**, 1–6.

White, M. J. D., Blackith, R. E., Blackith, R. M., and Cheney, J. (1967). Cytogenetics of the *viatica* group of morabine grasshoppers. The "coastal" species. *Aust. J. Zool.* **15**, 263–302.

White, M. J. D., Key, K. H. L., André, M., and Cheney, J. (1969). Cytogenetics of the *viatica* group of morabine grasshoppers. II. Kangaroo Island populations. *Aust. J. Zool.* **17**, 313–328.

Whitten, M. J. (1969). Automated sexing of pupae and its usefulness in control by sterile insects. J. Econ. Entomol. **62**, 272–273.

Whitten, M. J. (1971a). Use of chromosome rearrangements for mosquito control. *In* "Sterility Principle for Insect Control or Eradication," pp. 399–411. Int. Atomic Energy Agency, Vienna.

Whitten, M. J. (1971b). Insect control by genetic manipulations of natural populations. *Science* **171**, 682–684.

Whitten, M. J., and Norris, K. R. (1967). "Booby-trapping" as an alternative to sterile males for insect control. *Nature (London)* **216**, 1136.

Whitten, M. J., and Taylor, W. C. (1970). A role for sterile females in insect control. *J. Econ. Entomol.* **63**, 269–272.

27

INTEGRATED CONTROL: A REALISTIC ALTERNATIVE TO MISUSE OF PESTICIDES?

Philip S. Corbet and Ray F. Smith

I. INTRODUCTION

Sections I to IV of this treatise contain a comprehensive review of the theory and practice of biological control and of the relative merits of biological and other types of pest suppression. From this review we may draw three conclusions: (1) continued reliance on simple, nonselective methods of chemical control, at least as these have been used in the past, will incur unacceptably high environmental costs; (2) the integration of biological and chemical or of other methods (integrated control) seems to offer the most practicable means of reducing these costs without incurring unacceptably high economic losses due to pests; and (3) biological control (as defined in Chapter 1) constitutes a central element in most, if not all, integrated control programs.

The preceding chapters in Section V each provide a detailed treatment of one component of a potential integrated control program, that is, of one weapon (or rather one class of weapons) that stands ready to be drawn upon in the armory of integrated control.

In this final chapter we discuss the implementation of integrated control, namely the combination or synthesis of these several weapons into a practical program of pest management. First, we review the definition and the objectives of integrated control, and discuss briefly the way in which it is conducted and its effectiveness assessed; next, we present examples of successful integrated control programs and draw conclusions from them; and finally, with such material as a background, we examine the prospects of integrated control during the next few decades.

II. DEFINITION AND OBJECTIVES OF INTEGRATED CONTROL

Integrated control has been defined in various ways (e.g., Smith and Reynolds, 1966, 1972; FAO, 1968; Smith and van den Bosch, 1967; Kennedy, 1968). A recent definition by van den Bosch *et al.* (1971) is as follows:

> Integrated control is a pest population management system that utilizes all suitable techniques either to reduce pest populations and maintain them at levels below those causing economic injury, or to so manipulate the populations that they are prevented from causing such injury. Integrated control achieves this ideal by harmonizing techniques in an organized way, by making the techniques compatible, and by blending them into a multifaceted, flexible system (Smith and Reynolds, 1966). In other words, it is an holistic approach aimed at minimizing pest impact while simultaneously maintaining the integrity of the ecosystem.

This definition is typical in that it is intended to convey the idea that, in a successful integrated control program, the desired control of a pest is achieved in the most intelligent and practical way that knowledge can devise (FAO, 1971).

Although attempts to produce an incisive definition of integrated control are fairly recent, the *concept* of integrated control is far from new. Indeed it is justifiable to regard it as embodying a common sense, ecological approach to pest suppression and, as such, it has had its distinguished proponents for almost a century and certainly long before synthetic organic pesticides came to be widely used (Smith, 1969). However, integrated control has received increasing attention during the last 10—15 years largely because of two related circumstances: (1) man has become increasingly reliant on chemical pesticides for crop protection, and (2) at the same time he has become increasingly aware that such pesticides can have harmful effects. This realization has caused attention to be focused once more on the principles that underlie ecologically sound pest management and (in the context of recent ecological knowledge) it has elicited thorough consideration of what is needed to achieve pest suppression by these means. Looked at in this way the *intent* of this recent interest would be realistically expressed by regarding integrated control as a system of pest management that avoids or decreases reliance on simple, nonselective methods of suppression, especially those methods involving toxic materials that the environment accumulates more rapidly than it can degrade (Corbet, 1973a).

As integrated control is now envisaged, it embodies two fundamental ideas, both of which can be stated in the form of maxims (Smith, 1968). The first of these is "Consider the ecosystem"; that is, pest control should be developed and applied in the context of the whole environment. The second is "Utilize economic injury levels"; this offers the two great benefits of avoiding unnecessary control measures and of encouraging the investigator to attempt a cost/benefit analysis. The latter activity can bring into focus some of the long-term costs and gains of different control strategies which are often liable to be overlooked (see Steiner, 1973). This said, it must be recognized that both these maxims, if followed, lead to actions difficult to evaluate incisively. No one, however conscientious, can claim to consider *all* aspects of the ecosystem or *all* possible costs and benefits of a given treatment. Furthermore, to recognize a cost is not necessarily to be able to express it in quantitative terms, especially if it is a social or a long-term cost. Nevertheless these considerations orient our thinking in a direction that can only lead us closer to sound policies of pest management.

From these and related principles of ecology and crop protection a set of guidelines has been derived for practitioners of integrated control (see Smith, 1968), some applying generally, and others only to certain crop protection situations.

1. Identify the environmental factors that permit a species to achieve pest status, and then try to manipulate these factors in an integrated way to prevent the pest from causing economic damage.

2. Try to reduce the annual effort devoted to pest management and the annual cost of crop protection, especially in so far as it draws on an energy subsidy.

3. Maintain flexibility in crop protection procedures to allow for local variations and for the inevitable evolution of crop production within the agroecosystem.

4. The establishment of integrated control programs in ecologically disrupted areas (for example, large areas devoted to monocultures or those treated heavily with broad-spectrum pesticides) should usually be developed gradually, and crop production and crop protection research should be coordinated closely with practical applications.

5. Primary sources or reservoirs of pest populations should be evaluated and managed carefully, especially when they are small in relation to the total infested area.

6. Pest control procedures should be directed to affect only a limited part of the agroecosystem; that is, they should be used selectively and have their greatest and, when possible, only impact on a target pest species.

7. Use of resistant varieties and of parasites, predators, and pathogens which have been determined to be of consequence in the regulation or limitation of pest species should be augmented or fostered.

8. Appropriate attention should be given to diversity in the ecosystem and to the ways in which it can be manipulated as part of a pest control strategy (see Southwood and Way, 1970).

9. Surveillance of pests and ecological change must be maintained to detect the need for reestablishment or modification of control procedures.

These guidelines indicate the main objectives underlying the application of integrated control. Also, to some extent, they provide a set of criteria by which the success of such programs can be evaluated or (to express this in another way) a means of deciding how realistic the theory of integrated control is in practice! In the next part of this chapter we shall look at some recent integrated control programs with these criteria in mind.

III. SOME SUCCESSFUL INTEGRATED CONTROL PROGRAMS

For purposes of this review we are regarding integrated control as a procedure that (apart from complying with one or more of the guidelines listed above) involves the integration of two or more methods (biological, chemical, or other), and, where chemicals are used, a progressive decrease in the amount of pesticide applied to the ecosystem.

Programs included here are only those concerned with agricultural pests and are listed in order of decreasing ecosystem diversity, from orchard via field crops to greenhouse. We shall consider programs that have not only proved successful in experiments and field trials but have also been adopted as standard procedures or have recently begun to be implemented by growers or "industry." Comprehensive reviews of such programs have recently been published elsewhere (Huffaker, 1971; Corbet, 1973b; Wildbolz and Meier, 1973; Wood, 1973); so here we shall include only a few examples of special interest.

A. Apple Pests in Eastern Canada

Only a year or two after DDT began to be widely used as an insecticide, A. D. Pickett and his colleagues in Nova Scotia laid the foundations of a system for integrated control of apple pests that has been used by growers in Nova Scotia and eastern Quebec for about 20 years (e.g., Pickett *et al.*, 1946). The basic approach was to identify the more important natural enemies of the key pests and then to conserve them by using selective pesticides and by timing the application of these chemicals carefully in relation to the phenology of the pests and of their parasites and predators (MacPhee and MacLellan, 1971a,b). A major advance was made when it became possible to achieve selective control of the codling moth, *Carpocapsa pomonella* (L.), by using the chemical ryania; since then effort has been concentrated on the apple maggot, *Rhagoletis pomonella* (Walsh), which remains a key pest. This, and a similar program in southern Quebec (e.g., Le Roux *et al.*, 1963), has resulted in reduced damage from pests and reduced application of pesticides.

Significant progress has also been made in the development of integrated control programs for apple in Europe, Australia, the United States, and other parts of the world (Gonzalez, 1973).

B. Cotton Pests in the United States

For over 25 years integrated control for cotton has existed in California. The evolution of cotton pest control in California and in other parts of the United States indicates some of the difficulties of implementation and the need for continuing research effort closely tied to on-going pest control programs in the crop. It is unrealistic to envision a long-term intensive research effort which would produce a complicated pest control system which then, in turn, would be inserted *in toto* into the crop production system. The integrated control system must be responsive to change in the agroecosystem both from season to season and over longer periods.

Integrated control of cotton in California at first simply depended on weekly population counts to limit the application of pesticides to those times when they were presumed to be needed. Gradually over the years many items have been added to the programs and the old elements have been refined many times. Some farmers were able to accept these changes as rapidly as they came along, others accepted only a portion of the changes and often more slowly, and some farmers accepted none of the integrated control elements. The proportion of farmers in this last category is now small. The California program in its more sophisticated version now involves better pest sampling and prediction methods, more precise economic injury levels, an improved understanding of how naturally occurring biotic mortality factors can be utilized, the manipulation of the pests and natural enemies through cultural and agronomic practices, e.g., alfalfa interplantings, correlation of plant physiological condition with risk of insect damage, and knowledge of impact of moon phase on pest activity (Falcon and Smith, 1973; Smith and Falcon, 1973).

The principal cotton growing areas of Texas differ significantly from those of California and from each other in topography, soil, climate, and associated natural and cultivated vegetation. Each agroecosystem also differs from others in the land area it occupies and in the degree of ecological stability. Hence, each cotton agroecosystem has its own unique features that have important effects on the types of pests and the types of problems associated with these pests. Nonetheless, these several cotton agroecosystems in Texas can be grouped into three general categories (Bottrell, 1973).

1. Low insecticide usage areas, typified by the High Plains of Texas, without perennially occurring key pests and with only occasional invasion by severe migrant pests, and where potential pests (e.g. *Heliothis* spp., and *Trichoplusia ni* (Hübner)) do not cause regular damage under conditions currently prevailing in the agroecosystem.

2. High insecticide usage areas, typified by the Trans-Pecos of Texas, also without serious key pests, but where prevailing conditions (obviously unwise management conditions in the case of the Pecos area) have unleashed damaging populations of potential or "man-made" pests.

3. Areas, local, or wide-scale, in the rest of the state of Texas where key pests occur every year, or nearly every year, at population densities above the economic injury level and cause severe damage unless actions are taken to curb this damage.

A more complex integrated control program which has enjoyed success has been developed for the lower Rio Grande Valley (category 3 above) in response to the threat presented by the development of a strain of *Heliothis virescens* (Fabr.) (budworms), which is resistant to all insecticides registered for use on cotton. The fundamental strategy in this program is to avoid inducing *Heliothis* outbreaks by the premature, or unwise, use of insecticides. This program utilizes cultural practices for control of the pink bollworm, *Pectinophora gossypiella* (Saunders), and is centered around a shortened growing season. The cultural control program is effective against both the pink bollworm and the boll weevil, *Anthonomus grandis* Boh. In addition to early stalk destruction, cotton farmers are encouraged to apply one or two insecticidal treatments during the harvest period to reduce numbers of diapausing boll weevils. All the farmers in the area are encouraged by legislative and social pressures to comply with this program. The economic threshold for initiating chemical control of the fleahopper has been increased from 25 to 50 insects per 100 plants since the damage inflicted by this level of fleabeetle is considerably less than that caused by *Heliothis virescens* when freed from natural control. The outbreaks of *H. virescens* mostly result from destruction of beneficial arthropods by insecticidal application for control of the fleahopper. Also, farmers are advised to use a low dosage of the least disruptive insecticide appropriate when fleahopper control is deemed necessary. This program requires careful and frequent checking of the crop for insect pest numbers and crop damage.

The implementation of this integrated control program has preserved commercial cotton production in the lower Rio Grande Valley of Texas during a period when the crop has almost disappeared from adjacent areas of Mexico. Yields have been maintained near normal levels, and the number of insecticidal treatments per year have been greatly reduced (by more than 50% in most years). Production costs are considerably less than in former years when cotton was treated with insecticides from time of planting to harvest (Adkisson, 1973).

Another integrated control program that has not been so successfully implemented by producers is the one developed by Pate *et al.* (1972) for the Pecos Valley region of Texas (category 2 above) where the boll weevil is not present and the pink bollworm is sporadic in appearance. The primary insect problem on cotton in this area is *Heliothis*. Outbreaks are induced by insecticidal treatments applied prematurely on the basis of egg numbers or first signs of crop damage. Pate and co-workers have shown that a program of supervised insect control combined with sensible fertilization and irrigation of the cotton, and maintenance of crop diversity, has almost eliminated the need for insecticidal treatment of cotton. In this program only one application of insecticide has been applied to the cotton grown in experimental plots over a 5-year period and yields have been maintained at a level slightly above the average for the area.

The degree of success in getting Pecos cotton growers to follow this program has not been encouraging. The average Pecos producer still applied from eight to twelve insecticidal treatments per year to his cotton. However, eventually many farmers began to question the wisdom of this practice. The cost of cotton production in the area was exceeding income from the crop and it was obvious that *Heliothis* was becoming considerably more difficult to control with chemicals. In 1972, a local extension entomologist was employed to work in this area. This entomologist, by providing intensive technical assistance, was able to convince several key leaders to implement the cotton production program of the Pecos Research Station. These producers were able to reduce greatly insecticidal control costs and the losses inflicted by *Heliothis*. As a result, there are indications that a great many more farmers will soon adopt the program.

An even more remarkable demonstration of the value of this program was made (Pate and Neeb, 1972) in the Bakersfield area of Pecos County in 1971. This production area is an isolated irrigation scheme consisting of approximately 1100 acres of cotton grown by five farmers. These farmers produced the 1971 crop without any insecticides when treatment decisions were based on frequent and careful assessment of *Heliothis* numbers and crop damage. Yields were almost doubled, averaging 494 pounds of lint per acre in 1971, compared to 263 in 1970, and the cost of insecticidal treatment was reduced from a high of $70 per acre to zero. The Bakersfield farmers implemented this program only because two professional entomologists, in whom they had great confidence, were checking their fields each week. However, it is obvious that this level of assistance cannot be provided to every cotton producer.

An analysis of why some integrated insect control programs for cotton have been successfully implemented while others have not shows certain characteristic points. One of the main features that emerges is that proper understanding and acceptance of the economic threshold is critical to the success of integrated control programs of insect pest control. Economic thresholds are absolutely essential to most programs but they are a feature that farmers often reject. This is because the determination of the economic threshold requires frequent pest population assessment, evaluation and prediction of crop damage, and a new decision almost weekly as to whether to apply an insecticide or not.

The pink bollworm program in South Texas has worked because there has been little decision making left to the farmer as to when he should carry out the cultural control practices. The Texas legislature has assumed much of the responsibility as it has divided Texas into control zones in which the farmer must plant, grow the crop, and destroy stalks within certain dates. A majority of the farmers are convinced that this is a good program and they exert considerable social pressure on the few farmers each year who do not comply with the law. Economic thresholds play a small role in this program and there is little decision making required on the part of the farmer.

The present program in the lower Rio Grande Valley requires constant population assessment of the boll weevil, fleahopper, and *Heliothis*. The farmers know that the old control practice of unilateral dependence on insecticides has failed them and that an improperly timed insecticidal treatment may mean great loss to the crop. Their current fear of the pests is so great that they have been willing to adopt measures that they would not have considered a few years ago.

The Pecos farmer has not even accepted supervised control because insecticides have provided "crop insurance" for him. He is unsure of his ability to correctly assess insect pest populations and crop damage in order to time insecticidal treatments properly. Thus, he has continued to go his old way even though over the long term his practices may be greatly to his disadvantage. However, recent events have shaken his faith in the old system. Fear of chemical control failures combined with economic pressures are forcing him to reevaluate his concepts of insect pest control. A few farmers are beginning to implement the proposed integrated system.

The Bakersfield farmer will accept a pest management program when the state of Texas provides a well-trained specialist to check his crop for insect pests and make control decisions. When this is not available the farmers lack confidence in their ability to check the crop for pests; thus, they have substituted *Trichogramma* releases for insecticides. This provides them with the feeling that they are taking positive action to protect the crop from *Heliothis* attack.

C. Tobacco Pests in North Carolina

Insecticidal use patterns on flue-cured tobacco historically have consisted of regularly timed preventive applications or applications based mainly on intuition. Through the years, many tobacco growers, in seeking the best possible yields and quality, have come to overestimate their insect problems. As a result, insecticide use on tobacco has been excessive and applications poorly timed. The perpetual overuse of insecticides magnifies many of the problems faced by tobacco growers. These include: unnecessary residues on their product, increased personal hazards, and lowered environmental quality. The goal of the North Carolina program was to integrate insect control actions with crop production practices to obtain economically sound crop protection with the least possible disruption of natural control and environmental quality (Ellis *et al.*, 1972). This program had two basic objectives: (1) to lower the mean level of pest abundance over a wide geographical area and thus reduce the frequency of pest outbreaks, and (2) to suppress pest outbreaks that did occur according to sound economic and ecological principles. The North Carolina project has been successful in meeting these objectives.

The Tobacco Pest Management Program in North Carolina was initiated in 1971 to demonstrate a simple management program for insect pests of tobacco foliage. The management program consists of: (1) effective sucker control, (2) early stalk destruction, and (3) the application of recommended insecticides only when pest infestations exceed the economic threshold. The first two practices serve to lower the mean level of abundance of the key pests. The last practice allows suppression of pest outbreaks on a field-to-field basis, but requires frequent field inspection to determine the presence and infestation level of various pests. The application of insecticides only when needed, based on known levels of insect infestation, is economically and ecologically sound. It is economically sound in that the number of applications can generally be reduced, resulting in less expense for the grower. It is ecologically sound in that it reduces chances for product and environmental residues and removes undue insecticidal pressure from beneficial insect populations.

Area-wide tobacco insect pest management efforts in 1972 involved 1680 farms and 8498 acres of tobacco. Thirty-nine full-time summer employees scouted 3339 tobacco fields each week during the 1972 growing season. Weekly records on each field included: counts, or ratings, of population levels of certain insect pests and natural enemies; hornworms, budworms, flea beetles, aphids, stilt bugs (egg predators), *Campoletis sonorensis* Cameron (an inchneumonid budworm parasite), *Apanteles congregatus* (Say) (a braconid hornworm parasite), and *Winthemia manducae* Sabrosky and DeLoach (a tachinid hornworm parasite); and information on plant size, plant damage, phenological events, cultural practices, pesticide application, and phenology of surrounding crops. Data were stored on computer tapes for subsequent summary and retrieval. Growers were furnished with a copy of insect counts which informed them when pest levels exceeded the economic thresholds. These economic levels were: 5 or more 4th or 5th instar hornworms per 50 plants; 5 or more plants with budworms per 50 plants; "heavy infestation" of flea beetles or aphids. County agents also advised growers as to recommended insecticides and application procedures. Growers who followed scout reports avoided unneeded, or poorly timed, chemical applications.

The sucker control and early stalk destruction coupled with buildup of natural enemies as a result of lowered insecticide application has so reduced the major insect pests of tobacco in the program that it is now doubtful if the scouting program is any longer needed, at least in some regions. The program of early stalk destruction has also reduced the nematode and tobacco mosaic problems significantly.

D. Sorghum Pests in the United States

In 1968, a major aphid problem emerged in parts of the midwestern and southwestern United States on grain sorghum, one of the major crops in that area. Extremely high populations of a new biotype of greenbug, *Schizaphis graminum* Rondani, developed and caused extensive damage to the crop. The pest has attacked grain sorghum repeatedly since 1968 and has become a serious agricultural problem in some areas. The High Plains of Texas, where about one-fourth of the grain sorghum crop of the United States is produced annually, serves to illustrate the acuteness of this problem (Cate *et al.*, 1973a,b, 1976).

Prior to 1968, grain sorghum in the High Plains had not been confronted with widespread, perennially occurring insect pests. During a few years in the early 1960's, the sorghum midge, *Contarinia sorghicola* (Coquillett), appeared in the area and temporarily caused serious crop losses. However, an area-wide practice of early, uniform planting since the mid-1960's has restricted damage by this insect to isolated, late-planted grain sorghum fields which bloom late in the season. The amount of acreage treated for sorghum midge and other insect pests prior to 1968 was relatively small compared to the total grain sorghum acreage in the area.

In contrast, a vast amount of insecticide has been applied to grain sorghum in the High Plains since 1968. Estimates indicate that $14 million was spent for control of grain sorghum pests in 1970. Most of these expenditures were for control of greenbug. However, large quantities of insecticide also have been used to control corn leaf aphid, *Rhopalosiphum maidis* (Fitch). The latter insect frequently attacked grain sorghum prior to 1968, and heavy infestations sometimes caused damage to small plants. However, this aphid had not been regarded as a serious threat in most years. The recent actions to control corn leaf aphid obviously reflect a change in attitude that has come about as a result of experience with the more severe greenbug problem.

J. R. Cate and colleagues (1973a,b, 1976) recently described the evolution of an integrated control system to meet this emergency situation. Since noninsecticidal control methods were at first not available, an insecticidal evaluation program was initiated to find materials to control the pest at a reasonable cost and to meet residue tolerance regulations on the crop. These promising application methods and materials were further evaluated, application methods which are not amenable to management strategies were eliminated, and dosages of materials were sought which are sufficient to keep the pest below damaging levels while having the least possible effect on natural enemies. During the course of these emergency measures, data were accumulated on economic injury levels and research was initiated to find and develop sources of plant material resistant to the aphid. A search was also initiated for natural enemies that might be used. Each segment of these findings was eventually integrated into an effective management scheme. These management practices have the goal that the cost: benefit ratio is brought into proper balance, and the likelihood of ecological upsets is reduced in the crop under direct consideration as well as in the total farming system.

In the sorghum growing area most greenbug control has been practiced on the basis of prevention rather than of actual need as determined by pest population levels. Most producers view as cheap insurance the cost ($4—6/acre) of a seedfurrow or broadcast treatment with granular insecticides which will provide season-long control of the greenbug. This preventive control, while not needed in most cases, not only eliminates the possibility that damaging populations may develop, but also the uncertainties facing the producer when he attempts to determine for himself the need for control. However, these heavy pesticide applications would appear to have created problems with *Oligonychus pratensis* (Banks), formerly a minor mite pest, and to have aggravated the situation in cotton with *Heliothis zea*, a secondary pest of cotton in this area.

Cate *et al.* have proposed an integrated control strategy based on selective treatment of fields based on a pest scouting program. This is entomologically and economically sound, but it is not yet convincing to individual producers in the large sorghum growing areas. The sorghum producer's investment in his grain sorghum crop is high and the potential profit from the crop is also high, especially if the crop has been produced on good land that is fertilized and irrigated. A pest like greenbug, with the habit of secreting honeydew and leaving behind numerous cast skins, is viewed as a potential economic threat. The risk that the greenbug will reduce the chances for realizing maximum profits is viewed as far greater than the possibility that it will not reduce these chances. Therefore he concludes that the greenbug can be controlled inexpensively and the insurance thus obtained will assure maximum profit.

E. Cucumber Pests in Greenhouses in Canada

As a sequel to many years of research, an economically viable program of control of greenhouse pests has recently been adopted in Ontario (McClanahan, 1972, 1973). Its cornerstone is the biological control of the greenhouse whitefly, *Trialeurodes vaporariorum* (Westwood), by mass release of the eulophid wasp, *Encarsia formosa* Gahan. This parasite was brought to Canada in 1928 and mass produced at Belleville for supply to growers until 1955 (McLeod, 1938, 1962); studies continued on factors influencing its efficiency as a biological control agent (Burnett, 1949, 1967) but it was little used between 1946 and 1969 because growers used DDT or a fumigant which killed the parasite (McClanahan, 1971, 1972). The integrated control program, developed at Harrow, Ontario, involves the combination of *E. formosa* and a selective chemical fungicide, quinomethionate (Morestan), which controls powdery mildew of cucumber and also the two-spotted spider mite, *Tetranychus urticae* (Koch). The parasites are introduced, and the fungicide is applied, according to a prescribed timetable; this allows the parasite to build up its numbers so that a satisfactory balance between whiteflies and parasites is maintained for the 6-month duration of the crop. Parasites are produced by the local Marketing Board and by a commercial enterprise. As the program was operating in 1972, it involved substantial savings in cost of materials, and in labor. *E Formosa* is also widely used as a biological control for whiteflies in greenhouse tomatoes in Ontario.

Considerable effort has been devoted to the integrated control of various greenhouse pests on cucumber and other crops in several countries. On the basis of this work this approach is considered to have considerable promise (see Hussey and Bravenboer, 1971; Bravenboer, 1974).

IV. THE PROSPECTS OF INTEGRATED CONTROL

From the foregoing survey of these and other pest control programs (see also Corbet, 1973a, b) certain general conclusions emerge that are relevant to our examination of the prospects of integrated control.

1. The great majority of our systems of pest control still involve the simple application of chemicals.

2. The main motivation for development of most integrated control programs has been the failure or impracticability of pesticidal methods.

3. Integrated programs that have been applied successfully are usually characterized by the integration of a few, straightforward techniques.

4. The approach that is most likely to be successful seems to be correlated with the type of ecosystem. In Canada, for example (see Corbet, 1973b), biological control by inoculation and conservation, respectively, have been the methods of choice in forests and orchards, cultural control in field-crop situations, and inundation as cultivation becomes cleaner and the ecosystem simpler.

5. Several programs exist, ready for use, which are technologically successful but which are making no contribution to integrated control because nonentomological obstacles prevent their application.

Most pests that man attempts to control consciously are today suppressed solely by chemicals, many of which are administered as preventive sprays, that is by the application of pesticides according to the calendar and without reference to current pest density. If we are to answer the question that forms the title of this chapter our main objective here must be to examine the likelihood that integrated control (as defined in the preceding section) will, by its progressive adoption, modify this situation to a significant extent. Accordingly we must look at the causes of the present situation, the probability of their changing, and the probability of such changes occurring soon enough to influence trends in agricultural practice.

The events and trends that have shaped existing patterns of pest control, at least in regions where intensive cultivation is practiced, are well known. Since man first began to select certain plants for cultivation he has, through the practice of agriculture, been steadily simplifying ecosystems. Working against the normal course of ecological succession, he has been holding ecosystems back in stages that are characterized by a high ratio of productivity to biomass and by an increased susceptibility to wide-amplitude fluctuations of constituent populations (E. P. Odum, 1971). Man's ability to maintain ecosystems in this condition has increased greatly during the last 50 years as he has made increasing use of energy derived from fossil fuels. Thus the stage has been reached when, in the opinion of Reichle and Auerbach (1972), "expansion of simplified man-made agricultural ecosystems has renewed the question of whether or not these systems are increasingly vulnerable to biological insults and increasingly costly to maintain in terms of direct energy expenditures or requirements for energy-dependent products such as fertilizers and insecticides."

The objective of any system of pest control, briefly, is to replace, or improve upon, the ecological regulating mechanisms that previously prevented wide-amplitude fluctuations by substitutes of man's devising. In articulating this objective we assume a priori that such substitutes will not cause long-term harm to the environment and will be based primarily on an understanding of a pest's ecology and behavior rather than on methods requiring a fossil-fuel energy subsidy (see Watt, 1973b).

When the prospects of advances within a discipline are being assessed, it must be remembered that social and economic factors outside that discipline may greatly influence and sometimes determine its progress. Geier (1970) has stressed the relevance of this consideration to pest control, an activity in which far-reaching decisions are often made for political or economic, rather than for ecological or entomological reasons.

Thus, if the prospects of integrated control are to be assessed realistically, factors outside as well as within the discipline must be considered.

Progress *within* the discipline, at a given level of funding, will depend largely on the approach that entomologists are adopting toward the solution of problems, and thus where research effort is being concentrated. As Rothman and Woodhead (1968) show, this tends to be toward the *newest* approaches. Though there is general agreement that the development of an integrated control program requires the exercise of judgment and the selective acquisition of much ecological information, opinions vary considerably regarding the extent to which mathematical and statistical techniques are needed to exploit this information. At opposite ends of the spectrum of opinion we see positions that, respectively, may be called the "quantitative" and the "qualitative" view. To so designate them involves an oversimplification, since each position necessarily contains a combination of both approaches; nevertheless, the attitudes they represent are sufficiently different that the future of integrated control will certainly be influenced according to which predominates.

The *quantitative* view holds that, in general, the ecosystems in which integrated control is needed are susceptible to detailed analysis to an extent that the relations between their main components can be understood and the interactions of these components predicted. That is to say: by a series of formal steps a mathematical model can eventually be prepared that simulates the quantitative interactions of the main entities of the ecosystem; and such a model can then be used to predict the result of changing the action of certain of these entities and so make possible pest management by conscious, orderly manipulation based on a prior knowledge of the outcome. To some ecologists such an approach constitutes an essential part of *systems analysis*, which is therefore seen not only as "the use of scientific method with conscious regard for the complexity of the object of study" but also as an application "distinguished by the use of advanced mathematical and statistical techniques and by the use of computers" (Dale, 1970). In its latter, more restricted sense, and as it applies to ecological problems, systems analysis comprises the four successive phases that characterize problem solving.

1. The lexical phase, in which a choice is made (by an ecologist rather than a mathematician) of the entities that compose the system.

2. The parsing phase, in which the relationships between selected entities are defined.

3. The modeling phase, in which an attempt is made to specify, by means of a model, the mechanisms by which changes in the system take place.

4. The analysis phase, which involves solution of the model, in some sense, and validation of the model outputs by comparing them to the real system outputs (Dale, 1970).

Two considerations, in particular, have encouraged the view that systems analysis, pursued in this way, will come to play a predominant role in the solution of ecological problems.

First, there is a growing recognition that the organization of even the simplest ecosystem is formidably complex (see Pimentel, 1966), so that now few people would disagree with Handler's statement (1970) that "the general problem of ecosystem analyses is, with the exception of sociological problems that are also ecological, the most difficult problem ever posed by man." Indeed, Watt (1966) sees this complexity of ecological processes as the central motive for using systems analysis.

Second, the advent of digital computers, able to process information many times more rapidly than was possible before, would seem to provide (for the first time) the tools whereby nonlinear interactions involving many variables can be simulated and thus by degrees the regulatory mechanisms in an ecosystem understood.

With these considerations in mind, some ecologists predict that systems analysis will contribute importantly to integrated control by enabling pest management strategies to be based on a quantitative understanding of the key components of an ecosystem. Others go further and contend that systems analysis will provide the future framework of all major pest control programs.

The *qualitative* view (in its most extreme form) contends that ecosystems are too complex to be analyzed *and* simulated, at least to an extent that would have practical utility in devising control programs in the near future. Thus there are practicing entomologists who maintain that successful control programs have owed (and will owe) their success primarily to a combination of several talents and circumstances: the intuitive judgment of experienced and perceptive ecologists; the determination or verification of key relationships or interactions by field observation; hard work; and luck (or, more properly, chance!). Different facets of this view find expression in the writings of several biologists. Krebs (1972), summarizing known cases where introduction of natural enemies has led to control of a target pest, remarks "We cannot adequately account for even one of the successes, nor can we explain why failure is so common. Biological control will remain an art until we can do this." From the vantage point of 30 years' experience in the field of biological control, Simmonds (1972) writes as follows:

> The introduction of a natural enemy into a new area is in fact, the final and crucial experiment and the outcome has certainly to date never been absolutely predictable, except in so far as harmful repercussions can be guarded against. For all the careful selection of promising species for introduction, and for all the knowledge of the ecological requirements of the behaviour of the selected species, it may, on introduction into a new area with a somewhat different climate, flora, and "ecology" react in an unpredictable way, as has been seen in the past on many occasions when carefully studied promising biological control agents have failed to live up to expectations, whereas apparently unlikely species have been very successful.

Since the two approaches that we have identified are seldom discrete, and thus seldom reflect the extreme viewpoints represented here, discussion of their merits should not contain the implication that they are necessarily *alternative* strategies but one should rather look at situations in which one or the other approach might usefully be given greater emphasis.

It might be contended that, if promise is to be gauged by past experience, the qualitative approach offers the better prospect, although this could be no more than an expression of its far longer history! However, we have already noted above that successful integrated-control programs are generally characterized by their simplicity; and, indeed, it would be unrealistic not to acknowledge the formidable problems of human organization and comprehension associated with the management of very complex systems. As Huffaker (1972) observes (of systems analysis), "No one has developed a clear case history of the use of such a system, even on a hypothetical input basis." This state of affairs is clearly not improved by the great difficulty that even leaders in the field have encountered in obtaining funds for the development of models of ecosystems of relevance to pest management (Watt, 1970). Moreover, when a large program based on the systems approach *does* receive realistic financial support, there is always a danger that this will be done at the expense of many smaller research programs, some of which may be urgently needed. Another consideration to be kept in mind when large programs are established is that they will have a strong tendency to be self-sustaining regardless of the progress they make; it is accordingly essential that each such program be planned with a built-in periodic review, that can provide for improvement and redirection as necessary.

It is in the planning and review that we must also recognize some of the far-reaching benefits that the systems approach has to offer which are not necessarily associated with "the use of advanced mathematical and statistical techniques." These are the benefits that accrue to any investigation or program from "the use of scientific method with conscious regard for the complexity of the object of study" and from "the precise statement of the problems and the discipline imposed by an ordered approach to the complexities of the real system" (Dale, 1970). Such practices are not, of course, peculiar to any one approach but are essential components of good science. The compelling virtue of systems analysis (as a formal approach) is that it makes these practices *unavoidable*. Thus it encourages investigators to focus on a common goal and on one of high priority, so that the available resources (always fewer than are needed) can be used more efficiently and the most important pest problems are attended to first. Because the systems approach requires the explicit statement of each participant's contribution to the common goal, there is a logical and useful flow of information which can only enhance the effectiveness of each component of the overall program. So although gifted scientists will continue to advance the science of pest management using a predominantly qualitative approach, there is no doubt that the systems viewpoint, mainly through its impact on planning and thinking, can have a highly beneficial effect on the future of any discipline (such as integrated control) that is concerned with complex systems. This said, it is important to recognize that the approaches that will be followed within the discipline of integrated control will be largely determined by factors *outside* the discipline; and these we shall now consider.

Figure 1 constitutes an attempt to identify the main components in the socioeconomic system of which the practice of pest control is a part and to illustrate some

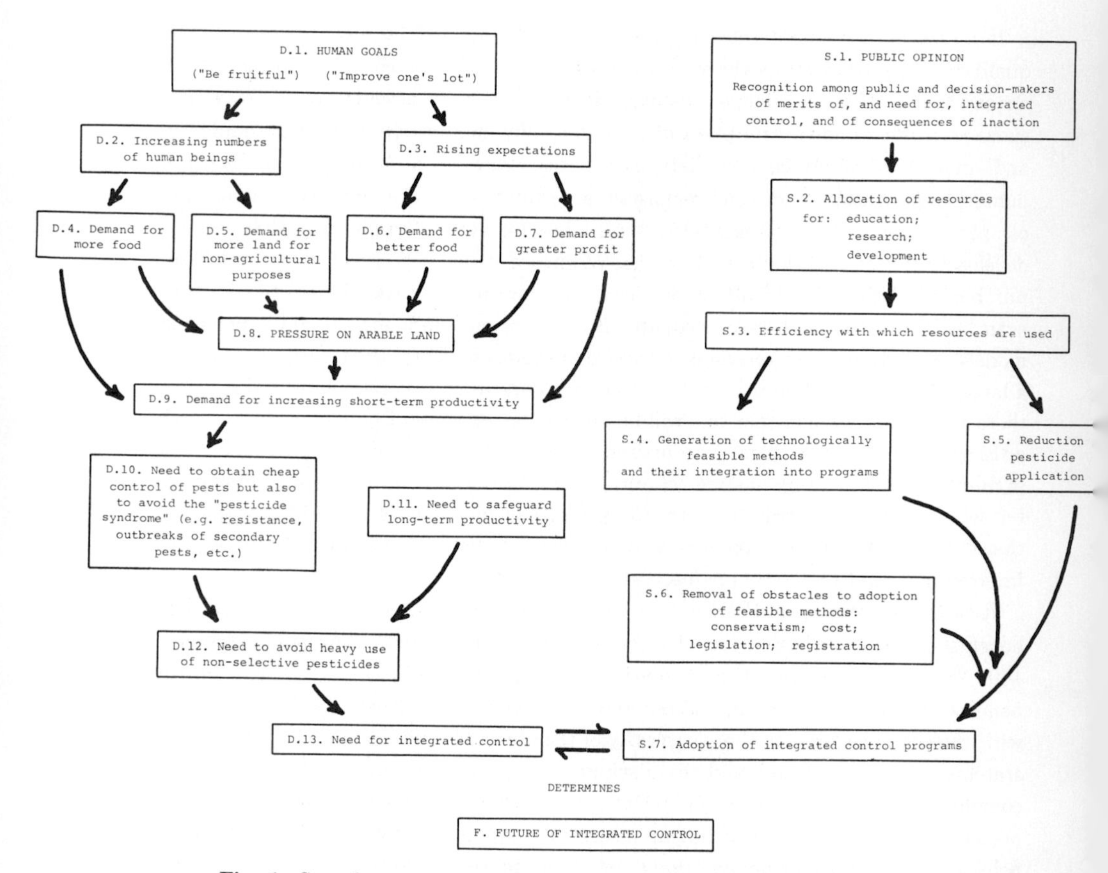

Fig. 1. Some key components that determine the prospects of integrated control.

of their interactions. This diagram is necessarily tentative and greatly simplified; it can, however, aid us in discussing some of the factors external to integrated control that are likely to determine its future.

The scheme illustrated applies only to agricultural pests. Comparable diagrams could be made to illustrate factors affecting integrated control of forest pests and of insect vectors of human disease, but neither of these categories has the same disproportionately great *short-term* relation to the major problems that man will encounter during the next few decades (see Corbet, 1970). (A scheme for disease vectors would, of course, expose the dilemma that resides in attempts to feed the world's growing human population, since success in the control of disease vectors contributes greatly to the growth of this population and thus, indirectly, to the increased severity of pest outbreaks on crops.)

In constructing the diagram we have assumed that the future of integrated control (F) will depend on the outcome of the interaction between factors that generate the need or *demand* for integrated control (D) and those that dictate the means for meeting this demand, i.e., the *supply* (S). Inasmuch as we wish S to equal or exceed D, factors designated "S" will be positive and those designated "D" will be negative. (Thus, within the restricted objective of improving the integrated control of crop pests, the suppression of disease vectors would receive the prefix "D".)

To make the main causal routes easier to follow in Figure 1, we have omitted the many feedback loops that connect them. Such loops are, of course, extremely important in affecting the rate at which processes occur. For example, any improvement in F (which reflects the relative strengths of D.13 and S.7) will generate a powerful positive feedback to S.1 and thus to all S factors, as increasing confidence in integrated control is generated among people who allocate resources (S.1), those who use them for research and development (S.2), and those who are in a position to accelerate the removal of obstacles that prevent adoption of integrated control programs (S.6). Equally important, an increasingly positive F will generate a positive feedback loop to D.10 and thus reduce the severity of several facets of the pesticide syndrome which makes the need for integrated control so pressing.

Time may be regarded as the most important single factor affecting the future of integrated control. Though not mentioned explicitly in Figure 1 (except perhaps in D.9 and D.11), it affects almost every pathway shown. Where pest management on food crops is concerned it is highly probable that time will prove to be the limiting resource (Corbet, 1970). As Watt (1973a) points out, the probability of an event occurring depends on the time available for it to occur!

Before summarizing the conclusions to be drawn from Figure 1 we shall comment on a few of the items and relationships it contains.

The fact that agriculture is an *industry* in most regions where integrated control is envisaged (D.7) has a large qualitative effect on the current attitude to land use (D.8,9) and to the economic acceptability of certain alternatives to pesticides (S.6); it combines with the exponential growth of human numbers (D.2) and the concomitant demand for food (D.4) to ensure that the "short-term" element identified in D.9 becomes progressively shorter. This, we may note, conflicts with a major objective of integrated control—to safeguard *long-term* yield (D.11).

The overriding importance of factors *outside* the discipline of pest management is clearly evident: only S.3–5 and perhaps aspects of D.10 and D.11 can be said to lie within it. The consequences of adopting a particular *approach* in research on integrated control will be encompassed by S.3.

Since there are many nonchemical methods of suppressing pests that, though known to be effective, have not been adopted, the importance of the factors listed in S.6 comes into sharp focus. A compelling, and not atypical, instance is provided by the methods now available for protecting tobacco in Ontario against its three major pests there: after many years of research, alternative methods of proved effectiveness have been developed for controlling all of them. That these methods are not being used is the result of, respectively, grower conservatism, difficulties of storing and marketing microbial insecticides, and the nonregistration of pathogens and (therefore) their high cost (see Corbet, 1973b). All such constraints are, of course, interdependent.

V. CONCLUSIONS

The analysis presented in Figure 1, albeit greatly simplified, emphasizes the great effect that current human demands have on the future of integrated control. So long as the pressure on arable land continues to grow, and so long as the goal of agriculture is first and foremost to increase short-term productivity, many growers will encounter strong incentives to continue practices (such as intensive monoculture or repeated, preventive application of nonselective chemicals) that make the adoption of integrated control increasingly necessary (in order to avoid progressive damage to the environment) and at the same time more difficult to implement (because of the diminished capacity of the agroecosystem to support natural enemies). It would seem then that the most rapid improvement in the prospects of integrated control is likely to come first, from the reduction of pressure on arable land; and second, from the orientation of agricultural policies toward improving sustainable yields rather than only increasing short-term profits.

If these two crucial changes could be effected, and agroecosystems could in this way be made receptive and responsive to integrated control practices, a determining question would then be: "What approaches and strategies *within* the discipline are most likely to be beneficial?"

Here we must keep in mind certain related considerations that are not evident in Figure 1. These concern the prospective availability of two finite resources: fossil fuels and mineral ores. As agriculture is now practiced, a capability to conduct coordinated research on pest management and particularly to test and develop certain manipulative techniques depends on a large subsidy from fossil fuels (which provide "cheap" energy) and mineral ores (which provide metals for machines). Since there exists a likelihood that both these resources will be critically depleted during the next 35 years or so (Cloud, 1969), we must anticipate the "postfossil fuel phase of agriculture" (Corbet, 1973a) and recognize that prospects for moving toward the goal of integrated control will be better the sooner the work can be done, and discontinuously improved if it can be accomplished during the next 35 years. This consideration becomes compelling if one reflects also that current agricultural *productivity* relies heavily on a subsidy from these two resources (H. T. Odum, 1971; Byerly, 1973; Pimentel *et al.*, 1973). Thus, unless the demand for food has diminished greatly by then, or substitutes for these two resources have become available, their withdrawal can only increase the pressure on arable land, encourage the use of broad-spectrum chemicals, and thus make integrated control more, rather than less, difficult to implement. Another consideration becomes evident from inspection of crop protection practices in countries which do not at present rely on large subsidies of this kind. The most feasible methods there are those that are simple to understand and apply, and that do not rely on an allochthonous input of expertise and materials. This point has recently been made persuasively by McClelland (1972) in relation to control of disease vectors, and by Watt (1973b) in relation to control of crop pests.

These facts would seem to show the approaches that are most likely to be rewarding within the discipline of pest management. Because time is a limiting resource, emphasis should be given to approaches that are known to be productive and environmentally acceptable, as distinct from those that, though intellectually appealing, remain untested. Although it is to be hoped that exponents of the discipline will always be receptive to new initiatives, the support that these receive should not be allowed to weaken programs based on approaches of known effectiveness. As Huffaker (1970) says "It could be catastrophic if disproportionate priority were given to new, relatively unproven *ideas*." Because of the need for integrated control, now or in the future, in the possible absence of heavy subsidies, attention should also be focused on methods that can be understood and applied effectively by indigenous personnel using locally made materials.

To summarize: in so far as a growing *need* will exist for integrated control, its prospects are excellent. Whether or not these prospects can be realized depends mainly on the rate at which pressure on arable land can be reduced which, in turn, will be influenced greatly by the speed with which the public, and legislators, can be made aware of the causal relationships shown in Figure 1. As such awareness comes to be reflected in formal policy (that is, as the needs identified in D.11–13 become *perceived* needs), a social and ecological environment may develop that can be responsive to the implementation of integrated control programs. Such programs, in turn, are most likely to be successful, and to benefit from the confidence and momentum that their success will generate, if effort is directed mainly to approaches that have already proved widely useful and are simple to apply. It is to be expected that biological control will play a prominent role among such approaches.

REFERENCES

Adkisson, P. L. (1973). The principles, strategies and tactics of pest control in cotton. *In* "Insects: Studies in Population Management" (P. W. Geier, L. R. Clark, D. J. Anderson, and H. A. Nix, eds.), pp. 274–283. Ecol. Soc. Australia (Mem. 1), Canberra.

Bottrell, D. G. (1973). Development of principles for managing insect populations in the cotton ecosystem: Texas. *Proc. Beltwide Cotton Conf. Phoenix, January, 1973.*

Bravenboer, L. (ed.). (1974). Integrated control in glasshouses. *Bull. West. Palearctic Reg. Sect. Int. Organ. Biol. Contr. 1973–1974,* 73 pp.

Burnett, T. (1949). The effect of temperature on an insect host-parasite population. *Ecology* **30**, 113–134.

Burnett, T. (1967). Aspects of the interaction between a chalcid parasite and its aleyrodid host. *Can. J. Zool.* **45**, 539–578.

Byerly, T. C. (1973). Towards abundance and good environments. *Ann. N.Y. Acad. Sci.* **217**, 238–243.

Cate, J. R., Bottrell, D. G., and Teetes, G. L. (1973a). Management of the greenbug on grain sorghum. I. Testing foliar treatments of insecticides against greenbugs and corn leaf aphids. *J. Econ. Entomol.* **66**, 945–951.

Cate, J. R., Bottrell, D. G., and Teetes, G. L. (1973b). Management of the greenbug on grain sorghum. II. Testing seed and soil treatments for greenbug and corn leaf aphid control. *J. Econ. Entomol.* **66**, 952–959.

Cate, J. R., Bottrell, D. G., and Teetes, G. L. (1976). Management of the greenbug on grain sorghum. III. Integration of insecticides and biological control. *Entomophaga*, in press.

Cloud, P. (ed.). (1969). "Resources and Man," 259 pp. Freeman, San Francisco, California.

Corbet, P. S. (1970). Pest management: objectives and prospects on a global scale. *In* "Concepts of Pest Management" (R. L. Rabb and F. E. Guthrie, eds.), pp. 191–204. North Carolina State University, Raleigh, North Carolina.

Corbet, P. S. (1973a). Application, feasibility and prospects of integrated control. *In* "Insects: Studies in Population Management" (P. W. Geier, L. R. Clark, D. J. Anderson, and H. A. Nix, eds.), pp. 185–195. Ecol. Soc. Australia (Mem. 1), Canberra.

Corbet, P. S. (1973b). Habitat manipulation in the control of insect pests in Canada. *Proc. Tall Timbers Conf. Ecol. Anim. Contr. Habitat Manage.* **5**, 147–171.

Dale, M. B. (1970). Systems analysis and ecology. *Ecology* **51**, 2–16.

Ellis, H. E., Ganyard, M. C., Singletary, H. M., and Robertson, R. L. (1972). *North Carolina Tobacco Pest Manage. 2nd Annu. Rep.*, 67 pp. North Carolina State University, Raleigh, North Carolina.

FAO (1968). *Rep. 1st Session FAO Panel Experts Integrated Pest Control*, 19 pp. FAO of the United Nations, Rome.

FAO (1971). "Integrated Pest Control," 27 pp. FAO of the United Nations, Rome.

Falcon, L. A., and Smith, R. F. (1973). Guidelines for integrated control of cotton insect pests. *FAO Publ. AGPP MISC* **8**, 92 pp.

Geier, P. W. (1970). Temporary suppression, population management, or eradication: how and when to choose. *In* "Concepts of Pest Management" (R. L. Rabb and F. E. Guthrie, eds.), pp. 170–189. North Carolina State University, Raleigh, North Carolina.

Gonzalez, R. H. (1973). Integrated control strategies on deciduous fruit trees. *FAO Plant Prot. Bull.* **21**, 56–64.

Handler, P. (1970). "Biology and the Future of Man," 471 pp. Oxford University Press, Oxford.

Huffaker, C. B. (1970). Summary of a pest management conference—a critique. *In* "Concepts of Pest Management" (R. L. Rabb and F. E. Guthrie, eds.), pp. 227–242. North Carolina State University, Raleigh, North Carolina.

Huffaker, C. B. (ed.). (1971). "Biological Control," 511 pp. Plenum, New York.

Huffaker, C. B. (1972). Ecological management of pest systems. *In* "Challenging Biological Problems. Directions towards their Solution" (J. A. Behnke, ed.), pp. 313–342. Oxford University Press, New York.

Hussey, N. W., and Bravenboer, L. (1971). Control of pests in glasshouse culture by the introduction of natural enemies. *In* "Biological Control" (C. B. Huffaker, ed.), pp. 195–216. Plenum, New York.

Kennedy, J. S. (1968). The motivation of integrated control. *J. Appl. Ecol.* **4**, 492–499.

Krebs, C. J. (1972). "Ecology," 694 pp. Harper and Row, New York.

Le Roux, E. J., Paradis, R. O., and Shreve, F. (1963). Major mortality factors in the population dynamics of the eye-spotted bud moth, the pistol case-bearer, the fruit-tree leafroller, and the European corn borer. *Mem. Entomol. Soc. Can.* **32**, 67–82.

McClanahan, R. J. (1971). *Trialeurodes vaporariorum* (Westwood), greenhouse whitefly (Homoptera: Aleyrodidae). *Commonw. Inst. Biol. Contr. Tech. Commun.* **4**, 57–59.

McClanahan, R. J. (1972). Integrated control of the greenhouse whitefly. *Can. Dep. Agr. Ottawa Publ.* **1469**, 1–7.

McClanahan, R. J. (1973). Integrated control of greenhouse pests. *Can. Agr.* **18**, 34–35.

McClelland, G. A. H. (1972). Control of *Stegomyia*: achievements and prospects. Unpublished Contribution to Symposium, 14th Int. Cong. Entomol., Canberra.

McLeod, J. H. (1938). Control of the greenhouse whitefly in Canada by the parasite *Encarsia formosa* Gahan. *Sci. Agr.* **18**, 529–535.

McLeod, J. H. (1962). Biological control of pests of crops, fruit trees, ornamentals, and weeds in Canada up to 1959. *Commonw. Inst. Biol. Control Tech. Commun.* **2**, 1–33.

MacPhee, A. W., and MacLellan, C. R. (1971a). Cases of naturally-occurring biological control in Canada. *In* "Biological Control" (C. B. Huffaker, ed.), pp. 312–328. Plenum, New York.

MacPhee, A. W., and MacLellan, C. R. (1971b). Ecology of apple orchard fauna and of integrated pest control in Nova Scotia. *Proc. Tall Timbers Conf. Ecol. Anim. Contr. Habitat Manage.* **3**, 197–208.

Odum, E. P. (1971). "Fundamentals of Ecology," 574 pp. Saunders, Philadelphia, Pennsylvania.

Odum, H. T. (1971). "Environment, Power and Society," 331 pp. Wiley, New York.

Pate, T. L., and Neeb, C. W. (1972). Unpublished report to Department of Entomology, Texas A & M University, College Station, Texas.

Pate, T. L., Hefner, J. J., and Neeb, C. W. (1972). A management program to reduce cost of cotton insect control in the Pecos area. *Texas Agr. Exp. Sta. Misc. Publ.* **1023**.

Pickett, A. D., Patterson, N. A., Stultz, H. T., and Lord, F. T. (1946). The influence of spray programs on the fauna of apple orchards in Nova Scotia. I. An appraisal of the problem and a method of approach. *Sci. Agr.* **26**, 590–600.

Pimentel, D. (1966). Complexity of ecological systems and problems in their study and management. *In* "Systems Analysis in Ecology" (K. E. F. Watt, ed.), pp. 15–35. Academic Press, New York.

Pimentel, D., Hurd, L. E., Bellotti, A. C., Forster, M. J., Oka, I. N., Sholes, O. D., and Whitman, R. J. (1973). Food production and the energy crisis. *Science*, **182**, 443–449.

Reichle, D. E., and Auerbach, S. I. (1972). Analysis of ecosystems. *In* "Challenging Biological Problems. Directions towards their Solution" (J. A. Behnke, ed.), pp. 260–280. Oxford University Press, New York.

Rothman, H., and Woodhead, M. (1968). Publication trends in biological control. *Nature (London)* **220**, 1053–1054.

Simmonds, F. J. (1972). Approaches to biological control problems. *Entomophaga* **17**, 251–264.

Smith, R. F. (1968). Recent developments in integrated control. *Pestic. News Sum.* **A14**, 201–206.

Smith, R. F. (1969). The new and the old in pest control. *Proc. Accad. Naz. Lincei, Rome* **366**, 21–30.

Smith, R. F., and Falcon, L. A. (1973). Insect control for cotton in California. *Cotton Grow. Rev.* **50**, 15–27.

Smith, R. F., and Reynolds, H. T. (1966). Principles, definitions and scope of integrated control. *Proc. FAO Symp. Integrated Pest Control Rome* **1**, 11–17.

Smith, R. F., and Reynolds, H. T. (1972). Effects of manipulation of cotton agro-ecosystems on insect pest populations. *In* "The Careless Technology" (M. T. Farvar and J. P. Milton, eds.), pp. 373–406. Natural History Press, New York.

Smith, R. F., and van den Bosch, R. (1967). Integrated control. *In* "Pest Control" (W. W. Kilgore and R. L. Doutt, eds.), pp. 295–340. Academic Press, New York.

Southwood, T. R. E., and Way, M. J. (1970). Ecological background to pest management. *In* "Concepts of Pest Management" (R. L. Rabb and F. E. Guthrie, eds.), pp. 6–29. North Carolina State University, Raleigh, North Carolina.

Steiner, H. (1973). Cost-benefit analysis in orchards where integrated control is practised. *OEPP/EPPO Bull.* **3**, 27–36.

van den Bosch, R., Leigh, T. F., Falcon, L. A., Stern, V. M., Gonzales, D., and Hagen, K. S. (1971). The developing program of integrated control of cotton pests in California. *In* "Biological Control" (C. B. Huffaker, ed.), pp. 377–394. Plenum, New York.

Watt, K. E. F. (ed.), (1966). The nature of systems analysis. *In* "Systems Analysis in Ecology," pp. 1–14. Academic Press, New York.

Watt, K. E. F. (1970). The systems point of view in pest management. *In* "Concepts of Pest Management" (R. L. Rabb and F. E. Guthrie, eds.), pp. 71–79 and discussion on p. 81. North Carolina State University, Raleigh, North Carolina.

Watt, K. E. F. (1973a). "The Principles of Environmental Science," 319 pp. McGraw-Hill, New York.

Watt, K. E. F. (1973b). The goals and means of resource management. *In* "Insects: Studies in Population Management" (P. W. Geier, L. R. Clark, D. J. Anderson, and H. A. Nix, eds.), pp. 45–51. Ecol. Soc. Australia (Mem. 1), Canberra.

Wildbolz, T., and Meier, W. (1973). Integrated control: critical assessment of case histories in affluent economies. *In* "Insects: Studies in Population Management" (P. W. Geier, L. R. Clark, D. J. Anderson, and H. A. Nix, eds.), pp. 221–231. Ecol. Soc. Australia (Mem. 1), Canberra.

Wood, B. J. (1973). Integrated control: critical assessment of case histories in developing economies. *In* "Insects: Studies in Population Management" (P. W. Geier, L. R. Clark, D. J. Anderson, and H. A. Nix, eds.), pp. 196–220. Ecol. Soc. Australia (Mem. 1), Canberra.

SECTION VI

APPENDIX

28

BIOLOGICAL CONTROL OF INSECT PESTS AND WEEDS BY IMPORTED PARASITES, PREDATORS, AND PATHOGENS

J. E. Laing and Junji Hamai

TABLE A-1

Scientific name	Common name	Crop	Location of infestation	Natural enemy: Scientific name
Plants (weeds)				
Alternanthera phylloxeroides (Mart.) Griseb.	alligator weed	waterways, canals	U.S.A. (southern)	*Agasicles hygrophila* Selman and Vogt
			U.S.A. (southern Florida)	*Vogtia malloi* Pastrana
Cirsium arvense (L.) Scop.	Canada thistle, creeping thistle		Canada (Ontario)	*Ceutorrhynchus litura* (Fab.)
Clidemia hirta (L.) D.Don	Koster's curse	rangeland	Fiji	*Liothrips urichi* Karny
			U.S.A. (Hawaii)	*Liothrips urichi* Karny
Cordia macrostachya (Jacq.) R. & S.	black sage	grass-lands aloe	Mauritius	*Schematiza cordiae* Barb.
				Eurytoma attiva Burks.
Emex spinosa (L.) Campd.	emex	pasture	U.S.A. (Hawaii)	*Apion antiquum* Gyll.
Emex australis (Steinh.)	spiny emex	pasture	U.S.A. [Hawaii (Kauai)]	*Apion antiquum* Gyll.
Eupatorium adenophorum Spreng	pamakani	rangeland	U.S.A. (Hawaii)	*Procecidochares utilis* (Stone)
	Crofton weed,		Australia and New Zealand	*Procecidochares utilis* (Stone)
Hypericum perforatum L.	Klamath weed, or St. Johnswort	range and arable land	U.S.A. (western)	*Chrysolina quadrigemina* (Suffr.)
				Chrysolina hyperici (Forst.)
				Agrilus hyperici (Creut.)
				Zeuxidiplosis giardi (Kieff.)
			New Zealand	*Chrysolina hyperici* (Forst.)
				Zeuxidiplosis giardi (Kieff.)

Biological Control Projects, Worldwide

Natural enemy				
Order Family	Type	Origin	Degree of success [a]	References
Col Chr		So. America, 1964	S	van den Bosch and Messenger, 1973; Andres, this vol., Chapter 19
Lep			P-S	Andres, personal communication
Col Cur		Europe, 1965–1967	P	Int. Org. Biol. Control Newsletter (March, 1972)
Thy P		Trinidad, 1930	C	Sweetman, 1958
Thy P		Trinidad	P	NAS, 1968
Col Chr		West Indies	C	Sweetman, 1958; van den Bosch and Messenger, 1973
Hy Eur		Trinidad	C	Simmonds, 1969
Col Cur		South Africa	S	Andres, this vol., Chapter 19
Col Cur		South Africa	C	Andres, this vol., Chapter 19
Dip Tr		Mexico prior to 1950	C	Sweetman, 1958; van den Bosch and Messenger, 1973
Dip Tr			P	Clausen, in press
Col Chr		France via Australia, 1944	C	Sweetman, 1958; DeBach, 1964, 1974
Col Chr		England via Australia, 1944	P (until displaced)	
Col Bup		France, 1950	P	
Dip Cec		France, 1950	P (until displaced)	
Col Chr		England via Australia, 1943	P-S	
Dip Cec			P-S	

TABLE A-1 *(continued)*

Scientific name	Common name	Crop	Location of infestation	Natural enemy: Scientific name
			Australia	*Chrysolina quadrigemina* (Suffr.)
				Chrysolina hyperici (Forst.)
				Agrilus hyperici (Creut.)
				Zeuxidiplosis giardi (Kieff.)
			Chile	*Chrysolina quadrigemina* (Suffr.)
				Chrysolina hyperici (Forst.)
			Canada (British Columbia)	*Chrysolina quadrigemina* (Suffr.)
				Chrysolina hyperici (Forst.)
Lantana camara L.	lantana		U.S.A. (Hawaii)	*Telonemia scrupulosa* Stål.
				Catabena esula (Druce)
				Syngamia haemorrhoidalis Guenée
				Hypena strigata F.
				Plagiohammus spinipennis (Thom.)
				Octotoma scabripennis (Guer.)
				Uroplata girardi Pic
				Epinotia lantana Busck
				Ophiomyia lantanae Frog.
				Orthezia insignis Dougl.
			Fiji	*Teleonemia scrupulosa* Stal.
			Australia (Norfolk Islands)	*Teleonemia scrupulosa* Stal.
			Uganda (Sereve)	*Teleonemia scrupulosa* Stal.

Natural enemy				
Order Family	Type	Origin	Degree of success [a]	References
Col Chr		France, 1937–1939	P	
Col Ch		England, 1934	P	
Col Brup		France, 1939–1940	P	
Dip Cec		France 1950 via U.S.A. (California), 1953	P	
Col Chr		U.S.A. (California), 1952	S	van den Bosch and Messenger, 1973
Col Chr				
Col Chr		U.S.A., 1950–1952	C	Andres, this vol., Chapter 19
Col Chr		U.S.A., 1950–1952	C	
Hemip Ting		Mexico, 1902	P-S	Andres and Goeden, 1971
Lep Noc		U.S.A. (California), 1954	P-S	
Lep Pyraus		U.S.A. (Florida), Cuba, 1954	P-S	
Lep Noc		Kenya and Rhodesia, 1975	P-S	
Col Cer		Mexico, 1953–1965	P-S	Clausen, in press
Col Chr		Mexico, 1953–1959	P-S	
Col Chr		Brazil, 1961	P-S	
Lep Ole		Mexico, 1902	S-C	
Dip Ag		Mexico, 1902	S-C	
Ho Coc			S-C	Andres, this vol., Chapter 19
Hemip Ti			P	Clausen, in press
Hemip Ti			C	Sweetman, 1958
Hemip Ti			S	Int. Org. Biol. Control Newsletter March, 1972

TABLE A-1 *(continued)*

				Natural enemy
Scientific name	Common name	Crop	Location of infestation	Scientific name
Leptospermum scoparium Forst.	manuka weed		New Zealand	*Eriococcus orariensis* Hoy
Linaria vulgaris Mill.	toadflax		Canada and U.S.A.	*Gymnaetron antirrhini* Paykull
				Brachypteroius pulicarius (L.)
Opuntia aurantiaca Lindl.	tiger pear		Australia	*Dactylopius* sp. near *confusus* Ckll.
			South Africa	*Cactoblastis cactorum* (Berg.)
Opuntia dillenii (Ker. Gawl.) Haw.	prickly pear		New Caledonia	*Cactoblastis cactorum* (Berg.)
			South India, Ceylon	*Dactylopius opuntiae* Licht. (= *tomentosus* Lam.)
			East Africa	*Dactylopius opuntiae* Licht.
Opuntia elatior Mill.			South India	*Dactylopius opuntiae* Licht.
			Ceylon	*Dactylopius opuntiae* Licht.
			Celebes	*Dactylopius opuntiae* Licht.
Opuntia imbricata (Haw.) D. C.	walking stick, cholla		Australia	*Dactylopius newsteadi* Ckll.
Opuntia inermis D. C.	prickly pear		Australia	*Cactoblastis cactorum* (Berg.)
				Chelinidea tabulata (Burm.)
				Dactylopius opuntiae Licht.
				Tetranychus desertorum Banks
Opuntia littoralis (Engel) Ckll.			U.S.A. (California)	*Dactylopius opuntiae* Licht.
Opuntia megacantha Salm-Dyck	mission prickly pear		South Africa	*Dactylopius opuntiae* Licht.
				Cactoblastis cactorum (Berg.)
			U.S.A. (Hawaii)	*Dactylopius opuntiae* Licht.
				Cactoblastis cactorum (Berg.)

Natural enemy				
Order Family	Type	Origin	Degree of success [a]	References
Ho Co		Australia	S-C	Andres, this vol., Chapter 19
Col Cur		Europe	P	NAS, 1968
Col Nit			P	CIBC, 1971
Ho Co		Argentina, 1932–1935		NAS, 1968
Lep Phy		Argentina, 1933	S	Sweetman, 1958
Lep Phy		U.S.A. via Australia, 1932, 1933	S–C	Sweetman, 1958; NAS, 1968
Ho Co		U.S.A.	S	NAS, 1968
Ho Co		U.S.A. via South Africa	S	
Ho Co		U.S.A. via Australia	P	Sweetman, 1958; NAS, 1968
Ho Co			P	
Ho Co		U.S.A. (Texas)	P	
Ho Co			C	
Lep Phy		Argentina	C	van den Bosch and Messenger, 1973; NAS, 1968
He Cor		U.S.A. (Texas), 1921–1923		
Ho Co		U.S.A.		
Ar Tet		U.S.A.		
Ho Co		Mexico via U.S.A. (Hawaii), 1951	S	Clausen, in press
Ho Co		Mexico via Australia	P	NAS 1968
Ho Co		Australia, 1932		
Ho Co		Mexico via Australia	S–C	Sweetman, 1958
Lep Phy		Argentina via Australia, 1951	S–C	DeBach, 1964

TABLE A-1 *(continued)*

Scientific name	Common name	Crop	Location of infestation	Natural enemy: Scientific name
			Australia	*Dactylopius confusus* Ckll.
				Cactoblastis cactorum (Berg.)
				Archlagocheirus funestus Thom.
Opuntia nigricans Haw.			North Celebes	*Dactylopius opuntiae* Licht.
Opuntia oricola Phil.			U.S.A. (California)	*Dactylopius opuntiae* Licht.
Opuntia streptacantha Lem.	white spined pear	range, timber, arable land	Australia	*Dactylopius confusus* Ckll.
				Cactoblastis cactorum (Berg.)
				Archlagocheirus funestus Thom.
Opuntia stricta Haw.	spiny pest pear	range, timber, arable land	Australia	*Cactoblastis cactorum* (Berg.)
				Dactylopius opuntiae Licht.
				Chelinidea tabulata (Burm.)
				Tetranychus desertorum Banks
Opuntia tomentosa Salm-Dyck	velvety tree pear	scrub and timberland	Australia	*Dactylopius opuntiae* Ckll.
				Archlagocheirus funestus Thom.
				Cactoblastis cactorum (Berg.)
Opuntia triacantha Sweet			West Indies	*Cactoblastis cactorum* (Berg.)
Opuntia tuna Mill.	prickly pear		Mauritius	*Dactylopius opuntiae* Licht.
				Cactoblastis cactorum (Berg.)

Natural enemy				
Order Family	Type	Origin	Degree of success [a]	References
Ho Co		U.S.A. (Florida)	C	NAS, 1968
Lep Phy		Argentina	S	NAS, 1968
Col Cer		Mexico	P	NAS, 1968
Ho Co		Australia	C	Sweetman, 1958
Ho Co		Mexico via U.S.A. (Hawaii)	P	Andres and Goeden, 1971
Ho Co		Mexico, 1927 U.S.A. (Florida)	S–C	NAS, 1968
Lep Phy		Argentina	S–C	Sweetman, 1958
Col Cor		Mexico	S–C	Sweetman, 1958
Lep Phy		Argentina	C	NAS, 1968; Sweetman, 1958
Ho Co		U.S.A.	C	
Hemip Cor		Mexico	C	
Ar Tet			C	
Ho Co		U.S.A.	P–C	Sweetman, 1958; NAS, 1968
Col Cer		Mexico, 1935	P	
Lep Phy		Argentina	P	
Lep Phy		South America via South Africa prior to 1958	S	van den Bosch and Messenger, 1973; Simmonds, 1969
Ho Co		U.S.A. via Ceylon, 1927, 1928	S	Sweetman, 1958; NAS, 1968
Lep Phy		Argentina via South Africa	S	

TABLE A-1 *(continued)*

Scientific name	Common name	Crop	Location of infestation	Natural enemy: Scientific name
Opuntia vulgaris Mill.	smooth tree pear	range and arable land	Australia	*Dactylopius ceylonicus* Green
			South India	*Dactylopius ceylonicus* Green
			Sri Lanka	*Dactylopius ceylonicus* Green
			South Africa	*Dactylopius ceylonicus* Green
Senecio jacobaea L.	tansy ragwort	rangeland	U.S.A. (California, Oregon)	*Tyria jacobaeae* (L.)
			Canada	*Tyria jacobaeae* (L.)
Tribulus terrestris L.	puncture vine		U.S.A. (Hawaii, Kauai)	*Microlarinus lypriformis* (Woll.)
				Microlarinus lareynii (Jacquelin-Duval)
Tiribulus cistoides L.	Jamaica fever plant		U.S.A. (Hawaii, Kauai)	*Microlarinus lypriformis* (Woll.)
				Microlarinus lareynii (Jacquelin-Duval)
			St. Kitts	*Microlarinus lypriformis* (Woll.)
				Microlarinus lareynii (Jacquelin-Duval)
Ulex europaeus L.	gorse	rangeland	Mauritius	*Apion ulicis* Forst.
			New Zealand	*Apion ulicis* Forst.
			U.S.A. (Hawaii)	*Apion ulicis* Forst.
			U.S.A. (California)	*Apion ulicis* Forst.
ORTHOPTERA				
Gryllotalpa africana P. deBeau.	African mole cricket	sugarcane	U.S.A. (Hawaii)	*Larra luzonensis* Roh.
Nomadacris septemfasciata (Serv.)	Red locust		Mauritius	*Acridotheres tristis* (L.) (mynah bird)
Oxya chinensis (Thunb.)	Chinese grasshopper	sugarcane	U.S.A. (Hawaii)	*Scelio pembertoni* Timb.

Natural enemy				
Order Family	Type	Origin	Degree of success [a]	References
Ho Co		Brazil, 1795 via India; via Ceylon, 1913	S-C S-C S-C	Sweetman, 1958
Ho Co			S-C	
Lep Arct		Europe	S	van den Bosch and Messenger, 1973
Lep Arct		Europe	S	Int. Org. Biol. Control Newsletter, March, 1972
Col Curc		Italy	S-C	Andres, this vol., Chapter 19
Col Curc		Italy	S-C	van den Bosch and Messenger, 1973
Col Curc			S-C	van den Bosch and Messenger, 1973
Col Curc			S-C	Andres, this vol., Chapter 19
Col Curc		Europe via U.S.A. (Hawaii)	S	Simmonds, 1969
Col Curc		U.S.A. (California)		
Col Curc		Europe	C	Sweetman, 1958
Col Curc		England	P	Clausen, in press
Col Curc		England via New Zealand	P	Clausen, in press
Col Curc			P	Clausen, in press
Hy Sp	para	Philippines, 1925	C P	Sweetman, 1958; DeBach, 1974
Pa St	pred	India, 1762	C	Sweetman, 1958; DeBach, 1974
Hy Sce	para	Malay peninsula, 1930	S	DeBach, 1964

TABLE A-1 *(continued)*

Scientific name	Common name	Crop	Location of infestation	Natural enemy: Scientific name
Periplaneta americana (L.)	American cockroach	household pest	U.S.A. (Hawaii)	*Ampulex compressa* (F.)
				Dolichurus stantoni (Ashm.)
Periplaneta australasiae (F.)	Australian cockroach	household pest	U.S.A. (Hawaii)	*Ampulex compressa* (F.)
				Dolichurus stantoni (Ashm.)
Sexava nubila (Stål)	coconut grasshopper	coconut	Celebes	*Leefmansia bicolor* Waterst.
			Bismark Archipelago	*Leefmansia bicolor* Waterst.
Sexava spp.		coconut	New Guinea	*Leefmansia bicolor* Waterst.
				Doirania leefmansi Waterst.
DERMAPTERA				
Forficula auricularia (L.)	European earwig	gardens and households	Canada (British Columbia)	*Bigonicheta setipennis* Fall.
			U.S.A. (Washington)	*Bigonicheta setipennis* Fall.
HEMIPTERA				
Nezara viridula (L.)	green tomato bug or southern green stink bug	vegetables, ornamentals, fruits, grains	Australia	*Trissolcus basilis* (Woll.)
			U.S.A. (Hawaii)	*Trissolcus basalis* (Woll.)
				Trichopoda pennipes (Fabr.)
			New Zealand	*Trissolcus basalis* (Woll.)
HOMOPTERA				
Acyrthosiphon pisum (Harris)	pea aphid	alfalfa	U.S.A. (California)	*Aphidius smithi* Sharma & Rao

Natural enemy				
Order Family	Type	Origin	Degree of success [a]	References
Hy Amp	para	New Caledonia, 1940	P	DeBach, 1964
Hy Amp	para	Philippines, 1917	P	
Hy Amp	para	New Caledonia, 1940	P	DeBach, 1964
Hy Amp	para	Philippines, 1917	P	
Hy En	para	Amboina, 1924 (Indonesia)	P	DeBach, 1964
Hy En	para	Amboina, 1929–1933	P	
Hy En	para	Moluccas (Indonesia), 1965–1969	S-C	Bennett *et al.*, this vol., Chapter 15
Hy Trich	para	Moluccas (Indonesia)	S-C	
Dip Ta	para	France via U.S.A. (Oregon), 1934	P	DeBach, 1964
Dip Ta	para	Europe, 1931–1935	P	
Hy Scel	para	Egypt, 1933 Pakistan	C	DeBach, 1974
Hy Scel	para	West Indies and Australia	C	
Dip Ta	para	West Indies, 1961–1962		
Hy Scel	para	Australia, 1949	S-C	
Hy Br	para	India, 1958	S	DeBach, 1964

TABLE A-1 *(continued)*

Scientific name	Common name	Crop	Location of infestation	Natural enemy: Scientific name
Aleurocanthus spiniferus (Quaint.)	spiny blackfly or orange spiny whitefly	citrus	Japan	*Prospaltella smithi* Silv.
			Guam	*Prospaltella smithi* Silv.
				Amitus hesperidum Silv.
Aleurocanthus woglumi Ashby	citrus blackfly	citrus	Cuba	*Eretmocerus serius* Silv.
			Barbados	*Eretmocerus serius* Silv.
				Prospaltella opulenta Silv.
			Jamaica	*Prospaltella opulenta* Silv.
				Eretmocerus serius Silv.
			Seychelles	*Eretmocerus serius* Silv.
			South Africa	*Eretmocerus serius* Silv.
			East Africa	*Eretmocerus serius* Silv.
			Costa Rica	*Eretmocerus serius* Silv.
			Haiti	*Eretmocerus serius* Silv.
			Panama	*Eretmocerus serius* Silv.
			Mexico	*Amitus hesperidum* Silv.
			Mexico	*Prospaltella opulenta* Silv.
			Mexico	*Prospaltella clypealis* Silv.
			Kenya	*Eretmocerus serius* Silv.
				Prospaltella opulenta Silv.
			U.S.A. (Hawaii)	*Eretmocerus serius* Silv.
			El Salvador	*Prospaltella* sp.
			Mexico	*Prospaltella opulenta* Silv.

Natural enemy				
Order Family	Type	Origin	Degree of success [a]	References
Hy Aph	para	China, 1925	C	DeBach, 1964
Hy Aph	para	Mexico, 1952	C	Sweetman, 1958; DeBach, 1964
Hy Pl	para			
Hy Aph	para	Southeast Asia, 1930	C	DeBach, 1964
Hy Aph	para	Jamaica, 1955	C	Simmonds, 1969
Hy Aph	para	Mexico	C	Bennett *et al.*, this vol., Chapter 15
Hy Aph	para	Mexico	C	Simmonds, 1969;
Hy Aph	para	Asia via Cuba	C	Bennett *et al.*, this vol., Chapter 15
Hy Aph	para	Jamaica	C	Simmonds, 1969; Bennett *et al.*, this vol., Chapter 15
Hy Aph	para	Kenya, 1960	C	
Hy Aph	para		C	
Hy Aph	para	Asia via Cuba	C	
Hy Aph	para		C	
Hy Aph	para		C	DeBach, 1964
Hy Pl		India and Pakistan, 1949	C	
Hy Aph	para	India and Pakistan, 1949	C	
Hy Aph	para	India and Pakistan, 1949	C	
Hy Aph	para		S	Bennett *et al.*, this vol., Chapter 15
Hy Aph	para		S	
Hy Aph	para	Asia via Cuba	S	
Hy Aph	para	Mexico	C	
Hy Aph	para	Mexico	C	Quezada, 1974

TABLE A-1 *(continued)*

Scientific name	Common name	Crop	Location of infestation	Natural enemy: Scientific name
Aleurodicus cocois Curt.	coconut whitefly	coconuts and other palms	Barbados	*Prospaltella* sp.
Aleurothrixus floccosus (Mask.)	woolly whitefly	citrus	U.S.A. (California)	*Amitus spiniferus* Brethes
				Eretmocerus paulistus Hempl.
			Mexico (Baja Calif.)	*Amitus spiniferus* Brethes
				Eretmocerus paulistus Hempl.
Antonina graminis (Mask.)	Rhodes-grass scale	forage grasses	U.S.A. (Texas and Florida)	*Neodusmetia sangwani* (Rao)
			Bermuda	*Neodusmetia sangwani* (Rao)
Aonidiella aurantii (Mask.)	Calif. red scale	citrus	U.S.A. (Texas and California)	*Aphytis lingnanensis* Comp.
				Aphytis melinus DeBach
				Comperiella bifasciata How.
				Chilocorus confusor (Casey)
				Prospaltella perniciosi Tower
			Greece	*Aphytis melinus* DeBach
				Aphytis melinus DeBach
			Australia	*Aphytis chrysomphali* (Mercet)
				Aphytis chrysomphali (Mercet)
				Aphytis melinus DeBach
				Comperiella bifasciata How.
Aonidiella citrina trina (Coq.)	yellow scale	citrus	U.S.A. (Texas and California)	*Comperiella bifasciata* How.
			Australia (Victoria)	*Aphytis melinus* DeBach
				Comperiella bifasciata How.

Natural enemy				
Order Family	Type	Origin	Degree of success [a]	References
Hy Aph	para	Trinidad	P	DeBach, 1964
Hy Pl	para	Mexico, 1967	S	Bennett *et al.*, this vol., Chapter 15
Hy Aph	para	U.S.A. (Mexico and Florida), 1967, 1968, 1969	S	
Hy Pl	para	U.S.A. and Mexico, 1969	S	
Hy Aph	para		S	
Hy En	para	India, 1959	S	DeBach, 1974
Hy En	para	India, 1968	S-C	Simmonds, 1974
Hy Aph	para	Southern China, 1947	S	DeBach *et al.*, 1971
Hy Aph	para	India, Pakistan, 1956–1957	P	
Hy En	para	Orient, 1941	P	
Col Coc	pred		P	
Hy Aph	para	Formosa, 1949	P	
Hy Aph	para	India, Pakistan, 1956–1957; 1961–1964	S-C	Bennett *et al.* this vol., Chapter 15
Hy Aph	para		S-C	DeBach and Argyriou, 1967
Hy Aph	para	China, 1905	P	DeBach, 1964
Hy Aph	para	U.S.A. (California), 1945	S-C	Bennett *et al.* this vol., Chapter 15
Hy Aph	para	U.S.A. (California), 1961	S-C	
Hy En	para	U.S.A. (California), 1942–1947	S-C	
Hy En	para	Japan, 1924–1925	S	DeBach, 1964
Hy Aph	para	U.S.A. (California), 1946	C	Bennett *et al.*, this vol., Chapter 15
Hy En	para		C	

TABLE A-1 *(continued)*

				Natural enemy
Scientific name	Common name	Crop	Location of infestation	Scientific name
Longiunguis (=*Aphis*) *sacchari* (Zhnt.)	sugar-cane aphid	sugar-cane	U.S.A. (Hawaii)	A complex of predators and parasites
				Eumicromus navigatorum (Brauer)
				Lysiphlebus testaceipes (Cres.)
				Coleophora inaequalis (Fabr.)
Aspidiotus destructor Sign.	coconut scale	coconut and other palms	Fiji	*Cryptognatha nodiceps* Marsh.
			Mauritius	*Chilocorus politus* Muls.
				Chilocorus nigritus Muls.
			Portuguese West Africa (Principe)	*Cryptognatha nodiceps* Marsh.
			Bali	*Aspidiotiphagus citrinus* (Craw)
			Tuamoto, Wallis	*Cryptognatha nodiceps* Marsh.
				Rhizobius pulchellus Montr.
Asterolecanium bambusae (Bdvl.);	bamboo scales		Puerto Rico, Haiti	*Chilocorus cacti* (L.)
Asterolecanium militaris (Bdvl.)				*Cladis nitidula* (F.)
Asterolecanium pustulans (Ckll.)	bamboo (pustule) scale	forest shade trees, ornamentals, oak, fig	Puerto Rico	*Chilocorus cacti* (L.)
Asterolecanium variolosum (Ratz.)	golden oak scale	oak	New Zealand	*Habrolepis dalmani* (Westw.)
			Tasmania	*Habrolepis dalmani* (Westw.)
			Chile	*Habrolepis dalmani* (Westw.)
Brevicoryne brassicae (L.)	cabbage aphid	cabbage	Australia	Unnamed sp.
Cavariella aegopodii (Scop.)	carrot aphid	carrot	Australia and Tasmania	*Aphidius* sp.

Natural enemy				
Order Family	Type	Origin	Degree of success [a]	References
		Various, 1900–1923	S	DeBach, 1964
Neur Her	pred	Australia, 1919	S	Sweetman, 1958
Hy Br	para	U.S.A. (California), 1923	S	
Col Coc	pred	Australia, 1894	S	
Col Coc	pred	Trinidad, 1928	C	DeBach, 1964
Col Coc	pred	Java, 1937	C	
Col Coc	pred	Ceylon, 1939	C	
Col Coc	pred	Trinidad, 1955	C	
Hy Aph	para	Java, 1934	P	
Col Coc	pred	Fiji, Trinidad	P	Cochereau, 1972
Col Coc	pred	New Caledonia New Hebrides	P	Cochereau, 1972
Col Coc	pred	Cuba, U.S.A. (Texas), 1937	S	Sweetman, 1958
Col Coc	pred			
Col Coc	pred	Cuba, Louisiana, U.S.A. (Texas), 1937	P	DeBach, 1964
Hy En	para	North America	S	DeBach, 1964
Hy En	para	New Zealand, 1931–1933	P	
Hy En	para	North America	P	Clausen, in press
Hy	para	Ceylon, 1907	P	DeBach, 1964
Hy Br	para	U.S.A.	C	van den Bosch and Messenger, 1973

TABLE A-1 *(continued)*

Scientific name	Common name	Crop	Location of infestation	Natural enemy Scientific name
Ceroplastes cirripediformis Comstock	barnacle scale	passion fruit	U.S.A. (Hawaii)	*Coccidoxenus mexicanus* Girault
Ceroplastes rubens Mask.	red wax scale	citrus persimmon tea	Japan (Honshu, Shikoku)	*Anicetus beneficus* Ishii & Yasumatsu
Chloropulvinaria (Pulvinaria) psidii (Mask.)	guava mealybug	guava and others	Bermuda	*Microterys kotinskyi* (Feld.)
				Azya luticeps (Muls.)
				Cryptolaemus montrouzieri Muls.
Chromaphis juglandicola (Kalt)	walnut aphid	walnut	U.S.A. (California)	*Trioxys pallidus* (Hal.)
Chrysomphalus dictyospermi (Morgan)	dictyospermum scale	wide host range	Greece	*Aphytis melinus* DeBach
			U.S.A. (California)	*Aphytis melinus* DeBach
			Sicily	*Aphytis melinus* DeBach
Chrysomphalus aonidum (L.) [= *C. ficus* (Ash.)]	Florida red scale	citrus	Israel	*Aphytis holoxanthus* DeBach
			Seychelles	*Chilocorus nigritus* (F.)
			Mexico	*Aphytis holoxanthus* DeBach
Edwardsiana froggatti (Baker) (=*australis* Frogg.)	apple leafhopper	apple	Tasmania	*Anagrus armatus nigriventris* Grlt.
Eriococcus coriaceus Mask.	blue gum scale	eucalyptus	New Zealand	*Rhizobius ventralis* (Erich.)
Eriosoma lanigerum (Hausm.)	woolly apple aphid	apple	Shillong (Assam)	*Aphelinus mali* (Hald.)
			Argentina	
			Australia	
			Brazil	
			Canada (Brit. Columbia)	

Natural enemy				
Order Family	Type	Origin	Degree of success [a]	References
Hy En	para	Trinidad Station CIBC	S	Bennett *et al.* this vol., Chapter 15
Hy En	para	Japan (Kyushu), 1948	C	DeBach, 1964
Hy En	para	U.S.A. (Hawaii), 1954–1955	S-C	Bennett *et al.* this vol., Chapter 15
Col Coc	pred		S-C	
Col Coc	pred		S-C	
Hy Br	para	France, Iran, 1959, 1960	S-C	van den Bosch, 1971
Hy Aph	para	India, 1961	C	Bennett *et al.*, this vol., Chapter 15
Hy Aph	para	India, 1961	C	Franz and Krieg, 1972
Hy Aph	para	India, 1964	S	Bennett *et al.*, this vol., Chapter 15
Hy Aph	para	Hong Kong, 1956–1957	C	Harpaz and Rosen 1971; Bennett *et al.*, this vol.
Col Coc	pred	India, 1938	S	DeBach, 1964
Hy Aph	para		C	Bennett *et al.*, this vol., Chapter 15
Hy Mym	para	New Zealand, 1935	P	DeBach, 1964
Col Coc	pred	Australia, 1905	C	DeBach, 1964
Hy Aph	para		C	Simmonds, 1969
		Uruguay, 1922	C	DeBach, 1964
		New Zealand, 1922–1923	S	
		Uruguay, 1923	S	
		Canada (Ontario), 1921	C	

TABLE A-1 *(continued)*

				Natural enemy
Scientific name	Common name	Crop	Location of infestation	Scientific name
			Chile	
			Columbia	
			Costa Rica	
			Cyprus	
			Germany	
			Germany	
			Israel	
			Italy	
			Japan	
			Kenya	
			Malta	
			New Zealand	
			Peru	
			Poland	
			South Africa	
			Spain	
			Tasmania	
			U.S.A. (Pacific northwest)	
			U.S.A. (California)	
			Uruguay	
			U.S.S.R.	
			Venezuela	
			Yugoslavia	
Eucalymnatus tessellatus (Sign.)	coconut scale	coconut	Seychelles	*Chilocorus distigma* Klug.
				Chilocorus nigritus Muls.
				Exochomus ventralis Gerst.
				Exochomus flavipes (Thun.)
Eulecanium coryli (L.)	European fruit lecanium	forest trees, ornamentals	Canada (British Columbia)	*Blastothrix sericea* (Dalm.)
Eulecanium persicae (F.)	vine scale	grape, plum, etc.	Australia (Western)	*Aphycus timberlakei* Ishii

Natural enemy				
Order Family	Type	Origin	Degree of success [a]	References
		U.S.A. and Uruguay, 1921–1922	C	
		U.S.A., 1933	C	
		U.S.A., 1933, 1936	C	
		England	C	
		Uruguay, 1923	P	
		Italy, 1926–1933		
		Egypt, 1935	S	
		Uruguay, 1921	S	
		France, 1921–1923		
		U.S.A., 1931	C	
		England, 1927–1928	S	
		Italy, 1933–1934	S	
		Northwestern U.S., 1921		
		U.S.A., 1922	S	
		England, 1928	P	
		U.S.A., 1921–1922	S	
		Italy, Uruguay, 1926	P	
		Australia, 1924	S	
		U.S.A. (New England), 1929–1931	C	
		U.S.A. (Pacific northwest), 1935–1939	S	
		U.S.A., 1921	C	
		Italy, England, 1930	S	
		U.S.A., 1941	P	
		Italy, 1930	S	
Col	pred		S	Sweetman, 1958
Coc			S	
Col Coc	pred		S	
Col Coc	pred		S	
Hy En	para	England, 1928–1929	S	DeBach, 1964
Hy En	para	U.S.A. (California), 1907	S	DeBach, 1964

TABLE A-1 *(continued)*

				Natural enemy
Scientific name	Common name	Crop	Location of infestation	Scientific name
Icerya aegyptiaca (Dougl.)	Egyptian mealybug or fluted scale	bread, fruit, avocado, citrus	Caroline Islands Marianas, and Marshall Islands	*Rodolia pumila* (Weise)
		citrus, figs	India	*Rodolia cardinalis* (Muls.)
Icerya montserratensis Riley and Howard	fluted scale	citrus	Ecuador	*Rodolia cardinalis* (Muls.)
Icerya palmeri Riley and Howard	fluted scale	citrus	Chile	*Rodolia cardinalis* (Muls.)
Icerya purchasi Mask.	cottony-cushion scale	citrus, ornamentals	U.S.A. (California)	*Rodolia cardinalis* (Muls.)
				Cryptochetum iceryae (Will.)
			Argentina	*Rodolia cardinalis* (Muls.)
			Bahamas	
			Bermuda	
			Chile	
			Cyprus	
			Egypt	
			Greece	
			Guam	
			Hawaii	
			India	
			Israel	
			Italy	
			Japan	
			Madeira	
			Malta	
			Morocco	
			New Zealand	
			Peru	
			Portugal	
			Puerto Rico	
			South Africa	
			Spain	
			Tripoli	
			Tunisia	
			Turkey	

Natural enemy				
Order Family	Type	Origin	Degree of success [a]	References
Col Coc	pred	Mariana Islands, 1947–1949	S	DeBach, 1964
Col Coc	pred	U.S.A. (California)	C	Clausen, in press
Col Coc	pred	U.S.A., 1941	S	DeBach, 1964
Col Coc	pred	U.S.A. (California), 1931	P	DeBach, 1964
Col Coc	pred	Australia, 1888–1889	C	DeBach, 1964
Dip Agr	para	Australia, 1888–1889		
Col Coc	pred	Uruguay, 1932	C	
		U.S.A., 1924	S	
		U.S.A., 1899	S	
		U.S.A., 1931	C	
		Egypt, 1938	C	
		U.S.A., 1892	C	
		France, Italy, 1927	C	
		U.S.A. (Hawaii), 1926	C	
		U.S.A. (California), 1890	C	
		Australia, 1928	C	
		Italy, 1912	C	
		Australia, 1899	C	
		Formosa, 1910	C	
		U.S.A., 1898	C	
		Italy, 1928	C	
		France, 1921	C	
		U.S.A., 1892	C	
		U.S.A., 1932	C	
		U.S.A., 1898	C	
		U.S.A., 1932–1933	C	
		U.S.A., 1891	C	
		Portugal, 1922	C	
		Italy, 1920	S	
		France	C	
		Egypt, Palestine, 1932	C	

TABLE A-1 *(continued)*

Scientific name	Common name	Crop	Location of infestation	Natural enemy: Scientific name
			Uruguay U.S.A. (except California and Hawaii) U.S.S.R. Venezuela	
Icerya seychellarum (Westw.)	Seychelles scale	fruit and timber trees	Seychelles	*Rodolia cardinalis* (Muls.)
		bread-fruit, etc.	Gambier Islands (Manareva Tuamotu Amanu, Hao)	*Rodolia cardinalis* (Muls.)
			Mariana Islands Caroline Islands Marshall Islands	*Rodolia pumila* (Weise)
Ischnaspis longirostris (Sign)	black thread scale	coconut palm	Seychelles	*Chilocorus nigritus* (F.)
				Chilocorus distigma (Klug.)
Lecanium tiliae (L.)	lecanium scale	ornamental trees	Canada	*Blastothrix sericea* (Dalm.)
Lepidosaphes beckii (Newm.)	purple scale	citrus	U.S.A. (California) Chile Peru Brazil Cyprus Greece, Crete Fiji, New Caledonia Turkey Israel Louisiana Florida Puerto Rico Jamaica Guadeloupe El Salvador Argentine Hawaii Australia U.S.A. (Texas)	*Aphytis lepidosaphes* Comp.

Natural enemy				
Order Family	Type	Origin	Degree of success [a]	References
		Portugal, 1916	C	
		U.S.A. (California)	C	
		Egypt, 1931	S	
		U.S.A., 1941	C	
Col Coc	pred	Mauritius, 1939	P	DeBach, 1964
Col Coc	pred	New Hebrides, Fiji	P	Cochereau, 1972
Col Coc	pred	Japan, 1928	S	Sweetman, 1958
Col Coc	pred	India, 1938	S	DeBach, 1964
Col Coc	pred	East Africa, 1936	S	
Hy En	para	England	S	CIBC, 1962
Hy Aph	para	Southern China and Formosa, 1948–1949	P	DeBach, 1964
			S-C	Bennett *et al.*, this vol. Chapter 15
			S-C	Bennet *et al.*, this vol., Chapter 15
			S-C	DeBach, 1974
			C	Bennett *et al.*, this vol., Chapter 15
			S-C	DeBach, 1974
			S-C	Bennett *et al.* this vol., Chapter 15
			S-C	
			S-C	
			S-C	
			S-C	
			S-C	
			S-C	
			S-C	
			S-C	
			S-C	
			S-C	
			S-C	
		U.S.A. (California), 1952	C	DeBach, 1974

TABLE A-1 *(continued)*

				Natural enemy
Scientific name	Common name	Crop	Location of infestation	Scientific name
			Mexico	
Lepidosaphes ficus (Sign.)	fig scale	fig	U.S.A. (Calif.)	*Aphytis mytilaspidis* (LeB.)
Lepidosaphes ulmi (L.)	oystershell scale	deciduous fruits, ornamentals	Canada (Br. Columbia)	*Hemisarcoptes malus* (Schimer)
Myzocallis annulata (Hart.)	oak aphid	oak	Tasmania	*Aphelinus flavus* (Nees)
Nipaecoccus (= *Pseudococcus*) *nipae* (Mask.)	avocado mealybug or coconut mealybug in Bermuda	guava, avocado, fig, mulberry, citrus, palms	U.S.A. (Hawaii)	*Pseudaphycus utilis* Timb.
Nipaecoccus vastator (Mask.) [= *Pseudococcus filamentosus* (Ckll.)]	Lebbeck mealybug	Lebbeck shade tree	Egypt	*Anagyrus aegytiacus* Moursi
				Leptomastix phenacocci Comp.
		various plants	U.S.A. (Hawaii)	*Anagyrus dactylopii* (How.)
Orthezia insignis Dougl.	greenhouse orthezia	coffee, ornamentals	Kenya	*Hyperaspis jocosa* Muls.
Parlatoria oleae (Colvée)	olive scale	olive, deciduous fruit trees, ornamentals	U.S.A. (California)	*Aphytis maculicornis* (Masi)
				Coccophagoides utilis Doutt
Perkinsiella saccharicida Kirk.	sugarcane leafhopper	sugarcane	U.S.A. (Hawaii)	*Tytthus mundulus* (Bredd.) (= *Cyrtorhinus*)
Phenacoccus (= *Pseudococcus*) *aceris* (Sign.)	apple mealybug	apple	Canada (British Columbia)	*Allotropa utilis* Mues.
Maconellicoccus (= *Phenacoccus*) *hirsutus* (Green)	hibiscus mealybug	many ornamentals	Egypt	*Anagyrus kamali* Moursi
				Achrysopophagus sp.

Natural enemy				
Order Family	Type	Origin	Degree of success [a]	References
		U.S.A. (California), 1954–1956	C	
Hy Aph	para	France, 1949	P	DeBach, 1964
Acar	pred	Canada (New Brunswick), 1917	P	DeBach, 1964
Hy Aph	para	England, 1937–1938	P	DeBach, 1964
Hy En	para	Mexico, 1922	S	DeBach, 1964
Hy En	para	Java, 1934–1939	P	DeBach, 1964
Hy En	para	Java, 1934–1939	P	
Hy En	para	Hong Kong, 1925	P	
Col Coc	pred	U.S.A. (Hawaii), 1948	S	DeBach, 1964
Hy Aph	para	Iran, 1951	C	DeBach, 1964; Huffaker and Kennett, 1966
Hy Aph	para	Pakistan, 1957	C	
Hemip Mi	pred	Australia, 1920	C	DeBach, 1964
Hy Pl	para	Canada (Nova Scotia), 1938	C	DeBach, 1964
Hy En	para	Java, 1934–1939	C	Clausen, in press
Hy En	para			

TABLE A-1 *(continued)*

Scientific name	Common name	Crop	Location of infestation	Natural enemy: Scientific name
Phenacoccus iceryoides (Green)	mealybug	coffee	Celebes	*Cryptolaemus montrouzieri* Muls.
Pineus laevis (= *borneri*) (Ann.)	pine chermid	Monterey pine	New Zealand	*Neoleucopis obscura* (Hal.)
Pinnaspis buxi (Bouché)	pine chermid	coconut and other palms	U.S.A. (Hawaii)	*Telsimia nitida* Chapin
			Seychelles	*Chilocorus nigritus* (F.)
Pinnaspis minor (Mask.)	cotton white scale	cotton	Peru	(A complex) *Aspidiotiphagus citrinus* (Craw.)
				Arrhenophagus chionaspidis (Auri)
				Microweisia (*Scymnus*) sp.
Planococcus (= *Pseudococcus*) *citri* (Risso)	citrus mealybug	citrus	U.S.A. (California)	*Cryptolaemus montrouzieri* Muls.
				Leptomastidea abnormis (Grlt.)
		citrus	Chile	*Cryptolaemus montrouzieri* Muls.
				Leptomastidea abnormis (Grlt.)
		citrus	U.S.A. (Hawaii)	*Leptomastidea abnormis* (Grlt.)
			South Africa	*Cryptolaemus montrouzieri* Muls.
Planococcus (= *Pseudococcus*) *kenyae* (Le Pelley)	coffee mealybug	coffee	Kenya	*Anagyrus nr. kivuensis* (Comp.)
Pseudococcus comstocki (Kuw.)	Comstock mealybug	apple	U.S.A. (eastern)	*Allotropa burrelli* Muls.
				Pseudaphycus malinus Gah.
		apple and others	U.S.S.R. (Uzbekistan)	*Pseudaphycus malinus* Gah.
Pseudococcus fragilis Brian (= *gahani* Green) (= *Citrophilus* Clausen)	citrophilus mealybug	citrus	U.S.A. (California)	*Coccophagus gurneyi* Comp.
				Hungariella pretiosa (Timb.) (= *Tetracnemus*)

Natural enemy				
Order Family	Type	Origin	Degree of success [a]	References
Col Coc	pred	Java, 1920	S	DeBach, 1964
Dip Cha	pred	Europe, 1932–1934	P	DeBach, 1964
Col Coc	pred	Guam, 1936	S	DeBach, 1964
Col Coc	pred	India, 1938	S	
Hy Aph	para	Barbados, Italy, Japan, U.S.A., 1904–1912	P P	DeBach, 1964 Sweetman, 1958
Hy En	para		P	
Col Coc	pred		P	
Col Coc	pred	Australia, 1891–1892	P	Bennett *et al.*, this vol., Chapter 15
Hy En	para	Sicily, 1914	P	
Col Coc	pred	U.S.A. (California), 1931	P	
Hy En	para	U.S.A. (California), 1931	P	DeBach, 1964
Hy En	para	U.S.A. (California), 1915	P	DeBach, 1964
Col Coc	pred		C	Bennett *et al.*, this vol., Chapter 15
Hy En	para	Uganda, 1938	C	DeBach, 1964
Hy Pl	para	Japan, 1939–1941	C	DeBach, 1964
Hy En	para			
Hy En	para	U.S.A., 1945	P-C	Rubtsov, 1957; Kobakhidze, 1965
Hy Aph	para	Australia, 1928	C	DeBach, 1964
Hy En	para	Australia, 1928	C	

TABLE A-1 *(continued)*

Scientific name	Common name	Crop	Location of infestation	Natural enemy: Scientific name
			Chile	*Coccophagus gurneyi* Comp.
			U.S.S.R.	*Coccophagus gurneyi* Comp.
Pseudococcus maritimus (Ehrh.)	grape mealybug	pears, grapes	U.S.A. (California)	*Acerophagus notativentris* (Gir.)
Pseudococcus adonidum (L.) (= *longispinus*) (Targ.)	long-tailed mealybug	citrus, avocado, bread-fruit, etc.	U.S.A., (California)	*Anarhopus sydneyensis* Timb.
				Hungariella peregrina (Timb.) (= *Tetracnemus*)
				Cryptolaemus montrouzieri (Muls.)
			Bermuda	*Hungariella peregrina* (Timb.) (= *Tetracnemus*)
Pseudococcus citriculus Green	Green's mealybug	citrus	Israel	*Clausenia purpurea* Ishii
			Palestine	*Clausenia purpurea* Ishii
Pseudaulacaspis (= *Diaspis*) *pentagona* (Targ.)	white peach scale	mulberry	Italy	*Prospaltella berlesei* (How.)
		mulberry, papaya, etc.	Puerto Rico	*Chilocorus cacti* (L.)
		oleander	Bermuda	*Aphytis diaspidis* (How.)
		mulberry	U.S.S.R. (Georgian, S.S.R.)	*Prospatella berlesei* (How.)
Pseudococcus spp.	mealybugs	citrus	Australia (western)	*Cryptolaemus montrouzieri* Muls.
Pulvinaria psidii Mask.	green shield scale	shade trees	Puerto Rico	*Cryptolaemus montrouzieri* Muls.
			Bermuda	*Microterys kotinskyi* (Full.)
				Cryptolaemus montrouzieri Muls.
				Azya luticeps (Muls.)

Natural enemy				
Order Family	Type	Origin	Degree of success [a]	References
Hy Aph	para	U.S.A. (California), 1936	S	
Hy Aph	para	U.S.A. (California), 1960	C	Simmonds, 1969
Hy En	para	1943	S	Sweetman, 1958
Hy En	para	Australia, 1933	P-S	DeBach, 1964, Bennett, *et al.*, this vol., Chapter 15
Hy En	para	Argentina, 1934		
Col Coc	pred			
Hy En	para	U.S.A (California), 1951	P	DeBach, 1964
Hy En	para	Japan, 1939	C	Bennett *et al.*, this vol., Chapter 15
Hy En	para	Japan, 1940	C	
Hy Aph	para	U.S.A., 1905	S	DeBach, 1964
Col Coc	pred	Cuba, 1938	S	
Hy Aph	para	Italy, 1924	S	DeBach, 1964
Hy Aph	para	Italy	S	van den Bosch and Messenger, 1973
Col Coc	pred	New South Wales, 1902 (Australia)	S	DeBach, 1964
Col Coc	pred	U.S.A. (California), 1911	P	DeBach, 1964
Hy En	para	U.S.A. (Hawaii and California), 1953–1955	C	DeBach, 1964; Simmonds, 1969
Col Coc	pred	U.S.A. (Hawaii and California), 1956	C	
Col Coc		U.S.A. (Hawaii and California), 1957	C	

TABLE A-1 *(continued)*

Scientific name	Common name	Crop	Location of infestation	Natural enemy
				Scientific name
Quadraspidiotus perniciosus (Comst.)	San Jose scale	deciduous fruit trees	U.S.A. (California)	*Prospaltella perniciosi* Tow.
			India (Kumaon region of Uttar Pradesh, and Kashmir)	*Prospaltella perniciosi* Tow.
Saccharicoccus (= *Trionymus*) *sacchari* (Ckll.)	pink sugarcane mealybug	sugarcane	U.S.A. (Hawaii)	*Anagyrus saccharicola* Timb.
Saissetia coffeae (Walker)	hemispherical scale		Peru	*Metaphycus helvolus* (Comp.)
Saissetia oleae (Olivier)	black scale	citrus, olive, ornamentals	U.S.A. (California), Australia	*Metaphycus helvolus* (Comp.)
				Scutellista cyanea Mots.
				Metaphycus lounsburyi (How.)
			Chile	*Metaphycus helvolus* (Comp.)
			Peru	*Metaphycus lounsburyi* (How.)
				Metaphycus helvolus (Comp.)
				Lecaniobius utilis Comp.
				Scutellista cyanea Mots.
			Greece	*Metaphycus helvolus* (Comp.)
Saissetia nigra (Nietn.)	nigra scale	ornamentals	U.S.A. (California)	*Metaphycus helvolus* (Comp.)
Siphanta acuta (Wlk.)	torpedo bug plant hopper	coffee, mango, citrus, others	U.S.A. (Hawaii)	*Aphanomerus pusillus* Perkins

Natural enemy				
Order Family	Type	Origin	Degree of success [a]	References
Hy Aph	para	U.S.A. (Georgia), 1933	P	DeBach, 1964
Hy Aph	para	China, U.S.A., U.S.S.R., 1960	S-C	Simmonds, 1969
Hy En	para	Philippines, 1930	S	DeBach, 1964
Hy En	para		C	Bennett *et al.*, this vol., Chapter 15
Hy En	para	South Africa, 1937	S	DeBach, 1964
Hy Pter	para	U.S.A. (California), 1904	S-C	
Hy En	para	South Africa, 1902	S-C	
Hy En	para	Peru, 1946	P	
Hy En	para	U.S.A. (California), 1936, 1961	S	
Hy En	para		S	
Hy Eup	para		S	
Hy Pter	para		S	
Hy En	para	U.S.A. (California), 1962	P	Franz and Krieg, 1972
Hy En	para	South Africa, 1937	S	DeBach, 1964
Hy Scel.	para	Australia, 1904, 1928	S	

TABLE A-1 *(continued)*

Scientific name	Common name	Crop	Location of infestation	Natural enemy: Scientific name
Steatococcus samaraius (Westw.)	Fluted scale	breadfruit	Palau Islands	*Rodolia pumila* Weise
Tarophagus (= *Megamellus*) *proserpina* (Kirk.)	taro leaf-hopper	taro	U.S.A. (Hawaii)	*Cyrtorhinus fulvus* Knight
			Guam	*Cyrtorhinus fulvus* Knight
			Ponape	*Cyrtorhinus fulvus* Knight
			Tahiti	*Cyrtorhinus fulvus* Knight
			Wallis, Futuna	*Cyrtorhinus fulvus* Knight
Therioaphis trifolii (= *maculata*) (Monell)	spotted alfalfa aphid	alfalfa	U.S.A. (California)	A complex of native predators, disease, and the imported parasites:
				Praon exsoletum Nees
				Trioxys complanatus Quilis
				Aphelinus asychis Walker
Trialeurodes vaporariorum (Westw.)	green-house whitefly	tomato, etc.	Australia	*Encarsia formosa* Gah.
			Tasmania	*Encarsia formosa* Gah.
			Canada	*Encarsia formosa* Gah.
COLEOPTERA				
Adoretus sinicus Burm.	Chinese rose beetle	many plants	U.S.A. (Hawaii)	*Composomeris marginella modesta* Smith
				Tiphia segregata Cwfd.
Anomala orientalis Waterh.	oriental beetle	sugar-cane	U.S.A. (Hawaii)	*Composomeris marginella modesta* Smith *Tiphia segregata* Cwfd.
Anomala sulcatula Burm.	oriental beetle	sugar-cane	Mariana Islands (Saipan)	*Composomeris annulata* Fabr.

Natural enemy				
Order Family	Type	Origin	Degree of success [a]	References
Col Coc	pred	Formosa, 1928	S	Sweetman, 1958
Hemip Mi	pred	Philippines	S	DeBach, 1964
Hemip Mi	pred	Hawaii, 1947	S	
Hemip Mi	pred	Guam, 1947	P	
Hemip Mi	pred	Hawaii, Philippines	P	Cochereau, 1972
Hemip Mi	pred	Samoa, Carolines, Guam	P	
			S	DeBach, 1964
Hy Br	para	Middle East, 1955–1956	S	
Hy Br	para	Middle East, 1955–1956		
Hy Aph	para	Middle East, 1955–1956		
Hy Aph	para	New Zealand, 1934	S	DeBach, 1964
Hy Aph	para	New Zealand, 1934	S	
Hy Aph	para	New Zealand, 1934	P	
Hy Scol	para	Philippines, 1916–1917	P	DeBach, 1964
Hy Ti	para	Philippines, 1916–1917	P	
Hy Scol	para	Philippines, 1916–1917	S	DeBach, 1964
Hy Ti	para	Philippines, 1916–1917	S	
Hy Scol	para	Philippines, 1940	C	DeBach, 1964

TABLE A-1 *(continued)*

Scientific name	Common name	Crop	Location of infestation	Natural enemy: Scientific name
Brontispa longissima (Gestro.) (=*froggatti* Sharp)	coconut leaf miner	coconut	Celebes	*Tetrastichus brontispae* (Ferr.)
			Russell Islands	*Tetrastichus brontispae* (Ferr.)
Brontispa mariana Spaeth	Mariana coconut beetle	coconut	Mariana Islands	*Tetrastichus brontispae* (Ferr.)
Cosmopolites sordidus (Germ.)	banana root borer (corn weevil)	banana	Fiji	*Plaesius javanus* (Erich.)
			Jamaica	*Plaesius javanus* (Erich.)
Criocerus asparagi (L.)	asparagus beetle	asparagus	U.S.A. (Washington)	*Tetrastichus asparagi* Cwfd.
Dysmicoccus boninsis (Kuw.)	grey sugarcane mealybug	sugarcane	U.S.A. (Hawaii)	*Aphycus terryi* (Full.)
				Cryptolaemus montrouzieri Muls.
Epilachna philippinensis (Dke.)	Philippine lady-beetle	various solanaceous vegetables	Guam	*Pediobius (Pleurotropis) epilachnae* (Roh.)
Phyrrhalta luteola (Muller) [=*Galerucella xanthomelaena* (Schr.)]	elm leaf beetle	elm	U.S.A. (California)	*Erynniopsis rondanii* Towns
				Tetrastichus brevistigma Gah.
Gonipterus scutellatus Gyll.	eucalyptus snout-beetle	eucalyptus	South Africa	*Patasson nitens* (Grlt.)
			New Zealand	*Patasson nitens* (Grlt.)
			Mauritius	*Patasson nitens* (Grlt.)
			Kenya	*Patasson nitens* (Grlt.)

Natural enemy				
Order Family	Type	Origin	Degree of success [a]	References
Hy Eu	para	Java, 1929	S	DeBach, 1964
Hy Eu	para		S	Bennett *et al.*, this vol., Chapter 15
Hy Eu	para	Java, Malaya, 1948	S-C	Bennett *et al.*, this vol., Chapter 15
Col Hi	pred	Java, 1913–1918	P-S	DeBach, 1964; Bennett *et al.*, this vol., Chapter 15
Col Hi	pred	Fiji, 1942	P	DeBach, 1964
Hy Eu	para	U.S.A. (Ohio), 1937	P	DeBach, 1964
Hy En	para	—	C	Sweetman, 1958
Col Coc	pred	—		
Hy Eu	para	Philippines, 1954	P C	DeBach, 1964; Clausen, in press
Dip Tach	para	France, 1939	P	DeBach, 1964
Hy Eu		U.S.A. (eastern), 1934	P	
Hy Mym	para	Australia, 1926	C	DeBach, 1964
Hy Mym	para	Australia, 1927	S	
Hy Mym	para	East Africa, 1946	C	
Hy Mym	para		S	

TABLE A-1 *(continued)*

Scientific name	Common name	Crop	Location of infestation	Natural enemy: Scientific name
			Madagascar	*Patasson nitens* (Grlt.)
Hypera postica (Gyll.)	alfalfa weevil	alfalfa	U.S.A. [California, Utah, eastern U.S.A. (in part)]	*Bathyplectes curculionis* (Thom.) *Microctonus aethiops* (Ness)
Oryctes tarandus Ol.	rhinocerus beetle	sugarcane	Mauritius	*Scolia oryctophaga* Coq.
Phyllophaga portoricensis (Smyth)	white grub	sugarcane	Puerto Rico	*Bufo marinus* (a toad)
Popillia japonica (Newm.)	Japanese beetle	turf, pasture, fruits	U.S.A. (eastern)	*Tiphia vernalis* Roh.
				Bacillus popillae Dutky
Promecotheca coeruleipennis (Blanch.) (=*reichei* Baly)	coconut leaf mining beetle	coconut	Fiji	*Pediobius parvulus* (Ferr.) (=*Pleurotropis*)
Promecotheca papuana Cziki	leaf miner	coconut	New Guinea, Bismark Archipelago	*Pediobius parvulus* (Ferr.)
Promecotheca sp.		coconut, other palms	Africa	*Pediobus parvulus* (Ferr.)
			Ceylon	*Pediobius parvulus* (Ferr.)
Rhabdoscelus obscurus (Bdv.)	New Guinea sugarcane weevil	sugar cane	U.S.A. (Hawaii)	*Lixophaga* (=*Microceromasia*) *sphenophori* (Vill.)
Syagrius fulvitarsis Pasc.	fern weevil	tree ferns	U.S.A. (Hawaii)	*Doryctes syagrii* Full.
LEPIDOPTERA				
Bedellia orchilella Walsingh.	sweet potato leaf miner	sweet potato	U.S.A. (Hawaii)	*Apanteles bedelliae* Vier.

Natural enemy				
Order Family	Type	Origin	Degree of success [a]	References
Hy Mym	para		S	
Hy Ich	para	Italy, 1911–1913	S-C	DeBach, 1964
Hy Br	para	Europe, 1958–1960	S	van den Bosch, 1971
Hy Sco	para	Madagascar, 1917	S	DeBach, 1964
An Bu	pred		S-C	Sweetman, 1958
Hy Tip	para	Korea, South China, 1924–1933	P	DeBach, 1964
	Disease			Sweetman, 1958
Hy Eu	para	Java, 1933	C	Bennett *et al.*, this vol., Chapter 15
Hy Eu	para	Fiji, 1938 Java, 1940	S	DeBach, 1964
Hy Eu	para	Fiji, New Hebrides	P-C	Cochereau, 1972
Hy	para	Fiji, New Hebrides	P-C	
Dip Tach	para	New Guinea, 1910	S	DeBach, 1964; Long and Hensley, 1972
Hy Br	para	Australia, 1921	P	DeBach, 1964
Hy Br	para	U.S.A. (Kansas), 1945	P	DeBach, 1964

TABLE A-1 *(continued)*

Scientific name	Common name	Crop	Location of infestation	Natural enemy
				Scientific name
Chilo infuscatellus Snellen	borer		Taiwan	*Lixophaga diatraeae* (Towns.)
Chilo sacchariphagus Boj.	sugarcane borer	sugarcane	Madagascar	*Diatreophaga striatalus* Tus.
Chilo suppressalis (Wlk.)	asiatic rice borer	rice	U.S.A. (Hawaii)	*Trichogramma japonicum* Ashm.
				Amyosoma chilonis Vier.
				Dioctes chilonis Cush.
Cnidocampa flavescens (Wlk.)	oriental moth	shade trees	U.S.A. (Massachusetts)	*Chaetexorista javana* B. and B.
Coleophora alcyonipennella (Koll.)	clover casebORER		New Zealand (Nelson Prov.)	*Bracon variegator* (Spin.)
Coleophora laricella (Hbn.)	larch casebORER	larch	U.S.A. (northeast) Canada	*Agathis pumila* (Ratz.)
Coleophora malivorella Riley	pistol casebORER	apple	Canada	*Chrysocharis laricinellae* (Ratz.)
Diatraea mauricella Wlk.	borer	sugarcane	Mauritius	*Xanthopimpla* sp.
Diatraea saccharalis (F.)	sugarcane borer	sugarcane	West Indies (Antigua; St. Kitts)	*Lixophaga diatraeae* (Towns.)
			St. Lucia	*Metagonistylum minense* Towns.
			Dominica	*Paratheresia claripalpis* (Wulp)
			Guadeloupe	*Paratheresia claripalpis* (Wulp)
				Lixophaga diatraeae (Towns.)
				Metagonistylum minense Towns.
			U.S.A. (Florida, Louisiana)	*Lixophaga diatraeae* (Towns.)
				Agathis stigmaterus (Cress.)

Natural enemy				
Order Family	Type	Origin	Degree of success [a]	References
Dip Ta	para	—	P	Chen and Hung, 1962
Dip Ta	para	—	P	Brenierre *et al.*, 1966
Hy Tr	para	Japan, Formosa	P	DeBach, 1964
Hy Br	para	China, 1928	P	
Hy Ich	para		P	
Dip Ta	para	Japan, 1929–1930	S	DeBach, 1964
Hy Br	para	European Sta., CIBC	S	CIBC, 1970
Hy Br	para	England, 1932–1937	S	CIBC, 1971
Hy Eu	para	England, 1939	S	CIBC, 1971
Hy Ich	para		S	Vinson, 1942
Dip Ta	para		S	DeBach, 1964
Dip Ta	para		S	
Dip Ta	para		S	
Dip Ta	para	Peru	C	Simmonds, 1960
Dip Ta	para		S	DeBach, 1964
Dip Ta	para		S	
Dip Ta	para	Cuba, 1915	P	
Hy Br	para	Peru, 1929	P	

TABLE A-1 *(continued)*

Scientific name	Common name	Crop	Location of infestation	Natural enemy Scientific name
			British Guiana,	*Metagonistylum minense* Towns.
			Barbados	*Apanteles flavipes* Cam.
			Virgin Islands (St. Croix)	*Lixophaga diatraeae* (Towns.)
Etiella zinckenella (Treit.)	lima-bean pod borer	pigeon peas	Mauritius	*Eiphosoma annulatum* Cress.
				Bracon thurberiphagae (Mues.)
				Bracon canji Mues.
				Apanteles etiellae Vier.
				Phanerotoma bennetti Meus.
				Perisierola sp.
				Apanteles flavipes (Cam.)
				Lixophaga diatraeae (Towns.)
Grapholitha (*Laspeyresia*) *molesta* (Busck.)	oriental fruit moth	peach	U.S.A. (other than New Jersey)	*Macrocentrus ancylivorus* Roh.
			Canada	*Macrocentrus ancylivorus* Roh.
Harrisinia brillians B. & McD.	western grape leaf skeletonizer	grape	U.S.A. (California)	*Apanteles harrisinae* Mues.
				Sturmia harrisinae Coq.
Homona coffearia Nietn.	tea tortrix	tea	Ceylon	*Macrocentrus homonae* Nixon
Laphygma exempta (Wlk.)	nutgrass army-worm	sugar-cane, pasture grasses	U.S.A. (Hawaii)	A complex
				Apanteles marginiventris (Cres.)
				Meteorus laphygmae (Vier.)
				Telenomus nawai (Ash.)
				Eucelatoria armigera (Coq.)

Natural enemy				
Order Family	Type	Origin	Degree of success [a]	References
Dip Ta	para		C	Cleare, 1939
Hy Br	para		C	Alam *et al.*, 1971
Dip Ta	para		P	Miskimen, 1962
	para	Trinidad, 1951 or 1952	P	Simmonds, 1969
Hy Br	para		P	
Hy Br	para		P	
Hy Br	para		C	
Hy Br	para		C	
Hy Bet	para		C	
Hy Br	para	Indian Sta., CIBC, 1966	C	Simmonds, 1970
Dip Ta	para		C	
Hy Br	para	U.S.A. (New Jersey)	P	DeBach, 1964
Hy Br	para	U.S.A. (New Jersey), 1929–1930	S	
Hy Br	para	U.S.A. (Arizona), 1950–1952	S	DeBach, 1964
			S	
Dip Ta	para			
Hy Br	para	Java, 1935–1936	C	DeBach, 1964
Hy Br	para	U.S.A. (California, Texas)	P	DeBach, 1964
Hy Br	para	Mexico	P	
Hy Sc	para		P	
Dip Ta			P	

TABLE A-1 *(continued)*

Scientific name	Common name	Crop	Location of infestation	Natural enemy: Scientific name
Laspeyresia nigricana (Steph.)	pea moth	legumes	Canada (British Columbia)	*Ascogaster quadridentata* (Wesm.)
				Glypta haesitator Grav.
Levuana iridescens B. B.	coconut moth	coconut	Fiji	*Bessa* (=*Ptychomyia*) *remota* Ald.
Lithocolletis messaniella Zell.	oakleaf miner	oak	New Zealand	*Apanteles circumscriptus* (Nees)
Macoleria octasema Meyr.	banana scab moth	banana	Fiji	*Chelonus striatigenas* Cam.
Malacosoma neustria L.			U.S.S.R.	*Teleas* (=*Telenomus*) *laeviusculus* Ratz.
				Telenomus ovulorum Bouché
Nygmia phaeorrhoea (Donov.)	brown-tail moth	decid. forest and shade trees	U.S.A. (northeast)	*Apanteles lacteicolor* Vier.
				Townsendiellomyia nidicola (Towns.)
			Canada	*Compsilura concinnata* (Mg.)
				Apanteles lacteicolor Vier.
				Meteorus versicolor (Wesm.)
Porthetria (=*Lymantria*) *dispar* (L.)	gypsy moth	forest and shade trees	U.S.A. (New England states)	*Compsilura concinnata* (Mg.)
				Blepharipoda scutellata (R-D)
				Apanteles melanoscelus (Ratz.) and others
			Spain	*Ooencyrtus kuwanai* (How.)
Proceras sacchariphagus Bojer	spotted borer	sugarcane	Taiwan	*Isotima javensis* (Rohw.)
Pseudaletia (=*Cirphis*) *unipuncta* (Haw.)	armyworm	sugarcane, etc.	U.S.A. (Hawaii)	A complex
				Euplectrus plathypenae How.
				Archytas cirphis (Cur.)

Natural enemy			Degree of success [a]	References
Order Family	Type	Origin		
Hy Br	para	England, 1937–1939	S	DeBach, 1964
Hy Ich	para		S	
Dip Ta	para	Malaya, 1925	C	DeBach, 1964
Hy Br	para	Europe, 1958–1959	C	Simmonds, 1969
Hy Br	para	Indonesia, 1960	P	Bennett *et al.*, this vol., Chapter 15
Hy Sc	para		S-C	Waters *et al.*, this vol., Chapter 13
Hy Sc	para		S-C	
Hy Br	para	Europe, 1905–1911	S	DeBach, 1964
Dip Ta	para	Europe, 1905–1911	S	
Dip Ta	para	Europe via U.S.A. (eastern), 1912–1915	C	
Hy Br	para		C	
Hy Br	para		C	
Dip Ta	para	Europe, 1905–1914	P	DeBach, 1964
Dip Ta	para	Japan, 1922–1923	P	
Hy Br	para		P	
Hy Enc	para		P	Franz and Krieg, 1972
Hy Ich	para	India, 1961–1963	P	Chen and Hung, 1962
Hy Eu	para para	Mexico, 1923–1924	P P	DeBach, 1964 Sweetman, 1958
Dip Ta	para		S	Sweetman, 1958

TABLE A-1 *(continued)*

Scientific name	Common name	Crop	Location of infestation	Natural enemy
				Scientific name
				Apanteles militaris Walsh
Ostrinia (=*Pyrausta*) *nubilalis* (Hbn.)	European corn borer	corn	U.S.A.	*Lydella stabulans grisescens* R-D
				Macrocentrus gifuensis Ashm. and others
Rhyacionia frustrana bushnelli (Busck.)	Nantucket pine tip moth	pine trees	U.S.A. (Nebraska)	*Campoplex frustranae* Cush.
Scirpophaga nivella F.	sugar-cane top borer	sugar-cane	Taiwan	*Isotima javensis* (Roh.)
Stilpnotia salicis (L.)	satin moth	forest trees	U.S.A. (New England, Pacific northwest regions)	*Apanteles solitarius* (Ratz.)
				Meteorus versicolor (Wesm.)
			Canada (British Columbia)	*Apanteles solitarius* (Ratz.)
				Meteorus versicolor (Wesm.)
				Compsilura concinnata (Mieg.)
			Maritime Prov.	*Apanteles solitarius* (Ratz.)
				Compsilura concinnata (Meig.)
Tirathaba complexa (Butler)	coconut spike moth	coconut	Fiji	*Apanteles tirathabae* Walk.
				Argyrophylax basifulva (Bezzi)
				Telenomus tirathabae Ferr.
				Venturia palmaris (Wlkn.)
Operophtera brumata (L.)	winter moth	forest shade and fruit trees	Canada (Nova Scotia)	*Cyzenis albicans* (Fall.)
				Agrypon flaveolatum (Grav.)
			Cyprus	*Agathis unicolor* (Schr.)
				Apanteles subandinus Blanch.

Natural enemy				
Order Family	Type	Origin	Degree of success [a]	References
Hy Br	para	U.S.A. (California), 1960	S	Clausen, in press
Dip Ta	para	Japan, France	P	DeBach, 1964
Hy Br	para	Italy, Korea, 1920–1928	P	DeBach, 1964
Hy Ich	para	U.S.A. (Virginia), 1925	P	DeBach, 1964
Hy Ich	para	India, 1962	P-S	Simmonds, 1969
Hy Br	para	Europe, 1927–1934	S	DeBach, 1964
Hy Br	para		S	
Hy Br	para	Canada (New Brunswick)	S	
Hy Br	para	U.S.A. (Massachusetts), 1929–1934	S	
Dip Ta	para		S	
Hy Br	para	U.S.A., 1933	S	CIBC, 1971
Dip Ta	para	U.S.A. 1912	P	DeBach, 1964
Hy Br	para	Java, 1930–1934	P	DeBach, 1964
Dip Ta	para		P	
Hy Sc	para		P	DeBach, 1964
Hy Ich	para		P	Clausen, in press
Dip Ta	para	Europe, 1955	C	van den Bosch and Messenger, 1973
Hy Ich	para		C	Franz and Krieg, 1972
Hy Br	para	South America (via India CIBC)	P	Simmonds, 1969
Hy Br	para		P	

TABLE A-1 *(continued)*

Scientific name	Common name	Crop	Location of infestation	Natural enemy Scientific name
				Copidosoma koehleri Blanch.
				Campoplex haywardi Blanch.
				Temelucha sp.
				Orgilus lepidus Mues.
Phthorimaea (=*Gnori-moschema*) *operculella* (Zell.)	potato tuber moth	potato	Cyprus	*Apanteles scutellaris* Mues.
			India (Mysore)	*Copidosoma koehleri* Blanch.
				Campoplex haywardi Blanch.
				Orgilus lepidus Mues.
				Orgilus parcus Turner
			Australia (Victoria)	*Copidosoma koehleri* Blanch.
			New South Wales (South Queensland)	*Apanteles subandinus* Blanch.
				Orgilus lepidus Mues.
				Campoplex haywardi Blanch.
Pieris rapae (L.)	imported cabbage worm	cruci-ferous crops	New Zealand	*Pteromalus puparum* (L.)
				Apanteles glomeratus (L.)
			Australia	*Pteromalus puparum* (L.)
				Apanteles glomeratus (L.)
			Tasmania	*Pteromalus puparum* (L.)
				Apanteles glomeratus (L.)
Plutella maculi-pennis (Curt.)	diamond-back moth	cruci-ferous crops	New Zealand	*Diadegma cerophaga* Grav.
			Australia	*Diadegma cerophaga* Grav.

Natural enemy				
Order Family	Type	Origin	Degree of success [a]	References
Hy En	para		P	
Hy Ich	para		P	
Hy Ich	para		P	
Hy Br	para		P	
Hy Br	para	North America	P	Simmonds, 1969
Hy En	para	South America	P	
Hy Ich	para		P	
Hy Br	para		P	
Hy Br	para		P	
Hy Br	para	U.S.A. (California)	P	
Hy Br	para	South America	P	
Hy Br	para	South America	P	
Hy Ich			P	
Hy Pter	para	North America, 1933	S	Franz and Krieg, 1972
Hy Br	para	Europe	P	DeBach, 1964
Hy Pter	para	New Zealand, 1941	P	
Hy Br	para	Canada, 1942	P	
Hy Pter	para	New Zealand, 1942	P	
Hy Br		Australia, 1949	P	
Hy Ich	para	England, 1936–1937	S	DeBach, 1964
Hy Ich	para	New Zealand, 1947	P	

TABLE A-1 *(continued)*

Scientific name	Common name	Crop	Location of infestation	Natural enemy: Scientific name
			Tasmania	*Diadegma cerophaga* Grav.
			Indonesia	*Diadegma cerophaga* Grav.
DIPTERA				
Anastrepha ludens Low.	Mexican fruit fly	citrus, mango	Mexico, U.S.A. (southern Texas, southern Calif.)	*Opius longicaudatus* (Ash.)
				Syntomosphyrum indicum Silv.
				Pachycrepoideus vindemmiae (Rond.)
Anastrepha suspensa (Loew)	West Indian fruit fly	citrus, mango	U.S.A. (Florida)	*Parachasma (Opius) cereum* (Gahan)
Ceratitis capitata (Wied.)	Mediterranean fruit fly	many fruits	U.S.A. (Hawaii)	*Opius tryoni* Cam.
				Opius fullawayi Silv.
			Central America	*Opius longicaudatus* (Ash.)
				Syntomosphyrum indicum Silv.
				Pachycrepoideus vindemmiae (Rond.)
Dacus cucurbitae Coq.	melon fly	melons, cucumber, squash	U.S.A. (Hawaii)	*Opius fletcheri* Silv.
Dacus dorsalis Hendel	oriental fruit fly	many fruits	U.S.A. (Hawaii)	*Opius oophilus* Full.
				Opius longicaudatus (Ash.)
				Opius incisi Silv.
Dacus passiflorae Frogg.	fruit fly	fruits	Fiji	*Opius oophilus* Full.
Dacus xanthodes Brown	fruit fly		Fiji	*Opius oophilus* Full.
Dasyneura mali Kieffer	apple leaf curling midge	apple	New Zealand	*Prosactogaster demades* (Walk.)

Natural enemy				
Order Family	Type	Origin	Degree of success [a]	References
Hy Ich	para	New Zealand, 1946–1947	P	DeBach, 1964
Hy Ich	para	New Zealand, 1950	S	Franz and Krieg, 1972
Hy Br	para		S	Bennett *et al.*, this vol., Chapter 15
Hy Eu	para		S	
Hy Pter	para		S	
Hy Br	para	Trinidad, 1967–1969	P-S	Bennett *et al.*, this vol., Chapter 15
Hy Br	para	Australia, 1913	P-S	DeBach, 1964
Hy Br	para	South Africa, 1913	P-S	
Hy Br	para		P	Bennett *et al.*, this vol., Chapter 15
Hy Eul	para		P	
Hy Pter	para		P	
Hy Br	para	India, 1916	P	DeBach, 1964
Hy Br	para	Malaya, India, Borneo, 1947	S	van den Bosch and Messenger, 1973
Hy Br	para		S	
Hy Br	para		S	
Hy Br	para	U.S.A. (Hawaii), 1951–1954	S	Bennett *et al.*, this vol., Chapter 15
Hy Br	para	U.S.A. (Hawaii), 1951–1954	S	
Hy Plat	para	France, 1925–1926	P	DeBach, 1964

TABLE A-1 *(continued)*

Scientific name	Common name	Crop	Location of infestation	Natural enemy: Scientific name
Dasyneura pyri (Bouché)	pear leaf curling midge	pear	New Zealand	*Prosactogaster demades* (Walk.)
Musca domestica L.	housefly		Mauritius	*Spalangia nigra* Latr.
				Muscidifurax raptor Gir.
Phytomyza ilicis (Curt.)	hollyleaf miner	English holly	Canada (British Columbia)	*Chrysocharis gemma* (Walk.)
				Opius ilicis Nixon
HYMENOPTERA				
Cephus pygmaeus (L.)	European wheat stem sawfly	wheat	Canada (Ontario)	*Collyria calcitrator* (Grav.)
				Pedobius beneficus (Gahan)
Diprion similis (Htg.)	introduced pine sawfly	pine	Canada (Montreal area)	*Monodontomerus dentipes* (Dalm.)
Gilpinia (=*Diprion*) *hercyniae* Htg.	European spruce sawfly	spruce	Canada	*Drino bohemica* Mesn.
				Exenterus vellicatus Cush.
				Exenterus confusus Ker.
				Borrelinavirus hercyniae
Neodiprion lecontei (Fitch)	red-headed pine sawfly		Canada (Ontario)	nucleopolyhedrosis virus plus nine parasitic species
Neodiprion sertifer (Geoff.)	European pine sawfly	pine	Eastern Canada and eastern U.S.A. (in Christmas tree plantations)	*N. sertifer* nuclear polyhedrosis virus, + four parasites
Neodiprion swainei Midd.	Swaine jack pine sawfly	pine	Canada	*Exenterus amictorius* Panzer

Natural enemy				
Order Family	Type	Origin	Degree of success [a]	References
Hy Plat	para	France, 1925–1926	S	DeBach, 1964
Hy Spal	para	W. I. Station, CIBC	P-S	Simmonds, 1969
Hy Pter	para			
Hy Eu	para	England, 1936–1939	P	DeBach, 1964
Hy Br	para		P	
Hy Ich	para	England, 1939–1940	S	CIBC, 1962
Hy Eu	para	Europe		DeBach, 1964
Hy Cal	para	Canada (Oakville, Ontario)	S	CIBC, 1962
Dip Ta	para	Europe	S-C	CIBC, 1971
Hy Ich	para	Japan, 1934, 1939		
Hy Ich	para			
	disease	Europe (accidental)		
	disease	Canada (Ontario) 1950	P	CIBC, 1971
	disease	Sweden, 1949	S	CIBC, 1971
Hy Ich	para		P	CIBC, 1971

TABLE A-1 *(continued)*

Scientific name	Common name	Crop	Location of infestation	Natural enemy Scientific name
				Pleolophus basizonus (Grav.)
Pristiphora erichsonii (Htg.)	larch sawfly	larch	Canada	*Borrelina* virus
				Mesoleius tenthredinis Morley
				Olesicampe benefactor Hinz
			U.S.A. (Maine)	*Olesicampe benefactor* Hinz
ARACHNIDA				
Latrodectus mactans (Fabr.)	black widow spider	household pest	U.S.A. (Hawaii)	*Baeus latrodecti* Doz.
MOLLUSCA				
Otala lactea Muller	snail		Bermuda	*Euglandia rosea* (Ferr.)
				Gonaxis quadrilateralis Preston
Achatina fulica Bow.	snail		U.S.A. (Hawaii and other Pacific Islands)	*Gonaxis kibweziensis* (Smith)

[a]C, complete success; S, substantial; P, partial.

TABLE A-2

Ratings of Biological Control of Insect Pests: Successes Listed for Only One Geographical Location for Each Species (The First Introduction Where Known)

Degree of success	Islands	Continents	Total
Complete	12	19	31
Substantial	32	41	73
Partial	25	28	53
Total	69	88	157

Natural enemy				
Order Family	Type	Origin	Degree of success [a]	References
Hy Ich	para			
	disease			
Hy Ich	para	England, 1910–1913	S	Franz and Krieg, 1972
Hy Ich	para	Europe, 1961	P	Simmonds, 1969
Hy Ich	para	Europe via Canada (Manitoba)	P	Simmonds, 1969
Hy Sc	para	U.S.A., 1939	S-C	Sweetman, 1958
Mol Pul	pred snail	Cuba via U.S.A. (Hawaii)	P-S	Simmonds, 1969
Mol Pul	pred snail	East Africa via U.S.A. (Hawaii), 1968		
Mol Pul	pred snail	Africa	C	Sweetman, 1958

TABLE A-3
Ratings of Biological Control of Insect Pests: Successes Listed for Every Location into which Importation Has Occurred

Degree of success	Islands	Continents	Total
Complete	35	67	102
Substantial	58	86	144
Partial	36	45	81
Total	129	198	327

TABLE A-4

Ratings of Biological Control of Weeds: Successes Listed for Only One Geographical Location for Each Species (The First Introduction Where Known)

Degree of success	Islands	Continents	Total
Complete	5	5	10
Substantial	7	7	14
Partial	0	5	5
Total	12	17	29

TABLE A-5

Ratings of Biological Control of Weeds: Successes Listed for Every Location into which Importation has Occurred

Degree of success	Islands	Continents	Total
Complete	7	6	13
Substantial	9	17	26
Partial	4	14	18
Total	20	37	57

REFERENCES

Alam, M. M., Bennett, F. D., and Carl, P. P. (1971). Biological control of *Diatraea saccharalis* (F.) in Barbados by *Apanteles flavipes* Cam. and *Lixophaga diatraeae* T.T. *Entomophaga* **16**, 151–158.

Andres, L. A., and Goeden, R. D. (1971). The biological control of weeds by introduced natural enemies. *In* "Biological Control" (C. B. Huffaker, ed.), pp. 143–164. Plenum, New York.

Brenierre, J., Pfeffer, P., Betbeder-Matibet, M., and Etienne, J. (1966). Tentative d'introduction a Madagascar et a la Reunion de *Diatraeophaga striatalis*, parasite de *Proceras sacchariphagus* "borer pronctue" de la canne a sucre. *Entomophaga* **11**, 231–238.

Chen, C. B., and Hung, T. H. (1962). Experimental results of introducing parasitic wasps and a fly for controlling sugarcane borers from India into Taiwan in 1961–1962. (English summary). *J. Agr. Ass. China* **40**, [N.S.] 63–67.

Clausen, C. P., in press. "Introduced Parasites and Predators of Arthropod Pests and Weeds: A World Review." *U.S. Dep. Agr. Handb.* **480**.

Cleare, L. D. (1939). The Amazon fly (*Metagonistylum minense* Towns.) in British Guiana. *Bull. Entomol. Res.* **30**, 85–102.

Cochereau, P. (1972). La lutte biologique dans le Pacifique. *Cah. ORSTOM, Ser. Biol.* **16**, 89–104.

Commonwealth Institute of Biological Control. (1962). A review of the biological control attempts against insects and weeds in Canada. *Commonw. Inst. Biol. Control Tech. Commun.* **2**, 216 pp.

Commonwealth Institute of Biological Control. (1971). Biological control programmes against insects and weeds in Canada 1959–1968. *Commonw. Inst. Biol. Control Tech. Commun.* **4**, 266 pp.

DeBach, P. (ed.). (1964). "Biological Control of Insect Pests and Weeds," 844 pp. Reinhold, New York.

DeBach, P. (1974). "Biological Control by Natural Enemies," 323 pp. Cambridge Univ. Press, Cambridge.

DeBach, P., and Argyriou, L. C. (1967). The colonization and success in Greece of some important *Aphytis* spp. (Hym. Aphelinidae) parasitic on citrus scale insects (Hom. Diaspididae). *Entomophaga* **12**, 325–342.

DeBach, P., Rosen, D., and Kennett, C. E. (1971). Biological control of coccids by introduced natural enemies. *In* "Biological Control" (C. B. Huffaker, ed.), pp. 165–194. Plenum, New York.

Franz, J. M. and Krieg, A. (1972). "Biologische Schadlingsbekampfung", 208 pp. Parey, Berlin. (In Ger.)

Harpaz, I., and Rosen, D. (1971). Development of integrated control programs for crop pests in Israel. *In* "Biological Control" (C. B. Huffaker, ed.), pp. 458–468. Plenum, New York.

Huffaker, C. B., and Kennett, C. E. (1966). Studies of two parasites of olive scale *Parlatoria oleae* (Colvée). IV. Biological control of *Parlatoria oleae* (Colvée) through the compensatory action of two introduced parasites. *Hilgardia* **37**, 283–335.

Kobakhidze, D. N. (1965). Some results and prospects of the utilization of beneficial entomophagous insects in the control of insects in Georgian SSR (U.S.S.R.). *Entomophaga* **10**, 323–330.

Long, W. H., and Hensley, S. D. (1972). Insect pests of sugar cane. *Annu. Rev. Entomol.* **17**, 149–176.

Miskimen, G. W. (1962). Studies on the biological control of *Diatraea saccharalis* F. (Lepidoptera: Crambidae) on St. Croix, U.S. Virgin Islands. *J. Agr. Puerto Rico Univ.* **46**, 135–139.

National Academy of Sciences. (1968). Weed control. Principles of plant and animal pest control, Vol. 2. *Nat. Acad. Sci. Nat. Res. Council, Publ.* **1597**, 471 pp.

Quezada, J. R. (1974). Biological control of *Aleurocanthus woglumi* (Homoptera: Aleyrodidae) in El Salvador. *Entomophaga* **19**, 243–254.

Rubtsov, I. A. (1957). État et problemes de l'etude et de l'utilisation en U.R.S.S. des entomophages dans la lutte biologique contre les insectes nuisibles. *Entomophaga* **2**, 125–128.

Simmonds, F. J. (1960). The successful biological control of the sugarcane moth-borer, *Diatraea saccharalis* F. (Lepidoptera, Pyralidae) in Guadeloupe B.W.I. *Proc. 10th Congr. Int. Soc. Sugarcane Technol.*, pp. 914–918.

Simmonds, F. J. (1969). Brief resumé of activities and recent successes achieved. *Commonw. Inst. Biol. Control., Commonw. Agr. Bur. Publ.* 16 pp.

Sweetman, H. L. (1958). "The Principles of Biological Control," 560 pp. Wm. Brown, Iowa.

van den Bosch, R. (1971). Biological control of insects. *Annu. Rev. Ecol. System.* **2**, 45–66.

van den Bosch, R., and Messenger, P. S. (1973). "Biological Control," 180 pp. Intext, New York.

Vinson, J. (1942). Biological control of *Diatraea mauriciella* Wlk. in Mauritius. I. Investigations in Ceylon in 1939. *Bull. Entomol. Res.* **33**, 39–65.

INDEX

For additional place names and other common names of insects (not included in this index) see Chapter 28 (Appendix), pp. 685–743.

C

D

G

J

K

L

M

N

O

P

Q

R

S

T

A 6
B 7
C 8
D 9
E 0
F 1
G 2
H 3
I 4
J 5